W0082707

Manual of
Rheumatology

Fifth Edition

Editors of Previous Editions

1st and 2nd Editions
Editor-in-chief
Prakash K Pispati
Associate editors
Nirupa E. Borges, SS Uppal

3rd Edition
Editor-in-chief
URK Rao
Editors
KM Mahendranath, Ramnath Misra, SJ Gupta, Rohini Handa, Annil Mahajan
Associate editor
Firdaus Fatima

4th Edition
Editor-in-chief
URK Rao
Editors
KM Mahendranath, Ramnath Misra, SJ Gupta, Rohini Handa, Annil Mahajan, Ved Chaturvedi, V Krishnamurthy, Binoy Paul, S Shankar, Firdaus Fatima

Manual of Rheumatology

Fifth Edition

Editor-in-Chief

Ved Chaturvedi MD, DM
Professor and Senior Consultant Rheumatologist
Department of Rheumatology and Clinical Immunology
Ganga Ram Institute for
Postgraduate Medical Education and Research (GRIPMER)
Sir Ganga Ram Hospital, New Delhi

Editors

Molly M Thabah MD
Associate Professor, Department of Medicine
Jawaharlal Institute of Postgraduate Medical Education and Research (JIPMER), Puducherry

Debashish Danda MD, DM, FRCP (Lond.) FACR, FAMS
Professor and Head
Department of Clinical Immunology and Rheumatology
Christian Medical College, Vellore

Rajiva Gupta MD, DNB FRCP (Glasg.) FRCP (Edin.)
Director and Head
Rheumatology and Clinical Immunology
Medanta, The Medicity, Gurgaon

Durga Prasanna Misra MD, DM
Assistant Professor
Department of Clinical Immunology
Sanjay Gandhi Postgraduate Institute of Medical Sciences, Lucknow

Vir Singh Negi MD, DM
Professor and Head
Department of Clinical Immunology
Jawaharlal Institute of Postgraduate Medical Education and Research (JIPMER)
Puducherry

Vinod Ravindran MD, MRCP (Rheumatology), FRCP (Edin.)
Centre for Rheumatology, Calicut
Department of Rheumatology
National Hospital, Calicut
Kerala

Vivek Vasdev MD, DNB Rheumatology
Head
Department of Rheumatology and Clinical Immunology
Army Hospital Research and Referral
New Delhi

CBS Publishers & Distributors Pvt Ltd

New Delhi • Bengaluru • Chennai • Kochi • Kolkata • Lucknow • Mumbai
Hyderabad • Jharkhand • Nagpur • Patna • Pune • Uttarakhand

Preface to the First Edition

Perhaps nowhere in the world as in India that the arthritic patient is vulnerable to influences and counter-influences of modem medicine, homoeopathy, ayurveda and unani systems of medicine, not to speak of acupuncture, copper bangles, magnetic therapy and what have you. In an ancient land such as India, a rheumatologist is competing with these modalities of alternate medicine all the time. Alternate medicine, if practiced in a scientific way may be justifiable, but under this garb outright quackery is practiced as well. Steroids are mixed on the sly with ayurvedic and homoeopathic preparations. Branding of joints with hot iron rods in the hope of curing arthritis still goes on rampantly. Some arthritic patients have lost teeth under the false belief that removal of "septic focus" would cure arthritis. Not surprisingly, even practitioners of modem medicine at times prescribe drugs of alternate medicine without necessarily understanding their role. The stark reality of the Indian situation is that there is such little understanding of rheumatic diseases in the minds of average practitioners. This is attributed to the fact that rheumatology is not taught in medical schools. Perhaps a few orthopaedic surgeons may refer to rheumatic diseases during their interaction with students. But is this enough?

There is thus a paramount need for a *Manual of Rheumatology* intended for practitioners, postgraduates in medicine and orthopaedics, orthopaedic surgeons who mainly treat rheumatic diseases day in and day out. Rheumatologists are so few in India that much of the arthritis care is delivered by these practitioners. Hence, this manual is mainly addressed to this target audience.

No doubt, the Indian Rheumatism Association (IRA) conducts CME programmes for practitioners throughout the country from time to time, but its reach is restricted. The Manual of Rheumatology aims at bringing recent diagnostic and therapeutic skills available in modem medicine to the practitioner. For example, there are few practitioners who realize that there are a hundred different types of arthritis. The manual aims at building their confidence in at least making a tentative diagnosis on clinical grounds. It sets out guidelines on laboratory investigations for arthritis, which to order and how to interpret the results. Radiology of arthritis is well illustrated. The reader should be able to discern between rheumatoid arthritis and seronegative spondyloarthropathies. The practitioner should be able to treat successfully a case of infective arthritis with cure as the outcome. The book emphasizes the wide prevalence of tuberculous arthritis in our country. It brings into focus the new diseases like HIV in arthritis. Complexities of vasculitis and antiphospholipid syndrome are explained in simple terms. Therapeutic modalities are described in a way that the practitioner should be able to choose a rational NSAID and not be afraid to prescribe DMARDs. Finally, the manual raises the challenge of patient education in the management of chronic diseases such as arthritis.

This Manual is a dream come true. It is a labour of love of us rheumatologists in India. It was indeed a privilege to write for this manual and to edit the various chapters and finally publish it within the stipulated time, for the new millennium. Initially, the print run is a modest five thousand. We hope that the sponsors M/s. Pfizer Ltd. will take up the challenge of reaching the practising community at large and multiply the print run several times. We pray that the manual will be received and read by the practitioners with enthusiasm in the hope of better care and treatment for the arthritic patient to whom the book is dedicated.

Prakash K Pispati
Editor-in-Chief

Nirupa E Borges
Associate editor

SS Uppal
Associate editor

Contributors

Aarthi Priya T MD, DM
Assistant Professor
Department of Rheumatology
Government Kilpauk Medical College and Hospital
Chennai

Abhishek Kumar MD, DNB Rheumatology
Senior Resident
Department of Rheumatology and Clinical Immunology
Army Hospital Research and Referral, New Delhi

Abhra Chandra Chowdury MD, DM
Consultant
Mission Hospital, Durgapur, West Bengal

Able Lawrence MD, DM
Professor
Department of Clinical Immunology
Sanjay Gandhi Postgraduate Institute of Medical Sciences
Lucknow

Ajay Wanchu MD, DM
Division of Arthritis and Rheumatic Disease
Oregon Health and Science University, Portland, OR, USA

Ajaz Kariem Khan MD
Director, Centre for Joint Diseases and Pain Management
SHIFA Medical Centre Srinagar

Alakendu Ghosh MD
Professor and Head, Department of Rheumatology
IPGMER & SSKM Hospital, Kolkata

Alladi Mohan MD, FRCP (Edin), FCCP, FAMS, FICP, FIACM
Chief
Division of Pulmonary, Critical Care and Sleep Medicine
Professor and Head, Department of Medicine
Sri Venkateswara Institute of Medical Sciences, Tirupati

Aman Sharma MD, FRCP (Lond.)
Additional Professor
Department of Internal Medicine
Post Graduate Institute of Medical Education and Research (PGIMER), Chandigarh

Amita Aggarwal MD, DM
Professor
Department of Clinical Immunology
Sanjay Gandhi Postgraduate Institute of Medical Sciences
Lucknow

Anand N Malaviya MD, FRCP (Lond.), FACP, FICP, FAMS, FNASc, ACR Master, APLAR Master
Consultant Rheumatologist, A&R Clinic
Visiting Senior Consultant Rheumatologist
Indian Spinal Injuries Centre Superspeciality Hospital
New Delhi

Annil Mahajan MD, FICP
Professor
Department of Medicine
Government Medical College, Jammu

Anuj Shukla MD, DM
Consultant Rheumatologist
Niruj Rheumatology Clinic, Maninagar, Ahmedabad

Anupam Wakhlu MD, DM
Professor
Department of Rheumatology, King George's Medical University, Lucknow

Anvesha M MSc
Consultant Nutritionist
Samarapan Health Centre
Bengaluru

S Aparna Reddy MD, DM
Assistant Professor
Division of Rheumatology
Department of Medicine
Sri Venkateswara Institute of Medical Sciences, Tirupati

Arun Hegde MD, DNB Rheumatology
Classified Specialist (Medicine and Rheumatology)
Army Hospital Research and Referral, Delhi Cantt

Arup Kumar Kundu MD, FICP, MNAMS
Professor of Medicine and In-charge, Rheumatology Clinic
IQ City Medical College and Narayana Multispeciality Hospital, Durgapur

Arvind Chopra MD
Director, Centre for Rheumatic Diseases, Pune

Arvind Ganapati MD
Senior Resident, Post-doctoral Fellow
Department of Clinical Immunology and Rheumatology, Christian Medical College, Vellore

Ashok Kumar MD, FRCP (Lond.)
Director and Head
Department of Rheumatology
Fortis Flt. Lt. Rajan Dhall Hospital, New Delhi

Ashish J Mathew DNB (Med), DM
Assistant Professor
Clinical Immunology and Rheumatology, Christian Medical College, Vellore

Ashwani Kumar MD, DM
Classified Specialist
Department of Rheumatology, Army Hospital Research and Referral, New Delhi

Aswin Nair MD, DM
Senior Resident
Department of Clinical Immunology and Rheumatology
Christian Medical College, Vellore

C Balakrishnan MD
Consultant Rheumatologist
PD Hinduja Hospital, Mumbai

Banwari Sharma MD, DM
Clinical Immunologist and Rheumatologist
Niramaya Health Care, Jaipur

Bidyut Kumar Das MD, PhD
Professor
Department of Medicine
Head of Clinical Immunology and Rheumatology
SCB Medical College, Cuttack, Odisha

Bimlesh Dhar Pandey MD, MRCP
Senior Consultant, Department of Rheumatology
Fortis Hospital, Noida

Binoy J Paul MD, PhD, DNB, FRCP (Edin.)
Professor of Medicine, Chief of Rheumatology
KMCT Medical College, Calicut

S Chandrashekara MD, DM
Director, ChanRe Rheumatology and Immunology Center
Bengaluru

P Damodaram MD, DM
Consultant Rheumatologist, Tirupati

Debashish Danda MD, DM, FRCP (Lond.), FACR, FAMS
Professor and Head
Department of Clinical Immunology and Rheumatology
Christian Medical College, Vellore

BG Dharmanand MD, DM
Senior Consultant and Head, Rheumatology, Sakra World Hospital, Bengaluru

Durga Prasanna Misra MD, DM
Assistant Professor
Department of Clinical Immunology
Sanjay Gandhi Postgraduate Institute of Medical Sciences
Lucknow

Firdaus Fatima MD
Consultant Rheumatologist
AB Rheumatology Center, Hyderabad

Gautam Dhar Choudhury MD
Professor of Medicine and Unit Chief Rheumatology
Command Hospital, Lucknow

Ghan Shyam Pangtey MD
Professor
Department of Medicine, Lady Hardinge Medical College
New Delhi

Harshini AS MD
Senior Resident
Clinical Immunology and Rheumatology, Christian Medical College, Vellore

Jeet Patel MD
Senior Resident
Department of Rheumatology and Clinical Immunology
Sir Ganga Ram Hospital
New Delhi

Jyoti Panwar MD (Radiology), FRCR
Professor
Department of Radiodiagnosis
Christian Medical College, Vellore

Jyoti Ranjan Parida MD, DM
Associate Professor and Head
Department of Immunology and Rheumatology
IMS and SUM Hospital, Bhubaneswar

Jyotsna Oak MD
Consultant, Rheumatology and Internal Medicine
Kokilaben Dhirubhai Ambani Hospital, Mumbai
Former Professor, Department of Medicine and In-Charge Rheumatology, LTMG Hospital & LTM Medical College
Sion, Mumbai

Kakarla Subbarao MD
Former Director
Nizam's Institute of Medical Sciences, Hyderabad
President, Krishna Foundation for Research
KIMS Hospital, Secunderabad

R Kaushik MRCP
Specialist Registrar, Rheumatology
Princess of Wales Hospital, Bridgend, UK

V Krishnamurthy MD, DM
Senior Consultant Rheumatologist
CEO Chennai Meenakshi Superspeciality Hospital
Chennai

Krishnan Shanmuganandan MD, DNB
Professor of Medicine and Rheumatology
Command Hospital, Lucknow

Lalit Duggal MD, FRCP (Glasg.)
Senior Consultant and Chairman
Department of Rheumatology and Clinical Immunology
Sir Ganga Ram Hospital, New Delhi

Latika Gupta MD, DM
Senior Resident
Department of Clinical Immunology
Sanjay Gandhi Post Graduate Institute of Medical Sciences
Lucknow

Liza Rajasekhar MD
Professor
Department of Rheumatology
Nizam's Institute of Medical Sciences, Hyderabad

KM Mahendranath MD, FRCP (Lond.)
Senior Consultant Rheumatologist
Samarapan Health Centre, Bengaluru

Manish Rathi MD, DM
Additional Professor
Department of Nephrology
Post Graduate Institute of Medical Education and Research (PGIMER), Chandigarh

Manjit Ahlawat MBBS
Resident in Medicine
Medical Division Command Hospital (Air Force)
Bangalore

HR Madhuri MD, DM
Consultant Rheumatologist
CARE Hospital, Hyderabad

Marcia Friedman MD
Division of Arthritis and Rheumatic Disease,
Oregon Health and Science University, Portland, OR, USA

Milind Nadkar MD, FICP
Professor
Department of Medicine
Chief of Rheumatology, Seth GS Medical College and KEM Hospital, Mumbai

Molly Mary Thabah MD
Associate Professor
Department of Medicine
Jawaharlal Institute of Postgraduate Medical Education and Research (JIPMER), Puducherry

Mohit Goyal MD
Consultant Rheumatologist
CARE Pain and Arthritis Centre, Udaipur

G Narasimulu MD, FICP, FIACM
Senior Consultant Rheumatologist
GVN Medical Centre, Hyderabad
Former Professor and Head, Department of Rheumatology
Nizams Institute of Medical Sciences, Hyderabad

Nagma Bansal MD
Senior Resident
Department of Rheumatology and Clinical Immunology,
Sir Ganga Ram Hospital, New Delhi

Natasha Negalur MD
Senior Resident
Department of Rheumatology and Clinical Immunology
Medanta, The Medicity
Gurgaon

Neha Arora DNB (Medicine)
Classified Registrar
Department of Rheumatology
Clinical Immunology and Allergies
Artemis Hospitals
Gurgaon

Nibha Jain MD
Senior Resident
Medicine and Rheumatology
VS Hospital, Ahmedabad

Nikhil Gupta MD
Senior Resident, Post-doctoral Fellow
Department of Clinical Immunology and Rheumatology
Christian Medical College, Vellore

Nutan Kamath MD
Professor
Department of Pediatrics
Kasturba Medical College, Mangalore

PD Rath MD
Associate Director and Head, Rheumatology
Max Super Speciality Hospital, New Delhi

Parasar Ghosh MD, DM
Professor
Department of Rheumatology
IPGMER and SSKM Hospital, Kolkata

Piyush Joshi MD, DNB Rheumatology
Consultant Rheumatologist
Hope Rheumatology Clinic, Bapunagar, Ahmedabad

Pooja Belani MD
Senior Resident
Department of Clinical Immunology
Jawaharlal Institute of Postgraduate Medical Education and Research (JIPMER), Puducherry

R Porkodi MD, DM
Consultant Rheumatologist
Dr Kamakshi Memorial Hospital, Chennai

Priyanka Kharbanda MD, DNB Rheumatology
Consultant Rheumatologist
Fortis Shalimar Bagh, New Delhi

Prakash Pispati MD, FRSM (Lond.), ICP, FACR, ACR Master, APLAR Master
Director of Rheumatology, Mentor Jaslok Hospital, Mumbai
Senior Consultant Rheumatologist, Saifee Hospital, Mumbai
Honorary Member, Past President APLAR
Past President, Indian Rheumatology Association
Editor-in-Chief, Voice of APLAR

Puja Srivastava MD, DM
Consultant Rheumatologist
STAR Clinics and VS Hospital, Ahmedabad

Puneet Mashru MD
Clinically Certified Densitometrist (ISCD)
Senior Registrar, Department of Rheumatology
Christian Medical College and Hospital, Vellore

V Rajamani MBBS, MRCP, FRCP
Consultant Rheumatologist
Kovai Medical Center and Hospital, Coimbatore

S Rajeswari MD, DM
Professor and Head
Institute of Rheumatology
Madras Medical College and
Rajiv Gandhi Government General Hospital, Chennai

Rajiva Gupta MD, DNB, FRCP (Glasg.) FRCP (Edin.)
Director and Head
Rheumatology and Clinical Immunology
Medanta, The Medicity, Gurgaon

Rajkiran Dudam MD, DM
Consultant Rheumatologist
Star Hospital, Hyderabad

Ramesh Jois MD, MRCP (UK)
Consultant, Department of Rheumatology
Fortis Hospital, Bannerughatta Road, Bengaluru

Ramnath Misra MD, FRCP (Lond.)
Professor and Head
Department of Clinical Immunology
Ex-Dean, Sanjay Gandhi Postgraduate Institute of Medical Sciences, Lucknow

Rasmi Ranjan Sahoo MD
Senior Resident
Department of Rheumatology
King George's Medical University, Lucknow

Ravichandran R MD, DM
Professor and Head, Department of Rheumatology
Government Kilpauk Medical College and Hospital
Chennai

Reena Gulati MD, DM Genetics
Associate Professor
Department of Pediatrics
Jawaharlal Institute of Postgraduate Medical Education and Research (JIPMER), Puducherry

Rohini Handa MD, DNB, FRCP (Glasg.)
Senior Consultant Rheumatologist
Indraprastha Apollo Hospitals, New Delhi

Rudra Prosad Goswami MD
Senior Resident, Department of Rheumatology
IPGMER and SSKM Hospital, Kolkata

Rupali Mathur MBBS
Resident Medicine
Kokilaben Dhirubhai Ambani Hospital, Mumbai

Sakir Ahmed MD
Senior Resident
Department of Clinical Immunology
Sanjay Gandhi Postgraduate Institute of Medical Sciences
Lucknow

Sanat Phatak MD
Senior Resident
Department of Clinical Immunology
Sanjay Gandhi Postgraduate Institute of Medical Sciences
Lucknow

Sanjeeb Kakati MD
Professor
Department of Medicine, In charge of Rheumatology Clinic
Assam Medical College, Dibrugarh, Assam

Sanjiv Kapoor MD, DM
Senior Consultant Rheumatologist
Indian Spinal Injuries Centre, Vasant Kunj, Delhi
Sanjivani Rheumatology and Infertility Centre, Delhi

Sapan Pandya MD, DM
Consultant Rheumatologist
VS Hospital and Vedanta Institute of Medical Sciences
Ahmedabad

Sarath Chandra Mouli Veeravalli MD, MRCP (UK)
Clinical Director
Department of Rheumatology and Clinical Immunology
Krishna Institute of Medical Sciences, Hyderabad

Sathish Kumar MD, DCH
Professor of Pediatrics
Christian Medical College, Vellore

Shankar Naidu MD, DM
Senior Resident
Department of Internal Medicine
Post Graduate Institute of Medical Education and Research (PGIMER), Chandigarh

Sharath Kumar MD (Ped), DNB Rheumatology
Consultant Rheumatologist
Columbia Asia Hospital, Bengaluru

V Shantaram MD
Consultant Physician
Former Professor and Head, Department of Medicine
Nizam's Institute of Medical Sciences, Hyderabad

Shefali Khanna Sharma MD
Associate Professor
Department of Internal Medicine
Rheumatology and Clinical Immunology Unit
Post Graduate Institute of Medical Education and Research (PGIMER), Chandigarh

J Shivanand MD
Consultant
Sri Deepti Rheumatology Centre, Hyderabad

Shweta Singhai MD
Consultant
Department of Rheumatology
Sakra World Hospital, Bengaluru

Shyamashis Das MD, DM
Consultant Rheumatologist
Institute of Neurosciences, Kolkata

SJ Gupta FRCP (Glasg.), FRCP (Edin.)
Senior Consultant Rheumatologist
Apollo Indraprastha and Sant Parmanand Hospitals
New Delhi

Siddharth Jain MBBS
Junior Resident
Department of Internal Medicine
All India Institute of Medical Sciences, New Delhi

Silas Supragya Nelson MD
Associate Professor, Medicine and
In-charge Rheumatology Unit
NSCB Medical College, Jabalpur

Sita Naik MD
Retired Professor of Immunology and Dean
Sanjay Gandhi Postgraduate Institute of Medical Sciences
Lucknow

MR Sivakumar MD, DM, PhD, FRCP, FACR, FAAN, FAHA
Program Director
Stroke Interventions and Neurocritical Care
GLB Hospitals and Acute Stroke Center, Chennai

Somnath Bhar MBBS, MRCP
Consultant Physician
GD Hospital and Diabetes Institute, Kolkata

Sonal Mehra MD, DM
Consultant Rheumatologist
Jaypee Hospital, Noida

Subhankar Haldar MD
Assistant Professor
Department of Rheumatology
IPGMER and SSKM Hospital, Kolkata

Subramanian Shankar MD, DNB FRCP (Glasg.)
Professor and Senior Advisor (Medicine and Clinical Immunology)
Command Hospital (Air Force), Bengaluru

Sujata Sawhney MD, MRCP (UK)
Senior Consultant
Pediatric and Adolescent Rheumatologist
Pediatric Rheumatology Division, Institute of Child Health
Sir Ganga Ram Hospital, New Delhi

Sukumar Mukherjee MD, FRCP, FRCPE, FICP
Ex-Professor and Head
Department of Medicine, Medical College, Kolkata
Consultant Physician
Calcutta Medical Research Institute
GD Hospital and Diabetes Institute
Kothari Medical Centre, Kolkata

Sumeet Agrawal MD, DM
Head
Department of Rheumatology
Clinical Immunology and Allergies
Artemis Hospitals, Gurgaon

Sundeep Grover MD, DM
Associate Professor
Department of Medicine, Subharti Institute of Medical Sciences, Meerut
Arthritis and Immunology Clinic, Meerut

Sundeep Upadhyaya MD, DM
Senior Consultant Rheumatologist
Department of Rheumatology, Indraprastha Apollo Hospitals, New Delhi
Associate Professor—AHERF

Surjit Singh MD, DCH (Lond.), FRCP (Lond.), FRCPCH (Lond.)
Professor of Pediatrics and
In-charge Allergy Immunology Unit
Advanced Pediatrics Centre
Post Graduate Institute of Medical Education and Research (PGIMER), Chandigarh

Taral Parikh MD, DNB Rheumatology
Consultant Rheumatologist
Columbia Asia Hospital and GCS Medical College
Ahmedabad

Uma Kumar MD
Professor and Head
Department of Rheumatology
All India Institute of Medical Sciences, New Delhi

Y Uday MD, DM
Associate Professor
Department of Haematology and
BMT Army Hospital (Research and Referral), New Delhi

URK Rao MD
Consultant Rheumatologist
Director, Sri Deepti Rheumatology Centre
Hyderabad

Varun Dhir MD, DM
Associate Professor
Department of Internal Medicine
Post Graduate Institute of Medical Education and Research (PGIMER), Chandigarh, India

Ved Chaturvedi MD, DM
Professor and Senior Consultant Rheumatologist
Department of Rheumatology and Clinical Immunology
Sir Ganga Ram Hospital, New Delhi

Velu Nair MD, FRCP (Lond.)
Director General, Medical Services (Army), New Delhi

VR Joshi MD, FRCP (Lond.)
Consultant Physician and Rheumatologist
Director Research, PD Hinduja Hospital and MRC, Mahim
Mumbai

Vijay KR Rao MBBS, MRCP (UK), FRCP
Manipal Hospital, Bengaluru

Vijaya Prasanna Parimi MD, DM
Consultant Rheumatologist
Krishna Institute of Medical Sciences, Hyderabad

Vinod Ravindran MD, MRCP, FRCP (Edin.)
Centre for Rheumatology, Calicut
Department of Rheumatology, National Hospital, Calicut
Kerala

Vikas Agarwal MD, DM
Professor
Department of Clinical Immunology
Sanjay Gandhi Post Graduate Institute of Medical Sciences
Lucknow

Vikram Londhey MD
Professor
Department of Medicine
Topiwala National Medical College and
BYL Nair Charitable Hospital, Mumbai

Vir Singh Negi MD, DM
Professor and Head
Department of Clinical Immunology
Jawaharlal Institute of Postgraduate Medical Education and Research (JIPMER)
Puducherry

Vishad Vishwanath MD, DM
Consultant Rheumatologist
Director, Institute of Rheumatology and Immunological Sciences, Thiruvanathapuram

Vivek Vasdev MD, DNB Rheumatology
Head, Department of Rheumatology and Clinical Immunology
Army Hospital Research and Referral
New Delhi

Yojana Gokhale MD
Professor
Department of Medicine and In-charge of Rheumatology
Lokmanya Tilak Medical College, Sion, Mumbai

Contents

SECTION II: INVESTIGATIONS OF RHEUMATIC DISEASES

SECTION III: THERAPEUTICS IN RHEUMATIC DISEASES

SECTION IV: RHEUMATOID ARTHRITIS

SECTION VII: VASCULITIS

SECTION VIII: CRYSTAL ARTHROPATHY AND OSTEOARTHRITIS

SECTION IX: OSTEOPOROSIS AND METABOLIC BONE DISEASES

SECTION X: JUVENILE IDIOPATHIC ARTHRITIS

SECTION XI: INFECTIONS AND ARTHRITIS

SECTION XII: MISCELLANEOUS

Chapter

1

Rheumatology Non-stop (A Short Story: History in the Making)

Prakash Pispati

Why history? When at school the subject I dreaded most was mathematics (probably drove me to medicine) and the subject I enjoyed the least was perhaps history. And now just as I enjoy visiting ancient temples, churches, castles, monuments, museums..., so do I enlighten myself dwelling into History of Medicine. Man's quest to conquer diseases is as ancient as mankind itself. Can you imagine a human being in his lifetime without complaining of aches and pains casual mostly, crippling at times? No wonder, medical scriptures are replete with accounts of what physicians in earlier times thought about 'gutta', 'rheuma', 'rheumatism' and now Rheumatology..... Are such words meaningful today? A young investigator may mock, yet didn't Winston Churchill write, "*The longer you look back, the further you look forward*".

Ancient Times and Thoughts

Greek physician Hippocrates (460–377 BC) developed a medical theory called humoralism, which held that four humors (liquids) coursing through the human body determined one's temperament and state of health. He used the term rheuma, which literally means "flowing", to describe an excess of the watery humor thought to flow down from the brain. The words rheuma and catarrhos ("flowing down") were used interchangeably by ancient Greeks to describe a variety of illnesses including joint problems.

In 13th century Europe any joint ailment was called gutta (Latin, "a drop") for a noxious humour falling drop by drop into the joint. Gout and gouty diathesis were terms used broadly to connote arthritis as we understand today. Thomas Sydenham (1624–1689), a London physician, who himself suffered from gout, distinguished the episodic disorder from acute arthritis that attacked children (probably rheumatic fever) and from chronic crippling arthritis [probably rheumatoid arthritis (RA)] that came to be called 'rheumatic gout'.

Not until 1642 was the term 'rheumatism' used by the French physician Ballonius, who sought to distinguish noxious humors.

In ancient Indian literature, Charaka Samhita (written in around 00–200 BC) also described a condition that describes pain, joint swelling and loss of joint mobility and function. In our own ancient Ayurveda science, the concept of vata, pitta, kapha (wing, bilious and phlegm) may seem parallel. Interestingly, while Charaka Samhita's texts suggest a joint disease rheumatoid like, it must have been inspired to prepare *Suvarna Bhasma* perhaps akin to 'gold oxide'. Remember injection sodium aurothiomalate ('Myocrisine'). Regretfully we stopped to use it even though its side effects were manageable, its cost... a fraction of today's biologicals.

Gout understandably is the most vividly described rheumatic disease, why not? Truly, its acute dramatic manifestations must have made "headline news" even in those days as victims included monarchs and ministers be they in office or in churches, authors or poets of romantic life-style. Hippocrates, himself a victim of gout described:

Eunchs do not take gout, nor become bald
A woman does not take gout unless her menses are stopped.
A young man does not take gout unless he indulges in coitus.
In gouty affection inflammation subsides in 40 days.
Hippocrates (5th Century BC, Greece)

Era and the Age of Reason, Experiments and Science

Such were powerful descriptions that any joint pain was thought to be 'gouty'. Yet multiple, painful chronic joint deformities were surely noticed by some physicians who correctly described these as "arthritis deformans". In 1858, Sir Alfred Baring Garrod insisted that the term be changed to Rheumatoid Arthritis. It did not make much of an impression. Sir William Osler had this to write:

"*Arthritis Deformans:* Once established, the disease is rarely curable. Too often it is a slow, but progressive, crippling of the joints, with a disability that makes the disease one of the most terrible of human afflictions (from Principles and Practice of Medicine, 1909)".

In 1923 a meeting of concerned physicians took place at the Royal Society of Medicine where Sir Archibald Edward Garrod (son of Sir Alfred Baring Garrod) strong plea carried weight and the name was formally changed from "Arthritis Deformans" to "Rheumatoid Arthritis". In 1941, the American Rheumatism Association followed suit and since then these two words seem here to stay permanently. Keen readers should ask themselves, what seems more apt, rheumatoid arthritis or rather Arthritis deformans? My vote for the latter.

Galen introduced the term 'Rheumatismus' in second century AD. Bannatyne described the radiological features of RA. It was in 1940 that Camroe coined the term 'rheumatologist'; and the term 'rheumatology' was coined by Hollander in 1949. Today we have subspecies like lupologists, osteologists, osteopologists, rheumatologists ...

The wars had minor salutary side effects. Reiter in the 19th century Russian war noticed that a triad of arthritis, urethritis, dermatitis christened Reiter's disease, a name that is now changed to Reactive Arthritis. Eponyms galore! 15,000 or more on record, a few are short listed.[1] The last significant eponym seems the "Hughes syndrome" synonymous with anti-phospholipid syndrome (Table 1.1).

SOLID FOUNDATIONS OF MICROBIOLOGY, IMMUNOLOGY AND RHEUMATOLOGY

Contemporary rheumatology is surely immunology based driven, inspired since the 18th century. A general practitioner, Edward Jenner in Yorkshire made a profound observation that those children who suffered from 'cowpox' never developed 'smallpox' (Fig. 1.1).[2]

Mere annotations were not enough. He experimented, injecting his son with extracts from

Table 1.1: Common eponyms in rheumatology

• Behçet's syndrome	• Hoffa's disease	• Sever's disease
• Caplan syndrome	• Hughes syndrome (antiphosholipid syndrome)	• Sjögren's syndrome
• Charcot's syndrome	• Jaccoud's syndrome	• Still's disease
• Churg-Strauss syndrome	• Kaschin-Beck disease	• Sudeck's disease
• Cogan's syndrome	• Kawasaki disease	• Sudeck's atrophy
• Crohn's disease	• Leri's pleonosteosis	• Sweet's syndrome
• de Quervain's tenosynovitis	• Libman-Sacks endocarditis	• Takayasu's disease
• Dupuytren's contracture	• Lyme arthritis	• Thiemann's disease
• Farber's disease	• Löfgren syndrome	• Tietze syndrome
• Felty's syndrome	• Morton's Metatasalgia	• Wegener's granulomatosis (granulomatosis with polyangiitis)
• François syndrome	• Osgood-Schlatter's disease	• Weil's disease
• Gaucher's syndrome	• Osler's nodes	• Werner's syndrome
• Handigodu syndrome	• Paget's disease	• Wilson's disease
• Heberden's nodes	• Pott's disease	• Whipple's disease
• Henoch-Schönlein purpura	• Reiter's disease	

Fig. 1.1: Edward Jenner. Portrait by John Raphael Smith[2]

cowpox. He was condemned and removed from the medical fraternity. Strangely, despite England and France going to war periodically, it was Napoleon Bonaparte who invited Jenner to vaccinate his troops! Later the US accepted this practice and then Britain had little choice but to honour Edward Jenner with Knighthood followed by Fellowship of the Royal Society and a reward of £30,000. This was laid the foundation of infection and immunity upon which the very edifice of immunology is carefully built on scientific experiments over centuries that followed.

A Dutch lensman van Leeuwenhoek who wanted to decipher the cause of coughs and common cold to answer everyone's question 'there must be something in the air', curiously led him to find out 'What's in the air' till he tumbled into moving creatures in a drop of water which came under his lens. Excitedly came the cry 'animalcules... eureka!' ...Archimedes repeated! (Fig. 1.2).

Fig. 1.2: Antonie van Leeuwenhoek discovered the single celled protozoa which he called animalcules

Microbiology and immunology were harmonised by Louis Pasteur and Robert Koch ... who knows it not? The Germ theory of disease was then thought sacrosanct. Question thus arose as to how come humans live that long when surrounded by millions of microbes throughout, outside and inside? What protects us? In his figment of imagination, Paul Ehrlich drew a pencil sketch, the hypothesis of antigens and antibodies (Fig. 1.3).[3] Experiments proved this to be very true. Every antibody was thought then to be protective till a couple of decades ago. We were then reminded that some unwanted antibodies without identified antigens were in fact damaging auto antibodies that led to the discovery of autoimmune diseases (Horror autotoxicus). So now we have autoimmune pathophysiology to explain, collagen vascular diseases, the main thrust of contemporary rheumatologists.

Foundations of Contemporary Science

Every simple observation in medicine has a deep-rooted complex message waiting to be decoded. Elie Metchnikoff's discovery that macrophages devour

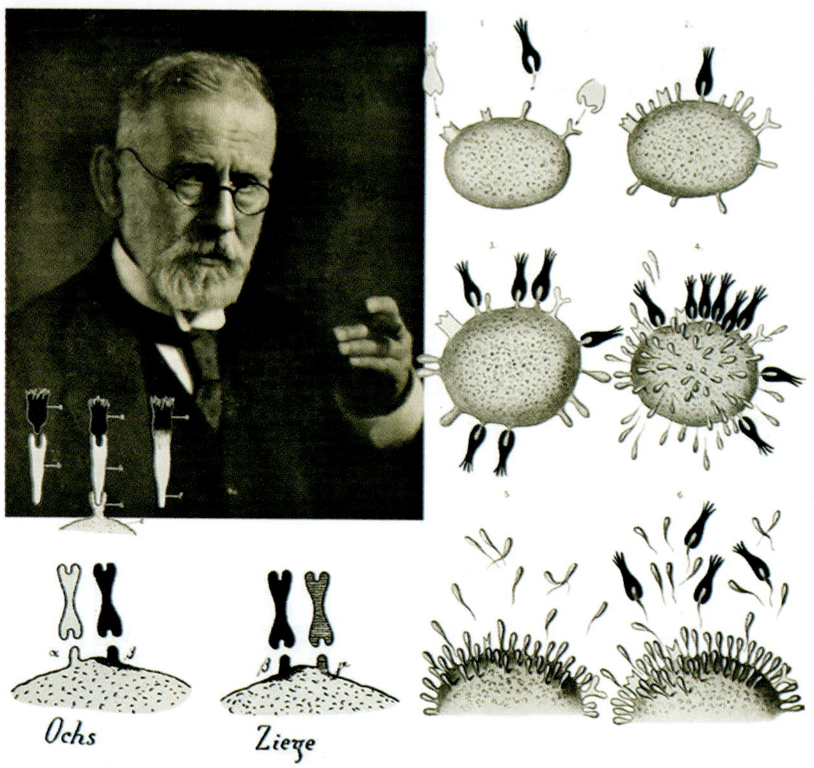

Fig. 1.3: Paul Ehrlich and his drawings of antibodies according to the side chain theory (Reprinted by permission from Macmillan Publishers Ltd: Nature Immunology. Kaufman SHE. Immunology's foundation: the 100-year anniversary of the Nobel Prize to Paul Ehrlich and Elie Metchnikoff. Nat Immunol 2008; 9(7):705–12)

invading bacteria as the defending soldiers of the body (Fig. 1.4).[3] His astute description of phagocytosis had a profound impact on every disease. These were laid the foundations of both cellular and humoral immunity challenging the sole germ theory of every disease. Later developments of immunology time line are depicted in Table 1.2. A mischief monger like a pathogen would fool, evade, enter and corrupt some of the soldiers. Thus, we have antigen presenting cells and T cells asking B cells to produce unwanted and damaging auto-antibodies, through signals produced by cell to cell talk. These are cytokines—some provoking, some inhibitory.

Investigations: Laboratory Diagnostic Approach

Diagnostic laboratory investigations evolved much later. Significantly perhaps, Rose and Waaler sheep's red blood cell agglutinate test changed the mindset to look for disease markers. Both were New York based, but never met (Waaler was Norwegian). 1948 recorded the first ever simple but significant lupus phenomenon (LE Cell test).[4] It got replaced by ANA immunofluorescence test invented by Henry Kunkel (1958). Since then, subsets of ANAs-ENAs, Anti-dsDNA antibody, ANCA and APLA are in vogue. Albeit, progress in diagnostic laboratory seems at turtle pace, isn't it?

1953 onwards Charles Plotz a famous rheumatologist from New York proved that antibodies could be antigenic and produce anti-antibodies (later called auto-antibodies) and that the latex fixation test could not only determine the presence of rheumatoid factor but that the technique could be used, with differing coatings, to determine the presence of auto-antibodies.

His first exposure to latex particles was to note their use as measures in electron microscopy. They were manufactured in very precise diameters by the Dow Chemical Company in Michigan. He tested various sizes and found that those of 0.9 micra in diameter worked best in the latex fixation test. Dow

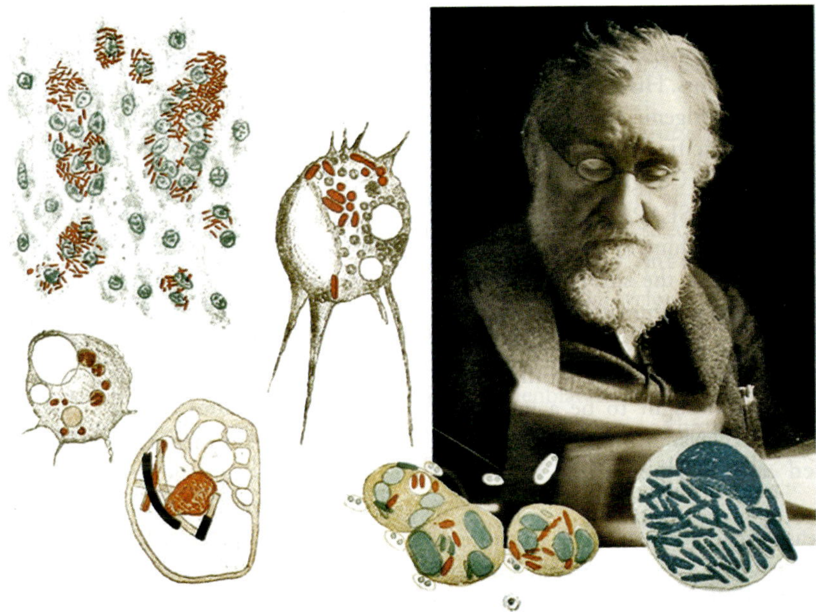

Fig. 1.4: Elie Metchnikoff and his drawings of bacterial phagocytosis by macrophages (Reprinted by permission from Macmillan Publishers Ltd: Nature Immunology. Kaufman SHE. Immunology's foundation: the 100-year anniversary of the Nobel Prize to Paul Ehrlich and Elie Metchnikoff. Nat Immunol 2008; 9(7):705–12)

Table 1.2: Immunology—rheumatology time-line

1798	Smallpox vaccination	Edward Jenner
1876	Validation of germ theory of disease discovering bacterial basis of anthrax	Robert Koch
1879	Chicken cholera vaccine development	Louis Pasteur
1890?	Discovery of Diphtheria "anti-toxin" in blood ShibasaburoKitasato	Emil Von Behring
1882	Isolation of tubercule bacillus	Robert Koch
1883	Delayed type of hypersensitivity	Robert Koch
1884	Phagocytosis: Cellular theory of immunity	Elie Metchnikoff
1891	Proposal that antibodies are responsible for immunity	Paul Erhlich
1891	Passive immunity	Emil Roux
1944	Immunological hypothesis of allograft rejection	Peter Medawar
1948	Demonstration of antibody production in plasma	Astrid Fagraeus
1949	Distinguishing self vs. non-self and its role in maintaining immunological unresponsiveness (tolerance) to self	Macfarlane Burnet & Frank Fenner
1953	Immunological tolerance hypothesis	Rupert Billingham, Leslie Brent, Peter Medawar & Milan Hasek
1954	Rheumatoid factor	Charles Plotz
1958	Identification of first autoantibody and first recognition of Autoimmune disease	Henry Kunkel
1968	Distinction of bone marrow and thymus derived lymphocyte populations: Discovery of T and B cell collaboration.	Jacques Miller & Graham Mitchell
1974	HLA-B27 predisposes to an autoimmune disease	Derek Brewerton
1983	Antiphospholipid (Hughes) syndrome	Graham Hughes
1990...	Discovery of TNF and Monoclonal antibodies	Feldmann and Maini

Finesse in stem cell biotechnology will actually set up functioning *tissue factories* discarding deranged and degenerate ageing cells and organs. *Nano technology* will offer bedside mini diagnostic devices replacing spread out laboratory floors housing space occupying machines. Electrostatic bio-waves will be captured on mobile phones displaying biomedical details, challenging and embarrassing our traditional medical personalities. Rheumatologists of tomorrow will be techno-savvy and I predict drastic changes in medical college curricula. Yet, traditional *'feel the pulse'* benign family doctor may well outweigh 'caught in between' superspecialist unable to cope with onslaught of biomedical technology.

Targeted and Tailored P4 Medicine

Now many hospitals in India offer genetic profiling in an attempt to predict later disease manifestations.

So far, our thrust is to diagnose diseases and treat as per their organ involvement. For example, angioplasty and coronary artery bypass surgery (CABS) for myocardial infarction. About 90% of our investment in medical research and treatment seems towards tackling fully manifested diseases rarely predicted well before they unfold (Fig. 1.6).

Evidenced Based Medicine (EBM) relies on data collection through a variety of studies localized or multicentric in specific, well defined diseases subsets of patients with elaborate criteria of inclusion and exclusion. Conclusions if any are then extrapolated into general recommendations presented commonly as 'guidelines', or algorithms. Such presumptions and premises are expected to be valid for any patient anywhere by the attending doctor who follows such guidelines to write standardized prescription, that

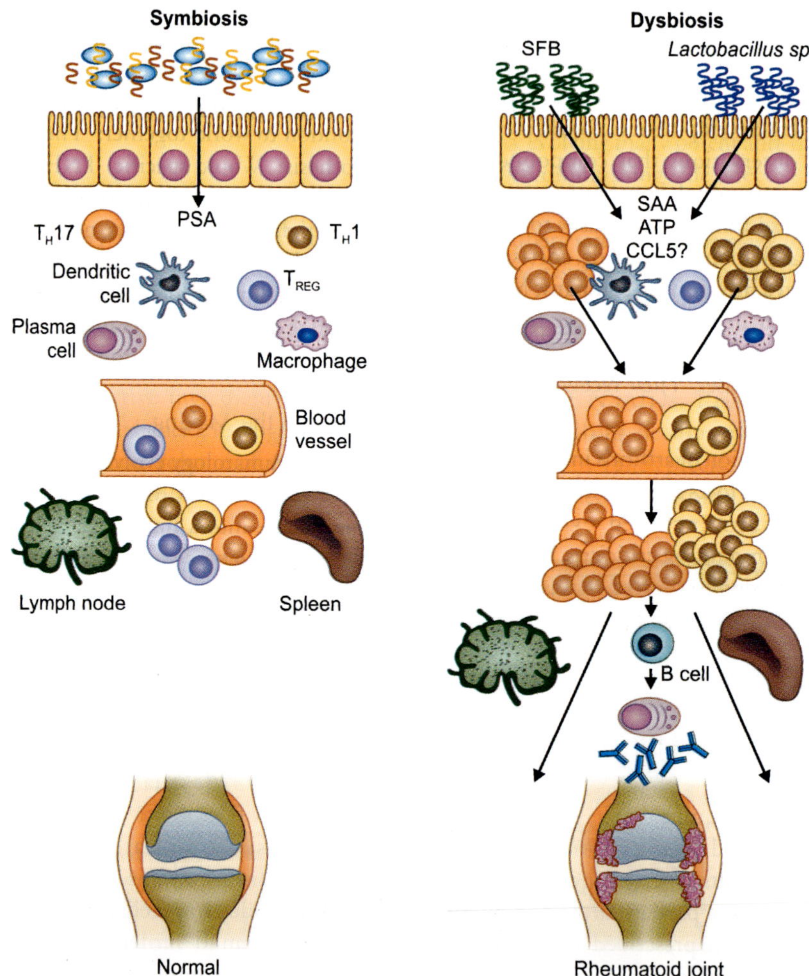

Fig. 1.6: Schematic representation of microbial symbiosis, dysbiosis, and rheumatoid arthritis (RA)

is how most of us are prone to practice medicine daily ... through generalized treatment regimes naturally with *highly variable treatment outcome.*

A distinguished immunologist Professor Leroy Hood, who won the prestigious Lasker Award in the US and many more the world over, propagates the concept of *P4 Medicine.*[12] He advocates that medicine should be *Predictive, Preventive, Participative and Personalized.* He further writes that new technologies available for studies in microbiome, genetics, metabolomics (metabolic end product of host microbe interaction) and meta-genomics can help predict diseases to come and install strategies to prevent, forecasts arrival of 'Wellness Clinics' with a thrust on preventive and a holistic approach to individuals with targeted and tailor-made treatments. He believes that this will result in not only high strike rates of treatment outcome but also heavy reduction in treatment budgets minimizing need of high-end, high cost procedures such as organ transplants, bypass or even joint replacement surgeries.

In the evolution of new galloping targeted therapy, we often forget and gloss over strange clinical symptoms and signs in changing environment whereby undiscovered diseases and syndromes wait to surface. A fine example is the vivid description of Antiphospholipid (Hughes) Syndrome. Rheumatology + Immunology = *Rheumanology* is the modern mantra. Let us be its disciples and advocates of the inevitable change from history to history in the making. Alvin Toffler's best-seller *Future Shock* is already upon us.[13] Are we ready? *Will we lag behind or ride the wave?*

REFERENCES

1. Pispati P. Confusion or clarity: Nomenclature, sematics, jargon, lingo, etymology and terminology. The Rheumatologist 2012; 6(10):10–13.
2. Janeway CA Jr, Travers P, Walport M, et al. Immunobiology: The Immune System in Health and Disease. 5th edition. New York: Garland Science; 2001. Figure 1.1, Edward Jenner. Available from: https://www.ncbi.nlm.nih.gov/books/NBK10779/figure/A36/
3. Kaufmann SH. Immunology's foundation: the 100-year anniversary of the Nobel Prize to Paul Ehrlich and Elie Metchnikoff. Nat Immunol. 2008; 9(7):705–12.
4. Pispati P. Seven wonders in the world of rheumatology: a short story of romance, reminiscences, and renaissance. Int J Rheum Dis 2008; 11:91–96.
5. Knight CJ. Antiplatelet treatment in stable coronary artery disease. Heart. 2003; 89(10):1273–8.
6. Milestones in rheumatology: Year Book 2001. New York, NY; the Parthenon Publishing Group: 1999.
7. Nanavati N. Rheumatoid arthritis. In: Pispati P, Editor. Manual of Rheumatology. Mumbai: Indian Rheumatology Association; 2002:72–82.
8. Weinblatt ME, Coblyn JS, Fox DA, Fraser PA, Holdsworth DE, Glass DN, et al. Efficacy of low dose methotrexate in rheumatoid arthritis. N Engl J Med 1985; 112:818–22.
9. Feldmann M, Maini RN. Lasker Clinical Medical Research Award. TNF defined as a therapeutic target for rheumatoid arthritis and other autoimmune diseases. Nat Med 2003; 9:1245–50.
10. Feldmann M, Maini RN. Perspectives From Masters in Rheumatology and Autoimmunity: Can We Get Closer to a Cure for Rheumatoid Arthritis? Arthritis Rheumatol 2015; 67:2283–91.
11. Sher JU, Abramson SB. The microbiome. A voyage to (our inner) Lilliput. The Rheumatologist 2011; 5(11):32–35.
12. Leroy H, Brogaard K, Price ND. A vision for 21st Century Healthcare. Bast Jr RC, Hong WK, Kufe DW, Hait WN, Pollock RE, Weichelsbaum RR, et al. In: Cancer Medicine. London, UK and Hoboken, USA: John Wiley & Co; 2017. pp 1–10.
13. Toffler A. Future Shock. New York: Random House. 1970.

Chapter 2

Simple Clinical Clues to Bedside Diagnosis

Prakash Pispati

It is an all-time truism that the British Medical Journal described Rheumatology as the last remaining bastion of clinical medicine. This primary lesson still holds water. Rheumatic diseases were described with substantial accuracy long before the advent of laboratory medicine and imaging technology. Of course we need these, in fact with newer innovations towards accurate and early diagnosis, with enhanced ratio of specificity and sensitivity. Therefore, mastering clinical approach to the arthritic patient with which any consultation begins is sacrosanct.

When patients approach us doctors with musculoskeletal complaints, then it is our turn to decide at once between 'soft tissue rheumatism' or 'arthritis'? Should it be arthritis, then let us ask ourselves these key questions:

- What is the presenting symptom?
- Is it primary or secondary to any other systemic disease?
- Which are cardinal, concomitant symptoms if any?
- Acute, subacute or of chronic onset?
- Monoarticular, pauciarticular or polyarticular?
- Symmetrical or Asymmetrical?
- Extra-articular features if any?
- Inflammatory, degenerative, metabolic, infective or post-infective?

Such key questions should help us proceed to subsequent algorithms[2], enabling the clinician to make a diagnosis with simplicity and clarity; add on minimum investigations, and we will probably hit the 'bull's eye' to initiate appropriate treatment: *therapeutic outcome in sight!*

As we drive our cars towards traffic signals don't we see green, amber or red lights? So also in the clinical approach to the rheumatic patients checking out green, amber, and red signals (Table 2.1).

Table 2.1: The green, amber, and red 'signals' to look for in a rheumatic patient

Green	*Amber*	*Red*
• Insidious onset • No joint involvement (e.g. soft tissue rheumatism) • One or few joints (<5) • Dull ache and pain, mainly localized • Normal ESR, negative CRP • No significant disability	• Chronic onset • Slow progression • Variable joint involvement • Regional pain, discomfort, moderate debility • Moderately elevated ESR, CRP	• Pyrexia • Loss of weight • Significant rash • Lymphadenopathy • Vital organ involvement • Unbearable pain and debility • Weakness and malaise • Vasculitic manifestations • Multisystem involvement • High ESR, CRP

Now let us check out the following seven masquerades, the devils in disguise:

- Depression
- Diabetes mellitus
- Drugs
- Anaemia
- Thyroid disease
- Urinary tract infection (UTI)
- Spinal dysfunction

All this is for doctors. Patients are certain to pose us multiple questions, relevant or irrelevant, so should we elicit history to purpose:

- What made you call up the doctor?
- Where exactly is the pain? Is this static or does it shift?
- What brought it on? Any specific event?
- Any injury, obvious or silent?
- Is this pain disturbing your sleep?
- How do you feel in the morning?
- As you move or exercise does pain increase? Are your movements difficult?
- Any fever? Any cough? Any skin rash?
- Do you feel tired, weak, fatigued, lethargic?
- Any sore throat?
- Any burning urine? What is its colour?
- Any loose motions?
- What is your marital/personal life?
- Any family history of arthritis?
- Have you suffered from episodes of fever preceding arthritis?
- What is your occupation, are you overworked? Is this aggravating your pain?
- Is your work disturbed by pain?
- Where do you live? Isolated or crowded locality?
- Do you travel frequently to any zones of endemic malaria, chikungunya or waterborne diseases?

To an experienced rheumatologist, posing such simplistic questions may seem naive, but there are two huge advantages. Firstly, it facilitates to zero on to the diagnosis, and not unimportantly, *enhance patient confidence* towards subsequent investigations and treatment that we may advise, eliciting *patient and family co-operation* and *therapeutic compliance* as well. The following two chapters on approach to monoarthritis and polyarthritis are certain to enhance your routine work for every patient.

Indeed, the clinical approach to a rheumatic patient is by far the first golden step towards effective arthritis management.

FURTHER READING

1. Davis J, Moder K, Hunder G. History and physical examination of the musculoskeletal system. In: Firestein GS, Budd RC, Gabriel S, McInnes IN, O'Dell JR, editors. Kelley's textbook of Rheumatology. Philadelphia, PA: Elsevier Saunders; 2009. pp. 515–31.
2. Pispati PK. Clinical approach to the rheumatic patient. In: Manual of Rheumatology, published by Indian Rheumatology Association (Second Edition, 2002). pp.1–13

By the way …

then … and now …

Chapter

3

Epidemiology of Rheumatic Diseases in India

Arvind Chopra

INTRODUCTION

Population and community-based epidemiological surveys are best suited to provide estimate of disease burden. Logistically and socio-economically the surveys are rather challenging. The sample size is statistically derived so as to capture the true extent of burden in the population. Though cumbersome, randomized methods are preferred for better representation. However, a well planned executed convenience sample survey may suffice. The success for a good survey depends upon the devotion and expertise of the investigators and honest and complete filing of data by the grass root paramedic field workers. Meticulous pre-survey planning is required to ensure survey completion in a systematic and timely fashion. While maintaining high scientific standards, the surveys must remain feasible and goal oriented.

Good epidemiologic data is hard to find in India for several reasons ranging from poor incentives to difficult logistics and socioeconomics. In our setting, surveys are often limited by shoe-string budgets. Also, a single survey is not enough to account for the intense diversity in India. Several surveys in time and place and using uniform standard protocols would be required to provide credible data for estimating national disease burden. However, the situation is considerably improved for musculoskeletal (MSK) and rheumatic disorders in India. In the last two decades or so, several surveys were carried out in different regions.[1–8]

WHO ILAR COPCORD SURVEYS

The maiden WHO ILAR (International League of Associations for Rheumatology) COPCORD (community oriented program for control of rheumatic diseases) India was launched in village Bhigwan (district Pune) in 1996.[4,5,7,9] COPCORD was conjointly launched by WHO and ILAR in 1981 to fill the void created by lack of MSK data from the developing countries.[10] Over 21 countries have completed COPCORD and in several instances with multiple surveys (Table 3.1). The prevalence of selected rheumatic and musculoskeletal disorders in the Asia-Pacific COPCORD surveys is shown in Table 3.1.[2,11-16] Detail descriptions of COPCORD are available on the COPCORD website.[2] The Bhigwan survey achieved a unique distinction in evolving a fast tract COPCORD model that was adopted by several countries and sites in India (Fig. 3.1).[2,7,17,18]

Two non-randomized surveys the Delhi study and the Pune-Bhigwan COPCORD have been frequently resourced to describe the Indian scenario.[3,4,7] The results of these surveys are not comparable because of significant methodology differences. The screening in the Delhi study targeted 'painful swollen joints' while in COPCORD the question was essentially 'pain in joints and/or MSK soft tissues (current or past)'. The Delhi survey by Malaviya and colleagues was primarily meant to collect data on systemic lupus erythematosus (SLE) and rheumatoid arthritis (RA). The survey covered nearly 40,000 semi-urban/rural population from the fringe villages of

Table 3.1: Point prevalence (%) of selected musculoskeletal (rheumatic) disorders from the WHO ILAR COPCORD surveys: global comparisons

Country	*Year*	*Sample size*	*Type*	*Design*	*MSK pain*	*RA*	*OA*	*OA knee*	*STR*	*AS*	*Gout*	*SpA*	*Low back pain*	*Knee pain*	*Remarks*
India Bhigwan	2001	4092	R	C,S	19.5	0.54	6.25	4.42	3.77	0.1	0.13	0.3	12.6	13.7	Adjusted for age-sex. India Census population 2001; except for RA, rest clinical diagnosis; Bhigwan IA-Un 0.76%; Pune IA-Un 0.32%
India Pune	2009	8145	U	C,S	11.5	0.19	4.01	3.41	1.2	0.07	0.06	0.27	4.6	6	
China Beijing	1987	4192	R	C,S	40.3	0.34*	NR	NR	NR	0.26	NR	0.26	35	29	Adjusted *1958 ARA criteria for RA, else clinical: SLE 0.07%
China Shanghai	2003	6584	U	R,S	13.3	0.28	NR	4.1	3.4	0.11	0.22	0.11	5.6	7	Adjusted to China national population
Lebanon	2012	3530	R	R,S	32.9	1	4	3	5.8	0.1	0.01	0.3	13.6	14.2	Adjusted to national population; ACR criteria when applicable: FM 1%
Egypt	2004	5120	R	C	16.2	0.29	8.5	NR	6.6	0.09	NR	0.15	4.9	9.1	Except RA, rest clinical diagnosis: FM 1.3%: CTD 0.15%
Indonesia	1992	4683	U+R	C	24	0.2	5.1	NR	15	NR	1.7	NR	15.1	12.2	Not standardized
Iran	2008	10291	U	R,S	41.9	0.33	16.6	15.3	4.6	0.12	0.13	0.23	21.7	25.2	Adjusted to national population; all diagnosis clinical including RA
Bangladesh Urban	2005	1259	U	C	27.9	0.2	NR	10.6	3.3	NR	NR	NR	18.4	15.8	Not standardized: clinical diagnosis; FM 4.4% rural, 3% urban
Bangladesh Rural	2005	2635	R	C	26.9	0.7	NR	7.5	2.6	NR	NR	NR	20.1	14	
Mexico	2011	19213	U	R,S	25.5	1.49	10.24	NR	3.8#	0.15	0.35	NP	5.8	NP	Adjusted to national population: ACR criteria when applicable; #Rheumatic regional pain syndrome: FM 0.68%: SLE 0.06%: Scl 0.02%

Abbreviations: COPCORD, community oriented program for control of rheumatic diseases; WHO, world health organization; ILAR, international league of associations for rheumatology; RA, rheumatoid; OA, osteoarthritis; STR, soft tissue rheumatism; AS, ankylosing spondylitis; SpA, spondyloarthritis; FM, fibromyalgia; SLE, systemic lupus erythematosus: Un, undifferentiated; IA, inflammatory arthritis: CTD, connective tissue disease; Scl, scleroderma; U, urban; R, rural; R, randomized sample; C, convenience sample; S, standardized/adjusted age-sex; NR, not reported/studied; *RA classified as per American College of Rheumatology (ACR) 1987 criteria unless shown otherwise (see remarks); See text for references (also available at www.copcord.org)

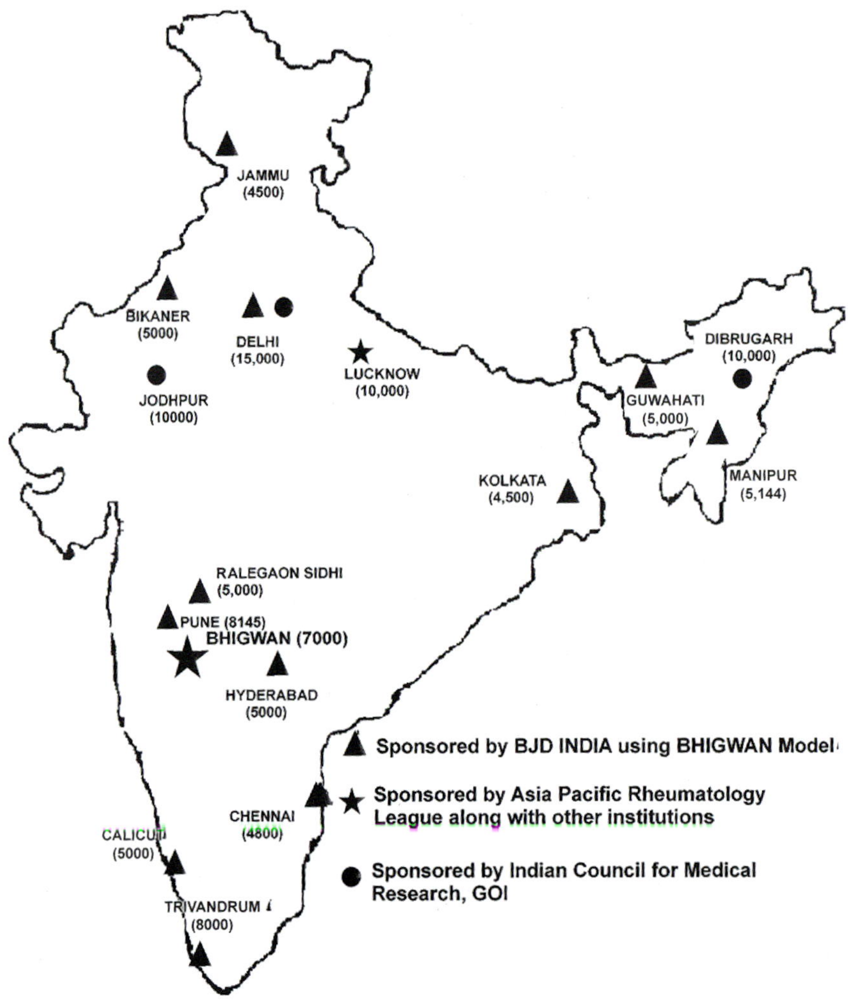

Fig. 3.1: COPCORD population survey sites in India

Delhi. For the study on prevalence of lupus the Delhi workers in addition included a larger number of blood samples from an on-going malaria program in Delhi.[19] Though point prevalence for several rheumatic disorders was published from the Delhi survey, important limitations were lack of appropriate screening methods and an unduly prolonged survey period (of three years).

In 1996 Chopra and colleagues completed the WHO-ILAR Bhigwan COPCORD rural survey (district Pune) in 5 weeks time. About 7000 villagers (children included) were screened. Overall the Bhigwan COPCORD study demonstrated for the first time that MSK rheumatic disorders are important health problems in the community. The community suffered mostly from soft tissue rheumatism, ill-defined MSK symptoms, degenerative arthritis, and about 10% individuals suffer from some form of inflammatory arthritis (including RA and seronegative spondyloarthritis).[5,7,9] The Bhigwan survey was continued as a prospective longitudinal observational follow up study for over 20 years till date. During the later period, the investigators provided free rheumatologic services and health education to the villagers and further carried out evaluation (including case-control nested cohort) to determine disease profile, therapy response, incidence cases and risk factors.[12,18,20–22] The Bhigwan data have also used by the World Health Organization (WHO) to project the burden of MSK illness in the South East Asia region.[23] The Bhigwan program is the longest running COPCORD in the world.[18]

Following the Bhigwan survey Das and colleagues completed a nearly 10,000 population COPCORD survey in Lucknow region and detail report is not yet published.[6,24]

Using the COPCORD Bhigwan model the Bone and Joint Decade India (BJD) and Indian Council of Medical Research (ICMR) launched population surveys at several sites during the period 2006–2011 to study about 90,000 population, both urban and rural.[8,25] Preliminary results were presented in the annual meeting of the American College of Rheumatology (ACR) 2012;[26] almost one-fifth of the Indian population suffered from some MSK pain and disorder and millions suffered from serious rheumatic diseases; about 5 million patients of RA was estimated.[26]

COPCORD was not designed to identify uncommon disorders like SLE which have been essentially described in the Indian scenario by case series. Traditionally medical camps which are popular in India have also contributed to MSK and rheumatic disease data although data collected from medical camps have many deficiencies from an epidemiological view point.[27,28]

Soft Tissue Rheumatism (STR)

STR encompasses a broad range of diverse conditions affecting different anatomical sites and tissues such as tendon, fascia, and bursae. The pathophysiological mechanisms of STR could be inflammatory or mechanical or traumatic and/or neurovascular. They are an extremely common cause of pain in the community, causing disability and absence from work. Data from the Bhigwan survey[7,9] show that 14% patients with non-specific arthralgias (NSA) and 11% STR cases are moderately to severe disabled as assessed by a health assessment questionnaire (HAQ) modified and validated for Indian use.[29]

Fibromyalgia is a condition with more widespread and diffuse musculoskeletal pain, well-defined tender points, and associated with undue fatigue and sleep disturbance. Several rural patients from the Bhigwan COPCORD with soft tissue and non-specific articular pains, in the absence of any obvious functional overlay also fulfilled the ACR criteria for fibromyalgia.[30] In the Lucknow Survey 3.8% and 4.5% of the urban and rural population respectively were found to suffer from fibromyalgia.[24]

Regional forms of STR (fascitis, bursitis, enthesitis, tendonitis, epicondylitis) are commonly encountered in clinical practice. Foot and heel pain due to local causes were common in rural surveys.[9] Painful tender swollen tendoachilles may be a forerunner of spondyloarthritis (SpA) and needs to be evaluated. Shoulder, knee, neck and back are common site for regional STR and NSA in COPCORD surveys and if not carefully evaluated may be misdiagnosed as inflammatory arthritis. Several cases of NSA and STR in COPCORD surveys were considered to be due to occupational overuse and trauma in a socioeconomically challenged community.

Osteoarthritis (OA)

Osteoarthritis is the second most common rheumatological problem in the community, and a leading cause of disability and poor quality of life beyond the fifth decade of life. OA commonly affects the knees and spine. Nodal and erosive forms of generalized OA are also seen. But it appears that OA hip is uncommon in the Indian population. In the Bhigwan survey almost 15–20% of population in the age range 60–80 years have knee OA (Table 3.2) and the pattern was similar in urban and rural communities.[23]

Rheumatoid Arthritis (RA)

The worldwide prevalence of clinical RA is about 1%.[12] In India the prevalence of RA varied considerably among several population surveys conducted by BJD India. Pooled age and sex

Table 3.2: Prevalence of knee osteoarthritis (OA) per 100,000 population in India (South East Asia region)[23]

Age bands	*15–24*	*25–34*	*35–44*	*45–59*	*60–69*	*70–79*	*80+*
Males	—	—	—	4644	15385	20000	6550
Females	—	—	2247	6587	14371	19608	14286

Diagnosis of knee OA is based on ACR clinical criteria.

Source: Chopra A, Bhigwan COPCORD study (India). Reproduced with permission from WHO-Publications.

adjusted (India census 2001) point prevalence for RA was reported to be 0.3% (95% confidence interval 0.08, 0.79) and this figure is likely to be a robust estimate for India.[26]

Contrastingly Malaviya and colleagues in 1994 from Delhi reported a much higher prevalence of RA (0.75%) and the patients diagnosed fulfilled the ACR 1987criteria.[3] In 1997 an urban population study from West Bengal using an index case linked cluster survey model reported a crude unadjusted clinical prevalence of 0.3%.[31] From the COPCORD Bhigwan project in 1996 we reported a clinical prevalence of 0.7% but on applying the 1987 ACR criteria and standardization (age, sex adjustment) the rate reduced to 0.55%.[5,7] The neighbouring urban COPCORD Pune (2004) reported an adjusted prevalence (ACR) of 0.3%.[7] The prevalence of RA in women aged 30–44 years in COPCORD Bhigwan was 1639/100,000 (Table 3.3), which was several times higher than that reported from other countries (300 in Norway and USA, 237 in China, 275 in Indonesia, and 1042 in West Africa; all rates are per 100,000 population in a recent WHO Technical Research Series.[23]

After a ten year follow-up in village Bhigwan (assuming geometric growth pattern; adult population about 4545), the period prevalence and incidence of RA was 1.17% and 0.044% (44/100,000) respectively.[12]

The natural history of RA in Indian patients is generally similar to Caucasians with a few differences.[32–34] In India patients with RA have (i) an earlier onset in younger women,[23] (ii) fewer extra-articular features, including subcutaneous nodules (less than 10%), (iii) and lower seropositivity for rheumatoid factor (RF) at 55–80%.

It is the author's contention that Indian RA patients suffer disproportionately from early severe crippling affection of the knee joints which is often made worse by the Indian habit of squatting and sitting cross legged on the floor.[35] The rheumatoid knee (and RA) is further worsened by the lack of basic toilet facilities in several Indian homes. There seems to be a difference in the MSK pain and disorders between the urban and rural settings.[7,35] Overall RA is likely to be more severe and functionally disabling in a rural setting.

In Delhi study 82% RA patients were positive for RF[3] which was surprisingly similar to that reported from hospital-based Indian series.[32,34] Seropositive RF is generally considered to be much less in community RA. Amongst clinical RA patients, 45% patients in rural COPCORD Bhigwan, 62% patients in urban COPCORD Pune and 57% patients in urban community referral practice (CRD, Pune) were seropositive for RF.[7,35] While 100% RA patients in COPCORD Pune and 60% RA patients in COPCORD Bhigwan were seropositive for anti-CCP (anti-cyclic citrullinated peptide) antibodies by second generation ELISA.[7,35] Also 54% of the clinical RA patients in COPCORD Bhigwan demonstrated radiological erosions.[35]

HLADRB1* alleles are considered to be an important factor genetic factor RA and may determine both the susceptibility and severity of disease.[36] The association of HLA DRB1* alleles (in particular *0401, *0404) or the shared epitope with RA is well established in reports from several parts of the world[36,37] and this has also been confirmed in hospital-based studies in India.[32,38] However, the rural COPCORD Bhigwan reported a modest association of HLA DRB1*1001 (odds ratio 1.85) with RA;[20] none was seen with *0404 and *0401. Subsequently, in a better designed larger sample study (both survey and incident cases) suffering from RA and undifferentiated inflammatory arthritis (IA) identified in the COPCORD Pune Bhigwan region(rural and urban),[39,40] a significant association of RA was shown for the shared epitope (including only HLADRB1*1001); other alleles significantly associated were HLA DQA1*0103, DQB1*0303, DPB1*0201 and DP

Table 3.3: Prevalence of rheumatoid arthritis (RA) per 100,000 population in India (South East Asia region)[23]

Age bands (years)	*15–29*	*30–44*	*45–59*	*60–74*	*75+*
Males	—	479	—	1136	—
Females	113	1639	1775	1914	3846

Diagnosis of RA based on ACR criteria.

Source: Chopra A, Bhigwan COPCORD study (India). Reproduced with permission from WHO publications

B1*0401. None of the haplotypes analysed showed significant association with RA in the later study.[40]

Tobacco use and smoking in particular is significantly associated with RA and adds to the risk of premature atherosclerosis and coronary artery disease.[41,42] COPCORD Bhigwan rural survey showed a rampant use of oral tobacco (called 'misri' in regional Marathi language) and showed a significant association with rheumatic MSK pain (odds ratio 1.71, 95% CI 1.44, 2.04). This was further supported by findings from urban COPCORD Pune[7] (odds ratio 3.16, 95% CI 2.88, 4.05) and rural COPCORD Ralegaon.[43] However, a significant association between tobacco use and RA could not be demonstrated in the Pune region study[43] because of a relatively smaller number of RA patients. Being an extremely important public health preventive measure, tobacco use in all its forms must be discouraged and stopped in patients suffering from RA.[35]

The use of steroids (and analgesics—NSAID) is rampant and often unsupervised. Often these drugs are sold mixed with herbal formulations under the false premise of Ayurvedic drugs by quacks. Patients often do not reveal their medications. All this makes the recognition of early RA and inflammatory arthritis difficult as the signs of inflammatory arthritis get masked, and this was the experience of the author in several COPCORD community surveys.[12,18]

RA is a recognized cause of premature atherosclerosis and coronary artery disease with its associated morbidity and mortality.[42] The author has reported mortality in RA in a crosssectional analysis of a case series of 60 RA patient who died. Infections (23%) and cardiac disorders (40%) were the leading causes of death;[35] other causes were respiratory (10%), hepatic (9%), stroke (8%), and miscellaneous (10%). At least 30% deaths could be due ischemic heart disease/myocardial infarction. Multi-organ failure and haemodynamic shock was the terminal event in patients dying of septicemia. Microbial culture reports of staphylococci and gram negative bacilli were available in 11 cases of septicemia (4 pneumonia, 1 total knee replacement, 1 dental surgery root canal, 1 pancreatitis, 1 fracture neck of femur with bed sores, 3 peripheral limb gangrene). Methotrexate associated severe hepatic dysfunction and pancytopenia was recorded in 3 cases of septicemia; none were diabetic or immunosuppressed due to any other cause. Disseminated intravascular coagulation complicated the terminal event in one patient of septicemia. Other clinically important contributory factors recorded in this cohort were: 5 diabetes, 5 hypertension, 1 cerebrovascular disease, 1 coronary artery bypass surgery, 2 gastrointestinal bleed, 2 peptic ulcer (probably non-steroidal anti-inflammatory drugs related), and 1 chronic renal failure. Though details of medication in the 3-month period prior to death could not be confirmed in the large majority, at least half of the cases were on long-term methotrexate and steroids.

Spondyloarthritis (SpA)

The commonest form of SpA reported in Indian patients is the undifferentiated type characterized by a dominant asymmetric lower limb oligoarthritis and often accompanied with enthesitis.[43–45] The latter are probably forme-fruste of the classical Reiter's syndrome and are believed to be triggered by a wide array of gut and genitourinary related infections.[45] The prevalence of ankylosing spondylitis (AS) in the Caucasian population is 0.1%. A large hospital-based population study from southern India reported a AS prevalence of 0.6%, with 87% patients being positive for HLAB27.[46] In a similar hospital-based study of AS from North India HLA B27 was positive in 94% patients.[47] The profile of AS and other SpA was similar to that seen in the Caucasians and association with HLA B27 varied in different SpA subsets at different investigation sites (Table 3.4).[48]

Psoriatic arthritis (PsA) is not too uncommon in the hospital setting. The Pune-Bhigwan and the Delhi survey did not report a single case (see above). During the ten year follow-up of about 5000 adult population, the author has diagnosed two cases (both males) of PsA in Bhigwan (unpublished). However, a retrospective study from southern India reported a hospital prevalence of 0.83% for PsA amongst the 16,293 rheumatology cases from 1981–94.[49] The pattern of arthritis in PsA reported was asymmetrical oligoarthritis in 63%, symmetrical polyarthritis in 32%, dominant distal interphalangeal (DIP) joint arthritis in 25%, and predominant axial SpA in 15%. Among these patients, the skin lesions preceded arthritis in over

Table 3.4: Frequency and relative risk of B27 locus antigen in Indian SSA patients[48]

Subset	*Present study*		*Delhi*		*Chennai*		*Caucasians*	
	f	*RR*	*f*	*RR*	*f*	*RR*	*f*	*RR*
AS	73	26	92	193	83	1091	90–100	90–118
RS	67	20	80	66	57	307	65–100	17–40
SpA-U	41	7	84	83	21.4	70	60–70	18
Healthy	9	—	6	—	0–2.5	—	9+	—

Abbreviations: f, frequency percent; R, relative risk; AS, ankylosing spondylitis; RS, Reiter's syndrome; SpA-U, Undifferented Spondyloarthritis.

60% patients, the skin and joints were simultaneously affected in 15% patients, and arthritis preceded the skin lesion in 19% patients, respectively.

Connective Tissue Disorders (CTD)

The extent of CTD in the Indian population is still unclear though many believe it to be uncommon. Twenty-one cases of CTD were diagnosed amongst 68,273 in patients (civilians and services) during 1982–1986 in a large Armed Forces referral-teaching hospital in Pune which gave an incidence of 0.03% in hospitalized patients.[50] Numerous hospital-based series of SLE,[40,41] scleroderma, mixed connective tissue disease, vasculitis are published and the descriptions are generally similar to the western series. The Delhi study reported a point prevalence of 3 per 100,000 for SLE.[19] This was much lower than that reported from the west varying from 12.5 per 100,000 adults in England to 124/100,000 in USA. The Pune–Bhigwan COPCORD follow-up study found a crude incidence rate for SLE of 1 per 25,000 person years, i.e. 4 per 100,000 populations per year. A large amount of referral non-standardized rheumatology/hospital based data on SLE collected from different regions has been recently published in the IRA (Indian Rheumatology Association) guidelines on SLE.[51,52] Neuropsychiatric SLE was reported in 63% and 32% patients of SLE from north and south India respectively.[52,53] In another hospital based in patient rheumatology emergency case series, 65% of the SLE patients were admitted with a neurological complication.[54]

Vasculitis

All forms of vasculitides have been described but the most frequent form is systemic necrotizing vasculitis.[55–60] In a retrospective study of 7700 patients attending a rheumatology referral out patient during the period 1979–1994 in a large referral hospital in Mumbai, 100 patients (1.3%) were found to suffer from vasculitis.[57] The majority was small vessel vasculitis (Henoch-Schönlein purpura being the commonest), followed by medium vessel vasculitis (polyarteritis nodosa, SLE) and large vessel disease. Interestingly not a single case of Wegener's granulomatosus (WG) was reported. This study also reported vasculitis in 0.9% of RA cases which was almost similar to that reported from hospital studies in Delhi and Chennai.[32,56] WG is reported from northern India.[59,60] The diagnosis of WG in these series was facilitated by ANCA and bronchoscopic biopsy of pulmonary tissue. Several cases treated for pulmonary TB were found to be actually WG.[59,60] Polyarteritis nodosa and Churg-Strauss syndrome have been reported.[58] Recently, vasculitis leading to avascular necrosis in the joints have been reported.[61]

Takayasu's arteritis is a rare disease, but unlike elsewhere in the world it is an important cause of renovascular hypertension in India.[62,63] From retrospective case series of 8600 rheumatology patients seen during the period 1992–96 in a large referral hospital in south India, five patients were diagnosed as Takayasu's arteritis.[62]

Kumar and colleagues described the clinical and immunogenetic features of 50 patients with Behçet's syndrome attending rheumatology outpatient clinic in AIIMS, Delhi during the period between 1978 and 2002.[64] The disease, at least in its complete form of oculomucocutaneous ulcers and neuro-vascular complications, is believed to be uncommon in India though HLA B5, a marker of Behçet's disease, is reported in almost 30% Indians.

Sjögren's Syndrome (SS)

In a retrospective analysis of 24,500 rheumatology cases seen over 20 years, only 3 patients were found to have primary SS;[65] 9 patients (RA in 8, SLE in 1) met the criteria for secondary SS. Twenty-six patients (21 women) of primary SS were reported over a 10-year period from a large tertiary rheumatology referral center in Lucknow.[66] The authors concluded that SS was an uncommon disease; presenting symptoms included dry eyes, dry mouth, and arthritis/arthralgias and extra-glandular manifestations were glomerulonephritis, vasculitis, renal tubular acidosis and peripheral neuropathy. The important laboratory abnormalities in the Lucknow cohort were hypergammaglobulinaemia, antinuclear antibodies, anti-La and anti-Ro, and in 60% cases a definitive diagnosis was provided by a minor salivary gland biopsy.

More recently in 2015 Danda and coworkers reported the clinical characteristics of 323 patients with confirmed primary SS.[67] They found that their cohort had younger age of onset, higher female to male ratio, paucity of cryoglobulinemia, Raynaud's phenomenon and hyperglobulinemia as compared to western series. Only 8% had sicca symptoms at presentation. In the author's opinion, secondary sicca in rheumatology outpatients is not uncommon, but is missed in population studies because of lack of awareness and poor facilities in field conditions for detail eye and oral examination and definitive investigation tests.

Hyperuricemia and Gout

It is generally stated that gout is probably less common in the Indians as compared to the Caucasians (Tables 3.1). In the ongoing Pune-Bhigwan COPCORD rural population (about 5000 adults) study, the author observed classical acute gout episodes in only 4 patients during the 10-year follow-up period (unpublished). In a retrospective study, 101 cases (95 males) of acute gout were diagnosed out of 20,000 patients attending rheumatology outpatients in a Government hospital in Chennai during the period 1989–2001.[68] Similar series were reported from Delhi and Kolkata.[69,70]

Overall the pattern of gout in Indian patients was similar to that seen in the Caucasians with perhaps a lesser prevalence of classical first metatarsophalangeal joint arthritis varying from 35 to 75%; maximum reported from Chennai and least reported from Kolkata versus about 90% in Caucasians. The reported prevalence of tophi in Indian patients ranged from 5 to 27%; maximum reported from Delhi and least reported from Kolkata versus about 35% in Caucasians. Almost 20% patients in the Chennai study had poly-articular onset; 20% patients in the Delhi study had renal calculi as compared to about 10% found amongst the Caucasians. All forms of gout, ranging from the classical monoarticular podagra to the polyarticular tophaceous gout were described. Interestingly, a large number of patients in the Delhi study had auricular tophi, which is uncommon in the experience of several Indian workers including the current author (unpublished). The majority of Indian patients, as reported by the above-mentioned studies, were under-excretors of uric acid. In the author's observation, asymptomatic hyperuricemia is over diagnosed and unnecessarily treated. We do see patients of gout who are strict vegetarians and have never consumed alcohol, and further, hail from a family lineage of vegetarians.

Infections

Chandrasekaran reported an extensive experience regarding infections in patients attending a rheumatology clinic in a Government hospital in Chennai during the period 1976–1992.[71] Their observations of infection related arthritis were septic arthritis usually due to *staphylococcus aureus* seen in 9 out of 24,500 patients. In almost 1.7% patients with post-diarrhea asymmetrical polyarthritis and lumbago, *Salmonella typhi* (not reported in the Western literature) and *Salmonella typhimurium* were ascertained to be the causative microbes. *Shigella shigae* was responsible for at least 14% of the cases with acute polyarthritis following diarrhea. Almost 1.9% patients with polyarthritis had significant Brucella titer; fever, hepatosplenomegaly and lymphadenopathy was reported in all. Almost a quarter of patients with monoarticular affection of the knee had tubercular etiology. Finally Poncet's disease, a form of acute painful arthritis and considered to be a hypersensitivity phenomenon to tubercular infection elsewhere, was diagnosed in 0.23% patients.

Several reports on infection and arthritis were recently published in a special supplement issue of the Indian Journal of Rheumatology.[72] In a retrospective analysis of 1,900 in patients over 2 years in a medical college in Chennai, 0.2% had septic arthritis, 0.68% had tuberculosis, 0.95% had reactive arthritis and 0.57% had post-viral arthritis.[73] In a report of infections among in-patient rheumatology patients from Hyderabad, almost one-fifth of all admissions (1,236 cases) in a rheumatology ward had some infection;[74] 84 had a diagnosis of SLE (*E. coli* was the commonest isolate); 72 suffered from RA and about one-third of patients had lower respiratory tract infection. A quarter of rheumatology patients in the Hyderabad study were complicated by tuberculosis infection.

In a report from Lucknow common infections in rheumatology practice were septic arthritis, tropical pyomyositis, tuberculosis, and leprosy associated arthritis;[75] MSK infections pose challenging diagnosis and community acquired methicillin resistant *Staphylococcus aureus* (with activated panton—valentine leucocidin gene) seems to be an important emerging pathogen.[76]

Tuberculosis (TB) arthritis: Usually less than one-tenth of all patients of TB in developing countries develop osteoarticular TB. Case series of TB arthritis and rheumatism were reported from Pune[77] and Kolkata[78] which included several atypical forms of articular and spinal TB that were identified early due to availability of MRI and newer myco-bacterium detection methods. In a retrospective cross sectional analysis of 23,651 patient records in central database in Pune from 2005 to 2009, TB associated arthritis was found in 0.35% cases; 0.4% cases of RA were complicated by concurrent TB. In this report 12 of the 16 cases with tubercular MSK infection were considered to have Poncet's disease, a hypersensitivity form of TB; 2 cases showed multifocal lesions which also included non-contiguous spine involvement.[77]

Leprosy: Leprous arthritis, usually a component of the lepra reaction, and mimicking seronegative RA or asymmetrical oligoarthritis, is sometimes encountered in rheumatology practice and needs a high index of clinical suspicion.[79] Several Indian workers have reported leprosy associated rheumatic complications.[80,81] Uncommonly, leprosy can present with a fairly acute onset and progressive rheumatic phenotype that can be mistaken for RA or SSA;[82] RF and other autoantibodies may be falsely positive. Uncommonly, leprosy may complicate patients with RA on long-term immunosuppressive therapy or be a manifestation of an immune reconstitution syndrome.

HIV infection: HIV is reported to present a varied spectrum of rheumatic syndromes – from reactive arthritis to polymyositis like CTD profile. A hospital-based case series from Mumbai reported that while 4% of HIV cases suffered from some form of rheumatism (commonly bone pains and arthralgias), almost 5% of rheumatology patients tested positive for HIV;[83] 46% of the 102 consecutive patients of HIV seen in an Armed Forces referral hospital had some form of rheumatic complaints, commonest being nonspecific myalgia and arthralgia.[84]

Chikungunya: The re-emergence of chikungunya virus (CHIKV) epidemic in 2006 in central and southern India was unprecedented.[85] Though several retrospective reports[86,87] emerged from hospital and/or community based studies, some rheumatology driven studies evaluated the long term MSK and rheumatism sequel.[88–93]

Following the acute illness several thousand patients suffered from chronic rheumatic MSK pain and rheumatism. Though the overall clinical phenotype was not different from classical textbook descriptions, there were several differences especially regarding complications and the magnitude of long-term sequel. This acute febrile arboviral illness, characterized by severe, excruciating musculo-skeletal aches and pains, and arthralgias, was overall benign and self-limiting as observed in a rural population study (Bavi, Sholapur) with almost two-third cases resolved completely within three weeks.[92] Chikungunya ought to be differentiated from dengue (also an arbovirus), which generally confines to a severe back ache, and may cause life threatening thrombocytopenia. The author described a wide spectrum of rheumatology symptoms occurring in arthritis-naïve patients following an acute CHIKV infection in a busy rheumatology outpatient setting.[88] The clinical manifestation ranged from a typical RA-like illness to a psoriatic illness, undifferentiated SpA, and speculated that no other virus had shown such a propensity for rheumatic disorders. Intriguingly, the frequency of RF and anti-CCP in patients with otherwise typical RA profile was low and the patients showed good therapeutic response to short-term steroids and supervised disease modifying antirheumatic drugs.[88] The Bavi[92] and Modnimb[89] (Sholapur) community studies also demonstrated a prolonged elevation of IL-6, anti-CHIKV IgM and IgG antibodies despite somewhat benign rheumatic sequel; only 4% cases had persistent MSK pains and arthralgias at 2 years and none suffered from significant persistent arthritis.

An intriguingly questionable report of 'erosive arthritis' following acute CHIKV infection ought to be validated by rheumatologists.[94] What is CHIKV arthritis? Is it a distinct entity or just a known phenotypic rheumatic disease precipitated by the CHIKV? Several Indian rheumatologists and the author believe that some of the CHIKV chronic arthritis may be unique in that it is RA-like but with absent RF and anti-CCP and a dominant periarticular soft tissue inflammation (prayer hand sign). Does the virus persist in the MSK tissues? The behaviour of the virus may be akin to that of its close cousin, the Ross River virus which is well known to cause chronic MSK sequel.[92] However, the virus seems to cause an autoimmune arthritis like clinical picture but the author did not observe an increase frequency of RF or other autoantibodies in community studies.[88,92] Will the Indian epidemic in 2006 lead to an increased incidence of rheumatic MSK disorders in years to come? Six years after the re-emergence, unlike the previous epidemics, the CHIKV continues to transmit and infect in several parts of India. There are several other epidemiological questions about the chronic CHIKV MSK disorders that need to be addressed. However, post-CHIKV arthritis and rheumatism continues to overload rheumatology outpatients and provide an excellent opportunity to probe the role of arboviruses in chronic inflammatory arthritis and other connective tissue syndromes.

Several other infections (bacterial, viral, fungal, parasitic) are also known to cause arthritis and sometimes are transient (e.g. respiratory viruses, chickenpox, measles, rubella, hepatitis).[71] Hepatitis C is known to cause several rheumatological syndromes but its extent in India is not yet reported but likely to be uncommon. Brucellosis, affecting spine and large joints, is reported.[95]

Childhood Arthritis and Juvenile Idiopathic Arthritis (JIA)

JIA is rather uncommon as compared to RA. The COPCORD Bhigwan survey (which included children as well) reported a point prevalence of 0.26% for JIA (an unusually high figure), and 0.09% rheumatic fever.[4]

Pediatric rheumatology is emerging as an important clinical service in India. There are several hospital/clinic outpatient based case series of JIA from Lucknow, Delhi, and Pune (Table 3.5).[96–98] The most common forms of JIA in these series was the polyarthritis and enthesitis related arthritis. A recent update review with a large clinical experience from Delhi and Mumbai also endorsed the latter observation.[99,100]

Table 3.5: Frequency distribution of juvenile idiopathic arthritis (JIA)—outpatients/clinic-based series

JIA subsets	*Lucknow* (n = 89)*	*Delhi* (n = 361)*	*Pune** (n = 272)*		
			JIA subset	*RF positivity (%)*	*ANA positivity (%)*
Systemic onset	14	24	9	0	27
Oligoarthritis	34	30	26	0	19
Persistent	—	—	20	0	19
Extended	—	—	6	0	0
Polyarthritis	52	46	31	29	26
Seronegative	—	—	22	0	16
Seropositive	—	—	9	100	10
Psoriatic arthritis	—	—	1	0	0
Enthesitis	—	—	27	0	7
Undifferentiated arthritis	—	—	6	88	41

*JRA subsets; **ILAR-JIA subsets. *Source*: References no. 96, 97, 98.

Khubchandani *et al* reported their experience with 1365 children examined in a pediatric rheumatology outpatient in corporate hospital in Mumbai between 2003 and 2011; 32% cases were JIA, 11% were CTD, 10% were vasculitis, 5% were pain syndromes and 1% was auto-inflammatory; several children also suffered from mechanical problems.[100]

Rheumatic fever and its cardiac sequel continue to be an important communicable and preventable disorder among children and adolescents in India.[101] Kawasaki's disease is also being increasingly recognized.[102] A case series of macrophage activation syndrome in patients with JIA (systemic onset) was recently reported.[103]

Miscellaneous

An unusual familial endemic form of crippling arthritis, usually affecting the lower spine and hips, in young adults called Handi-Godu disease was reported from Malnad area of Karnataka.[104] The overall picture suggested a primary spondyloepiphyseal dysplasia presenting as a degenerative disorder. Endemic fluorosis in central parts of India continues to produce disabling musculoskeletal axial and dental complications and can in early stages present as non-specific rheumatic syndrome.[105] Thorn prick arthritis which is non-infective, inflammatory, and usually in fingers has been observed by the author during the rural COPCORD Bhigwan. Kumar and coworkers found significantly low vitamin D levels among patients with non-specific MSK pains and arthralgias and who improved symptomatically with vitamin D supplements.[106] Vitamin D is being increasingly reported to be low in the Indian community and its MSK profile needs to be evaluated validated in controlled community based studies.

Conclusion

Indian population suffers from a large burden and wide variety of rheumatic MSK disorders as clearly shown in COPCORD population surveys. We still do not have enough data on their socioeconomic impact which by all logic and standards is bound to be immense. MSK disorders are bound to adversely impact national productivity and health. This paves way for the national health policy makers to recognize the disease burden and rheumatology as an important superspecialty. The time is ripe to include rheumatic disorders in public health programs, as the control of rheumatic diseases is likely to raise the standard of living.[107]

REFERENCES

1. Rheumatologyindia.org [Internet]. Pune: Center for Rheumatic Diseases (CRD); [cited 2017 Aug 2017]. Available from http://www.rheumatologyindia.org/
2. copcord.org [Internet]. Pune: COPCORD; [cited Aug 30 2017]. Available from http://copcord.org/
3. Malaviya AN, Singh RR, Kapoor SK, Sharma A, Kumar A, Singh YN. Prevalence of Rheumatic diseases in India: Results of a population survey. J Indian Rheumatol Assoc. 1994; 2:13–17.
4. Chopra A, Patil J, Billampelly V, Relwani J, Tandle HS. The Bhigwan (India) COPCORD: methodology and first information report. APLAR J Rheumatol 1997; 1:145–54.
5. Chopra A, Patil J, Billampelly V, Relwani J, Tandale HS. Prevalence of Rheumatic diseases in a rural population in Western India: A WHO-ILAR COPCORD Study. J Assoc Physicians India 2001; 49:240–46.
6. Das SK, Kumar P, Srivastava R, Srivastava S, Alok R, Bhattacharya D, et al. Distribution of various rheumatological diseases in rural and urban population of Lucknow. [O17 abstract] Indian J Rheumatol 2005; 13 (Suppl).
7. Joshi VL, Chopra Arvind. Is there an urban rural divide? Population surveys of rheumatic musculoskeletal disorders in the Pune region of India using the COPCORD Bhigwan model. J Rheumatol 2009; 36:614–22.
8. Chopra A, Ghorpade R, Sarmukkadam S, Joshi VL, Mathews A, Gaur L, et al. A Staggering Burden of Pain and Rheumatic Disorders in India: A National Bone and Joint Decade India Community Oriented Program for Control of Rheumatic Disease Survey 2006–2011. [abstract] Arthritis Rheum 2012; 64 Suppl 10:S23.
9. Chopra A, Saluja M, Patil J, Tandle H. Pain and disability, perceptions and beliefs of a rural Indian population: A WHO-ILAR COPCORD study. J Rheumatol 2002; 29:614–21.
10. Muirden KD. The origins, evolution and future of COPCORD. APLAR J Rheumatol 1997; 6:44–8.
11. Wigley RD. Rheumatic problems in the Asia pacific region. In: Wigley RD, editor. The primary prevention of rheumatic diseases. New York: The Parthenon Publishing Group Inc; 1994. pp. 21–25.

12. Chopra A, Abdel-Nasser A. Epidemiology of rheumatic musculoskeletal disorders in the developing world. Best Pract Res Clin Rheumatol 2008; 22:583–604.

13. Haq SA, Rasker JJ, Darmawan J, Chopra A. WHO ILAR-COPCORD in the Asia-Pacific: the past, present and future. Int J Rheum Dis 2008; 11:4–10.

14. Davatchi F, Jamshidi A, Banihashemi AT, Gholami J, Forouzanfar MH, Akhlaghi M, et al. WHO-ILAR COPCORD Study (Stage 1, Urban Study) in Iran. J Rheumatol 2008; 35:1384.

15. Chopra A. The WHO ILAR COPCORD Latin America: consistent with the world and setting a new perspective. J Clin Rheumatol 2012; 18:167–69.

16. Chopra A. The COPCORD world of musculoskeletal pain and arthritis. Rheumatology (Oxford) 2013; 52: 1925–8.

17. Chopra A. The WHO ILAR COPCORD Bhigwan (India) model, Foundation for a future COPCORD design and data repository. Clin Rheumatol 2006; 25:443–7.

18. Chopra A. Community rheumatology in India. COPCORD driven perspective. Indian J Rheumatol 2009; 4:119–26.

19. Malaviya AN, Singh RR, Singh Y, Kapoor SK, Kumar A. Prevalence of systemic lupus erythematosus in India. Lupus 1993; 2:115–18.

20. Chopra A, Poulton K, Silman A, Thomson W. HLA DRB1 associations in a community based study of inflammatory arthritis in India. [abstract] Arthritis Rheum 2000; 43:9:S71.

21. Saluja M, Chopra A. Bare Foot Applications of HAQ, and Identification of Some Risk Factors in WHO-ILAR COPCORD Bhigwan (India) Stages II and III: an Ongoing Longitudinal Population Based Study 1996–2004. In: Nilganuwong Surasak, editor. Proceedings of the 10th Asia Pacific League of Associations for Rheumatology Congress; 2002; Bangkok, Thailand. Bangkok: Supjaroon Printing Co Ltd; 2002. p. 335–40.

22. Chopra A, Chaturvedi S, Saluja M, Joshi VL, Sarmukadam S. Oral tobacco use is significantly associated with rheumatic musculoskeletal (RMSK) pain: result of a WHO ILAR COPCORD (Community Oriented Program for Control of Rheumatic Diseases) rural population study in India. [abstract] Arthritis Rheum. 2010; 62 Suppl 10: 1551.

23. Report of a WHO scientific group. The burden of musculoskeletal conditions at the start of the new millennium. Geneva (Switzerland): World Health Organization 2003. WHO Technical Report Series 919.

24. Das SK, Srivastava S, Kumar P, Srivastava R, Bhattacharya D, Agarwal S, et al. High prevalence of fibromyalgia in both rural and urban areas of Lucknow. [Abstract P57] Indian J Rheumatol 2006; 1(3):166.

25. bjdindia.org [Internet]. Pune: Bone and Joint Decade India; [cited 2017 Aug 30]. Available from www.bjdindia.org

26. Chopra A, Ghorpade R, Sarmukkadam S, Joshi VL, Mathews A, Gaur L et al. 5 Million Patients and Not 0.34% Is Worrisome: Burden of Rheumatoid Arthritis in India Based On a Bone and Joint Decade India Community Oriented Program for Control of Rheumatic Disease. Arthritis Rheum 2012; 64 (10) (suppl): S 23.

27. Chopra A. Rheumatology: Made in India (Camps, COPCORD, HLA, Ayurveda, HAQ, WOMAC and Drug Trials). J Indian Rheumatol Assoc 2004; 12:43–53.

28. Chopra A, Pispati P, Sancheti KH, et al. Medical camps in 'arthritis and rheumatism': a community oriented prospective study, with emphasis on diagnosis and education. J Indian RheumatolAssoc 1991; 5:31–37.

29. Chopra A, Saluja M. Validation and usefulness of Indian version (CRD Pune) Health Assessment Questionnaire: Drug Trials, Community practice and COPCORD Bhigwan population study (1994–2012). Indian J Rheumatol 2012; 7(2):74–82.

30. Wolfe F, Smythe HA, Yunus MB, Benett RM, Bombardier C, Goldenberg et al: The American college of Rheumatology 1990 criteria for the classification of fibromyalgia. Report of the multi-center criteria committee. Arthritis Rheum 1990; 33:1602–72.

31. Ganguly KS, Roy M. An epidemiological study of rheumatoid arthritis in an urban population in a district of West Bengal. J Indian Rheumatol Assoc 1997; 5:91–100.

32. Malaviya AN, Mehra NK, Dasgupta B, Khan MK, Rao UR, Agarwal A, et al. Clinical and immuno-genetic studies in rheumatoid arthritis in India. Rheumatol Int 1983; 3:105–08.

33. Chopra A, Raghunath D, Singh A, Subramanian AR. Pattern of rheumatoid arthritis in the Indian Population: A Prospective study. Br J Rheumatol. 1988; 27(6):454–56.

34. Chandrasekaran AN, Radhakrishna B, Porkodi R, Madhavan R, Parthiban M. Spectrum of clinical and immunological features of systemic rheumatic disorders in a referral hospital in South India: Rheumatoid Arthritis. J Indian Rheumatol Assoc 1994; 2(1):18–26.

35. Chopra A. Epidemiology of rheumatoid arthritis. In: Mukherjee S, Ghosh A, editors. Monograph on Rheumatoid Arthritis. Indian College of physicians. Academic Wing of API. Kolkata: Marksman Media Service; 2012. pp.1–9.

36. Gregersen PK, Silver J, Winchester RJ. The shared epitope hypothesis: an approach to understanding the molecular genetics of susceptibility to rheumatoid arthritis. Arthritis Rheum 1987; 30:1205–13.

37. Ghodke Y, Joshi K, Chopra A, Patwardhan B. HLA and disease. Eur J Epidemiol. 2005; 20(6):475–88.

38. Taneja V, Mehra NK, Chandrasekaran AN, Ahuja RK, Singh YN, Malaviya AN .HLA-DR4-DQw7 haplotypes occur in Indian patients with rheumatoid arthritis. Rheumatol Int 1992; 11:251–55.

39. Anuradha V, Bharadwaj R, Chopra A. HLA Class II associations Differ in Indian (Asian) patients suffering from RA: Regional population (urban and rural) surveys using COPCORD Bhigwan model. [Abstract] Arthritis Rheum 2010; 62 (Suppl 10); S684.

40. Anuradha V. Defining the immunogenetic markers in patients with inflammatory arthritis in a rural population of western Maharashtra. Ph.D dissertaion, University of Pune, 2010.

41. Harel-Meir M, Sherer Y, Shoenfeld Y. Tobacco smoking and autoimmune rheumatic diseases. Nat Clin Pract Rheumatol 2007; 3(12):707–15.

42. Gabriel SE, Michaud K. Epidemiological studies in incidence, prevalence, mortality, and comorbidity of the rheumatic diseases. Arthritis Res Ther 2009; 11(3); 229.

43. Malaviya AN, Mehra NK, Adher GC, Jindal K, Bhargava S , et al. The clinical spectrum of HLA B27 related rheumatic diseases in India. J Assoc Physicians India 1979; 27: 487–92.

44. Mahadevan R, Chandrasekaran AN, Parthiban M, Achutan, Porkodi R, Rajendran CP. HLA profile of seronegative spondarthropathies in a referral hospital in South India. J Indian Rheumatol Assoc 1996; V4; 3:91–95.

45. Aggarwal A, Misra R, Chandrasekara S, Prasad KN, Dayal R, Ayyagari A. Is undifferentiated seronegative spondyloarthropathy a formefruste of reactive arthritis? Br J Rheumatol 1997; 36:1001–04.

46. Chandrasekaran AN, Porkodi R, Achutan K, Madhavan R, Parthiban M. Spectrum of clinical and immunological features of systemic rheumatic disorders in a referral hospital in South India: Primary Ankylosing Spondylitis. J Indian Rheumatol Assoc 1994; 2(4):149–52.

47. Prakash S, Mehra NK, Bhargava S, Vaidya MC, Malaviya AN. Ankylosing Spondylitis in North India: A clinical and immunogenetic study. Ann Rheum Dis 1984; 43:381–5.

48. Chopra A, Raghunath D, Singh A. A spectrum of seronegative spondarthritis (SSA) with special reference to HLA profiles. J Assoc Physicians India 1990; 38:351–55.

49. Chandrasekaran AN, Ramakrishnan S, Radhakrishnan B, Parthiban M. Spectrum of clinical and immunological features of systemic rheumatic disorders in a referral hospital in South India: Psoriatic Arthropathy. J Indian Rheumatol Assoc 1995; 3(1):7–13.

50. Jagadish TK, Kasthuri AS, Chopra A, Gupta MM, Kumaravelu S, Dham SK, et al. Clinical profile of connective tissue diseases in a referral service hospital. J Assoc physicians India 1988; 36:602–05.

51. Kumar A. Indian Guidelines on The Management of SLE. J Indian Rheumatol Assoc 2002; 10:80–96.

52. Patwardhan M, Pradhan V, Rajadhyaksha A, Umare V, Rajendran V, Surve P, et al. Clinical and serological features of male Systemic Lupus Erythematosus patients from Western India. Indian J Rheumatol 2012; 7:204–08.

53. Chandrasekaran AN, Porkodi R, Radhakrishanan B, Ramakrishnan S, Mahadevan R, et al. Neuropsychiatric manifestations of systemic lupus erythematosus. J Indian Rheumatol Assoc 1994; 2(2): 90–3.

54. Bichile LS, Vishwanath, Rajdhyaksha. Rheumatologic emergencies in Indian patients: Experience from a rheumatology unit. APLAR J Rheumatol 2000; 4: 114–16.

55. Kumar A, Malaviya AN, Bhat A, Misra R, Banerjee S, Sindhwani R et al. Clinicopathological profile of vasculitides in India. J Assoc Physicians India 1985; 33:694–98.

56. Chandrasekaran AN, Venugopal B, Porkodi R, Ramakrishnan S, Rajendran CP, Radhakrishnan B, et al. Spectrum of clinical and immunological features of systemic rheumatic disorders in a referral hospital in South India: Vasculitis. J Indian Rheumatol Assoc 1995; 3:84–92.

57. Samant R, Vaidya S, Nadker M, Borges NE. Spectrum of clinical features of Vasculitides in a referral hospital from Western India. J Indian Rheumatol Assoc 1997; 5: 3: 69–74.

58. Gupta R, Kumar A, Malviya AN. Outcome of polyarteritis nodosa in northern India. Rheumatol Int 1997; 17:101–03.

59. Bamberry P, Sakhuja V, Bhusnurmath SR, Jindal SK, Deodhar SD, Chugh KS. Wegener's granulomatosis: clinical experience with eighteen patients. J Assoc Physicians India 1992; 40:597–600.

60. Kumar A, Pandhi A, Menon A, Sharma SK, Pande JN, Malaviya AN. Wegener's granulomatosis in India: Clinical features, treatment and outcome of 25 patients. Indian J Chest Dis Allied Sci 2001; 43:197–204.

61. Mittal G, Mangat G, Samant R, Pathan E, Sule A, Rajdhyaksha S, et al. Avascular Necrosis in systemic rheumatic diseases: An analysis of 29 cases. J Indian Rheumatol Assoc. 2000; 8:91–96.

62. Chandrasekaran AN, Rajasekhar G, Porkodi R, Rajendran CP, Madhavan R, Parthiban M. Spectrum of clinical and immunological features of systemic rheumatic disorders in a referral hospital in South India: Takayasu's Arteritis. J Indian Rheumatol Assoc 1996; 4(3): 77–82.

63. Uppal SS, Kakker R, Malviya AN, Sharma S. Clinical aspects of Takayasu arteritis: Current concepts. J Indian Rheumatol Assoc 2000; 8:34–41.

64. Kumar A, Malaviya AN, Grover R, Marwaha V, Talwar D, Kanga U, et al. Clinical and Immunogenetic features of Behçet's disease in India. In Nilganuwong Surasak, editor. Proceedings of the 10th Asia Pacific League of Associations for Rheumatology Congress; 2002; Bangkok, Thailand. Bangkok: Supjaroon Printing Co Ltd; 2002. p. 207–11.

65. Chandrasekaran AN, Rajendran CP, Porkodi R, Ramakrishana S, Venugopal B, Thenmozhi Valli PR, et al. Spectrum of clinical and immunological features of systemic rheumatic disorders in a referral hospital in South India: Sjögren's Syndrome. J Indian Rheumatol Assoc 1994; 2(2):71–75.

66. Misra R, Hissaria P, Tandon V, Aggarwal A, Krishnani N, Dabadghao S. Primary Sjögren's syndrome: rarity in India. J Assoc Physicians India. 2003; 51:859–62.

67. Sandhya P, Jeyaseelan L, Scofield RH, Danda D. Clinical Characteristics and Outcome of Primary Sjögren's Syndrome: A Large Asian Indian Cohort. Open Rheumatol J 2015; 9:36–45.

68. Porkodi R, Parthiban M, Rukmangatharajan S, Kanakarani P, Rajendran CP. Clinical spectrum of Gout in South India. J Indian Rheumatol Assoc 2002; 10:61–63.

69. Kumar A, Malviya AN, Singh YN, Chaudhary K, Tripathy S. Clinical profile, therapeutic approach and outcome of gouty arthritis in northern India. J Assoc Physicians India 1990; 38:400–2.

70. Gupta SK. Primary gout in an orthopedic practice in Calcutta. J Indian Rheumatol Assoc 1994; 2:153–56.

71. Chandrasekaran AN. Infections and arthritis: our experience – a retrospective and prospective study. J Indian Rheumatol Assoc 1993; 1(1):7–20.

72. Chopra A. Bugs and we: who is chasing whom? Indian J Rheumatol 2011; 6(1):1–4.

73. Porkodi R, Rathnasamy R. Infections in a rheumatology setting-Chennai experience. Indian J Rheumatol 2011; 6(1):31–35.

74. Irlapati RVP, Nagaprabu VN, Suresh K, Agrawal S, Gumdal N. Infections in rheumatology practice: an experience from NIMS, Hyderabad. Indian J Rheumatol 2011; 6(1):25–30.

75. Agarwal V, Aggarwal A. Acute and chronic bacterial infections in rheumatology practice. Indian J Rheumatol 2011; 6(1):69–74.

76. Gupta N, Shetty A, Soman R, Raodrigues C. Musculoskeletal infections: selected cases from a referral hospital in Mumbai. Indian J Rheumatol 2011; 6(1):55–58.

77. Chopra A, Joshi VL, Joshi V, Rane R, Anuradha V. Old enemy on a difficult terrain: tuberculosis in rheumatology practice with focus on uncommon extrapulmonary affection. Indian J Rheumatol 2011; 6:44–54.

78. Haldar S, Ghosh P, Ghosh A, Tuberculous arthritis-the challenges and opportunities: observations from a tertiary center. Indian J Rheumatol 2011; 6:62–68.

79. Samant R, Nadkar MY, Vaidya SS, Chhugani SJ, Balaji R, Borges NE. Leprosy: a close mimic in a rheumatology clinic. J Assoc Physicians India 1994; 47:576–79.

80. Prasad S, Misra R, Aggarwal A, Lawrence A, Haroon N, Wakhlu A, et al. Leprosy revealed in a rheumatology clinic: a case series. Int J Rheum Dis. 2013; 16:129–33.

81. Sarkar RN, Phaujdar S, Banerjee S, Siddhanta S, Bhattacharyya K, De D, et al. Musculoskeletal involvement in leprosy. Indian J Rheumatol 2011; 6(1):20–24.

82. Sheetal S, Arvind C. Lest we forget Hansen's disease (leprosy): an unusual presentation with an acute onset of inflammatory polyarthritis and the rheumatology experience. Int J Rheum Dis 2009; 12:64–69.

83. Vaidya S, Samant RS, Nadkar MY, Koppikar GV, Kulkarni MG, Wadhva SL et al. HIV infections and rheumatological disorders. J Indian Rheumatol Assoc 1996; 4(3):83–87.

84. Achuthan K, Uppal SS. Rheumatological manifestations in 102 cases of HIV infection. J Indian Rheumatol Assoc 1996; 4:43–47.

85. Lahariya C, Pradhan SK. Emergence of chikungunya virus in Indian subcontinent after 32 years: A review. J Vector Borne Dis 2006; 43:151–60.

86. Ray P, Ratagiri VH, Kabra SK, Lodha R, Sharma S, Sharma BS. Chikungunya infection in India: results of a prospective hospital based multi-centric study. PLoS One 2012; 7(2):e3002.

87. Nagpal BN, Saxena R, Srivastava A, Singh N, Ghosh SK, Sharma SK, et al. Retrospective study of chikungunya outbreak in urban areas of India. Indian J Med Res 2012:351–58.

88. Chopra A, Anuradha V, Lagoo-Joshi V, Kunjir V, Salvi S, Saluja M. Chikungunya virus aches and pains: an emerging challenge. Arthritis Rheum 2008; 58: 2921–22.

89. Chopra A, Venugopalan A. Persistent rheumatic musculoskeletal pain and disorders at one year post-chikungunya epidemic in south Maharashtra-a rural community based observational study with special focus on naïve persistent rheumatic musculoskeletal cases and selected cytokine expression. Indian J Rheumatol 2011; 6(1):5–11.

90. Paul BJ, Pannarkady G, Moni SP, Thachil EJ. Clinical profile and long-term sequel of Chikungunya fever. Indian J Rheumatol 2011; 6(1):12–19.

91. Ganu MA, Ganu AS. Post-chikungunya chronic arthritis-our experience with DMARDs over two years follow up. J Assoc Physicians India. 2011; 59:83–86.

92. Chopra A, Anuradha V, Ghorpade R, Saluja M. Acute Chikungunya and persistent musculoskeletal pain following the 2006 Indian epidemic: a 2-year prospective rural community study. Epidemiol Infect 2012; 140:842–50.

93. Mathew AJ, Goyal V, George E, Thekkemuriyil DV, Jayakumar B, Chopra A; Trivandrum COPCORD Study Group. Rheumatic-musculoskeletal pain and disorders in a naïve group of individuals 15 months following a Chikungunya viral epidemic in south India: a population based observational study. Int J Clin Pract 2011; 65:1306–12.

94. Manimunda SP, Vijayachari P, Uppoor R, Sugunan AP, Singh SS, Rai SK et al. Clinical progression of chikungunya fever during acute and chronic arthritic stages and the changes in joint morphology as revealed by imaging. Trans R Soc Trop Med Hyg 2010; 104: 392–99.

95. Pandit D. Brucella arthritis-an update. Indian J Rheumatol 2011; 6(1):75–79.

96. Agarwal A, Misra R. Juvenile chronic arthritis in India. Is it different from that seen in Western countries. Rheumatol Int 1994; 15:53–56.

97. Seth V, Kabra SK, Semwal OP, Jain Y. Clinico-immunological profile in juvenile rheumatoid arthritis – an Indian experience. Indian J Pediatr 1996; 63:293–300.

98. Kunjir V, Venugopalan A, Chopra A. Profile of Indian patients with juvenile onset chronic inflammatory joint disease using the ILAR Classification Criteria for JIA: A Community-Based Cohort Study. J Rheumatol 2010; 37:1756–62.

99. Sawhney S. Juvenile idiopathic arthritis: Classification, clinical features, and management. Indian J Rheumatol 2012; 7(1):11–21.

100. Khubchandani RP, Hasija RP. Spectrum of paediatric rheumatologic disease: The Mumbai experience. Indian J Rheumatol 2012; 7 (1):7–10.

101. Viswanathan V. Acute rheumatic fever. Indian J Rheumatol 2012; 7(1):36–43.

102. Gupta A, Singh S. Kawasaki disease-A preventable cause of cardiac morbidity. Indian J Rheumatol 2012; 7(1):87–91.

103. Singh S, Chandrakasan S, Ahluwalia J, Suri D, Rawat A, Ahmed N, et al. Macrophage activation syndrome in children with systemic onset juvenile idiopathic arthritis: clinical experience from northwest India. Rheumatol Int. 2012; 32:881–6.

104. Krishnamachari KA, Bhat RV. Endemic familial arthritis of Malnad an outbreak in southern India. Trop Geogr Med. 1978; 30:33–7.

105. Rawlani S, Rawlani S, Rawlani S. Assessment of Skeletal and Non-skeletal Fluorosis in Endemic Fluoridated Areas of Vidharbha Region, India: A Survey. Indian J Community Med. 2010; 35(2): 298–301.

106. Kumar A, Gopal H, Khamkar K, Prajapati P, Mendiratta N, Misra A, et al. Vitamin D deficiency as the primary cause of musculoskeletal complaints in patients referred to rheumatology clinic: A clinical study. Indian J Rheumatol 2012; 7(4):199–203.

107. Wigley R, Chopra A, Wigley S, Wigley A. Does control of rheumatic disease raise the standard of living in developing countries? Clin Rheumatol 2009; 28:621–22.

Chapter

4

Musculoskeletal System Examination

Milind Nadkar, Vikram Londhey

INTRODUCTION

Detail history and examination is the cornerstone of diagnosis in rheumatology. Since one of the most common symptoms in rheumatology are joint pain and swelling, one must be well versed with examination of the individual joints of the musculoskeletal (MSK) system.[1] Connective tissue disorders (CTD) may present with constitutional and systemic symptoms referable to the heart, kidneys, brain, etc. hence apart from joint examination systemic examination must also be done. The clinical approach to monoarthritis, polyarthritis, CTDs is discussed in Chapters 5, 6, and 8, respectively. In this chapter we will focus on how to perform MSK system examination.

Paul Dieppe and Mike Doherty developed a musculoskeletal (MSK) screening examination called "GALS" which stands for gait, arms, legs and spine.[2,3] This approach is good and helps one to quickly assess a patient's MSK problem. Regional Examination of the MSK System (REMS) is a more detailed method of examination.[4] The methods of doing GALS examination and its interpretation is available in the Arthritis Reseach UK website arthritisresearchuk.org as a free hand book download.[3]

EXAMINATION

Joint Examination

The MSK system consists of joints, tendons, muscles and bursa. Joints can be peripheral joints (such as wrists, elbows, knees, etc.) and the axial (spinal) joints. Hip and shoulder are root joints often considered part of the axial skeleton. A few basic principles apply when doing MSK examination so that findings will not be missed.

1. Inspection, palpation should be applied for each part of MSK system.
2. Examiner should always compare examination on the both sides of the body. Subtle changes can only be made by comparing bilaterally.
3. Follow a systematic approach. In other words, start from distal (hands) and go proximal toward the shoulders in upper limb. The same applies to lower limb.
4. This is followed by spine and central joints such as sternoclavicular, pubic symphysis, sacroiliac joints, etc.
5. Do not forget to examine the temporomandibular joints.

INSPECTION

Look for any swelling, changes in colour of the skin which may occur in gout. Look for any discharge from over the joint surface. This may occur in tubercular arthritis, septic arthritis, or osteomyelitis. With the joints and limbs in resting position note if there are any deformities, flexion contractures, or malalignment such as genu valgum or genu varum. The patient can be asked to move the joints while watching the patient's face for wincing to know whether the joint movement is painful or not. Look for erythema of the joint surface.

PALPATION

On palpation the presence of warmth, erythema, tenderness, presence of effusion in the joint, decreased range of joint movements should be duly noted. All these features occur in arthritis. Two or more components should be present to call it arthritis. Presence of bogginess or doughy feel suggests synovial proliferation and synovial thickening. This is called synovitis. There can be inter-examiner variation when examined at the same time or different time based on the experience of the examiner.

In early arthritis synovitis or joint swelling may not be overtly present. Nowadays subtle synovial thickening, synovitis, joint effusion is picked up by musculoskeletal ultrasound or MRI.

Joint Line Tenderness

It is important to differentiate arthritis (pain originating from the joint) from periarthritis (pain arising from surrounding structures). Palpation helps in this aspect. If there is tenderness along the joint line associated with restriction in range of motion in all the planes, the pathology is within the joint. If the tenderness is away from the joint, periarticular structures are likely to be involved.

Range of motion: Check whether the range of movement is complete or restricted in a single plane or all the planes. Also check whether the patient is able to move the joint actively or only passive movements are possible.

Other points to note during palpation: Look for any bony nodules which may indicate osteoarthritis (OA) of the hands. Heberden's nodes are hard bony overgrowth over the DIP joints in nodal OA. Similarly Bouchard's nodes are hard bony overgrowth over the PIP joints. The presence of nodules over the elbows should arouse the suspicion of tophus or rheumatoid nodules.

Manipulation

It gives more clues to diagnose some diseases and excludes others, for example, deformities in rheumatoid arthritis (RA) are non-reducible (fixed) while those of Jaccoud's arthropathy are reducible by manipulation. Similarly, stability of joints and abnormal movements, painful or painless range of motion, crepitus on motion and special tests for specific muscle or group of muscles (e.g. test for rotator cuff, test for hamstring muscles, subscapularis lift off test, etc.) should be carried out gently.

Once an abnormality has been detected either through the history or through the screening GALS, regional examination should be carried out as discussed below.

EXAMINATION OF SMALL JOINTS OF HANDS AND WRISTS

Examination of the small joints of hands includes distal interphalangeal (DIP), proximal interphalangeal (PIP), metacarpophalangeal (MCP), and wrist joints. We need to start with inspection and follow principles as outlined above.

DIP and PIP joints: To palpate DIP and PIP joints we use the four-finger method (Fig. 4.1). The examiner places his/her thumb and index finger of one hand on side by side of the joint, while the other hand's thumb and index finger should be kept antero-posteriorly of the patient's joint. The patient's fingers should be kept in horizontal positions while examiner will try to feel for bogginess or try to evoke pain.

MCP joints: The MCP joints should be examined by two-finger method. Here, examiner puts two thumbs anteriorly and two index fingers posteriorly to feel bogginess while patient's MCP joints should be in position of 30 degree flexion (Fig. 4.1).

Wrist joints: To palpate the wrist the examiner places his/her two thumbs in between radius and ulna and at the level of radial styloid process while remaining fingers should be on ventral side of forearms (Fig 4.2). Apply pressure gently to check for tenderness.

Palpation of hand joints should be followed by gentle manipulation to check for range of motion, particularly at wrist joint which may be the only finding suggestive of inflammatory arthritis. Normally the wrist can be dorsiflexed to 70 degrees and palmar flexed up to 80 or 90 degrees.

Metatarsal squeeze test for MTP joints and metacarpal squeeze test for MCP joints enables one to check for MTP or MCP joint arthritis.[4] The patient's hand is squeezed transversely, if there is pain the patient winces and the test is considered

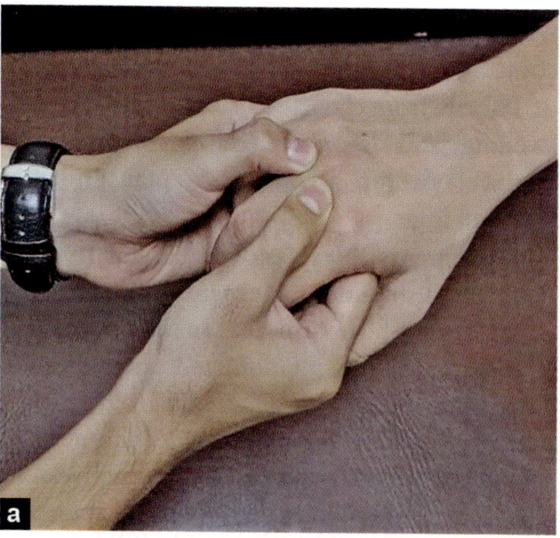

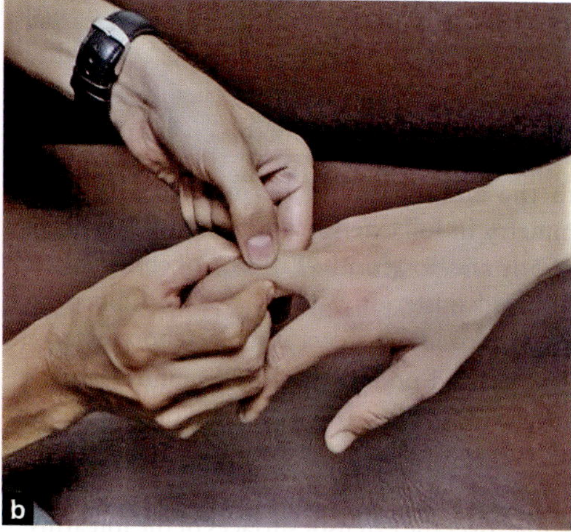

Fig. 4.1: (a) Two-finger method for examination of MCP joint. (b) Four-finger method for PIP/DIP joint examination

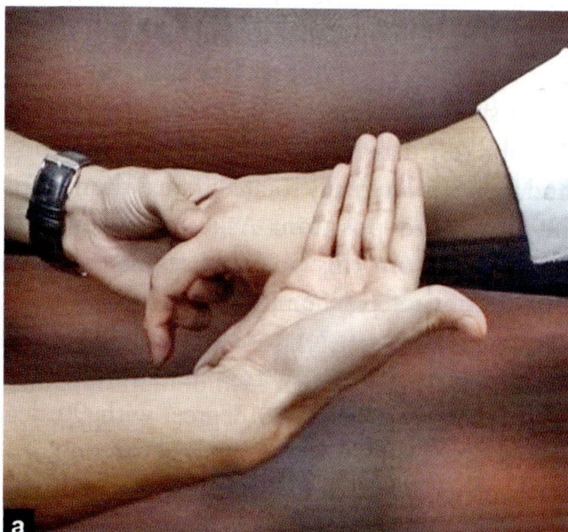

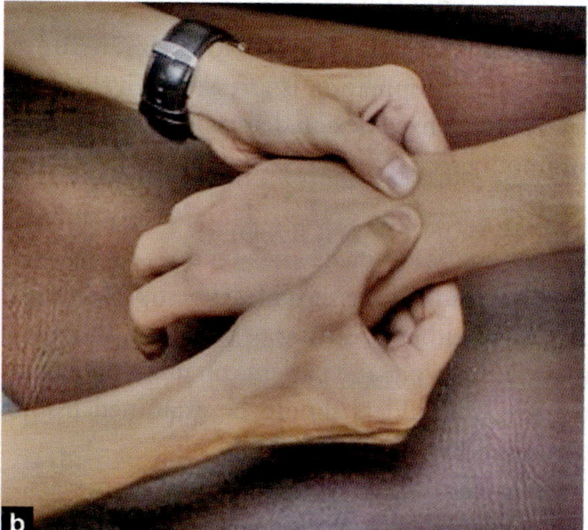

Fig. 4.2: (a) Palpation at wrist by ulnar dorsal surface of hand. (b) Wrist palpation by two thumbs in midway between radio-ulna, in parallel of radius styloid process

positive (Fig. 4.3). The foot is squeezed in a similar manner. This is easy to perform, reliable and reproducible test to check for tenderness of the MCP and MTP joints.

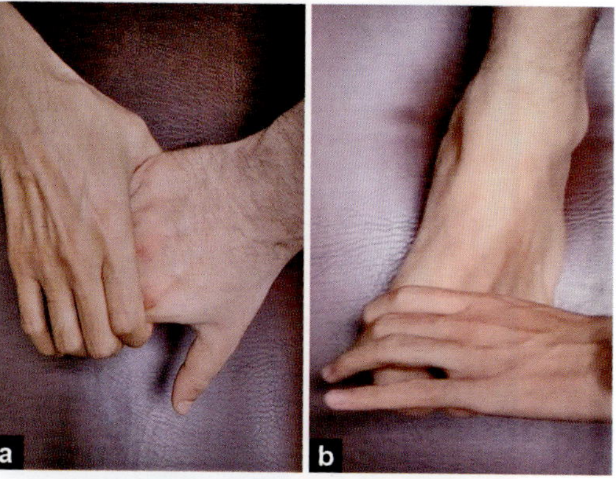

Fig. 4.3: (a) MCP squeeze test. (b) MTP squeeze test

Finkelstein's test is used to detect de Quervain's tenosynovitis.[5] To elicit Finkelstein test, the patient is instructed to make a fist with the thumb inside the fist, then deviate the fist toward the little finger, which will elicit pain if there is tenosynovitis of the tendons contained in the first dorsal compartment of the wrist, i.e. the extensor pollicis brevis and abductor pollicis longus tendon.

Finally, the palms should be palpated to look for trigger finger, diabetic cheiroarthropathy (prayer

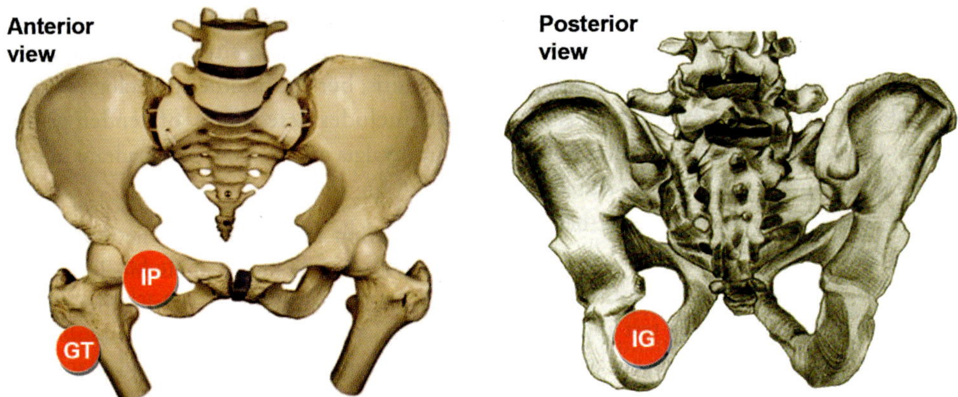

Fig. 4.6: Location of hip bursae. Greater trochanteric (GT) bursitis, iliopsoas (IP) bursitis, and ischeogluteal (IG) bursitis (posteriorly)

Special Tests of Hip Joint

FABER test: FABER stands for Flexion, **AB**duction, and External Rotation described in section of sacroiliac (SI) joint examination (Fig. 4.7).

FADIR test: FADIR stands for Flexion, **AD**dcution, and Internal Rotation. It is used to check femoro-acetabular impingement syndrome.

Thomas test is used to assess flexion contracture of hip. In supine position, limb opposite the test limb is flexed over abdomen to remove compensated lumbar lordosis and then test limb is allowed to extend normally. Failure of hip to lie flat over table or bed is suggestive of fixed flexion deformity.

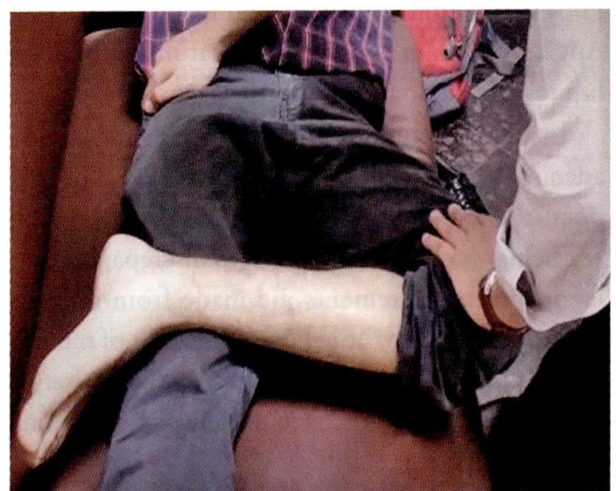

Fig. 4.7: FABER test

EXAMINATION OF THE KNEES

Knee is the largest joint in the body.

Inspect from anteriorly, posteriorly, and from the side preferably in weight bearing position for swelling, erythema, wasting of muscles (particularly vastus medialis and quadriceps femoris), and scars. Inspect knee for angulation. Angulation medially is genu valgum (knock knee). Angulation laterally is genu varum (bow knee) while angulation posteriorly is called genu recurvatum.

Palpation is done to note temperature, tenderness, synovial thickening, and effusion. *Tenderness* is assessed for hamstrings, quadriceps, patella, patellar tendons, tibial tubercle, femoral epicondyles, and anserine bursa. Palpate the joint line for synovial thickening (Fig. 4.8). Absence of dimple on medial side suggests synovitis. Synovial thickening is noted as a swelling of doughy consistency. Thickened synovial edge is reflected in the suprapatellar pouch and noted as a longitudinal ridge approximately of 4 to 5 cm above the upper border of the patella.

Effusion: A subtle joint effusion is detected by 'bulge' test also called the cross fluctuation test. In this test, examiner pushes fluid from medial gutter into suprapatellar pouch from one hand and then pressure is applied from lateral side of the knee from proximally to distally. Presence of any bulge in medial recess is considered a positive 'bulge test'. If the effusion is of moderate size the patellar tap test or ballottement test is positive. In this test the examiner milks fluid from suprapatellar pouch, apply one hand firmly over upper border of patella and with other hand, presses the patella down to feel any ballottment (Fig. 4.9).

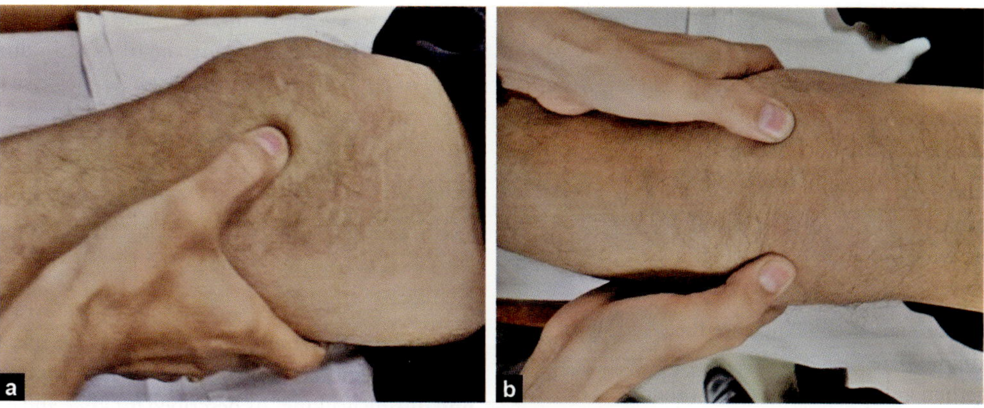

Fig. 4.8: (a) Palpation of medial and (b) lateral joint line at knee joint

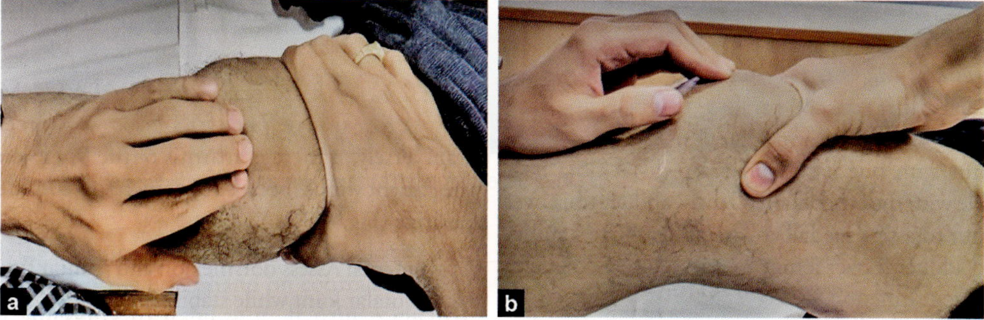

Fig. 4.9: Patellar tap test. (a) The patients knee is held firm at upper border of knee joint and (b) tap over middle part of patella to feel bounce of synovial fluid

Popliteal fossa is examined for any swelling, mass and popliteal artery pulse. Fullness over the popliteal fossa and pain suggests baker's cyst.

Movement: Normal range of motion of knees varies from 0 to 115–135 degree. Flexion and extension should be checked actively and passively. When there is flexion contracture, full extension of the knee will not be possible. Flexion contracture can happen as a sequela of inflammatory arthritis (Fig. 4.10).

Knee crepitus is a feature of osteoarthritis. It is checked by placing the hands lightly over patella and ask the patient to flex and then extend knees.

Provocative test: These are used to check for ligament injuries and instabilities. For anterior cruciate ligament (ACL) assessment tests described are Lachman test and anterior drawer test. Posterior cruciate ligament (PCL) is assessed by posterior sag test.

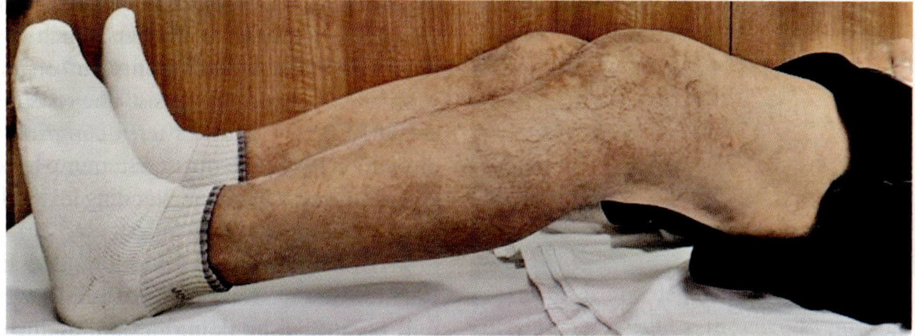

Fig. 4.10: Flexion contracture at both knee joints in a patient of rheumatoid arthritis

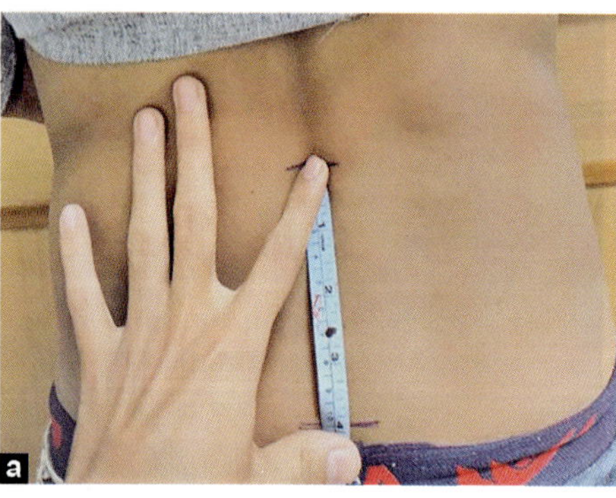

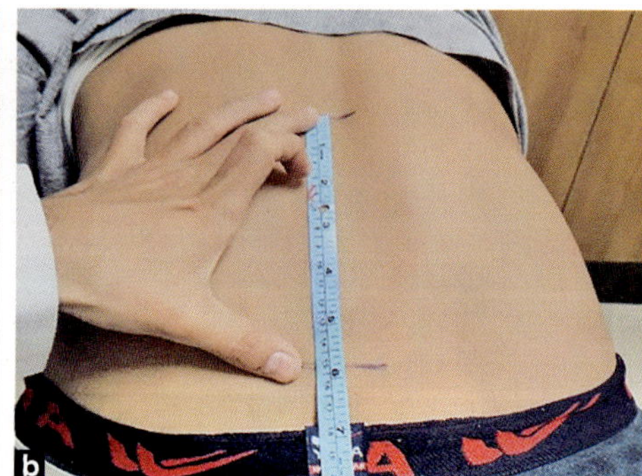

Fig. 4.13: Schöber test: (a) Mark a horizontal line at the level of dimple of Venus which corresponds to 5th lumbar spine or posterior superior iliac spine (PSIS). (b) Mark a second vertical line 10 cm above the 1st line, then ask the patient to bend forward maximally. Measure the distance between 2 lines. Normally on forward flexion this line should increase to >15 cm

Pelvic Rock Test

This test is done to find out stability of the SI joints and or pelvis. With the patient in supine position the examiner puts hands on patient's iliac crest with thumbs on ASIS. Both the hands are pushed towards each other to stress the SI joints. Any pain or discomfort, decrease or increase movements suggest SI joint or pelvis pathology.

Conclusion

Finally, as discussed in the chapter on approach to CTDs, look beyond the joints. Skin examination is important to diagnose SLE, scleroderma, dermatomyositis. Nodules over the tibial shins or around the ankles can be erythema nodosum. These can be easily confused with cellulitis. Or a beginner may mistake pain due to cellulitis for arthritis. The opposite is true when podagra can mimic cellulitis. Look for rash in hidden sites like behind the pinna, periumbilical region, anal cleft as they can be easily missed in psoriasis. Bruises, ecchymosis, livedo reticularis are signs of APS and cryoglobulinemia. Clubbing should arouse suspicion of hypertrophic pulmonary osteoarthropathy and/or ILD. The eyes are examined for uveitis, scleritis, episcleritis. Endocrine disorders and malignancies can present with joint pains.

MSK examination is not taught routinely in undergraduate curriculum. But there is a definite need to teach a medical student and rheumatology fellows on how to examine the joints. In addition to the MSK system in rheumatology we have to *look beyond the joints.*

Key Points

1. GALS an acronym for gait, arms, legs and spine helps one to quickly assess whether a patient has MSK problem or not.
2. This should be followed by detailed REMS (regional examination of MSK system)
3. The same principles of clinical medicine are followed in rheumatology. The joints are inspected, palpated, and check for range of movements.
4. In general inspection look for gait, symmetry, and posture.
5. Start systematically from distal (hands) and go proximal toward the shoulders in upper limb. The same applies to lower limb. This is followed by examination of spine, central joints such as sternoclavicular, pubic symphysis, and SI joint. Always compare both sides.
6. Do not forget to examine the temporomandibular joints and chest wall for costochondral joints.

REFERENCES

1. American college of Rheumatology Ad Hoc committee on clinical Guidelines. Guidelines for the initial evaluation of the adult patient with acute musculoskeletal symptoms. Arthritis Rheum 1996; 39:1–8.
2. Doherty M, Dacre J, Dieppe P, Snaith M, The "GALS" locomotor screen. Ann Rheum Dis 1992; 51:1165–9.
3. arthritisresearchuk.org[Internet]. Nottingham, UK.Arthritis Research UK. The musculoskeletal examination: GALS.(Cited 30 Aug 2017). Available from http://www.arthritisresearchuk.org/health-professionals-and-students/student-handbook/the-msk-examination-gals.aspx

4. Coady D, Walker D, Kay L. Regional Examination of the Musculoskeletal System (REMS): A core set of clinical skills for medical students. Rheumatology 2004; 43:633–9.
5. Finkelstein H. Stenosing tendovaginitis at the radial styloid process. J Bone Joint Surg Am 1930; 12: 509–40.
6. Dalton SE. Clinical examination of the painful shoulder. Baillieres Clin Rheumatol 1989; 3:453–74.
7. Lea RD, Gerhardt JJ. Range of the motion measurements. J Bone Joint Surg Am 1995; 77:784–98.
8. Russel AS, Maksymowych W, LeClercq S. Clinical examination of the sacroiliac joints: A prospective study. Arthritis Rehum 1981; 24:1575–7.

FURTHER READING

1. Wilson CH. The Musculoskeletal Examination. In: Walker HK, Hall WD, Hurst JW, editors. Clinical Methods: The History, Physical, and Laboratory Examinations. 3rd edition. Boston: Butterworths; 1990. Chapter 164. Available from: https://www.ncbi.nlm.nih.gov/books/NBK272/

Chapter

5

Approach to Monoarthritis

SJ Gupta

INTRODUCTION

Joint inflammation that starts and remains at a single joint is termed monoarthritis; inflammation at more than one but less than 5 joints is termed oligoarthritis and when inflammation occurs at 5 or more joints, it is termed polyarthritis. Monoarthritis can occur in a previously normal joint or might occur in a joint that is abnormal due to previous joint disease. Monoarthritis (MA) may be: (a) Acute MA—which is of recent onset and less than 6 weeks duration, or (b) Chronic MA—which has been present for at least 6 weeks. The evaluation of a patient presenting with MA involves a systematic clinical approach, keeping in mind specific clinical features in the individual patient. This chapter will aim to provide an overview of the suggested clinical approach to a patient with monoarthritis.

DIFFERENTIAL DIAGNOSIS

There can be many possible causes of MA and in order to determine the cause in an individual patient, it is important to be aware of these. Specific conditions may present with and exhibit differing clinical features. Therefore, one should bear in mind the diagnostic possibilities depending upon the presentation of MA; that is, is the MA acute or chronic and if is there any history of previous joint disease (Tables 5.1 and 5.2).[1,2] Though many causes are not immediately serious or potentially dangerous, the evaluation of a patient with MA needs to be completed fairly soon, as infection is one of the possible causes and if not recognised and treated adequately, can result in joint damage and considerable morbidity. It is also important to remember that occasionally a polyarticular or systemic disease may present with an initial MA involvement.

A. ACUTE MONOARTHRITIS (MA)

As the term suggests, acute MA implies the development of inflammation at a single joint fairly rapidly or suddenly. The patient will complain of the development of pain, swelling and stiffness at the affected joint over a matter of hours or at most, a few days, with the duration being less than 6 weeks (Fig. 5.1).

Acute MA constitutes a medical emergency, considering the rapid development of symptoms that can often incapacitate a patient in a short span of time. Prompt and rapid recognition of the cause in any given patient is important to institute correct treatment and alleviate the clinical symptoms. Infection (septic arthritis) is the most important cause of acute MA, as rapid treatment is essential in order to prevent joint damage and morbidity. Other possible common causes of acute MA that need to be kept in mind and excluded are: Crystal induced synovitis; osteoarthritis (OA); trauma; and mono-articular presentation of a polyarticular or oligoarticular disease (Table 5.1).

1. Septic Arthritis

Septic arthritis indicates joint infection/sepsis, caused by pathogenic bacteria, either by the

Table 5.1: Possible causes of acute/chronic monoarthritis

Acute monoarthritis	*Chronic monoarthritis*
Infection Bacterial, mycobacterial, fungi, viral, and lyme disease **Crystals** MSU (Gout); CPPD (pseudogout); BCP (destructive arthropathy); calcium oxalate **Trauma** Fracture, internal derangement, hemarthrosis **Osteoarthritis** Presenting as monoarthritis—1st CMC joint, MTP joint, hip or knee **Foreign body synovitis** **Ischaemic necrosis** **Tumour** Metastases, pigmented villonodular synovitis, osteoid osteoma ***Systemic disease presenting as monoarthritis***	**Infection** **Plant thorn (foreign body) synovitis** **Sarcoidosis** **Enteropathic arthritis** (Crohn's disease predominantly) **Pigmented villonodular synovitis** **Synovial chondromatosis** **Synovial sarcoma** **Monoarticular presentation of a polyarticular disease**

MSU—mono sodium urate; CPPD—calcium pyro-phosphate dihydrate;
BCP—basic calcium phosphate hydroxy-apatite

Table 5.2: Causes of monoarthritis related to presence or absence of previous joint disease

Previously normal joint (no previous joint disease)	*Abnormal joint (previous joint disease or damage)*
Septic arthritis **Juvenile idiopathic arthritis** **Crystal synovitis** (gout or pseudogout) **Monoarticular presentation of oligo/polyarthritis (initial)** Spondarthritis (reactive/psoriatic); erythema nodosum; rheumatoid arthritis; Juvenile idiopathic arthritis **Trauma** (haemarthrosis) **Foreign body reaction**	**Pseudogout** in association with osteoarthritis **Avascular necrosis** or subchondral collapse or fracture **Cartilage tear** **Haemarthrosis** **Septic arthritis**

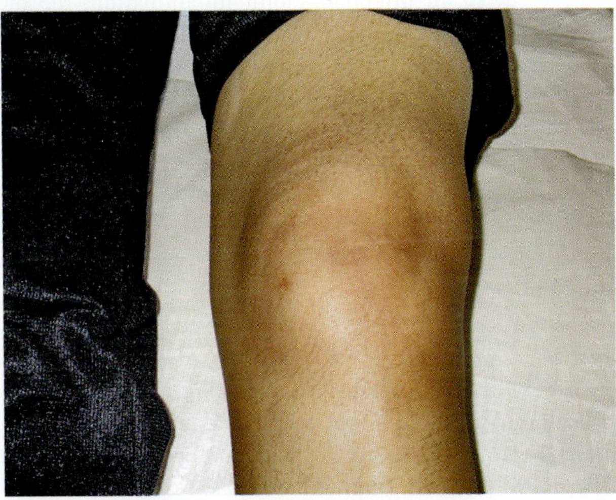

Fig. 5.1: Acute monoarthritis of the left knee: Note the obvious effusion at the joint

haematogenous route or by direct inoculation of the joint (such as following trauma); this distinguishes septic arthritis from reactive arthritis (ReA) where the joint inflammation is due to an immunological reaction to the organism, which may be elsewhere in the body. A patient with joint infection will typically present with a painful, hot and swollen joint, with restricted movements. The symptoms develop rapidly over a period of days and within about 2 weeks.[3] Delayed treatment of septic arthritis can lead to irreversible joint damage and in addition, there is a significant mortality, with an estimated case fatality rate of 11%.[4,5] It is therefore crucially important that the diagnosis of septic arthritis should be made as rapidly as possible. However, the diagnosis can be difficult and may be delayed, even in the hands of experienced clinicians. Some clinical situations when associated with the presentation of acute MA, should raise the index of suspicion for septic arthritis as the cause and these include: (a) The presence of existing joint disease such as rheumatoid arthritis (RA) or osteoarthritis (OA); (b) Acute inflammation in an artificial or prosthetic joint; (c) Low socio-economic status; (c) Intra-venous drug abuse; (d) Chronic alcoholism; (e) Other associated medical conditions such as diabetes mellitus; (f) Previous intra-articular injections; and (g) Ulcers or infection of the skin overlying the affected joint.[6] Usually, the presentation is at a large joint (knee or hip) with acute, pain, swelling, overlying warmth and restricted movement at the affected joint. Occasionally, the presentation may be with more than one joint being affected, so an oligo or polyarticular presentation does not necessarily mean that septic arthritis can be safely excluded. Staphylococcus and Streptococcus are the most common organisms that cause septic arthritis and account for about 90% of cases. Gram-negative organisms are more common in immuno-compromised or older patients and anaerobic bacteria may be responsible for septic arthritis following penetrating injuries.[5]

Once septic arthritis is suspected clinically, *prompt aspiration and analysis of synovial fluid from the affected joint is extremely important.* A Gram's stain and routine bacterial culture of the fluid would enable confirmation of septic arthritis and examination for crystals under polarising light microscopy would enable the diagnosis of crystal-induced arthritis.[7] Estimation of the white blood cell (WBC) count in the synovial fluid can also be helpful in identifying septic arthritis. WBC counts of >100,000/mm^3 are more likely to be associated with joint infection and counts of <50,000/mm^3 are less likely to be due to infection and more likely to be inflammatory in nature.[8] The synovial fluid sample should be transported as rapidly as possible to the laboratory and it would be prudent to warn the microbiologist that a sample from a suspected case of septic arthritis is being sent for analysis, beforehand. In order to improve the diagnostic yield from culture, it has been suggested that the synovial fluid should be inoculated into blood culture bottles in addition to direct culture on agar plates.[9] Blood cultures may help to isolate the causative organism in some cases and the serum WBC count, ESR and CRP levels may help to monitor response to treatment. There are no typical diagnostic imaging techniques to identify septic arthritis; however, isotope scanning may help to distinguish osteoarthritis from sepsis (but will not necessarily distinguish sepsis from other inflammatory causes) and magnetic resonance imaging (MRI) may help to identify appearances consistent with infection.[3] The diagnosis of septic arthritis therefore rests upon appreciating the clinical features, prompt joint aspiration with appropriate synovial fluid analysis, together with other relevant investigations.

2. Crystal Induced Synovitis

Crystal induced synovitis is an important cause of acute MA. A number of crystals can cause acute joint inflammation, the most common of these being mono sodium mono urate (MSU) crystals that cause gout. Recognition of gout as the cause of acute MA depends upon recognising the typical clinical features: Acute gout is eight times more common in men than women and the first attack commonly occurs between the third to sixth decades. *It is extremely uncommon in women before the menopause.* The commonest joint to be affected is the first metatarso-phalangeal (1st MTP) joint, also referred to as podagra. A typical attack is sudden in onset, develops rapidly with peak symptoms developing within 24 hrs and even within 6 to 8 hrs and may often awaken the patient from sleep. The joint becomes red, hot, and swollen, with shiny overlying skin, which may desquamate later. The joint is exquisitely painful and tender and the patient is usually unable to weight bear on it. The attack is not uncommonly misdiagnosed as local infection. During an acute attack there may be associated constitutional symptoms such as anorexia, nausea and fever. Leucocytosis and an elevated ESR may also be present. An elevated serum uric acid is usually (but not invariably) present.[10] These clinical features will often allow a diagnosis of acute gout to be made with fair confidence. Occasionally, the initial attack may be polyarticular and the 1st MTP joint may be spared. To establish the diagnosis of gout beyond doubt, synovial fluid should be aspirated from an affected joint whenever possible and examined under polarising light microscopy to look for typical MSU crystals. These crystals will appear as needle shaped, negatively birefringent crystals (Fig. 5.2). Recurrent acute attacks may result in chronic tophaceous gout, with tophi being found typically in the periarticular tissues, cartilaginous helix of the ear, bursae and tendon sheaths. The tophi appear as periarticular swellings, filled with a soft, 'cheesy' or 'chalky' material (Fig. 5.3). The presence of tophi in a patient presenting with acute MA and a compatible history is a sure clinical sign confirming the diagnosis of gout and the tophaceous material will reveal numerous MSU crystals under polarising light microscopy.

Some calcium containing crystals may also cause acute episodes of synovitis. Calcium pyrophosphate dihydrate (CPPD) and basic calcium phosphate (BCP) are the common calcium crystals. CPPD crystals can cause an acute attack of synovitis, which may be indistinguishable from an attack of gout. This is termed pseudogout. The deposits can often be seen as a fine rim of calcification overlying the articular cartilage of the affected joint (chondrocalcinosis) shown in Fig. 5.4; and the crystals of CPPD can be identified on polarising light microscopy, confirming the diagnosis.[11]

Fig. 5.2: Needle shaped crystals of MSU seen under polarising light microscopy

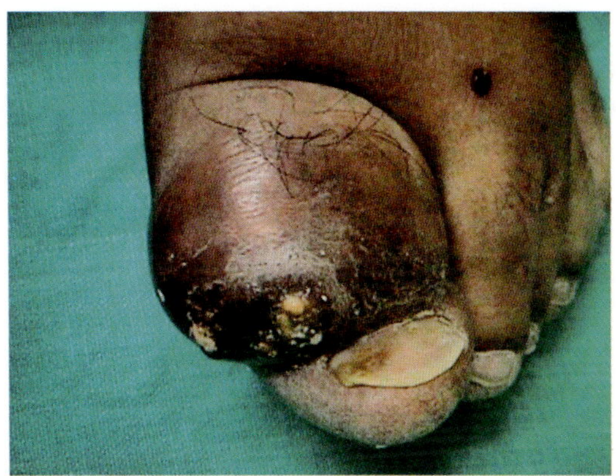

Fig. 5.3: Acute synovitis of 1st MTP joint with a large overlying tophus and typical 'cheesy' material extruding from the tophus

The BCP crystals include hydroxyapatite, octacalcium phosphate and tricalcium phosphate. BCP crystals most often cause a periarthritis or calcific tendonitis, but may also cause a synovitis of a large joint such as the knee or shoulder (Milwaukee shoulder). Though this tends to be a more chronic synovitis, it may occasionally present as an acute MA and ought to be kept in mind in the evaluation of a patient with acute MA.[11]

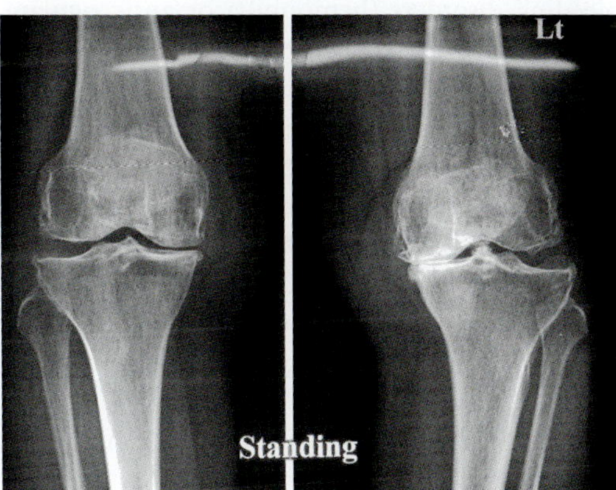

Fig. 5.4: Chondrocalcinosis, seen as a thin line of calcification over the articular cartilage of the right knee lateral compartment. Also demonstrates osteoarthritis changes at both knees, (left > right) with reduced joint space, osteophyte formation and varus deformities

3. Osteoarthritis

Osteoarthritis (OA) is the most common joint disease and therefore, should always be borne in mind as a possible cause of joint symptoms including acute MA. Though usually a chronic, slowly progressive disease, OA can occasionally present with acute worsening of symptoms at a joint, with associated swelling.[12] This is often seen at the knee joint, where mild trauma, overuse of the joint or chondrocalcinosis may cause an acute increase in symptoms with clinical signs of inflammation.[1] However, these acute symptoms would be part of the overall clinical spectrum of OA, which commonly affects multiple joints (hands, feet, spine, and hips) and when associated with bony swelling at the DIP joints of the hands (Heberden's nodes) is referred to a Primary Generalised Osteoarthritis (PGOA) or 'Nodal' OA. Occasionally, OA may affect single large joints (knees or hips)[13]. The diagnosis of OA is not too difficult, keeping in mind the clinical picture and can be easily confirmed by radiology (Fig. 5.4).

4. Trauma, Haemarthrosis and Foreign Body Reaction

Trauma to a joint may cause acute pain due to internal derangement (particularly at the knee), haemarthrosis or fracture. This may be mistaken for acute synovitis. Milder episodes of trauma may cause a transient (reactive) synovitis at the joint and should be borne in mind in the evaluation of a patient with MA.[14] Patients with coagulation or bleeding disorders (hemophilia or anticoagulant therapy) may present with bleeding into a joint, which clinically present quite like an episode of acute MA. Joint aspiration will however confirm the diagnosis. Penetrating injuries from thorns (plant thorn synovitis), wood fragments and other foreign particles (foreign body synovitis) may cause an acute synovitis, due to a foreign body reaction, which may appear shortly after the injury or even a few weeks or months later.[15] A careful history, eliciting possible trauma or even occupation of the patient (such as gardening) is therefore important in reaching an accurate diagnosis.

5. Monoarticular Presentation of a Polyarticular Disease

Many systemic or polyarticular diseases may initially present with an episode of acute MA. These conditions include: Rheumatoid arthritis, systemic lupus erythematosus, the arthritis of inflammatory bowel disease, psoriatic arthritis, Behçet's disease, Reiter's disease and other forms of reactive arthritis. Most of these conditions will however not persist as MA and in addition, the associated clinical features of the underlying disorder will enable an accurate diagnosis to be made.

B. CHRONIC MONOARTHRITIS

Inflammatory MA persisting beyond 6 weeks may have many causes (Table 5.1). As the duration of synovitis increases, infection (particularly chronic infection such as TB) becomes more likely and a synovial biopsy should be undertaken to confirm or exclude infection and also to exclude primary synovial conditions (such as villonodular synovitis)[2]. However, should there be associated radiological features suggesting the possibility of infection, such as periosteal reaction, osteopenia or bony erosion, the synovial biopsy should not be delayed. Some cases of chronic MA may eventually evolve to a polyarticular disease (OA or RA) but about 1/3rd may remain undiagnosed, particularly when the knee is involved.[2]

APPROACH TO THE EVALUATION OF A PATIENT WITH MONOARTHRITIS

The process of evaluation of a patient presenting with joint symptoms, really commences even before the formal consultation, by careful observation of the patient as he/she enters the consulting room. The gait of the patient, how ill the patient appears, any visible extra-articular features (such as a rash) may all suggest possible diagnoses to an astute clinician (Table 5.3).[16]

History

The next and certainly the most important step in evaluation, is to take a detailed and full history; this is of course true for the evaluation of all medical problems, but particularly so for the evaluation of musculoskeletal complaints and acute MA. Onset of symptoms is important in order to determine acute from chronic MA and often, patients may have some difficulty in recalling exactly when their symptoms began. Patients may not associate their present acute symptoms with pain at other joint(s)

Table 5.3: Clues to a possible cause of arthritis based on initial clinical observation

Clinical observation	*Possible diagnosis*
1. Middle aged man with an acute, swollen, painful foot (MTP joint) *OR* Elderly lady, single, swollen and painful joint, on diuretics	Crystal synovitis/gout
2. Young male. Knee or ankle pain and enters with stiff neck and back	Spondarthritis/ankylosing spondylitis
3. Elderly, ill looking individual with an acute, hot, swollen, painful joint	Septic arthritis

or even previous episodes of joint pain; however this would be important to elicit, even by specific questioning, as a history of previous similar episodes would suggest a non-infectious cause and the possibility of crystal arthritis. A history of fever; insect or tick bites; sexual contact; pre-existing systemic disease; intravenous drug use; or travel, might all raise the possibility of infection. The presence of associated features, such as skin rash, ocular complaints, urethritis or genital ulcers and diarrhoea, would suggest the possibility of Reiter's disease. History of trauma might suggest the possibility of traumatic synovitis or fracture. A history of concurrent medical problems might be relevant such as the presence of psoriasis would make one consider psoriatic arthritis. History of drug therapy would also be important; recent introduction of diuretics might suggest gout as a possibility. A family history of gout, osteoarthritis or other inflammatory joint diseases should be sought and might often lead to a diagnosis.

Examination

The most crucial aspect of the physical examination would be to distinguish true synovitis (joint inflammation) from bursitis, cellulites or tendonitis. The presence of a joint effusion and synovial thickening, tenderness over the joint line and restricted active and passive movements at the joint, indicate joint inflammation. By and large, this is not difficult and one can recognise joint inflammation clinically, particularly at certain joints such as the knee joint. However, some clinical situations pose specific problems: Olecranon bursitis may sometimes be difficult to differentiate from synovitis at the elbow; cellulitis around the ankle or knee may simulate synovitis; and deep seated joints such as the hip and sacroiliac joints can be difficult to evaluate.

Once the primary question is settled that one is dealing with a MA, one should attempt to look for specific clues that might suggest an underlying diagnosis on further examination (Table 5.4).

Table 5.4: Diagnostic clues on examination

Findings on examination	*Possible diagnosis/think of*
Significant lymphadenopathy	Infection (tubercular)
Oral ulcers	Behçet's disease; Reiter's syndrome; or systemic lupus erythematosus (SLE)
Cutaneous tophi	Gout
Nail pitting or onycholysis	Psoriatic arthritis
Specific skin rash:	
• Erythematous facial rash	• SLE
• Psoriatic patches (natal cleft, behind ears or at umbilicus)	• Psoriatic arthritis
• Erythema nodosum	• Sarcoidosis, inflammatory bowel disease, SLE
• Keratoderma blennorrhagicum	• Reiter's disease
• Pyoderma gangrenosum	• Inflammatory bowel disease
• Ulcers	• Infection
Sacroiliitis	Spondarthritis

Investigations

The most important clinical investigation in the evaluation of a patient with MA is synovial fluid examination (particularly if infection is suspected) and virtually every patient presenting with a MA should have the joint aspirated with this objective.[1] Joint aspiration is a simple, safe and easy bedside procedure with virtually no complications, providing a sterile no-touch technique with appropriate disposable equipment is used.[17] The synovial fluid should be examined for the following: (1) Gram's stain; (2) Routine culture; (3) WBC count (total and differential); and (4) Crystals under PLM. Virtually any joint can be aspirated easily and only a few drops of fluid are needed for these tests. Some joints such as the hip and sacroiliac joints are deep seated and are best needled under image guidance.

Synovial biopsy obtained by needle aspiration or arthroscopy, might be indicated for the evaluation of chronic MA rather than in the acute situation. Often, the biopsy reveals "non-specific inflammatory" changes on histology and this is particularly so for the non-infective, inflammatory conditions. Synovial biopsy would be of specific help in identifying conditions such as amyloidosis, sarcoidosis, pigmented villonodular synovitis or other synovial tumours. Some recent advances in the processing of synovial biopsy specimens, by polymerase chain reaction (PCR) and immuno-electron microscopy, might help to identify tiny fragments of the aetiological agent and have been found to be useful for the diagnosis of Lyme disease, gonococci and chlamydiae.[1] These new techniques may not be easily available but certainly rapid and appropriate synovial fluid analysis should be regarded as a sine-qua-non in the evaluation of MA.

Other investigations might be helpful depending upon the suspected clinical diagnosis. Leucocytosis might suggest infection and cultures of the blood and urine might also help in identifying the causative organism in some situations. Radiology of the affected joint will seldom identify the cause of MA, but might help to identify fracture (suggesting trauma), chondrocalcinosis, sacroiliitis or changes of osteoarthritis. Magnetic resonance imaging is helpful in identifying possible infective changes at deep-seated joints (such as the sacro-iliac) and can also help to detect bone and soft-tissue involvement.

Finally, one should keep in mind that MA may occasionally be the first clinical sign of a systemic disease and therefore one may need to consider specific investigations depending upon the clinical features, Table 5.5 enumerates the systemic conditions that might present as an acute MA.

Table 5.5: Possible systemic diseases presenting as acute monoarthritis

Clinical presentation	*Possible systemic disease(s)*
Acute monoarthritis	Psoriasis; Reiter's disease; inflammatory bowel disease
Septic arthritis	Hypogammaglobulinaemia; bacterial endocarditis; bacterial pneumonia; HIV/AIDS
Pseudogout	Haemochromatosis; acromegaly; hypoparathyroidism

Suggested algorithm for the evaluation of MA

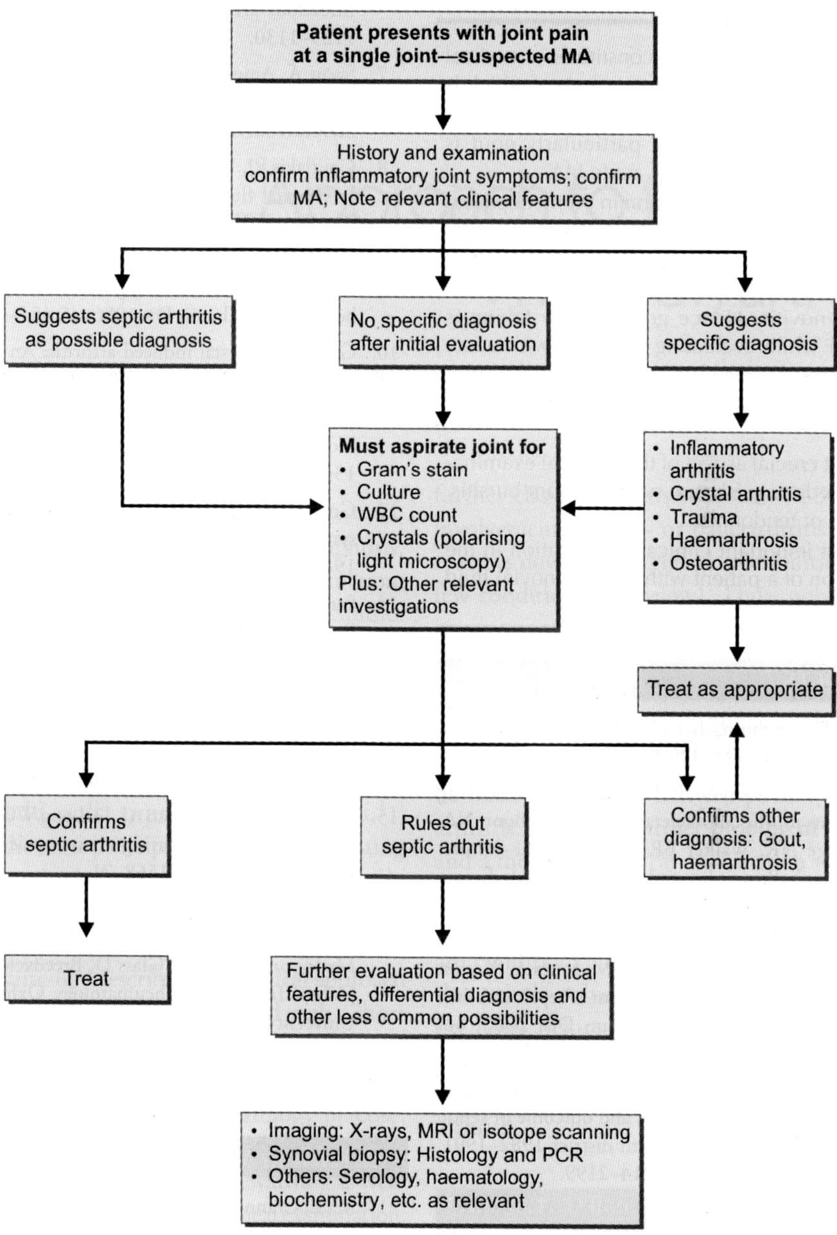

IS IT OLIGO-PAUCI ARTHRITIS?

As mentioned above, the involvement of 2 to 4 joints is called oligo-/pauciarthritis. Its separation from polyarthritis serves the purpose of narrowing down the diagnostic possibilities. Typically, pauci-/oligoarthritis is seen in patients with spondyloarthritis (SpA) that presents with lower extremity large joint involvement with prominent asymmetry and associated enthesopathy. Table 6.2 gives a short list of conditions that commonly present with oligoarthritis. It is to be noted that almost all forms of inflammatory arthritides may evolve through the oligoarticular stage but 'true' oligoarthritis would persist as oligoarticular over time.

Table 6.2: Common conditions causing oligoarthritis

- Spondyloarthritis
- Psoriatic arthritis
- Löfgren's syndrome (a form of acute sarcoidosis)
- Poncet's disease
- A certain subset of juvenile chronic arthritis (JIA)

POLYARTHRITIS: CLINICAL ANALYSIS FOR REACHING A DIAGNOSIS

The following points need to be ascertained:

1. Is it acute or chronic?
2. Is it inflammatory or noninflammatory?
3. The past, personal and family history
4. Review of systems and extra-articular manifestations
5. Pattern and topography of the joint involvement

IS POLYARTHRITIS ACUTE OR CHRONIC?

By definition involvement of more than 4 joints of less than 6-week duration is classified as acute polyarthritis. Patients presenting with acute polyarthritis pose a major diagnostic challenge as this category includes almost every disease of importance in rheumatology some of whom could be potentially life-threatening and associated with high morbidity. Yet, in others, it could simply be a benign self-limiting condition, e.g. a post-viral arthritis that resolves completely within 6 weeks. In the majority of acute polyarthritis cases definitive diagnosis is usually not possible at the outset. The reason is that by-and-large clinical rheumatology is a science of recognising pattern of joint involvement and its evolution and that still remains central for making a rheumatological diagnosis. Therefore, in acute polyarthritis the symptoms may not have yet evolved into a recognisable pattern for a definitive diagnosis. For example, acute dactylitis of a toe in a young person is strongly suggestive but not diagnostic of psoriatic arthritis. Further evolution of the disease over time and/or history of psoriasis or psoriatic arthritis in a blood relative would confirm the diagnosis. However, in other patients, no such clue may be forthcoming in the beginning and the diagnosis is possible only over time. Therefore, most experienced rheumatologists realise that despite their best efforts only further *evolution of pattern of symptoms over time* may give a clue to diagnosis. It may be difficult for a young in-training rheumatologist to accept this 'defeatist' attitude but in clinical rheumatology, it is no shame to wait-and-watch till either the symptoms resolve themselves (benign self-limiting conditions) or evolve over time to become recognisable for making a definitive diagnosis. It may, however, be noted that future may be much brighter. Rapid advances in diagnostic techniques based on advanced technologies may be able to provide a diagnostic clue not only in the early stages of polyarthritis but even in preclinical stages. The best example is the presence of anti-cyclic citrullinated peptides (CCP) antibody in patients with rheumatoid arthritis.[12] The presence of high titres of these antibodies in a patient with acute polyarthritis is now considered a strong clue that it would evolve into rheumatoid arthritis. The American College of Rheumatology (ACR) took cognisance of this new research finding (and several additional new advances) to propose re-revised classification criteria for rheumatoid arthritis in 2010 that is specifically suited for making an early diagnosis.[13] Using these classification criteria now, most rheumatologists would not wait for 6 weeks before embarking upon specific treatment for RA in such a patient. Thus, it is obvious that clinical rheumatology is an exciting and rapidly evolving field. The prevailing diagnostic paradigm may not hold true for too long. However, for the time being rheumatologists must follow the time-honoured methods of making a diagnosis.

In some patients, a more detailed history including past history, family history, review of systems, history of the similar disease in the community, physical examination and diagnostic evaluation may give a clue to the diagnosis. Table 6.3 gives common causes of acute polyarthritis.

IS THE POLYARTHRITIS INFLAMMATORY IN NATURE?

Possibly the most crucial point related to making a diagnosis in polyarthritis (or for that matter in all forms of arthritis) is to evaluate whether it is inflammatory or non-inflammatory in nature.[8] This clinical point is pivotal as it determines the management and prognosis of the patient. A beginner or a non-rheumatologist clinician may not realise the importance of this distinction or mistakenly believe that presence or absence of time honoured easily recognisable features of acute inflammation (namely red, hot, tender, swollen and non-functional) would be easily recognisable in joints for classifying polyarthritis as inflammatory or non-inflammatory. Unfortunately, the joints may be tender, swollen and non-functional even in non-inflammatory conditions, e.g. osteoarthritis (OA) or traumatic arthritis. On the other hand, redness and local heat may not always be present or difficult to perceive even in inflammatory joint disease. For this reason, rheumatologists have devised an entirely different set of clinical features that distinguishes inflammatory from non-inflammatory joint diseases with a high degree of sensitivity and specificity. Table 6.4 lists the contrasting clinical features that distinguish inflammatory from non-inflammatory joint diseases. It is strongly recommended that clinicians get used to properly quizzing the patients about these symptoms by putting open-ended questions. For example, a leading question *'Are you stiff in the morning'*, would elicit a wrong response as against an open-ended question *'How do you feel when you get up in the morning?'* If the clinical history suggests inflammatory polyarthritis, physical examination may reveal a warm joint. Erythema or redness over the inflamed joints is usually not perceptible especially in those with dark skin. However, in those with intense inflammatory arthritis as seen in the periankle-ankle involvement of acute sarcoidosis (Löfgren syndrome) and acute gouty arthritis involving joints in the feet, the overlying skin would show indurated inflammatory oedema with peeling off the outer layer of skin as the inflammation subsides. However, in most of the common forms of inflammatory polyarthritides (RA, PsA) features of local inflammation may only be minimal or imperceptible.

Table 6.3: Causes of acute polyarthritis

- RA, SLE, PsA, other CTDs, systemic vasculitides
- Septic (mainly gonococcal) arthritis, arthritis seen in infective endocarditis
- Reiter's and other 'classic' or 'non-classic reactive arthritides' (e.g. acute rheumatic fever, Poncet, leprosy-related)
- Viral arthritis: Chikungunya, rubella, hepatitis-associated arthritis, parvovirus-B19, others
- IBD-related arthritides
- Sarcoidosis (Löfgren's syndrome-type presentation)
- Erythema nodosum and other panniculitis-related arthritis
- Serum sickness
- Haematological causes: Sickle cell anaemia, leukaemic arthritis, coagulation disorders
- Palindromic rheumatism

(*Abbreviations*: SLE—systemic lupus erythematosus; PsA—psoriatic arthritis; CTDs—connective tissue diseases; IBD—inflammatory bowel disease)

Table 6.4: Clinical features to distinguish inflammatory from noninflammatory joint disease

1. Early morning stiffness that takes at least 30 minutes for maximum improvement, more often it takes about 1 hour for maximum improvement.
2. Activity (gentle movement) improves the symptoms but inactivity does not (inactivity may actually increase the stiffness).
3. Spontaneous 'flares' with fluctuation in the severity of symptoms over time without any obvious extraneous cause.
4. Constitutional symptoms: Fatigue and tiredness, feverish feeling or actual fever, night sweats, loss of appetite and weight.

Note: Presence of one, more or all of these manifestations would strongly suggest the inflammatory nature of the joint problem with increasing probability

PAST, PERSONAL AND FAMILY HISTORY

The past history may have the clue for diagnosing polyarthritis. A transient episode of heel pain (Achilles tendon enthesitis, plantar fasciitis) at the juvenile age or an episode of uveitis in the past may be the clue to the present problem of inflammatory low back pain due to ankylosing spondylitis/ spondyloarthritis (AS-SpA) group. History in the recent past, of a febrile episode with or without exanthema, sexual contact or frank venereal disease, an episode of diarrhoea, or acute conjunctivitis, may give a clue to the diagnosis of reactive arthritis, gonococcal arthritis or AS related peripheral arthritis. Quite often, past history may also unmask a diagnostic mistake made in the distant past, e.g. history of 'fever-of-unknown-cause' in the past that was mistakenly either treated as typhoid or tuberculosis may actually have been an early manifestation of systemic lupus erythematosus that became overt in due course of time. Three most common diagnostic mistakes seen in day-to-day rheumatology practice involve tuberculosis, gout and a rheumatic fever. It is quite common that a child or juvenile was given treatment for tuberculosis or rheumatic fever (long-acting penicillin) for months in the past for a mono- or oligoarthritis that evolved into rather typical enthesitis-related arthritis (ERA) over time. The other common mistake is to diagnose and treat the patient as 'gout' (only based upon trivial increase in serum uric acid) over years while the patient has been evolving into any of the common polyarthritides. The importance of personal history has sharply come under focus in the recent years. Thus, it has now been proven that cigarette smoking may be a trigger for most inflammatory arthritides (RA, SpA, PsA, SLE) as well as accelerate the cartilage damage.[14–18] There may be several other personal factors yet to be discovered. Family history is equally important in making a diagnosis of polyarthritis. Similar disease (e.g. RA), or closely related autoimmune disease (thyroid, connective tissue disease, inflammatory bowel disease (IBD), other autoimmune conditions) in the family members could be very useful in reaching a diagnosis. A so-called 'seronegative inflammatory polyarthritis' may actually be PsA (even in the absence of psoriasis in the patient) if a blood relative is known to have psoriasis. The significance of family history is more obvious in frank hereditary conditions, e.g. AS and gout where invariably a positive history is elicited.

REVIEW OF SYSTEMS AND EXTRA-ARTICULAR MANIFESTATIONS

Inflammatory polyarthritis has bidirectional association with other organ systems of the body.

On the one hand, it may affect different body parts and organs secondarily. On the other hand, by itself it may be part of a multisystem disease. Therefore, a detailed review of systems along with a careful general and systemic physical examination is mandatory for making an accurate diagnosis.

History of *skin, mucosal and subcutaneous lesions* is highly relevant for making a diagnosis in a large number of inflammatory polyarthritides so much so that a rheumatologist has to be a good dermatologist. Thus, diagnosis of an otherwise undiagnosed 'seronegative inflammatory polyarthritis' becomes 'PsA' on discovering a psoriatic patch anywhere in the body, however small it may be. Even a family history of psoriasis may help similarly. RA may have lesions of pyoderma gangrenosum (a form of neutrophilic dermatosis, also seen in IBD that itself may also be associated inflammatory polyarthritis, see below), subcutaneous nodules, and vasculitic lesions. Facial 'butterfly' rash of systemic lupus erythematosus (SLE) along with mucosal ulcers, alopecia and vasculitic skin lesions is diagnostic of the disease. Oral ulcers are seen in several other polyarthritides including Behçet's disease (very painful commonly associated with genital ulcers), IBD-related arthritis, and SpA. Gottron's papules, heliotrope-facial rash, 'mechanic's hands', 'shawl sign' along with periungual vasculitic infarcts are typical of dermatomyositis. Raynaud's phenomenon associated with scleroderma skin changes including telangiectasia are typical of the disease and must be distinguished from scleredema of Bushke (no Raynaud's phenomenon). Sarcoid skin lesions including lupus pernio may come handy in diagnosing the cause of otherwise undiagnosed 'seronegative inflammatory polyarthritis'. Urticarial vasculitis can be suspected on typical skin lesions. Erythema nodosum and several other forms of inflammatory lesions of the subcutaneous tissue (panniculus) is seen in association with a number of inflammatory arthritides the best known being

Löfgren syndrome (acute sarcoidosis), already mentioned above, Behçet's disease (the lesion often ulcerates) and IBD. Subcutaneous nodules in association with polyarthritis are seen with several diseases including RA, SLE, OA (Heberden's and Bouchard's nodules), gout (gouty tophi), sarcoidosis, rheumatic fever, polyarteritis nodosa, dislipidaemias, and others. Mucosal ulcers in SLE (moderately painful), SpA (little pain), and Behçet's disease (severely painful) are helpful clinical features for diagnosis. Palpable as well as non-palpable purpuric lesions are seen in several polyarthritides including SLE, Sjögren's syndrome (including hyper-gammaglobulinaemic purpura of Waldenström) and others. Leg ulcers (vasculitic or part of neutrophilic dermatosis, e.g. pyoderma gangrenosum) are seen in RA, SLE (subcutaneous inflammation that may ulcerate e.g. lupus profundus), polyarteritis-nodosa (including cutaneous polyarteritis), Behçet's disease and others. Keratoderma blenorrhagicum seen with Reiter's syndrome is a typical lesion that helps in making a definitive diagnosis. Recognition of gonococcal skin lesions could come in handy for diagnosing an otherwise idiopathic migratory inflammatory polyarthritis. Gross clubbing in a smoker presenting with painful swelling of distal joints in the extremities (wrists, small joints in the hands, ankles, joints in the feet) should immediately bring in the possibility of carcinoma lung-related hypertrophic pulmonary osteoarthropathy. The list is rather big to be covered here but suffice it would be to note that skin-subcutaneous tissue-mucosa (including nails, scalp) should be carefully evaluated in any patient with inflammatory polyarthritis.

The association of *eye and ear-nose-throat-mouth* with polyarthritis is possibly equally important. Sicca symptoms are well-known features of Sjögren's syndrome, but also occur secondary to RA, SLE, scleroderma and others. Episcleritis-scleritis is common in RA. Iridocyclitis-uveitis is typically seen in association with SpA group including Reiter's syndrome. Retinal lesions are seen in SLE, systemic vasculitides and others, retrobulbar lesions (causing proptosis) may be seen in granulomatosis with polyangiitis (GPA, Wegener's granulomatosis), the involvement of rectus muscles may occur as part of inflammatory myositis. Inflammation of cartilage in the pinna and bridge of the nose (with similar cartilage lesions at other sites) is typically seen in relapsing polychondritis causing flail or deformed pinna giving the appearance of a 'boxer's ear' and 'saddle-nose' deformity. The latter is also seen in granulomatosis with polyangiitis (GPA, Wegener's granulomatosis) that usually destroys the cartilage at the bridge of the nose. Involvement of middle and inner ear is typically seen in GPA (Wegener's granulomatosis).

Inflammatory polyarthritis also has a close association with *genitourinary* and *gastrointestinal* systems. A urethritic form of 'reactive arthritis', the most dramatic subset being Reiter's syndrome, is a well-known cause of seronegative inflammatory polyarthritis. Gonococcal infection with urethritis, cervicitis and migratory inflammatory polyarthritis is a well-recognised association. IBD also has a dual association with inflammatory polyarthritis. Thus, several patterns of inflammatory polyarthritis have been described in association with IBD including peripheral polyarthritis and erythema nodosum (mentioned above). Another form is that of spondyloarthritis some of whom may evolve into frank ankylosing spondylitis. Subclinical/asymptomatic small segments of lesions of IBD have been demonstrated by endoscopic examination in a significant proportion of patients with spondyloarthritides.[19] Behçet's disease is the other condition where IBD-type lesions may be seen. In fact, there are patients diagnosed with IBD who on detailed clinical evaluation may show features of milder forms of Behçet's disease including orogenital ulcers, erythema nodosum lesions, eye inflammation.

Renal disease and inflammatory polyarthritis combination are among the most serious, often life-threatening forms of arthritic disorders. Urinary sediment abnormality indicative of glomerular inflammation (microscopic haematuria, a variety of casts and WBC), sub-nephrotic or nephrotic range proteinuria, frank nephrotic syndrome, renal arterial involvement with renal hypertension, rapidly progressive glomerulonephritis, and end-stage renal disease are the manifestations of several diseases that may have associated inflammatory polyarthritis. This group mainly includes SLE and its overlap with other connective tissue diseases, antiphospholipid syndrome, and anti-neutrophil cytoplasmic antibody (ANCA)-associated small vessel systemic vasculitides [GPA (Wegener's granulomatosis), microscopic polyangiitis, eosinophilic granulomatosis with polyangiitis (Churg-Strauss syndrome) and the medium vessel vasculitis, namely the classic polyarteritis nodosa].

Peripheral, as well as *central nervous system* involvement and inflammatory polyarthritis, is another potentially serious, often life-threatening group of arthritides. These include SLE, antiphospholipid syndrome, systemic vasculitides of several types (mononeuritis multiplex is a common manifestation in some of them), and Behçet's disease.

Lung involvement is commonly seen with RA, connective tissue diseases especially in patients with inflammatory myositis, systemic vasculitides, sarcoidosis, relapsing polychondritis, and AS. There is an important clinical subset of pulmonary-renal syndrome where kidney and lung are affected in addition to an inflammatory polyarthritis. SLE, scleroderma and other connective tissue diseases, ANCA-associated vasculitides, Goodpasture's syndrome, pauciimmune necrotising and crescentic glomerulonephritis (CGN), Henoch-Schönlein's purpura (HSP), and essential mixed cryoglobulinaemia (EMC) are some of its causes.[20]

Heart involvement is typically seen in antiphospholipid syndrome, SLE, scleroderma, dermatomyositis, and in a form of systemic vasculitis seen in paediatric age group (Kawasaki disease). Rheumatic fever is too well known to be reminded of but, it must be noted that persistent inflammatory polyarthritis is not a feature of this disease. Moreover, it is a disease of paediatric age group. Polyserositis with inflammatory polyarthritis may be seen in SLE, overlap connective tissue diseases and a relatively uncommon condition in our country, namely Familial Mediterranean Fever (FMF). Pleuropericarditis is typically seen in SLE but may be present in RA and other systemic conditions.

Obstetric and gynaecological problems may occur in several diseases with inflammatory polyarthritis, mainly due to joint deformities causing biomechanical/physical issues. But, the antiphospholipid syndrome is the most notorious for causing foetal wastage and related problems. Increased risk during pregnancy and purpureum is characteristic of SLE.

Haematological abnormalities are universal in inflammatory polyarthritis and include high erythrocyte sedimentation rate (ESR), high platelets, anaemia of chronic disease and occasionally, leucocytosis that could be extreme on occasions. However, some haematological abnormalities may specifically point towards a certain diagnosis. Thus, low platelets may point towards the diagnostic possibility of SLE and antiphospholipid syndrome. Rare cases of leukaemic arthropathy in adults may have not only low platelets but also other abnormalities in the blood that may be a clue to the diagnosis. The hypercoagulable state with widespread thromboembolic manifestations is rather typical of the antiphospholipid syndrome that is often associated with SLE.

WHAT IS THE PATTERN OF JOINT INVOLVEMENT?

The next point to note in polyarthritis is the pattern of evolution of the joint involvement. Basically there are 3 patterns, of the joint involvement namely *additive, intermittent,* and *migratory.*

In *additive* pattern (increasing number) joints are progressively recruited with many joints getting affected over time. Rheumatoid arthritis is a typical example of such a pattern of involvement.

In *intermittent* pattern the joint involvement appears and then disappears till the next such episode with completely asymptomatic period in between these episodes. In its inflammatory form it is typically seen in palindromic rheumatism, acute gouty arthritis, relapsing polychondritis and relapsing seronegative symmetrical synovitis with pitting oedema (RS3PE). The last mentioned condition is associated with marked joint stiffness and symmetrical synovitis in the joints of the hands and feet. A condition called intermittent hydrarthrosis is an idiopathic noninflammatory intermittent arthritis.

In *migratory* pattern, the joints become symptomatic then subside while different joints get involved. The tempo of migration may differ, for example, it could be only hours between one to the next joint involvement in rheumatic fever. On the other hand, the tempo of 'migration' may be much slower (days) in gonococcal arthritis. These patterns are not mutually exclusive but the dominant pattern may indicate a specific diagnosis.

The *topography or distribution of joint involvement* (which joints are affected in what distribution) usually helps in narrowing down the diagnostic possibilities.[21] For describing the *topography* of the joint involvement the following anatomical regions are taken in account: *Peripheral joints* (in the extremities), *axial joints* (traditionally sacroiliac joints are considered part of axial joints), *root joints* (that overlap between peripheral and axial joints, namely shoulder and hip joints), and *upper segment* and *lower segment* of the body (above and below waist). In peripheral joints *symmetry* or *asymmetry* is important to note. Table 6.5 gives some of the common causes of polyarthritis classified as inflammatory versus noninflammatory further categorised on the basis of the *topographic pattern* of joint involvement. Figures 6.1 to 6.5 depict some of the typical topographic patterns of the joint involvement in different forms of inflammatory and noninflammatory arthritides. Caines A, et al[21] had diagrammatically depicted the pattern of the distribution of pain in different musculoskeletal diseases. Thus, inflammatory synovitis of small as well as large joints in the extremities in a symmetrical distribution on both sides of the body, equally distributed in upper and lower segment with

Table 6.5: Common causes of polyarthritis classified as inflammatory vs. noninflammatory further categorised on the topographic pattern of the joint involvement (Modified from Ref. 10)

1. Inflammatory polyarthritis

a. Peripheral polyarthritis, mostly symmetrical in distribution
- i. RA
- ii. SLE and other connective tissue diseases
- iii. Wegener's granulomatosis and other systemic vasculitides
- iv. PsA

b. Peripheral oligoarthritis or polyarthritis, often asymmetrical
- i. PsA*
- ii. Peripheral involvement in SpA including Reiter's disease and other forms of 'classical' reactive arthritis.*
- iii. Enteropathic arthritis*
- iv. Behçet's disease
- v. Polyarticular gout and pseudogout
- vi. Infective endocarditis
- vii. Relapsing polychondritis
- ix. Sarcoidosis
- xi. Amyloid arthropathy

(Those marked by '*' may have axial disease as well)

(*Note*: 1. PsA as well as sarcoid joint disease have several different patterns. Therefore, their names may appear in different lists of oligo or polyarthritides. 2. For labelling the condition as oligoarthritis (2–4 joints) midfoot and wrist is taken as a single joint. 3. Several conditions mentioned above as 'oligoarthritis' may have a subset that presents mainly as oligoarthritis, in others it may only be a phase in the evolution of a polyarthritis).

2. Noninflammatory polyarthritis (generic term osteoarthritis)

a. Hereditary factors
- i. Primary nodular OA of the hands (distal and proximal interphalangeal joint involvement)
- ii. Generalised primary OA

b. Environmental factors and age-related osteoarthritis: Life-style, occupational, trauma
- i. First metatarsophalangeal joint (bunion), the knee joints, the distal interphalangeal joints (including Heberden's nodes), proximal interphalangeal joints (including Bouchard's nodes), hip joints, lumbar and cervical spine—usually an obese elderly person
- ii. OA following local injury (including surgery) in a joint
- iii. OA in strenuous exercise-related cartilage damage (chondromalacia)

c. Metabolic conditions, some uncommon diseases
- i. Hypothyroidism
- ii. Acromegaly
- iii. Haemochromatosis
- iv. Ochronosis
- v. CPPD deposition disease

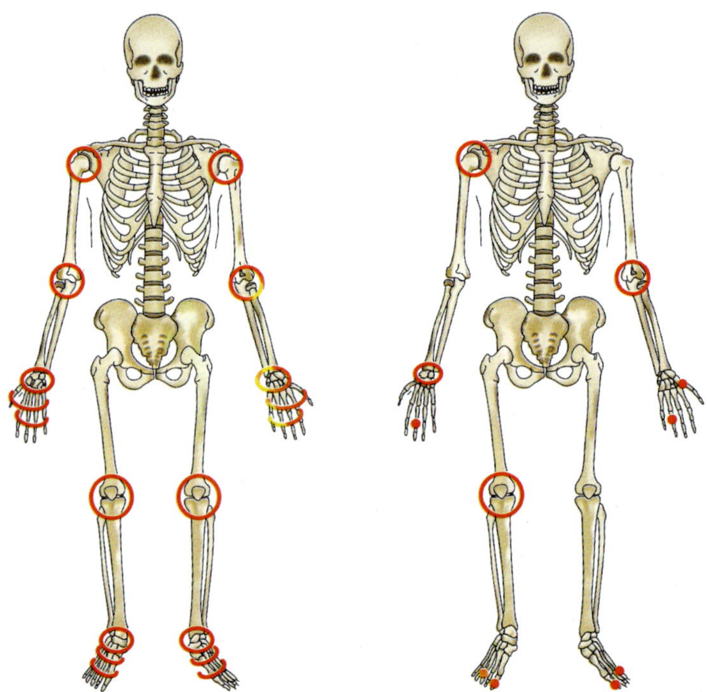

Fig. 6.1: Pattern of joint involvement in rheumatoid arthritis (RA) versus psoriatic arthritis (PsA). Note the polyarticular involvement in RA (left) but aymmetrical pattern in PsA (right)

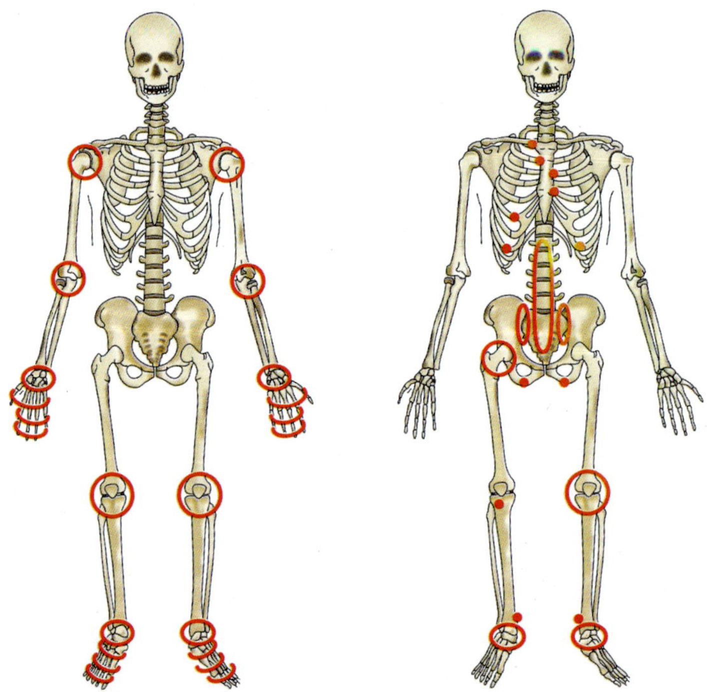

Fig. 6.2: Pattern of joint involvement in rheumatoid arthritis (RA) versus spondyloarthritis (SpA). Note the polyarticular involvement in RA (left). In SpA there is asymmetrical lower limb, sacroiliac joint, and spine involvement (right). Enthesitis is depicted as red dots

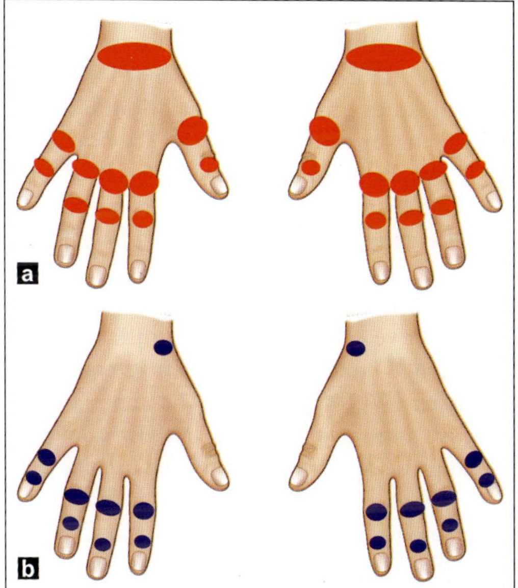

Fig. 6.3: Pattern of hand joints involvement in rheumatoid arthritis (RA) depicted as red circles (panel a); versus osteoarthritis (OA) (blue circles) panel b. Both have symmetrical involvement of the PIPs, but the DIPS are spared in RA

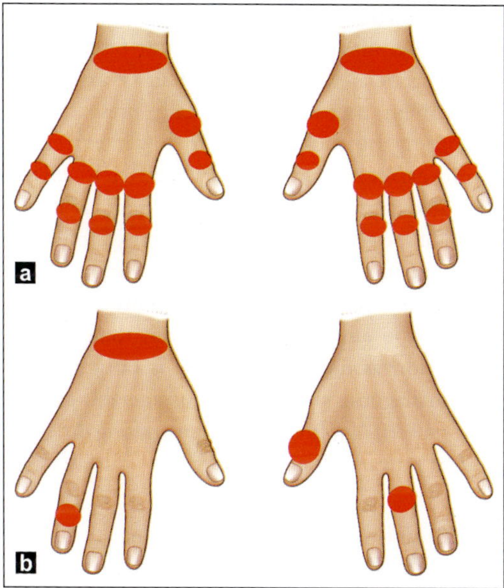

Fig. 6.4: Pattern of hand joints involvement in rheumatoid arthritis (RA) depicted as red circles (panel a); versus PsA (panel b). RA is symmetrical, but PsA is asymmetrical with DIP joint involvement

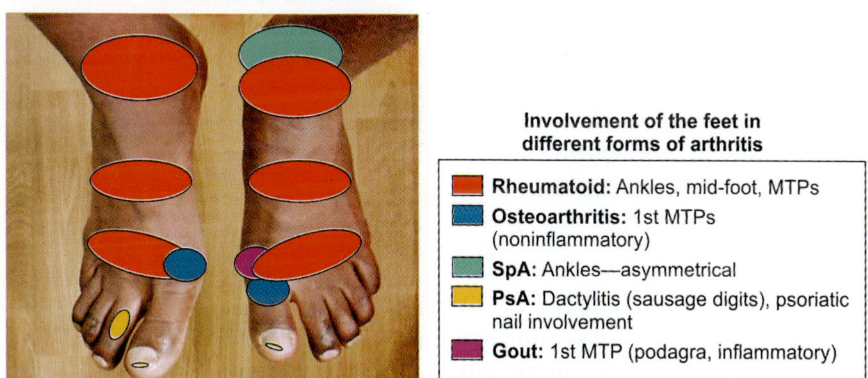

Fig. 6.5: Pattern of involvement of the feet in different forms of arthritis

predominant involvement of joints of the hands, wrists and feet, with sparing of the distal interphalangeal (DIP) joints, is rather typical of RA. Then, there are certain joints that, if involved in isolation or predominantly, may be a clue to diagnosis. Thus, involvement of DIP joints without synovitis or inflammation is typically seen in primary nodular OA but with synovitis, it is often seen in some subsets of PsA, enteropathic arthritis and sarcoidosis. Noninflammatory involvement of the first carpometacarpal joint (floor of the 'snuff-box') is typically seen in primary OA. Similarly, noninflammatory first metatarsophalangeal joint involvement is most commonly seen in OA. In acute gouty attack the same joint is often the first one to be affected with severe acute inflammatory response (podagra). Periankle-ankle inflammatory arthritis, (as mentioned earlier) usually associated with erythema nodosum is seen in acute sarcoid arthritis (Löfgren syndrome). Acute *dactylitis* has already been mentioned to be strongly suspicious of psoriatic arthritis. *Enthesitis* (inflammation at the sites of insertion of ligaments, tendons to the bone such as Achilles tendonitis, plantar fasciitis and others) associated with inflammatory axial joint involvement, sacroiliitis (classically presents with alternating buttock pain) and lower extremity asymmetrical inflammatory arthritis, often associated with root joint involvement, is rather typical of spondyloarthritis (SpA) group. Prominent involvement of metacarpophalangeal joints without synovitis is typically seen in haemochromatosis. It needs to be emphasised, however, that pattern recognition as mentioned above is only a rough guide to diagnosis since there is considerable overlap in the pattern of joint involvement among different types of polyarthritis.

OLIGOARTHRITIS

Involvement of 2 to 4 joints is called oligo-/pauci arthritis. Its separation from polyarthritis serves the purpose of narrowing down the diagnostic possibilities. Typically, inflammatory peripheral arthritis seen in the various forms of SpA is oligoarticular in nature with involvement of joints of the lower segment of the body with prominent asymmetry and associated enthesopathy. Table 6.3 has already been referred to above providing a short list of conditions that commonly present with oligoarthritis. It is to be noted that almost all forms of inflammatory arthritides may evolve through the oligoarticular stage but 'true' oligoarthritis would remain oligoarticular over time.

JOINT COUNT

It has been documented that formal joint count is not include in most visits of most patients with rheumatoid arthritis to most rheumatologists.[22] However, rheumatologists need to be reminded that with rapid advancements in drug therapy of patients with a variety of musculoskeletal diseases and especially that of inflammatory arthritides, some sort of objective assessment of the disease activity, the simplest method being joint count, must be carried out in every patient at the first encounter.[23] Irrespective of what methodology is adopted, it should not be ignored. One may try to adopt the most suited method for the type of clinical set-up one works in. Original Disease Activity Score (DAS), its modification (DAS28), more recently devised and validated Clinical Disease Activity Index (CDAI) and Simplified Disease Activity Index (SDAI) are widely recommended for use in RA and other inflammatory arthritides.[24] Bath Ankylosing Spondylitis Disease Activity Index (BASDAI), Bath Ankylosing Spondylitis Functional Index (BASFI), Bath Ankylosing Spondylitis Metrology Index (BASMI) are available for objective assessment of different aspects of inflammatory spinal arthritis. The methodology is also available for objective assessment of noninflammatory diseases like osteoarthritis Western Ontario and McMaster (WOMAC) index. The importance of objective assessment of disease activity cannot be overemphasised. Over the years it has become clear that tight control of disease activity in inflammatory arthropathies prevents joint damage and disability and reduces morbidity and mortality. But, tight-control of disease activity with 'treat-to-target' approach[25] is possible to achieve only if it is measured routinely. Treatment of diabetes mellitus or hypertension is good example to understand this argument. If blood glucose or blood pressure readings are not documented regularly they cannot be controlled satisfactorily. Unfortunately, most clinicians would treat inflammatory polyarthritis but without knowing the degree of disease activity to guide them. Such a patient is likely to develop increasing damage and disability over time with increased morbidity and mortality.

INVESTIGATIONS

As mentioned above, the most crucial point towards making a diagnosis in polyarthritis is to differentiate between inflammatory and noninflammatory conditions. History and physical examination provide good idea whether the condition is inflammatory in nature. In most situations the inflammatory nature of polyarthritis can be confirmed by routine laboratory investigations for non-specific markers of inflammation, namely raised levels of acute phase proteins. Table 6.6 gives the list of commonly available tests for confirming inflammation. Some exceptions, however, must be pointed out. In certain diseases ESR may not truly reflect the inflammatory state. This is often true of SLE, scleroderma and AS where ESR may be disproportionately high or low. Similarly, platelet counts may be low in SLE as part of its clinical

Table 6.6: Commonly used nonspecific laboratory markers of inflammation (acute phase reactants)

1. Increased in
- a. ESR
- b. Platelet count
- c. Total serum globulins with reversal of albumin/globulin ratio
- d. Leukocyte count (often but not always)
- e. CRP
- f. Alkaline phosphatase

2. Decreased in
- a. Haemoglobin (normocytic normochromic anaemia of chronic disease)
- b. Serum albumin (with reversal of albumin/globulin ratio)

manifestation, as also in a relatively uncommon condition in adults, namely leukaemic arthropathies. Examination of *synovial fluid* in polyarthritis only for confirming inflammatory nature of the joint disease is generally not recommended. This is because inflammatory condition can be reasonably differentiated from noninflammatory condition on history and physical examination that can be further confirmed by tests for acute phase reactants. However, on rare occasions there could be some confusion because of poor history, no definite clue in physical examination and inconclusive or borderline results of acute phase reactants. In such cases joint fluid examination (if joints have effusion) for leukocyte count can help. A leukocyte count of >2000/cmm is a definite sign of inflammatory synovitis. Additional tests including gram staining, bacterial and mycobacterial culture, and examination under polarised light microscopy for crystals would also help in making a definitive diagnosis. *Imaging examination* of the joints and other routine imaging investigations do not distinguish between inflammatory and noninflammatory conditions. They may, however, detect typical pattern of *radiographic abnormalities* associated with certain diseases (e.g. typical erosions of RA or characteristic juxta-articular new bone formation {appearing as ill-defined ossification near the joint margins, but excluding osteophyte formation} on plain radiograph of the hand or foot seen in PsA, typical 'scooped-out' lesion of chronic gouty joint, sacroiliitis of SpA, aggressive destructive lesions of joint tuberculosis and other infections, chondrocalcinosis in calcium pyrophosphate dihydrate {CPPD} crystal deposition disease, etc.). This may indirectly give a clue to the inflammatory nature of the joint disease. *Magnetic resonance imaging* (MRI), on the other hand, can demonstrate bone marrow oedema that has been demonstrated to be the earliest sign of inflammatory nature of the joint disease in recent studies.[26] Being an expensive investigation it cannot be recommended for routine use for confirming the inflammatory joint disease. *Histopathology* is another investigation that may give a definitive diagnosis in difficult cases of polyarthritis. However, only in rare circumstances, a synovial biopsy may be necessary for making a diagnosis. Thus, it may be required in suspected polyarticular infection (mycobacterial infection, fungal), villonodular synovitis, etc. On the other hand, skin or subcutaneous lesions may often yield useful diagnostic information (e.g. sarcoidosis, erythema nodosum, vasculitis). Specific tissue biopsy may be indicated in specific diseases, e.g. sural nerve biopsy in systemic vasculitis, muscle biopsy in inflammatory myositis. Once a provisional diagnosis has been made, in the second-step of diagnostic work-up, more *specific investigations* are usually required. Thus, in RA, one may like to have some idea of disease prognosis. Quantitative measurement of rheumatoid factor (RF) and antibodies against anti-cyclic citrullinated proteins (ACPA) are useful in this respect. In suspected SLE and related connective tissue disorders screening for immunofluorescent antinuclear antibody (F-ANA) is of great importance. F-ANA has a very high sensitivity for SLE and other connective tissue diseases but it has rather a low specificity. This makes F-ANA a very useful test *for excluding the diagnosis of these diseases;* if F-ANA test is negative it would be most unlikely that the patient would have SLE or related connective tissue disease. Immunofluorescent anti-neutrophil cytoplasmic antibody (ANCA) test is similarly a very useful screening test for patients suspected of systemic vasculitides; a negative immunofluorescent ANCA test would go strongly against the diagnosis of ANCA-associated vasculitides. ANCA titres have also been shown to have some degree of correlation with disease activity. For routine *monitoring of drug toxicities* standard complete blood counts, liver enzymes, renal parameters, and routine urine examination would suffice in most cases. Baseline electrocardiogram, chest radiograph and stool

Chapter

8

Approach to a Patient with Suspected Connective Tissue Disorder

Anupam Wakhlu, Rasmi Ranjan Sahoo

INTRODUCTION

Connective tissue disorders (CTDs) are a heterogeneous group of diseases affecting the connective tissues of the body. These tissues are found in all structures in the body and are composed of two important proteins, collagen and elastin. Sometimes, the term "collagen vascular disease" is used interchangeably. The CTD's are classified into two groups: Inherited (e.g. Marfan's syndrome, Ehlers-Danlos syndrome, osteogenesis imperfecta) and acquired (e.g. systemic lupus erythematosus (SLE), rheumatoid arthritis (RA), systemic sclerosis, Sjögren's syndrome, idiopathic inflammatory myopathies). Most clinicians do not include systemic necrotizing vasculitides, e.g. polyarteritis nodosa (PAN), eosinophilic granulomatosis with polyangiitis (EGPA) and granulomatosis with polyangiitis (GPA) in the category of CTD.[1] The acquired CTDs are characterized by immune responses against self-antigens resulting in the production of autoreactive T and B cells. Thus, they are also called systemic autoimmune diseases and can affect any organ system.

The CTDs are uncommon/rare, but potentially life-threatening diseases. Young females of child bearing age are most frequently affected. The spectrum of clinical manifestations is wide with skin and joints affected most frequently. A definitive diagnosis of CTD can be made based on the history and physical examination alone in 80–90% of patients. Further, 'classification criteria' for CTDs are not meant for diagnosing diseases in individual cases, as many patients do not satisfy requisite criteria early in the disease course or may do so sequentially.[2] Thus, it is important to know the common manifestations and basic diagnostic work up for CTDs, which will help in avoiding unnecessary investigations and allow timely management.

Patients with CTDs can present with either non-specific or organ-specific features. The common manifestations of CTDs are listed in Table 8.1.

The presence of one or more of these specific features warrants further investigation for CTDs.

Table 8.1: Common manifestations of CTDs

Non-specific	*Specific*
Fever	Skin rash
Myalgia	Photosensitivity
Arthralgia	Palpable purpura
Malaise	Subcutaneous nodule
Fatigue	Mucocutaneous ulcer
Weight loss	Alopecia
	Arthritis
	Skin thickening
	Raynaud's phenomenon
	Dry eyes/mouth
	Pleurisy, ILD
	Carditis, PAH
	Glomerulonephritis
	Myositis
	Neuropathy
	Gangrene of fingers/toes
	Vascular event at an early age (CVA, MI)

ILD: Interstitial lung disease; PAH: Pulmonary arterial hypertension; CVA: Cerebrovascular accident; MI: Myocardial infarction

Skin Rash (Table 8.2)

There are numerous causes of skin rash; however, certain characteristics may aid in the diagnosis of CTDs, e.g. morphology (discoid rash of SLE), distribution (heliotrope rash of dermatomyositis (DM), association with photosensitivity.

The malar rash of SLE (Fig. 8.1) is not pathognomonic and can also be seen in DM (**Fig. 8.2a**) and acne rosacea. However, sparing of the nasolabial folds is characteristic of SLE (Fig. 8.1). Typical discoid lesions of SLE are characterized by papules or plaques with hyperpigmented erythematous borders and atrophic centers with follicular plugging. The heliotrope rash of DM (Fig. 8.2) is seen in less than 50% of cases (specially in dark-skinned individuals) and is purplish in color, located in the periorbital area, particularly over the upper eyelids.[4] The Gottron papules (Fig. 8.3b) are erythematous or violaceous lesions located over the bony prominences. The Gottron's sign refers to erythematous macules over the same distribution area as the Gottron papules. Lupus pernio, diagnostic of chronic sarcoidosis is a maculopapular, erythematous or purple rash involving the nose, the area beneath the eyes and cheeks.

Photosensitivity

Photosensitivity is defined as an exaggerated response to UV light, inducing symptoms such as burning, itching, and redness. It is a relatively sensitive indicator of SLE and is seen in up to 69% of patients.[5] Photosensitivity is also seen in DM. Although it is seen in many primary cutaneous disorders including polymorphic light eruption, solar urticaria, actinic prurigo, its association with other specific features may favour a diagnosis of CTD.

Palpable Purpura

Palpable purpura is a manifestation of small vessel vasculitis, which can be either primary, e.g. Henoch-Schönlein purpura (HSP), cryoglobulinemic vasculitis, ANCA-associated vasculitis or secondary, e.g. CTDs, drugs, infections and malignancy.

Subcutaneous Nodules

CTDs like RA, sarcoidosis can present with subcutaneous nodules and it has to be differentiated from other causes (Table 8.3). Rheumatoid nodules can occur at a number of locations in the body but are most commonly located on the extensor surface of the forearms (Fig. 8.4).

Table 8.2: Mucocutaneous manifestations in CTDs

SLE[3]	*Dermatomyositis[4]*
Acute • Malar rash (Fig. 8.1) • Bullous lesions • Maculopapular rash • Photosensitivity • Toxic epidermal necrolysis • Subacute (psoriasiform, annular polycyclic lesions) **Chronic** • Discoid rash • Hypertrophic (verrucous) rash • Lupus panniculitis • Lupus tumidus • Mucosal lupus • Chilblain lupus • Discoid lupus/lichen planus overlap	**Pathognomonic** • Heliotrope rash (periorbital region) • Gottron papules (over the bony prominences, e.g. MCP, PIP and DIP joints of the hands) **Less specific** • Gottron's sign • Photosensitivity • "V" sign (anterior part of the chest) (Fig. 8.2) • Pruritic scalp involvement • "Shawl" sign (nape of the neck and upper part of the back) (Fig. 8.2) • "Holster" sign (lateral surfaces of the thighs) • Nail-fold capillary changes with cuticular overgrowth • Mechanic's hand (Fig. 8.3a)
Sarcoidosis	*Systemic sclerosis*
• Maculopapular rash • Lupus pernio • Erythema nodosum • Subcutaneous nodule	• Sclerodactyly • "Salt and pepper" appearance (Fig. 8.5) • Fixed flexion contractures of the fingers • Telangiectasia • Calcinosis cutis

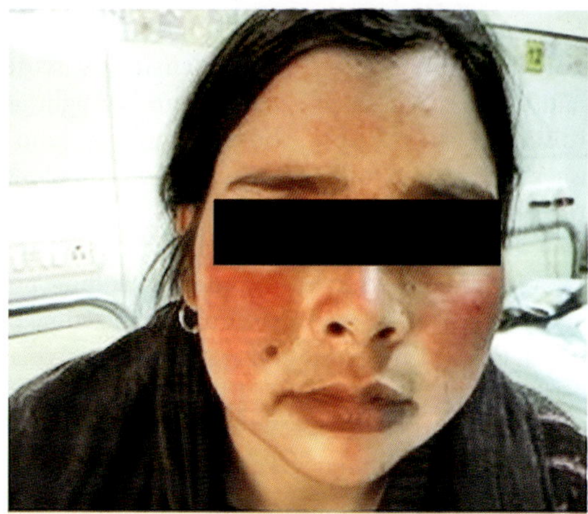

Fig. 8.1: Malar rash in a patient with SLE. Note that the nasolabial folds are spared

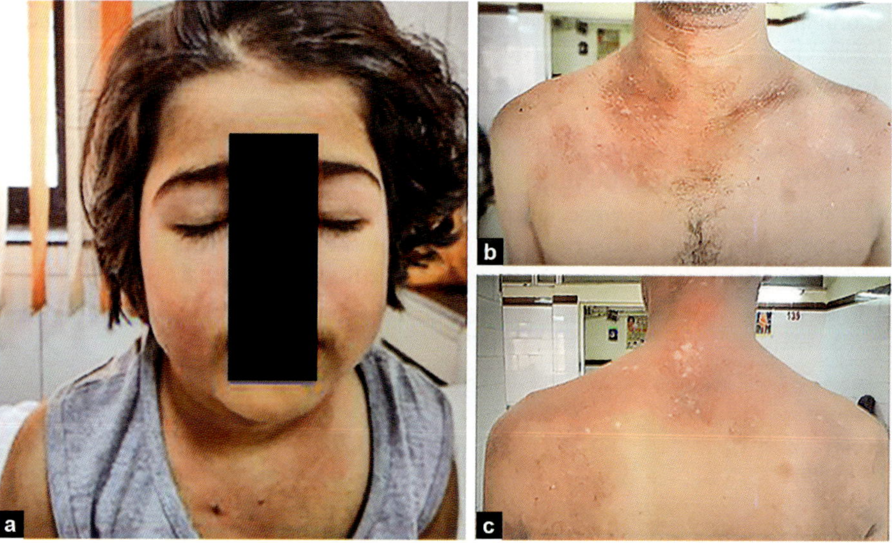

Fig. 8.2: Skin manifestations of dermatomyositis. Heliotrope and malar rash (a); V sign (b); Shawl sign (c)

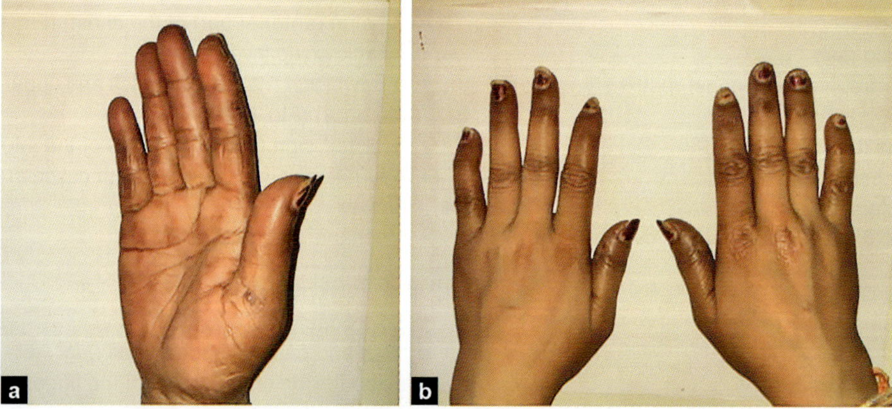

Fig. 8.3: The hands in dermatomyositis. Mechanic's hands (a); Gottron's papules (b) note the erythematous lesions over the bony prominences

Table 8.3: Differential diagnosis of subcutaneous nodules[6]

Basic pathogenesis	*Disease*
Inflammatory	Erythema nodosum Weber-Christian disease
Infections	Acute rheumatic fever HBV infection Hansen's disease Syphilis Coccidioidomycosis Sporotrichosis
Metabolic	Diabetes mellitus Hypercholesterolemia
Crystal arthritis	Gout
Storage diseases	Farber disease Multicentric reticulo-histiocytosis
Others	Amyloidosis

Mucocutaneous Ulcer

The presence of mucocutaneous ulcers may favour a diagnosis of CTD. However, infections like HIV, Herpes simplex, malignancy, vitamin B_{12} deficiency, inflammatory bowel disease and reactive arthritis must be ruled out. Painful ulcers are usually seen in oral cavity in SLE. Recurrent, painful oral and genital ulcers are seen in Behçet's disease.

Alopecia

Alopecia may be seen in dermatological conditions like alopecia areata, telogen effluvium, etc. Alopecia has been included in the 2012 SLICC/ACR classification criteria for SLE.[3] Both non-scarring and scarring alopecia are seen in SLE (Fig. 8.1) of which scarring alopecia is considered to be more specific.

Arthritis

Duration of arthritis, presence and duration of EMS, number of joints involved and their pattern of involvement are important considerations. It is also important to determine whether the source of pain is articular or extra-articular. RA is characterized by symmetrical polyarthritis affecting small and large joints. Bone erosions and deformity are features of chronic RA. The classical deformities described in RA are swan-neck deformity (hyperextension at PIP joint and flexion at DIP joint), boutonnière deformity (flexion at PIP joint and hyperextension at DIP joint), Z-line deformity (subluxation of 1st MCP joint with hyperextension of 1st interphalangeal joint). Other CTDs like SLE, MCTD and Sjögren's syndrome may also present with polyarthritis which is usually nonerosive and nondeforming.

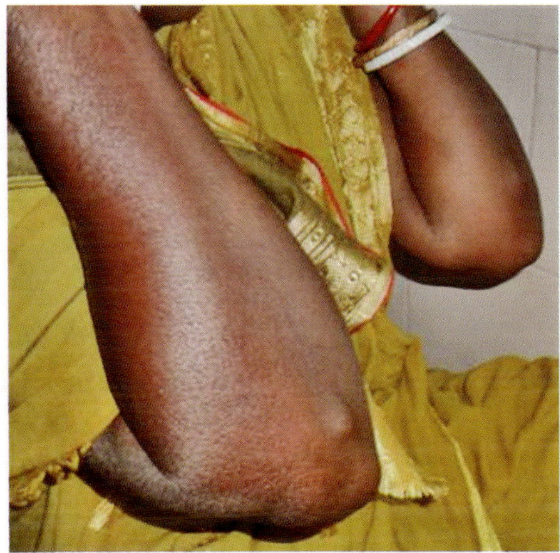

Fig. 8.4: Rheumatoid nodules over extensor surfaces of the forearms

Septic arthritis, gout, pseudogout, reactive arthritis and heamarthrosis are important causes of acute monoarthritis, while tuberculosis, osteoarthritis, chronic traumatic, benign tumors, e.g. PVNS and neuropathic are causes of chronic monoarthritis. Polyarticular gout, polyarticular septic arthritis and diffuse osteoarthritis would be uncommon causes of polyarthritis.

Skin Thickening

The thickening of skin is a primary feature of systemic sclerosis which can be of two types: Limited cutaneous and diffuse cutaneous. In limited cutaneous, the skin thickening is restricted with involvement distal to elbows and knees with or without face, whereas in diffuse cutaneous, it is generalized. The causes of skin thickening are listed in Table 8.4. Patients with systemic sclerosis often have mask-like face due to tautness of skin over the face (Fig. 8.5a). They also may have gangrene of the fingers (Fig. 8.5b), and salt and pepper appearance of the skin (Fig. 8.5c).

Raynaud's Phenomenon

The Raynaud's phenomenon (RP) is classically defined as a triphasic colour change, beginning with

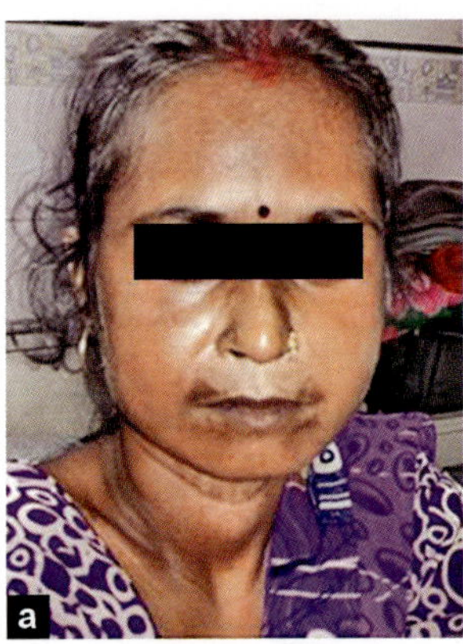
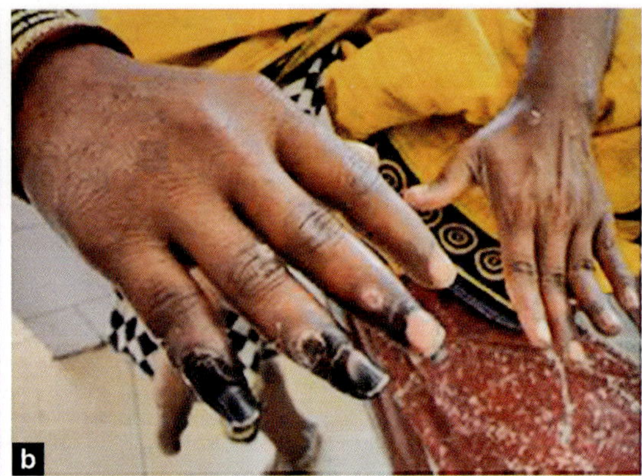
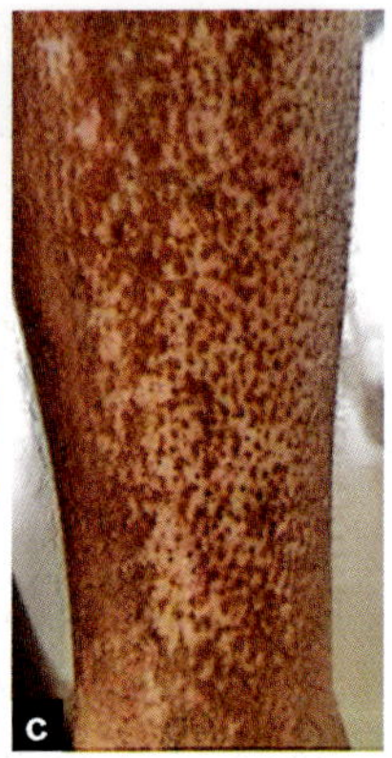

Fig. 8.5: Systemic sclerosis: Mask like face (a); gangrene of fingers (b); salt and pepper appearance of skin (c)

Table 8.4: Conditions associated with scleroderma like induration[7]

Systemic sclerosis	Localized scleroderma
Limited cutaneous	Morphea: Guttate/diffuse/ pansclerotic
Diffuse cutaneous	Linear scleroderma
	Scleredema
	Scleromyxedema
	Nephrogenic fibrosing dermopathy
	Chronic graft-versus-host disease
	Eosinophilic fasciitis
	Eosinophilia-myalgia syndrome

blanching of the affected part (pallor), followed by cyanosis and then rubor. At least a biphasic (pallor and cyanosis) change in skin colour of the digits is required for the diagnosis of Raynaud's phenomenon.[8] It may either be primary when no cause is identifiable (also called Raynaud's disease) or secondary, usually associated with a CTD (Table 8.5).[9] The factors associated with secondary RP are enumerated in Table 8.6. One study reported that 37.2% of 3029 persons of suspected Raynaud's disease subsequently developed a CTD.[10] Raynaud's phenomenon is seen in almost all patients of systemic sclerosis and mixed connective tissue disorder (MCTD). It is included in the 2013 ACR/ EULAR classification criteria for scleroderma.[11]

Dry Eyes/Mouth

Dry eyes/mouth are primarily seen in Sjögren's syndrome (SS) and are included in the Revised International Classification Criteria for SS. Sjögren's syndrome can be either primary or secondary. CTDs associated with Sjögren's syndrome are RA, SLE, scleroderma and MCTD. Dry eyes/mouth may have several other causes, e.g. HIV and HCV infection, anticholinergics and antihistaminics, past head and neck radiation treatment, graft-versus-host disease, etc.

Table 8.5: CTDs associated with Raynaud's phenomenon[9]

CTD	*Prevalence (%)*
Systemic sclerosis	90
MCTD	85
Sjögren's syndrome	33
SLE	10–45
Polymyositis/dermatomyositis	20
RA	10–45
Cryoglobulinemia/cryofibrinogenemia	10

Table 8.6: Factors associated with secondary Raynaud's phenomenon[9]

Clinical
- Older age at onset (>35 years)
- Recurrence of chilblains as adult
- Vasospasm all year round
- Asymmetric attacks
- Sclerodactyly or any other sign of CTD
- Digital ulceration
- Finger pulp pitting scars

Laboratory
- High inflammatory markers
- Detection of autoantibodies
- Increased level of von Willebrand factor antigen

Nail-fold microscopy
- Abnormal vessels

Gangrene of Fingers and Toes

The presence of gangrene in the extremities may be due to CTDs, vasculitis or other causes. Raynaud's phenomenon associated with CTDs, e.g. systemic sclerosis can present with digital ischemia, ulcer and gangrene (Fig. 8.5B). Antiphospholipid syndrome can present with arterial or venous thrombosis leading to gangrene. RA, SLE, MCTD, Sjögren's syndrome associated with secondary vasculitis, can have similar manifestations.

However, conditions other than CTDs must be considered in the differential diagnosis of gangrene, e.g. atheroembolic lesions, bacterial endocarditis, valvular heart disease, sepsis and DIC, procoagulant state, etc.

Myopathy and Neuropathy

Proximal muscle weakness is a manifestation of idiopathic inflammatory myopathies (IIM), e.g. polymyositis (PM), dermatomyositis (DM) and inclusion body myositis (IBM). The presence of skin rash suggests a diagnosis of DM. The characteristic feature of muscle weakness in IIM is sparing of extraocular and facial muscles. Myositis can also be associated with SLE, SS and MCTD. Proximal myopathy may have several other causes (Table 8.7).

Neuropathy can be a manifestation of CTDs, either because of disease per se (demyelinating or axonal) or associated with vasculitis. The different types of neuropathy seen in CTDs are distal symmetric sensory or sensorimotor, mononeuropathy, mononeuritis multiplex, autonomic neuropathy and cranial neuropathy.

Table 8.7: Other causes of proximal myopathy

Drugs	Statins, fibrates, corticosteroids, alcohol
Endocrine	Thyroid, parathyroid, adrenal or pituitary dysfunction
Metabolic	Osteomalacia
Hereditary and congenital myopathies	Muscular dystrophies
Malignancy	Paraneoplastic myopathy/ myositis

Laboratory Investigations

Laboratory tests must be ordered and interpreted in light of the clinical context

1. **Hemogram**: Anemia is usually multifactorial. Anemia due to chronic inflammation is a common finding in CTDs. Other possible etiologies are haemolytic, megaloblastic or iron deficiency anaemia. Leucocyte count may be low in SLE or RA with Felty's syndrome, whereas leucocytosis may suggest a vasculitis, Still's disease or an infection. Leukocytosis is a common accompaniment of steroid use and should not be confused with infection. Lymphopenia is an important feature of SLE but may be seen in HIV infection as well. Thrombocytopenia may be seen in SLE and thrombocytosis in vasculitis.

2. **Acute phase reactants:** Elevation of erythrocyte sedimentation rate (ESR) and C-reactive protein (CRP) are usually seen in CTDs which indicate persistent inflammation. CRP levels in SLE are either normal or only modestly elevated. Marked CRP elevation in SLE indicates infection.[12] It is possible for only one of the inflammatory markers to be selectively elevated in a particular disease; that marker should be repeated during follow-up.

3. **Urine routine microscopy** and **renal function tests** should always be ordered in suspected CTD patients. The presence of active sediments (RBCs, WBCs, casts) and/or proteinuria indicates renal involvement in the form of glomerulonephritis, interstitial nephritis and nephrotic syndrome. Renal involvement is seen in 30–50% of SLE patients. Hence a urine microscopic examination under a trained eye is mandatory.

4. **Muscle enzymes:** In suspected case of myositis, SGOT, SGPT, lactate dehydrogenase (LDH) and creatinine phosphokinase (CPK) should be ordered. Once a definitive diagnosis is made, these investigations are repeated to monitor response to treatment, relapse of disease and for possible side-effects to disease modifying therapy. Hematological malignancies may closely mimic a CTD and a raised serum LDH may be indicative of a hematoproliferative disorder.

5. Estimation of **serum uric acid** (UA) is one of the most misused tests in rheumatology. Mild elevations of UA do not explain vague arthralgias, aches and pains and often erroneously take the attention of the clinician away from a diagnosis of a more serious CTD. Asymptomatic hyperuricemia almost never needs treatment. Serum UA will almost never be significantly raised in a premenopausal woman or a child unless a secondary cause is present and hence routine testing of UA in these categories is not warranted.
6. **Rheumatoid factor** (RF): The sensitivity of RF in diagnosing RA depends upon the clinical suspicion and the prevalence of the disease in the population. The sensitivity and specificity of RF are 50–85% and 80–95% respectively, and depend upon the population studied. As RF can be negative in 20% of patients and early in the disease course (up to 40%), negative RF does not rule out a diagnosis of RA. RF in significant titers may be seen in other conditions like HIV, tuberculosis, syphilis, infective endocarditis and malignancy and so a positive RF does not always indicate RA.
7. **Anti-citrullinated peptide antibody** (ACPA): ACPA has a higher specificity (97%) and similar sensitivity compared to RF. It can be detected in the early stages of the disease and usually correlates with bone erosions. Hence, ACPA may help in confirming or ruling out the diagnosis of RA in RF negative patients.
8. **Antinuclear antibody(ANA):** The ANA test is most reliably done using indirect immunofluorescence technique to detect autoantibodies that bind to several nuclear antigens. The sensitivity and specificity of ANA test varies among different CTDs. ANA is detected in more than 95% of SLE patients, whereas its specificity is only 57%, which indicates that the presence of the antibody is not necessarily diagnostic of SLE.[12] The test can also be positive in 5% of healthy population, albeit in low titers. Therefore, ANA should not be routinely ordered in patients who present with joint pain to exclude or diagnose SLE as false positive results are particularly common in the elderly age group.[13] ANA titers of values above 1:160 are considered very significant.[14] Although high titers of ANA are diagnostic, they do not correlate with disease activity. Different autoantibodies manifest as different ANA patterns (Table 8.8).
9. **ENA (extractable nuclear antibody):** The extractable nuclear antibodies target small ribonuclear proteins, viz. Sm, U1RNP, SS-A/Ro and SS-B/La. These tests are highly specific and should only be ordered in ANA positive patients with clinical features suggestive of a CTD or in ANA negative patients with known or suspected CTD.[13] They are intended for diagnostic confirmation and do not correlate with disease activity (Table 8.9).

Table 8.8: Different ANA patterns and clinical associations[15]

Pattern	*Antigen target*	*Clinical correlate*
Homogenous *Or* Diffuse	Nucleosomes Histones dsDNA	SLE, drug-induced lupus, juvenile idiopathic arthritis
Fine speckled	SS-A/Ro SS-B/La Mi-2 Ku	SLE, dermatomyositis, Sjögren's syndrome
Coarse speckled	U1 RNP Sm RNA polymerase III	MCTD, SLE, SSc
Homogenous nucleolar	PM/Scl Th/To	SSc, SSc/PM overlap
Centromere *Or* Kinetochore	CENP-A/B(C)	Limited cutaneous SSc Primary biliary cirrhosis
PCNA like	PCNA	SLE

Source: Chan EKL, et al. Front Immunol 2015; 6:412

Table 8.9: ENA's and associated conditions

Anti-dsDNA	SLE Correlates with disease activity and severity of nephritis
Scl-70 (Topoisomerase)	Diffuse cutaneous SSc
Centromere	Limited cutaneous SSc
Anti SS-A/Ro Anti SS-B/La	Sjögren's syndrome SCLE Neonatal lupus
U1RNP	MCTD
Pm/Scl	SSc/PM overlap

10. **Complements (C3, C4, CH50) and anti-dsDNA:** Low complement levels in SLE usually indicate active disease and correlate with severity of nephritis. Anti- dsDNA antibodies in SLE have a sensitivity of 57% and specificity of 97% and high titres are associated with severity of nephritis.
11. **Lupus anticoagulant (LAC), dilute Russell viper venom time (dRVTT), β2 glycoprotein1, and anticardiolipin antibodies (IgM and IgG)** are tested in patients with suspected antiphospholipid syndrome having either recurrent arterial or venous thrombosis and/or pregnancy loss.

Key Points

- The diagnosis of a CTD is mainly based on the history and physical examination.
- Fulfilling classification criteria for CTDs is not mandatory for diagnosis in individual patients.
- ANA and specific autoantibodies must be tested in an appropriate clinical circumstance.

REFERENCES

1. Kumar TS, Aggarwal A. Approach to a Patient with Connective Tissue Disease. Indian J Pediatr 2010; 77:1157–64.
2. Gaubitz M. Epidemiology of connective tissue disorders. Rheumatology 2006; 45:iii3–iii4.
3. Petri M, Orbai A-M, Alarcón GS, Gordon C, Merrill JT, Fortin PR, et al. Derivation and Validation of Systemic Lupus International Collaborating Clinics Classification Criteria for Systemic Lupus Erythematosus. Arthritis Rheum 2012;64(8): 2677–86.
4. Oddis CV, Ascherman DP. Clinical features, classification, and epidemiology of inflammatory muscle disease. In: Hochberg MC, Silman AJ, editors. Rheumatology. Philadelphia: Elsevier; 2015. P. 1224–36.
5. Chong BF, Werth VP. Skin disease in cutaneous lupus erythematosus. In: Wallace DJ, Hahn BH, editors. Dubois' Lupus Erythematosus and Related Syndromes. Philadelphia: Elsevier; 2013. P. 319–32.
6. Moore CP, Wilken's RF. The subcutaneous nodule: Its significance in the diagnosis of rheumatic disease. Semin Arthritis Rheum 1977; 7(1):63–79.
7. Varga J. Systemic sclerosis (Scleroderma) and Related Disorders. In: Kasper DL, Fauci AS, editors. Harrison's Principles of Internal Medicine. New York: McGraw-Hill; 2015. pp. 2154–65.
8. Wigley FM, Flavahan NA. Raynaud's phenomenon. N Engl J Med 2016; 375(6):556–65.
9. Baines C, Kumar P, Belch JJF. Raynaud phenomenon. In: Hochberg MC, Silman AJ, editors. Rheumatology. Philadelphia: Elsevier; 2015. pp. 1213–18.
10. Pavlov-Dolijanovic SR, Damjanov NS, Vujasinovic Stupar NZ, Baltic S, Babic DD. The value of pattern capillary changes and antibodies to predict the development of systemic sclerosis in patients with primary Raynaud's phenomenon. RheumatolInt 2013; 33:2967–73.
11. van den Hoogen F, Khanna D, Fransen J, Johnson SR, Baron M, Tyndall A, et al. 2013 classification criteria for systemic sclerosis: an American college of rheumatology/ European league against rheumatism collaborative initiative. Ann Rheum Dis. 2013; 72:1747–55.
12. Ter Borg EJ, Horst G, Limburg PC, van Rijswijk MH, Kallenberg CG. C-reactive protein levels during disease exacerbations and infections in systemic lupus erythematosus: a prospective longitudinal study. J Rheumatol 1990; 17:1642–8.
13. Habash-Bseiso DE, Yale SH, Glurich I, Goldberg JW. Serologic testing in Connective Tissue Diseases. Clin Medicine Res. 2005; 3:190–93.
14. Tozzoli R, Bizzaro N, Tonutti E, Villalta D, Bassetti D, Manoni F, et al. Italian Society of Laboratory Medicine Study Group on the Diagnosis of Autoimmune Diseases. Guidelines for the laboratory use of autoantibody tests in the diagnosis and monitoring of autoimmune rheumatic diseases. Am J Clin Pathol 2002; 117:316–324.
15. Chan EKL, Damoiseaux J, Carballo OG, Conrad K, de Melo Cruvinel W, Francescantonio PL, et al. Report of the First International Consensus on Standardized Nomenclature of Antinuclear Antibody HEp-2 Cell Patterns 2014–2015 [Internet]. Front Immunol 2015; 6:412.Available from: http://journal.frontiersin.org/article/10.3389/fimmu.2015.00412

FURTHER READING

1. Kumar TS, Aggarwal A. Approach to a Patient with Connective Tissue Disease. Indian J Pediatr 2010; 77:1157–64.
2. Satoh M, Vázquez-Del Mercado M, Chan EK. Clinical interpretation of antinuclear antibody tests in systemic rheumatic diseases. Mod Rheumatol. 2009; 19:219–28.

Chapter

9

Approach to Vasculitis

Ramnath Misra

INTRODUCTION

Vasculitides are a group of heterogeneous diseases with diverse aetiology characterized by inflammation of blood vessels of any size or any site. The clinical manifestations are varied depending the size, site and number of vessels involved in the process. It may be confined to one organ or vessel type or to multiple organs and vessels and may be clinically insignificant or rapidly fatal unless recognized and prompt treatment instituted. The major manifestations are consequent to ischaemia and inflammation of the tissues affected by the vasculitic process. Vasculitis can be primary or secondary to infections, drugs, malignancies or to connective tissue diseases like rheumatoid arthritis, systemic lupus erythematosus, etc. The following description shall pertain to the description of primary vasculitides.

EPIDEMIOLOGY

Vasculitides are rare diseases. Epidemiological studies have been done mostly from Western Europe countries and USA involving the white Caucasian population. Epidemiologic studies from England, Spain and Scandinavia show an overall annual incidence of primary systemic vasculitis of approximately 20 per million population. Vasculitis appears more common with advancing age and the prevalence varies from one geographic and ethnic population to the other. Giant cell arteritis is the commonest vasculitis in North America and Western Europe with an incidence of 178 per million described in Minnesota in adults aged 50 or more. Takayasu's arteritis is more common in eastern countries and in females. Classical polyarteritis nodosa is a rare disease with an estimated annual incidence of 2 to 9 per million. But its incidence is more common in Kuwait (45 per million) and in Alaska. (77 per million). A study from a predominantly English county showed that the annual incidence of Wegener's granulomatosis is 8.5 per million.[1] Kawasaki disease and Henoch-Schönlein purpura are vasculitic diseases predominantly of the childhood. The annual incidence of Kawasaki's disease was 900 per million children younger than 5 years in Japan in 1991–92 and that of Henoch-Schönlein purpura in UK is 204 per million in children.[2]

In India, Henoch-Schönlein purpura[3, 4] is fairly common in pediatric population and Takayasu's arteritis[5] in adults. There are reports of small case series of patients with Wegener's granulomatosis[6,7], Polyarteritis nodosa[8], Kawasaki[9,10] disease from our country. This is attributable to under reporting of vasculitis due to lack of awareness and diagnostic facility. Polymyalgia rheumatica and giant cell arteritis usually occur in the elderly population and since only 5% of our population is above 65 years we tend to see less of these vasculitis.

CLASSIFICATION

The classification is based on vessel size, histology and serology. The American College of Rheumatology (ACR) proposed a classification system in 1990[11] used both vessel size and the type of inflammatory infiltrate. It classified vasculitides into seven types which included polyarteritis nodosa (PAN), Churg-Strauss syndrome (CSS), Wegener's granulomatosis (WG), hypersensitivity vasculitis, Henoch-Schönlein purpura, giant cell arteritis and Takayasu's arteritis. The sensitivity and specificity varied considerably ranging from 71 to 95.3% and 78.7 to 99.7% respectively. ACR 1990 criteria has two drawbacks, it did not recognize microscopic polyangitis as a distinct entity and did not include antibodies to neutrophil cytoplasmic antigen (ANCA). In 1994, the Chapel Hill Consensus Conference (CHCC) proposed definition for ten types of vasculitis.[12] In contrast to the 7 ACR subtypes, they removed hypersensitivity vasculitis and defined microscopic polyangiitis, Kawasaki disease and essential cryoglobulinemic vasculitis. The two-classification system does not necessarily classify individual patients into same category. It had also recognized role of ANCA in diagnosis. In 2012, a revised CHCC nomenclature of vasculitides[13] was proposed to include additional categories of vasculitis (Table 9.1).

CLINICAL FEATURES OF VASCULITIS

The clinical feature of different types of vasculitis depends upon the size of the vessel that is predominantly affected. Table 9.2 shows the common clinical features associated with large, medium and small size vessel involvement.

Table 9.1: Nomenclature of systemic vasculitides according to International Chapel Hill Consensus Conference (2012)[13]

Type of vasculitis	*Diseases*
Large vessel vasculitis (LVV)	Takayasu's arteritis (TAK) Giant cell arteritis (GCA)
Medium vessel vasculitis (MVV)	Polyarteritis nodosa (PAN) Kawasaki disease (KD)
Small vessel vasculitis (SVV)	**ANCA associated vasculitis (AAV)** Microscopic polyangiitis (MPA) Granulomatosis with polyangiitis (Wegener's) (GPA) Eosinophilic granulomatosis with polyangiitis (Churg-Strauss) (EGPA) **Immune complex small vessel vasculitis** Anti-glomerular basement membrane (anti-GBM) disease Cryoglobulinemic vasculitis (CV) IgA vasculitis (IgAV) (Henoch-Schönlein) Hypocomplementemic urticarial vasculitis (HUV)
Variable vessel vasculitis (VVV)	Behcet's disease (BD) Cogan's syndrome (CS)
Single-organ vasculitis (SOV)	Cutaneous leukocytoclastic angiitis Cutaneous arteritis Primary central nervous system vasculitis Isolated aortitis Others
Vasculitis associated with systemic disease	Lupus vasculitis Rheumatoid vasculitis Sarcoid vasculitis Others
Vasculitis associated with probable aetiology	Hepatitis C virus-associated cryoglobulinemic vasculitis Hepatitis B virus-associated vasculitis Syphilis-associated aortitis Drug-associated immune complex vasculitis Drug-associated ANCA-associated vasculitis Cancer-associated vasculitis Others

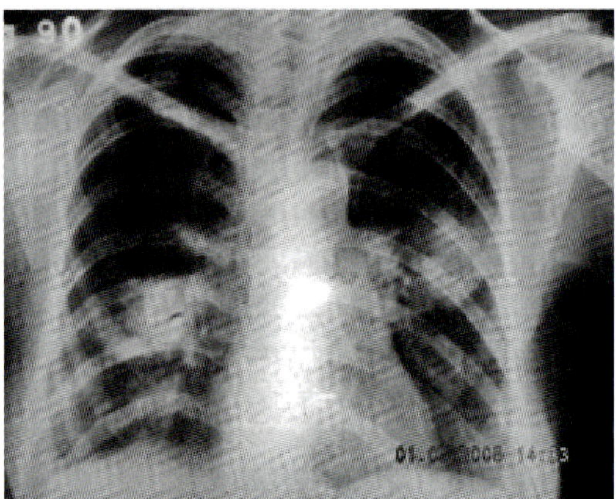

Fig. 9.4: X-ray chest showing nodular consolidation in right lung in Wegener's granulomatosis

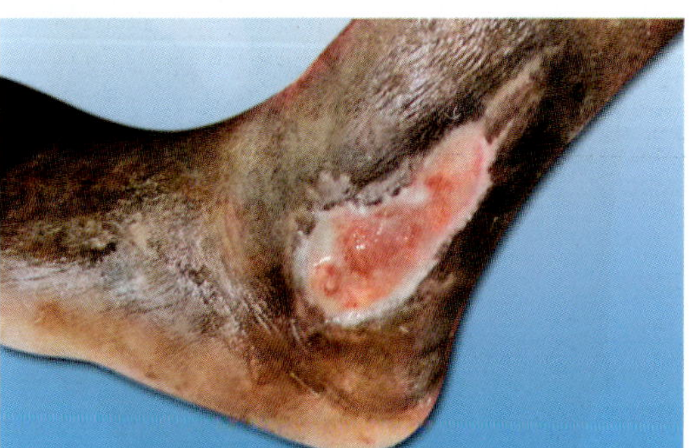

Fig. 9.5: Non-healing ulcer in a patient with systemic necrotizing vasculitis

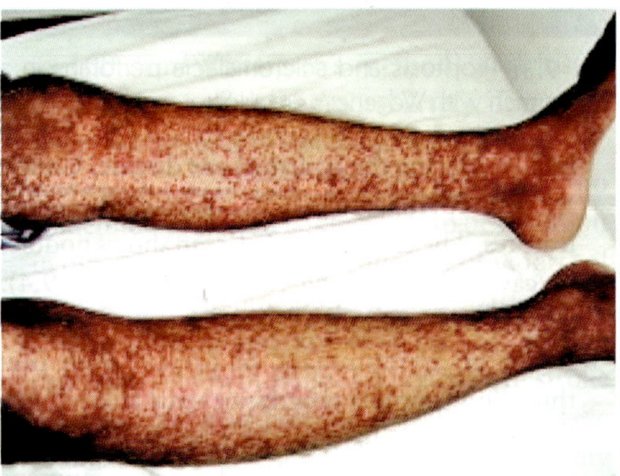

Fig. 9.6: Palpable purpura in the lower limbs in a patient with Henoch-Schönlein purpura

adults. Typically the rashes are present on the buttocks and lower limbs. Besides skin rash, there occurs arthalgias/arthritis, microscopic haematuria/proteinuria, haemoptysis and gastrointestinal bleeding. The prognosis is excellent in majority of cases; 10–30% of older children and adults may progress to chronic renal failure. The diagnosis is confirmed by demonstration of IgA immunoglobulin deposits in the skin and kidney on immunofluorescence.

To renew ACR criteria using details from cases collected worldwide including India, the DCVAS group will provide weighted criteria based on sum of features present towards a diagnosis. Currently only criteria for AAV (ANCA associated small vessel vasculitis is finalized.

Investigation and Approach to Diagnosis

It is mandatory to exclude infections and other vaso-occlusive diseases before diagnosing a primary vasculitis, the vasculitis mimickers (Table 9.3). The investigations are directed to confirm the diagnosis, extent of organ involvement and assessing the disease activity. Table 9.4 summarizes the salient features in establishing the diagnosis and assessing both activity and extent of disease. The type and extent of organ involvement helps in determining the type of vasculitis, urgency of initiating treatment and planning diagnostic tests. Hepatitis B and C and human immunodeficiency virus serologies should be obtained since these can be associated with vasculitis.

Antineutrophil Cytoplasmic Antibodies

Antineutrophil cytoplasmic antibodies (ANCA) are useful diagnostic markers for the primary systemic necrotizing vasculitis in AA small vessel vasculitis to separate MPA from GPA. These are detected by

Table 9.3: Vasculitis mimickers

Infections	Subacute infective endocarditis *Neisseria meningitides* Viral infections
Malignancies	Metastatic carcinoma
Connective tissue diseases	Systemic lupus erythematosus Rheumatoid arthritis
Occlusive vascular disease	Atrial myxoma Cholesterol embolism Infections (mycotic) Antiphospholipid antibody syndrome Ergotism

Table 9.4: Assessing extent of disease and salient investigations for the diagnosis of various vasculitides

Diseases	*Routine tests and serologies*	*Radiology/ultrasound*	*Electro-physiology*	*Biopsy*
Temporal arteritis	ESR, CRP	Ultrasound PET CT	—	Superficial temporal Art.
Takayasu's arteritis	ESR, CRP	Angiography Aorta and branches PET CT	—	—
PAN	ESR, TLC, Creat.	Angiography abdominal vessels	NCV	Sural nerve/ muscle/ testis/ skin
MPA	Urinalysis, Creat. pANCA (MPO)	Chest X-ray	—	Renal
WG	TLC, Urinalysis, creatinine, cANCA (PR3)	X-ray chest, PNS CT thorax	—	Kidney Lung Nasal mucosa
CSS	Blood eosinophilia, pANCA (MPO)	X-ray chest	NCV	Sural nerve
HSP	Normal platelet count	—	—	Immunofluorescence (IgA deposits)

indirect immunofluorescence which are reported as cytoplasmic or cANCA, perinuclear or pANCA and atypical or xANCA (Fig. 9.7). cANCA is directed against proteinase-3 and pANCA is directed predominantly against myeloperoxidase. cANCA is present in at least 90% of patients with active, diffuse GPA Wegener's granulomatosis and in up to 70% of patients with airways limited disease. pANCA is typical of MPA (50–60%) and Churg-Strauss disease (25–30%). Because vasculitides associated with ANCA are rare disorders, the likelihood of false-positive result (ulcerative colitis) of ANCA is greatly increased if not used in appropriate clinical setting. Enzyme linked immunoassays (EIA) were developed for detection of anti-PR3 and anti-MPO antibodies.

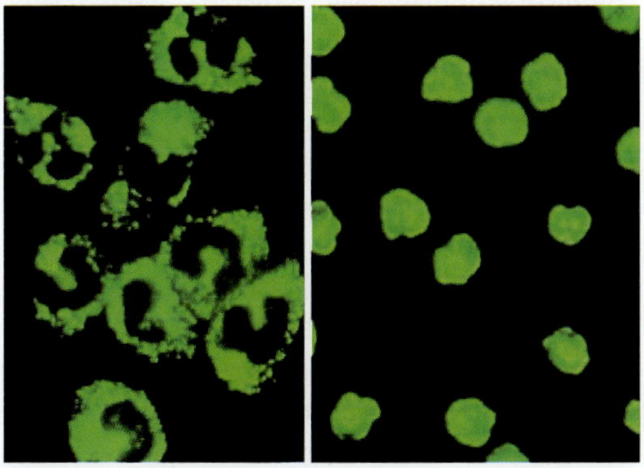

Fig. 9.7: Cytoplasmic (left) and peripheral (right) ANCA pattern by indirect immunofluorescence

Like cANCA antibodies to PR3 is both specific and sensitive for the diagnosis of WG and similarly antibodies to MPO are specific for microscopic polyangiitis. But EIA was found to have its own problems. Hence, it is advisable to order for both IIF and EIA to confirm the presence and type of ANCA. Indiscriminate ordering of ANCA should be avoided as it would lead to misdiagnosis. It is of value only when there is a high pretest probability of vasculitis. ANCAs of a low titer including borderline PR3 and MPO do not necessarily indicate vasculitis.

Treatment

Treatment of vasculitis depends upon the type of vasculitic syndrome and the seriousness of involvement of organs by the disease. It is imperative to assess disease activity and damage before initiation of immunosuppressive therapy. Composite clinical indices, radiology and serological parameters are useful.[15]

The medical treatment of TA is directed against improving systemic symptoms and controlling hypertension. Assessment of disease activity there is no valid assessment of disease activity in TA, the Indian Rheumatology Association Vasculitis (IRAVAS) Group have recently come out with a validated instrument the Indian Takayasu Activity Score (ITAS) which is a composite disease activity scores.[15] A combination of oral prednisolone starting with 1 mg/kg of body weight along with methotrexate (15–25 mg/week), azathioprine (2 mg/kg/day) or mycophenolate mofetil (1.5–3 g/day) generally are employed to remit disease activity. Prednisolone is to be tapered within 2–3 months to a maintenance of 0.15 mg/day. Tocilizumab (IL-6 receptor blocker) has also been found to be effective in small case series to lessen disease activity such as ITAS score. Surgery or percutaneous transluminal angioplasty are important considerations for severely occluded vessels. In GCA, oral prednisolone in a dose of 40–60 mg/day for 6 weeks is given followed by maintenance for 6–12 months with a dose of .15 mg/kg/day. Various agents such as azathioprine, cyclophosphamide, dapsone and cyclosporine have been tested as steroid sparing drugs in GCA; however, all were tested in too small or uncontrolled series to permit conclusion.

PAN is considered to be a monophasic illness. Corticosteroids improved the 5 year survival rate of PAN patients from 10 to 55%. Survival was further prolonged by adding cytotoxic agents, cyclophosphamide or azathioprine, attaining a 5-year survival of 80% for patients given steroids and cyclophosphamide. Patients of PAN who are HBsAg positive are also treated with interferon and lamivudine.

It is imperative to assess disease activity and damage before In ANCA associated vasculitis (GPA and MPA) disease activity is assessed by clinical scoring instruments. The most widely used is Birmingham vasculitis activity score (BVAS), measures clinical features of 9 categories , systemic and 8 organ system and is used in drug trials to measure activity. It is simple to use, though is generally not used in day-to-day clinical practice. Damage in vasculitis is attributable to disease process and secondary to complication of treatment. Evaluation of damage is important for prognostication and avoidance of overt reatment. Vasculitis damage index (VDI) is most widely used. It is an objective assessment based on eleven organ systems. It records long-term consequences of disease and its therapy.

GPA with disease having arthritis, sinusitis, epistaxis or solitary pulmonary lesions is treated with methotrexate at dose of 7.5 to 25 mg/week and low-dose prednisolone 7.5 mg daily. The duration of treatment will normally be of 3–3 years.

Widespread and serious life-threatening GPA or MPA having pulmonary haemorrage, nephritis, neuropathy or tissue infarcts or a combination of these manifestations require aggressive treatment as the mortality is 90% in untreated cases by the end of 2 years. Treatment is generally induced with either oral (2 mg/kg) or IV cyclophosphamide (15 mg/kg IV bolus fortnightly for 3 doses and then three weekly for 3 more).[16] Rituximab 500 mg IV weekly for 4 weeks or 2 doses of 1 g each given at fortnightly intervals) is a suitable alternative to cyclophosphamide.[18,19] Both cyclophosphamide and retuximab is given along with prednisolone 1 mg/kg of body weight for at least 4 weeks and tapered down to 20 mg/daily by 8 weeks. Thereafter, reduced to 7.5 mg/daily by 12 to 16 weeks. In those with rapidly progressive renal disease or massive hemoptysis suggesting diffuse pulmonary vasculitis or recent onset motor neuropathy or gangrene treatment is initiated with IV methylprednisolone 1 g daily for 3 consecutive days. Plasma exchange has shown to be effective with patients with hemoptysis suggestive of diffuse alveolar haemorrage or whose serum creatinine at presentation is above 4 mg/dl.

Remission is usually achieved by 4 to 6 months following which maintenance with daily azathioprine is administered at a dose of 2 mg/ kg/day along with prednisolone (0.15 mg/kg/day) for a period of one year. Some cases prolonged maintenance for a period of 18 months may be needed. The French group has shown rituximab administered 6 monthly provides effective can be used as almost 50% of GPA patients relapse after remission is achieved while during maintenance or after completion of maintenance therapy. Severe relapse is managed with cyclophosphamide or retuximab as mentioned above. Mycophenolate mofetil at a dose of 2 g/daily can be used as an alternative to cyclophosphamide. Recently it has shown by the French group that rituximab is effective in maintaining remission given 6 monthly.[20]

Treatment of Henoch-Söhonlein purpura is largely supportive. NSAIDs help joint pains and do not worsen purpura. Steroids have been shown to be of benefit in managing painful oedema. It is also used for GI involvement although its efficacy in this regard is unproven. Treatment of renal involvement is highly controversial. Various treatment modalities have included pulse corticosteroids, cyclophosphamide, plasmapheresis, cyclosporine and azathioprine. Cryoglobulinemic vasculitis has a wide manifestation ranging from mild symptoms to life threatening conditions. The clinical course can be characterized by remissions and exacerbations. Serological parameters do not correlate with disease activity and hence cannot be relied upon for therapeutic decisions. Treatment should therefore not focus upon lowering cryoglobulin levels or normalizing complements but rather on controlling manifestations. Low to moderate dose steroids are sufficient for most minor symptoms but larger doses may be needed for managing renal involvement, neuropathy or serositis. Plasma exchange with or without immunosuppressive may be needed for life-threatening conditions. Interferon has been used with good success for inducing clinical and biological remission but the disease relapses in most following cessation. Anti-viral treatment combined with steroid and immunosuppression appears most effective.

Conclusion

Primary systemic vasculitides are a group of rare diseases. The aetiology and pathogenesis are not well-understood. The discovery of ANCA has significantly helped us to further our knowledge with regards to pathogenesis and diagnosis. Immunosuppressive treatment has helped to reduce mortality but morbidity is a major concern in ANCA associated vasculitis. B cell depleting therapy has proven to an effective in inducing and maintaining remission in AAV.

REFERENCES

1. Watts R, Carruthers D, Scott D. Epidemiology of systemic vasculitis: changing incidence or definition? Semin Arthritis Rheum 1995; 25:28–34.
2. Gardner-Medwin JM, Dolezalova P, Cummins C, Southwood TR. et al. Incidence of Henoch-Schönlein purpura, Kawasaki's disease and rare childhood vasculitides in children of different ethnic origins. Lancet 2002; 360:1197–1202.
3. Bagga A, Kabra SK, Srivastava RN, Bhuyan UN. Henoch-Schönlein syndrome in northern Indian children. Indian Pediatr. 1991; 28:1153–57.
4. Kumar L, Singh S. Vasculitis in children. Indian J Pediatr. 1996; 63:323–34.
5. Jain S, Kumari S, Ganguly NK, Sharma BK. Current status of Takayasu arteritis in India. Int J Cardiol. 1996; 54 (Suppl: 111–16).
6. Malaviya AN, Kumar A, Singh YN, Singh RR, Dash SC, Khare SD. et al. Wegener's granulomatosis in India: not so rare. Br J Rheum 1990; 29:499–500.
7. Bambery P, Sakhuja V, Bhusnurmath SR, Jindal SK, Deodhar SD, Chugh SK. Wegener's granulomatosis: clinical experience with eighteen patients. J Assoc Physicians India 1992; 40:597–600.
8. Handa R, Wali JP, Gupta SD, Dinda AK, Aggarwal P, Wig N, et al. Classical polyarteritis nodosa and microscopic polyangiitis a clinicopathologic study. J Assoc Physicians India 2001; 49:314–19.
9. Pendse RN, Bhandari H, Vats AK, Bhandari B. Kawasaki disease Indian perspective. Indian J Pediatr 2001; 68:775–77.
10. Bagyaraj B, Krishnan U, Farzana F. Kawasaki disease in India. Indian J Pediatr 2003; 70:919–22.
11. Fries JF, Hunder GG, Bloch DA, Michel BA, Arend WP, Calabrese LH, et al. The American college of Rheumatology criteria for the classification of vasculitis: summary. Arthritis Rheum 1990; 33: 1135–36.
12. Jeanette JC, Falk RJ, Andrassy K, Bacon PA, Churg J, Gross WL, et al. Nomenclature of systemic vasculitides. Proposal of an international consensus conference. Arthritis Rheum 1994; 37:187–92.
13. Jennette JC, Falk RJ, Bacon PA, Basu N, Cid MC, Ferrario F, et al. 2012 Revised International Chapel Hill Consensus Conference Nomenclature of Vasculitides. Arthritis Rheum 2013; 65:1–11.
14. Hoffman GS, Kerr GS, Leavitt RY, Hallahan CW, Lebovics RS, Travis WD, et al. Wegener's granulomatosis: an analysis of 158 patients. Ann Intern Med 1992; 116:488–98.
15. Sharma A, Misra R. Assessment of disease activity and damage. In: Sharma A, editor. Textbook of Systemic vasculitis. New Delhi: The Health Science Publisher; 2014. pp.232–37.
16. Misra R, Danda D, Rajappa SM, Ghosh A, Gupta R, Mahendranath KM, et al; Indian Rheumatology Vasculitis (IRAVAS) group. Development and initial validation of Indian Takayasu Activity score (ITAS 2010). Rheumatology (Oxford) 2013; 52:1759–801.
17. de Groot K, Harper L, Jayne DR, Flores Suarez LF, Gregorini G, Gross WL, et al. Pulse versus daily oral cyclophosphamide for induction of remission in antineutrophil cytoplasmic antibody-associated vasculitis: a randomized trial. Ann Intern Med 2009; 150:670–80.

18. Jones RB, Tervaert JW, Hauser T, Luqmani R, Morgan MD, Peh CA, et al; European Vasculitis Study Group. Rituximab versus cyclophosphamide in ANCA-associated renal vasculitis. N Engl J Med 2010; 363:211–20.

19. Stone JH, Merkel PA, Spiera R. Seo P, Langford CA, Hoffman GS, et al; RAVE-ITN Research Group. Rituximab versus cyclophosphamide for ANCA-associated vasculitis. N Engl J Med 2010; 363: 221–32.

20. Pugnet G, Pagnoux C, Terrier B, Perrodeau E, Puéchal X, Karras A, et al. French Vasculitis Study Group. Rituximab versus azathioprine for ANCA-associated vasculitis maintenance therapy: impact on global disability and health-related quality of life. Clin Exp Rheumatol. 2016; 34 (3 Suppl 97): S54–9.

Chapter

10

Outcome Measures in Rheumatic Diseases

INTRODUCTION AND GENERAL PRINCIPLES

Sapan Pandya

Why the need arose? A peek into the History

Old is gold but new is welcome. Which is better? Especially with relevance to therapeutics. Individual health care givers would have their own experiences, coupled with patient heterogeneity, different brands, socio-economic issues, habits of patients, health care givers own core competence—there are innumerable variables that could affect the decision of choice as regards which is better—the old or the new? Outcome assessment is the measurement of these in a standardized and validated manner—the way by which consequences of diseases and health care management decisions are evaluated.[1] Process measures include histological, biochemical or imaging procedures, whereas outcome measures reflect the suffering or loss of health experienced by patients as a consequence of the disease.

As newer armamentarium gathered for treating RA, reports kept piling up on the use of these by several centers and rheumatologists from the world on their utility and results—mainly in the 1980s.[2] There was marked variability in the methodology each used both to carry out the trial and producing the results. The measures used lacked validity and were insensitive to change.

Thus arose the need for a group that brought together rheumatologists of the world and allied specialists, statisticians to uniformly lay out measures that should be used to assess response and disease status. This gave birth to the OMERACT—initially called outcome measures in RA clinical trials but now is the outcome measures in rheumatology.[3] The first OMERACT meet was held in Maastricht in 1992 under the auspices of the WHO, ILAR, EULAR and several other rheumatology bodies.[4] This is where the 'core set' of outcome measures of the WHO/ILAR and later adopted by the ACR took birth. The meet also emphasized on the need for developing measurement methodology. Subsequently the OMERACT has been meeting every two years. Other diseases like osteoarthritis, gout, connective tissue diseases also have been included since. It was after about the 5th OMERACT meet that the significance of having patients in the groups was realized. Fatigue is a classic example of an outcome measure that only a patient can figure out and it was later included in the list.[5]

Outcome measures can also be divided into two categories—observer dependent (rated by the evaluator, e.g. swollen joint count, tender joint count, grip strength, etc.) or observer independent (self-reported, e.g. visual analogue scales VAS for pain, global disease, etc.) or qualitative/quantitative. The various domains within the core set defined for different diseases relied on the Fries' paradigm (5Ds—Death, Disability, Discomfort, Drug and Dollar). Put simply, they define disease consequence in terms of mortality or survivorship and morbidity or impact of disease and the symptom severity on

Table 10A.1: Various joint counts used in rheumatoid arthritis

	Ritchie index	*ACR joint count*	*44 joint counts*	*28 joint counts*
Year	1968	1965	1992	1989
No. of joints assessed	78	68/66	46/44	28
Graded	Yes	No	No	No
Joints involved				
DIPs		+		
PIPs	+	+	+	+
MCPs	+	+	+	+
CMCs				
Carpus				
Wrist	+	+	+	+
Elbow	+	+	+	+
Shoulder	+	+	+	+
DIP (feet)				
PIP (feet)		+	+	
MTP	+		+	
Tarso-metatarsal	+			
Tarsus	+	+	+	
Ankle	+	+	+	
Knee	+	+	+	+
Hip	+	+		
Acromio-clavicular	+	+	+	
Sternoclavicular	+	+	+	
Temporomandibular	+	+	+	
Cervical spine	+			

ACR: American College of Rheumatology
DIP: Distal inter-phalangeal joint
PIP: Proximal inter-phalangeal joint
MCP: Metacarpo-phalangeal joint
CMC: Carpometacarpal joint
MTP: Metatarso-phalangeal joint

Adapted from: Aletaha D, Smolen JS. The definition and measurement of disease modification in inflammatory rheumatic diseases. Rheum Dis Clin North Am 2006; 32:9–44.

(DAS-3 and DAS28-3). The cut-off points between high, moderate, low disease activity and remission are 5.1, 3.2, and 2.6, respectively for DAS 28.[3] Table 10A.3 shows the complex mathematical formulae used to calculate DAS. Complex computation is one of the drawbacks of DAS28, which is addressed by other simplified indices.

- *Simplified disease activity index (SDAI)*: It is the simple sum of 5 core domains and does not use complex and weighted mathematical formula like DAS (Table 10A.3). SDAI has been widely validated, and definitions of states of remission and of low, moderate, and high disease activity have been standardized.[4]
- *Clinical disease activity index (CDAI)*: This is purely clinical instrument and does not require the availability of an acute-phase reactant (CRP or ESR), as unavailability of these laboratory results frequently precludes immediate assessment of disease status. It consists of linear sum of 4 core domains (Table 10A.3) and correlates well with the disease activity.[5]

Table 10A.2: Instruments to assess outcome in rheumatoid arthritis

Disease parameters	*Core domains*	*Composite indices/instruments*
1. Disease activity	Tender joint count (TJC) Swollen joint count (SJC) Pain (VAS) Patient global Physician global Acute phase reactant (ESR/CRP)	DAS28 ESR/CRP DAS-3 CDAI SDAI **Patient reported outcomes (PROs)** RADAR RADAI RAPID 3/4/5
2. Improvement criteria	Tender joint count (TJC) Swollen joint vount (SJC) Pain (VAS) Patient global Physician global ESR/CRP HAQ-DI	ACR 20, 50, 70 Hybrid ACR EULAR response criteria
3. Functional improvement	Patient reported Questionnaires	HAQ-DI SF-36 AIMS
4. Radiographic damage	X-rays of hands and/or feet	Larsen method Sharp score Sharp van der Heijde (SvdH) score

Abbreviations: DAS 28, disease activity score based on 28 joint counts; CDAI, clinical disease activity index; SDAI, simplified disease activity index, RADAR, rapid assessment of disease activity in RA; RADAI, RA disease activity index; RAPID 3/4/5, routine assessment of patient index data 3/4/5; HAQ-DI: health assessment questionnaire-disability index; ACR, American College of Rheumatology; EULAR, European League Against Rheumatism; SF-36, Short Form 36; AIMS, arthritis impact measurement scale

Table 10A.3: Composite disease activity indices for rheumatoid arthritis

Index	*Formula*	*Remission*	*Low*	*Moderate*	Severe
DAS28-ESR	$0.56 \times \sqrt{(TJC28)} + 0.28 \times \sqrt{(SJC28)} + 0.70 \times \log_{nat}(ESR) + 0.014 \times GH$	<2.6	2.6–3.1	3.2–5.1	>5.1
DAS28-CRP	$0.56 \times \sqrt{(TJC28)} + 0.28 \times \sqrt{(SJC28)} + 0.36 \times \log_{nat}(CRP + 1) + 0.014 \times GH + 0.96$	<2.6	2.6–3.1	3.2–5.1	> 5.1
SDAI	SJC28 + TJC28 + PGA + EGA + CRP	<3.3	3.4– <11	11.1 – <26	>26
CDAI	SJC28 + TJC28 + PGA + EGA	<2.8	2.9– <10	10.1 – <22	≥22

Abbreviations: DAS 28, Disease Activity Score based on 28 joint counts; ESR, erythrocyte sedimentation rate; CRP, C-reactive protein; SDAI, simplified disease activity index; CDAI, clinical disease activity index.

- *Patient reported questionnaires (PROs)*: Multiple patient reported measures of disease activity have been developed (Table 10A.2), but are rarely used in clinical practice. Examples: RADAR—six-point questionnaire, RADAI—five item questionnaires, RAPID3, RAPID4 and RAPID5.

2. Disease improvement criteria: American College of Rheumatology (ACR) criteria are widely used in clinical trials of RA. The ACR improvement criteria[6] require 20% improvement in SJC and TJC and in three of the five remaining core set variables (Table 10A.2). Minimum of 20% improvement from baseline is called ACR20. This criterion has also been applied to study more profound responses such as ACR50 and ACR70 for 50% and 70% response, respectively. ACR criteria do not consider the starting level of a patient's disease activity, it only provides us a dichotomas "yes/no" result. To overcome the limitation of dichotomy, a modified numerical ACR-N has been developed which has a continuous scale ranging from 0 to 100%. It has the capability to readout the smallest relative improvement in three measures: SJC, TJC, and the median of the five remaining core set variables.[7] EULAR response criteria is based on change in DAS28 measurement. Moderate to good response is used as marker of efficacy in RA clinical trials.

3. Physical function assessment: Common measures used to assess this parameter include Short Form 36 (SF-36) and Health Assessment Questionnaire (HAQ). While the former score allows the comparison in quality of life parameters of patients with different diseases and normal population, the latter helps in assessment of quality of life of patients with different diseases and different patients with same disease.

Full HAQ includes assessment of disease related discomfort, side effects of drugs, economic burden of disease, acquired disability and mortality associated with the disease. However, it is time consuming and difficult to perform on every outpatient visit. Hence a modified version, called HAQ Disability Index (HAQ-DI) has been devised, which tests patients' ability to perform activities of daily living with the help of a questionnaire.[8] It includes 20 questions divided into eight categories: Arising, hygiene, walking, dressing, eating, reach, grip and usual activities. Each question is weighted on Likert scale from no difficulty to inability to perform (0 to 3). Maximum score in each of the eight categories is divided by eight to calculate the HAQ score (range 0 to 3). HAQ is determined by both disease activity and damage. Minimum change in HAQ should be $\geq$0.22, to observe any clinically significant difference in quality of life of a given patient.

4. Assessment of radiographic damage: This is done using simple X-rays of hands and feet. Different authors have given different scores.[9] Commonly used ones are Larsen score, Sharp score, SvdH score. These measures assess disease severity in terms of radiographic damage (joint space narrowing and erosions), but only sharp score and its modifications are sensitive to change.

Conclusions

Outcome assessment in RA is complex, in view of systemic nature of the disease. Disease activity, radiographic joint damage and functional impairment are three main areas of outcome assessment. All three are interlinked, with one influencing the other and thus, they serve as the determinants of the most important outcome measure, i.e. patient's quality of life.

REFERENCES

1. Solomon DH, Bitton A, Katz JN, Radner H, Brown E, Fraenkel L. Treat to Target in Rheumatoid Arthritis: Fact, Fiction or Hypothesis? Arthritis Rheumatol. 2014; 66:775–82.
2. Prevoo ML, van't Hof MA, Kuper HH, Van Leeuvan MA, Van de Putte LBA, van Riel PL. Modified disease activity scores that include twenty-eight joint counts. Development and validation in a prospective longitudinal study of patients with rheumatoid arthritis. Arthritis Rheum 1995; 38:44–8.
3. Aletaha D, Ward MM, Machold KP, Nell VP, Stamm T, Smolen JS. Remission and active disease in rheumatoid arthritis: defining criteria for disease activity states. Arthritis Rheum 2005; 52:2625–36.
4. Smolen JS, Breedveld FC, Schiff MH, Kalden JR, Emery P, Eberl G, et al. A simplified disease activity index for rheumatoid arthritis for use in clinical practice. Rheumatology 2003; 42:244–57.
5. Aletaha D, Nell VP, Stamm T, Uffman M, Pflugbeil S, Smolen JS. Acute phase reactants add little to composite disease activity indices for rheumatoid arthritis: validation of a clinical activity score. Arthritis Res 2005; 7:R796–806.
6. Felson DT, Anderson JJ, Boers M, Bombardier C, Furst D, Goldsmith C, et al. American College of Rheumatology preliminary definition of improvement in rheumatoid arthritis. Arthritis Rheum 1995; 38:727–35.
7. Siegel JN, Zhen B-G. Use of the American College of Rheumatology N (ACR-N) index of improvement in rheumatoid arthritis. Argument in favor. Arthritis Rheum 2005; 52:1637–41.
8. Kumar A, Malaviya AN, Pandhi A, Singh R. Validation of an Indian version of the Health Assessment Questionnaire in patients with rheumatoid arthritis. Rheumatology 2002; 41:1457–59.
9. Ory PA. Interpreting radiographic data in rheumatoid arthritis. Ann Rheum Dis 2003; 62:597–604.

10B. OUTCOME MEASURES IN SYSTEMIC LUPUS ERYTHEMATOSUS

Taral Parikh

Introduction

Systemic lupus erythematosus (SLE) is a prototype immune complex mediated disease. It affects multiple organ systems and has a relapsing remitting course, thus making assessment of disease activity and damage more difficult than other rheumatic diseases.

Four core domains are assessed in SLE trials routinely: Disease activity (reversible features), damage (irreversible features), health related quality of life (HRQOL) and adverse events.[1]

The US Food and Drug Administration (USFDA) in the 2010 guidelines have included the following as outcome measures for SLE trials—disease activity, damage, SLE flares, concomitant corticosteroid use, patient-reported outcomes, and biomarkers.[1]

Outcome Measures

Thorough history, physical examination, and laboratory findings help to assess SLE disease activity and efficacy of treatment in the clinic.

Laboratory tests like presence of anaemia, leucopoenia, thrombocytopenia, hypocomplementemia, elevated erythrocyte sedimentation rate, a rise in anti-double-stranded DNA antibody levels, or any combination of these features may be associated with increased SLE disease activity. Proteinuria, hematuria, urinary casts, and a rise in creatinine levels indicate active renal involvement.

Disease Activity Indices

The following disease activity indices (Table 10B.1) and its modifications are used in trials and clinical practice. They capture organ specific or global disease activity. Disease activity indices measure reversible clinical features amenable to therapy and not damage which is usually irreversible.[1,2]

Disease Damage Indices

As SLE has a relapsing remitting nature, damage in various systems accumulate over time, both from the disease and concomitant use of immuno-

Table 10B.1: Composite disease activity measures in SLE

Disease activity measure	
The Systemic Lupus Erythematosus Disease Activity Index (SLEDAI)	24 questions assessing the physical findings and laboratory values of SLE Weighted across organ systems
1. SLEDAI	SLEDAI measures manifestations over the past 10 days
2. Safety of Estrogens in Lupus Erythematosus National Assessment (SELENA) 3. SLEDAI 2000 (SLEDAI-2K)	SELENA-SLEDAI and SLEDAI-2K score manifestations over the past 28 to 30 days
4. SLEDAI-2K Responder Index 50% (S2KRI-50)	Is found to reliably detect 50% improvement in the 24 descriptors
The British Isles Lupus Assessment Group (BILAG) Index 1. BILAG classic 2. BILAG 2004 modification	Measures disease activity in individual organ systems based on an intent-to-treat principle The BILAG and BILAG 2004 are the only indices that scores features as: Improved, the same, worse, or new rather than as present or absent Evaluates activity in 9 organ systems occurring within the past 4 weeks versus the previous month Not generally used in clinical practice
The Systemic Lupus Activity Measure (SLAM) SLAM-R modification	Based on signs and symptoms observed over the past 4 weeks, with weighting of more severe clinical manifestations. It is the only measure which scores patient-reported symptoms of fatigue
The European Consensus Lupus Activity Measure (ECLAM)	Scores clinical and laboratory manifestations of SLE over the past 4 weeks

suppression.[1,2] Damage accrual is measured using the SDI (Table 10B.2).

Table 10B.2: Measuring damage in SLE

Damage index	
The Systemic Lupus International Collaborating Clinics (SLICC)/American College of Rheumatology Damage Index (SDI)	12 organ systems are assessed to capture damage that has been present for ≥6 months

Health Related Quality of Life (HRQoL)

HRQoL is the impact of disease on the physical, social, psychological and mental aspects of health. Various patient reported outcome measures are shown in Table 10B.3.

Composite Responder Indices[1]

These composite indices (Table 10B.4) have been developed as endpoints for new treatment molecules in SLE. These indices include multiple outcome measures as defined by each trial, thus it increases the power of the study.

Conclusions

- Measuring disease activity, damage and HRQoL is essential in SLE.
- Clinical examination and laboratory tests remain basic to assessing disease activity.
- Damage accrual is major problem in SLE and should be monitored carefully.
- Indices used in clinical trials, help access the treatment efficacy of new drugs.

Table 10B.3: Patient reported outcome measures (HRQoL)

Generic The medical outcomes short form 36-item (SF-36) survey	It is important to assess HRQoL in SLE, especially since it has clearly been shown that disease activity, damage, and HRQoL are independent of each other and thus reflect different domains affected by disease.
SLE specific Lupus quality of life tool (Lupus QoL) Quality of life in SLE (L-QoL) Systemic lupus erythematosus Quality of life (SLE-QoL) Lupus patient-reported outcome tool (Lupus PRO) Lupus impact tracker	Both generic and disease-specific measures have been used in RCTs in SLE.

Table 10B.4: Responder indices used in randomized controlled trials

Responder index Responder index for DHEA trials	Responder indices define patients who have improved with treatment versus those who have not. It usually includes multiple outcome measures.
Response Index for Lupus Erythematosus (RIFLE) SLE Responder Index (SRI)	Apart from the specific components defined within each index, "responses" first require that patients are not "treatment failures", as defined prospectively in each trial.
British Isles Lupus Assessment Group-based Composite Lupus Assessment (BICLA) endpoint	

REFERENCES

1. Strand V, Chu AD. Assessing disease activity and outcome in systemic lupus erythematosus. In: Hoschberg MC, Silman AJ, editors. Rheumatology (6th edition). Philadelphia: Elsevier, Mosby; 2015. p. 1093–1098.
2. Romero-Diaz J, Isenberg D, Ramsey-Goldman D. Measures of Adult Systemic Lupus Erythematosus. Arthritis Care Res. 2011; 63 (S11):S37–S46.

10C. OUTCOME MEASURES IN SCLERODERMA

Taral Parikh

Measuring Disease Activity and Severity

Scleroderma or systemic sclerosis (SSc) is a connective tissue disease which can affect multiple organ systems.

Since disease activity[1] is defined as those features which are variable over time or has potential to reverse spontaneously or with therapy (e.g. tendon friction rubs, acute-phase reactants, inflammatory polyarthritis, inflammatory myositis), it warrants treatment. While, damage is generally irreversible and increases as the disease progresses (e.g. calcinosis, end-stage pulmonary fibrosis, deformities), here immunosuppression may be less effective.

Thus, assessment of disease activity and severity[2,3] (Table 10C.1) is important in SSc as in any other rheumatic disease, as treatment options are limited and also they carry significant toxicity.

Both activity and damage contribute to disease severity; early in SSc, activity is prominent, but as the disease advances, damage is more likely to accumulate.

Outcome Measures in SSc

Although many novel outcome measures are currently in development those outcome measures that are feasible, reliable, and valid in clinical trials and routine clinical care are listed in Table 10C.2.

Conclusions

These outcome measures, developed for each organ system, helps to decide whether a particular organ is worsening requiring a close surveillance and need for scaling up therapy. For example, if %FVC is less than 70% or HRCT extent of ILD is more than 20%, it indicates need for treatment of the ILD or rapid rise in MRSS would suggest rapidly progressing disease and need for active intervention.

Table 10C.1: Outcome measures to assess—disease activity and severity

A. Disease activity in SSc	*B. Disease severity in SSc*
The European Scleroderma Study Group* Composite Index[2]	Disease severity includes both disease activity and damage.
Includes (clinical examination, patient assessment of activity—last month, laboratory measures, and percent predicted diffusing capacity of the lung for carbon dioxide—%DLCO)	The revised Medsger severity index 3 identifies 9 organ systems (general, peripheral vascular, skin, joint/tendon, muscles, gastrointestinal tract (GIT), lung, heart, and kidney).
It is scored on a 0 (no activity) to 10 (severe activity) basis, with the greatest weight assigned to deterioration of the relevant organ system as evaluated by the patient with respect to the previous month.	Each system is scored from 0 (uninvolved) to 4 (end-stage disease).

*The European Scleroderma Study Group Composite Index, can be used to assess disease activity in practice, but still awaits validation.

Table 10C.2: Instruments for SSc outcome measures[4]

Instruments	*Remarks*
A. Skin	
Modified Rodnan skin score (MRSS)	Is a measure of skin thickness in SSc Used as the primary outcome measure in clinical trials Surrogate measure of disease severity and mortality in patients with diffuse cutaneous SSc May improve with time even without treatment is a limitation
Durometer	Hand held device—measures hardness of skin Costly

(*Contd.*)

Table 10C.2: Instruments for SSc outcome measures[4] (*Contd.*)

Instruments	*Remarks*
B. Musculoskeletal	
Tendon friction rubs	Associated with a higher likelihood of the development of diffuse cutaneous SSc and more severe disease
Tender joint count	Active arthritis
Serum creatine phosphokinase (CPK)	Myositis
Cochin hand function scale	3 scales—dexterity, rotational movement, and flexibility of the first three fingers
Hand mobility in SSc	9 items that assess hand function
Mouth handicap in systemic sclerosis scale	
C. Cardiac	
Echocardiogram with Doppler imaging	Most widely used for diagnosis and staging of PAH, evaluation of cardiac dimensions and valvular abnormalities Has disadvantages in using for diagnosis, may be used as screening tool
Right-sided heart catheterization (RHC)	RHC remains the gold standard for diagnosing PAH, for trials and clinical practice Helps identify the cause of PAH and response to therapy
6-Minute walk test (6MWT)	6MWT is currently the most widely used primary endpoint for studies investigating SSc-related PAH. Pain and musculoskeletal involvement can influence the 6MWT, and it is not always solely reflective of changes in cardiopulmonary involvement when used in patients with SSc.The 6MWT is not sensitive to change in SSc-ILD. It should not be used as a screening measure but may provide prognostic information in patients with SSc-PAH.
Borg dyspnoea index	
D. Pulmonary	
Pulmonary function test with diffusion capacity (PFT)	% FVC has been used as the main parameter of restrictive lung disease and %DLCO for pulmonary vascular disease % FVC used in clinical trials % DLCO is sensitive but not specific for SSc-ILD or SSc-PAH and can decline in both diseases
High-resolution computed tomography (HRCT)	HRCT of the lungs has two key roles in SSc: 1. Detection and staging of baseline severity 2. As a surrogate endpoint or more accurate measure of serial change.
Validated measure of dyspnoea	The Mahler dyspnoea index is a patient-reported outcome measure and has been used in SSc trials to assesses the level of dyspnoea
Breathing VAS from the SSc HAQ	Allows patients to assess their degree of difficulty in performing daily activities because of shortness of breath on a continuous 100 mm scale

(*Contd.*)

Table 10C.2: Instruments for SSc outcome measures[4] (*Contd.*)

Instruments	*Remarks*
E. Gastrointestinal	
Gastrointestinal VAS from the SSc HAQ	GIT VAS assesses interference in daily activities on a continuous 100-mm scale
UCLA Scleroderma Clinical Trial Consortium Gastrointestinal Tract 2.0	A 34-item instrument, is a validated patient reported outcome measure to assess GIT symptoms and health-related quality of life (HRQOL) in patients with SSc
Body mass index	
F. Renal	
Estimated creatinine clearance	Outcome measures for diagnosis and assessing outcomes in patients with scleroderma renal crisis (SRC)
Systolic and diastolic blood pressure	
Serum creatinine	
G. Raynaud's phenomenon	
Raynaud's condition score (RCS)	RCS is an outcome used in clinical trials and is calculated by summing the daily score over a period of 1 or 2 weeks
Number of attacks as reported by the patient	
Duration of attacks	
Raynaud's VAS from the SSc HAQ	
H. Digital ulcer	
Active digital tip ulcer count on the volar surface	
Digital ulcer VAS from the SSc HAQ	
I. Biomarkers	
Acute-phase reactants	Erythrocyte sedimentation rate and C-reactive protein are associated with disease activity and predict mortality
Serum BNP/NT-Pro BNP	NT-Pro BNP has recently been shown to predict incident cases of PAH
Vascular (von Willebrand factor [vWF]), T-cell (soluble IL-2 receptor [sIL- 2R]), B-cell (autoantibodies), and fibroblast (type III procollagen N-terminal peptide pro-peptide – PIIINP)	Have been proposed as biomarkers for SSc
Transforming growth factor-β-regulated and interferon-regulated genes	
Novel markers include myofibroblast staining, four-gene marker	
J. Patient reported outcome measures	
Short form 36 (SF 36)	
Health Assessment Disability Index (HAQ DI)	
UK Functional Score (UKFS)	
Patient-Reported Outcomes Measurement Information System	
Preference-based measures for health-related quality of life: The short form 6D	

REFERENCES

1. Symmons DP. Disease assessment indices: activity, damage and severity. Baillieres Clin Rheumatol. 1995; 9:267–85.
2. Valentini G, Della Rossa A, Bombardier S, Bencivelli W, Silman AJ, D'Angelo S, et al. European multicentre study to define disease activity criteria for systemic sclerosis. II. Identification of disease activity variables and development of preliminary activity indexes. Ann Rheum Dis 2001; 60:592–8.
3. Medsger TA Jr, Silman AJ, Steen VD, Black CM, Akesson A, Bacon PA, et al. A disease severity scale for systemic sclerosis: development and testing. J Rheumatol 1999; 26:2159–67.
4. Khanna D. Assessing disease activity and outcome in systemic sclerosis (scleroderma). In: Hoschberg MC, Silman AJ, editors. Rheumatology. Philadelphia: Elsevier, Mosby; 2015. p. 1159–64.

10D. OUTCOME MEASURES IN SPONDYLOARTHRITIDES

Anuj Shukla

INTRODUCTION

The outcome measure of spondyloarthritis (SpA) can be broadly divided into activity-measure (for example, BASDAI/ASDAS) or damage-measure (for example, BASMI). Arthritic diseases particularly affect the functions of an individual due to pain and limitation of joints, thus demanding for a third important outcome-measure called the functional-outcome-measures (FOMs).

Activity measures a component of disease that can improve with an effective treatment and thus is the best immediate measurement of therapy-responsiveness. Damage measures the components that cannot improve with a treatment and thus cannot assess the immediate effects of the therapy. But it can still assess long-run effects as a measure of accruing damage. Thus, damage measures provide a hard indicator of failure or success of a therapy. Function depends on activity as well as damage, both contributing to the limitation of movement. Thus FOMs partially improve with an effective treatment (activity component) but the damage component does not improve leading to limitation of functions. Thus it cannot precisely assess the response to the treatment. This makes BASMI, a FOM, just a measure of damage after the activity is completely controlled with proper treatment.

The outcome measures can be a single-component-measure or a composite-measure in order to give a single comprehensive value to the disease outcome. Outcome measures have four facets that is patient-based (pain, EMS), physician-based (TJC, SJC, range of movement), laboratory-based (ESR, CRP) and imaging-based (X-rays and MRI). For example, BASFI and BASMI are respectively patient-based and physician-based composite FOMs. Patient-based outcome measures in addition to disease process are affected by individual perception of pain and emotional well-being, which varies among patients. These are soft, subjective and may not exactly reflect the improvement in disease process with an effective therapy but on the other hand cannot be ignored, as the ultimate goal of therapy is patient satisfaction. On the other side, physician-based, lab-based and imaging based measures are increasingly hard, objective measures precisely reflecting the effect of therapy on disease process but may not exactly translate into patient improvement. Thus a good composite outcome measure should have a fine balance of all these four-facets of outcomes. While using all these four-facets of outcomes, the final composite outcome measure must still have good validity, reliability, responsiveness and feasibility (The OMERACT filter) for clinical and research utility.

The spondyloarthritis are currently classified as axial and peripheral. Axial includes classical ankylosing spondylitis (AS) with X-ray evidence of sacroiliac joint involvement and limitation of lumbar and thoracic spine and a major new group of non-radiographic axial-SpA (axSpA) (based on MRI and HLA-B27 with other features). Peripheral SpA (pSpA) includes patients suffering from peripheral arthritis with either HLA-B27 positivity or other typical clinical features of SpA, the major group being undifferentiated spondyloarthritis in the earlier classification. Arthritis associated with psoriasis and inflammatory bowel disease although have many clinical overlapping features, are better classified separately as Psoriatic arthritis (PsA) and enteropathic arthritis. The vast majority of assessment tools have been developed for AS and PsA.

Outcome measures in SpA will be discussed in three categories:

1. Axial spondyloarthritis (axSpA)
2. Non-psoriatic peripheral SpA (pSpA)
3. Psoriatic arthritis (PsA)

Outcome Measure in axSpA

Disease activity in axSpA is a composite measure of clinical features, laboratory parameters and overall impression of patient and the treating physician. The challenge of including all these measures in a single composite index is complicated by the need for ensuring proper validity, reliability, responsiveness and practical feasibility of calculating the index in a busy outpatient department. The Assessment of SpondyloArthritis International Society (ASAS) founded in 1995 has published many landmark papers on outcome measures in SpA.[1,2,3] Major outcome measures of axSpA are summarized in Table 10D.1 and Box 10D.1.

Table 10D.1: Outcome measures for axial-spondyloarthritis (axSpA)

Outcome measure domains	*Outcome measures*	*Year*	*Improvement*	*Issues*
Activity	BASDAI	1994	Time tested, valid, reliable composite index for assessment of disease activity in AS. Does not require any laboratory tests. BASDAI scores are the current employed criteria used to decide for biological drug therapy in AS.	Since they are patient self-reported parameters so scores are affected by patient mood and education, measures only part of disease activity, lacks specificity to inflammatory process, does not weigh individual components.
	ASDAS (ASDAS-CRP is preferred over ESR)	2009	Better construct validity and discrimination than BASDAI, includes lab parameters with less redundancy between individual components and are weighed variably	Requires ESR, CRP. Requires scientific calculator or internet connection*
	SASDAS	2012	Simplified version of ASDAS (summed variables) no need of calculator, same parameters as ASDAS, includes ESR (which is less expensive)	45% patients classified by ASDAS-CRP as moderate disease activity were classified differently by SASDAS. Requires validation in larger cohorts.
Functional	BASFI	1994	Largely supplanted earlier functional indices, underwent rigorous psychometric analysis and has shown good reproducibility	Measures only physical functions, performance for peripheral SpA is poorer compared to HAQ-S
	ASAS-HI	2011	A linear composite patient-reported measure containing 17 dichotomous queries addressing pain, emotional functions, sleep, sex, mobility, self-care, community life and employment	Further studies of its psychometric properties are required prior to its use in clinics and trials It assesses functioning based on objective description and does not assess the subjective appraisal of the problems
Quality of life	ASQoL	2003	A well validated composite patient-reported measure containing 18 dichotomous queries Has excellent scaling and psychometric properties	QoL perspective is based on subjective appraisal of the impact of the disease on life, does not give objective idea of whole range of common difficulties faced by patients
Spinal mobility (Activity + damage)	BASMI	1995	A composite measure including cervical rotation, tragus to wall distance, lumbar side and forward flexion and intermalleolar distance; BASMI-linear (2008) showed better responsiveness than both BASMI-10 step and BASMI-2 step score in the golimumab trial	Not a pure measure of spine mobility as includes inter-malleolar distance, does not include chest expansion. There is room to develop new composite scores that may differ from current BASMI

(*Contd.*)

Structural damage (radiological)	mSASSS	2005	Quantification of structural spinal changes in AS using lateral X-ray of cervical and lumbar spine	Very low sensitivity to change, does not include thoracic spine and does not assess inflammation activity. Newer scores being developed are the RASSS which also includes assessment of lower thoracic spine and only osteoprolifera-tive changes are scored, and MRI-based scoring-system.

Abbreviations: AS—Ankylosing spondylitis, SpA—spondyloarthritis, ASAS—Assessment of Spondyloarthritis international Society, ASDAS—AS disease activity score, SASDAS—Simplified ASDAS, BASDAI—Bath AS disease activity index, BASMI—Bath AS metrological index, BASFI—Bath AS functional Index, ASQoL—AS quality of life, ASAS-HI—ASAS health index, mSASSS—Modified Stokes AS Scoring system, HAQ-S Health assessment questionnaire modified for SpA, RASSS—Radiographic AS scoring system

*ASDAS can be calculated on line http://www.asas-group.org/clinical-instruments/asdas_calculator/asdas.html

Box 10D.1 Disease activity slabs, improvement and flare scores for axial-spondylo-arthritis (axSpA)

Disease activity scores

ASDAS-CRP scores

- Inactivity disease—<1.3 (defining remission better than ASAS partial response as it included BASFI)
- Moderate disease activity—1.3 to <2.1
- High disease activity—2.1 to <3.5
- Very high disease activity >3.5

Improvement scores

- Clinically important improvement—at least 1.1 unit change in ASDAS-CRP
- Major improvement—at least 2 units change in ASDAS-CRP
- Other improvement criteria used to define improvement are ASAS20, ASAS40, BASDAI50 and delta BASDAI of at least 2 units

Definition of Flare in axSpA

- An increase of at least 1.3 in ASDAS-CRP or 2.1 in BASDAI has been defined as a flare in axial SpA

Outcome Measures for Non-psoriatic Peripheral Spondyloarthritis (pSpA)

ASDAS-CRP and BASDAI as well as single item measures like Patient Global Assessment (PGA) and Physician Global Assessment (PhGA) all perform equally well and better than the other single measures like ESR and CRP, in detecting disease activity in pSpA and differentiating treatment arm from the placebo arm.[4] But it has been shown that these composite measures fail to assess different aspects of disease in pSpA, for example, enthesitis and dactylitis. This is shown by their poor correlation with PGA and PhGA suggesting that aspects of disease activity considered important by physicians or patients are not captured in these composite measures. This calls for derivation of a new composite measure specific for pSpA excluding PsA.

The ABILITY-2 trial of adalimumab in pSpA used a new disease specific response criteria termed Peripheral SpondyloArthritis Response Criteria (PSpARC-40) as the primary end point. Both response criteria, disease specific PSpARC-50/70 and disease non-specific ACR-20 response criteria performed best in terms of discriminative ability in both the trials of anti-TNF therapy in pSpA. The performance of axial-specific response criteria ASDAS and BASDAI50 was worse compared to the above two. This is likely because the cut-off levels of improvement in ASDAS response and BASDAI50 has been tested in axSpA and not in pSpA and thus failed to capture the improvement.

Outcome Measures for Psoriatic Arthritis (PsA)

Assessment of PsA is complicated by the fact that the disease has heterogeneous manifestations from peripheral arthritis to involvement of skin, spine, enthesis, dactylitis and nail involvement.[5] The most important question that arises is whether to include all these in a single outcome measure or to assess them separately. The variation is not only limited to manifestations but these clinical features also respond varyingly to various immunomodulation. Thus making a single outcome measure including peripheral arthritis, skin and spine involvement increases heterogeneity of responsiveness and decreases discriminative power of the score while assessing different therapies. This makes a case in favor of assessing peripheral arthritis, spine involvement and skin separately to allow evaluating therapies that are effective for certain but not other manifestations of this heterogeneous disease.

The Group for Research and Assessment of Psoriasis and Psoriatic Arthritis (GRAPPA) and *Outcome measures in Rheumatology* (OMERACT) has published 6 mandatory core outcome measures to be included in all clinical trials for PsA. These 6 single variables are peripheral joint activity, patient global and pain assessment, physical function, skin disease and quality of life. Spinal assessment, dactylitis, enthesitis, acute phase reactants, nail changes, fatigue, structural damage and imaging were not included in the core set of obligatory assessment and were included in outer circle and as research agenda.

Most of the current validated and frequently used composite disease activity state measures and response criteria are borrowed from other more common arthritides, for example, *Disease Activity Score-28* (DAS-28) and *European League Against Rheumatism* (EULAR) response criteria from RA. Peripheral arthritis in PsA is also heterogeneous varying from polyarthritis like RA to oligoarthritis, predominant DIP arthritis and monoarthritis. This makes a case against the use of RA based outcome measures using 28-joints involved commonly in RA. Another composite index initially developed for reactive arthritis is *The Disease Activity Index for PSoriatic Arthritis* (DAPSA). It uses 66/68 joint count and is a summation of five disease activity variables: Tender and swollen joint count (TJC, SJC), patient global and pain assessment on visual analogue scale 0–10, and CRP. The other two specific disease activity indices are *The Composite Psoriatic Disease Activity Index* (CPDAI) and *The Psoriatic Arthritis Disease Activity Score* (PASDAS). CPDAI has 5 domains: 66/68 joint count, dactylitis, enthesitis, skin disease and spine disease. Each domain is assessed for activity and quality of life or health assessment questionnaire and given a score of 0 to 3. The single domains are then summarized to give a final score from 0 to 15. In a modified version of CPDAI, spine assessment has been dropped, thus resulting in a maximum score of 12. Another version of CPDAI includes joint count, enthesitis and dactylitis (CPDAI-JED), but excludes skin assessment. PASDAS includes physician and patient global assessment, 66/68 SJC/TJC, physical component score of Survey-Short Form-36, enthesitis, dactylitis as well as CRP. Comparison of various indices is shown in Table 10D.2.

Among the response criteria for PsA, *European League Against Rheumatism* (EULAR) criteria based on DAS-28 improvement performed better than 20/50/70 ACR criteria and both were better than more disease specific *Psoriatic Arthritis Response Criteria* (PsARC). PsARC includes 66/68 SJC/TJC, physician and patient global assessment both on a 0–5 Likert rating scale. 30% improvement in joint count and 1-point reduction in Likert scale is defined as response. PsARC excludes acute phase reactants and pain assessment. Response criteria help to assess change in activity from baseline, thus

Table 10D.2: Outcome measures (disease activity) for psoriatic arthritis (PsA)

Outcome measures	*Advantages*	*Disadvantages*
Disease activity score-28 (DAS-28)	Proven to be highly responsive and discriminative instrument in PsA trials	Does not include joints which are commonly involved in PsA, for example, DIP, foot and ankle
Disease activity for psoriatic arthritis (DAPSA)	Applies 66/68 swollen and tender joint count, exhibits good validity responsiveness and discrimination for PsA patients, shows correlation with ultrasound assessed synovitis	Does not include enthesitis, dactylitis, spine and skin frequently involved in PsA
Composite psoriatic disease activity index (CPDAI)	Disease specific includes all the aspects of PsA, for example, spine and skin involvement	Needs validation in trials
Psoriatic arthritis disease activity score (PASDAS)	PsA specific compound score does not include skin and spinal assessment	Needs validation in trials

improvement with therapy but do not allow the quantification of disease activity.

Indian Perspective for Outcome Measures in SpA and PsA

There are multiple problems with using these outcome measures in Indian routine clinical practice.

1. Low education level with less health related awareness makes it extremely difficult to scale the disease on visual analogue scale of 0–10 by our patients. In fact most of the patients usually scale it as 0, 5 or 10 with 5 being the most common score. This markedly affects the assessment of therapeutic responsiveness of most of these patient-oriented-scores including ASDAS and BASDAI. Patient assessment of global disease activity and pain on visual analogue scale is often the same, as patients are hardly aware of swollen joints or fatigue. Early morning stiffness term is often difficult to understand, which is again confused with pain.
2. SpA in comparison to rheumatoid arthritis (RA) is complicated as it has many components to assess the disease, for example, entheseal assessment, axial and peripheral joint assessment, so it is difficult to make a single composite measure for different aspects of disease activity like DAS-28 in RA. Measuring different aspects of the disease with different outcome measures is difficult in busy Indian outpatient departments.
3. Complicated more for PsA where Psoriasis Area and Severity Index (PASI) for skin involvement is time consuming. Physician training for skin assessment is limited and there is poor interaction between dermatologists and rheumatologists. Due to socio-cultural influences, examining ladies is often incomplete due to reluctance in undressing.
4. In addition to outcome measures there are many other problems related to diagnosis of Axial-SpA, for example, marked delay in diagnosis, use of costly tests like HLA-B27 (PCR) and MRI in diagnosis with limited availability, unsatisfactory management due to cost of anti-TNF therapy. It becomes futile to keep measuring outcome measures in every follow up OPD without giving appropriate therapy.

Altogether there are multiple challenges in diagnosis, management and assessment of outcome measures in SpA and PsA. The classification criteria have recently evolved, the definition of optimum treatment and their targets are likely to evolve with more therapeutic options in the future and so the need for better outcome measures with more construct validity, discriminative capabilities in research trials and feasibility in routine clinical practice.

REFERENCES

1. Machado P, van der Heijde D. How to measure disease activity in axial spondyloarthritis? Curr Opin Rheumatol. 2011; 23:339–45.
2. Braun J, Kiltz U, Baraliakos X, van der Heijde D. Optimisation of rheumatology assessments—the actual situation in axial spondyloarthritis including ankylosing spondylitis. Clin Exp Rheumatol. 2014; 32:S-96–104.
3. Zochling J. Measures of symptoms and disease status in ankylosing spondylitis: Ankylosing Spondylitis Disease Activity Score (ASDAS), Ankylosing Spondylitis Quality of Life Scale (ASQoL), Bath Ankylosing Spondylitis Disease Activity Index (BASDAI), Bath Ankylosing Spondylitis Functional Index (BASFI), Bath Ankylosing Spondylitis Global Score (BAS-G), Bath Ankylosing Spondylitis Metrology Index (BASMI), Dougados Functional Index (DFI), and Health Assessment Questionnaire for the Spondyloarthropathies (HAQ-S). Arthritis Care Res (Hoboken). 2011; 63 (Suppl11):S47–58.
4. Turina MC, Ramiro S, Baeten DL, Mease P, Paramarta JE, Song I, et al. A psychometric analysis of outcome measures in peripheral spondyloarthritis. Ann Rheum Dis. 2016; 75:1302–7.
5. Schoels M. Psoriatic arthritis indices. Clin Exp Rheumatol. 2014; 32:S-109–12.

10E. OUTCOME MEASURES IN PRIMARY SJÖGREN'S SYNDROME

Nibha Jain, Sapan Pandya

Primary Sjögren's syndrome (pSS) is a systemic disorder extending beyond the exocrine glands. The disease spectrum includes sicca symptoms alone or in combination with systemic extra glandular manifestations.[1]

Clinical features include:

1. Sicca, pain and fatigue affecting almost all
2. Systemic manifestations affecting 20–40%

The natural course of pSS is assessed with respect to three aspects:

1. Disease activity (potentially reversible with intervention) with flares in between
2. Disease damage (irreversible and can increase with time)
3. Subjective findings (patient's perception of the symptoms of the disease).

The need for outcome measures in pSS arose out of interest in conducting clinical trials, including RCTs.[2] Assessments of disease activity was essential for inclusions in clinical trials especially when using biologic or other immunosuppressive therapies, and determining end points was needed. At the same time it ought to be useful for daily clinical practice.

1. Instruments for Measuring Disease Activity (Table 10E.1)

Activity indices are designed for patients with systemic complications of primary pSS. They are based on physician's judgment and include multiple domains (i.e. organ systems). The EULAR Sjögren's syndrome disease activity index (ESSDAI) had domains which encompassed all organ systems involved in the disease, and was agreed upon by a large number of experts.[3] Each domain is rated according to degree of activity with final score calculated as sum of all weighted scores. Domains include fatigue, constitutional symptoms, arthritis, muscle, gland swelling, and skin, pulmonary, renal, neurological and hematological domains.

Constitutional: After excluding infectious causes, symptoms of fever night sweats and weight loss graded according to severity.

Lymphadenopathy: Exclusion of infection followed by assessment of lymph node involvement in any nodal region or inguinal region and/or splenomegaly (clinical or proven by imaging). Current B-cell malignant proliferative disorders are taken as high grade.

Glandular: After exclusion of calculi or infection, salivary gland swelling limited to parotids or submandibular with lacrimal swelling.

Articular: Exclusion of osteoarthritis, involvement of arthralgia or inflammatory arthritis with 28 joint counts.

Cutaneous: Presence of erythema multiforme, cutaneous vasculitis including urticarial vasculitis, or purpura, or subacute cutaneous lupus or ulcers related to vasculitis.

Pulmonary: Evidence of interstitial lung disease with any involvement of breathlessness or cough and lung function tests.

Renal: In the form of tubular acidosis or glomerular involvement with proteinuria, hematuria and reduced GFR or cryoglobulinemia.

Muscular: Assessment of myositis with clinical grading of power, EMG, biopsy and elevated CK levels.

Peripheral nervous system (PNS): Cranial nerve involvement of peripheral origin (except trigeminal (V) neuralgia), pure sensory axonal polyneuropathy shown by NCS, sensory neuropathy with presence of cryoglobulinemic vasculitis, ganglionopathy with symptoms restricted to mild/moderate ataxia, chronic inflammatory demyelinating polyneuropathy (CIDP) with mild functional impairment.

CNS: Rated as cranial nerve involvement of central origin, optic neuritis or multiple sclerosis like syndrome with symptoms or proven cognitive impairment, cerebral vasculitis with cerebrovascular accident or transient ischemic attack, seizures, transverse myelitis, lymphocytic meningitis.

Hematological: In the form of autoimmune cytopenias.

Biological: This domain takes into account the clonal component and/or hypocomplementemia (low C4 or C3 or CH50) and/or hypergammaglobulinemia or high IgG level and presence of cryoglobulinemia.

In 2016 a clinical index (ClinESSDAI) without the biologic domains was devised (Table 10E.1) which was found to correlate well with the ESSDAI.[4] This could be useful especially in the practice setting where patients do not always get investigations done.

2. Measuring Damage in pSS (Table 10E.2)

To assess the accumulated permanent damage in various organs and systems the Sjögren's syndrome damage index (SSDI) was developed.[5] Damage indices become relevant as the duration of illness gets longer due to increasing damage accrual as seen with lupus. The SSDI was developed by the British Sjögren's Group and was based on the SLICC/ACR for lupus. It included 3 main domains—ocular, oral and systemic with the last broken into further organ systems as shown in Table 10E.2.

3. Subjective Measure (Table 10E.3)

The original subjective symptom outcome measure PROFAD[6] (had 8 domains which were scaled on a Likert scale) has been now reduced to just 3 domains (fatigue, dryness and musculoskeletal pain) called the ESSPRI (Table 10E.3 and Fig. 10E.1).[7] Patients with pSS have fatigue and discomfort as major components which can be disabling. Subjective measures take into account this component without

Table 10E.1: Activity index of pSS

Activity index	*Remarks*
SCAI (SS clinical activity index) Exhaustive but too complicated to use in clinical practice	• 10 domains: Fatigue, constitutional symptoms, arthritis, muscle, gland swelling, skin, pulmonary, renal, neurological and hematological domains • Items are recorded as 0(absent) 1(improving) 2(same) 3(worse) and 4(new)
SSDAI (Sjögren's syndrome disease activity index) Simple test but lacks exhaustiveness	• 8-domain global score • Final score >5 suggestive of active disease
(ESSDAI) EULAR SS disease activity index (2009)[3]	• Based on physician's judgment • 12 domains, each is divided in 3–4 levels according to their degree of activity • The final score between 0 and 123 • 0 indicating no disease activity • Any feature stable for 12 months should be taken as damage • Validated in large independent cohort; reproducible • Correlates well with biomarkers like BAFF levels • Thresholds for disease activity and minimal clinically important improvement defined (e.g. ESSDAI ≥5, moderate disease and ≥3 needed for MCII)
ClinESSDAI[4]	• Exclusion of the biological domain of ESSDAI • Comparable ESSDAI with slight lower sensitivity to change • Avoids collinearity of data in biological studies • Can be used in absence of laboratory investigations

Table 10E.2: Measuring damage in pSS

Damage index	*Remarks*
SSDI (Sjögren's syndrome damage index)[5]	• Assess the combined permanent systemic global score on 3 separate scales (ocular, oral and systemic damage). Maximum score was 1 for each item and total items were 27
SSDDI (SS disease damage index)	• A global score including exocrine and non-exocrine features • Not very sensitive to change and generally not used in practice

Table 10E.3: Subjective measure of Sjögren's syndrome

Subjective measure	*Remarks*
PROFAD (Profile of fatigue and discomfort)[6]	• Psychometric instrument • Somatic and mental fatigue domain with the fatigue, arthralgia and vascular dysfunction
PROFAD-SSI (Profile of fatigue and Discomfort: Sicca symptoms inventory)	• 64 questions in eight 'domains' • Scored on an 8-point (0–7) on Likert scale. • Shorter version with 19 questions available.
Xerostomia inventory (XI)	• 11 questions that address xerostomia symptoms in daily activities
ESSPRI (EULAR Sjögren's syndrome patient reported index) (2011)[7]	• Scores 1 to 5 with final score range 11–55 • Patient-administered questionnaire • Evaluates subjective symptoms • Sensitive to change and used in trials—validated in prospective international cohort • Correlates well with quality of life measures • Poorly correlated to ESSDAI

1. How severe has your dryness been during the last 2 weeks?

No dryness	0	1	2	3	4	5	6	7	8	9	10	Maximal imaginable dryness

2. How severe has your fatigue been during the last 2 weeks?

No fatigue	0	1	2	3	4	5	6	7	8	9	10	Maximal imaginable fatigue

3. How severe has your pain (joint or muscular pains in your arms or legs) been during the last 2 weeks?

No pain	0	1	2	3	4	5	6	7	8	9	10	Maximal imaginable pain

Fig. 10E.1: The EULAR Sjögren's syndrome patient reported index (ESSPRI)

any measure of damage. ESSPRI and ESSDAI are complementary as outcome measures in clinical trials and hence a clear definition of which component is being tested is needed while designing them.

4. Responder Indices (Table 10E.4)

Sjögrens syndrome responder index (SSRI) was devised in the TEARS trial (Tolerance and Efficacy of Rituximab in Sjögren's syndrome) and is similar to the SLE responder index.[8] This is particularly useful when assessing treatment response to costly treatments as biologics.

5. Measure of Health Related Quality of Life (HRQoL)

Measure of HRQoL is an important but difficult issue with several factors contributing to the impairment of the HRQoL in pSS. The disease activity, accumulated damage with disease-specific issues such as dryness, chronic pain and fatigue all contribute to it. Since no disease-specific HRQoL index exists, the most widely used tool in pSS has been the short form 36 (SF-36).

Conclusion

Since Sjögren's syndrome is one of the diseases with a lot of subjective overlay with the objective symptoms, use and application of outcome measures which can gauge disease activity (both symptoms and signs) versus damage become very relevant in today's era where more and more biologic therapies are being used early in diseases. At the moment, the ESSPRI and ESSDAI are the most useful measures and both are recommended to be used in clinical trials as they complement each other. In the practice setting, ESSPRI and ClinESSDAI can

Table 10E.4: Responder indices

Responder index	*Remarks*
SSRI (SS responder index)[8]	• Composite index with five outcome measures: VAS fatigue score, VAS oral dryness score, VAS ocular dryness score, unstimulated whole saliva flow and ESR • Includes both subjective and objective measures • SSRI-30 response is defined as a 30% improvement in at least two of the five outcome measures

be easily used. We recommend the latter as a bare minimum in our country if we were to contribute to global Sjögren's data from the subcontinent.

REFERENCES

1. Seror R, Theander E, Bootsma H, Bowman SJ, Tzioufas A, Gottenberg JE, et al. Outcome measures for primary Sjögren's syndrome: a comprehensive review. J Autoimmun 2014; 51:51–6.
2. Price EJ, Rigby SP, Clancy U, Venables PJ. A double blind placebo controlled trial of azathioprine in the treatment of primary Sjögren's syndrome. J Rheumatol. 1998; 25: 896–9.
3. Seror R, Bowman SJ, Brito-Zeron P, Theander E, Bootsma H, Tzioufas A, et al. EULAR Sjögren's syndrome disease activity index (ESSDAI): a user guide. RMD Open. 2015; 1(1):e000022.
4. Seror R, Meiners P, Baron G, Bootsma H, Bowman SJ, Vitali C, et al: EULAR Sjögren Task Force. Development of the ClinESSDAI: a clinical score without biological domain. A tool for biological studies. Ann Rheum Dis. 2016; 75:1945–50.
5. Krylova L, Isenberg D. Assessment of patients with primary Sjögren's syndrome—outcome over 10 years using Sjögren's Syndrome Damage Index. Rheumatology (Oxford) 2010; 49:1559–62.
6. Bowman SJ, Hamburger J, Richards A, Barry RJ, Rauz S. Patient-reported outcomes in primary Sjögren's syndrome: comparison of the long and short versions of the Profile of Fatigue and Discomfort—Sicca Symptoms Inventory. Rheumatology (Oxford). 2009; 48:140–3.
7. Seror R, Ravaud P, Mariette X, Bootsma H, Theander E, Hansen A, et al; EULAR Sjögren's Task Force. EULAR Sjögren's Syndrome Patient Reported Index (ESSPRI): development of a consensus patient index for primary Sjögren's syndrome. Ann Rheum Dis. 2011; 70:968–72.
8. Cornec D, Devauchelle-Pensec V, Mariette X, Jousse-Joulin S, Berthelot JM, Perdriger A, et al. Development of the Sjögren's Syndrome Responder Index, a data-driven composite endpoint for assessing treatment efficacy. Rheumatology (Oxford). 2015; 54:1699–708.

10G. OUTCOME MEASURES IN OSTEOARTHRITIS

Piyush Joshi

Osteoarthritis (OA) is the most common arthritis worldwide. Degenerative nature of the disease leads to a poor response to treatment from any of the drugs including hyaluronic acid injections or slow acting symptomatic osteoarthritic drugs (diacerein, glucosamine sulfate, chondroitin sulfate, etc.), making it necessary to find a good outcome measure which can evaluate treatment responsiveness in osteoarthritis.

Understanding the derivation and validation of outcome measures and minimal clinically important differences (MCID) helps clinicians interpret data from published RCTs (randomized control trials). OARSI (Osteoarthritis Research Society International) and OMERACT (outcome measures in rheumatology) both advocated the use of core outcome measures which access pain and function in people with OA. Both patient-reported and performance-based outcome measures have been used to assess physical function but currently there is no gold standard measure in the assessment of OA. Absence of a consensus on the best performance-based tests makes it difficult to select the most feasible and reliable outcome measures for clinical and research purposes.[1]

Outcome measure included by OMERACT should show truth, reliability and discrimination.[2] Reliability is measured by evaluating the extent to which the same or similar scores are assigned the same valuation over multiple replications, reported as internal consistency coefficients (ICCs) (kappa coefficient value of >0.8 suggests good test retest reliability). "Truth" is assessment of an instrument according to its content, criterion, and construct validity. Content validity is extent to which instrument measures what it should with meaningful information. Criterion validity is how instrument fare to gold standard and construct validity is the extent to which an instrument correlates with other measures with which it should be related (convergent evidence) and is able to distinguish between groups (e.g. *between patients with or without a condition*—discriminate evidence). "Discrimination" is the ability of an instrument to detect changes over time.[1, 3]

For commonly used outcome measures, it is useful to understand whether improvements represent clinically important differences perceptible to patients or not. Determination of MCID (minimal clinically important difference) helps in setting goals and sample size calculations.

Table 10G.1 shows all the outcome measures for osteoarthritis available, while in the following discussion only important ones are discussed further. Outcome measures can be divided among patient reported and performance based outcome measures.

Patient Reported Outcome Measures

1. **Pain assessment:** Pain is often considered the most important symptom of OA. Pain can be accessed based on VAS (Visual analogue scale) (0–100 mm); Likert scale (5 point scale) and SF 36 (short form 36) bodily pain subscale can be used for objective measurement of pain. VAS scales are potentially more sensitive to change. People also advocate the use of a numeric rating scale for patients with language, cultural, or cognitive difficulties comprehending VAS scales. Short form-36 (SF-36) bodily pain subscale is considered as a brief measure of pain severity because of its reliability, validity, and the extensive normative data available for this instrument.
2. **Global disease activity by patient:** Patient global assessment question can be answered by the use of a Likert scale (very good, good, fair, poor, or very poor), or a 100-mm VAS. In general, VAS scores are more sensitive to change, physician's global assessment is not part of the obligatory OMERACT III core set, in part because of the disparity between patient and physician reported outcomes, but is recommended only as a supportive measure.
3. **Physical function assessment**

FOR KNEE AND HIP OA

WOMAC (Western Ontario and McMaster Osteoarthritis Index)[4, 5]: The WOMAC includes 24 questions in the following three sections: pain (five questions), stiffness (two questions), and

Table 10G.1: Outcome measures in osteoarthritis

	Outcome measures
Pain scale	1. Visual analogue scale (VAS) 2. Likert scale 3. SF-36 bodily pain subscale
Physical function scales	**Hip and Knee:** 1. Western Ontario and McMaster Osteoarthritis Index (WOMAC) 2. Hip injury and osteoarthritis score (HIOS) 3. Knee injury and osteoarthritis score (KIOS) 4. Oxford hip score 5. Oxford knee score 6. Le Quesne **Index Shoulder:** 1. American Shoulder and Elbow Surgeons (ASES) Standardized Shoulder Assessment Form **Hand OA:** 1. Australian/Canadian (AUSCAN) Osteoarthritis Hand Index 2. Disabilities of the Arm, Shoulder, and Hand (DASH) 3. Dreiser and Cochin Indices
General measures	1. Patient global assessment (PGA) 2. Health assessment questionnaire (HAQ) 3. Arthritis Impact management scale (AIMS and AIMS-2) 4. Osteoarthritis Global Index 5. Patient Reported Outcome Measurement Information System (PROMIS-29) 6. Short form-36 (SF-36) 7. Health related quality of life (HRQoL) 8. Work limitation questionnaire (WLQ)

physical function (17 questions). The WOMAC is completed using Likert 5-point (none, mild, moderate, severe, or extreme) or VAS (0 [no difficulty] to 100 mm [extreme difficulty]) scales and has been well validated. Test-retest reliability is 0.8 for physical function and 0.7 for pain. Overall truth, reliability, and discrimination were well demonstrated for the pain and physical function subscales across patient groups and types of interventions. Completion of the WOMAC requires approximately 5 minutes.

Hip Injury and Osteoarthritis Score (HIOS)

It is a WOMAC modification for hip. It is a 40 item self-report questionnaire with 5 subsets (pain, stiffness, QoL, daily activities and sports) on Likert scale.

Knee Injury and Osteoarthritis Score (KIOS)

It is a modification of WOMAC score for knee. A 42 item self-administered assessment of five outcomes: Knee related QoL (quality of life), activities of daily living, sports and recreation function, symptoms and pain on Likert scale 0–4.

Oxford Hip Score

12 point score on patient's pain and function on scale of 0–5, making a total score of 60.

Oxford Knee Score

12 point score on patient's pain and function on scale of 0–5, making a total score of 60.

Le Quesne Index[6]

The Le Quesne Index measures pain (five questions) and physical function, specifically maximal walking distance and activities of daily living (four items). In post arthroplasty patients, on comparing WOMAC with Le Quesne Index, WOMAC was more responsive. Total WOMAC and pain subscale scores had SRMs (Standardized response means) of 2.0 (knee) and 2.4 (hip) compared with 1.5 and 2.1, respectively.

Chapter

11

Synovial Fluid Analysis

Sanjeeb Kakati

BACKGROUND

Synovial joints consist of joint capsule, cartilaginous articular surface, synovial fluid and synovium. Synovial fluid is regarded as body fluid lying within the tissue bounded by synovium. It is often discussed in similar terms to those used for lining and contents of other body cavities.[1,2]

Synovial fluid is composed of glycoprotein and hyaluronate in addition to small filtrate molecules from plasma. Passage of larger molecules like fibrinogen results from increased vascularity, i.e. injury or inflammation. Lymphatics are the drainage path of these molecules back to plasma and plays a least role in disease condition.

The main function of synovial fluid is to provide lubrication, joint stability and supply of nutrition to the cartilage and it also has antibacterial properties.[3,4]

Synovial fluid analysis consists of a series of test done on synovial fluid with the aim of diagnosing joint disease.

INDICATIONS FOR SYNOVIAL ANALYSIS[5,6]

Diagnostic

- Septic arthritis
- Crystal deposition disease

Therapeutic

Large joint effusion with pain and swelling.

Suspected septic arthritis—the American College of Rheumatology (ACR) clinical guideline suggest that unexplained inflammatory fluid, be assumed to be infected until proven otherwise by appropriate culture.[7] According to the ACR synovial fluid analysis should be performed in a febrile patient with an acute flare of established arthritis (e.g. rheumatoid arthritis, osteoarthritis) to rule out superimposed septic arthritis.[7]

Repeated aspiration and analysis may be used to monitor response to treatment in septic arthritis and may also be valuable for diagnosis of some cases of gout in which initial aspiration does not have detectable crystals.[8]

Timing of the arthrocentesis is of utmost importance. Institution of antibiotic therapy may alter the finding of synovial fluid and hence delay or never establish the diagnosis in septic arthritis. In crystal-induced arthritis especially the level of monosodium urate or hydroxyapatite crystals may reduce rapidly and hence the synovial fluid analysis may become inconclusive. Therefore, once the decision of synovial fluid analysis is taken, it should be carried out promptly.[9]

Contraindication of Arthrocentesis

There is no absolute contraindication for synovial fluid aspiration. However, relative contraindications are: (1) Bleeding diathesis, e.g. hemophilia, anticoagulant therapy, thrombocytopenia, etc. (2) Cellulitis overlying the joint.[2]

Selection of Site

Knee, elbow, ankle, shoulder, wrist including olecranon, or prepatellar bursae are preferred site for synovial fluid aspiration. Arthrocentesis of the hip needs ultrasonographic or fluoroscopic guidance.[1,2,10]

Procedure

Synovial fluid aspiration or arthrocentesis should be carried out under aseptic precautions. A topical antiseptic preferably povidone iodine is to be applied over the area. Use of sterile gloves is mandatory. 25 gauze needle is preferred for administration of local anesthetic (e.g. 1% lidocaine). With the help of a 20–30 cc syringe and 18 gauze needle, aspiration should be carried out.[10]

Use of talc-free gloves is recommended when preparing synovial fluid for a crystal search, as contamination of slide with birefringent talc particles may make the microscopic examination for pathogenetically important crystals difficult.[11]

Complications of Arthrocentesis

Infection (<1 in 10,000), pain, haemarthrosis, cartilage injury can occur during the procedure of joint fluid aspiration.

Causes of Dry Tap

Very thick fluid, obstruction of the needle lumen with debris, deposition of thick layer of fat in chronically inflamed synovium, obese patient with thick layer of medial fat in the knee joint and faulty technique are the common causes of dry tap.[10] Dry taps may still have fluid within the needle, which may be sufficient for the most critical test. Such specimen should be submitted with the needle still on the syringe and its tip stuck into a sterile cork.

Collection and Transportation

Inflamed synovial fluid tends to clot. Therefore, it is to be collected in sodium heparin container to prevent clotting. For culture and sensitivity testing the synovial fluid should be sent in the syringe itself. While transporting the syringe it should be wrapped with a sterile plastic cap. The fluid should be analyzed at the earliest otherwise there may be reduction in the cell count, decrease in the number of crystals, or rarely, appearance of artifactual crystals.[1,10]

EXAMINATION OF SYNOVIAL FLUID

Joint effusion can be due to inflammatory or non-inflammatory in origin (Table 11.1). Evaluation of synovial fluid for the following will help in narrowing the differential diagnosis.

Table 11.1: Causes of joint effusion[1,3,7]

Non-inflammatory	*Inflammatory*
Osteoarthritis	Rheumatoid arthritis
Trauma	Crystal induced arthritis
Internal derangement	Psoriatic arthritis
Osteochondritis dissecans	Reactive arthritis
Charcot's arthropathy	Whipples disease
Villonodular synovitis	Polymyalgia rheumatica
Myxoedema	Polychondritis
Acromegaly	Sarcoidosis
Haemochromatosis	Behçet's syndrome
Ochronosis	Ankylosing spondylitis
Sickle cell disease	Lyme disease
Milwaukee shoulder	Leukaemia

- **Physical appearance**—an evaluation of the appearance of the fluid.
- **Microscopic examination**—cells and crystals that may be present are counted and identified by type under a microscope.
- **Chemical tests**—detect changes in the chemical constituents of the fluid.
- **Infectious disease tests**—detect and identify microorganisms, if present.

Normal synovial fluid is a clear, yellowish fluid and transparent enough to read newsprint through.

Physical Characteristics

Volume, clarity, color and viscosity of synovial fluid may give an indication of the underlying cause. Normal fluid is either colorless or straw color and transparent. In septic arthritis fluid is either translucent, or opaque. Physical characteristic of three groups of fluid, viz. non-inflammatory, inflammatory and septic arthritis are shown in Table 11.2.

Traumatic tap—uneven distribution of blood during arthrocentesis although pale yellow xanthochromia is may be difficult to distinguish from normal; a red brown colour following centrifugation is good evidence of pathologic hemarthrosis.

A shimmering, oily appearing specimen suggests an abundance of cholesterol crystal which may grossly resemble pus. A ground pepper appearance from pigmented cartilage fragment may be the result of a metabolic disorder (i.e. ochronosis).

Table 11.2: Characteristics of synovial fluid in normal and various disease conditions[7,21,22]

Test	*Normal*	*Non-inflammatory*	*Inflammatory*	*Sepsis*	*Crystal*	*Hemarthrosis*
Appearance	Clear	Slightly turbid	Turbid	Turbid	Turbid	Bloody
Color	Clear	Straw	Straw to opalescent	Opaque	Yellow-milky	Red-brown
Viscosity	Very high	High	Low	Variable	Low	Low
Mucin clot	Firm	Firm to friable	Friable	Friable	Friable	Friable
Clotted	No	Occasional	Occasional	Often	Occasional	Yes
WBC/mm³	0–200	200–2000	2000–50,000	2000–>50,000	200–>50,000	50–10,000
% Polys	10	<20	20–70	>70	<90	<50
Glucose difference	0–10	0–10	0–40	20–100	0–80	0–20
Crystals	Absent	Absent	Absent	Absent	Present	Absent
Culture	Sterile	Sterile	Sterile	Positive	Sterile	Sterile
Diseases and differential diagnosis		Osteoarthritis, osteochondritis, pigmented villonodular synovitis, sickle cell disease, neuropathic joint	Rheumatoid arthritis, SLE, Reiter's syndrome, ankylosing spondylitis, ulcerative colitis, psoriasis, rheumatic, fever, gout psudogout	Bacteria, tubercular arthritis, gonococcal arthritis, fungi	Gout, pseudogout	Trauma, haemophilia, heman-gioma, pigmented villonodular synovitis, anticoagulant therapy, tumours

MICROSCOPIC ANALYSIS OF SYNOVIAL FLUID

Specimen requirement for cell count and differential is transferred to an EDTA tube. Leucocytes are counted in the same way as it is done in peripheral blood but normal saline is used as diluent instead of acetic acid. White blood cell (WBCs) are normally less than 200/mm^3 with fewer than 25% neutrophils. A WBC count of 2,000/mm^3 and a neutrophil count of 75% serve as useful cutoff points to distinguish inflammatory from non-inflammatory disease. However, there is much overlap in synovial fluid WBCs between the inflammatory, crystal-induced, and sepsis categories. WBC is greater than 50,000/mm^3 in 70% of patients with septic arthritis, 15% with gout, 10% with pseudogout, and 4% with rheumatoid arthritis.

Extremely elevated synovial WBCs >100,000/mm^3 are not always due to infection. Sterile processes that can cause this degree of leukocytosis, sometimes termed as "pseudoseptic arthritis, include reaction to intracellular injection (e.g. hyaluronans), flares of rheumatoid arthritis, leukemic infiltration and gout.[12]

Under light microscopy abnormal cells like Reiter's cells, i.e. macrophages with phagocytosed neutrophils, or ragocytes, i.e. neutrophils with large intracytoplasmic inclusion bodies can be detected.[1] These cells with large inclusion bodies consist of complement and immunoglobulin are found in rheumatoid arthritis which are poor responder to therapy. The joint fluid leucocyte count and its interpretation is shown in Table 11.2. Joint fluid eosinophilia is observed in—hypereosinophilic syndrome, Lyme disease, parasitic arthritides, psoriatic arthritis, rheumatoid arthritis, etc.

Presence of apoptotic polymorphs, erythrophagocytic mononuclear cells and mast cells distinguish seronegative arthropathy from rheumatoid arthritis (RA) as the latter does not have cytophagocytic cells in the synovial fluid.[2] Joint fluid may be haemorrhagic either due to trauma during arthrocentesis, or other disease conditions, viz. bleeding diathesis, tumor, joint prosthesis, hemangioma, AV fistula, etc. In case of trauma during arthrocentesis the fluid will be non-homogenously stained and clump rapidly. Demonstration of crystals in the joint fluid is the mainstay of diagnosis of crystal induced arthritis (Table 11.3).

Crystals like monosodium urate (MSU) can be identified under light microscope, but if they are too small they may be missed. Compensated polarized light microscope helps in the identification of variety of crystals. Many times the mere presence of crystals in synovial fluid does not suggest disease, but the identification of intracellular crystals either MSU or calcium pyrophosphate (CPP) crystals is diagnostic of gout and pseudogout respectively. Synovial fluid analysis of previously inflamed, asymptomatic knees or first metatarsophalangeal joints is a simple bedside procedure for the diagnosis of intercritical gout. Such an approach may facilitate the diagnosis of gout and help to avoid unnecessary diagnostic procedures.[13,14]

Other crystals like basic calcium phosphate/calcium hydroxyapatite which is the pathological agent in "Milwaukee shoulder disease" is seen only on electron microscopy. They form clump and appear like shiny coin and stain with Alizarin red S.

Biochemical Tests

Biochemical tests for synovial fluid typically includes:

- Glucose—typically a bit lower than blood glucose levels; may be significantly lower with joint inflammation and infection.
- Protein—increased with bacterial infection.
- Lactate dehydrogenase—increased LD (LDH) level may be seen in rheumatoid arthritis, infectious arthritis, or gout.
- Uric acid—increased with gout

Synovial fluid protein concentration is usually 25% of serum (1–3 g/dl). Synovial fluid glucose is normally 10 mg/dl lower than the corresponding serum glucose (Table 11.4). Neither synovial fluid glucose nor protein provides much useful diagnostic information. Glucose levels below 20 mg/dL suggest infection and also occur in rheumatoid arthritis. Normal synovial fluid contains extremely low concentration of lipid. Synovial fluid lipid abnormalities include (1) rare cholesterol-rich pseudochylous effusions typically associated with RA; (2) lipid droplets, usually the result of trauma; and (3) extremely rare chylous effusions seen in association with RA, SLE, filariasis, pancreatitis and trauma.[15] Together with elevated 2-hydroxyl fatty acids, elevated *sn*-1 lysolipids, sphingomyelins, and subsequent lipid metabolites in synovial fluid may be biomarkers of injury.[16] Rheumatoid factor may be positive in synovial fluid even if it is negative in serum but it is not disease specific. Synovial fluid biochemical and immunological assay have been used widely as research tool but not advocated for clinical use.[17] Currently, there is interest in new type of biomarkers, e.g. cartilage oligometric matrix protein considered to be raised in osteoarthritis.[18,19] The total amount of protein in synovial fluid is approximately 20 gm/L. Electrophoretic pattern is similar to the serum. Investigations of synovial proteins are largely concentrated in inflammatory joint diseases. Increased level of immunoglobulin with lowered level of complement is found in RA. However, their level does not correspond to the disease activity. Complement level remains normal in most non-rheumatoid situations.[20,21] Rheumatoid

Table 11.3: Differences between gout and pseudogout crystals

	Gout	*Pseudogout*
Composition	Monosodium urate (MSU)	Calcium pyrophosphate dihydrate (CPPD)
Birefringence	Needle shaped	Rhomboid, or rectangular
Shape	Negative	Positive
Colour	Yellow	Blue

Table 11.4: Biochemical properties of normal synovial fluid

Test	*Normal result*
Protein	Less than 3 g/dl (approx. 20 gm/L)
Glucose	Less than 10 mg/dl of blood level
Uric acid	Parallels serum level
RF	Parallels serum level
ANA	Parallels serum level
C3	Parallels serum level

factor is found in synovial fluid of about 60% of RA. Antinuclear antibodies (ANA) are found in the synovial fluid in about 70% of patients with SLE and 20% of patients with RA.

Test for Infectious Diseases

Gram stain along with both aerobic and anaerobic cultures of synovial fluid are indicated whenever infection is suspected in a joint. Gram stain allows for the direct observation of bacteria or fungi under a microscope. There should be no microorganisms present in normal synovial fluid. If there is strong suspicion for infection, culture must be done even if the fluid appears clear. Concentration methods including cytocentrifugation may increase the sensitivity of the Gram stain. Bacterial cultures require 2–3 mL of fluid in a tube. If bacteria are present, susceptibility testing against certain antibiotics can be performed to guide antimicrobial therapy. If there are no microorganisms present, it does not rule out an infection; they may be present in small numbers or their growth may be inhibited because of prior antibiotic therapy. Culture sensitivity ranges from 75 to 95% for non-gonococcal joint infections in patients who have not received antibiotics. For patients with gonococcal arthritis, the sensitivity is only 10–50%.[22] In partially treated patients, the use of resin-containing blood culture bottles for culturing synovial fluid may improve isolation and identification of the responsible organism. When anaerobic bacteria is suspected, anaerobic blood culture bottles should be used.

In suspected case of tuberculous arthritis, Ziehl-Neelsen stain supported by culture for mycobacteria should be carried out. At times, PCR test for mycobacteria is helpful for early diagnosis of joint tuberculosis.[23] The use of PCR with universal primers to detect bacterial DNA is helpful, particularly for the more fastidious, uncultivable pathogens such as *Borrelia burgdorferi*, *Chlamydia* spp. *Mycoplasma* spp.[24] Viruses are often associated with acute infectious arthritis, depending on the putative virus serology, viral culture and detection of viral DNA by nucleic acid amplification should be performed. Coexistence of septic and crystal arthritis is possible, therefore one must look for crystals even after diagnosis of septic arthritis.[25,26]

Key Points

1. Synovial fluid analysis is an important tool for diagnosis in joint diseases, especially in case of monoarthritis.
2. Methodical collection followed by physical analysis including volume, viscosity, color should be noted.
3. Prompt laboratory investigations should be carried out subsequently. These include:
   a. Cell count
   b. Detection of crystals
   c. Gram stain
   d. Culture and sensitivity
4. Presence of intracellular crystals in the synovial fluid confirms the diagnosis of gout or pseudo-gout.

REFERENCES

1. Lawrence HJ. Joint fluid. In: Isenberg DA, Madison PJ, Woo P, Glass DN, editors. Oxford Text Book of Rheumatology. Oxford: Oxford University Press; 1993. p. 677–85.
2. Freemont A. Role of cytological analysis of synovial fluid in diagnosis and research. Ann Rheum Dis 1991; 50:120–23.
3. Simkin PA. Physiology and Pathophysiology. In: Gattar RA, Schuwacher HR, editors. Practical Hand Book of Joint Fluid Analysis (2nd edition). Philadelphia: Lee and Febiger; 1991, p. 8–23.
4. Gruber BF, Miller BS, Onnen J, Welling RD, Wojtys EM. Antibacterial properties of synovial fluid in the knee. J Knee Surg 2008; 21:180–5.
5. Dieppe P, Campion G, Doharty M. Mixed crystal deposition. Rheum Dis Clin North Am 1988; 14: 415–26.
6. Cohen AS, Brandt KD, Krey PR. Synovial fluid. In: Cohen AS, editor. Laboratory Diagnostic Procedures in the Rheumatic Diseases. Second edition. Boston: Little Brown and Company; 1975, p. 1–62.
7. Guidelines for the initial evaluation of the adult patient with acute musculoskeletal symptoms. American College of Rheumatology Ad Hoc Committee on Clinical Guidelines. Arthritis Rheum 1996; 39:1–8.
8. Schumacher HR, Jimenez SA, Gibson T, Pascual E, Traycof R, Dorart BB, et al. Acute gouty arthritis without urate crystals identified on the initial examination of synovial fluid: report of 9 patients. Arthritis Rheum 1975; 18:603–12.

9. Kerolus G, Clayburne G, Schumacher HR Jr. Is it mandatory to examine synovial fluid promptly after arthrocentesis? Arthritis Rheum 1998; 32:271–8.
10. Gattar RA, Schymacher HR. A practical handbook of joint fluid analysis. Philadelphia: Lea and Febiger: 1991; p. 14–23.
11. Dieppe P, Swan A. Identification of crystals in synovial fluid. Ann Rheum Dis 1999; 58:261–3.
12. Krey PR, Bailen DA. Synovial fluid leukocytosis. A study of extremes. Am J Med 1979; 67:436–42.
13. Pascual E, Batlle-Gualda E, Martínez A, Rosas J, Vela P. Synovial fluid analysis for diagnosis of intercritical gout. Ann Intern Med 1999; 131:756–9.
14. Kohn NN, Hughes RE, Mccarty DJ Jr, Faires JS. The significance of calcium phosphate crystals in the synovial fluid of arthritic patients: the "pseudogout syndrome" II. Identification of crystals. Ann Intern Med 1962; 56:738–45.
15. Wise CM, White RE, Agudelo CA. Synovial fluid lipid abnormalities in various disease states: Review and classification. Semin Arthritis Rheum 1987; 16:222.
16. Leimer EM, Pappan KL, Nettles DL, Bell RD, Easley ME, Olson SA, et al. Lipid profile of human synovial fluid following intra-articular ankle fracture. J Orthop Res 2017; 35:657–66.
17. Schumacher HR. Reproducibility of Synovial fluid analysis. Arthritis Rheum 1986; 29:770–74.
18. Swan A, Amer H, Dieppe P. The value of synovial fluid assay in the diagnosis of joint disease: a literature survey. Ann Rheum Dis 2002; 61:493–8.
19. Chaturvedi V, Handa R, Rao DN, Wali JP. Estimation and significance of serum and synovial fluid malondialdehyde levels in Rheumatoid arthritis. Indian J Med Res 1999; 109:170–74.
20. Bunch TW. Hunder GG, Offord K, McDuffie FC. Synovial fluid complement: usefulness in diagnosis and classification of rheumatoid arthritis. Ann Intern Med 1974; 81:32–5.
21. Cracchiolo A 3rd, Barnett EV. The role of immunological tests in routine synovial fluid analysis. J Bone Joint Surg Am 1972: 54:828–40.
22. Shmerling RH. Synovial fluid analysis. A critical reappraisal. Rheum Dis Clin North Am 1994; 20:503–12.
23. Negi SS, Gupta S, Lal S. Comparison of various microbiological tests including PCR for diagnosis of osteoarticular tuberculosis. Indian J Med Microbiol 2005; 23:245–48.
24. Nocton JJ, Dressler F, Rutledge BJ, Rys PN, Persing DH, Steere AC. Detection of *Borrelia burgdorferi* DNA by polymerase chain reaction in synovial fluid from patients with Lyme arthritis. N Engl J Med 1994; 330:229–34.
25. Baer PA, Tenenbaum J, Fam AG, Little H. Coexistent septic and crystal arthritis. Report of four cases and Literature review. J Rheumatol 1986; 13:604–7.
26. Pispati PK, Rao URK. Laboratory investigations for Arthritis. J Assoc Physicians India 1999; 47:298–304.

FURTHER READING

1. Dinneen A, Guyot A, Clements J, Bradley N. Synovial fluid white cell and differential count in the diagnosis or exclusion of prosthetic joint infection. Bone Joint J 2013; 95-B(4):554–7.
2. Kumar A. How to investigate new-onset polyarthritis? Best Pract Res Clin Rheumatol 2014; 28:844–59.

Chapter

12

Autoantibodies in Rheumatology

Sita Naik

The pathogenesis of most rheumatic diseases involves autoimmune processes, which is a consequence of the breakdown of self-tolerance. While we understand little about the events that initiate this breakdown, the autoimmune or anti-self processes itself involves both antibody mediated and cellular arms of the immune response. While the cellular responses are difficult to demonstrate, autoantibodies are easier to detect and are the *sine qua non* of these diseases. They may be directed to any constituent of cell, namely the cell membrane, DNA or other proteins in the nucleus. While their presence *per se*, does not indicate any disease, and their pathogenic role is not clear, they are valuable markers for the diagnosis and sometimes, follow-up of these diseases. This is because from experience we have learnt of their frequent association with specific diseases. However, the **absence of an auto-antibody does not exclude a disease.**

COMMON ANTIBODIES USED IN THE CLINICAL PRACTICE (Table 12.1)

Table 12.1: Autoantibodies that are commonly used and their disease association

Antibody	*Disease association*
Rheumatoid factor	Rheumatoid arthritis, Sjögren syndrome
Anti-nuclear antibody	Systemic lupus erythematosus, systemic sclerosis, inflammatory myositis, Sjögren syndrome
Anti-dsDNA	Systemic lupus erythematosus
Anti-proteinase 3	Wegener's granulomatosis
Anti-myeloperoxidase	Microscopic PAN, Churg-Strauss syndrome
Anti-cardiolipin	Anti-phospholipid syndrome, SLE

Rheumatoid Factor

Rheumatoid factor (RF) is an antibody directed against the Fc portion of IgG and is usually of IgM type. RF is detected by latex agglutination, nephelometery or ELISA and the results should always be expressed in IU for ease of comparison between different laboratories or methods. Presence of RF in a patient with joint disease increases the probability of diagnosis of RA. It is present in 75–85% of patients with RA and in 10% of children with juvenile arthritis. Those with a positive RF test are classified as seropositive RA. The test has a sensitivity of ~70% and specificity of ~80% for the diagnosis of RA. In a patient with established RA its presence suggests a more aggressive disease with extra-articular features and is predictive of poor prognosis. It is also present with lesser frequency in other chronic inflammatory conditions (Sjögren syndrome, SLE, other connective tissue disease, sarcoidosis, autoimmune hepatitis, primary biliary cirrhosis) some chronic infections (malaria, leprosy and tuberculosis, hepatitis C), and even rarely malignancies. The prevalence of rheumatoid factor increases with age and geriatric population have a prevalence of approximately 20 per cent. Thus, in elderly patients with vague joint pains, presence of RF should not lead to a diagnosis of RA. RF titres correlate poorly with activity of disease in RA and therefore there is no utility in repeating the test.

Antibodies to Citrullinated Peptides (ACPA)

ACPA are a set of antibodies directed against citrullinated self-peptides and are highly specific for a diagnosis of RA. Anti-cyclic citrullinated peptide (anti-CCP) is the most often used test and has a specificity of 95% and a sensitivity of 65–70%. ACPA is detected in around one-third of patients who are negative for RF. In contrast to RF it is rarely seen in other connective tissue diseases and infections. In early arthritis presence of anti-CCP antibodies increases the likelihood of RA.

Antinuclear Antibodies (ANA)

ANA is a heterogeneous group of autoantibodies directed against components of nucleus including dsDNA, RNA associated proteins (nRNP, Sm, La, Ro), nucleolar proteins, centromere, etc.

The 'gold standard' for ANA detection is indirect immunofluorescence (IIF) assay using Hep2 cells. The results of the IIF assay are reported in titers, starting from a 1:40 dilution. The dilution then is taken through a series of additional steps, creating tubes of 1:80, 1:160, 1:320, and 1:640 dilutions, respectively. Labs vary in their standards for "positive," e.g. some labs will report any titer above 1:160 as positive.

The recently developed ELISA tests are inferior to IIF both in sensitivity and specificity. The IIF test also has the advantage that the fluorescence pattern can give an indication regarding the target antigen against which the antibody is directed. For instance, rim pattern suggests antibodies to dsDNA and is suggestive of systemic lupus erythematosus (SLE), while a centromere pattern is typical of patients with limited systemic sclerosis. ANA is used as a screening test and its detection increases the likelihood of the diagnosis of a connective tissue disease in a patient with multi-system involvement (Fig. 12.1). Further investigations, like anti-dsDNA antibodies, anti-Sm antibodies are needed to establish the specificity of the ANA present in the sera. ANA is also positive in other conditions such as in infection, malignancy, RA and juvenile idiopathic arthritis.

Antibodies to dsDNA

Antibodies to dsDNA are a hallmark of systemic lupus erythematosus (SLE). They are detected by IIF using *Crithidia lucillae*, Farr assay (liquid phase radio-immunoassay) or ELISA. *Crithidia lucillae* is a protozoan parasite of the *Leishmania* family which has a prominent kinetoplast made of dsDNA. In the Farr assay, bound and free DNA are separated by precipitating immunoglobulins with 50% saturated ammonium sulphate. Bound radioactive DNA precipitates with the immunoglobulins whereas the free DNA remains in the supernatant.

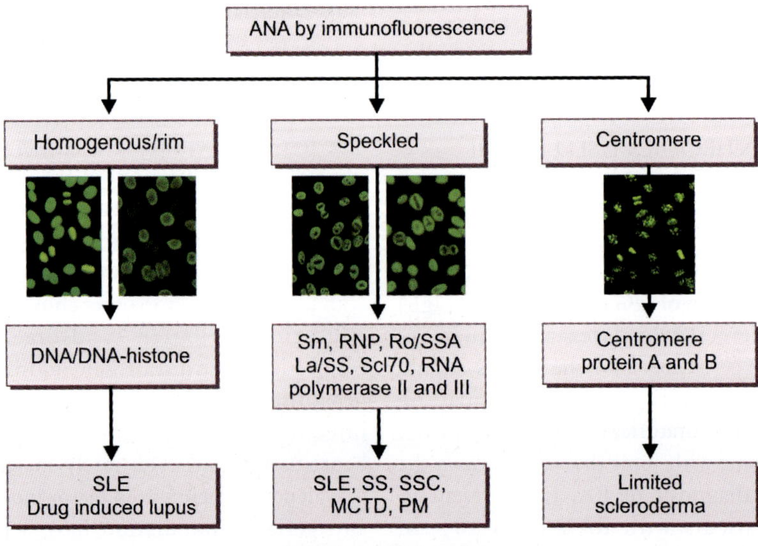

Fig. 12.1: ANA by IIF

A nitrocellulose filter allows free DNA to pass through but dsDNA-antibody complexes do not and the retained radioactivity on the filter is proportional to the serum anti-dsDNA antibody concentration. This method is highly reproducible, but may miss low-avidity anti-dsDNA antibodies. However, due to ease of performance, the ELISA assay is most widely used and is useful in follow-up of patients.

Antibodies to Extractable Nuclear Antigens

Antibodies to extractable nuclear antigens (ENA) comprise antibodies to various RNA associated proteins. These are anti-Sm, (for SLE), anti-RNP (for MCTD), anti-La and anti-Ro (for Sjögren's), anti-Scl70 (for scleroderma), anti-Jo (for inflammatory myositis). This test should be ordered only if the ANA is positive. Each antibody has a specific disease phenotype associated with it and has variable prevalence (Table 12.2). ENAs are detected using ELISA or immunoblotting. Since antibody titres do not correlate with disease activity, qualitative assays like immunoblot or line assays are as useful as quantitative assays like ELISA.

Antiphospholipid Antibodies

Presence of antibodies to phospholipids is mandatory for the diagnosis of anti-phospholipid (APL) syndrome. Cardiolipin is the most commonly used substrate. IgG and IgM antibodies are measured by ELISA and results are expressed as mild, moderate and marked elevation. IgA antibodies to cardiolipin can also be detected, but have less diagnostic value.

Anti-beta 2-glycoprotein 1 antibodies which are directed against a co-factor required for binding of cardiolipin are more specific for diagnosis and are detected by ELISA. Detection of anti-prothrombin antibodies may be used as a substitute for lupus anticoagulant assay but have low sensitivity. Antibodies against other phospholipids like phosphotidyl serine, annexin-V are not useful clinically.

Antibodies to Neutrophil Cytoplasmic Antigens (ANCA)

Anti-neutrophilic cytoplasmic antibodies (ANCA) is a serological marker for severe necrotising vasculitis affecting small vessels. It is positive in patients with granulomatosis with polyangitis (GPA) previously known as Wegener's granulomatosis, eosinophilic granulomatosis with polyangitis (EGPA) previously known as Churg-Strauss syndrome and microscopic polyarteritis (PAN). ANCA is detected using IIF assay with human neutrophils as substrate. A cytoplasmic pattern (cANCA) and a perinuclear pattern (pANCA) of staining are recognized (Fig. 12.2). The antigenic target of cANCA is proteinase 3 (PR3) which has specificity of 85–90% for diagnosis of GPA. pANCA is mainly directed against myeloperoxidase (MPO) and is present in 60% of patients with microscopic PAN. pANCA or an atypical staining pattern which may be positive in other diseases like ulcerative colitis, rheumatoid arthritis, SLE etc. are less useful for diagnosis as their antigenic targets are variable (cathepsin G, lactoferritin, BPI, elastase, etc.). Antibodies to MPO and PR3 can be detected by ELISA and quantitative measurement is beneficial for monitoring disease activity in patients with vasculitis. ANCA are not useful in a patient with classical polyarteritis nodosa, Takayasu's arteritis, etc.

Table 12.2: Antibodies to nuclear target antigens

Antibody	*Disease*	*Prevalence*
Anti-Sm	SLE	30%
Anti-nRNP	SLE	40%
	Mixed connective tissue disease	100%
	Rheumatoid arthritis	15–20%
Anti-Ro	Primary Sjögren syndrome	60%
	Systemic lupus erythematosus	30%
Anti-La	Primary Sjögren syndrome	30%
	Systemic lupus erythematosus	15%
Anti-centromere	Limited systemic sclerosis	50%
Anti-Scl 70	Diffuse systemic sclerosis	45%

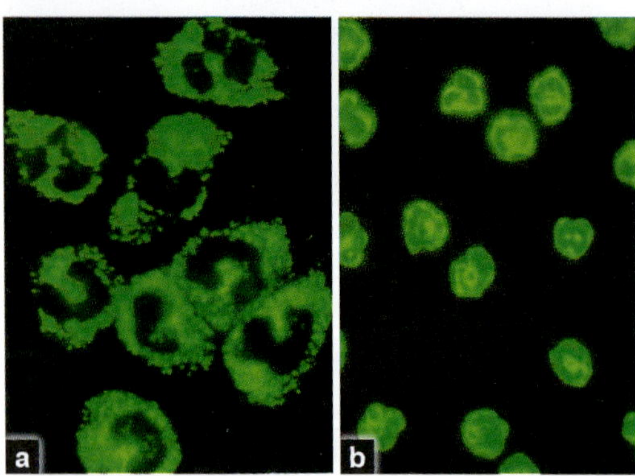

Fig. 12.2: ANCA patterns; cANCA (a), pANCA (b)

AUTOANTIBODIES IN SPECIFIC DISEASES

Systemic Lupus Erythematosus (SLE)

The IIF-ANA test is a good screening test for the diagnosis of SLE, since most patients (more than 95% of individuals) with the disease will test positive. A negative ANA test is helpful in excluding the diagnosis. However, the test has a poor specificity for SLE, since only about 11–13% of persons with a positive ANA test have lupus. Also, up to 15% of completely healthy people may also have positive ANA test and as in the case of RF, positivity among normal persons increases with age, with 10–35% of those over 65 years testing positive. A homogenous pattern of fluorescence is suggestive of antibodies to DNA. A positive ANA test mandates the ordering of anti-dsDNA, which has a high specificity (about 60%) for SLE. It also has a good correlation with presence of lupus nephritis. Serial measurement can help in the monitoring of disease activity. However, a rise in the antibody titre alone does not warrant a change in treatment which should be made in the context of clinical parameters. A number of anti-nuclear antibodies other than anti-DNA are also positive in SLE (Table 12.2) and these are associated with different clinical aspects of the disease. Anti-Ro positivity is seen in subacute cutaneous lupus and both anti-Ro and anti-La positivity is seen in neonatal lupus. Anti-Sm antibodies have a high specificity, but are positive in only 10% of cases.

Systemic Sclerosis

More than 95% of cases of systemic sclerosis are ANA positive. A nucleolar pattern of staining on IIF, is suggestive of antibodies to anti-topoisomerase1, and it is present in 50% of cases with diffuse cutaneous disease, where it has 100% specificity. Patients with this antibody frequently develop peripheral vascular disease, pulmonary fibrosis and cardiac involvement. Anti-centromere antibodies are found in 75% of patients with limited disease.

Sjögren's Syndrome

Seventy percent of patients have a positive ANA on IIF and it usually shows a speckled pattern, indicating presence of anti-SSA or anti-SSB antibodies. Line blot assays to confirm the specificity shows a 70% positivity for SS-A and 40% for SS-B antibodies and these have a high specificity for the disease.

Inflammatory Muscle Disease

Inflammatory myositis is a feature of overlap syndromes and scleroderma, and the myositis-associated autoantibodies, anti-U1-RNP, anti-U3-RNP (fibrillarin), anti-PM-Scl and anti-Ku, may be positive in these cases. The myositis-specific autoantibodies (MSAs) are found only in a small percentage of cases of polymyositis (PM) and dermatomyositis (DM). They target either nuclear or cytoplasmic components involved in gene transcription, protein translocation and anti-viral responses. Autoantibody against the cytoplasmic aminoacyl-tRNA synthetase (ARS) enzymes is the most frequently detected, in about 30–40% of adult patients with myositis. Although antibodies to individual enzymes can be detected by the method of immunoprecipitation, this is not widely available and hence utility in clinical practice is limited. The detection of the specific serological type is, however, beneficial in diagnosing the specific clinical syndrome due to the close association of some of these antibodies with the syndromes.

ORGAN-SPECIFIC AUTOANTIBODIES

The characteristic autoantibodies found in the organ-specific autoimmune diseases along with their prevalence is given in Table 12.3. Most of these can be detected by IIF assays using appropriate tissue substrates or by ELISA. Quantitative titres are not useful in the management of these conditions and the IIF tests have adequate sensitivity to assist in diagnosis. In many of these diseases, detection of antibody only suggests an autoimmune aetiology and is not required for diagnosis. For instance, the diagnosis of Graves' disease is based on

Table 12.3: Organ-specific autoantibodies

Disease	*Antibodies*	*Prevalence*
Hashimoto's thyroiditis	Thyroid microsomal antibodies, antibodies to TPO	85–90%
Graves' disease	Antibodies to TSH receptor	90–95%
Type 1 diabetes	Antibodies to islet cells, insulin, GAD65, IA2	75%
Pernicious anaemia	Antibodies to parietal cells	85–90%
Autoimmune hepatitis (AIH)	Antibodies to smooth muscle (actin) Antibodies to liver-kidney microsomes	Type I AIH Type II AIH
Primary biliary cirrhosis	Antibodies to mitochondria (pyruvate dehydrogenase)	90%
Celiac disease	Anti-endomysial antibodies, antibodies to tissue transglutaminase	95%
Immune thrombocytopaenia	Antibodies to platelets (GpIIb/IIIa)	80–90%
Autoimmune haemolytic anaemia	Antibodies to RBCs (Rh antigen)	95%
Myasthenia gravis	Antibodies to acetylcholine receptor	70–75%
Goodpasture's syndrome	Antibodies to basement membrane	60–70%

Abbreviations: TPO, thyroid peroxidase; TSH, thyroid stimulating hormone; GAD65, glutamic acid decarboxylase 65; IA2, islet antigen 2

clinical findings and hormone levels. Antibodies to insulin, GAD65 and IA-2 are seen in patients with insulin-dependent diabetes mellitus prior to development of IDDM but are not useful in diagnosis.

Key Points

1. Autoantibodies are useful in diagnosis of autoimmune diseases.
2. ACPA are more specific than RF for a diagnosis of RA.
3. Presence of ANA increases the likelihood of connective tissue disease in a patient.
4. ANCA are useful in diagnosis of small vessel vasculitis like microscopic PAN, granulomatosis with polyangitis.
5. Mere presence of antibody does not make a diagnosis of connective tissue disease.

FURTHER READING

1. Aggarwal A. Role of autoantibody testing. Best Pract Res ClinRheumatol. 2014; 28:907–20.
2. Conigliaro P, Chimenti MS, Triggianese P, Sunzini F, Novelli L, Perricone C, et al. Autoantibodies in inflammatory arthritis. Autoimmun Rev. 2016; 15:673–83.
3. Fritzler MJ. Choosing wisely: Review and commentary on anti-nuclear antibody (ANA) testing. Autoimmun Rev. 2016; 15:272–80.
4. Kumar A. How to investigate new-onset polyarthritis. Best Pract Res Clin Rheumatol. 2014; 28:844–59.
5. Manual of Molecular and Clinical Laboratory Immunology. Detrick B, Hamilton RG, Folds JD, Editors. Seventh edition. Published by American Society of Microbiology 2006.

Chapter

13

Conventional Radiology in Rheumatological Disorders

URK Rao, Kakarla Subbarao, J Shivanand

Conventional radiographs are the cornerstone in the evaluation of patients with various rheumatological disorders. Plain radiographs are inexpensive, widely available and can be easily obtained. Hence, imaging of rheumatological disorders should start with plain radiographs. Well performed good quality radiographs depict the extent and progression of the disease and are helpful in patient follow up and assessment.[1] They include bone erosions, joint space narrowing, subluxation, misalignment and ankylosis of the joints. The disadvantages of conventional radiography include insensitivity in detecting early bone disease or soft tissue manifestations.[2] Today digital radiography has become popular as the resolution is much better than conventional radiography. The following mnemonic is easy to remember and helpful in the evaluation of a joint on a plain radiograph.

A Alignment
B Bone density
C Cartilage
D Distribution of lesions
E Erosions/enthesopathy/eburnation
S Soft tissue evaluation

Appropriately selected radiographic projections would serve the purpose of delineating the lesions in various joints (Table 13.1). In general the symptomatic joint should be radiographed in two different positions. The second radiograph taken at an angle of 90° to the first view can show some abnormalities. On occasion, special radiographs in different positions are taken.

Table 13.1: Commonly requested radiographic projections

Hand and wrist	Posteroanterior (PA) Anteroposterior (AP). Oblique
Elbow	AP, lateral
Shoulder	AP (with internal and/or external rotation of humerus)
Hip	Pelvis AP [internal rotation and/or external rotation of leg (frog leg)]
Knee	AP (standing), lateral (AP in semi flexion-tunnel view and sunrise) view
Ankle	AP (lateral/oblique)
Foot	AP (lateral/oblique)
Sacroiliac joints	Pelvis AP (SI oblique)
Lumbar spine	AP, lateral (oblique, coned views)
Thoracic spine	AP, lateral
Cervical spine	AP, lateral (flexion, extension and open mouth view)
Chest	PA, lateral

(The projections mentioned in brackets are for special circumstances)

Chest radiograph (CXR) should be included among the investigations requested for a patient with rheumatic disease. Despite various advances in the field of radiology, radiograph of chest continues to be an important screening modality, especially in Indian patients. CXR is taken in erect position in postero anterior (PA) view.

The first step in interpreting any radiograph is to check whether it has been labeled properly—name, age, gender, position, medical record number

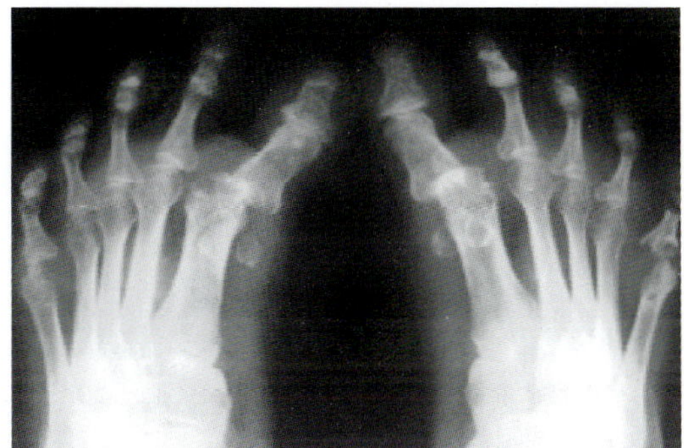

Fig. 13.9: Feet (AP): RA—dorsi flexion deformities of IP joints and hallux varus deformities of big toe with erosions noted at 1st and 5th MTP joints

within the acetabulum. There are varying degrees of joint space narrowing (Fig. 13.10) with erosions and synovial cyst formation seen in head of femur. It may need surgical intervention if it is advanced and disabling (Fig. 13.11).

Shoulder involvement is seen in 60% of patients with RA. Radiograph shows uniform narrowing of all compartments of shoulder joints (glenohumeral, acromiohumeral and acromioclavicular joint). Except for upper cervical spine, axial skeleton is seldom involved in RA. Subluxation at atlanto-axial (C1–C2) joint due to progressive erosive RA affecting the odontoid process is seen in some of the patients with late RA (Table 13.3 and Fig. 13.12).

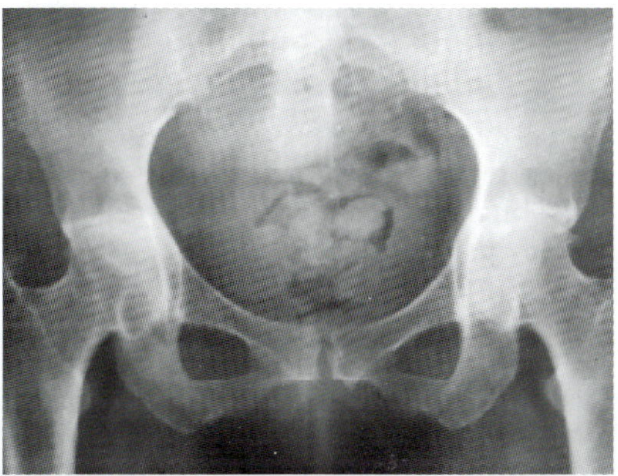

Fig. 13.10: Pelvis (AP): RA—bilateral uniform hip space narrowing with erosions

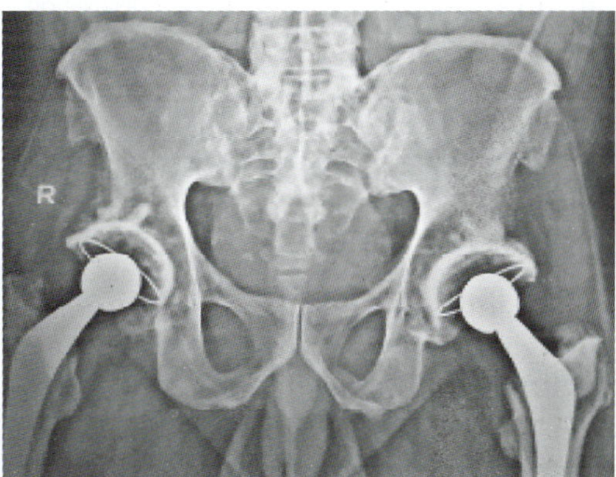

Fig. 13.11: Pelvis (AP): RA—bilateral hip replacement

Table 13.3: Deformities in rheumatoid arthritis

Site	*Abnormality*	*Name*
Hand		
PIP and DIP joints	Flexion PIP, extension DIP	Boutonniere deformity
	Extension PIP, flexion DIP	Swan-neck deformity
MCP joints	Ulnar deviation, Volar subluxation	
Wrist		
Radio carpal joint	Radial deviation	
Hip	Acetabular wall medial to ilioischial line	Protrusioacetabuli
Knee	Outward deviation of knee	Genu varus
	Inward deviation of knee	Genu valgum
Feet		
First MTP joint	Lateral deviation	Hallux valgus
	Deviation away from 2nd toe	Hallux varus
PIP and DIP	Hyperextension and dorsal	
	Subluxation of MTP	Cock-up toe
	Hyperflexion PIP or DIP	Hammer toe
Cervical spine		
Atlanto-axial joint	Atlanto-odontoid space > 3 mm	Atlanto-axial subluxation
All levels	Subluxation of apophyseal joints	Stair-step deformity

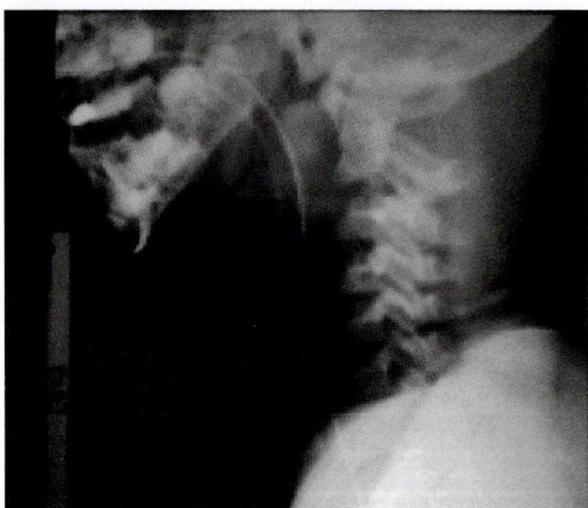

Fig. 13.12: Cervical spine (lateral view): Subluxation at atlanto-axial joint

Juvenile Idiopathic Arthritis

Juvenile Idiopathic Arthritis (JIA) is most common type of arthritis in childhood. Radiologically, soft tissue swelling around an affected joint is typical with para-articular osteoporosis. Erosions are late finding in JIA (Fig. 13.13), since erosions in the cartilaginous epiphysis will not be visible until ossification is complete. In the advanced disease, complete cartilage loss with bony ankylosis occurs. In the cervical region, there may be fusion of segments of C2 to C4 vertebrae. Atlanto-axial subluxation is common in seropositive, poly-articular JIA. Abnormalities in growth and maturation may lead to advanced skeletal maturation, enlargement of secondary centers of ossification, increased longitudinal growth, shortening of small tubular bones, undergrowth of mandibular condyles and facet joint ankylosis in cervical spine. In the knees, enlargement of epiphyses and the meta-physeal ends of the shafts are characteristic findings.

Spondyloarthropathy

The spondyloarthropathies (SpA) are a group of multi-system inflammatory joint disorders that share a variety of genetic, clinical and radiological features. They include ankylosing spondylitis (AS), psoriatic arthritis (PsA), reactive arthritis (ReA), arthritis associated with inflammatory bowel disease and unclassified SpA. The hallmark of SpA group of diseases is involvement of the sacroiliac joint and the spine resulting in sacroiliitis and spondylosis (Fig. 13.14). Inflammation at the bony insertions of ligaments and tendons (enthesopathy) is radiologically characterized by bony erosions followed by bony proliferation and eventually ankylosis of the adjacent bones.

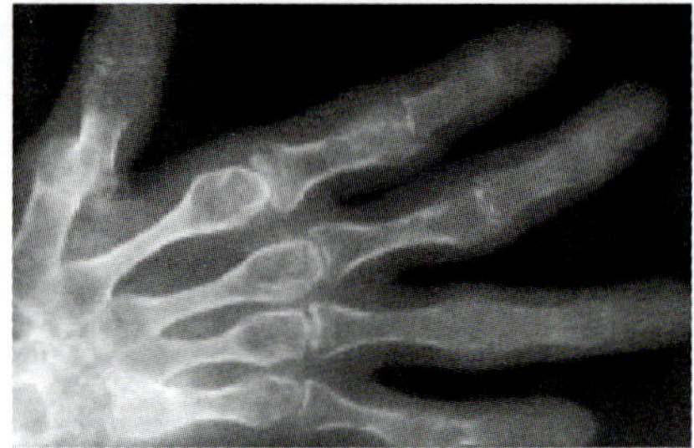

Fig. 13.13: Hand (AP): JIA—diffuse osteoporosis. Erosions noted 2nd and 3rd PIP joints

Ankylosing Spondylitis

Ankylosing spondylitis (AS) is the prototype of SpA, primarily affecting the axial skeleton. The disease starts in the sacroiliac (SI) joints with sub-chondral bony erosions on the iliac side of the joint with "saw-teeth" appearance. The erosions are followed by bony proliferation and sclerosis. The changes first appear in the lower third (synovial part) of the SI joint. Initially, the joint may look widened. The changes gradually affect whole of the SI joint and the joint eventually fuses. Once the fusion is complete, sclerosis disappears. The sacroiliitis in AS is usually symmetrical, and should be differentiated with osteitis condensans ilii (OCI). OCI is a benign disorder characterized radiologically by presence of triangular shaped sclerosis localized to both the ilium (Fig. 13.15). In AS the spine is initially involved at lumbosacral and thoracolumbar segments. In PsA and ReA no such orderly spread is seen and paravertebral ossifications tend to appear in a random fashion.

The enthesopathy in the spine manifests as small erosions surrounded by bony sclerosis at the vertebral corners seen as "shiny corners". This is followed by ossification of the outer annular fibers to form marginal and symmetric syndesmophytes (Fig. 13.16). They are different from osteophytes seen with degenerative disc disease in that syndesmophytes are parallel to the spinal axis, whereas in degenerative spine disease the osteophytes are perpendicular to the spine (grow outward). Squaring of the vertebral bodies is characteristic of early AS caused by the corner erosions and anterior apposition of periosteal new bone formation. Squaring is easier to appreciate in the lumbar spine. The complete fusion of the syndesmophytes is the so called "bamboo-spine" which is a late feature (Figs 13.17 and 13.18). Erosions are seen in the apophyseal and costovertebral joints followed by ankylosis of these joints. Following the spinal fusion, the spine becomes osteoporotic and is vulnerable to stress. Fractures in AS occur most commonly at the thoracolumbar and cervicothorax junctions. It may proceed to non-union with pseudoarthrosis, an appearance resembling disc space infection (Andersson lesion) (Fig. 13.19). Enthesopathy of ischial tuberosities is described as "Whiskering".

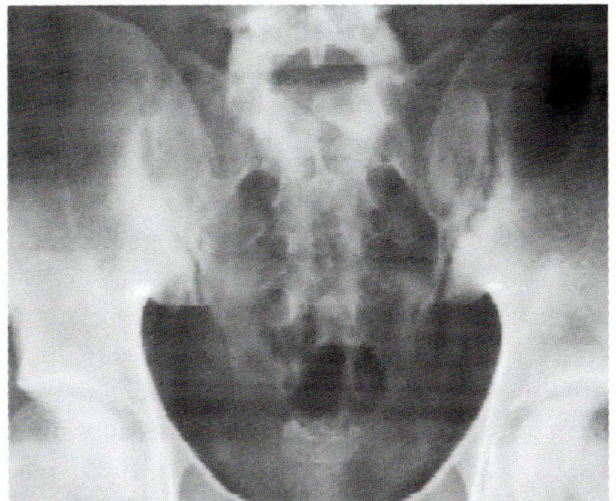

Fig. 13.14: Pelvis (AP): Sacroiliitis—sclerosis and erosions of both SI joints

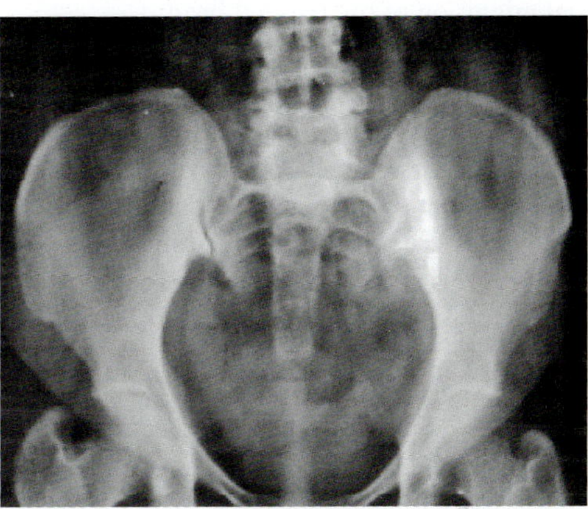

Fig. 13.15: Pelvis (AP): Osteitis condensans ilii—triangular homogenous sclerosis involving the iliac side of left SI joint

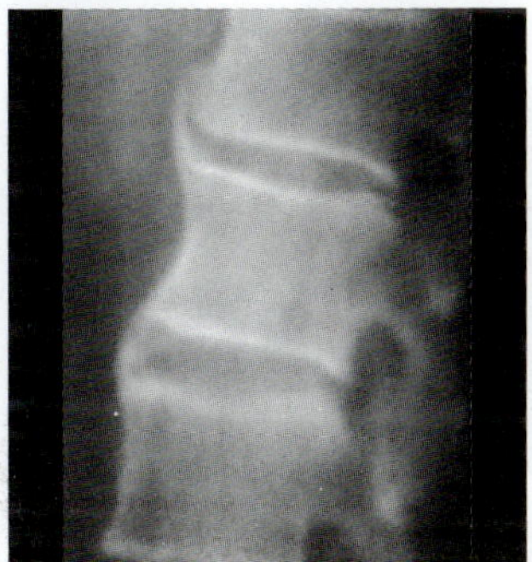

Fig. 13.16: Lumbar spine lateral view: Ankylosing spondylitis—early syndesmophytes noted bridging L2–L3, L3–L4

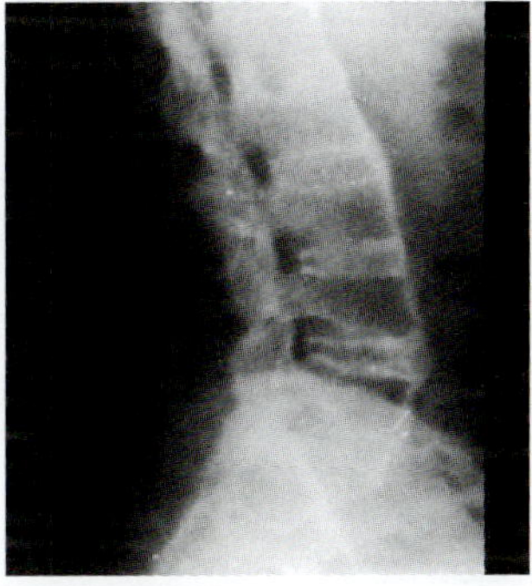

Fig. 13.17: Lumbar spine (Lateral): Syndesmophytes joining anterior vertebral margins in ankylosing spondylitis

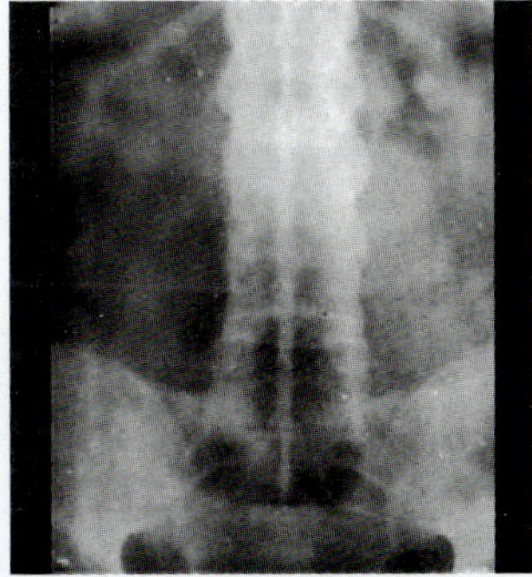

Fig. 13.18: Lumbar spine (AP): AS with bamboo-spine

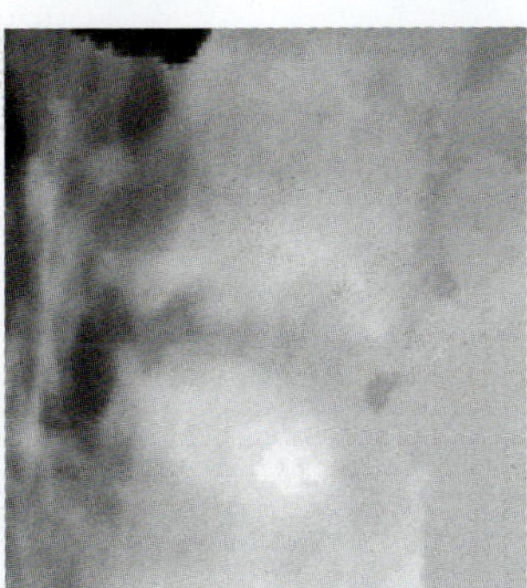

Fig. 13.19: Dorso lumbar spine (lateral view): AS with pseudoarthrosis D11–D12 (Andersson lesion)

The changes of AS are less severe and delayed in appearance in women with AS, and in children with AS changes in appendicular joints are noted more frequently. Peripheral arthritis occurs in about 30 percent of the patients with AS. They include hips, shoulders and knees. Hip disease in AS is characterized by uniform joint space narrowing and axial migration of the femoral head.

Psoriatic Arthritis (PsA) and Reactive Arthritis (ReA)

SI joint and spine involvement occur in 20 to 40% of patients with PsA. The sacroiliitis in PsA and ReA may or may not be symmetrical and spondylitis occurs in a more random fashion. It is associated with presence of coarse, symmetric paravertebral ossification (non-marginal syndesmophytes). The joint involvement in psoriasis can be variable (Table 13.4).

Erosions in the phalanges are seen in the bare areas with associated fluffy bone formation (mouse ears) and sometimes followed by fusion. The distal erosions (Fig. 13.20) can evolve into the so-called 'pencil and cup' appearance (Fig. 13.21). In the feet, erosions appear in metatarsophalangeal (MTP) and interphalangeal joints, especially in great toe. Tuftal proliferation of the digit results in a shaggy dense distal phalanx (ivory phalanx). Periosteal reaction can be seen early in the disease along with shafts of the phalanges and metatarsals and is associated in the feet with sausage digits. ReA typically affects the lower extremities. Inflammatory spurs commonly seen in ReA have a fluffy appearance (Fig. 13.22a) which distinguishes them from degenerative calcaneal spurs (Fig. 13.22b).

Table 13.4: Modes of presentation in psoriatic arthritis

- Distal interphalangeal joint involvement
- Proximal interphalangeal and metacarpophalangeal joint involvement that resemble RA
- Asymmetric oligoarthritis affecting large joints
  Destructive form of arthropathy (arthritis mutilans)
  Seronegative spondyloarthropathy

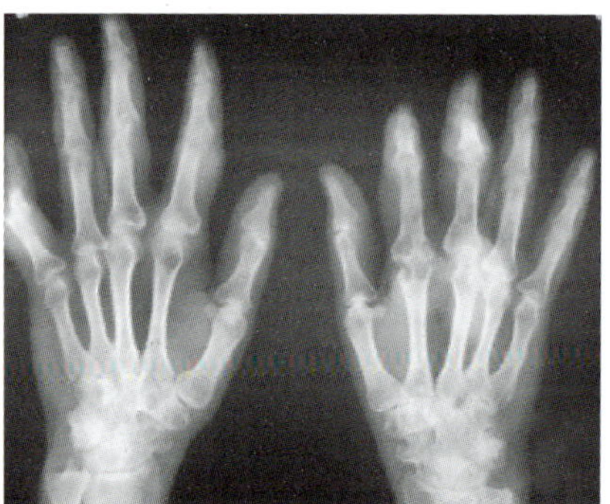

Fig. 13.20: Hands (PA): Left 2nd, 3rd and right 5th DIP joint involvement in PsA

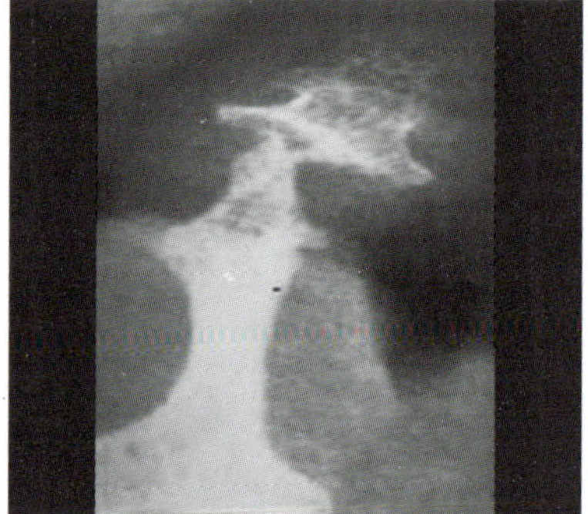

Fig. 13.21: Distal interphalangeal joint (DIP) AP view showing pencil in cup appearance

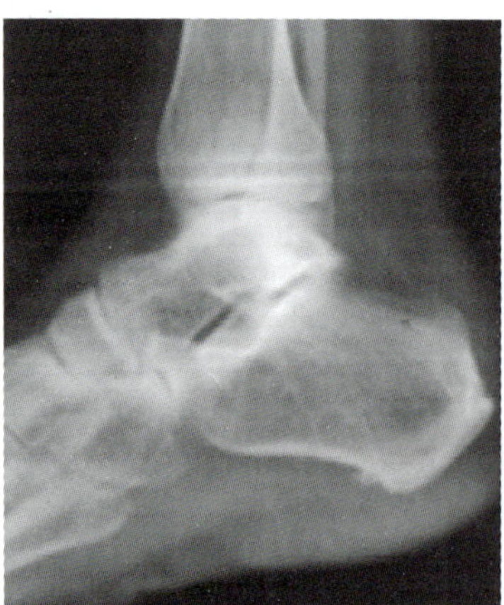

Fig. 13.22a: Ankle (lateral); Reactive arthritis: Fluffy appearance of calcaneal spur

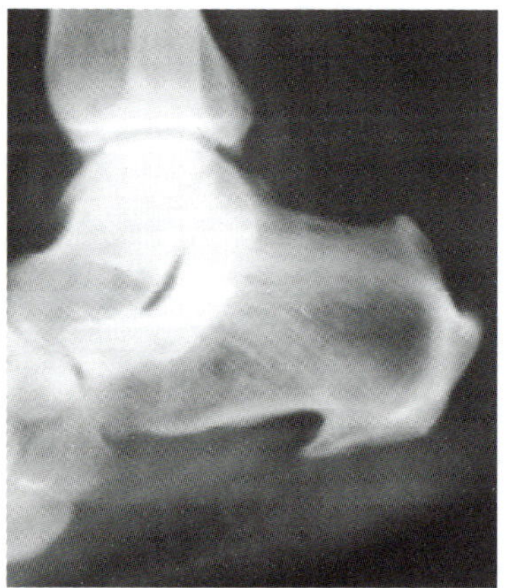

Fig. 13.22b: Ankle (lateral): Degenerative calcaneal spur

Osteoarthritis (Degenerative Joint Disease)

The common joints involved in osteoarthritis (OA) are knees, hips, DIP, first carpometacarpal (CMC) joints, first MTP joint and spine (cervical, lumbar and lumbosacral). The radiographic appearance of OA in different joints varies, being dependent on anatomic relationships and on the stress to which each joint is subjected. The most typical pattern is the presence of reparative changes in both the "stressed and non-stressed" segments of the joints. In the stressed segment, there is narrowing of joint space due to thinning of cartilage with the development of erosions and eburnations. Subchondral bone sclerosis with formation of cystic lesions follows. In non-stressed segment, reparative changes lead to osteophytosis. Osteophytes are classified as marginal, central, periosteal and capsular. Degeneration of surrounding tissues such as ligaments and capsule may lead to malalignment and subluxation of the joint. In advanced OA, fragmentation of the cartilaginous or osseous surface can lead to the formation of intra-articular "loose bodies" (Table 13.5).

Hip

Asymmetric joint space narrowing is the most reliable sign of OA in the hip joint. The femoral head displacement can be categorized into three different migration patterns—superior migration, medial migration and axial migration. The superior migration pattern in turn is of 2 types—superolateral and superomedial.

Knee

Weight bearing AP view of the knee is better for accurate assessment. This view shows the extent of joint space loss and the degree of angulation (varus or valgus) and subluxation. The knee joint is divided into three compartments—the medial femorotibial, lateral femorotibial and patellofemoral. Most frequently, the medial compartment is involved in primary OA of knees (Fig. 13.23). Both or only lateral compartment are/is involved in secondary OA knees due to inflammatory arthritis such as RA. The patellofemoral compartment (Fig. 13.24) is affected usually in association with medial compartment disease. The best position to take AP-view of the knees is in standing position.

Table 13.5: Features of OA
• Narrowing of joint space
• Osteophytosis
• Subarticular sclerosis
• Deformities

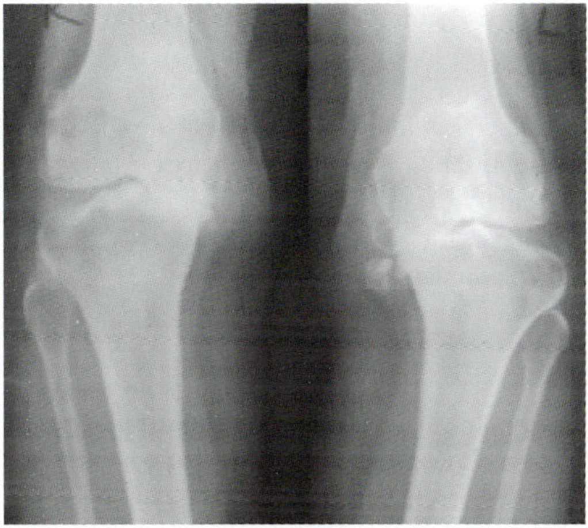

Fig. 13.23: Knees (AP): Osteoarthritis knees with obliteration of medial compartments K-L grade IV

Conventional radiography is one of the parameters useful in the assessment of outcome measurements in RA, PsA, OA and SpA and other autoimmune diseases.[3–7]

The cardinal features of OA are asymmetric joint space narrowing accompanied by subchondral cysts and sclerosis. Marginal osteophytes are frequent finding at the femoral and tibial margins of the joints in advanced cases at the intercondylar

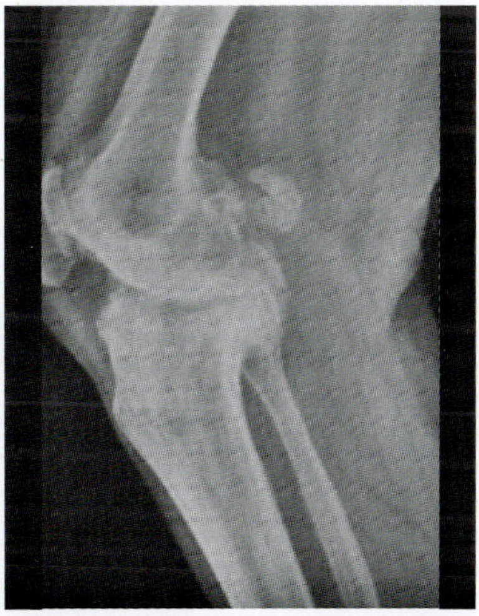

Fig. 13.24: Knees (lateral): Patellofemoral joint space narrowing with loose bodies in knee OA

tubercles. Features of patellofemoral OA include joint space loss with deformity at the patellar articular surface and in severe cases formation of osteophytes. Additional signs of OA are angulation and subluxation with varus (primary OA) or valgus (secondary OA) deformities, small joint effusions and presence of intra-articular bodies (loose bodies), called secondary osteocondromatosis.

Spine

In the axial skeleton, degenerative joint disease is mainly observed at the sites of maximum weight (Fig. 13.25) bearing and movement, namely the cervical C5–C6, C6–C7, and lumbosacral spine at L4–L5, L5–S1 (Fig. 13.26).

In intervertebral (disc) spaces, marginal osteophytes are observed early. As the disease advances, thinning of the disc space with sclerosis of subchondral bone takes place with subluxation of various degrees leading to instability and pain. The hyaline cartilage of the facet joints may also be involved by the degenerative process.

Hands and Feet

OA of hands occur primarily in DIP joints (Fig. 13.27), first CMC joint (Fig. 13.28) and first MTP joints (Fig. 13.29). The erosive and nodal osteoarthritis may mimic RA or PsA. However, subarticular osteoporosis is not encountered in OA.

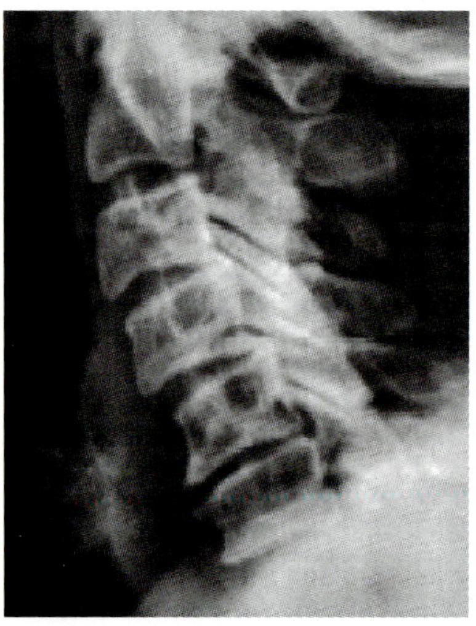

Fig. 13.25: Cervical spine (lateral): Cervical spondylosis with narrow C5–6 intervertebral space and osteophytosis

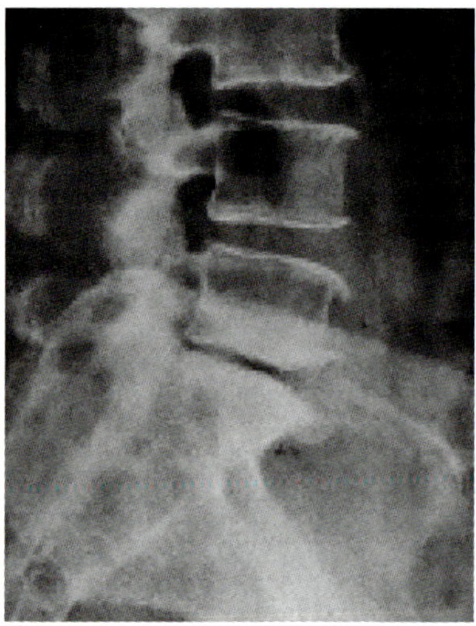

Fig. 13.26: Lumbosacral spine (lateral): Lumbar spondylosis with narrow L5–S1 and anterior osteophytes noted at L5 with vacuum sign L5–S1

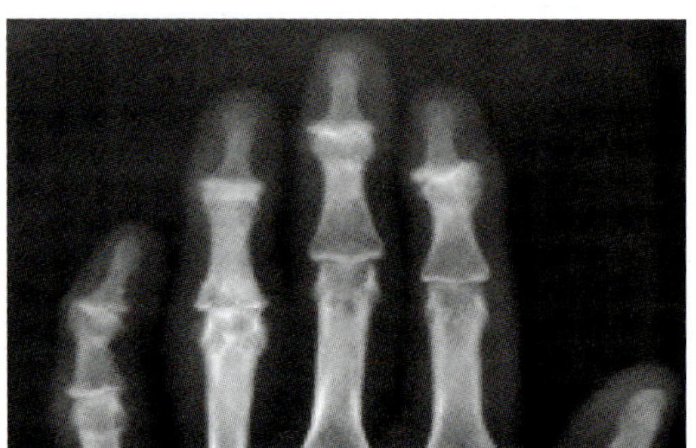

Fig. 13.27: Hand (PA): OA DIP joints with narrow joint space, subarticular sclerosis and erosions

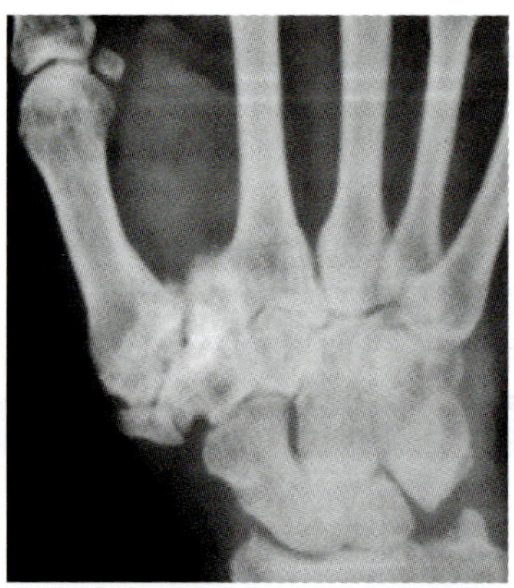

Fig. 13.28: Wrist (AP): OA first CMC joint with narrow joint space, subarticular sclerosis and osteophytes

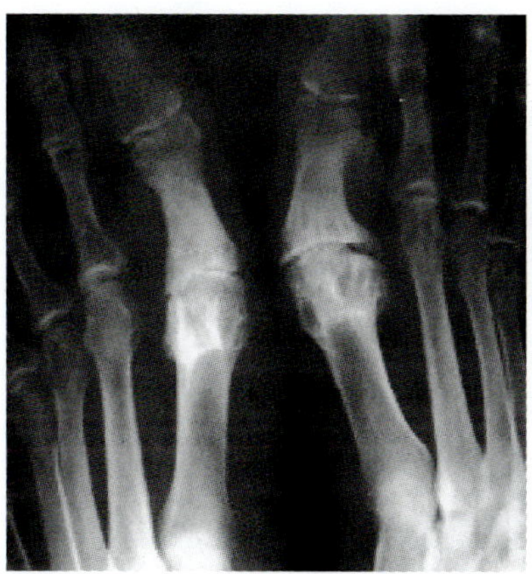

Fig. 13.29: Feet (AP): First MTP joints with narrow joint space, subarticular sclerosis and osteophytes (OA)

Crystal Arthropathies

The crystal induced arthropathies are a heterogeneous group of biochemical disorders that share the common features of crystal deposition in and around joints. The common crystals are monosodium urate (MSU), calcium pyrophosphate dihydrate (CPPD) and calcium hydroxyapatite (CHA). Radiography often demonstrates characteristic changes in these disorders and they delineate the extent of articular and periarticular disease.

Gout

The radiographic findings of acute gout are nonspecific. Soft tissue swelling, osteopenia or a joint effusion may be seen. In chronic tophaceous gout, the radiological features include loss of joint space, preservation of bone density leading to osteopenia in late stages and osseous erosions (Table 13.6).

Table 13.6: Radiographic features of gout

- Asymmetrical involvement
- Eccentric soft tissue prominence
- Relative preservation of joint space
- Relative preservation of bone density
- Punched out erosions
- Overhanging margins
- Intra-osseous calcification/lytic lesions

Though lesion is more common in first metatarsophalangeal joint, similar lesions are also seen in ankles, elbows, wrists and hands (Fig. 13.30). Intra-articular erosions tend to involve the joint margins before extending centrally. An elevated margin of the bone may extend outside the erosions as "overhanging edge" which is very suggestive of gout (Fig. 13.31). Intraosseous MSU crystals deposition and tophus formation may give rise to osteolytic lesions that simulate neoplastic disease.

Calcium Pyrophosphate Deposition (CPPD) Disease

Articular calcification related to CPPD is most frequent in the knee followed by symphysis pubis, wrist, elbow, shoulder and hip. Calcification in the hyaline cartilage and/or fibrocartilage (chondrocalcinosis) is present (Fig. 13.32a and b). Fibrocartilage calcification is coarse and irregular, commonly occurring within the meniscus of the wrist and the pubic symphysis. Hyaline cartilage

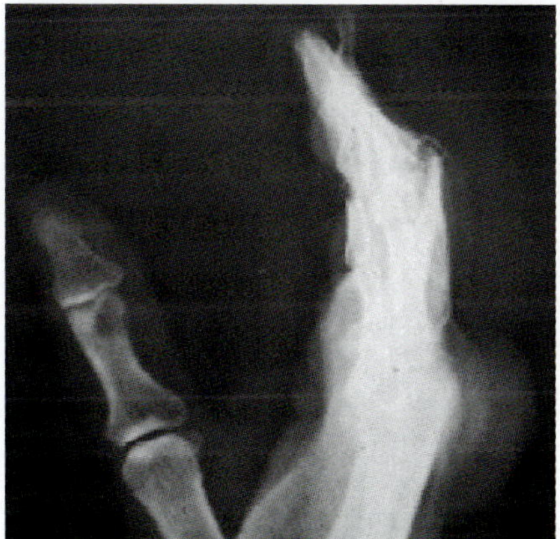

Fig. 13.30: Hand (oblique): Gouty tophus at 2nd MCP joint, with mouse ears (overhanging margins) at interphalangeal joint

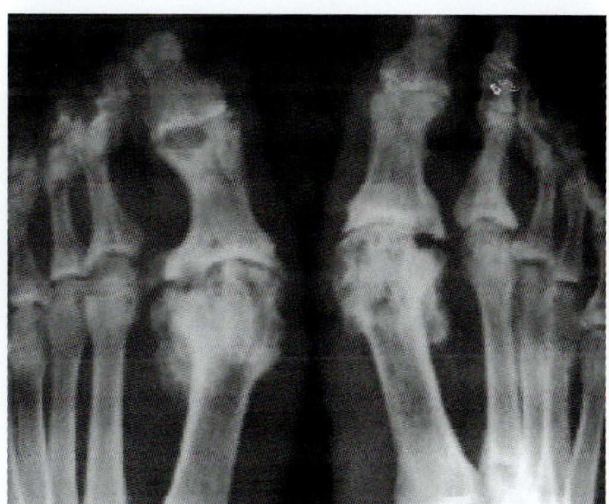

Fig. 13.31: Feet (AP): Interosseous osteolysis at both 1st MTP joints with overhanging margins. Punched out erosions are also noted near IP joints

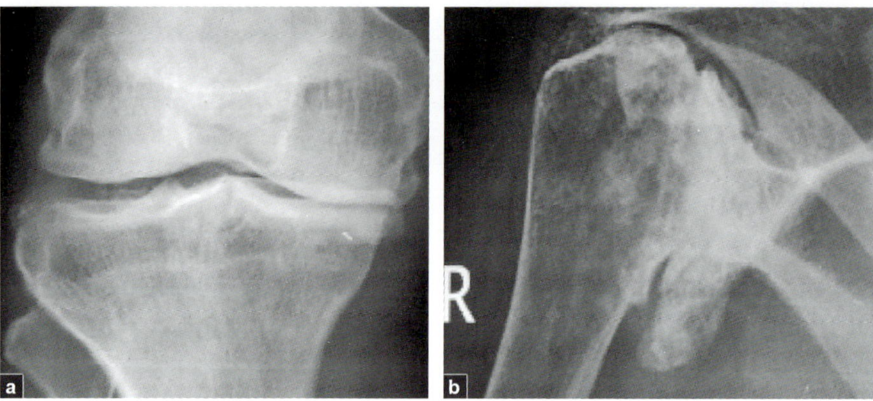

Fig. 13.32: Knee AP view, linear calcification seen in hyaline cartilage at lateral compartment indicating CPPD disease (a). Shoulder AP view, there is loss of the gleno-humeral joint space with osteophyte formation, subchondral sclerosis and cyst formation (CPPD arthropathy) (b)

calcification is fine and linear following the contour of underlying subchondral bone. Apart from the cartilage, CPPD deposition may also occur in the capsule, synovium, ligaments, and soft tissue mostly appearing as curvy linear calcifications.

Calcium hydroxyapatite: CHA crystal depositon commonly occurs in shoulder and is called peritendinitis calcarea. But it may also be seen in hip, elbow and wrist joints. Radiographically amorphous calcific deposits are noted (Fig. 13.33).

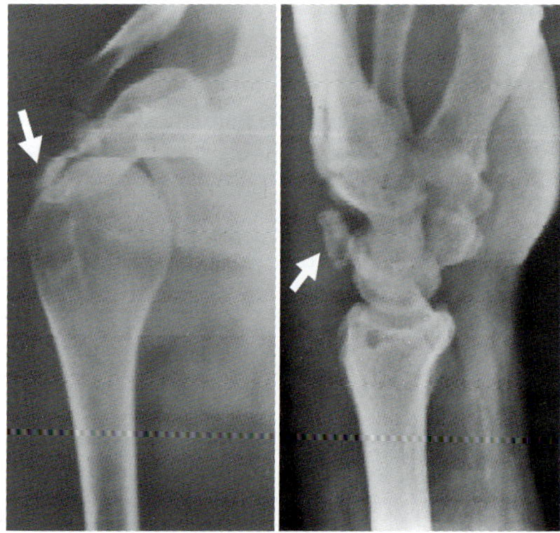

Fig. 13.33: Shoulder AP and Wrist oblique view showing calcium hydroxyapatite (CHA) crystal depositions

Connective Tissue Diseases

Systemic lupus erythematosus (SLE): The radiological features of SLE may mimic early RA, but erosions are uncommon in SLE. However, deforming arthropathy can occur due to soft tissue involvement. Osteonecrosis may occur in patients with SLE especially on glucocorticosteroids (GCs) (Fig. 13.34).

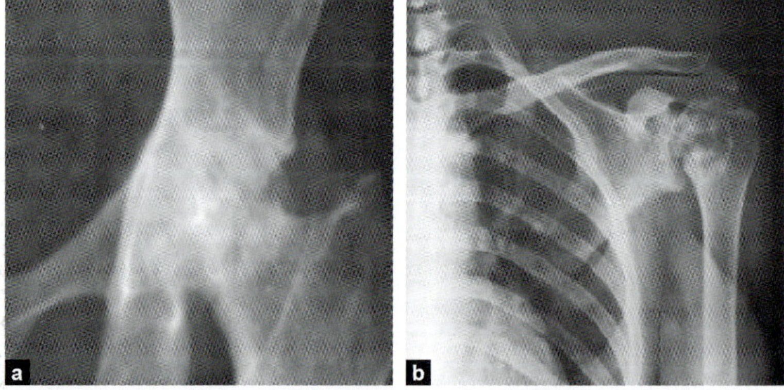

Fig. 13.34: Hip AP view, glucocorticoid induced osteonecrosis of the hip in a patient with SLE (a). Shoulder AP view showing osteonecrosis of humeral head in a patient with SLE on glucocorticoids (b)

Systemic sclerosis (scleroderma): Articular or non-articular features are commonly seen in scleroderma especially with Raynaud's phenomena (Fig. 13.35) (Table 13.7). CREST syndrome in localized scleroderma includes calcinosis, Raynaud's phenomenon, esophageal hypomotility, sclerodactyly and telangiectasia.

Inflammatory myositis: The most characteristic soft tissue abnormality in dermato/polymyositis is calcification in subcutaneous tissue and intermuscular fascia.

Miscellaneous Disorders

Osteonecrosis

The most common sites of osteonecrosis (avasular necrosis) are the femoral heads, femoral condyles, humeral head (Fig. 13.35), talus, scaphoid and lunate. In children, the idiopathic form of osteonecrosis of the femoral head is called Legg-Calvé-Perthes disease. Secondary osteonecrosis is commonly seen in adults. Radiographic changes are not evident in early epiphyseal osteonecrosis. As the disease progresses, arc like radiolucent subchondral band (crescent sign) and the formation of patchy subchondral lucent and sclerotic foci are seen. Fragmentation and collapse of the articular surface follows. The joint space is usually preserved until late in the course of the disease when secondary osteoarthritis may develop (Fig. 13.34).

Table 13.7: Radiographic features in systemic sclerosis
• Soft tissue resorption of the terminal fingers
• Soft tissue calcification—subcutaneous tissue, joint capsule, tendons or ligaments
• Osteolysis—resorption of terminal phalanges (acro-osteolysis)
• Erosive articular disease

Fluorosis

Excessive ingestion of fluoride through water or food causes toxicity resulting in a debilitating disease called 'fluorosis'. Excessive intake during pre-eruptive stage of teeth leads to dental fluorosis and further continued ingestion over years and decades causes bony or skeletal fluorosis. Patient complains of pains in the small joints of the limbs and back, which are often mistaken for RA or AS. Radiological findings show calcification and ossification of various bones (Fig. 13.36), joints, spine (Fig. 13.37), related ligaments and tendinous attachments to the bones and interosseous membranes.

Diffuse Idiopathic Skeletal Hyperostosis (DISH)

DISH radiologically resembles fluorosis. Since it is seen in older individuals the density of bone is not like that of flourosis. There is ossification of the anterior longitudinal ligaments of the spine, which gives the appearance of "flowing wax"on plain radiography (Fig. 13.38).

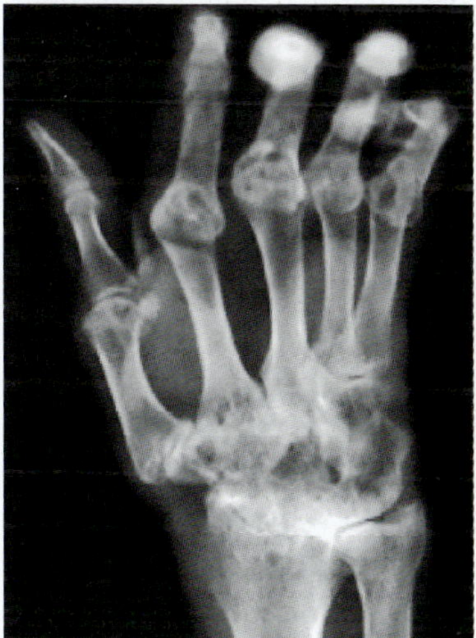

Fig. 13.35: Hand (AP): Acro-osteolysis, subcutaneous calcifications of fingers, with erosive articular disease of the wrist in scleroderma

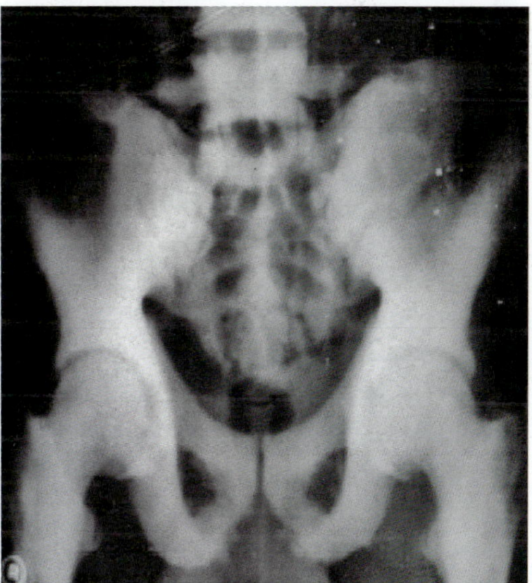

Fig. 13.36: Pelvis (AP): Dense bones in fluorosis

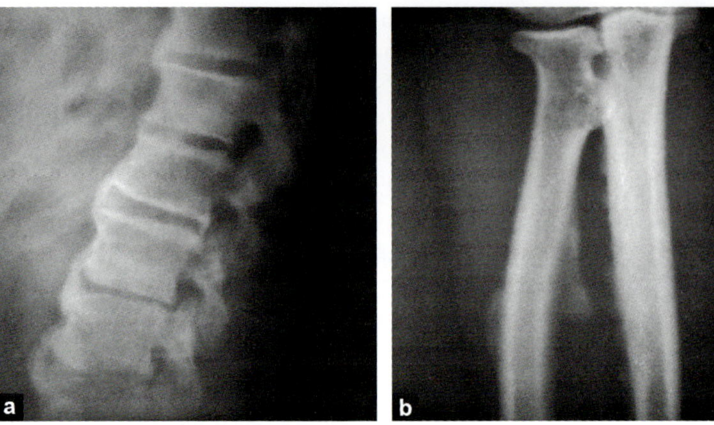

Fig. 13.37: Flourosis—Lumbar spine lateral view shows dense vertebrae with bridging osteophytes in a patient with fluorosis (a). Forearm AP view shows calcification of interosseous membrane (b)

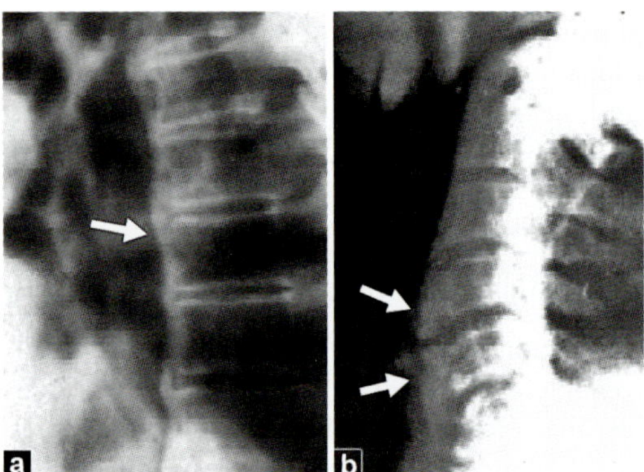

Fig. 13.38: Lumbar spine lateral view (a) and cervical spine (b) showing flowing wax appearance in diffuse idiopathic skeletal hyperostosis (DISH)

Infectious Arthritis

Infectious arthritis is a form of joint inflammation caused by live or inactive degraded organism, which may be bacteria, or virus, or fungus. Infection of the joints usually occurs as a secondary infection elsewhere in the body. Usually only one joint is involved, though sometimes more joints can become infected. Mostly, infectious arthritis affects the large joints (shoulders, hips, knees), but smaller joints (fingers, ankles) can also be involved (Fig. 13.39).

Osteoporosis and Osteomalacia

Osteoporosis is a systemic skeletal disease with low bone mass, increase in bone fragility and susceptibility to fracture. Radiologically, reduced bone density is observed in vertebrae, wrist, hips, humerus and tibial bones. The bones have prominent vertical striations resulting from a relatively greater loss of the horizontally oriented trabeculae (Fig. 13.40). In severe osteoporosis

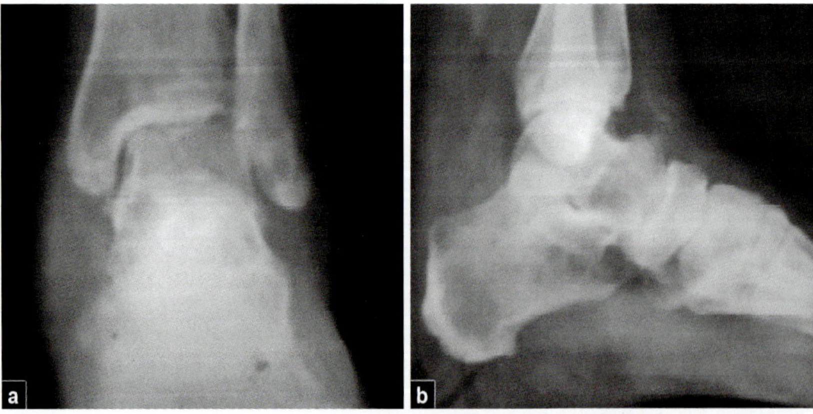

Fig. 13.39a and b: Ankle (lateral): Infective arthritis

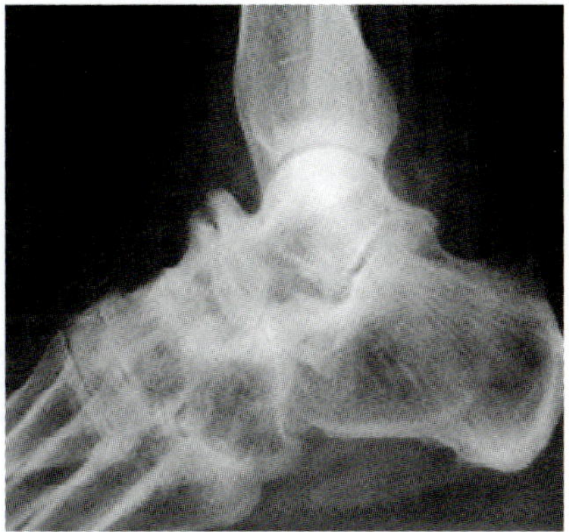

Fig. 13.40: Ankle (lateral): Osteoporosis with vertical striations at calcaneal bone

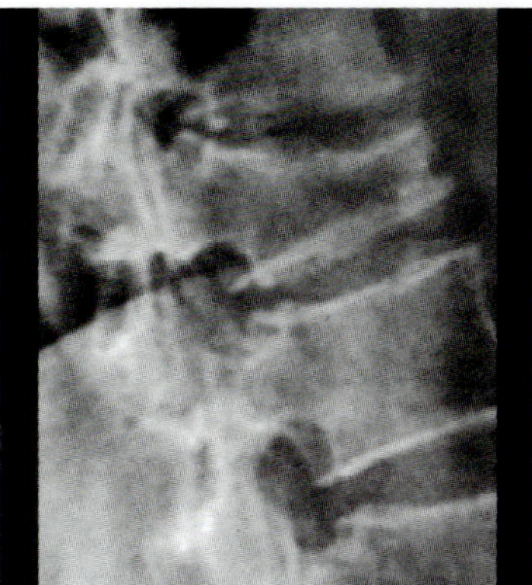

Fig. 13.41: Dorsal spine (lateral): Wedge compression at D10 due to osteoporosis

vertebral fractures (wedge compression) can be seen (Fig. 13.41).

Defective mineralization of the organic bone matrix in adults leads to osteomalacia. Most common clinical features are bone pain and tenderness in spine, shoulders, ribs and pelvis. The characteristic features of osteomalacia is ground glass appearance of bone with loss of differentiation between cortex and spongiosa. Looser's zone or psudofractures are commonly seen in the pelvis (Fig. 13.42), upper femur, and scapulae.

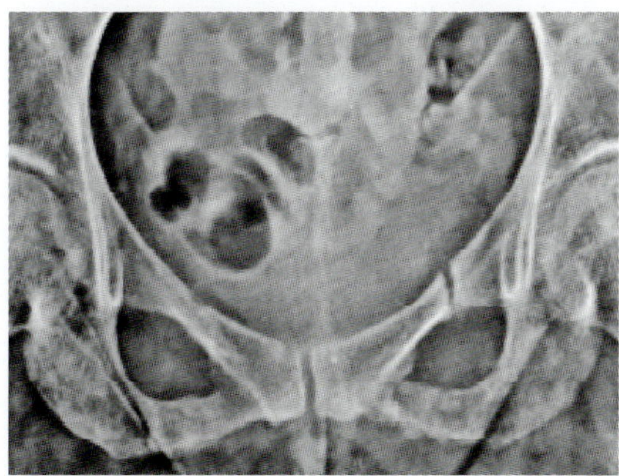

Fig. 13.42: Pelvis (AP): Osteomalacia with pseudofractures at superior ramus of pubic bone on both sides

Key Points

1. Radiographs are requested after obtaining the patient's complete history and physical examination.
2. Plain radiography remains the core imaging technique. It is simple, convenient and economical.
3. Clinical indications will guide which joints are to be investigated.
4. Radiographs are helpful in the diagnosis, prognosis and management of rheumatological disorders.
5. Conventional radiography is one of the parameters useful in the assessment of outcome measurements of RA, PsA, OA and SpA and other autoimmune diseases.

FURTHER READING

1. Subba Rao K. Arthritides (disorder of joints). In: Subba Rao K, Banerjee S, Aggarwal S, Bhargava S (editors). Diagnostic radiology and imaging. New Delhi: Jaypee Bro; 1997. pp 269–71.
2. Ostergaard M, Ejbjerg B, Szkudlarek M. Imaging in early rheumatoid arthritis: roles of magnetic resonance imaging, ultrasonography, conventional radiography and computed tomography. Best Pract Res Clin Rheum 2005; 19:91–116.
3. van der Heijde D. How to read radiographs according to the Sharp/van der Heijde method. J Rheumatol 2000; 27:261–23.
4. van der Heijde D. Quantification of radiological damage in inflammatory Arthritis: RA, PsA, and AS. Best Pract Res Clin Rheumatol 2004; 18:847–60.
5. Hellio Le Graverand MP, Mazzuca S, Duryea J, Brett A. Radiographic grading and measurement of joint space width in Osteoarthritis. Rheum Dis Clin North Am 2009; 35:487–502.
6. Pandya S. Outcome assessment in Rheumatic diseases. In: Rao URK, editor. Manual of Rheumatology. Mumbai: National Book Depot; 2014. p. 35–41.
7. Ostergaard M, Lambert RGW, Grassi HJW. Imaging in Rheumatic diseases. In: Firestein GS, Budd RC, Gabriel SE, McInnes IB, O'Dell JR, editors. Kelley and Firestein textbook of Rheumatology. Philadelphia: Elsevier Saunders; 2016. p. 858–907.

Chapter

14

Ultrasound in Rheumatology

Ved Chaturvedi, Priyanka Kharbanda

Introduction

The practice of musculoskeletal ultrasound (US) in rheumatology is a rapidly developing field, and forms an important part of curriculum of rheumatology training programs in Europe and the USA. In 1972 Daniel McDonald and George Leopold first described the use of a contact B-mode scanner to differentiate Baker's cysts from thrombophlebitis.[1] Musculoskeletal (MSK) ultrasonography (US) in rheumatology was usually done to diagnose Baker's cyst. But now MSK US is done routinely to diagnose and prognosticate rheumatoid arthritis (RA). MSK US is commonly performed to differentiate between arthralgia and arthritis, to look for erosions in early arthritis which is not visible on plain radiographs, for enthesopathies, to image blood vessels, etc. The US joint probe is likened as a "stethoscope" for the rheumatologist.[2]

Over time MSK US in rheumatology became popular because clinical examination of deeper joints especially the hip and sacroiliac joint was difficult. MSK US is an inexpensive and easy to examine these joints. Another good reason of the popularity of MSK US is the accuracy with which intra-articular injections can be given. The other advantages of MSK US are lack of radiation, the ability to look at tissue perfusion and inflammation all in real time. The disadvantages are that it is highly operator dependent, and it takes time to acquire the necessary skills, and that skill needs to be maintained by regular performance. The American College of Radiology (ACR) recommends that US trainees undertake 500 supervised scans in order to achieve an acceptable standard.[4] The European league against rheumatism (EULAR) conducts ultrasound course at three levels (basic course, intermediate course and advanced course) on an annual basis as part of its educational activity.[5]

1. Principles of MSK US: US images are produced by high frequency sound waves, typically above 20 kHz which are inaudible to the human ear. The machine consists of a computer processing unit and a transducer. The transducer is the most important part of the machine, it converts one type of energy into another. Based upon the **pulse-echo** principle occurring with **ultrasound piezoelectric crystals**, ultrasound transducers convert electricity into sound = pulse, and sound into electricity = echo. Pulse of sound is sent to soft tissues, the sound interaction with soft tissue is called bioeffects. Pulsing is determined by the transducer or probe crystal(s) and is not operator controlled. Echo produced by soft tissues are received by the transducer crystals, echoes are interpreted and processed by the ultrasound machine to produce images. Reflected echo intensity is proportionate to the amount of difference between tissue impedances at that interface, mainly due to differing water content of the tissues which provide those signals that finally define the edges of images. **Intense echoes** such as from bone appear **whiter or "hyperechoic"**, whereas **weaker echoes** from fluid or muscle produce darker more **"hypoechoic"** image. **Water** is the least reflective body material and it appears black **"anechoic"** on screen.

US images are presented in "black and white" or gray scale where the brightness of the white dot corresponds to the intensity of the reflected wave. This is known as B-mode or brightness modulated ultrasound. Doppler imaging is superimposed on the B-mode image. There are two main types of Doppler US: Colour flow Doppler (CFD) and power Doppler scan (PDS). Both produce a similar colour spectral map superimposed onto the gray scale image. CFD is an estimate of the mean Doppler frequency shift and relates to ***velocity*** and ***distribution of red blood cells***. Whereas PDS denotes the amplitude of the Doppler signal, which is determined by the volume of blood present. CFD is better suited for evaluating high velocity flow in large vessels like the carotids, whereas PDS is better suited for assessing low velocity flow in small vessels like in the synovium.

Points to consider when selecting an US machine

1. PDS is theoretically more sensitive to detect low velocity flow. But this sensitivity has disappeared in the newer high-end machines where the trend is that CFD now is more sensitive than PDS. In less expensive equipment, PDS has the highest sensitivity. The choice between CFD and PD depends on the equipment.[7]
2. Pulse radiofrequency (PRF) is the Doppler sampling frequency of the transducer and is reported in Hz. In rheumatology, we wish a high sensitivity to any flow and therefore we select a low PRF because the machine then will apply the lowest possible filters.
3. Colour priority must be maximised because as rheumatologist we evaluate vessels that are not visible on grey-scale US. Colour priority is a function in the machine where when colour information is obtained, the machine has to decide whether to show grey scale or colour.
4. Filters: Every Doppler instrument has high-pass filters, which eliminate the lowest Doppler shifts from the display. The filters should be kept at their lowest setting for use in rheumatology.
5. The gain setting determines the sensitivity of the system to flow. To get a acceptable trade-off between excess noise and picking up weak flow signals we should set the Doppler gain by turning it up until random noise is encountered and then lowering it until the noise disappears.
6. Focus positioning: Doppler study is highly dependent on focus positioning. In most machines once Doppler is activated the machines move the focus point to the colour box. Focus should be positioned where highest sensitivity is required. In longitudinal studies the focal point should be consistent.

Ultrasound transducers are commonly referred to by the operating, resonant or main frequency. Transducer frequency ranges used in medical ultrasound imaging are 2–20 MHz. There is improved resolution with high frequency transducers, at the cost of loss of depth of penetration, therefore making it unsuitable to image deep structures. Low frequency transducers, on the other hand, have full depth penetration, with poorer resolution. Therefore, low frequency 2.5 to 5 MHz is used for abdominal imaging, whereas for musculoskeletal imaging the frequency chosen is 7 MHz and above. Modern transducers have multiple frequencies, so it is not necessary to have multiple transducers with different frequencies. Therefore, a single transducer can have a frequency with range from 4 to 12.5 MHz. The transducer can be linear array, annular array or radial array based on the arrangement of the piezoelectric crystals. In MSK US the transducer is linear array whereby the piezoelectric crystals are arranged in parallel to one another, the surface is flat and produces rectangular image.

Scanning Technique

The patient must be positioned comfortably, and the area under investigation must be completely relaxed; otherwise, tension in the muscles and tendons will produce slight tremor resulting in movement artefacts.

The transducer is held in the hand between the thumb and index finger close to the imaging surface while anchoring the transducer to the patient with the small finger or the heel of the hand. This will allow one to make fine adjustments in transducer position and control the amount of transducer pressure on the patient. Very little pressure should be applied by the transducer. Use generous amounts of scanning gel with visible gel between the transducer and skin to ensure light pressure. False findings of absence of flow may occur if the examiner presses too hard on the tissue with the transducer, thereby blocking the flow.

The resulting image is then optimized by adjusting the depth of the image to bring the area of interest to view. If the US machine has adjustable focal zones, these also should be moved to the depth of interest to optimize resolution. The gray scale gain then is adjusted for brightness of the image. When describing anatomic structures at US, one refers to the imaging plane relative to the structure itself, such as transverse and longitudinal, rather than the imaging plane relative to the body.

Artifacts in MSK US

Anisotropy occurs when the region of interest is not perpendicular to the transducer, due to either the angle of transducer or the angle of object. If the US beam is not perpendicular to the tissue being scanned, the sound waves are scattered rather than being reflected back to the transducer. This causes the structure to appear darker than it should and can result in wrong diagnosis.

Refractile shadowing: This is reflection away from the transducer when the US beam hits a structure with different acoustic impedance, at an oblique angle. This most commonly occurs when scanning tendons, when the edge of the tendon and tendon sheath may look erroneously hypoechoic.

Acoustic shadowing occurs when the US beam hits a highly refractile surface like bone. The region beyond the refractile surface appears hypoechoic or anechoic as very few sound waves can reach it.

UTILITY OF ULTRASOUND IMAGING IN RHEUMATOLOGY

Diagnosis of Synovitis, Joint Effusion in Early Arthritis

Normal synovium is not seen on US. Synovitis is diagnosed on US by the presence of abnormally thickened hypoechoic intra-articular tissue. Synovitis can be distinguished from effusion, as in effusion the fluid is compressible and displaceable. US is superior to clinical examination in detection of joint inflammation and effusion.[5] It is a sensitive tool to detect presence of synovitis and effusion in joints which are normal on clinical examination in patients with very early synovitis (<3 months duration) and arthritis only in a single joint (Fig. 14.1).[6] This study has shown that presence of synovitis in early undifferentiated arthritis predicted the progression to RA.

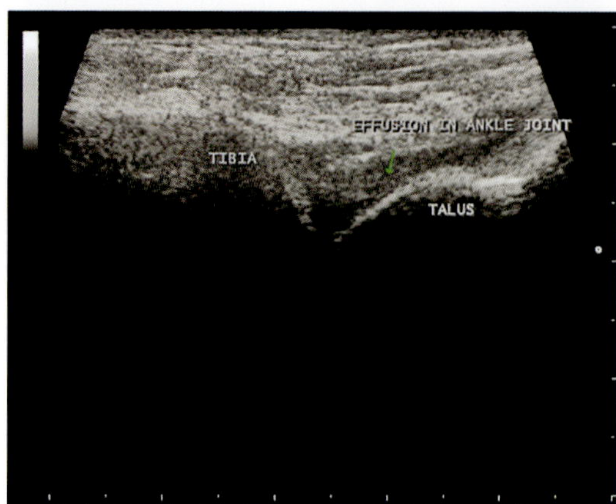

Fig. 14.1: Longitudinal scan of the ankle showing effusion in the ankle joint (not otherwise appreciable by clinical examination)

US Joint Counts—How many Joints to Scan

Filer *et al* evaluated 58 patients with clinically apparent synovitis of at least one joint and symptom duration of ≤3 months with clinical, laboratory, radiographic and 38 joint ultrasound assessments and followed these patients prospectively for 18 months.[6] At the end of the study 16 patients had resolution of synovitis, 13 developed non-RA persistent disease, and 29 developed RA by 1987 criteria. US demonstrated subclinical wrist, elbow, knee, ankle and metatarsophalangeal joint (MTP) involvement in patients who developed RA. Large joint and proximal interphalangeal joint (PIP) ultrasound variables had poor predictive ability, whereas US erosions lacked specificity. The most important finding of this longitudinal study is that both grey scale US and PDS of metacarpophalangeal (MCP) joints, wrists and MTP joints is the "optimum minimal ultrasound data" required to accurately predict development of RA.

In daily practice, it is not feasible to scan all joints and on the other hand scanning only the clinically involved joints is pointless. Based on the results of a recently published systematic review and on the spectrum of joints most frequently involved in early RA, whenever one uses US for diagnosis of early RA, it is recommended to scan at the minimum the wrists, MCP, and MTP bilaterally using PDS; PIP joints could be also included.[7] Time taken to scan depends on the experience of the person performing the US. For an experienced person the average time to scan small joints of hands and feet is 20 minutes.

Power Doppler Scan (PDS) in the Evaluation of Early Inflammatory Arthritis

As mentioned above Doppler US is a technique used to assess blood flow in normal or inflamed tissues and is widely used in cardiology to evaluate blood flow across heart valves. While colour flow Doppler (CFD) encodes direction and velocity of blood flow, Power Doppler scanning (PDS) detects the strength (amplitude) of blood flow. PDS is sensitive to detect even small increases in synovial perfusion. PDS is useful in arthritis evaluation, because it can measure blood flow at the microvascular level. The presence of Doppler hyperemia clinically means hypervascularity and inflammation. PDS is preferred to CFD for the assessment of disease activity because of greater sensitivity for low velocity blood flow. Thus, PDS is a very important tool in the diagnosis of early RA and evaluation of the remission of RA.

Freeston *et al*, evaluated 30 rheumatoid factor (RF) and/or ACPA negative individuals with inflammatory hand symptoms with or without clinical synovitis. In this study, they scanned 12 joints; bilateral MCPs, wrists, and flexor tendons of fingers.[8] The presence of US evidence of synovitis at baseline in addition to clinical parameters increased the pretest probability of 6% to 94% post-test for the progression to inflammatory arthritis at 12-month follow up at patient level. Other studies have also included the shoulder, elbow, PIPs, knees, and MTP joints in addition to the involved joints.[7]

A prospective cohort study from the Netherlands of 192 ACPA and/or RF positive arthralgia patients who were evaluated by US and PDS for joint effusion, synovitis, and PD signals in the synovial membrane of the joints, and adjacent and contralateral joints.[9] After a mean of 11 months, 45 (23%) developed arthritis. A significant association was found between patient's arthritis development and US abnormalities. Therefore, US and PDS can diagnose subclinical arthritis in seropositive arthralgia patients.[9]

A word of caution here: Signs of inflammation can be seen in patients without arthritis by only grey scale (GS) US.[8] Therefore, whenever feasible evaluation of arthritis patients by US should always be done with a PDS.

There is as yet no uniform US criteria for diagnosis remission of rheumatoid arthritis. But for all practical purposes the absence of synovitis according to PDS can be considered as the main criterion of remission.[10,11] Intervals between two US examinations should be at least 3 months.

A patient of early inflammatory arthritis with mild subclinical synovitis on US should be followed up, as presence of US evidence of synovitis among patients with inflammatory hand symptoms (with or without synovitis) increased the probability of progression to inflammatory arthritis at 12-month follow-up.[8] Other connective tissue disease must be ruled out. In these patients DMARDs can be offered. Patient has to be followed up, since up to 25% of patients with undifferentiated arthritis (and fulfil the ACR/EULAR criteria for RA) develop RA by the 1987 criteria at 1 year.[12]

Bone Erosions in RA

Erosion is defined as an interruption of the bone surface in two planes perpendicular to each other (Fig. 14.2). It is well known that the finding of erosion on plain radiograph is a very important outcome measure in RA. It is shown that US can detect more erosions than plain radiographs in early RA.[13]

Osteoarthritis (OA) and Bursal Disease

US shows the presence of effusion in a bursa as anechoic or hypoechoic lesion between two hyperechoic lines. The most common bursa imaged by US is the Baker's cyst (semimembranosus bursa) (Fig. 14.3). By high resolution US it is possible to

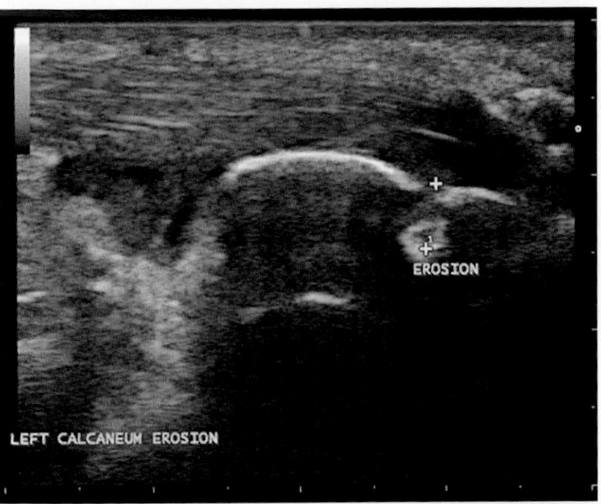

Fig. 14.2: Erosion of the calcaneum in a patient with RA

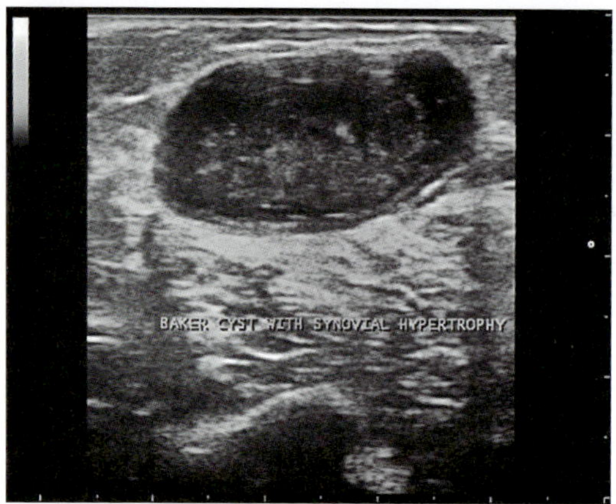

Fig. 14.3: Ultrasound scan of a patient with RA showing baker cyst with gross synovial hypertrophy in the bursal cavity

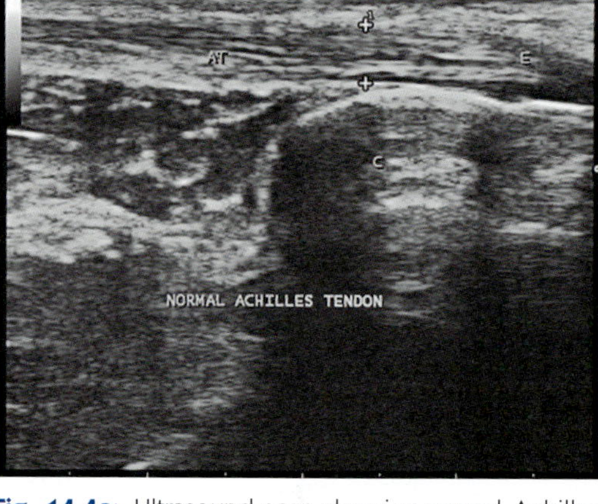

Fig. 14.4a: Ultrasound scan showing normal Achilles tendon

look for cartilage thinning, osteophytes, effusion, synovial proliferation, popliteal cyst, and meniscal protrusion in knee OA.[14] Similarly patients with hand OA patients who have synovitis by US have more severe radiological damage.[15]

Tendon Disease

Normal tendons fibrillary pattern along long axis and stippled in transverse axis by US (Fig. 14.4a). In tenosynovitis fluid will be seen adjacent to the tendon, and PDS will show increase in vasculature. Later the tendon becomes thicker and loses its normal fibrillar characteristic (Fig. 14.4b). Tendinosis is characterized by the presence of hypoechoic areas within the tendon substance. While if there is complete rupture of tendon there will be absence of signal between the free edges of the tear and lack of movement of the tendon on dynamic imaging (Fig. 14.4c). Plantar fasciitis a manifestation of ankylosing spondylitis can be appreciated by ultrasound as thickness of the fascia (Fig. 14.5).

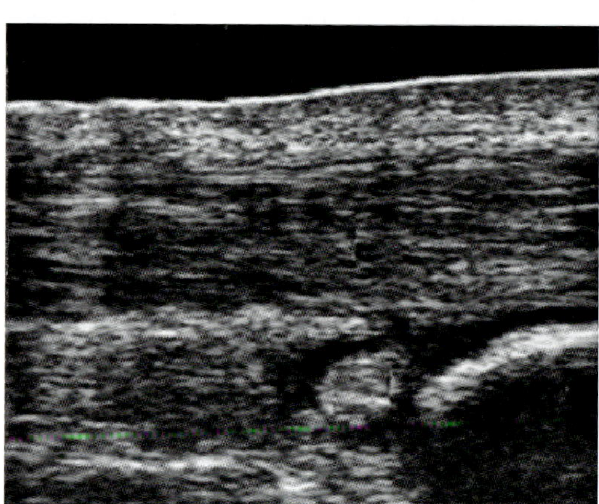

Fig. 14.4b: Achilles tendinitis and retrocalcaneal bursitis. Ultrasound detects diffuse thickening and loss of the normal fibrillar echotexture of the Achilles tendon

Enthesitis

Enthesitis the hallmark of the spondyloarthropathy group of disease, may also be secondary to mechanical stress. By US enthesitis is characterized by abnormally hypoechoic appearance of the entheses, and/or increased thickness of tendon or ligament at its bony attachment.[16,17] Other findings are enthesophyte, erosions, calcifications, and adjacent bursitis. US is again more sensitive than clinical examination at detecting enthesitis.

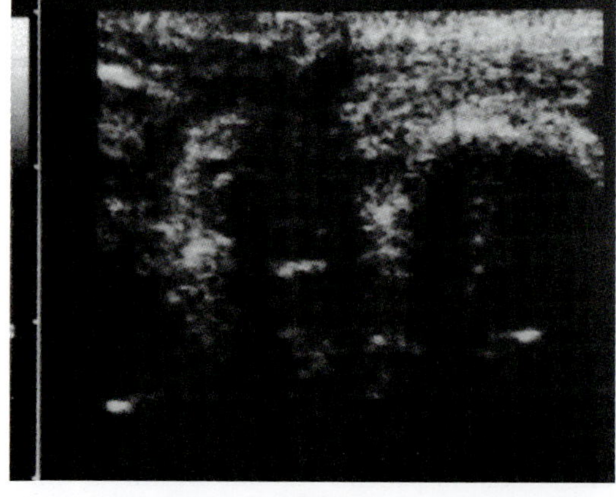

Fig. 14.4c: Achilles tendon rupture

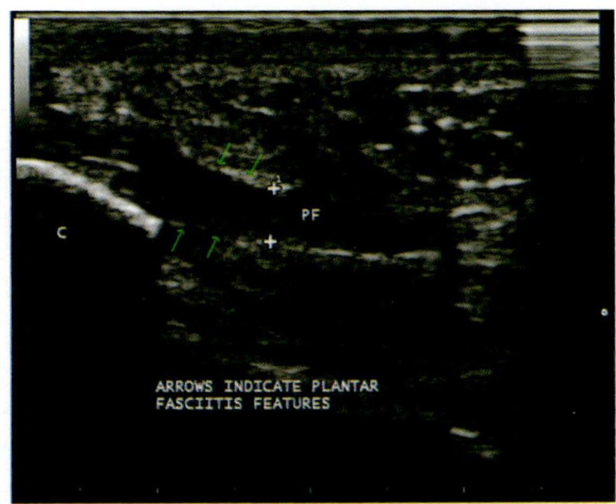

Fig. 14.5: Longitudinal scan of the foot showing plantar fasciitis (green arrows)

Sacroilitis

US Doppler may be useful to detect active sacroiliitis, though with difficulty due to deep location and angulation of the joint.[18] Contrast-enhanced ultrasound (CEUS) technique using second-generation microbubbles allows for the detection of active sacroiliitis, by showing deep contrast enhancement into the SI joints not detectable in inactive joints of patients or controls.[19] PDS US has been used to monitor response to anti-TNF (infliximab) therapy, and it was observed that blood flow signals by PDS reduced or even disappeared after infliximab therapy.[20] It may also be used to direct needles into the SI joint space to deliver local therapy, biopsy and/or to aspirate fluid.

Crystal Arthropathies

Normal cartilage is anechoic. US of the cartilage is useful in diagnosis of crystal deposition disease. A characteristic US finding in gout is the 'double contour' sign, whereby an echogenic line (representing MSU crystal) can be detected parallel to the cortex of bone (for example, a metatarsal head) with an anechoic region between, representing hyaline cartilage (Fig. 14.6).[21] In gout the crystals appear as hyperechoic or mixed echogenicity areas which sit on superficial articular surface, whereas in chondrocalcinosis the hyperechoic lesions are seen in deeper areas of the cartilage. Some authors reported the double contour sign to have a sensitivity of 36.8% and a specificity of 97.3% for the diagnosis of gout.[22]

Large Vasculitis

Colour flow Doppler is an important imaging modality in giant cell arteritis (GCA). Characteristic features of GCA include a hypoechoic swollen artery wall with surrounding oedema ('halo sign') and an irregular narrowed lumen.[23] One study found this sign is found to have a high negative predictive value, that means absence of the halo sign can practically rule out GCA.[24] The halo sign is also found to correspond with a pronounced inflammatory infiltrate of the temporal artery on biopsy.[25] A meta-analysis showed that the sensitivity and specificity of the halo sign is 68% and 91% respectively.[26] The usefulness of US also lies in guiding biopsy of involved vessels, particularly in GCA which is characterized by 'skip lesions'.

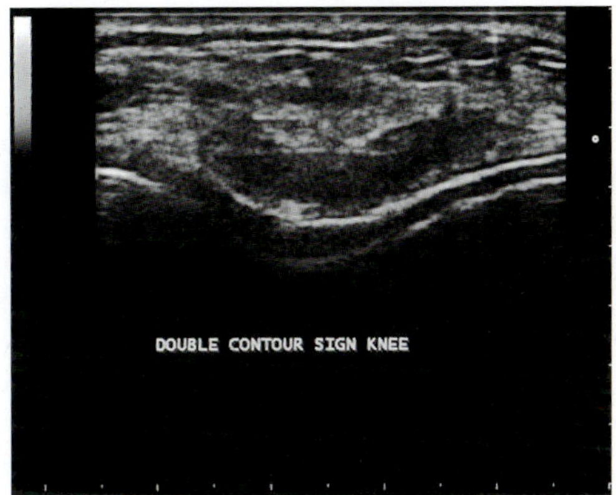

Fig. 14.6: Ultrasound of the knee shows the "double contour sign" in a patient with gout. An echogenic line (representing MSU crystal) is seen parallel to the cortex of bone (knee joint) with an anechoic region between, representing hyaline cartilage

Muscle Disease

Presently MRI is more sensitive than US in detecting muscle edema in inflammatory myositis and in guiding muscle biopsy. The usefulness of US lies in detecting muscle tears, localization and aspiration of muscle abscess in pyogenic myositis, to detect calcifications and myositis ossificans. However, a recent small study of 7 patients with dermatomyositis PDS was useful to detect fasciitis, whereby increased blood flow signals were observed along the fascia by PDS and this correlated with histologically confirmed fasciitis.[27]

Nerve Entrapments

In rheumatology practice the most common nerve entrapment syndrome is the carpal tunnel syndrome (CTS). By US the median nerve can be easily identified from tendons as it is hyperechoic and speckled in transverse section and has a hypoechoic fascicular pattern in longitudinal section. The cause of nerve compression in the carpal tunnel can also be identified, for example, tenosynovitis, tendon effusion, increased fatty tissue, ganglion cyst etc. An increase in the cross sectional area of the median nerve of greater than 9 mm^2 at the level of pisiform bone, aids in confirming the clinical diagnosis of CTS.[28] Local injections can be given with US guidance, so that the needle tip is visualized continuously and the needle is placed in the desired location, thus avoiding the risk of nerve damage.

Systemic Sclerosis (SSc)

The availability of high frequency transducers (13–20 MHz) has enabled evaluation of skin thickness in scleroderma and to assess the severity of skin disease and response to treatment. More recently shear wave elastography by US is found to be more sensitive than the modified Rodnan skin score (mRSS) to detect skin tightness.[29] Even at the sites where mRSS was normal the mean elasticity was higher in SSc compared with control. Therefore, US assessment of skin fibrosis by shear wave elastography is potentially useful tool to detect early skin fibrosis in SSc.

INTERVENTIONAL MSK US

The common bedside procedures in rheumatology are aspiration of joint fluid and intra-articular injections which have always been performed blind. In 1993, Jones *et al* studied the accuracy of 109 injections into various joints, by mixing the depot steroid with a radiographic contrast medium.[30] They found that approximately one-third of knee and ankle injections were extra-articular, only half of the wrist injections were definitely intra-articular, with even less accuracy for shoulder injections. Musculoskeletal US can be used to guide aspirations of joint fluid and intra-articular injection, this way it will ensure correct placement of needle and that medication will reach the desired location. This is particularly important when injecting the hip joint, shoulder joint and sacroiliac joint. Needle visualization is usually good by US because of the reflective nature of the needle. US can also be used for cyst decompression (e.g. Baker's cyst). US is definitely useful to direct needles into the joint space to deliver local therapy, biopsy and/or to aspirate fluid.

Limitations of ultrasound: First of all being highly operator dependent one has to ensure that the operator has been trained adequately, and studies from the OMERACT group show that with standardization, operator dependency can be diminished. To become an experienced sonographer requires long period of training, dedication and practice. Also since US beam do not penetrate bone, edema of bone marrow cannot be visualized.

Ultrasonography Abnormalities in Normal Population

Hypoechoic rims within joints and around tendons, and fluid in bursae are found in normal healthy individuals. Erosions in the humeral head can also be seen as a normal finding. However, normal healthy individuals should not have erosions in the MCP and PIP joints. For a detail description on what can be normal findings on MSK US, readers can read the elegant paper by WA Schmidt.[31]

Conclusion

MSK US is an inexpensive, non-invasive, and sensitive aid to diagnosis in rheumatology practice. Images are acquired in real time. An initial hurdle to overcome is to procure a good machine, and allotting a specific amount of time for learning. A sound knowledge of anatomy is essential, one needs to scan more normal joints so that the abnormal findings can be easily appreciated.

Key Points

1. Power Doppler Scan (PDS) is a very important tool in the diagnosis of early arthritis, because it can measure blood flow at the microvascular level. Increase in blood flow indicates synovitis.
2. Whenever one uses US to evaluate patients with early undifferentiated arthritis, the minimum set of joints to be scanned by PDS are both wrists, bilateral MCPs, bilateral MTPs.
3. Increase in PDS signals indicate active disease in rheumatoid arthritis, while absence of synovitis by PDS signals is considered as a criterion for remission.
4. US guided joint injections ensure accurate needle placement and correct delivery of the medication.
5. Colour flow Doppler (CFD) is useful to diagnose giant cell arteritis.
6. Shear wave elastography appears a promising tool to detect early skin fibrosis in systemic sclerosis.

Acknowledgement

Figures Courtesy: Dr Sajjan Shenoy MD DM
KMC, Mangalore

REFERENCES

1. McDonald DG, Leopold GR. Ultrasound B-scanning in the differentiation of Baker's cyst and thrombophlebitis. Br J Radiol 1972; 45:729–32.
2. Meenagh G, Filippucci E, Kane D, Taggart A, Grassi W. Ultrasonography in rheumatology: developing its potential in clinical practice and research. Rheumatology (Oxford) 2007; 46:3–5.
3. American College of Radiology. ACR standard for performing and interpreting diagnostic ultrasound examinations. In: Standards. Reston, VA American College of Radiology, 1996:235–6.
4. eular.org [Internet]. EULAR ultrasound course. Zurich, Switzerland. [Cited May 30, 2017]. Available from https://www.eular.org/edu_course_ultrasound.cfm
5. Kane D, Balint PV, Sturrock RD. Ultrasonography is superior to clinical examination in the detection and localization of knee joint effusion in rheumatoid arthritis. J Rheumatol 2003; 30:966–71.
6. Filer A, De Pablo P, Allen G, Nightingale P, Jordan A, Jobanputra P, et al. Utility of ultrasound joint counts in the prediction of rheumatoid arthritis in patients with very early synovitis. Ann Rheum Dis 2011;70:500–7.
7. Ten Cate DF, Luime JJ, Swen N, Gerards AH, De Jager MH, Basoski NM, et al. Role of ultrasonography in diagnosing early rheumatoid arthritis and remission of rheumatoid arthritis—a systematic review of the literature. Arthritis Res Ther 2013; 15(1):R4.
8. Freeston JE, Wakefield RJ, Conaghan PG, Hensor EM, Stewart SP, Emery P. A diagnostic algorithm for persistence of very early inflammatory arthritis: the utility of power Doppler ultrasound when added to conventional assessment tools. Ann Rheum Dis 2010; 69:417–19.
9. van de Stadt LA, Bos WH, Meursinge Reynders M, Wieringa H, Turkstra F, van derLaken CJ,et al. The value of ultrasonography in predicting arthritis in auto-antibody positive arthralgia patients: a prospective cohort study. Arthritis Res Ther. 2010;12(3):R98. doi: 10.1186/ar3028.
10. Foltz V, Gandjbakhch F, Etchepare F, Rosenberg C, Tanguy ML, Rozenberg S, et al. Power Doppler ultrasound, but not low-field magnetic resonance imaging, predicts relapse and radiographic disease progression in rheumatoid arthritis patients with low levels of disease activity. Arthritis Rheum 2012; 64:67–76.
11. Balsa A, de Miguel E, Castillo C, Peiteado D, Martín-Mola E. Superiority ofSDAI over DAS-28 in assessment of remission in rheumatoid arthritis patients using power Doppler ultrasonography as a gold standard. Rheumatology (Oxford). 2010; 49(4):683–90.
12. Krabben A, Abhishek A, Britsemmer K, Filer A, Huizinga TW, Raza K, van Schaardenburg DJ, van der Helm-van Mil AH. Risk of rheumatoid arthritis development in patients with unclassified arthritis according to the 2010 ACR/EULAR criteria for rheumatoid arthritis. Rheumatology (Oxford) 2013; 52:1265–70.
13. Wakefield RJ, Gibbon WW, Conaghan PG, O'Connor P, McGonagle D, Pease C, et al. The value of sonography in the detection of bone erosions in patients with rheumatoid arthritis: a comparison with conventional radiography. Arthritis Rheum 2000; 43:2762–70.
14. Razek AA, El-Basyouni SR. Ultrasound of knee osteoarthritis: inter observer agreement and correlation with Western Ontario and McMaster Universities Osteoarthritis. Clin Rheumatol. 2016;35(4): 997–1001.
15. Mancarella L, Magnani M, Addimanda O, Pignotti E, Galletti S, Meliconi R. Ultrasound-detected synovitis with power Doppler signal is associated with severe radiographic damage and reduced cartilage thickness in hand osteoarthritis. Osteoarthritis Cartilage. 2010;18:1263–8.
16. Gandjbakhch F, Terslev L, Joshua F, Wakefield RJ, Naredo E, D'Agostino MA; OMERACT Ultrasound Task Force. Ultrasound in the evaluation of enthesitis: status and perspectives. Arthritis Res Ther. 2011; 13(6):R188.
17. Kehl AS, Corr M, Weisman MH. Review: Enthesitis: New Insights Into Pathogenesis, Diagnostic Modalities, and Treatment. Arthritis Rheumatol. 2016; 68:312–22.
18. Unlü E, Pamuk ON, Cakir N. Color and duplex Doppler sonography to detect sacroiliitis and spinal inflammation in ankylosing spondylitis. Can this method reveal response to anti-tumor necrosis factor therapy? J Rheumatol 2007; 34:110–6.
19. Klauser AS, De Zordo T, Bellmann-Weiler R, Feuchtner GM, Sailer-Höck M, Sögner P, Gruber J. Feasibility of second-generation ultrasound contrast media in the detection of active sacroiliitis. Arthritis Rheum 2009; 61:909–16.
20. Jiang Y, Chen L, Zhu J, Xue Q, Wang N, Huang Y, et al. Power Doppler ultrasonography in the evaluation of infliximab treatment for sacroiliitis in patients with ankylosing spondylitis. Rheumatol Int. 2013; 33:2025–9.
21. McQueen FM, Doyle A, Dalbeth N. Imaging in gout—what can we learn from MRI, CT, DECT and US? Arthritis Res Ther 2011;13:246.
22. Lai K, Chiu Y. Role of ultrasonography in diagnosing gouty arthritis. Journal of Medical Ultrasound 2011, 19:7–13.

23. Schmidt WA, Kraft HE, Vorpahl K, Völker L, Gromnica-Ihle EJ. Color duplex ultrasonography in the diagnosis of temporal arteritis. N EnglJ Med 1997; 337:1336–42.
24. Nesher G, Shemesh D, Mates M, Sonnenblick M, Abramowitz HB.The predictive value of the halo sign in color Doppler ultrasonography of the temporal arteries for diagnosing giant cell arteritis. J Rheumatol 2002; 29:1224–6.
25. Schmidt D, Hetzel A, Reinhard M, Auw-Haedrich C. Comparison between color duplex ultrasonography and histology of the temporal artery in cranial arteritis (giant cell arteritis). Eur J Med Res 2003; 8(1):1–7.
26. Arida A, Kyprianou M, Kanakis M, Sfikakis PP. The diagnostic value of ultrasonography-derived edema of the temporal artery wall in giant cell arteritis: a second meta-analysis. BMCMusculoskeletDisord 2010; 11:44.
27. Yoshida K, Nishioka M, Matsushima S, Joh K, Oto Y, Yoshiga M, et al. Brief Report: Power Doppler ultrasonography for detection of increased vascularity in the fascia: apotential early diagnostic tool in fasciitis of dermatomyositis. Arthritis Rheumatol. 2016; 68:2986–91.
28. Duncan I, Sullivan P, Lomas F. Sonography in the diagnosis of carpal tunnel syndrome. Am J Roentgenol 1999; 173:681–684.
29. Wakhlu A, Chowdhury AC, Mohindra N, Tripathy SR, Misra DP, Agarwal V. Assessment of extent of skin involvement in scleroderma using shear wave elastography. Indian J Rheumatol 2017; 12 (3): ahead of print DOI:10.4103/injr.injr_41_17
30. Jones A, Regan M, Ledingham J, Pattrick M, Manhire A, Doherty M. Importance of placement of intra-articular steroid injections. Br MedJ 1993;307: 1329–30.
31. Schmidt WA, Schmidt H, Schicke B, Gromnica-Ihle E. Standard reference values for musculoskeletal ultrasonography. Ann Rheum Dis 2004; 63:988–94.

FURTHER READING

1. Iagnocco A, Naredo E, Bijlsma JW. Becoming a musculoskeletal ultrasonographer. Best Pract Res ClinRheumatol. 2013; 27:271–81.
2. Plaza M, Nowakowska-Plaza A, Pracoń G, Sudol-Szopińska I. Role of ultrasonography in the diagnosis of rheumatic diseases in light of ACR/EULAR guidelines. J Ultrason. 2016; 16:55–64.

Chapter

15

Extremity Magnetic Resonance Imaging (eMRI) in Rheumatology Practice

Ashish J Mathew, Harshini AS, Jyoti Panwar, Debashish Danda

INTRODUCTION

Advanced imaging has effectively enabled early diagnosis, understanding of pathogenesis, monitoring therapy and outcome measurement in clinical trials for patients with inflammatory arthritis. Magnetic resonance imaging (MRI) has evolved over the past two decades as an integral part of management in patients with inflammatory arthritis. Lack of ionizing radiation, 3D imaging, excellent resolution and better soft tissue contrast are a few of the areas where MRI scores over other advanced imaging modalities. Recent office-based extremity MRI scanners have taken care of patient inconvenience caused by conventional MRI scanners, albeit with lower resolution.[1] Newer imaging techniques like whole body MRI (WBMRI), dynamic contrast-enhanced MRI, and sequences like delayed gadolinium enhanced MRI of cartilage (dGEMRIC) and T2 mapping will greatly enhance the application of MRI in inflammatory arthritis.[2–5] There is growing body of evidence indicating the role of MRI as an outcome tool in clinical trials, especially for peripheral arthritis. As the debate on tapering as opposed to discontinuation of biologic therapy rages, MRI may find its utility in defining and classifying patients who would benefit by tapering biologic agents once disease activity is optimally controlled.[6–10] This chapter provides a bird's-eye view of the role of MRI in rheumatologic conditions, focusing on peripheral arthritis, and discusses recent advances based on published literature.

Parts of a MR Scanner (Fig. 15.1)

a. Main magnet: Generates a constant, strong external magnetic field.
b. Three gradient coils lie concentric to each other within the main magnet, representing the three orthogonal directions (x, y and z). Each coil generates a magnetic field in the same direction as the external magnetic field (B_0), but with a strength that shifts with position along the three orthogonal directions, depending on the coil used. The main magnetic strength varies along the direction of the applied gradient field.
c. Radiofrequency (RF) coils are mounted inside the gradient coils, lying concentric to them and to each other. These coils transmit RF energy to the tissues of interest and receive induced RF signal back from them.[11,12]

Principles and Sequences in MRI

A sound understanding of the basic physical principles of MRI acquisition is pivotal for a clinician interested in interpreting images. MRI utilizes the principles of nuclear magnetic resonance to enhance contrast between tissues in the body.

Understanding the Physics of Magnetic Field

Hydrogen nucleus, consisting of a single proton carrying positive electrical charge is the primary source of all MR signals. A proton constantly spins around its own axis, generating a magnetic field called magnetic moment. Under normal biological state magnetic moments from various protons in the human body are randomly oriented. The main

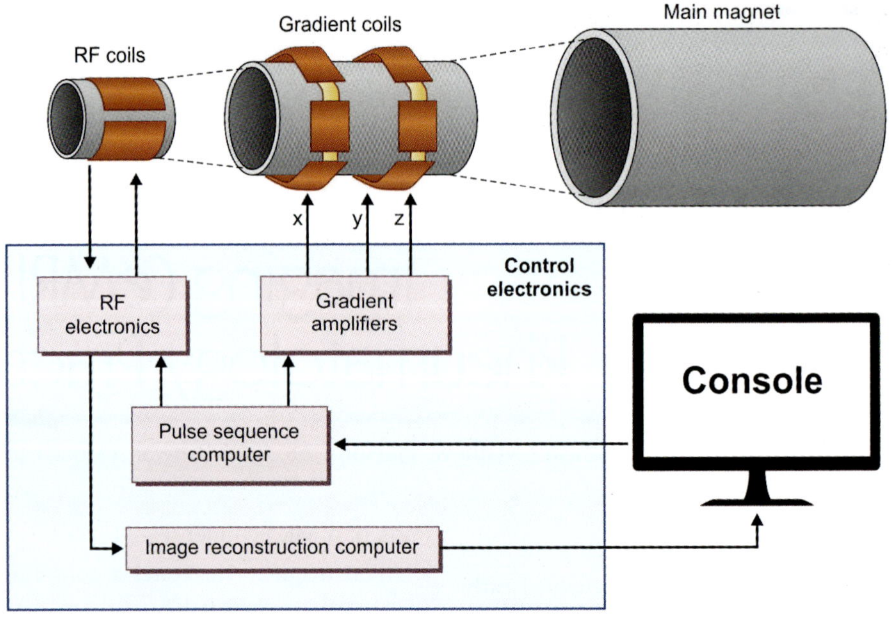

Fig. 15.1: Parts of a MR scanner

[(*Adapted from*: JL Prince, JM Links (2014). Medical Imaging Signals and Systems (2nd Edition): Pearson Education)]

magnet generates a constant, external magnetic field (B_0). When the human body is brought in close proximity to this external magnetic field, these magnetic moments align either with (parallel) or against (antiparallel) it (Fig. 15.2a). More protons align parallel to B_0, as this alignment requires the least energy. The difference between parallel and antiparallel alignment of protons depends on the strength of B_0, which is measured in units of Tesla (T). Once the protons in the human body generate a sum magnetic field or sum magnetization (longitudinal magnetization) after getting exposed to the main magnetic field, RF pulses are switched on and off. RF pulse transfers its energy to the protons and disturbs their alignment with B_0. The RF pulse causes protons to move together in the same direction (transverse magnetization) in line with the precessing protons (Fig. 15.2b). In order to make the protons fall out of alignment with B_0 the RF pulse should have the same frequency as the precession frequency of protons. This phenomenon is called resonance. If a conductive receiver coil is placed in proximity to the object being imaged, an alternating voltage can be induced across

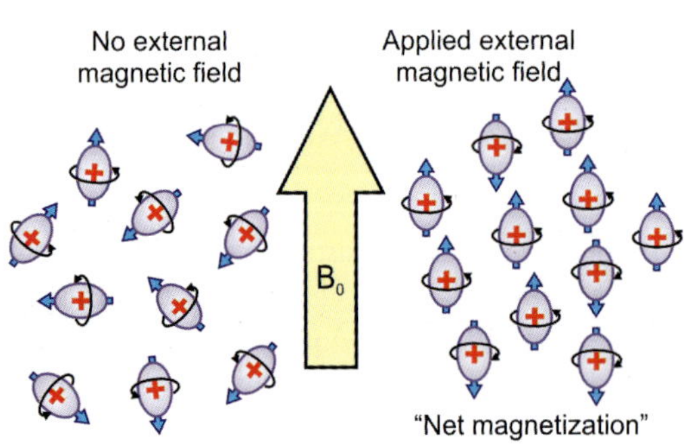

Fig. 15.2a: Alignment of protons in external magnetic field

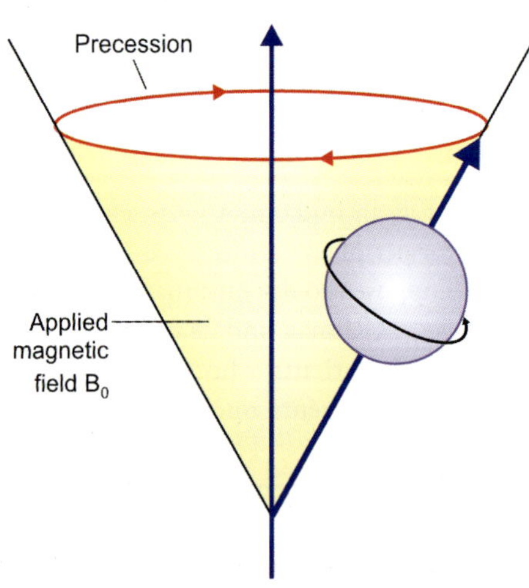

Fig. 15.2b: Precessional frequency

it. This gives rise to an electric current, which can be picked up as an MR signal and processed or reconstructed to obtain 3D gray-scale MR images. On switching off the RF pulse, protons fall out of phase with each other and return to a lower energy state, thereby, realigning parallel with B_0 again. This is called relaxation.

What are T1 and T2 in MRI?

This realignment can either occur by relaxation of transverse magnetization (T2) or by return of longitudinal magnetization to its original value (T1). The RF pulse is repeated at predetermined intervals called repetition time (TR). The interval between repetition of RF pulses and response signal is depicted as echo delay time (TE). By adjusting the TR and TE MR images can be made to contrast different tissues in the body. T1 and T2 weighted MR images are generated predominantly by manipulating the TR and TE, respectively.[13–16]

Core Variables Assessed in Inflammatory Arthritis

These include **erosions, synovitis, osteitis (bone marrow edema) and tenosynovitis**. Erosions are best visualised in T1W sequences. Post intravenous (IV) gadolinium (Gd) T1W images identify synovitis and tenosynovitis with good resolution. T2W fat suppressed sequences can be used for identifying osteitis, synovitis and tenosynovitis.[17] The outcome measures in rheumatology (OMERACT) has suggested imaging in two planes with T1W images before and after IVGd contrast and T2W fat saturated sequence or a short tau inversion recovery (STIR) sequence for imaging inflammatory and destructive changes in inflammatory arthritis. For optimal visualization, the slice thickness should be ≤1 mm and pixel size ≤ 0.5 × 0.5 mm. For detecting bone erosions, bone edema and/or bone proliferation, IV Gd can be omitted. Detection of inflammatory changes (synovitis and tenosynovitis) without IV Gd contrast in inflammatory arthritis has been shown to have lower sensitivity and greater inter-reader variability, especially in smaller joints of fingers.[18–21]

Definitions of Important Joint Pathologies[18–20]

- *Synovitis*: Area in the synovial compartment showing above mentioned normal post-Gd enhancement greater than the width of a normal synovium. Synovitis and joint fluid are generally very difficult to differentiate on non-contrast MRI. Heavily T2W images or STIR images sometimes can be used to differentiate these variables, as synovitis has a lower signal intensity than joint effusion and is more irregular (Fig. 15.3).[22] Acute synovitis enhances rapidly and intensely after IV Gd contrast. However, it lasts only for up to 5 minutes. Hence, it is important to image within 5 minutes of IV contrast injection to accurately delineate the extent of synovitis.[23]
- *Bone erosion*: Sharply marginated juxta-articular bone lesion, with typical signal characteristics, which is visualized in two planes. A breach of the cortex should be seen at least in one plane. Identification of erosions in two planes is important to avoid over-estimation of lesions (Fig. 15.4).
- *Bone edema*: Lesion with ill-defined margins and signal characteristics consistent with increased water content within the trabecular bone (Fig. 15.5). Histologic studies have shown that bone marrow edema corresponds to inflammatory cellular infiltrates in the bone marrow, representing osteitis.[24,25] Osteitis may be seen alone or surrounding bone erosions. It is considered to be "responsive to therapy".[26]

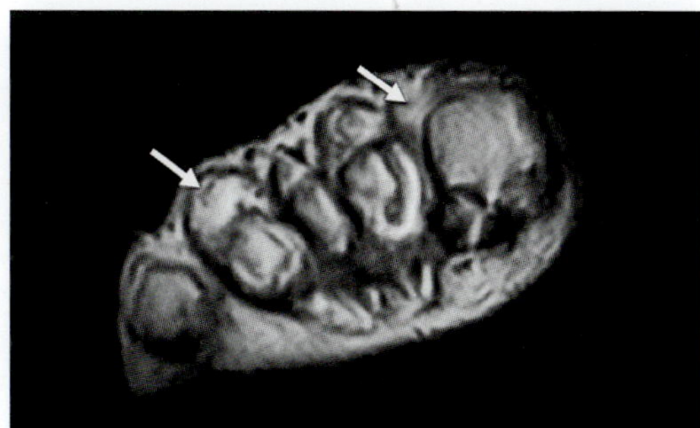

Fig. 15.3: e-MRI of left foot showing synovitis of 2nd and 4th MTP joints (white arrows)

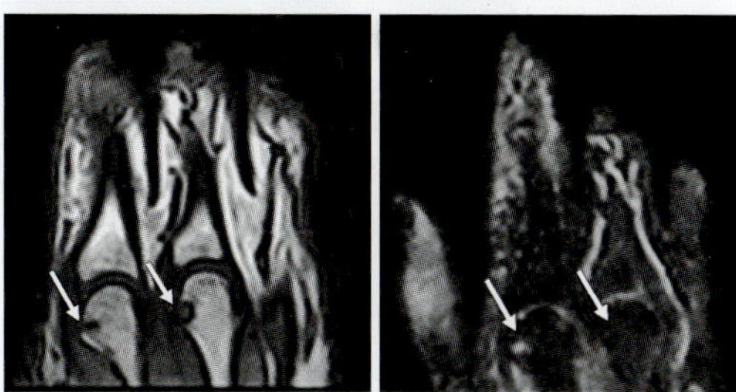

Fig. 15.4: 3DT1 and STIR Cor sequences showing erosions—acute (2nd MC head) and chronic (3rd MC head)

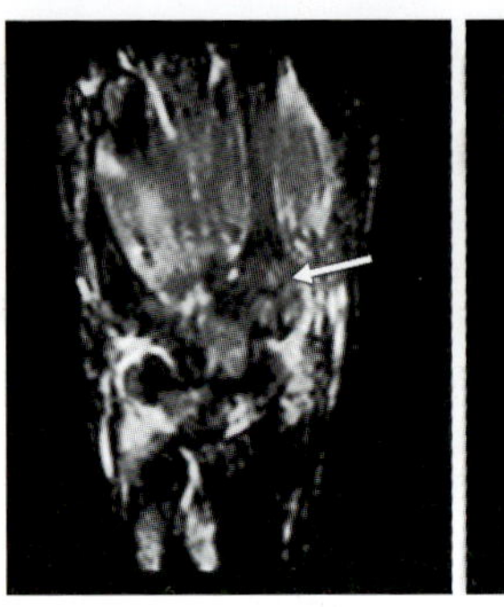
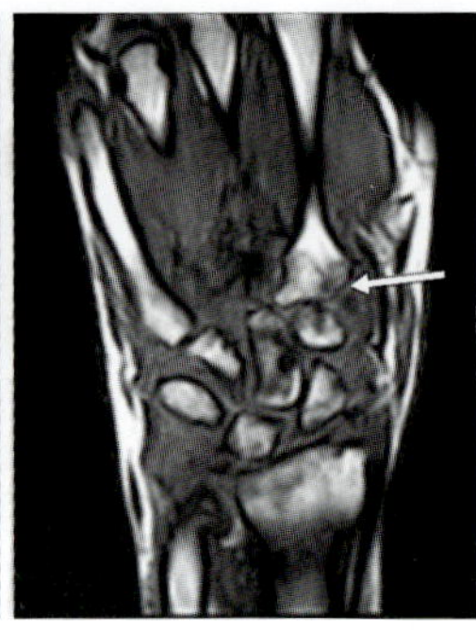

Fig. 15.5: STIR Cor image showing increased signal intensity and 3D T1 Cor image showing low signal intensity representing osteitis

- *Tenosynovitis*: Signal characteristics in T2W fat-saturated or STIR images are consistent with increased water content adjacent to a tendon in an area with a tendon sheath; however, it appears as low signal intensity on T1W images (Fig. 15.6). The fluid around the tendon sheath is considered abnormal only if its diameter exceeds that of the corresponding tendon.[27]
- *Periarticular inflammation*: Signal characteristics are consistent with increased water content at extra-articular sites like periosteum and entheses, but not tendon sheaths (Fig. 15.7).
- *Bone proliferation*: Abnormal formation of bone in pericarticular region like entheses and across the joint.
- *Peritendonitis*: Similar to tenosynovitis, but in an area without a tendon sheath.
- *Tendonitis*: Abnormal thickening and/or signal characteristics consistent with water content inside a tendon.

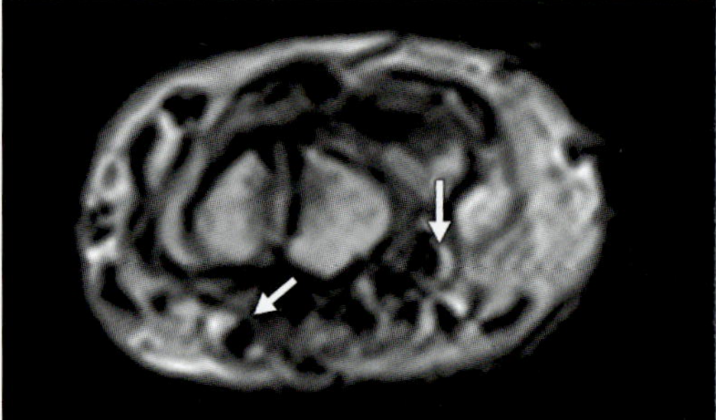
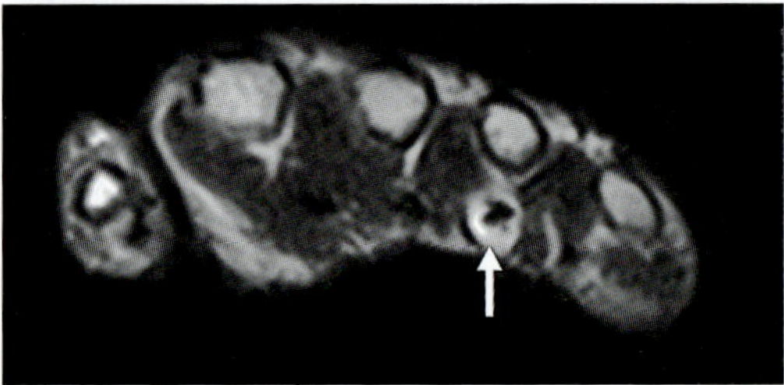

Fig. 15.6: T2W axial images showing flexor tenosynovitis at wrist (above) and 4th MCP (below)

In-office Extremity MRI Scanners (eMRIs)

Extremity MRIs have gained popularity over the past two decades owing to their portability, ease of installation and maintenance, as well as patient convenience, as compared to conventional MRI scanners. Poor quality image, especially in the STIR sequence secondary to a worsened signal–noise ratio, reduced field of vision (FOV) and less spectral separation between water and fat are the major trade-offs.[28] The imaging task force assigned by the American College of Rheumatology (ACR) in 2006 had highlighted the importance of investigating eMRI as a potential tool for outcome measure monitoring in inflammatory arthritis.[29] Over the

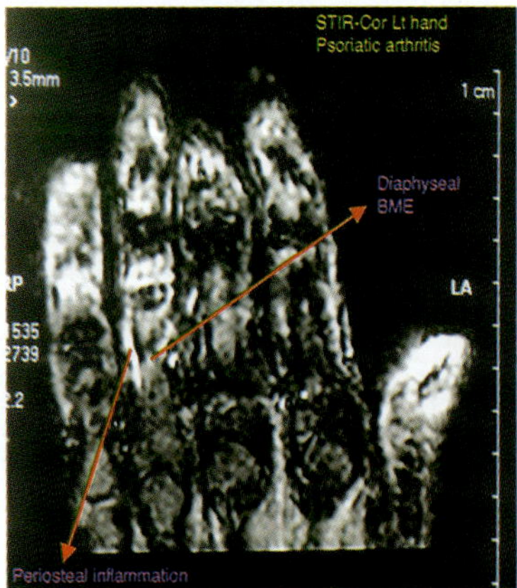

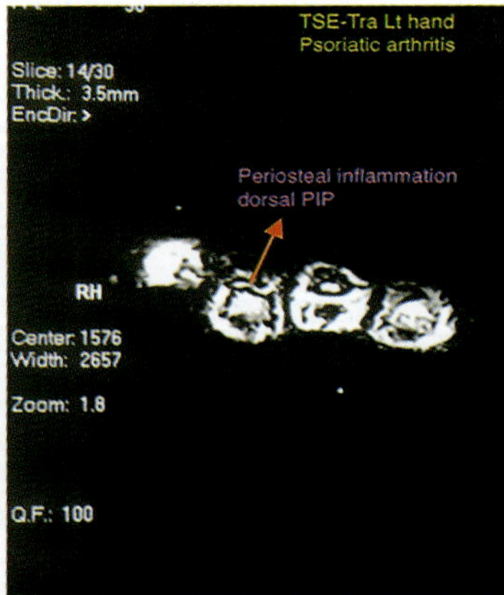

Fig. 15.7: STIR Cor and axial images of left hand showing periarticular inflammation (arrows)

high specificity, but loses out on sensitivity, when compared to reference standards.[93] Thus, it may find utility as a good test in ruling out OA. MRI has established its niche as the most widely used modality in clinical research in OA to evaluate risk factors, determine predictors of disease progression and monitor therapeutic changes. Table 15.2 summarizes the prominent semi-quantitative scoring systems for knee OA.

Role in Early Diagnosis

Epidemiological and genetic studies specify pre-OA disease states that can be modified. Diagnosis of pre-OA can potentiate therapeutic trials aiming to alter the preliminary course of OA, thus preventing future degenerative changes. Regional subchondral bone attenuation has been shown to have strong association with cartilage loss in the same region.[98] Quantitative MRI (qMRI), which includes volumetric measurement and physiologic MRI is useful for characterizing pre-OA. qMRI three dimensional sequences like double-echo steady state and fast low-angle shot detect cartilage volume and minimal thickness changes precisely.[99] dGEMRIC, T2 and T1 rho mapping and ultrashort echo-time (UTE) enhanced T2 mapping are various qMRI techniques that have been applied in pre-OA cohorts for better characterization.[100] dGEMRIC index parallels proteoglycan content in the cartilage and is decreased in pre-OA, suggesting tissue glycosaminoglycan loss.[101] Lower dGEMRIC index at baseline predicts disease in patients with pre-radiographic OA.[102] T2 and T1rho are sensitive to tissue hydration and collagen matrix architecture. T2 is focally increased in pre-OA patients. UTE-T2 measures short T2 signals from meniscus and deep layers of articular cartilage. Bone marrow lesions (BML) are seen in areas of cartilage loss and meniscal tears. The presence and extent of BMLs as indicated by MRI can be used for evaluating pre-OA states. BML associates closely with cartilage degeneration and can be used as an imaging marker for future OA development.[103] BML also strongly correlates with development of subchondral bone erosions.[104] Synovial biopsies in patients with pre-OA have shown evidence of active synovitis. MRI, both contrast-enhanced and non-contrast types are sensitive modalities to detect synovitis in early stages.

Table 15.2: Semi-quantitative whole joint MRI scoring systems for knee OA

Scoring system (year of publication)	*Scoring features and grades*
WORMS (2004)[94]	Cartilage (0–6); BMLs (0–3); subchondral cysts (0–3); bone attrition (0–3); effusion and synovitis (0–3); periarticular cysts
KOSS (2005)[95]	Cartilage size and depth (0–3); BMLs (0–3); subchondral cysts (0–3); osteophytes (0–3); effusion (0–3); meniscal tear (0–3); meniscal extrusion (0–3); popliteal cysts (0–3); synovial thickening (0–1)
BLOKS (2008)[96]	Cartilage size and depth (0–3, plus extent of any cartilage loss at specified point); BMLs (0–3, for each lesion); osteophytes (0–3); effusion (0–3); meniscal extrusion (0–3); synovitis (in Hoffa's fat pad 0–3 and at 5 additional sites 0–1); meniscal status (0–1 for intrameniscal signal, tears, maceration, meniscal cyst, each scored individually); ligaments (0–1); periarticular cysts/bursitis (0–1); loose bodies (0–1)
MOAKS (2011)[97]	Cartilage size and depth (0–3); BMLs (0–3, for each lesion); osteophytes (0–3); effusion-synovitis (0–3); Hoffa synovitis (0–3); meniscal extrusion (0–3); meniscal status (0–1, for intrameniscal signal, tears, maceration, meniscal cyst, hypertrophy; scored individually); ligaments (0–1); periarticular cysts/bursitides (0–1, scored individually); loose bodies (0–1)

WORMS: Whole organ magnetic resonance imaging score; KOSS: Knee osteoarthritis scoring system; BLOKS: Boston-Leeds osteoarthritis knee score; MOAKS: MRI osteoarthritis knee score

**Adapted from*: Guermazi A, et al. Nat Rev Rheumatol 2013; 9:236–51

Role in Measuring Treatment Response

BMLs have been shown to be reversible with therapy and are sensitive markers of therapeutic response. Rise in dGEMRIC index following therapy can be measured objectively. Many of the clinical trials on disease modifying osteoarthritis drugs (DMOADs) have utilized MRI end points as outcome to show therapeutic response.[105–107] Recent studies have addressed synovitis as a treatment target.[108–109]

Predictive Value in OA

Of late, the focus has been on prediction of structural progression as regards symptomatic progression. The MOST study has shown significant association between higher WORMS synovitis score (Table 15.2), superolateral Hoffa's fat pad hyperintensity with incident knee OA.[110] Change in thickness of cartilage as detected in MRI is a robust biomarker for disease progression in knee OA. The Strontium Ranelate trial in OA reported BML as an effective predictor of change in width of joint space.[111] Severe cartilage damage, BML, synovitis in MOAKS, baseline BLOKS and WORMS cartilage scores can predict future knee replacement.[112,113] High baseline T2 values in knee cartilage increase the likelihood of knee OA.[114] Bone shape may evolve as a potential MRI predictor of OA development. Changes in bone area and shape using 3D MR images were shown to be associated with radiographic and pain progression over 4 years in a recent study.[115]

Pitfalls in Extremity MRI

Considering the high sensitivity of MRI in detecting joint pathologies, clinicians should be well acquainted with the pitfalls to minimize false positive reporting.

Artifacts, misrepresentation of normal findings, slice thickness, field of view (FOV), normal anatomical variations, bone erosion, osteitis, synovitis and tenosynovitis are a few of the potential pitfalls, which an unwary reporter may overlook.[116]

Partial Voluming

Partial voluming is one of the common artifacts in which a structure is contained only partially within a voxel. This occurs mostly in areas where two tissues of different signal intensity lie adjacent to each other, mimicking erosion. Viewing the image in another plane is of paramount importance in such cases.

Artifacts

Susceptibility artifacts are seen following fluctuations in the local magnetic field due to variations in external magnetic field depending on the degree to which tissues become magnetized. Presence of metal in the region being imaged can distort the magnetic field and is the best example of susceptibility artifact. When two tissues of very different composition are imaged against one another, adjacent lines of hyperintensity and hypointensity appear at the tissue interface. These are called chemical shift artifacts. MR image clarity is significantly affected by motion. For patients with active arthritis, maintaining same position for prolonged periods can be a major challenge. Movement artifacts are seen mostly in T2 weighted sequences, as they warrant prolonged imaging periods. In homogeneity of fat suppression can at times result in appearance akin to bone edema.

Misinterpreting Normal Anatomical Variants

Normal anatomical features can often be misinterpreted. Interosseous ligaments attachments at the wrist may simulate erosions. Normal subjects can have small erosion-like lesions in around 2% of metacarpal and wrist bones. These lesions do not enhance post-contrast.[117] Nutrient foramina in carpal bones may be apparent on certain sequences, and may be mistaken as small erosions.

Technical Cognizance

Optimal slice thickness and FOV for imaging a hand with inflammatory arthritis are important factors to be considered. The minimum thickness recommended is 3 mm or less, with a gap of 0.5 mm between the slices. However, certain small erosions do get missed out even with this thickness. Hence, in 3D T1 sequences, a thickness of 0.9 mm is deemed necessary to include all erosions. Signal noise ratio (SNR), however, decreases with reduction in thickness. A FOV of 10 cm or less is ideal for good spatial resolution.

Distortion due to Disease Process

Distortion of anatomy due to the disease process can cause many pitfalls in MRI. MCP subluxation in RA patients makes erosion assessment at metacarpal heads more difficult. Advanced disease result in fusion of bones at the carpus, which can cause confusion in identification. Irregular margins of small carpal bones with adjacent synovitis can mimic pannus within small erosion. Erosions can at times appear similar to osteitis, especially when they contain pannus. Erosions have well defined margins and a clear cortical break. Normal joints have a thin layer of synovial membrane, which can at times enhance in post-contrast images. For this reason definition of synovitis includes both the degree of enhancement following contrast injection and thickness of synovial membrane. Tenosynovitis can be confused with synovitis as both these pathologies look alike in T2 weighted sequences. A sound knowledge of anatomy is pivotal to avoid such doubts.

Conclusion

MRI is increasingly finding its utility in rheumatology, both in research as well as in clinical practice. Future initiatives to device algorithms for cost-effective use of MRI in clinical practice are essential. Novel MRI sequences are being discovered which would aid in better understanding of the pathogenesis of inflammatory arthritis. Incorporating MR imaging with clinical and omics markers to device a composite biomarker score for diseases like psoriatic arthritis will be an ideal way to diagnose pre-clinical disease and modify the treat-to-target concept, even in preclinical disease. In clinical trials, conventional radiography may give way to MRI in the light of ample evidence from studies suggesting that MRI can discriminate therapeutic efficacy of different therapies within 6 months. Utility of MRI while tapering or discontinuing biologic therapy need to be refined further. With evolving technology and quantification, MRI can find better and more efficient utility in the field of rheumatology, both in terms of clinical research and routine patient care.

Key Points

- MRI is the gold standard for imaging soft tissue and joint pathology in inflammatory arthritis.
- Core variables in MRI for assessing inflammatory arthritis include erosions, synovitis, osteitis and tenosynovitis
- MRI has shown utility in diagnosis, understanding of pathogenesis, outcome measurement in clinical trials and determining remission in inflammatory arthritis.
- Patient friendly, low maintenance office-based e-MRI machines for peripheral joint assessment have evolved over time and the latest 1.5T high-field magnet machine overcomes a major limitation of earlier machines with better signal–noise ratio.
- The OMERACT group defined and validated scores for imaging in RA (RAMRIS) and PsA (PsAMRIS), which have been successfully used in clinical trials and cross-sectional studies.
- In early diagnosis of osteoarthritis, MRI has been shown to have high specificity, with limited sensitivity and thus, useful in ruling out the disease.
- Pitfalls, including partial voluming, potential artefacts, misinterpretation of normal anatomical variants and distortion of normal anatomy due to disease process should be borne in mind while reporting.

Acknowledgement

The authors would like to thank Dr Priya Mathew for all hand-drawn figures.

REFERENCES

1. Mathew AJ, Crues JV III, Danda D. Office eMRI: viewing joints from the inside. Int J Rheum Dis 2014; 17:706–09.
2. Poggenborg RP, Pedersen SJ, Eshed I, Sorensen IJ, Moller JM, Madsen OR, et al. Head-to-toe whole-body MRI in psoriatic arthritis, axial spondyloarthritis and healthy subjects: first steps towards global inflammation and damage scores of peripheral and axial joints. Rheumatology. 2015; 54:1039–649.
3. Cimmino MA, Barbieri F, Boesen M, PaparoF, Parodi M, Kubassova O, et al. Dynamic contrast-enhanced magnetic resonance imaging of articular and extraarticular synovial structures of the hands in patients with psoriatic arthritis. J Rheumatol. 2012; 89:44–8.
4. Miesse F, Buchbender C, Scherer A, Wittsack HJ, Specker C, Schneider M, et al. Molecular imaging of cartilage damage of finger joints in early rheumatoid arthritis with delayed gadolinium—enhanced magnetic resonance imaging. Arthritis Rheum 2012; 64:394–99.

5. Herz B, Albrecht A, Englbrecht M, Welsch GH, Uder M, Renner N, et al. Osteitis and synovitis, but not bone and erosion, is associated with proteoglycan loss and microstructure damage in the cartilage of patients with rheumatoid arthritis. Ann Rheum Dis 2014; 73:1101–06.
6. Mathew AJ, Danda D, Conaghan PG. MRI and ultrasound in rheumatoid arthritis. Curr Opin Rheumatol 2016; 28:323–29.
7. McGonagle D, Ash ZR, Hodgson RJ, Emery P, Radjenovic A. MRI for the assessment and monitoring of RA—what can it tell us? Nat Rev Rheumatol 2011; 7:185–89.
8. Baker JF, Conaghan PG, Emery P, Baker DG, Ostergaard M. Relationship of patient-reported outcomes with MRI measures in rheumatoid arthritis. Ann Rheum Dis 2017; 76:486–90.
9. Haavardsholm EA, Lie E, Lillegraven S. Should modern imaging be part of remission criteria in rheumatoid arthritis? Best Pract Res Clin Rheumatol 2012; 26:767–85.
10. Gandjbakhch F, Haavardsholm EA, Conaghan PG, Ejbjerg B, Foltz V, Brown AK, et al. Determining an magnetic resonance imaging inflammatory activity acceptable state without subsequent radiographic progression in rheumatoid arthritis: results from a follow up MRI study of 254 patients in clinical remission or low disease activity. J Rheumatol 2014; 41:398–406.
11. Currie S, Hoggard N, Craven IJ, Hadjivassiliou M, Wilkinson ID. Understanding MRI: basic MR physics for physicians. Postgrad Med J 2013; 89:209–23.
12. Jacobs MA, Ibrahim TS, Ouwerkerk R. AAPM/RSNA physics tutorials for residents: MRI imaging: brief overview and emerging applications. Radiographics 2007; 27:1213–29.
13. Tall MA, Thompson AK, Greer B, Campbell S. The pearl and pitfalls of magnetic resonance imaging of the lower extremity. J Orthop Sports Phys Ther 2011; 41:873–86.
14. Lisbona MP, Maymo J, Carbonell J. Magnetic resonance of the hand in rheumatoid arthritis. Review of methodology and its use in diagnosis, monitoring and prognosis. Rheumatol Clin 2007; 3:126–36.
15. McMohan KL, Cowin G, Galloway G. Magnetic resonance imaging: the underlying principles. J Orthop Sports Phys Ther 2011; 41:806–19.
16. Amrami KK. Radiology corner: Basic principles of MRI for hand surgeons. J American SocSurg Hand 2005; 5:81–86.
17. Mathew AJ, Bird P. Utility of in-office extremity magnetic resonance imaging in rheumatology. Indian J Rheumatol 2015;10:140–6.
18. Ostergaard M, Edomonds J, McQueen F, Peterfy C, Lassere M, Ejbjerg B, et al. An introduction to the EULAR-OMERACT rheumatoid arthritis MRI reference image atlas. Ann Rheum Dis 2005; 64(Suppl 1):i3-i7.
19. Ostergaard M, McQueen F, Wiell C, Bird P, Boyesen P, Ejbjerg B, et al. The OMERACT psoriatic arthritis magnetic resonance imaging scoring system (PsAMRIS): Definitions of key pathologies, suggested MRI sequences, and preliminary scoring system for PsA hands. J Rheumatol 2009; 36:1816–24.
20. Ostergaard M, Peterfy C, Conaghan P, McQueen F, Bird P, Ejbjerg B, et al. OMERACT rheumatoid arthritis magnetic resonance imaging studies. Core set of MRI acquisitions, joint pathology definitions, and the OMERACT RA-MRI scoring system. J Rheumatol 2003; 30:1385–6.
21. Stomp W, Krabben A, van der Heijde D, Huizinga TW, Bioem JL, Ostergaard M, et al. Aiming for a simpler early arthritis MRI protocol: can Gd contrast administration be elimintated? Eur Radiol 2015; 25:1520–27.
22. Narváez JA, Narváez J, Roca Y, Aguilera C. MR imaging assessment of clinical problems in rheumatoid arthritis. Eur Radiol 2002; 12(7): 1819–28.
23. Østergaard M, Klarlund M. Importance of timing of post-contrast MRI in rheumatoid arthritis: what happens during the first 60 minutes after IV gadolinium DTPA? Ann Rheum Dis 2001; 60(11):1050–54.
24. Jimenez-Boj E, Nöbauer-Huhmann I, Hanslik-Schnabel B, Dorotka R, Wanivenhaus AH, Kaiberger F, et al. Bone erosions and bone marrow edema as defined by magnetic resonance imaging reflect true bone marrow inflammation in rheumatoid arthritis. Arthritis Rheum 2007; 56(4):1118–24.
25. McQueen FM, Gao A, Ostergaard M, King A, Shalley G, Robinson E, et al. High-grade MRI bone oedema is common within the surgical field in rheumatoid arthritis patients undergoing joint replacement and is associated with osteitis in subchondral bone. Ann Rheum Dis 2007; 66(12):1581–87.
26. McQueen FM, Benton N, Perry D, Crabbe J, Robinson E, Yeoman S, et al. Bone edema scored on magnetic resonance imaging scans of the dominant carpus at presentation predicts radiographic joint damage of the hands and feet six years later in patients with rheumatoid arthritis. Arthritis Rheum 2003; 48(7):1814–27.
27. Rubens DJ, Blebea JS, Totterman SM, Hooper MM. Rheumatoid arthritis: evaluation of wrist extensor tendons with clinical examination versus MR imaging—a preliminary report. Radiology 1993; 187(3):831–38.

28. Ghazinoor S, Crues III JV, Crowley C. Low-field musculoskeletal MRI. J MagnReson Imaging 2007; 25:234–44.

29. American College of Rheumatology Extremity Magnetic Resonance Imaging Task Force. Extremity magnetic resonance imaging in rheumatoid arthritis: report of the American College of Rheumatology Extremity Magnetic Resonance Imaging Task Force. Arthritis Rheum. 2006; 54:1034–47.

30. Taouli B, Zaim S, Peterfy CG. Rheumatoid arthritis of the hand and wrist: comparison of three imaging techniques. Am J Roentgenol. 2004; 182:937–43.

31. Savnik A, Malmskov H, Thomson HS, Bretlau T, Graff LB, Nielsen H, et al. MRI of the arthritic (small joints: comparison of extremity MRI (0.2T) vs high field MRI (1.5T). EurRadiol. 2001;11:1030–1038.

32. Ejbjerg BJ, Narvestad E, Jacobson S, Thomasen HS, Ostergaard M. Optimized, low cost, (low field dedicated extremity MRI is highly specific and sensitive for synovitis and bone erosions in rheumatoid arthritis wrist and finger joints: comparison with conventional high field MRI and radiography. Ann Rheum Dis. 2005; 64:1280–7.

33. Bird P, Ejbjerg B, Lassere M, Ostergaard M, McQueen F, Peterfy C, et al. A multireader reliability study comparing conventional high-field magnetic resonance imaging with extremity low field MRI in rheumatoid arthritis. J Rheumatol. 2007; 34: 854–56.

34. Schirmer C, Scheel AK, Althoff C, Schink T, Eshed I, Lemboke A, et al. Diagnostic quality and scoring of synovitis, tenosynovitis and erosions in low-field MRI of patients with rheumatoid arthritis: a comparison with conventional MRI. Ann Rheum Dis. 2007; 66:522–29.

35. Suzuki T, Ito S, Handa S, Kose K, Okamoto Y, Minami M, et al. A new low-field extremity magnetic resonance imaging and proposed compact MRI score: evaluation of anti-tumor necrosis factor biologics on rheumatoid arthritis. Mod Rheumatol. 2009; 19:358–65.

36. Krabbe S, Eshed I, Pedersen SJ, Boyessen P, Moller JM, Therkildsen F, et al. Bone marrow edema assessment by magnetic resonance imaging in rheumatoid arthritis wrist and metacarpophalangeal joints: the importance of field strength, coil type and image resolution. Rheumatology. 2014; 53:1446–51.

37. Freeston JE, Olech E, Yocum D, Hensor EMA, Emery P, Conaghan PG. A modification of the OMERACT RA MRI score for erosions for use with an extremity MRI with reduced field of view. Ann Rheum Dis. 2007; 66:1669–71.

38. Strube H, Becker-Gaab C, Saam T, Reiser M, Schewe S, Schulze-Koops H, et al. Feasibility and reproducibility of the PsAMRIS-H score for psoriatic arthritis in low field strength dedicated extremity magnetic resonance imaging. Scan J Rheumatol. 2013; 42:379–82.

39. Hetland ML, Stengaard-Pedersen K, Junker P, Ostergaard M, Ejbjerg BJ, Jacobsen S, et al. Radiographic progression and remission rates in early rheumatoid arthritis e MRI bone edema and anti-CCP predicted radiographic progression in the 5-year extension of the double-blind randomized CIMESTRA trial. Ann Rheum Dis. 2010; 69:1789–95.

40. Conaghan PG, Peterfy C, Olech E, Kaine J, Ridley D, Dicarlo J, et al. The effects of Tocilizumab in osteitis, synovitis and erosion progression in rheumatoid arthritis: results from the ACT-RAY MRI substudy. Ann Rheum Dis. 2014; 73:810–16.

41. Axelsen MB, Eshed I, Horslev-Petersen K, Stengaard-Pedersen K, Hetland ML, Moller J, et al. A treat-to-target strategy with methotrexate and intra-articular triamcinolone with or without adalimumab effectively reduces MRI synovitis, osteitis and tenosynovitis and halts structural damage progression in early rheumatoid arthritis: results from the OPERA randomized controlled trial. Ann Rheum Dis. 2015; 74:867–75.

42. Ostergaard M, Peterfy C, Conaghan P, McQueen F, Bird P, Ejbjerg B, et al. OMERACT rheumatoid arthritis magnetic resonance imaging studies. Core set of MRI acquisitions, joint pathology definitions and the OMERACT RA-MRI scoring system. J Rheumatol. 2003; 30:1385–86.

43. McQueen F, Lassere M, Edmonds J, Conaghan P, Peterfy C, Bird P, et al. OMERACT Rheumatoid arthritis magnetic resonance imaging studies. Summary of OMERACT 6 MR imaging module. J Rheumatol 2003; 30:1387–92.

44. Ostergaard M, McQueen F, Wiell C, Bird P, Boyesen P, Ejbjerg B, et al. The OMERACT psoriatic arthritis magnetic resonance imaging scoring system (PsAMRIS): definitions of key pathologies, suggested MRI sequences and preliminary scoring system for PsA hands. J Rheumatol. 2009; 36:1816–24.

45. Haavardsholm EA, Ostergaard M, Ejbjerg BJ, Kvan NP, Uhlig TA, Lilleas FG, et al. Reliability and sensitivity to change of the OMERACT rheumatoid arthritis magnetic resonance imaging score in a multireader, longitudinal setting. Arthritis Rheum 2005; 52:3860–67.

46. Haavardsholm EA, Ostergaard M, Ejbjerg BJ, Kvan NP, Kvien TK. Introduction of a novel magnetic resonance imaging tenosynovitis score for rheumatoid arthritis: reliability in a multireader longitudinal study. Ann Rheum Dis 2007; 66:1216–20.

locus associated with RA. HLA-DRB1*01 and HLADRB1*04 alleles containing QKRAA, QRRAA and RRRAA amino acid motifs at third hypervariable region of DRβ-chains (amino acid 70 through 74) called shared epitope (SE), confer an increased risk. RAA sequence at position 72 to 74 of shared epitope is the key component, predisposing to increased susceptibility and it is modulated by amino acids at position 70 and 71.[5] Certain HLA-DR04 alleles are associated with increased risk and severity of RA (DRB1*04:01, *04:02, *04:04, *04:05, *04:08), whereas DRB1*04:03, *04:06 and *04:07 are low risk alleles. Others like DRB1*1301, *1302 and *1304 alleles linked to DERAA motif confer protection against development of RA probably by affecting the peptide binding properties of the MHC molecule.

There are significant ethnic differences in MHC Class II association with RA. Amongst the Japanese, DRB1*04:05 of DRB1*04 allele is most prevalent. RA in the native American populations is associated with DRB1*14:02. DRB1*04:05 and *04:04 frequency is increased among Chinese patients. Within India, susceptibility genes differ in different regions. In north and west India, HLA-BRB1*04 was associated with increased susceptibility while DQB1*03:01-DRB1*04:03 conferred protection against RA. Whereas, in South Indian Tamils, DRB1*10 was the major risk allele while DRB1*13 and DRB1*14 were protective.[6] Among Indian migrants to South Africa, HLA-DRB1*10 was found to be the major susceptibility allele.[7] Thus, different HLA-DR alleles containing the same SE appear to have different effect in different ethnic populations, suggesting the role of genes and amino acids outside SE region in disease pathogenesis.

HLA and Disease Phenotype

There is an association of SE with extra-articular disease manifestations. Individuals with SEs contained in DR4+ alleles (HLA-DRB1*04:01, *04:04, *04:08) have the highest risk of developing extra-articular disease including rheumatoid vasculitis, rheumatoid lung disease, and Felty's syndrome. Among other HLA class genes, HLA-G is shown to be a disease modifier, modifying the clinical phenotype and autoantibody production.

HLA Genes and Disease Prognosis

Presence of SE has been linked to ACPA positivity, which at disease onset is the strongest predictor of later development of radiographically demonstrable damage to joints. Association of SE with erosive disease is now well-established. SE alleles are also associated with increased risk of cardiovascular disease and mortality. Presence of two copies of SE alleles is associated with premature mortality and increased cardiovascular mortality in various studies. This effect appears to be independent of ACCP status.

Non-MHC Genes Associated with RA

Various non-MHC susceptibility genes in RA have been studied by either investigating candidate genes whose functions suggest a possible role in disease pathogenesis, or by screening the entire human genome using genome-wide association studies. Candidate gene association studies have shown that polymorphisms of *PTPN22* gene in Caucasians, *PADI4* (peptidyl arginine deiminase type 4) in Asians and *CTLA4, STAT4* are associated with risk of RA. Some Indian studies have reported an association of *IRF5* and *NKG2D* polymorphism with increased risk of developing RA and with disease severity. In north Indian population, the risk of RA was high with *KIRDS1* and *KIRDS2* polymorphism whereas *KIRDL1, KIRDL2* and *KIRDL3* protected against the disease. Extra-articular manifestations were common in patients with *KIRDS1* and *KIRDS3*. In recent genome-wide association studies, new genetic risk factors (*TRAF1-C5, ARL15*) have been described for RA.

Axial Spondyloarthritis (SpA)

Worldwide prevalence of SpA varies in different populations. High prevalence seen in Whites parallels a high prevalence of HLA-B27 in this population. Amongst genetic factors, HLA-B27 is known to have the strongest association with the disease. HLA-B27 accounts for approximately 25% of disease heritability, and is, thus thought to have a role in disease pathogenesis.

Pathogenic mechanisms of HLA-B27 associated spondyloarthritis

a. Arthritogenic peptide hypothesis based on molecular mimicry mechanism. It is postulated that presentation of certain microbial peptides by HLA-B27 to CD8⁺T cells leads to reactivity of lymphocytes against self-antigens.

b. *HLA-B27 misfolding hypothesis.* Improperly folded HLA-B27 molecules accumulate in endoplasmic reticulum (ER) causing ER stress and subsequent inflammatory response with activation of autophagy and IL-23/IL-17 pathway.

c. *Heavy chain homodimer hypothesis.* HLA-B27 heavy chains can form stable dimers without beta-2-microglobulin, such dimers are thought to engage receptors on NK and T cells, and activate them.

The latest WHO-ILAR COPCORD surveys (2009, 2015) report the prevalence of SpA in India to be 0.3%. Prevalence of HLA-B27 in one study on 1,170 unrelated healthy individuals in western India was 9.6% and HLA-B27 frequency was 90% in Indian SpA patients similar to that in Caucasians.[8] HLA-B27 subtyping identified B*2702 (1.43%), B*2704 (14.29%), B*2705 (70%), B*2707 (12.86%) and B*2718 (1.43%), respectively. Different HLA-B27 subtypes are associated with susceptibility to disease in different populations -B*2705 in Caucasians, B*2704 in Chinese and Japanese, whereas B*2706 and B*2709 are not associated with the disease.

Non-HLA Genes

Various GWAS studies in European and Chinese populations have identified at least 48 genome-wide risk loci. Endoplasmic reticulum aminopeptidases *ERAP1* and *ERAP2*, genes involved in the IL-23/IL-17 pathway, TNF receptor gene family (*lymphotoxin beta-LTBR and TNFRSF1A*) are also seen to be playing an important role in disease pathogenesis.

Table 16.2: Association of HLA class II alleles with antibody subsets in patients with SLE

HLA class II allele	*Associated antibodies*
HLA-DR2/DX	Anti –Sm
HLA-DR3/DX, DQ2	Anti-Ro and Anti-La
HLA-DR2/DR3	Anti-Ro, Anti-La, Anti-Sm, Anti-dsDNA
HLA-DR3/DR3	Anti-Sm
HLA-DR4/DQ8	Anti-cardiolipin, lupus anticoagulant, anti-β2 glycoprotein I
HLA-DR4	Anti-Sm, anti-RNP
HLA- DR5	Absence of anti-Sm and anti-RNP

Modified from: Kelley's Textbook of Rheumatology, 9th edition

Genetics of Systemic Lupus Erythematosus (SLE)

The prevalence of lupus in India is reported to be 0.01–0.03% in general population.[9, 10] MHC class II and class III are implicated in disease pathogenesis. Amongst HLA alleles, HLADRB1*03 and HLADRB1*15 are reported to be associated with SLE. There is strong association between HLA class II alleles, specifically HLA-DR and DQ, with different antibody subsets in SLE patients are summarized in Table 16.2.

Fifty different polymorphic loci have been implicated by genome-wide association studies till date. Amongst non-MHC genes, the highest risk is conferred by the deficiency of classic complement pathway components C1q, C4A, B, and C2. Non-MHC susceptibility genes found to be involved in pathogenesis of SLE are given in Table 16.3.

Table 16.3: Non-MHC susceptibility genes involved in pathogenesis of SLE

Genes related	*Genes*	*Mechanism*
Innate immune system	*IRAK1, STAT 4, IRF5, TNFAIP3, TLR7*	Most of these genes are associated with increased IFN α production.
Adaptive immune system T cell signalling	*IRAK1, STAT4, HLA-DR, PTPN22, PDCD1*	PTPN22 is associated with persistent activation of auto-reactive T cells.
Adaptive immune system B cell signalling	*BANK1, BLK, LYN*	Lead to auto-reactivity of B cells and persistent auto-antibody production.
Immune complex clearing	*C1q, C2, C4, ITGAM, FCRG2A, TREX1, CRP*	Defective clearance of apoptotic debris by CRP may predispose to auto-immunity. TREX 1 is the major intracellular DNAase that cleaves ssDNA and dsDNA. TREX1 gene polymorphism leads to inadequate clearance of macromolecules leading to autoimmunity.

Pertaining to Indian scenario, a few studies have reported genetic risk alleles associated with SLE. *CTLA4, TLR9* and *FcāR IIB* have been reported to be related to increased disease susceptibility and severity in South Indian Tamil population.

Juvenile Idiopathic Arthritis (JIA)

Most of the genetic predisposition to JIA is by MHC alleles (Table 16.4).

Non-MHC genes associated with JIA include *PTPN22, MIF, SLC11A6, WISP3, TNF, CCR5, IL-6-174G allele, NIF gene, IL-2/IL-21 and IL-2RA*, the chain of the IL-2 receptor.

GENETICS OF OTHER RHEUMATIC DISEASES

A number of immune related genes are implicated in the pathogenesis of a large number of auto-immune/inflammatory rheumatic disorders. They are summarized in Table 16.5.

Benefits of Genetics Studies in Rheumatic Diseases

Genetic testing, in general, is not recommended for diagnosis or prognostication of autoimmune rheumatic diseases, although it may provide certain critical clues in specific scenarios. More importantly, it provides information pertaining to ethnic

Table 16.4: MHC gene association in juvenile idiopathic arthritis (JIA)

HLA	*Disease group*	*Increased risk*	*Protective*
MHC class I	JIA-ERA	HLA-B27	
	Oligoarticular JIA	HLA-A*0201	
	Persistent oligoarticular	HLA-C*0202	
MHC class II	All types of JIA	DRB1*08, DQA1*0401-DQB1*0402	DRB1*0401, *0701 and *1501
	Oligoarticular JIA	DRB1*0801 and 11, DRB1*1301	
	Distinguishes persistent from extended oligo-articular JIA	DRB1*1301-DQA1*01, DQB1*06 haplotype	
	Rheumatoid factor (RF)-positive polyarticular JIA	HLA-DRB1*0401	
	Polyarticular	DRB1*0801 and DRB1*1401 DRB1*0101	
	Evolution from oligo-articular to polyarticular, erosive disease	DRB1*0101	
	Systemic onset JIA	DRB1*11-DQA1*05-DQB1*03	
	Predisposes to eye disease	DRB1*1104	

Table 16.5: Genetic association in other rheumatic disease

Disease	*MHC gene polymorphisms*	*Non-MHC genes*
Sjögren's syndrome	DRB1*0301and DRB1*1501	*IRF-5, TLR-4, 7, 9, STAT4, EBF1 TNFSF4*
Scleroderma	HLA DRB1*1104, DQA1*0501, and DQB1*0301	*PTPN22, NLRP1, IRF5, STAT4, BANK1, TNFSF4, T-bet*
Behçet's	HLA-B51	MICA6 allele; perth block (PERB); B (NOB); and TAP
Idiopathic inflammatory myositis	In whites: HLA-DRB1*0301 and HLA-DQA1*0501 In African Americans-HLA-DRB1*08	*TNF, TNFR1, IL-1, C4, C2* gene mutations
ANCA vasculitis	- anti-PR3 ANCA-HLA-DP - anti-MPO ANCA-HLA-DQ	anti-PR3 ANCA-SERPINA1, PRTN3

variation and disease pathogenesis, which may, in future, direct for the development of new drugs targeting these pathways.

Future Perspective

Newer advances are being made towards diagnosis of 'Pre-clinical RA', 'Pre-clinical SLE', 'serologically active but clinically quiescent disease', so as to identify and prevent/treat the disease at its preclinical stage itself. Early identification of pre-clinical disease requires thorough knowledge of genetic risk factors in a specific population to better understand disease pathogenesis and the individuals at risk for the disease. Using these strategies, it may be possible to develop novel therapeutics to prevent the disease in future.

Key Points

- Genetics plays an important role in disease pathogenesis of rheumatic diseases.
- Understanding of genetics may help in determining the target molecules for treatment.
- Recent advances in genetics, points to the importance of yet unknown risk factors which need to be identified- for disease prevention and treatment.

REFERENCES

1. Prahalad S, Ryan MH, Shear ES, Thompson SD, Glass DN, Giannini EH. Twins concordant for juvenile rheumatoid arthritis. Arthritis Rheum. 2000; 43:2611–2.
2. Deapen D, Escalante A, Weinrib L, Horwitz D, Bachman B, Roy-Burman P, et al. A revised estimate of twin concordance in systemic lupus erythematosus. Arthritis Rheum 1992; 35:311–8.
3. Mcgonagle D, Aydin SZ, Gül A, Mahr A. "MHC-I-opathy"—unified concept for spondyloarthritis and Behçet disease. Nat Rev Rheumatol 2015; 51:1–10.
4. MacGregor AJ, Snieder H, Rigby AS, Koskenvuo M, Kaprio J, Aho K, et al. Characterizing the quantitative genetic contribution to rheumatoid arthritis using data from twins. Arthritis Rheum. 2000; 43:30–7.
5. Raychaudhuri S, Sandor C, Stahl EA, Freudenberg J, Lee H-S, Jia X, et al. Five amino acids in three HLA proteins explain most of the association between MHC and seropositive rheumatoid arthritis. Nat Genet. 2012; 44:291–6.
6. Mariaselvam CM, Fortier C, Charron D, Krishnamoorthy R, Tamouza R, Negi VS. HLA class II alleles influence rheumatoid arthritis susceptibility and autoantibody status in South Indian Tamil population. HLA. 2016; 88:253–8.
7. Mody GM, Hammond MG. Differences in HLA-DR association with rheumatoid arthritis among migrant Indian communities in South Africa. Br J Rheumatol. 1994; 33:425–7.
8. Chhaya SU. HLA-B27 polymorphism in Mumbai, Western India. Tissue Antigens. 2005; 66:48–50.
9. Chopra A. Disease burden of rheumatic diseases in India: COPCORD perspective. Indian J Rheumatol. 2015; 10:70–7.
10. Malaviya AN, Singh RR, Singh YN, Kapoor SK, Kumar A. Prevalence of systemic lupus erythematosus in India. Lupus. 1993; 2:115–8.

FURTHER READING

1. Deng Y, Tsao BP. Genetic susceptibility to systemic lupus erythematosus in the genomic era. Nat Rev Rheumatol. 2010; 6:683–92.
2. Fernando MMA, Stevens CR, Walsh EC, De Jager PL, Goyette P, Plenge RM, et al. Defining the Role of the MHC in Autoimmunity: A Review and Pooled Analysis. Fisher EMC, editor. PLoS Genet. 2008; 4(4):e1000024.
3. Mcgonagle D, Aydin SZ, Gül A, Mahr A. "MHC-I-opathy"—unified concept for spondyloarthritis and Behçet disease. Nat Rev Rheumatol 2015; 51:1–10.

THERAPEUTICS IN RHEUMATIC DISEASES

Chapter

17

Patient Education in Rheumatic Diseases

KM Mahendranath, Anvesha M

1. What is Patient Education?

Clinical Practice is shifting from traditional paternalistic model to shared decision making.[1] This process involves an exchange of information in order to prepare patients to participate in taking treatment decisions with their physicians. This process is patient education and in rheumatology, it is as important as the treatment.

2. Why Patient Education is Important in Rheumatic Diseases?

To offer patient centered therapy (treatment) rather than disease centered. To improve patient care by getting an insight and to innovate. To achieve sustained remission by complete compliance with painfree, disability free and good quality of life.

Educating the patient with rheumatic disease should be a lifelong commitment for the rheumatologist, and the family.

Educating the patient is the process which should begin with the very first visit and followed through and updated with subsequent consultations. It is important to keep in check, the impact it has made on the patient as well as the family.

Unfortunately, nobody teaches us about patient education. It is acquired by years of experience, by keen observation of patient and family, by the adaptation and innovation of skills and by the ability to empathize.

3. Patient's Initial Response to the Disease or Illness

At the onset of the disease, the patient's emotional response to their afflictions may interfere with their understanding and acceptance of the disease. Feelings of bewilderment, despair, apprehension (of medical care and medications) and fear of disability often undermines patient care by fostering denial, apathy, non-compliance or reliance on unproven remedies.

4. Rheumatologist as an Educator

Early in the disease, the rheumatologist should educate the patient as to the certainty of the diagnosis, the short-term prognosis (what to expect in the next few months), the types of treatments, the benefits and risks of therapies and the need for ongoing professional care and evaluation.

Rheumatologist should encourage the involvement of other members of the family during the consultations and reinforce the therapeutic plan and transfer of information. This involvement will initiate the formation of an emotional and educational support network. More than communication skills, doctor's time is needed because we spend very little time with patients. The rheumatologists should encourage the involvement of other members of the family and reinforce the therapeutic plan and transfer of information.

This involvement will initiate the information of an emotional and educational support network. Each interaction with the patient and the family is an opportunity to further educate them. The patient and the family should be directed to the educational material available from libraries, bookstores, arthritis foundations, reliable internet sources and patient support groups.

5. What to Educate?

Education is a continuous and a lifelong process like a Continuing Medical Education (CME) program and it is a lifelong commitment.

Questions most patients and relatives most commonly ask is:

a. *Is it curable?*

No. But it is treatable and the patient can lead a near normal life if diagnosed and treated early.

b. *Is it life threatening?*

Though living pain free and a good quality of life might not be their immediate concern but one of the major concerns of the patient and the family is about the disease and its impact like deformities, affliction of other organs like heart, kidney, etc.

They have to be made aware that the disease will not kill directly but the involvement of organs like the cardiovascular system, bones, etc. may impair the quality of life and shorten the lifespan of the patient.

c. *Will it affect others in the family?*

They want to know whether others can be affected by the contact or sharing things. The would also worry about the side-effects of the medicines particularly non-steroidal anti-inflammatory drugs (NSAIDs), disease modifying anti-rheumatic drugs (DMARDs) and steroids.

d. *About special issues like Conceiving/ Pregnancy/ Motherhood?*

Pregnancy is not contraindicated in most of the rheumatic diseases. Some of them even go into remission during pregnancy. Pregnancy can be complicated if the disease is not well controlled and they should be aware of the medications to be avoided during conception and breastfeeding. It should be a team effort of the rheumatologist and obstetrician who has experience in high-risk pregnancies.

6. Modalities of Education

a. One on one/person to person education: It can be considered the most crucial form of education. Other forms being important, it ultimately comes down to the rheumatologist spending time with the patient to explain the disease and the treatment, answer questions and encourage compliance.
b. Printed and electronic media: Access to printed media in the form of pamphlets, brochures and books should be given to patients to help them understand the disease and its management.
c. Audio-visual media in the form of CDs form the most convenient mode of patient education, at least for the urban population.
d. The need for appropriate exercises, rest and diet need to be stressed in any education material.
e. Patient groups: Interaction among patients from various backgrounds to share their experiences has proven to be useful, as it might reduce anxiety and fear among other patients and encourage compliance.
f. Public forums: Public forums like Rotary or Lions Clubs, schools, professional organizations have played an important role in educating people.

7. How to Measure your Efforts to Educate are Succeeding?

There are a few indicators which will help in measuring the success of the rheumatologist's efforts. Most important pointers are:

a. Improvement in compliance
b. Regular follow-up
c. Good disease control
d. Patients stop shopping around for doctors or rheumatologists and trying unproven remedies

8. Difficulties we Face in India

a. Huge patient population and acute shortage of rheumatologists.
b. Very little teaching of musculoskeletal diseases at undergraduate as well as postgraduate levels.
c. Non-existence of rheumatology services even in major towns.
d. Total absence of rheumatology services in rural areas.
e. Failure of the health planners and educators to understand the importance of disease burden of musculoskeletal problems and its impact on the society.
f. Shortage of physiotherapists, occupational therapists and non-existence of counselors in rheumatic diseases.

The biggest problem is the huge patient population and the acute shortage of rheumatologists. We need to reach the patients at peripheral levels. Some rheumatology services should be available to everyone. The only way is to increase the number of rheumatologists, which is difficult. Some good basic rheumatology training, both in theory and in practice to manage day-to-day rheumatological problems should be compulsory at the final year of undergraduation and during postgraduation. Patient education should be part of the curriculum to doctors at every level.

9. Other Aspects of Patient Education

a. Educate, empathise, empower

b. Importance of nutrition

   Some patients do have a very sensitive stomach. Year after year use of NSAIDs and medicines can make their stomach wall lining very sensitive. In spite of using gastro-protective drugs they can still have severe gastritis or stomach upsets, when they consume very spicy foods and chillies. Some advice on avoiding hot and spicy foods can help them in dealing with severe gastritis or bleeding. Some patients who have co-morbidities like diabetes should be advised to avoid high carbohydrate foods and not to gain weight and also make more honest efforts in reducing obesity.

   A balanced nutritious diet with adequate proteins, carbohydrates, fats and minerals and vitamins supplements whenever needed must be advised.

c. Importance of lifestyle modification: Patient should be encouraged to stop smoking, reduce alcohol intake, and to lose weight.

d. Importance of exercise.

10. Conclusion

Patient education provides insight to the patient and family and inspires them to take control of the disease and effectively manage it. It also helps the rheumatologists to innovate and improve their strategies to handle difficult problems.

REFERENCE

1. Zong JY, Leese J, Klemm A, Sayre EC, Memetovic J, Esdaile JM, et al. Rheumatologists' Views and Perceived Barriers to Using Patient Decision Aids in Clinical Practice. Arthritis Care Res (Hoboken). 2015; 67:1463–70.

Chapter

18

Non-pharmacological Therapy in Rheumatic Diseases

Shweta Singhai, Abhra Chandra Chowdhury, BG Dharmanand

INTRODUCTION

Despite advances in pharmacological therapy for rheumatic diseases, many patients continue to experience some measure of ongoing disease activity. Most rheumatic diseases are chronic; hence have a huge impact on the quality of life. Patients have to make long-term changes in their lifestyle. They have to adapt and accustom to living with chronic illness. Non-pharmacological interventions contribute to better compliance and thus help in disease control. A comprehensive management program includes

a. Patient education
b. Psychosocial interventions
c. Lifestyle modification
d. Nutritional and dietary intervention

A. PATIENT EDUCATION AND COUNSELING

In the modern era, its easy for the patient to get intimidated by the 'internet' information about the disease manifestations as well as the side effects of the pharmacological therapy. As most of the rheumatic diseases are life long, counseling plays an important role in acceptance of the disease by the patient, compliance and regular follow-up. Many patients have misconceptions about the nature of the disease and its cause, communicability of disease, effects on married life and childbearing. Correcting these misconceptions helps in establishing good long-term relationship between the clinician and the patient.

Patient education can be provided by the physician or a trained educator. Inputs from the physiotherapist, clinical psychologist, nutritionist and nurse educator help in overall management. The general theme of education could include:

1. Explaining the nature of the illness and how it occurs.
2. Treatment options available with pros and cons of each treatment.
3. Importance of blood tests and why some of the blood tests need to be done periodically during the treatment.
4. Importance of compliance with treatment.
5. Role of diet, exercise and fitness.
6. Education of the spouse or the caregiver is equally important.

Patient education, though not proven to reduce disease activity or pain' can help patients and their family to understand and cope with the difficult rheumatic diseases.[1]

Self Management Strategies

These help the patients to be better prepared during their consultations and also in day-to-day management of their diseases.

- Patients should write a list of the top three issues they wish to discuss before each visit.
- Remain open-minded and supportive about use of unproven, but non-harmful complementary and alternative medicines (CAM) therapies. At a minimum, they may evoke a placebo response and be conducive to maintaining a therapeutic alliance.

- Realistic expectation regarding drugs.
- Reassure about non-progressive nature of diseases such as fibromyalgia and stress potential serious and unpredictive nature of connective tissue diseases.
- Teach about minor flare management like using local heat or cold, analgesics and NSAIDs.
- Instruct about warning signs and major flares and what to do when they occur.

B. PSYCHOSOCIAL INTERVENTIONS

The chronic, unpredictable course of rheumatic diseases brings out a variety of emotions in these patients including anxiety, guilt, helplessness and depression. Contributing factors in depression are socio-economic status, social support, educational status and intrinsic factors of patients themselves. Psychosocial interventions improve coping and reduce psychological distress.

These can be divided into:

1. Educational programs
2. Coping skills training (CST)
3. Cognitive behavioral therapy (CBT)

1. Educational programs are included in above section on patient education and counseling.
2. Coping skills training (CST) enables the patient to face the pain and disability and thus helps to reduce the distress because of the disease. Coping can be defined as the process of responding and managing or contending with life stresses and difficulties arising out of stressful situations, which are beyond the control of individual patient. Coping strategies are:
   a. *Active coping:* Problem-based coping including learning and enacting behaviors designed to face the problems.
   b. *Passive coping:* Avoiding situations or giving up situations.
   c. *Catastrophizing:* Exaggerating the perceived threat and focusing on the worst that can happen.

   Identifying passive coping, catastrophizing early and addressing them will help the patients to deal with their illness more efficiently.
3. Cognitive behavioral therapy (CBT) was originally developed to target mood issues in adults but has been adapted to treat individuals with chronic pain.[1] CBT is a structured self-management intervention that includes
   - cognitive (distraction, guided imagery, cognitive restructuring) and
   - behavioral (activity pacing, pleasant activities, relaxation training) strategies

   Coping skills are taught by clinician psychologist instruction, guided practice in either individual or group based sessions (usually 8–12 sessions).

   Coping skills learnt in CBT, for example, relaxation strategies, may also be taught using biofeedback,[1] which uses computer assisted instruction in reducing sympathetic arousal, for example, reducing muscle tension or increasing peripheral body temperature, with the purpose of increasing relaxation and reducing pain.

C. LIFESTYLE MODIFICATION

This involves altering long-term habits, typically of eating or physical activity and maintaining the new behavior for months to years. These may include:

1. **Rest:** Since fatigue is a frequent complaint, the performance of day-to-day activities becomes difficult. Resting an inflamed joint as well as the entire body by taking a nap, may therefore be beneficial. These frequent rest periods can alternate with exercise.
2. **Exercise:** Pain and stiffness often leads patients to avoid using affected joints. The lack of use can result in loss of joint motion, contractures and muscle atrophy, thus decreasing the joint function.

   Range-of-motion exercises help preserve joint function. Exercises to increase muscle strength (such as isometric, isotonic, isokinetic) performed even once or twice a week improve function and do not worsen disease activity.

   Regular aerobic exercises (such as walking, swimming, cycling and supervised cardio-respiratory aerobic conditioning) improves muscle function, joint stability, aerobic capacity and physical performance over short term and can result in overall pain control. Physical activity has also been shown, in a systematic review and meta-analysis, to decrease the level of fatigue in RA patients. Preliminary evidence suggests aerobic weight-bearing exercise may help prevent glucocorticoid associated osteoporosis in RA.

Exercise programs should be prescribed by a physical therapist and tailor-made for each patient according to her disease severity, body build and previous activity level.

Physiotherapy and occupational therapy are discussed in detail elsewhere in this book.

3. **Weight reduction:** Obesity (BMI > 30 kg/m^2) is a significant health problem globally. Earlier thought to be a simple fat store, it is now recognized that adipose tissue and adipocytes are metabolically active and contribute to systemic inflammatory processes.

   Adipocytes release proinflammatory adipokines such as leptin, resistin and visfatin. Leptin produces pro-inflammatory cytokines such as tumor necrosis factor (TNF), interleukins (IL) such as IL-1β, and IL-6 increases the endothelial expression of adhesion molecules such as intercellular adhesion molecule (ICAM), vascular cell adhesion molecule-1 (VCAM1), monocyte chemoattractant protein-1(MCP1). Leptin increases proliferation of Th1 cells, and inhibits proliferation of Th2 and Treg cells.

   Resistin is produced by mononuclear cells and adipocytes, increases production of TNF, IL-6, and IL-1β and increases the production of adhesion molecules.

   Visfatin, produced by lymphocytes and adipocytes, has similar pro-inflammatory effects including induction of IL-8, IL-6, IL-1β and TNF, increased endothelial expression of ICAM-1, VCAM-1 and matrix metalloproteinases (MMPs), and enhancement of B cell differentiation. IL-6 enters the systemic circulation and increases C-reactive protein (CRP) and serum amyloid A production by the liver.

   Adipocytes also produce anti-inflammatory adiponectin, which inhibits TNF induced adhesion molecule expression and NF-κB (nuclear factor-κB). It reduces production of TNF and IL-6 and increases production of anti-inflammatory cytokines IL-10 and IL-1RA. It also increases number of Tregs.

   The net result of obesity is thus an inflammatory state (Fig. 18.1).[2] The obese patient should be encouraged to lose weight, as even mild excess weight increases the stress upon joints involved with synovitis, potentially hastening joint destruction. Referral to a dietician can assist in weight reduction.

4. **Cardiovascular risk reduction:** As chronic rheumatic diseases such as RA are associated with increased risk of coronary atherosclerosis, efforts should be made to modify risk factors such as cigarette smoking, alcohol intake, dyslipidemia, hypertension and sedentary life-style.

D. NUTRITIONAL AND DIETARY INTERVENTION

Nutrition plays a role in the management of most chronic diseases like diabetes. Although dietary counseling is a part of management of gout, the role of nutrition in other rheumatic diseases such as RA is less well accepted. Despite a widespread lack of conviction among rheumatologists about the role of nutrition, many patients believe food plays an important role in their symptom severity and more than 50% would have tried some dietary manipulation. A balanced diet involving less carbohydrates, more of natural vegetables and fruits, adequate proteins and healthy fats is advocated to all patients.

NON-PHARMACOLOGICAL THERAPY IN RA

Complementary and Alternative Medicine (CAM) Therapies

It is important to advise patients not to replace conventional medical therapy with unproven CAM therapy. CAM may be added to treatment regimens to control pain and inflammation, improve physical function or to cope with the disease or side effects of treatment. This includes:

a. *Mind body therapies:* Meditation, relaxation and Taichi are most common mind body therapies used. A recent preliminary study found that a type of meditation called "mindfulness based stress reduction" had positive effects on psychological aspects of RA (e.g. depression and coping ability) but did not improve symptoms.

   Taichi which involves specific movements, postures and breathing techniques.[3] Performance of Taichi show improvement in mood, quality

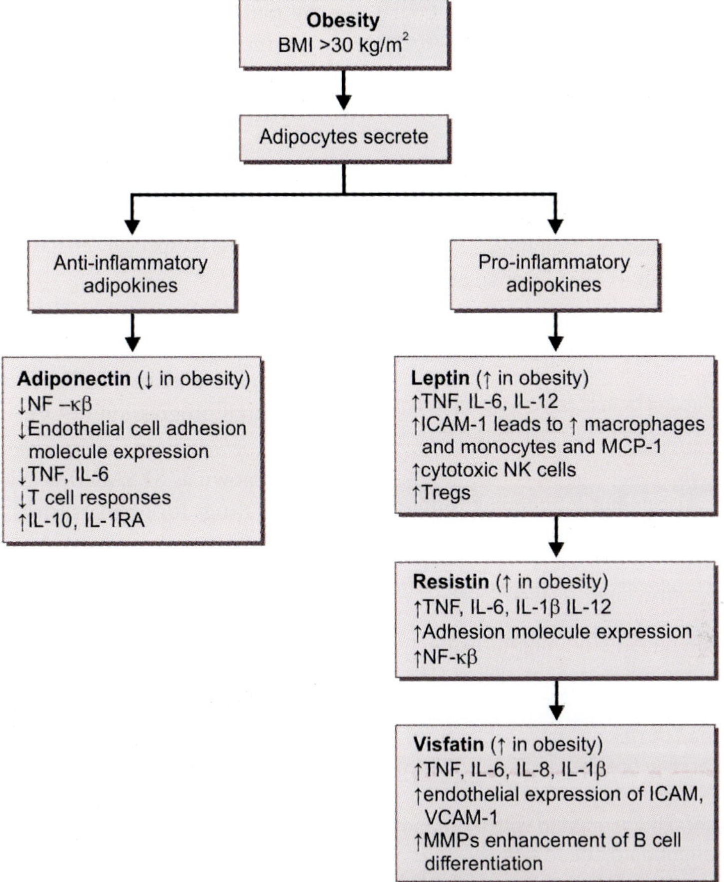

Fig. 18.1: Obesity and inflammation

of life and overall physical function, but is not effective for joint pain, swelling and tenderness.

b. *Role of diet and Dietary Supplements in RA:*

   i. Fish oil contains high amounts of omega-3-fatty acids which are anti-inflammatory. Types of fish high in omega-3-fatty acids include herring, mackerel, salmon and tuna.

      A recent study compared high dose fish oil to low dose fish oil, 5.5 or 0.4 gm/day respectively, of omega-3-fatty acid, in DMARD naïve patients with early RA.[4] Fish oil was used in combination with a treat to target predetermined intensive combination DMARD protocol. During 12 months of treatment, patients on high dose fish oil were more likely to achieve ACR remission (HR, 2.17; 95% CI, 1.07 to 4.42; P = 0.03) and triple therapy with methotrexate, sulfasalazine and hydroxychloroquine (HR, 0.28; 95% CI, 0.12 to 0.63; P = 0.002) was less likely to fail.[4] Table 18.1 summarizes the advantages and role of fish oils in RA.

   ii. Gamma linolenic acid (GLA) is an omega-6-fatty acid found in plant seeds. It is converted to anti-inflammatory substances in the body. Studies have shown that there was reduction in symptoms such as joint pain, stiffness and tenderness and decreased NSAID usage.

   iii. Turmeric (*Curcuma longa*)

   iv. Research on other supplements like thunder god wine, boswellia (*Boswellia serrata*), ginger (*Zingiber officinale*), green tea is still in initial stages.

   v. Other CAM therapies include acupuncture and balneotherapy (treatment with hot water baths). Table 18.2 summarizes the role of diet in the management of RA.

Table 18.1: Potential benefits of fish oil supplements in rheumatoid arthritis

Omega 3 fatty acid (fish oil supplement)
• Decreases NSAID requirement • Decreases cardiovascular risk • Decreases risk of RA • Large doses required • Delay in onset of action by 3 months • Bottle fish oil better than 10–15 capsules • Best taken with food

NSAIDs, Non-steroidal anti-inflammatory drugs; RA, Rheumatoid arthritis;
Source: Kelley and Firestein's Textbook of Rheumatology, 10th ed.

Table 18.2: Role of diet in rheumatoid arthritis

Role of diet in RA
• Alcohol may decrease risk of RA (especially ACPA positive RA) Limit alcohol in patients on methotrexate • Red meat consumption increases risk of RA • Coffee increases RA risk Tea decreases RA risk • Vitamin D may or may not decrease risk of developing RA • Low Vitamin C intake associated with increased risk of RA; no effect on disease activity • Vitamin E has no association with onset of RA; no effect on disease activity • Obesity increases risk of developing RA • Fasting, vegetarian/vegan and elimination diets are difficult to sustain, and its difficult to predict which patients will respond

RA, Rheumatoid arthritis; ACPA, Anti-cyclic citrullinated peptide antibody;
Source: Kelley and Firestein's Textbook of Rheumatology, 10th ed.

NON-PHARMACOLOGICAL THERAPY IN OSTEOARTHRITIS (OA)

A. Exercise and Physical Therapy

For knee and hip OA, trials have shown consistently that exercise lessens pain and improves physical function.

- Exercise regimens are aerobic or resistance training (focusing on strengthening muscles around joint).
- Low impact exercises, including water aerobics and water resistance training are better tolerated by patients than impact training exercise like treadmill or running.
- Taichi may be effective for knee OA.[3]
- Physicians should reinforce the necessity for exercise at each visit and help the patient recognize the barriers to ongoing exercise.
- Combination of exercise with calorie reduction and weight loss is especially effective in reducing pain.

B. Role of Diet

- Studies suggest that n-3 fatty acid supplementation in low dose may be beneficial in preventing structural progression in OA.[5]
- *GAGs (glucosamine sulfate, chondroitin sulfate):* Also known as SYSADOAs (Symptomatic slow acting drugs for osteoarthritis)

  A recent study aimed to investigate the effectiveness and safety of glucosamine, chondroitin, the two in combination, or celecoxib in the treatment of knee OA. Pubmed, Embase and Cochrane library were searched from inception to February 2015. A total of 54 studies covering 16427 patients were included. Celecoxib (back transformed difference was –0.59 units, 95% CI was –0.74 units to –0.46 units), glucosamine (–0.40 units, –0.61 units to –0.19 units), and the combination of glucosamine and chondroitin (–0.48 units, –0.80 units to –0.17 units), all showed a significantly better effect on function improvement compared with the placebo group. Combination of glucosamine and chondroitin was second best treatment (72%), celecoxib was the best (92%) and placebo ranked last (1%).[6]
- *Hyaluronic acid (HA):* Recent analysis of 29 studies (n = 4866) of intra-articular (IA) HA injections versus placebo was done. Compared to saline controls, standardized mean difference (SMD) with IA HA was maintained at 0.38 for knee pain and 0.32 for knee function at weeks 14–26 d ($p < 0.001$) which equates to true but moderate effect. A therapeutic trajectory of IA HA versus placebo found that IA HA is efficacious by 4 weeks (ES:0.31; 95% CI:0.17–0.45), reaching a peak in effectiveness at 8 weeks (ES:0.46, 95% CI:0.28–0.65) and with a residual detectable effect for knee OA pain at 6 months post-intervention (ES:0.21,95% CI: 0.10–0.31). In addition, IA HA induces longer lasting pain control compared with IA corticosteroids.[7]

- Methionine with active form—S adenosyl methionine (SAMe), has antioxidant properties and prevents degradation and promotes anabolic processes of cartilage. Found to be equally effective as NSAID.[5]
- *Curcumin (an extract from turmeric):* In a recent study, literature searches were conducted using 12 electronic databases, including PubMed, Embase, Cochrane Library, Korean databases, Chinese medical databases, and Indian scientific database.[8] A pain visual analogue score (PVAS) and Western Ontario and McMaster Universities Osteoarthritis Index (WOMAC) were used for the major outcomes of arthritis. Initial searches yielded 29 articles, of which 8 met specific selection criteria. Three among the included RCTs reported reduction of PVAS (mean difference: –2.04 [–2.85, –1.24]) with turmeric/curcumin in comparison with placebo ($p < 0.001$), whereas meta-analysis of four studies showed a decrease of WOMAC with turmeric/curcumin treatment (mean difference: –15.36 [–26.9, –3.77]; $p < 0.009$). Furthermore, there was no significant mean difference in PVAS between turmeric/curcumin and pain medicine in meta-analysis of five studies. Eight RCTs included in the review exhibited low to moderate risk of bias. In conclusion, these RCTs provide scientific evidence that supports the efficacy of turmeric extract (about 1000 mg/day of curcumin) in the treatment of arthritis. However, more rigorous and larger studies are needed to confirm the therapeutic efficacy of turmeric for arthritis.[8]
- Botanical extracts such as Avocado/soy unsaponifiable (an extract from avocado) and soyabean oleoresin (extracted from Boswellia serrata), phytoflavanoids, bioflavanoids and ginger require further research to better understand their role.[5]

C. CAM

Various randomized clinical trials reveal no benefit of TENS, therapeutic ultrasound, laser or needle acupuncture over education and exercise for knee OA.

NON-PHARMACOLOGICAL THERAPY IN ANKYLOSING SPONDYLOSIS

a. **Exercise:** A recent meta-analysis summarized the available scientific evidence on effectiveness of physiotherapy interventions in the management of AS.[9] Some recommendations are:
   - Regular exercise program to start immediately on diagnosis.
   - Personal home exercises are more effective than no exercise.
   - Intensity of exercise should be as per activity and stage of disease of patient.
   - Posture training while walking, sitting and sleep should be taught.
   - Swimming and hydrotherapy are most effective methods to achieve all physiotherapy targets.

b. **Role of diet:** Low starch diet is found to be beneficial in AS, however, further research is warranted.

NON-PHARMACOLOGICAL THERAPY IN GOUT

Link between gout and diet has been recognized for centuries. Sustained reduction of serum uric acid to <6 mg/dl, or <5 mg/dl if tophi are present, is critical for management of gout. This can be achieved by a combination of urate lowering therapy and dietary/lifestyle modification. It is advised to restrict intake of purine rich food and drinks which are thought to precipitate gout flare, e.g. meat, seafood, beer/wine. Such diet is also high in saturated fats and carbohydrates, and may add to the risk of metabolic syndrome, often associated with gout. The role of diet in gout is summarized in Table 18.3.

It has been suggested that cherries may prevent gout flares. In a study of 633 individuals, cherry intake during a two-day period was associated with a 35% lower risk of gout attack (OR, 0.65; 95%CI, 0.5 to 0.85). Furthermore, when cherry intake was combined with allopurinol treatment, risk of gout was 75% lower than during periods without either. (OR, 0.25; 95%CI, 0.15 to 0.42).[10]

In obese patients, weight loss may reduce serum uric acid. In one study, 13 males with gout lost weight during a 16-week period. Their pre-study BMI was 30.5 ± 8.1 kg/m^2 while the post-study BMI was 27.8 ± 7.9 kg/m^2. Serum uric acid decreased in all participants from a baseline of 9.6 ± 1.7 mg/dl to 7.9 ± 1.5 mg/dl, $p = 0.001$. Also a decrease in frequency of gouty attacks was seen from 2.1 ± 0.8 attacks per month to 0.6 ± 0.7 attacks per month, $p = 0.002$.

Table 18.3: Dietary modification in management of gout

Increases risk of gout	*Decreases risk of gout*
• Alcohol consumption other than moderate wine increases risk of gout, with beer conferring higher risk than liquor	• All vegetarian food is recommended
• Shell fish increases serum uric acid levels	• Low fat dairy decreases risk
• Red meat especially organ meat	• Cherries and vitamin D supplementation decrease serum urate levels
• Fructose, found in corn syrup, sugar sweetened soft drinks and fruit juices	• Coffee decreases risk
• Increased BMI >30 kg/m^2	• High intake of water decrease risk

BMI: Body mass index
Source: Kelley and Firestein's Textbook of Rheumatology, 10th ed.

NONPHARMACOLOGICAL THERAPY FOR FIBROMYALGIA (FM)

Mind and body therapy

- CBT plays an important role in management of FM and has been discussed above.
- Aerobic exercise 2–3 times a week shows a sustained effect on patients at 1–2 years of treatment.[1] Aquatic exercise (in waist deep water) is found to improve pain, stiffness, muscle strength as compared to no exercise.[11]
- "Qigong" refers to cultivation of "qi" (life energy). The goal is to improve the flow of "qi" throughout the body. Daily self-practice matters significantly and can lead to long-term benefits (>6 months).[12]
- A meta-analysis in 2014 of seven studies evaluating taichi for fibromyalgia showed improvement in symptoms of pain and sleep quality.[13]
- *Massage therapy:* In a 2014 systemic review and meta-analysis, Li *et al* included randomized control trials conducted after 2006; the authors examined fatigue, anxiety, depression and sleep disturbance.[14] They concluded that massage therapy of more than 5 weeks showed significant improvements in pain, anxiety and depression. Myofascial release techniques were most beneficial.
- *Acupuncture:* A recent Cochrane review showed a low to moderate level evidence that acupuncture improves pain and stiffness in fibromyalgia.[15]

Conclusion

Rheumatic diseases are typically chronic, and hence diet and nutrition play an important role in optimization of long-term health. Although patients as individuals can make their own individual choices, it is beneficial for the treating rheumatologist to be adequately informed regarding the evidence for frequently considered non-pharmacological therapies. A convincing, informed advice plays an important role in management of rheumatic diseases and helps patients avoid interventions that are worthless, time consuming, expensive and in some cases harmful and can distract the patient from effective measures.

Key Points

- Patient education and counseling lay the foundation of the long-term doctor–patient relationship.
- Psycho-social interventions enable the patient to face the pain and disability and enhance the day-to-day functioning.
- Obesity is a pro-inflammatory state. Patient should be encouraged to reduce weight.
- Due to an increased risk of coronary atherosclerosis and associated morbidity and mortality, efforts should be made to modify risk factors such as cigarette smoking.
- There is a definite role of diet in the management of gout.
- Omega-3-fatty-acid decreases the need for NSAIDs in RA.
- Low starch diet is said to play a role in ankylosing spondylitis.
- Cognitive behavioral therapy (CBT) and exercise play an important role in the management of fibromyalgia.
- Taichi and yoga can help to improve physical functioning, balance, and reduce pain in fibromyalgia, osteoarthritis, and rheumatoid arthritis.

REFERENCES

1. Cunningham NR, Zuck SK. Non-pharmacological treatment of pain in rheumatic diseases and other musculoskeletal pain conditions. Curr Rheumatol Rep. 2013; 15:306.
2. Stofkova A. Resistin and visfatin: regulators of insulin sensitivity, inflammation and immunity. Endocr Regul: 2010; 44:25–36.
3. Wang C. Tai chi and rheumatic diseases. Rheum Dis Clin North Am. 2011; 37:19–32.
4. Proudman S, James M, Spargo L. Reduction in cardiovascular risk factors with long term fish oil treatment in early rheumatoid arthritis. Ann RheumDis.2015; 74:89–95.
5. Castrogiovanni P, Trovato FM, Loreto C, Nsir H, Szychlinska MA, Musumeci G. Neutraceutical Supplements in the Management and Prevention of Osteoarthritis, Review. Int J of Mol Sci.2016; 17:2042.
6. Zeng C, Wei J, Li H, WangY, Xie D, Yang T, et al. Effectiveness and safety of Glucosamine, Chondroitin, the two in combination, or celecoxib in the treatment of osteoarthritis of the knee. Sci Rep. 2015; 5:16827.
7. Maheu E, Rannou F, Reginster JY. Efficacy and safety of hyaluronic acid in the management of Osteoarthritis: Evidence from real life setting trials and surveys. Semin Arthritis Rheum 2015; 528–33.
8. Daily JW, Yang M, Park S. Efficacy of turmeric extracts and curcumin for alleviating the symptoms of joint arthritis: a systemic review and meta-nalysis of randomized clinical trials. J Med Food. 2016; 19:717–29.
9. Pécourneau V, Degboé Y, Barnetche T, Cantagrel A, Constantin A, Ruyssen-Witrand A. Effectiveness of exercise programs in ankylosing spondylitis: A meta-analysis of randomized controlled trials. Arch Phys Med Rehabil. 2017 [epub ahead of print]
10. Zhang Y, Neogi T, Chen C. Cherry consumption and decreased risk of recurrent gout attacks. Arthritis Rheum 2012:64:4004–4011.
11. Bidonde J, Busch AJ, Webber SC, Schachter CL, Danyliw A, Overend TJ, et al. Aquatic exercise training for fibromyalgia. Cochrane Database Syst Rev. 2014; 2014 (1) CD0011336.
12. Sawynok J, Lynch M. Qigong and fibromyalgia: randomized controlled trials and beyond. Evid Based Complement Alternat Med 2014; 379715.
13. Raman G, Mudella S, Wang C. How effective is Taichi-min body therapy for fibromyalgia; a systematic review and metanalysis. J Altern Complement Med. 2014; 20: A66–6.
14. Li YH, Wang FY, Feng CQ, Yang XF, Sun YH. Massage therapy for fibromyalgia: a systematic review and meta-analysis of randomized controlled trials. PLoS ONE 2014; 9, e89304 [Internet] cited on May 2017. Available from http://journals.plos.org/plosone/article?id = 10.1371/journal.pone.0089304.
15. Deare JC, Zheng Z, Xue CC, Liu JP, Shang J, Scott SW, et al. Acupuncture for treating fibromyalgia. Cochrane Database Syst Rev. 2013; (5):CD007070.

FURTHER READING

1. Cunningham NR, Kashikar-Zuck S. Nonpharmacological treatment of pain in rheumatic diseases and other musculoskeletal pain conditions. Curr Rheumatol Rep. 2013; 15:306.
2. Uhlig T. Tai Chi and yoga as complementary therapies in rheumatologic conditions. Best Pract Res Clin Rheumatol. 2012; 26: 387–98.
3. Wang C. Role of Tai Chi in the treatment of rheumatologic diseases. Curr Rheumatol Rep. 2012;14: 598–603.

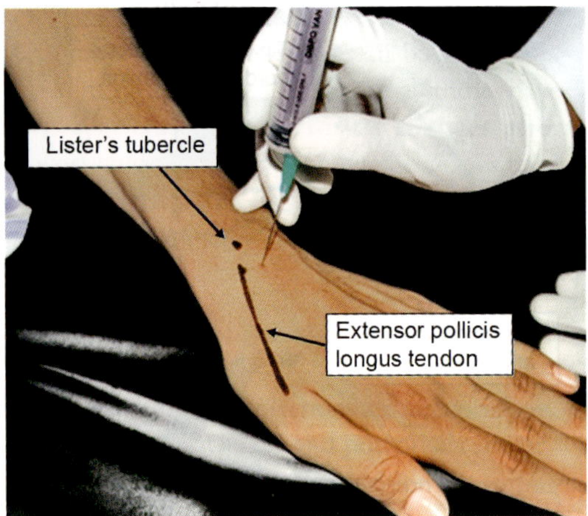

Fig. 19.4: Approach to wrist joint injection

the patella, thus facilitating multiple approaches for injection or synoviscocentesis. Anterior and posterior approaches will be discussed here. For either approach patient should be laying supine with knees extended and relaxed quadriceps. For the medial approach, the needle should be inserted just posterior to the medial portion of patella in slight posterior and inferior direction (Fig. 19.5). In the lateral approach to knee since there is no muscle bulk the needle is inserted laterally at the junction of the middle and upper thirds of patella in slight posterior and inferior direction (Fig. 19.6).[13]

Ankle Joint

The ankle joint is covered by many tendons and ligaments and there are nerves and arteries passing over the joint. The ankle joint can be entered through an anteromedial approach. Patient should be placed in supine position, the ankle should be kept flexed at an angle of 90 degrees to the leg. The space between tibia (medial malleolus) and talus should be palpated by gently moving the ankle to and fro (Fig. 19.7). The needle is then inserted just medial to the tibialis anterior tendon or medial to the tendon of the extensor hallucis longus (the neurovascular structures, i.e. deep peroneal nerve and dorsalis pedis vessels lie just lateral to the extensor hallucis longus) and lateral and distal to

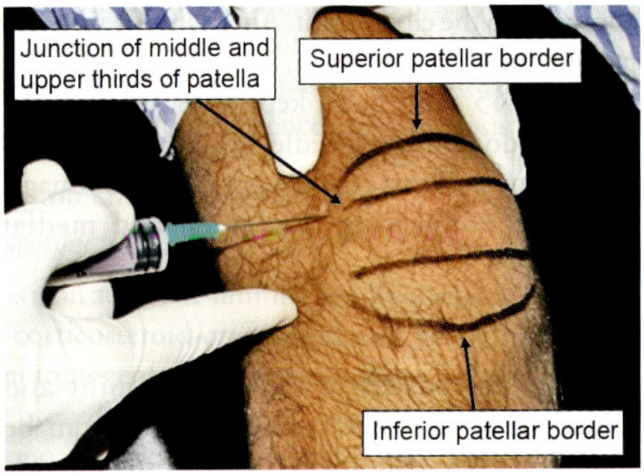

Fig. 19.5: Medial approach of injecting the knee joint

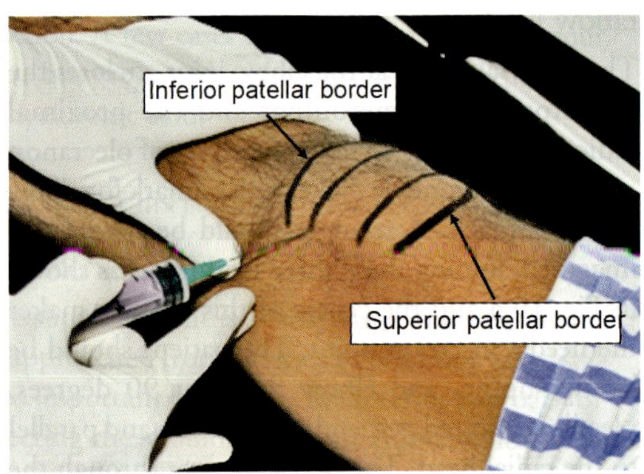

Fig. 19.6: Lateral approach of injecting the knee joint

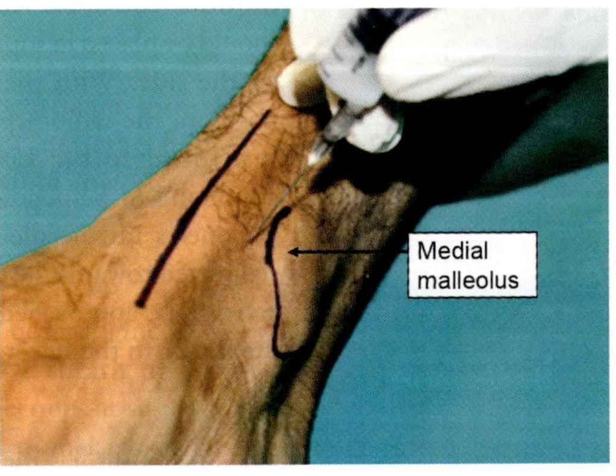

Fig. 19.7: Approach to ankle joint

the lower margin of medial malleolus, in posterior and lateral direction.[12,14]

Hip Joint

The hip joint is generally aspirated or injected under ultrasonographic (USG) guidance, using a long needle (usually a lumbar puncture needle), approaching the joint capsule from an anterolateral approach taking care to avoid the neuro-vascular bundle. Needle is inserted just inferior and anterior to greater trochanter and advanced along the femoral neck. Position of needle inside the capsule should be ascertained by USG. Hip joint can also be approached from anterior aspect by inserting the needle at a point 2.5 cm lateral to femoral artery and 2.5 cm inferior to inguinal ligament. Needle is advanced medially and proximally.[15]

Sacroiliac Joints

The inflammation of sacroiliac joints is the hallmark of axial spondyloarthritis and it is also found in a significant proportion of other subclasses of spondyloarthropathies like psoriatic spondyloarthropathy and reactive arthritis. Being a deep-seated joint overlaid by a thick cover composed of subcutaneous fat, muscles and other periarticular structures it is not approachable for a blind procedure easily and presence of important anatomical structure like sciatic nerve in vicinity discourages clinicians from using the unguided approach. Hence so far, clinicians use fluoroscopy or other imaging modalities of guidance for injections in SI joints. However, there have been many proponents of unguided technique to SI joint approach also.[16] The authors have also been using this technique successfully over the years and more than 800 SI joints have been injected at author's centre using unguided technique. Accuracy of this technique has been confirmed by the clinical improvement in patient symptoms as well as confirmation of needle's position in SI joints by 3D processing of CT images. The author is sharing his personal technique of unguided approach to SI joint injection.

With patient lying in prone position, draw a horizontal line connecting the centres of two dimples of Venus. Draw another perpendicular line in the midline along the spinous process of lumbar vertebrae. Point of needle insertion is at the junction of 1 cm below the horizontal line and 1 cm lateral to the perpendicular line (Fig. 19.8). Needle should be advanced in outward and downward direction at an angle of about 75 degrees (Fig. 19.9).

Conclusion

Intra-articular injections form a keystone of day-to-day rheumatology practice. However, appropriate technique with asepsis must always be used. Always consider septic arthritis as a differential while aspirating a single inflamed joint, whether a monoarthritis or a monoarticular flare of prior rheumatic disease.

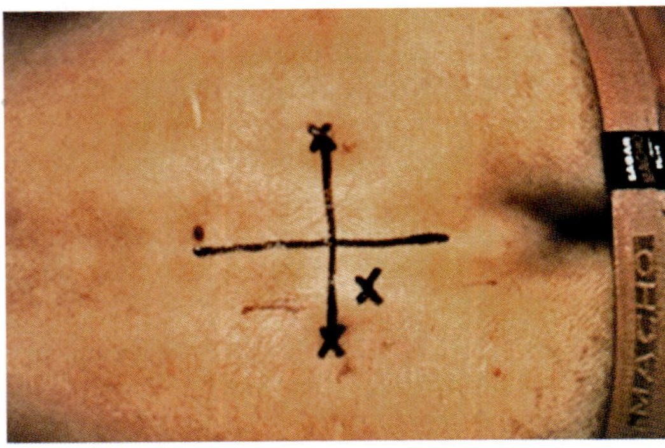

Fig. 19.8: Point of insertion of unguided SI joint injection

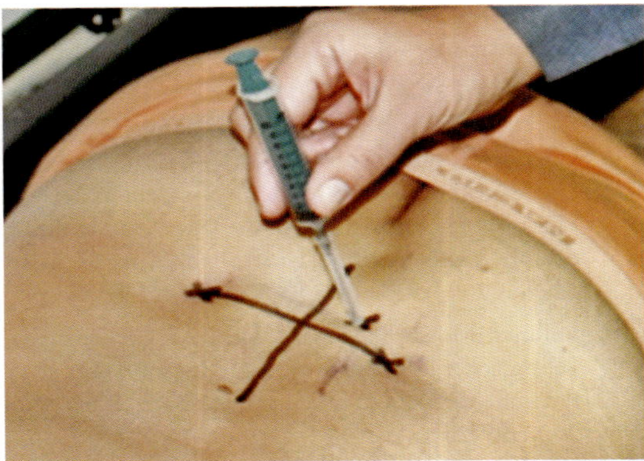

Fig. 19.9: Direction of needle insertion in unguided approach to SI joint

Key Points

1. Learn and use the appropriate technique (with asepsis) while aspirating or injecting any joint.
2. Always keep the possibility of septic arthritis when a single joint is inflamed, before deciding to inject glucocorticoids.
3. Keep in mind that every time a joint is injected or aspirated, there exists a definite risk of introducing infection into the joint.

REFERENCES

1. Gatter RA, Schumacher HR Jr. Joint aspiration: Indications and technique. In: Gatter RA, Schumacher HR Jr, Editors. A Practical Handbook of Synovial Fluid Analysis. Philadelphia: Lea and Febiger; 1991, p.14.
2. Shmerling RH, Delbanco TL, Tosteson AN, Trentham DE. Synovial fluid tests. What should be ordered? JAMA 1990; 264:1009.
3. Ahmed I, Gertner E. Safety of arthrocentesis and joint injection in patients receiving anticoagulation at therapeutic levels. Am J Med 2012; 125:265–9.
4. Kirschke DL, Jones TF, Stratton CW, Barnett JA, Schaffner W. Outbreak of joint and soft-tissue infections associated with injections from a multiple-dose medication vial. Clin Infect Dis 2003; 36: 1369–73.
5. Wittich CM, Ficalora RD, Mason TG, Beckman TJ. Musculoskeletal injection. Mayo Clin Proc. 2009; 84(9):831–6.
6. Bhagra A, Syed H, Takahashi PY. Efficacy of musculoskeletal injections by primary care providers in the office: a retrospective cohort study. Int J Gen Med 2013;6:237–43.
7. Behrens F, Shepard N, Mitchell N. Alterations of rabbit articular cartilage by intra-articular injections of glucocorticoids. J Bone Joint Surg Am 1975; 57:70.
8. Raynauld JP, Buckland-Wright C, Ward R, Choquette D, Haraoui B, Martel-Pelletier J. Safety and efficacy of long-term intra-articular steroid injections in osteoarthritis of the knee: a randomized, double-blind, placebo-controlled trial. Arthritis Rheum 2003; 48 (11):370–7.
9. Pullman-Mooar S, Mooar P, Sieck M, Clayburne G, Schumacher HR. Are there distinctive inflammatory flares after hylan intra-articular injections? J Rheumatol 2002; 29:2611.
10. Owen DS. Aspiration and injection of joints in soft tissue. In: Kelley WN, Editor. Textbook of Rheumatology. St. Louis, MO: Saunders; 1993, p.545.
11. George LV, Elizabeth G. The Elbow. In: George LV, Hans KJ, Hawker GA, Editors. Musculoskeletal examination and joint examination techniques. Second edition. Philadelphia: Mosby Elsevier; 2010, pp. 21–28.
12. Naredo E, Rull M. Aspiration and injection of jointsand periarticular tissue and intralesional therapy. In: Hochberg MC, Silman AJ, Smolen JS, Weinblatt ME, Weisman MH, editors. Rheumatology. Philadelphia: Mosby Elsevier; 2015. pp.546–548.
13. Hans KJ, Hawker GA. The knee. In: George LV, Hans KJ, Hawker GA, et al., Editors. Musculoskeletal examination and joint examination techniques. Second edition. Philadelphia: Mosby Elsevier; 2010, pp. 65–88.
14. Stephen DJG, Choy WG, Fam GA. The Ankle and Foot. In: George LV, Hans KJ, Hawker GA, et al (Editors). Musculoskeletal examination and joint examination techniques. Second edition. Philadelphia: Mosby Elsevier; 2010. pp. 89–101.
15. Hans JK, Dana J. The Hip. In: George LV, Hans KJ, Hawker GA, et al., Editors. Musculoskeletal examination and joint examination techniques. Second edition. Philadelphia: Mosby Elsevier; 2010, pp. 45–63.
16. Sadreddini S, Noshad H, Molaeefard M, Ardalan MR, Ghojazadeh M, Shakouri SK. Unguided sacroiliac injection: Effect on refractory buttock pain in patients with spondyloarthropathies. Presse Méd. 2009; 38: 710–16.

FURTHER READING

1. Courtney P, Doherty M. Joint aspiration and injection and synovial fluid analysis. Best Pract Res Clin Rheumatol 2013; 27(2):137–69.

Chapter

20

Non-steroidal Anti-inflammatory Drugs

Annil Mahajan, Vishal Tandon

INTRODUCTION

Nonsteroidal anti-inflammatory drugs (NSAIDs) are among the most widely prescribed medications for reducing pain and inflammation in patients with various rheumatological disorders. More than 70 million prescriptions for NSAIDs are written each year in United States and more than 30 billion over the counter tablets are sold annually.[1] NSAIDs are a necessary choice in pain management and symptomatic relief of inflammatory conditions.[2] This chapter will provide a brief overview of NSAIDs and their usefulness in rheumatology practice.

Mechanism of action: NSAIDs work by interfering with the cyclo-oxygenase pathway, which involves the conversion of arachidonic acid, by the enzyme cyclo-oxygenase (COX) to prostaglandins (PGs) (Fig. 20.1). There are two types of COX, i.e. COX-1 and COX-2. The COX-1 enzyme is constitutive, meaning that its concentration in the body remains stable. It is present in most tissues and it generates 'good' PGs for 'housekeeping' functions such as gastric mucosal integrity, platelet homeostasis and regulation of renal blood flow. In contrast, the COX-2 enzyme is induced dramatically by the action of macrophages, the scavenger cells of the immune system. COX-2 is involved in producing 'bad' PGs involved in inflammatory reactions (i.e. arthritis, pain, hyperalgesia). This hypothesis suggest that the anti-inflammatory actions of NSAIDs are due to inhibition of COX-2 enzyme, whereas the unwanted side effects of conventional NSAIDs such as gastrotoxicity, are due to the inhibition of the constitutive COX-1 isoenzyme based on this selective action on COX enzyme, NSAIDs can be classified in to non-selective NSAIDs, preferential COX-2 inhibition, and selective COX-2 inhibition (Table 20.1).

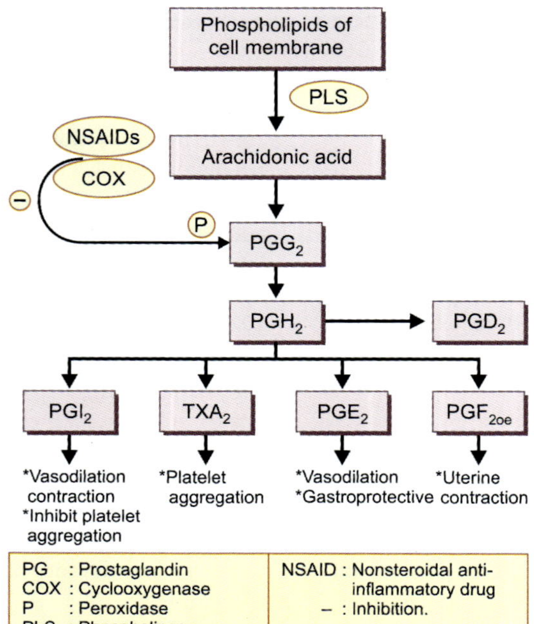

Fig. 20.1: COX dependent mechanism of action[4,5]

NSAIDs may also act by mechanisms independent of COX enzyme inhibition as elaborated in Table 20.2.

Table 20.1: Recent classification of NSAIDs on the basis of COX selectivity[3,4]

Non-selective NSAIDs	*Average adult dose/ frequency*
Aspirin	600–900 mg 4–6 hrly
Diclofenac	50 mg 8 hrly
Aceclofenac	100 mg 12 hrly
Indomethacin	50 mg 8 hrly
Ketorolac	10 mg 6 hrly
Sulindac	150–200 mg twice daily
Ibuprofen	600–800 mg 8 hrly
Flurbiprofen	100 mg 8 hrly
Naproxen	500 mg 12 hrly
Mefenamic acid	500 mg 8 hrly
Piroxicam	20 mg once daily
Tenoxicam	20 mg once daily
Preferential COX-2 (shows preferential COX-2 inhibition over COX-1)	
Nimesulide	100 mg 12 hrly
Nabumetone	200–400 mg once daily
Etodolac	7.5–15 mg twice daily
Meloxicam	7.5 mg once daily
Selective COX-2 inhibitors	
Celecoxib (FDA alert)	100–200 mg 12 hrly
Etoricoxib	60–120 mg once daily
Lumiracoxib	400 mg once daily
Parecoxib	10 mg/kg daily dose (injectable)
Zaltoprofen	80 mg thrice daily

COX-3: A new isoform of the COX family has been discovered very recently, which is expressed abundantly in cerebral cortex and heart and it is thought to contribute for central mechanism by which paracetamol decrease fever. The relevance of COX-3 in humans is still debatable.[3]

Apart from anti-inflammatory effects, NSAIDs causes anti-pyretic, analgesic and other beneficial effects (Table 20.3).

Toxicity

The selective inhibition of COX-2 causes CVS toxicity while inhibition of both COX-1 and COX-2 causes renal toxicity.[3–8]

Gastrointestinal (GI) toxicity: 35 million people consume NSAIDs on a daily basis, and about 30% of these users may develop GI toxicity. Approximately 107,000 patients are hospitalized annually for NSAID-related GI complications, and at least 16,500 NSAID-related deaths occur each year among arthritis patients alone.[10]

Topical preparations of NSAIDs are acidic in nature of which promotes the accumulation of ionised molecules (ion trapping) within the mucosal

Table 20.2: COX independent mechanisms of NSAIDs[4,5,9]

- Inhibitor effects on lipooxygenase products (sulindac, diclofenac)
- Inhibition of superoxide formation (indomethacin, piroxicam)
- Inhibition of neutrophil aggregation, adhesion and enzyme release (salicylates, indomethacin, aceclofenac)
- Depression of lymphocyte transformation (salicylates)
- Inhibition of rheumatoid factor production (piroxicam)
- Inhibition of cytokine production by inhibiting factor kappa B (salicylates)
- Suppression of proteoglycan production in cartilage (salicylates, piroxicam, ibuprofen, fenoprofen)
- Inhibit mononuclear cell phospholipase C activity and inhibit inducible nitric oxide synthase (aspirin)
- Stimulation of adenosine receptors (salicylates)
- Peroxisome proliferators activated receptor (PPAR) activation
- COX independent apoptosis and angiogenesis (COX-2 inhibitors)
- Inhibition of IL-beta and TNF in inflammatory cells and stimulation of synthesis of the extracellular matrix of human articular cartilage (aceclofenac)

Table 20.3: Other beneficial pharmacological actions of NSAIDs[4,5]

- Anti-inflammatory effects (Utilizes various mechanism but main is PGs inhibition)
- Analgesic effects (most effective against inflammation associated pain)
- Antipyretic effect (by inhibiting synthesis of PGE2 in hypothalmus)
- Anti-platelet effect (low dose aspirin selectively inhibits TXA2)
- Closure of ductus arteriosus (indomethacin by inhibiting PGE2 and I2)
- Cancer chemoprevention (familial adenomatous polyposis, colorectal cancer/adenoma)
- Alzheimer's disease (epidemiological data supports unknown mechanism)

cells, thereby causing GI toxicity. Systemically, GI toxicity is mediated through (COX-1) inhibition and a subsequent decrease in gastroprotective prostaglandins.

NSAID induced GI injury includes superficial damage such as mucosal haemorrhages and erosions, endoscopically documented non-symptomatic ('silent') ulcers and symptomatic ulcers causing complications such as GI haemorrhage.

Renal toxicity : There is no difference between COX-2 selective and non-selective NSAIDs with regard to renal toxicity.[11] Locally synthesized PGI_2, PGE_2 and PGD_2 cause vasodilatation, decrease vascular resistance and enhance perfusion with redistribution of blood flow from the renal cortex to nephrons in the juxtamedullary region. Moreover, PGE_2 and $PGF_{2\alpha}$ cause diuresis by inhibiting the transport of Na and Cl in the thick ascending limb of loop of Henle and collecting ducts, thus their inhibition by NSAIDs can cause renal toxicity. Nephrotoxicity with NSAIDs includes acute tubular necrosis, acute tubulo-interstitial nephritis, glomerulonephritis, renal papillary necrosis, salt and water retention, chronic renal failure, hypertension, hyperkalemia.[12]

Cardiovascular toxicity: The CLASS trial,[13] VIGOR trial,[14] a case-control study,[15] adenomatous Polyp Prevention on Vioxx (APPROVe) trial,[16] clinical trial (APC) for prevention of colorectal adenoma continued exposure to celecoxib[17] and another recent study,[18] demonstrated that rofecoxib, celecoxib, valdecoxib and parecoxib increase the incidence of thrombotic events such as myocardial infarction, unstable angina, cardiac thrombus, resuscitated cardiac arrest, sudden or unexplained death, ischemic stroke, and transient ischemic attacks. The mechanisms by which selective COX2 inhibitors cause cardiovascular toxicity is summarised in Table 20.4. Because of this increase in cardiovascular events valdecoxib and rofecoxib have been withdrawn from the market.

Traditional NSAIDs have varying associations with risk for myocardial infarction (MI). In nested case control study diclofenac was shown to increase the risk of MI.[19] In another population based study, a significant increase (24–55%) in risk of MI was observed in patients receiving rofecoxib, diclofenac and ibuprofen.[20] In a recent meta-analysis, among the seven NSAIDs (naproxen, ibuprofen, diclofenac, celecoxib, etoricoxib, rofecoxib, lumiracoxib) analyzed, naproxen seemed least harmful for cardiovascular safety.[21] Cardiovascular toxicity of COX-2 inhibitors can be prevented to some extent by combining essential fatty acids (EFAs) eicosapentaenoic acid.[22]

Table 20.4: Mechanisms of cardiovascular toxicity[3–8]

- COX-2 inhibitors decrease vascular prostacyclin (PGI_2) production and may affect the balance between prothrombotic and antithrombotic eicosanoids (thromboxane A2)
- COX-2 may also play a role in destabilizing plaques
- Inhibition of COX-2 cause an increase in the flux of arachidonate through the lipooxygenase (LO) pathways, which is important in the setting of inflammation in the atheromatous plaque. Generated 12-LO and 15-LO contribute to LDL oxidation
- COX-2 has been recognized cardioprotective in ischemia-reperfusion injury and has antiproliferative effects toward vascular smooth muscle cells in conjunction with NO
- COX-2 inhibitors have also been shown to increase blood pressure (BP), due to alterations in the renin angiotensin pathway, change in sodium and water retention by the kidney, inhibition of vasodilating PG's and production of various vasoconstricting factors, including endothelin-1

Hepatic: NSAIDs can cause transient rise in liver enzymes. Reye's syndrome has been reported with aspirin, whereas cholestasis is seen with sulindac and diclofenac therapy and fulminant hepatitis with the use of nimesulide in children has been reported.[4,5]

Bone: NSAIDs also interfere with and suppress bone repair and remodeling.[4,5]

Sleep pattern: The drugs can decrease sleep.[4,5]

Endocrinological effects: Aspirin and ibuprofen augment insulin secretion.[4,5]

Psychiatric: NSAIDs have also been reported to induce adverse psychiatric reactions like depressive disorders and suicidal tendencies.[4,5]

Central nervous system (CNS): Headache, aseptic meningitis, drowsiness, altered mood, tinnitus, etc. are commonly reported with NSAIDs therapy.[4,5]

Hematologic, dermatologic and respiratory: Phenylbutazone or oxyphenbutazone can cause aplastic anemia and thrombocytopenia. However, COX-2 specific inhibitors do not affect platelet functions. Macular eruptions, urticaria, pseudoporphyria, erythema multiforme have been reported with their use. Asthma can be precipitated with NSAIDs especially aspirin.[4,5]

Factors to be Considered While Prescribing NSAIDs

Drug Factors

Efficacy: Nonspecific NSAIDs or COX-2 specific NSAIDs show comparable efficacy in different rheumatological conditions.[1,4]

Safety: Greater the potency more is the chance to cause side effects. Long-term use of nonspecific NSAIDs is a known risk factor for gastrointestinal (GI) toxicity unlike the COX-2 inhibitors. Nimesulide remains safe therapeutic choice in patients with asthma; however, nimesulide should not be used in children less than 12 years.[23] Sulindac has lower incidence of renal side effects. Increased risk of acute MI has been associated with selective COX-2 inhibitors and treatment with diclofenac and ibuprofen, but not with naproxen.[24]

Pharmacokinetics: The shorter the half life, more the frequency of administration. The compliance is better with drugs like piroxicam (once daily) or naproxen (twice daily) than drugs like ibuprofen having shorter half life. It is advantageous to start with a high dose of a short-life drug (e.g. ibuprofen) and then adjust the dose downward when analgesic efficacy has been achieved. For management of chronic pain, administration of NSAIDs with long half-lives (naproxen and COX-2 inhibitors) has clear advantages in allowing once or twice a day dosing.[2,4]

Dosage/formulation: Diclofenac sodium softgel has also been shown to provide a very rapid onset of analgesic activity and prolonged analgesic duration compared with conventional diclofenac potassium and better tolerated.[2,4]

Concomitant treatment:[4,5,25] The likelihood of NSAID-induced GI toxicity is increased with concomitant use of corticosteroids. NSAIDs can decrease the clearance of methotrexate and lithium. Warfarin competes with NSAIDs for protein binding site and can increase the risk of GI bleed. NSAIDs can also decrease antihypertensive efficacy of betablockers, diuretics, angiotensin converting enzyme inhibitors, calcium channel blockers, etc.

Patient's Factors

Disease: Indomethacin is the traditional choice in gout. Ankylosing spondylitis (AS) is better treated with indomethacin or naproxen. RA is better treated with drug having long half life such as piroxicam and naproxen.[4,5,25]

Target symptom: Morning stiffness may be better controlled with a long half life drugs such as piroxicam or a nighttime dose of indomethacin.[4,5,25]

Concomitant disease: It is best to avoid NSAIDs in the presence of renal dysfunction. If a patient has no risk factors for GI bleed (i.e. no prior GI bleed, no concomitant glucocorticoids or anticoagulation use or antithrombotics) and is younger than 70 years, then nonselective NSAIDs as appropriate can be used. In patients with prevalent GI risk factors, cotherapy with a PPI should be used. In the presence of few (average) CV risk factors either a nonselective NSAID or a coxib can be used. Naproxen is preferred in patients with high CV risk. However, in patients with multiple GI risk factors and high CV risk NSAIDs should be avoided.[26]

Old age: NSAIDs may be able to protect neurons directly by reducing cellular response to glutamate and have potential to reduce the risk of Alzheimer's disease in old patients.[27] However, the chances of drug interaction and serious adverse effects are more.[4,5,25,26]

Pregnancy:[28] High dose aspirin should be avoided during all stages of pregnancy, particularly in last trimester, as it can inhibit uterine contractility and prolong labour. It can cause premature closure of ductus arteriosus, kernicterus, haemorrhage and renovascular complications. *Non-selective COX-inhibitors:* Can be used in the first and second trimester of pregnancy, if necessary to control maternal disease. Paracetamol may be considered as safe analgesic. COX-2 *Selective Inhibitors:* theoretically the effect of COX-2 selective inhibitors on duration of labour and clotting should be less, whereas effect on fetal vasculature and renal function should be equal. However, there is lack of data on these drugs, hence should be avoided as far as possible.

Lactation: Most NSAIDs are safe during breast feeding as amount of drug excreted in breast milk is not significant, although there is potential risk of jaundice and kernicterus. There is insufficient data on COX-2 inhibitors and lactation. Drugs like indomethacin and aceclofenac should be avoided in lactation.[28]

NSAIDs in Rheumatology

Role in RA: NSAIDs are perhaps the most effective adjunctive therapy in RA, providing both analgesic and anti-inflammatory benefits. However, they do not cure or alter the course of the disease and articular cartilage damage may continue despite symptomatic relief. In a meta-analysis it was shown that only naproxen showed significant decrease in CRP, whereas lumiracoxib increased CRP.[29]

Gout: NSAIDs are preferred in patients with uncomplicated gout.[30] Indomethacin is the traditional choice. An initial dose of 50 to 75 mg followed by 50 mg every 6 to 8 hours with maximum of 200 mg in the first 24 hours and then tapering is recommended.[30] Oral naproxen, ibuprofen, sulindac, piroxicam, ketoprofen as well as intramuscular ketorolac are also effective.[4,5]

Osteoarthritis: NSAIDs are cornerstone in the management of osteoarthritis (OA) and are considered for patients who do not obtain adequate pain relief with paracetamol and have associated swelling and inflammation, however, they have no structural or disease modifying effect.[31] Opioids (codeine and tramadol) and paracetamol in combination provide better analgesia than paracetamol alone.

Low back pain: NSAIDs prescribed at regular intervals provide effective pain relief for simple acute low back pain, but do not affect return to work.[4,5]

Juvenile idiopathic arthritis (JIA): Naproxen (15–20 mg/kg/day twice daily) and ibuprofen (35 mg/kg/day four times daily) are the approved drugs. An unusual toxic result of naproxen and other NSAID has been occurrence of pseudoporphyria. The response of NSAIDs varies among children with JIA but satisfactory treatment of arthritis is achieved in about 50%.[4]

Anklyosing spondylitis: NSAIDs are first line agents in the management of axial spondyloarthritis (SpA), both non-radiographic SpA and ankylosing spondylitis (AS). There is no clear efficacy of one NSAID over the other. However, clinical trial data is available with celecoxib, and diclofenac sodium. In everyday practice clinicians often prescribe indomethacin, or diclofenac, or etoricoxib.

A study by Wanders et al in 2005 found that the continuous use of celecoxib, in contrast with on-demand use, was associated with less radiographic progression in AS.[32]

An RCT (2008) by Sieper and colleagues, randomized active AS subjects (n = 458) into three treatment arms: celecoxib 200 mg once a day, celecoxib 200 mg twice a day, and diclofenac SR 75 mg twice a day. Data from this RCT suggest high NSAID intake is better than low dosage NSAID to reduce AS disease activity.[33]

Data from the German Early Spondyloarthritis Inception Cohort (GESPIC) shows that patients with AS with a high NSAID intake over 2 years demonstrated lower likelihood of significant radiographic progression in the spine compared to patients with low NSAID intake.[34] In AS NSAIDs are given in full dose for at least 3 months before considering treatment with anti-TNF agents. However, long term data on retardation of radiographic progression are still not available yet.

Undifferentiated SpA and reactive arthritis: With the advent of anti-TNF alpha therapy effective treatment of inflammation of both peripheral joints and the axial skeleton is now possible. Nevertheless, first line treatment for all these patients is still NSAIDs. NSAIDs provide maximum suppression of pain and inflammation within 2 weeks.[4,5,25]

Psoriatic arthritis: The treatment of psoriatic arthritis is directed at controlling the inflammatory process. Initial treatment is NSAIDs for joint disease.[4,5,25]

SLE: In patients with mild symptoms such as arthralgia administration of NSAIDs may provide adequate relief. Ibuprofen, sulindac can cause aseptic meningitis. Through their effects on renal prostaglandin all NSAIDs can reduce GFR and renal blood flow especially in patients who have associated nephritis. Salt retention secondary to NSAIDs use in SLE patients may elevate blood pressure and cause pedal edema. Finally patients with SLE have a higher incidence of hepatotoxicity. Thus, SLE patients treated with NSAIDs should be monitored regularly for renal, GI and hepatic side effects.[4,5,25]

Others: The anti-inflammatory effects of NSAIDs have been demonstrated in acute rheumatic fever. Although not as rigorously proven their efficacy is also widely accepted in the treatment of acute and chronic bursitis, tendinitis, radiation synovitis and sport injuries.[4,5,25]

A stepped care approach has been recommended in chronic pain disorders. It is recommended to start with simple analgesics (acetaminophen or nonsteroidal anti-inflammatory drugs) followed by tricyclic antidepressants (if neuropathic, back or fibromyalgia pain) or gabapentin, duloxetine or pregabalin if neuropathic pain and cyclobenzaprine, pregabalin, duloxetine, or milnacipran for fibromyalgias.[35]

FDA Expert Advisory Committee and Multidisciplinary European Expert Panel Recommendations[36,37]

- COX-2 inhibitors and NSAIDs should be prescribed with the lowest effective dose and for the shortest duration.
- They should not be prescribed for high risk patients. Two or more NSAIDs in combination should not be given.
- COX-2 selective inhibitors alone and non-selective NSAID plus PPI should be preferred for patients with high GI risk and low cardiovascular risk.
- Naproxen plus PPI is favoured in patients with high cardiovascular risk.
- For the combination of high gastrointestinal/high cardiovascular risk the use of any NSAID is discouraged; if needed, naproxen plus PPI or a COX-2 inhibitor plus PPI could be considered.

Nitric oxide (NO) NSAIDs: Nitric oxides releasing nonsteroidal anti-inflammatory drugs are a novel group of drugs synthesized by the ester linkage of an NO releasing moiety to conventional (NSAID) such as NO aspirin, NO-iburbiprofen, NO-naproxen and NO ibuprofen.[2,3,6,8,33] NO-NSAID cause markedly diminished GI and hepatic toxicity. NO is a potent inhibitor of platelet activation and adhesion to the vessel wall. NO-NSAIDs theoretically, are expected to exhibit a greater degree of inhibition of platelet function. In addition, the vasodilator activity of NO may assist in counteracting the vasoconstriction caused by platelet-derived mediators such as TXA2. Thus they may prove to be cardio-protective over COX2 inhibitors. However, the existing data is from preclinical studies clinical trials yet to prove these roles.

NSAIDs and Cancer Protection

Recent studies suggest that, NSAID use is associated with a reduced colorectal cancer risk,[39] prostate cancer risk,[40] pancreatic cancer risk[41] and breast cancer risk on long-term use.[42] In rheumatology practice, patients requiring NSAIDs for longer duration may have some blessing effects in the form of chemo protection (Table 20.3).

Recent Causations and Advances

Frequent use of NSAIDs but not aspirin seemed to be associated with increased absolute incidence of Crohn disease and ulcerative colitis.[43] Recently, thiazolo [3, 2-b]-1, 2, 4-triazole-5 (6H) - derivatives of ibuprofen has been shown to improve analgesic and anti-inflammatory activities and lower ulcerogenic risk.[44] Stable gastric pentadecapeptide BPC 157 is an anti-ulcer peptidergic agent, has been proposed to be used as an antidote against NSAIDs induced toxicity.[45]

Key Points

- NSAIDs are among the most widely prescribed for pain and inflammation in various rheumatological disorders.
- Logical therapeutic approach is to choose low dose of NSAID for shorter duration.
- During this initial stage an analgesic such as paracetamol may be added to supplement analgesic activity.
- However, the choice of NSAID for a particular patient is based upon a number of factors, including relative efficacy, toxicity, concomitant drug use, concurrent disease states, the patient's age, renal function, cost and to a certain extent, on the prescriber's preference.
- It is important to give each NSAID an appropriate therapeutic trial before an alternative is tried.

REFERENCES

1. Genovese MC, Harris ED. Treatment of rheumatoid arthritis. In: Harris ED, Budd RC, Genovese MC, Firestein GS, Sargent JS, Sledge CB,Ruddy S, editors. Kelley's Textbook of Rheumatology. Philadephia, PA: Elsevier Saunders; 2005. pp. 1079–1100.
2. Ong C, Lirk P, Tan C, Seymour R. An Evidence-Based Update on Nonsteroidal Anti-Inflammatory Drugs. Clin Med Res. 2007;5:19–34.
3. Kulkarni S, Jain N. Coxibs: The new super aspirins or unsafe pain killers? Indian J Pharmacol. 2005;37: 86.
4. Robert LJ, Morrow D. Analgesic-antipyretic and anti-inflammatory agents and drugs employed in the treatment of gout. In: Robert LJ, Morrow D, Hardman JG, Limbird LE, Gilman AG, editors. Goodman & Gilman's The Pharmacological basis of Therapeutics. New York: McGraw-Hill; 2005. pp. 687–731.
5. Crofford LJ. Nonsteroidal anti-inflammatory drugs. In: Crofford LJ, Harris ED, Budd RC, Genovese MC, Firestein GS, Sargent JS, Sledge CB. Kelley's Textbook of Rheumatology. Philadephia, PA: Elsevier Saunders; 2005.pp.839–59.
6. Mahajan A, Sharma R. COX-2 Selective Nonsteroidal Anti-inflammatory Drugs: Current Status. J Assoc Physicians India. 2005; 53:200–4.
7. Mahajan A, Sharma R. Cox-2 Inhibitors: Cardiovascular Safety. JK Science.2005;7(2): 61–62.
8. Mahajan R, Sharma R, Singh JB, Hussain J. Non-steroidal Anti-inflammatory Drugs: Controversies. J Indian Academy Clin Med 2007; 8(1):65–72.
9. Collier DH. Nonsteroidal anti-inflammatory drugs (NSAIDs). In: West SG, editor. Rheumatology Secrets. Philadelphia, PA: Hanley &Belfus, Elseviers; 2003. pp. 561–71.
10. Graumlich J. Preventing gastrointestinal complications of NSAIDs. Postgrad Med. 2001;109:117–128.
11. Schneider V, Levesque L, Zhang B, Hutchinson T, Brophy J. Association of selective and conventional non-steroidal anti-inflammatory drugs with acute renal failure: a population-based, nested case-control analysis. Am J Epidemiol 2006; 164:881–89.
12. Ejaz P, Bhojani K, Joshi VR. NSAIDs and Kidney. J Assoc Physicians India. 2004; 52:632–40.
13. White W, Faich G, Whelton A, Maurath C, Ridge N, Verburg K et al. Comparison of thromboembolic events in patients treated with Celecoxib, a Cyclooxygenase-2 specific inhibitor, versus Ibuprofen or Diclofenac. Am J Cardiol 2002; 89:425–430.
14. Bombardier C, Laine L, Reicin A, Shapiro D, Burgos-Vargas R, Davis B, et al. Comparison of Upper Gastrointestinal Toxicity of Rofecoxib and Naproxen in Patients with Rheumatoid Arthritis: VIGOR study group. N Engl J Med 2000; 343:1520–28.
15. Graham D, Campen D, Hui R, Spence M, Cheetham C, Levy G,et al. Risk of acute myocardial infarction and sudden cardiac death in patients treated with cyclo-oxygenase 2 selective and non-selective non-steroidal anti-inflammatory drugs: nested case-control study. The Lancet. 2005;365:475–481.
16. Antman E, DeMets D, Loscalzo J. Cyclooxygenase Inhibition and Cardiovascular Risk. Circulation. 2005;112:759–70.
17. Solomon S, McMurray J, Pfeffer M, Wittes J, Fowler R, Finn P, et al. Cardiovascular risk associated with celecoxib in a clinical trial for colorectal adenoma prevention. N Engl J Med 2005; 352:1071–80.
18. Nussmeier NA, Whelton AA, Brown MT, Langford RM, Hoeft A, Parlow JL, et al. Complications of the COX-2 inhibitors parecoxib and valdecoxib after cardiac surgery. N Engl J Med. 2005;352:1081–91.
19. Hippisley-Cox J, Coupland C. Risk of myocardial infarction in patients taking cyclo-oxygenase-2 inhibitors or conventional non-steroidal anti-inflammatory drugs: population based nested case-control analysis. BMJ. 2005;330(7504):1366.
20. Johnsen SP, Larsson H, Tarone RE, McLaughlin JK, Nørgård B, Friis S, Sørensen HT. Risk of Hospitalization for Myocardial Infarction Among Users of Rofecoxib, Celecoxib, and Other NSAIDs. A population-based case-control study. Arch Intern Med 2005; 165:978–84.
21. Trelle S, Reichenbach S, Wandel S, Hildebrand P, Tschannen B, Villiger PM, et al. Cardiovascular safety of non-steroidal anti-inflammatory drugs: network meta-analysis. BMJ. 2011;342:c7086.
22. Das UN. Can COX-2 inhibitor-induced increase in cardiovascular disease risk bemodified by essential fatty acids? J Assoc Physicians India 2005; 53:623–7.
23. Rainsford K. Nimesulide—a multifactorial approach to inflammation and pain: scientific and clinical consensus. Curr Med Res Opin. 2006;22:1161–70.
24. Singh G, Wu O, Langhorne P, Madhok R. Risk of acute myocardial infarction with nonselective non-steroidal anti-inflammatory drugs: a meta-analysis. Arthritis Res Ther 2006; 8: R153.
25. Woodfork KA, Dyke KV. Anti-inflammatory and antirheumatic drugs. In: Craig CR, Stitzel RE, editors. Modern Pharmacology with clinical applications. Philadelphia :Lippincott Williams and Wilkins; 2004.pp. 423–39.
26. Chan FK, Abraham NS, Scheiman JM, Laine L; First International Working Party on Gastrointestinal and Cardiovascular Effects of Nonsteroidal Anti-inflammatory Drugs and Anti-platelet Agents. Management of patients on nonsteroidal anti-inflammatory drugs: a clinical practice recommendation from the First

International Working Party on Gastrointestinal and Cardiovascular Effects of Nonsteroidal Anti-inflammatory Drugs and Anti-platelet Agents. Am J Gastroenterol 2008; 103:2908–18.

27. Imbimbo BP, Solfrizzi V, Panza F. Are NSAIDs useful to treat Alzheimer's disease or mild cognitive impairment? Front Aging Neurosci. 2010 May 21;2. pii:19. doi: 10.3389/fnagi.2010.00019. eCollection 2010.
28. Temprano K, Bandlamudi R, Moore T. Antirheumatic drugs in pregnancy and lactation. Semin Arthritis Rheum 2005; 35:112–121.
29. Tarp S, Bartels EM, Bliddal H, Furst DE, Boers M, Danneskiold-Samsøe B, et al. Effect of nonsteroidal anti-inflammatory drugs on the C-reactive protein level in rheumatoid arthritis: a meta-analysis of randomized controlled trials. Arthritis Rheum. 2012;64:3511–21.
30. Khanna D, Khanna PP, Fitzgerald JD, Singh MK, Bae S, Neogi T, et al; American College of Rheumatology. 2012 American College of Rheumatology guidelines for management of gout. Part 2: therapy and antiinflammatory prophylaxis of acute gouty arthritis. Arthritis Care Res (Hoboken). 2012;64: 1447–61.
31. Hochberg MC, Altman RD, April KT, Benkhalti M, Guyatt G, McGowan J et al. American College of Rheumatology 2012 recommendations for the use of nonpharmacologic and pharmacologic therapies in osteoarthritis of the hand, hip, and knee. Arthritis Care Res 2012; 64:465–74.
32. Wanders A, Heijde D, Landewe R, Béhier JM, Calin A, Olivieri I, et al. Nonsteroidal anti-inflammatory drugs reduce radiographic progression in patients with ankylosing spondylitis: a randomized clinical trial. Arthritis Rheum 2005; 52: 1756–65.
33. Sieper J, Klopsch T, Richter M, Kapelle A, Rudwaleit M, Schwank S, et al. Comparison of two different dosages of celecoxib with diclofenac for the treatment of active ankylosing spondylitis: results of a 12-week randomised, double-blind, controlled study. Ann Rheum Dis 2008;67: 323–9.
34. Poddubnyy D, Rudwaleit M, Haibel H, Listing J, Märker-Hermann E, Zeidler H, et al. Effect of non-steroidal anti-inflammatory drugs on radiographic spinal progression in patients with axial spondyloarthritis: results from the German Spondyloarthritis Inception Cohort. Ann Rheum Dis 2012; 71: 1616–22.
35. Kroenke K, Krebs EE, Bair MJ. Pharmacotherapy of chronic pain: a synthesis of recommendations from systematic reviews. Gen Hosp Psychiatry 2009; 31: 206–19.
36. Young D. FDA labors over NSAID decisions: panel suggests COX-2 inhibitors stay available. Am J Health Syst Pharm 2005; 62: 668–72.
37. Burmester G, Lanas A, Biasucci L, Hermann M, Lohmander S, Olivieri I, et al. The appropriate use of non-steroidal anti-inflammatory drugs in rheumatic disease: opinions of a multidisciplinary European expert panel. Ann Rheum Dis 2011; 70: 818–22.
38. Keeble J, Moore P. Pharmacology and potential therapeutic applications of nitric oxide-releasing non-steroidal anti-inflammatory and related nitric oxide-donating drugs. Br J Pharmacol. 2002;137:295–310.
39. Ruder E, Laiyemo A, Graubard B, Hollenbeck A, Schatzkin A, Cross A. Non-Steroidal Anti-Inflammatory Drugs and colorectal cancer risk in a large, prospective cohort. Am J Gastroenterol 2011; 106:1340–50.
40. Mahmud S, Franco E, Turner D, Platt R, Beck P, Skarsgard D, et al. Use of Non-Steroidal Anti-Inflammatory Drugs and Prostate Cancer Risk: A Population-Based Nested Case-Control Study. PLoS One. 2011; 6(1):e16412.
41. Bradley M, Hughes C, Cantwell M, Napolitano G, Murray L. Non-steroidal anti-inflammatory drugs and pancreatic cancer risk: a nested case-control study. Br J Cancer. 2010; 102:1415–21.
42. Brasky T, Bonner M, Moysich K, Ambrosone C, Nie J, Tao M, et al. Non-steroidal anti-inflammatory drug (NSAID) use and breast cancer risk in the Western New York Exposures and Breast Cancer (WEB) Study. Cancer Causes & Control. 2010;21:1503–12.
43. Ananthakrishnan A. Aspirin, nonsteroidal anti-inflammatory drug use, and risk for Crohn disease and ulcerative colitis. Ann Intern Med 2012; 156:350–9.
44. Uzgören-Baran A, Tel BC, Sar?göl D, Oztürk E?, Kazkayas? I, Okay G, et al. Thiazolo[3,2-b]-1,2,4-triazole-5(6H)-one substituted with ibuprofen: novel non-steroidal anti-inflammatory agents with favorable gastrointestinal tolerance. Eur J Med Chem. 2012; 57:398–406.
45. Sikiric P, Seiwerth S, Rucman R, Turkovic B, Rokotov DS, Brcic L, et al. Toxicity by NSAIDs. Counteraction by stable gastric pentadecapeptide BPC 157. Curr Pharm Des. 2013; 19:76–83.

FURTHER READING

1. Díaz-González F, Sánchez-Madrid F. NSAIDs: learning new tricks from old drugs. Eur J Immunol 2015; 45:679–86.
2. Hunt RH, Choquette D, Craig BN, De Angelis C, Habal F, Fulthorpe G, Stewart JI, et al. Approach to managing musculoskeletal pain: acetaminophen, cyclooxygenase-2 inhibitors, or traditional NSAIDs? Can Fam Physician 2007; 53:1177–84.

Chapter

21

Glucocorticoid Therapy in Rheumatic Diseases

Vinod Ravindran

INTRODUCTION

Glucocorticoids (GCs); also called corticosteroids or more loosely, "steroids" are one of the most widely used groups of medicines. They have a special place in the management of autoimmune diseases such as rheumatoid arthritis (RA) and vasculitides. After the extremely successful initial clinical trial published in 1948, wherein cortisone was used to in a patient with RA to dramatic improvement, GCs rapidly consolidated their position as a viable therapeutic option in the management of RA.[1] Since then, the uses of GCs in many more autoimmune rheumatic diseases have been established. However, therapeutic advances in the management of these conditions and concerns regarding the safety of GCs have led to continuous re-appraisal of their relevance. At the same time, effective strategies to deal with adverse effects of GCs have been developed. Newer GC molecules hold promise for optimum use of GCs in future. In this review, an up-to-date summary of some of these issues is presented.

MECHANISMS OF ACTION

The efficacy of GCs in alleviating inflammatory disorders results from the pleiotropic effects of the glucocorticoid receptor (GCR) on multiple signalling pathways. Evidence indicates that the GCs inhibit inflammation through all three following mechanisms: Direct and indirect genomic effects and non-genomic mechanisms.

Genomic Pathways

At physiological or low GC doses after easily passing through the cell membranes into the cytoplasm, cortisol binds with high affinity to a cytoplasmic GCR-alpha (cGCRα). Within the cell, cortisol acts in two ways. First, the cortisol-GCR complex moves to the nucleus and activates genomic signalling by interacting with GC response elements (GREs). The resulting complex recruits either coactivator (for transcriptional activation or transactivation) or corepressor (for transcriptional repression or transrepression) proteins that modify the structure of chromatin, thereby facilitating or inhibiting assembly of the basal transcription machinery and the initiation of transcription by RNA polymerase II.[2] Broadly speaking, transrepression leads to anti-inflammatory effects of GCs and transactivation leads to many of the recognised side-effects of GCs. This distinction is not clear-cut, however, and exceptions include osteocalcin and osteoprotegerin, proteins that are critical to new bone formation and are repressed.

Second, regulation of other GC-responsive genes involves interactions between the cGCR complex and other transcription factors, such as nuclear factor-kB (NF-kB).[3]

Non-genomic Pathways

At higher concentrations GCs are thought to interact with membrane-bound receptors or directly with the membrane to induce non-genomic (i.e. not mediated by changes in gene expression) effects. The exact mechanisms of non-genomic signalling remain unclear but are probably responsible for very rapid effects of high-dose GC (e.g. pulsed intravenous methylprednisolone) such as inhibition of arachidonic acid release.[4]

Therefore, GCs have a mixture of therapeutic effects depending on the dose used (Table 21.1). This has important implications for consideration of the potential for adverse effects, as a simple extrapolation from effects observed in high-dose regimens to the expectation of similar but lesser effects at lower doses may not be appropriate if the adverse effect is related to mechanisms which are not employed at lower doses.

DEFINITIONS OF "DOSES"

The recognition of different mechanisms of action also provides an underlying rationale to the clinically recognisable differences in effect for different doses, and for the definition of 'low', 'medium' and 'high' doses of GC. In the clinical setting, GC doses can be clustered using the following scheme: Low doses are <7.5 mg of prednisone-equivalent doses per day; medium doses are between 7.5 and 30 mg per day; high doses >30 mg and up to 100 mg prednisone-equivalent per day; very high doses >100 mg prednisone-equivalent per day; finally, when doses are >250 mg of prednisone-equivalent per day for a few days, usually no more than 5 days, it is termed as pulse therapy.[5] It is recommended that a treatment with ≤7.5 mg prednisolone equivalent a day should be considered as low dose GC treatment because this dose occupies <50% of the GC receptors, this dose is often used for the maintenance therapy for many rheumatic diseases requiring GCs and relatively few adverse effects are expected. The recognition that a relative hypocortisolism is present in chronic inflammatory conditions such as RA lead observers to regard low dose GC treatment as a means of replacement therapy for reduced adrenal production. Therefore most authors now regard ≤7.5 mg prednisolone/day to be near the physiologic range.

Table 21.1: Key differences between genomic and non-genomic actions of glucocorticoids

Characteristics	Mechanism	
	Genomic	*Non-genomic*
Glucocorticoid dose	Low	High
Onset of action	Hours to days	Minutes
Efficacy	Moderate	High
Adverse effects	Low	High

TYPES OF GLUCOCORTICOIDS AND THEIR POTENCY

It has been suggested that equivalent GC doses should be calculated using prednisone as the comparator of the group regarding the different potencies. The traditional ways to measure GC and mineralocorticoid (MC) potencies of steroids refer to anti-inflammatory (rat ear oedema test, McKenzie vasoconstriction test) or metabolic (liver glycogen assay) effects and sodium retention. Depending on the testing conditions, absolute potency values vary considerably. It is, therefore, extremely difficult, if not impossible, to arrange the individual results from different research groups to compile a comprehensive list (Table 21.2). Some of the recent studies using more standardised ways of comparing MC and GC properties of different steroids have found clinically relevant differences in the potencies such as for deflazacort (DFZ).[6] Deflazacort, was initially thought to be as effective as prednisone while inducing fewer adverse events (less bone wasting, less significant alteration in the pattern of growth hormone (GH) secretion, no negative on the affect the overall amount of GH secreted, and on the carbohydrate metabolism). Overall, as the evidence for some of these beneficial effects comes mainly from the animal models[7] together with potency issues and availability of studies of relatively poor quality using very small number of patients, DFZ usage in clinical rheumatology in particular in the management of RA appears to offer no superiority compared to routinely used GCs.[8,9]

Table 21.2: Potency comparison of selected therapeutic glucocorticoids

Agent	*GC potency*	*MC potency*
Hydrocortisone	1	0.002
Deflazacort	4	No data
Prednisolone	4	0.002
Methylprednisolone	5	0.0013
Betamethasone	25	0
Dexamethasone	25	0

GC; glucocorticoid, MC; mineralocorticoid

Newer preparations like modified-release prednisolone are now available, which when ingested while going to sleep, release GC into the bloodstream in the early hours of the morning, mimicking the circadian rhythm and associated with potentially lesser hypothalamopituitary axis suppression.

GLUCOCORTICOID USE IN RHEUMATOLOGICAL CONDITIONS

In rheumatology, GCs are utilized in several different ways and by various routes (Table 21.3). Prior to discussing some of the more specific indications it would be useful to consider the following EULAR evidence based recommendations[10] on the management of systemic GC therapy in rheumatic diseases:

1. The adverse effects of GC therapy should be considered and discussed with the patient before GC therapy is started. This advice should be reinforced by giving information regarding GC management. If GCs are to be used for a more prolonged period of time, a "glucocorticoid card" is to be issued to every patient, with the date of commencement of treatment, the initial dosage and the subsequent reductions and maintenance regimens.
2. Initial dose, dose reduction and long-term dosing depend on the underlying rheumatic disease, disease activity, risk factors and individual responsiveness of the patient. Timing may be important, with respect to the circadian rhythm of both the disease and the natural secretion of GCs.

Table 21.3: Glucocorticoid utilization in rheumatic diseases

Routes	*Strategies*
Oral	Short-term, high dose Medium term, low to moderate dose Long-term, low dose
Intramuscular	At initiation and for flares
Intravenous	Pulse therapy
Intra-articular	Single or more affected joints, refractory joint
Intra-lesional	Epicondylitis, bursitis, tendinosis, etc.

3. When it is decided to start GC treatment, comorbidities and risk factors for adverse effects should be evaluated and treated where indicated; these include hypertension, diabetes, peptic ulcer, recent fractures, presence of cataract or glaucoma, presence of (chronic) infections, dyslipidemia and co-medication with non-steroidal anti-inflammatory drugs.
4. For prolonged treatment, the GC dosage should be kept to a minimum, and a GC taper should be attempted in case of remission or low disease activity; the reasons to continue GC therapy should be regularly checked.
5. During treatment, patients should be monitored for body weight, blood pressure, peripheral oedema, cardiac insufficiency, serum lipids, blood and/or urine glucose and ocular pressure depending on individual patient's risk, GC dose and duration.
6. If a patient is started on prednisone >7.5 mg daily and continues on prednisone for more than 3 months, calcium and vitamin D supplementation should be prescribed.
7. Antiresorptive therapy with bisphosphonates to reduce the risk of GC-induced osteoporosis (GIOP) should be based on risk factors, including bone-mineral density measurement.
8. Patients treated with GCs and concomitant non-steroidal anti-inflammatory drugs (NSAIDs) should be given appropriate gastroprotective medication, such as proton pump inhibitors or misoprostol, or alternatively could switch to a cyclo-oxygenase-2 selective inhibitor.
9. All patients on GC therapy for longer than 1 month, who will undergo surgery, need perioperative management with adequate GC replacement to overcome potential adrenal insufficiency.
10. GCs during pregnancy have no additional risk for mother and child.
11. Children receiving GCs should be checked regularly for linear growth and considered for growth-hormone replacement in case of growth impairment.

Rheumatoid Arthritis

Though many patients now receive disease-modifying antirheumatic drugs (DMARDs) and biologics, GCs are commonly used in the treatment of RA with recent registry based studies showing, in the UK that the long-term use of GCs for RA between 1991 and 2008 did not decrease[11] and in Germany that the percentage of GC treated patients with RA remained fairly constant over the years at about 56%, but with an increase in the use of GC dosages of ≤7.5 mg/prednisone equivalent per day.[12] Several meta-analyses have confirmed the therapeutic advantages of short-term, medium-term and long-term GC therapy in RA in combinations with DMARDs.[13–15] The rapidity of onset of symptom relief with GCs results in their near-universal use as a "bridge" therapy until the action of DMARDs sets in. However, in early RA (disease duration of < 2 years), the use of low-dose GCs is based not solely on disease symptoms, but also on joint sparing effects on the long-term, as GCs can be categorized as DMARDs; based on category IA level of evidence presently available.[16] The current recommendations are therefore to "use GCs added at low to moderately high doses to synthetic DMARD monotherapy (or combinations of synthetic DMARDs) to provide benefit as initial short-term treatment, but should be tapered as rapidly as clinically feasible".[17] Scenarios of GC use in RA are presented in Box 21.1.

Small and Medium Vessel Vasculitis

In vasculitis the use of high-dose GCs is an important part of remission induction therapy.[18] There are no direct clinical trials examining the role of GC therapy but every clinical trial or cohort study conducted has used GC therapy in combination with immunosuppressive therapy. It is common practice to commence prednisolone or prednisone at 1 mg/kg/day as also seen in recent clinical trials. Current recommendations state that the initial high dose should be maintained for 1 month, and should not be reduced to <15 mg/day for the first 3 months.[18] The GC dose should then be tapered to a maintenance dose of 10 mg/day or less during remission. When a rapid effect is needed, intravenous pulsed methylprednisolone may be used in addition to the oral prednisolone as part of remission induction therapy. Furthermore, GC in low doses has been recommended in the remission-maintenance therapy in combination with immunosuppressive therapy such as azathioprine, leflunomide or methotrexate.[18]

Box 21.1: Indications for the possible use of glucocorticoids in RA

- Bridge therapy for patients who have experienced a severe functional decline with limitations interfering with necessary daily living or vocational activities.
- Patients with NSAID contraindication who have acute inflammation unlikely to respond rapidly enough to second-line agents.
- Rheumatoid vasculitis or other serious life-threatening or organ damaging manifestation.
- Men or women not at reduced risk of bone loss, for whom GCs may be reasonable in combination therapy regimens with a single daily dose of <7.5 mg/day.
- Pregnant or lactating women with active RA.

Temporal Arteritis and Polymyalgia Rheumatica

The signs and symptoms of polymyalgia rheumatica (PMR) and temporal arteritis (TA) overlap and can often be confused, but both are treated with GCs. Early intensive therapy with high-dose GC induces remission in patients with large vessel vasculitis.[19] The initial dose of prednisolone is 1 mg/kg/day (maximum 60 mg/day) and the initial high-dose should be maintained for a month and tapered gradually. The taper should not be in the form of alternate day therapy, as this is more likely to lead to a relapse of vasculitis. At 3 months, the GC dose in clinical trials has been between 10–15 mg/day. The duration of GC therapy for patients with TA is variable and can extend to several years, but some patients may not be able to tolerate discontinuation of GC therapy due to recurrent disease or secondary adrenal insufficiency. The high doses of GC required initially and the duration of treatment can be associated with GC side effects, and epidemiological studies have shown a high incidence of fractures. These risks can be substantially reduced with concurrent osteoporosis prophylaxis and this should now be routine practice.

Systemic Lupus Erythematosus

In the treatment of SLE without major organ manifestations, GCs are indicated.[20] Concomitant use of anti-malarials and immunosuppressive drugs may help keep daily doses of prednisone <7.5 mg/day, or even to withdraw it. In patients with severe lupus activity higher dose of GC (1 mg/kg/day) is being used, though recent evidence suggest that significantly lower doses (0.5 mg/kg/day) might be equally effective and have less adverse effects. In patients with proliferative lupus nephritis, GCs in combination with immunosuppressive agents are effective against progression to end-stage renal disease.[20] With regard to pulse methylprednisolone, a recent prospective study reported similar efficacy of 500 mg/day during three consecutive days vs. 1000 mg/day for three to five consecutive days in treating severe lupus flares.[21] The lower dose group suffered a significant lower number of serious infections and it is likely that lower dose of GCs in "pulse therapy" would become a standard practice in near future. However in SLE, GC usage is a major determinant of accrued damage.[22] Pioneering recent work from UK in treating lupus nephritis without oral GC leads to hope of treating SLE without long-term GC.[23]

Gout

GCs (and in the past ACTH) have been used in gout mainly in treating an acute attack. In acute gout, GC can either be given systemically, intra-venous, intra-muscular or intra-articular routes if one or two joints are affected. GCs are a good alternative where NSAID and/or colchicine cannot be used (e.g. in patients with chronic kidney disease or liver disease) or in refractory cases. A recent randomized control trial (RCT) involving 90 patients compared the analgesic efficacy of oral prednisolone (30 mg daily for 5 days) plus paracetamol vs. oral indomethacin (50 mg three times daily for 2 days and 25 mg three times daily for 3 days) plus paracetamol.[24] Pain reduction was similar but the steroid group experienced fewer adverse events (AEs). Similar results were found in a randomized double-blind trial demonstrating that 35 mg of prednisolone gives equivalent pain relief to naproxen 500 mg twice a day.[25] Glucocorticoids may have fewer AEs than other acute treatments when used for short-term, particularly in the elderly.

ADVERSE EFFECTS OF GC USE

GCs may cause a variety of adverse effects. Recently the evidence has emerged regarding two patterns of GC associated AEs namely "threshold" and "linear".[26] For cushingoid phenotype, ecchymosis, leg oedema, mycosis, parchment-like skin, shortness of breath and sleep disturbance, a "linear" increase in the frequency with increasing dose has been noted. The term "threshold" pattern describes an elevation in the frequency of health problems beyond a certain threshold value. The threshold for the increase in glaucoma, as well as depression/listlessness and an increase in blood pressure was a dosage of more than 7.5 mg/day.[26] Dosages of 5 mg/day or more was identified to be relevant for epistaxis and weight gain. A very low threshold was seen for eye cataract (<5 mg/day).[26]

Whereas the clinical benefits of low-dose GC and their ability to inhibit the progression of radiological damage is generally accepted in RA, the balance of benefits and risks associated with prolonged use is less certain because of concerns over toxicity. A comprehensive review of four prospective trials in RA and the COBRA study concluded that definitive associations of low-dose GC with many adverse effects remain elusive.[27] In general, the precise data on the frequency and severity of AEs of GCs and methods of assessment are not well-established. By using patient withdrawal as a marker of adverse events, a recent meta-analysis of six randomised controlled trials utilizing low-dose GC lasting? 2 years found no significant difference in toxicity compared to placebo.[15] Nevertheless, the long-term use of GC in RA continues to generate debate and in this regard standardization of assessment of AEs may also enable aetiological and casual inferences.[28]

Evidence-based recommendations for the safer use of systemic GC therapy in rheumatic diseases have been published recently.[10] These include patient education, monitoring for AEs and use of concomitant therapy to reduce unwanted side effects and special safety advice. Adherence to these guidelines might reduce the AEs in patients on GC therapy.

Table 21.4 lists some of the important adverse effects of GCs. A special mention has to be made of GC induced osteoporosis (GIOP). There is no doubt that long-term GC use even in moderate doses is associated with reduced bone mineral density, particularly loss of trabecular bone, and increased fracture risk at the hip and spine. The mechanisms of GIOP are multiple but include disturbed calcium homeostasis, osteoblast dysfunction and osteocyte apoptosis. Although

Table 21.4: Toxicity of glucocorticoids

Organ system	*Dose*		
	Probably with high dose only	*High or low dose*	
		Probably	*Definitely*
General	Infections		Weight gain Redistribution of body fat Facial dema Withdrawal symptoms
Musculoskeletal	Osteonecrosis	Osteoporosis	
Gastrointestinal	Pancreatitis Bowel rupture	Peptic ulcer	
Cardiovascular	Hypertension Dyslipidemia Atherosclerosis		
Reproductive	Foetal wastage		
Neurological	Psychosis		
Cutaneous	Delayed wound healing		Purpura Skin atrophy Acne Easy bruising
Eye			Glaucoma Cataract
Endocrine		HPA dysfunction	Diabetes

fracture risk rises soon after GCs are commenced it also rapidly declines when treatment is stopped and this is independent of effects on bone mineral density. Chronic inflammatory conditions such as RA are themselves associated with an increased fracture risk and bone density loss, and therefore distinguishing disease effects from GC effects at low doses is complicated. Nevertheless several studies including from India have confirmed much higher prevalence of reduced bone mass in RA.[29]

Guidelines in several countries and by several different organisations have now been produced for managing GIOP. The Indian Rheumatology Association (IRA) has produced guidelines for the management of GIOP which has also been endorsed by the Endocrine Society of India and Indian Society for Bone and Mineral Research.[30] The key recommendations are as follows:

1. Use minimum dose of oral GC for shortest period of time.
2. Use GC sparing immunosuppressive drugs like methotrexate, azathioprine whenever indicated.
3. Use inhalational route in patients with asthma and enema in ulcerative colitis.
4. Use supportive therapy like balanced nutrition, adequate calcium, regular exercise, avoidance of smoking and alcohol.
5. Assess fracture risk at initiation of therapy by using FRAX tool.
6. Start primary prevention in high-risk group (post-menopausal women, men >65 years and those with previous fragility fracture).
7. In those without high-risk factors do a baseline BMD if T-score <–1.5 start primary prevention.
8. If baseline T-score >–1.5, follow-up and assess BMD after 1–2 years if therapy with GC is being continued.
9. Assess fracture risk at initiation of therapy by using FRAX tool (FRAX tool for Indians is available and can be applied to our country: http://www. sheffield.ac.uk/FRAX).

Conclusion

From the foregoing discussion, it is apparent that GCs have several desirable clinically beneficial effects. At the same time, ever growing concerns regarding their toxicity combined with issue of GC

Box 21.2 Key learning points

- Glucocorticoids (GCs) are important therapeutic option in several rheumatic diseases.
- Although no dose of GC is totally without adverse effects, low-dose and high-dose GC have different mechanisms of action, therapeutic and adverse effects.
- Better understanding of GC mechanism of actions (genomic and non-genomic) helps in understanding issues with GC adverse effects observed at a particular dose.
- A disease modifying effect of GC therapy in RA has become established.
- Several newer GCs molecules have the potential to enhance the current role of GCs by selective augmentation of the beneficial effects.
- Adherence to relevant guidelines such as for GC induced osteoporosis is mandatory in minimizing the GC adverse effects.

resistance have driven research in to newer GC molecules. A logical approach to diminish the adverse effects of GCs is the development of novel drugs that induce transrepression but stimulate very little, or virtually no, transactivation.[31, 32] Novel drug delivery systems and combinations are also being experimented.[32]

GCs remain an important treatment option in many conditions in rheumatology, and new understanding of disease-specific and dose-specific effects has resulted in renewed effort to make full use of their potential. Box 21.2 summarises some important points presented in this review.

REFERENCES

1. Hench PS, Kendall EC, Slocumb CH, Polley HF. The effect of a hormone of the adrenal cortex (17-hydroxy-11-dehydrocorticosterone: compound E) and of pituitary adrenocorticotropic hormone on rheumatoid arthritis. Mayo Clin Proc 1949; 24: 181–97.
2. Hebbar PB, Archer TK. Chromatin remodelling by nuclear receptors. Chromosoma 2003; 111:495–504.
3. Nissen RM, Yamamoto KR. The glucocorticoid receptor inhibits NF-κB by interfering with serine-2 phosphorylation of the RNA polymerase II carboxy-terminal domain. Genes Dev 2000; 14:2314–29.
4. Rhen T, Cidlowski JA. Anti-inflammatory actions of glucocorticoids-new mechanisms for old drugs. N Engl J Med 2005; 353:1711–23.
5. Buttgereit F, da Silva JA, Boers M, Burmester GR, Cutolo M, Jacobs J, et al. Standardised nomenclature for glucocorticoid dosages and glucocorticoid treatment regimens: current questions and tentative answers in rheumatology. Ann Rheum Dis 2002; 61:718–22.
6. Grossman C, Scholz T, Rochel M, Bumke-Vogt C, Oelkers W, Pfeiffer AFH, et al. Transactivation via the human glucocorticoid and mineralocorticoid receptor by therapeutically used steroids in CV-1 cells: a comparison of their glucocorticoid and mineralo-corticoid properties. European Journal of Endocrinology 2004; 151: 397–406.
7. Latta K, Krieg Jr RJ, Hisana S, Veldhuis JD, Chan JCM. Effect of the synthetic glucocorticoid, deflazacort, on body growth, pulsatile secretion of GH and thymolysis in the rat. European Journal of Endocrinology. 1999; 140:441–46.
8. Mithun MC, Gonzalez-Perez O, Ramos-Remus C, Ravindran V. Deflazacort in rheumatology: Where does it stand? Indian J Rheumatol 2014; 9; 161–162.
9. Kroggsgaard MR, Thamsborg G, Lund B. Changes in bone mass during low-dose corticoid treatment in patients with polymyalgia rheumatica: a double blind, prospective comparison between prednisolone and deflazacort. Ann Rheum Dis 1996; 55:143–6.
10. Hoes JN, Jacobs JWG, Boers M, Boumpas D, Buttgereit F, Caeyers N, et al. EULAR evidence-based recommendations on the management of systemic glucocorticoid therapy in rheumatic diseases. Ann Rheum Dis 2007; 66:1560–7.
11. Fardet L, Petersen I, Nazareth I. Prevalence of long-term oral glucocorticoids prescriptions in UK over the last 20 years. Rheumatology 2011; 50:1982–90.
12. Buttgereit F. Do the treatment with glucocorticoids and/or the disease itself drive the impairment in glucose metabolism in patients with rheumatoid arthritis? Ann Rheum Dis 2011; 70:1881–83.
13. Gotzsche PC, Johansen HK. Short-term low dose corticosteroids vs. placebo and non-steroidal anti-inflammatory drugs in rheumatoid arthritis. Cochrane Database Syst Rev 2005; (1):CD000189.
14. Criswell LA, Saag KG, Sems KM, Robinson V, Shea B, Wells G et al. Moderate-term, low-dose corti-costeroids for rheumatoid arthritis. Cochrane Database Syst Rev 1998 ;(3):CD001158.
15. Ravindran V, Rachapalli S, Choy EH. Safety of medium-to long-term glucocorticoid therapy in rheumatoid arthritis: a meta-analysis. Rheumatology (Oxford) 2009; 48:807–11.
16. Kirwan JR, Bijlsma JWJ, Boers M, Shea BJ. Effects of glucocorticoids on radiological progression in rheumatoid arthritis. Cochrane Database Syst Rev 2007;(1):CD006356.

17. Smolen JS, Landewé R, Breedveld FC, Dougados M, Emery P, Gaujoux-Viala C, et al. EULAR recommendations for the management of rheumatoid arthritis with synthetic and biological disease-modifying antirheumatic drugs. Ann Rheum Dis 2010; 69: 964–97.
18. Mukhtyar C, Guillevin L, Cid MC, Disrupt B, de Grout K, Gross W, et al. EULAR recommendations for the management of primary small and medium vessel vasculitis. Ann Rheum Dies 2009; 68:310–17.
19. Mukhtyar C, Guillevin L, Cid MC, Dasgupta B, de Groot K, Gross W, et al. EULAR recommendations for the management of large vessel vasculitis. Ann Rheum Dis 2009; 68:318–23.
20. Bertsias G, Ioannidis JPA, Boletis J, Bombardieri S, Cervera R, Dostal C, et al. EULAR recommendations for the management of systemic lupus erythematosus. Report of a Task Force of the EULAR Standing Committee for International Clinical Studies Including Therapeutics. Ann Rheum Dis 2008;67: 195–05.
21. Kong KO, Badsha H, Lian TY, Edwards CJ, Chng HH. Low-dose pulse methylprednisolone is an effective therapy for severe SLE flares. Lupus 2004; 13:212–3.
22. Bruce IN, O'Keeffe AG, Farewell V, Hanly JG, Manzi S, Su L, et al. Factors associated with damage accrual in patients with systemic lupus erythematosus: results from the Systemic Lupus International Collaborating Clinics (SLICC) Inception Cohort. Ann Rheum Dis 2015; 74:1706–13.
23. Condon MB, Ashby D, Pepper RJ, Cook HT, Levy JB, Griffith M, et al. Prospective observational single-centre cohort study to evaluate the effectiveness of treating lupus nephritis with rituximab and mycophenolate mofetil but no oral steroids. Ann Rheum Dis 2013;72:1280–1286.
24. Man CY, Cheung I, Cameron I, Peter A, Rainer TH. Comparison of oral prednisolone/paracetamol and oral indomethacin/paracetamol combination therapy in the treatment of acute gout like arthritis: a double-blind, randomized, controlled trial. Ann Emerg Med 2007; 49:670–7.
25. Janssens HJ, Janssen M, Van de Lisdonk EH, van Riel PL, van Weel C. Use of oral prednisolone or naproxen for the treatment of gout arthritis: a double-blind, randomised equivalence trial. Lancet 2008; 371:1854–60.
26. Huscher D, Thiele K, Gromnica-Ihle E, Hein G, Demary W, Dreher R, et al. Dose-related patterns of glucocorticoid-induced side effects. Ann Rheum Dis 2009; 68:1119–24.
27. Da Silva JA, Jacobs JW, Kirwan JR, Boers M, Saag KG, Inês LB, et al. Safety of low dose glucocorticoid treatment in rheumatoid arthritis: published evidence and prospective trial data. Ann Rheum Dis 2006; 65:285–93.
28. van der Goes MC, Jacobs JW, Andrews T, Blom-Bakkers MA, Buttgereit F, et al. Monitoring adverse events of low-dose glucocorticoids therapy: EULAR recommendations for clinical trials and daily practice. Ann Rheum Dis 2010; 69:1913–9.
29. Jain S, Ravindran V, Mathur DS. Detection of low bone mass using quantitative ultrasound measurements at calcaneus: comparative study of an Indian rheumatoid arthritis cohort. Int J Rheum Dis. 2008; 11:414–20.
30. Krishnamurthy V, Sharma A, Aggarwal A, Kumar U, Amin S, Rao URK, et al. Indian rheumatology association guidelines for management of glucocorticoid-induced osteoporosis. Ind J Rheumatol 2011; 6:68–75.
31. Baudy AR, Reeves EKM, Damsker JM, Heier C, Garvin LM, Dillingham BC, et al. ?9,11 Modification of glucocorticoids dissociates nuclear factor—B inhibitory efficacy from glucocorticoid response element-associated side effects. JPET 2012; 343: 225–32.
32. Ravindran V. Newer glucocorticoids: Overcoming mechanistic hurdles. Indian J Rheumatol 2012; 7:221–225.

FURTHER READING

1. Krishnamurthy V, Sharma A, Aggarwal A, Kumar U, Amin S, Rao URK, et al. Indian rheumatology association guidelines for management of glucocorticoid-induced osteoporosis. Ind J Rheumatol 2011; 6:68–75.
2. Rhen T, Cidlowski JA. Antiinflammatory actions of glucocorticoids-new mechanisms for old drugs. N Engl J Med 2005; 353:1711–23.
3. van der Goes MC, Jacobs JW, Andrews T, Blom-Bakkers MA, Buttgereit F, et al. Monitoring adverse events of low-dose glucocorticoids therapy: EULAR recommendations for clinical trials and daily practice. Ann Rheum Dis 2010; 69:1913–9.

Chapter

22

Disease-modifying Antirheumatic Drugs

Ashok Kumar

INTRODUCTION

Disease-modifying antirheumatic drugs (DMARDs) are the key therapeutic agents for RA (Table 22.1). They act at some crucial steps in the aetiopathogenetic mechanisms thereby retarding the progression of disease to negligible or low level. Steady progress in therapeutics of rheumatoid arthritis (RA) in recent years has culminated in a radical change in our approach to the treatment of this disease. Today, the goal of drug therapy of RA is complete remission or low disease activity. 'Drug free' remissions have also been reported, though infrequently.[1] To be designated a DMARD, a drug must change the course of RA for at least 1 year as evidenced by sustained improvement in physical function, decreased inflammatory synovitis, and slowing or prevention of structural joint damage. Low doses of systemic steroids have also been shown to have a disease-modifying activity in RA.[2] Steroids rapidly reduce the burden of inflammation and facilitate induction of remission with routinely used DMARDs. The latter are divided into conventional and biological DMARDs. The biological DMARDs represent a very important component of therapeutic armamentarium of modern rheumatology and will be discussed in a separate chapter. The present chapter deals with so-called conventional synthetic DMARDs, i.e. non-biological DMARDs.

Table 22.1: 'Conventional' disease-modifying antirheumatic drugs

1. Methotrexate
2. Sulphasalazine
3. Antimalarials (hydroxychloroquine, chloroquine)
4. Leflunomide
5. Minocycline
6. Cyclosporine
7. Gold (sodium aurothiomalate, auranofin)
8. Miscellaneous group (levamisole, mycophenolate mofetil, tacrolimus)

RELEVANT CLINICAL PHARMACOLOGY OF DMARDs

Methotrexate

Methotrexate is currently the sheet anchor of treatment of RA. Therapeutic dose ranges from 7.5 to 25 mg orally, once a week. Depending on the grade of disease activity, the starting dose could be 7.5 mg/week to 15 mg/week. Dose may be escalated by 2.5 mg increments at intervals of 4 weeks with careful monitoring of disease activity score (DAS28) and drug toxicity. Absorption falls by at least 30% when the weekly dose is 15 mg or more.[3] Therefore, parenteral administration (subcutaneous or intramuscular) is preferred for doses in the range of 15–25 mg/week. Beneficial effect is appreciable in 1–2 months and peaks at 6 months. Most importantly, patients show long-term adherence to methotrexate monotherapy: >70% continue to use the drug for 5 years. Such high retention rates were not known with older DMARDs such as injectable gold salts. Anorexia and nausea in the 24 hours after the dose is the commonest side effect. Other side effects include diarrhea, oral ulcers, stomatitis, elevated liver enzymes, hepatic fibrosis, skin rash, methotrexate nodulosis, methotrexate pneumonitis and risk of opportunistic infections. Gastrointestinal side effects can be reduced by administering folic acid (5–15 mg) per week or ondansetron. The drug is contraindicated in pregnancy, renal insufficiency and liver disease. Caution is needed in obesity, diabetes and alcohol abuse. In frail, elderly patients, slow and careful dose escalation is recommended because cytopenias occur more frequently in this age group. At the same time, the elderly must not receive suboptimal treatment. Methotrexate can be safely continued in the perioperative period in patients undergoing orthopaedic surgery.[4] Concerns have been raised about possible increased risk of lymphoma with long-term use of methotrexate. However, the results of one large retrospective study do not support this possibility.[5] Non-steroidal anti-inflammatory drugs can reduce the renal clearance of methotrexate thereby increasing its toxicity. Hydroxychloroquine when combined with methotrexate, enhances the effect of methotrexate by increasing the 'area under the curve', i.e. plasma concentration × time.

Baseline evaluation before starting methotrexate comprises blood counts, serum creatinine, liver function tests and chest X-ray. Patients at risk should also be screened for hepatitis B and C and HIV serology. A urine pregnancy test may be done where appropriate, to rule out pregnancy. Periodic monitoring includes blood counts and liver function tests every 4–8 weeks and serum creatinine every 6 months. If liver enzymes are increased less than twice the upper normal value, it is recommended to repeat these tests in 2–4 weeks. If liver enzymes are increased 2–3 times the upper normal value, it is recommended to decrease the dose and repeat measurements every 4 weeks. If ALT and AST concentrations are greater than 3 times the upper limit of normal on 2 occasions, methotrexate should be stopped.[6] The possibility that enzyme elevation may be due to other medications (e.g. NSAIDs, antituberculous drugs), alcohol, viral hepatitis or autoimmune hepatitis should not be overlooked and carefully ruled out before abandoning methotrexate. In India, non-alcoholic fatty liver disease (NAFLD) is very common cause of mild but persistent elevation of transaminases.[7] In this situation, methotrexate therapy can be started after obtaining a careful evaluation by gastroenterologist and instituting appropriate measures for management of NAFLD.

Sulphasalazine

This drug is a poorly absorbed sulphonamide, which splits into 5-aminosalicylic acid (5-ASA) and sulphapyridine after reaching the colon. Sulphapyridine is responsible for efficacy in RA while 5-ASA is the component, which benefits ulcerative colitis. Sulphasalazine is started at 500 mg/day and then increased by 500 mg weekly to 2–3 gm daily. Side-effects include nausea, abdominal pain, skin rashes, CNS disturbances, reversible oligospermia, discoloration of urine and sweat, blood dyscrasias and elevated liver enzymes. It is generally a well-tolerated drug with mild side-effects. Its efficacy is lower than that of methotrexate. Blood counts and LFTs should be regularly monitored every 4 weeks for 3 months and then every 3 months. If the results are normal during first year, monitoring may be done 6 monthly during second year. Thereafter, the monitoring may be discontinued.[8]

Antimalarials

Antimalarials are used either in mild RA or in combination with other DMARDs. Both chloroquine (CQ) and hydroxychloroquine (HCQ) are equally efficacious for the treatment of RA. HCQ is preferred because ocular toxicity is rare with its use. Both have very long half-lives and get accumulated in tissues (steady state reached only after 3 to 4 months). Recommended dose for CQ is 5 mg/kg/day and that for HCQ is 6.5 mg/kg/day. Maintenance dose is 50% of initial dose (some patients may require higher dose). Length of time to onset of benefit varies from 1 to 6 months. Retinal toxicity is rare at doses below 6.5 mg/kg/day. American Academy of Ophthalmology recommends a baseline screening within first year of use and annual check-up after 5 years of use of HCQ.[9] Recommended screening procedures include ocular examination, automated visual field examination and in addition, one of the 3 others: SD-OCT (spectral domain ocular coherence tomography), mfERG (multifocal electroretinogram) and FAF (fundus autofluorescence). Amsler grid and colour testing may be used only as adjuncts. Other side-effects of HCQ include nausea, indigestion, pruritus, skin rash, hyperpigmentation of exposed parts and rarely, hypoglycaemia.

Leflunomide

Leflunomide inhibits dihydro-orotate dehydrogenase, an enzyme involved in *de novo* pyrimidine synthesis. It is comparable to methotrexate (MTX) and sulphasalazine (SSZ) in efficacy and retards the development of erosions. The practice of administering a loading dose of 100 mg/day for 3 days is no longer recommended. Leflunomide is started directly at 20 mg/day, orally. Beneficial effect starts appearing after 4–6 weeks. The drug has a good safety profile. Liver toxicity, nausea, diarrhoea, alopecia, weight loss (sometimes impressive) and occasionally hypertension are known to occur. Many of these side-effects tend to decrease with continued use. The drug tends to accumulate in the body and may not get eliminated till 3 years. It was found to be highly teratogenic in animal studies and is, therefore, contraindicated in patients who are pregnant or who can become pregnant. Similar restrictions apply to males who may be contemplating fathering a child. Monitoring for this drug is done on the same lines as for methotrexate. In case of severe liver toxicity and bone marrow toxicity, cholestyramine (8 gm TDS for 11 days) is administered for elimination of the drug. The goal should be to achieve plasma concentrations below 0.02 mg/L. The drug can be used in patients with CKD.[10]

Minocycline

Minocycline has been shown to be beneficial in patients with early and mild RA. The antirheumatic effect of this semi-synthetic derivative of tetracycline is related to its immunomodulatory and anti-inflammatory, rather than to its antibacterial properties. Its dose is 100 mg BD. The common adverse effects include GI side effects, dizziness, rash and headaches. Persistent skin and mucosal hyperpigmentation, lupus-like syndrome and acute hepatic injury are less common.

Cyclosporine

Cyclosporine is a potent immunosuppressant approved for use at a dose of 3–5 mg/kg/day in RA. It selectively blocks the release of IL-1 from monocytes and IL-2 from helper T cells. Its absorption is usually incomplete and reduced by food. 'Go slow and go low' is the key to use of this drug. Complete blood counts, liver enzymes, serum creatinine and urine analysis should be done weekly until stable dose is reached and then monthly.

Rise in serum creatinine and blood pressure are common adverse effects, particularly at doses higher than 3 mg/kg/day. Elevation of serum creatinine >30% over baseline is an indication to stop the drug. Other adverse effects include hyperkalemia, gum hyperplasia, hypertrichosis and hepatotoxicity. Cyclosporine is efficacious but expensive as well as toxic. It is recommended for patients refractory to methotrexate. Most patients are unable to continue the drug beyond 2–3 years because of renal toxicity.

Gold Salts

Gold compounds are of 2 types: Water-soluble thiolates and the fat-soluble phosphine derivatives. Gold is eliminated very slowly from the body. Significant tissue concentrations have been found even 20 years after the last dose. Injectable gold is started at a test dose of 10 mg intramuscularly and then 50 mg weekly until a cumulative dose of 1 gm has been administered. If there is no improvement, the drug is stopped. If there is significant improvement, the frequency of injections is reduced to once in a fortnight for next 3 months, once in 3 weeks for 3 months and then continued at once a month. Oral gold (auranofin) is given at a dose of 3–6 mg/day.

Very few patients are able to continue gold beyond 5 years. Adverse reactions to gold are the major reason for discontinuation. These include skin rashes, oral ulcers, proteinuria, thrombocytopenia and pulmonary toxicity. Aplastic anaemia is rare but potentially fatal. Eosinophilia may predict adverse reactions. Monitoring of blood counts and urine protein should be done before each injection for 4–8 weeks and then before every 3rd or 4th injection. Oral gold is much less toxic but dose-related diarrhoea is a common side-effect.

Practical Approach to use of DMARDs

The current strategy of treatment of RA is 'treat-to-target'. The treatment is targeted at achieving remission/low disease activity. According to 2015 update of ACR recommendations, DMARD therapy should commence as soon as the diagnosis of RA is made.[11] Thus, both early RA (<6 months) and established RA (>6 months) qualify for DMARD therapy. If treatment is delayed, beneficial

Table 22.2: Summary of clinical pharmacology of conventional DMARDs

Drug	*Dosage*	*Adverse effects*	*Monitoring*
Methotrexate	7.5–25 mg per week (orally or subcutaneously)	GI, cytopaenias, ↑ AST and ALT, alopecia, stomatitis, infection, pneumonitis	CBC, LFT every 4–8 weeks
Sulphasalazine	2–3 gm daily (build up the dose slowly, over 1 month)	GI, cytopaenia, ↑ AST and ALT, photosensitivity	CBC, LFT every 4 weeks for 3 months, then 3 monthly; if stable for 2 years, may stop monitoring
Chloroquine	150 mg (base) daily for 3 months, then alternate day	GI, hyperpigmentation, pruritus, retinopathy	Fundoscopy and visual field testing, 6 monthly
Hydroxychloroquine	200 mg BD for 3 months, then OD	Retinopathy is very rare	Baseline evaluation (as per American Academy of Ophthalmology) and then annual evaluation after 5 years of use
Leflunomide	20 mg OD, orally (loading dose not needed)	GI, alopecia, mild weight loss, ↑ AST and ALT	CBC, LFT every 4–8 weeks
Minocycline	100 mg BD	GI side effects, dizziness, rash, pigmentation, lupus-like syndrome, headache	Monitoring not required
Cyclosporine	3–5 mg/kg/day, orally	↑ BP, ↑ hair, hypertrophy of gums, impaired renal function and LFT	CBC, LFT, creatinine, urinalysis weekly until dose is stable, then monthly; monitor BP
Gold thiomalate (Inj.)	10 mg IM stat (test dose) then 50 mg per week; frequency reduced after 1 gm total dose	Stomatitis, oral ulcers metallic taste, rashes, photosensitivity, pruritus, cytopaenias, proteinuria	CBC and urine protein before each injection for 1 month and then before every 4th injection
Auranofin (oral gold)	3–9 mg OD	Diarrhoea, stomatitis, rash	CBC and urine protein every month for 3 months and then every 3 months

effect will still occur but it may not be possible to avoid some irreversible joint damage. In fact, the new ACR-EULAR classification criteria for RA have been formulated primarily to facilitate early diagnosis of RA and obtain best outcome of treatment by instituting it early.

Two important parameters, which guide the treatment are: The level of disease activity and presence or absence of poor prognostic factors. Although various scores are available to quantify disease activity, DAS28 is the most commonly used in clinical practice. It requires 4 parameters: Tender joint count (out of 28), swollen joint counts (out of 28), patient's global assessment (0–100, VAS) and ESR. A DAS28 calculator can be downloaded for computing the score. DAS28 value (range: 0–10) categorises the disease activity into remission (<2.6), low (2.6–3.1), moderate (3.2–5.1) and high (>5.1). Poor prognostic factors include functional limitation (using HAQ DI or similar valid tools), extra-articular disease (rheumatoid nodules, rheumatoid vasculitis, Felty's syndrome), positive rheumatoid factor or anti-citrullinated peptide antibodies and bony erosions by radiograph. Presence of one or more, poor prognostic features, even with moderate disease activity is an indication for combination DMARD therapy in early RA. EULAR guidelines recommend monotherapy with methotrexate in a DMARD-naive patient or when poor prognostic factors are absent.[12] Systemic steroids may be used as "bridge therapy" to cover the lag period before beneficial effect of DMARDs appears. EULAR recommends short-term steroid treatment when initiating or changing DMARD therapy. Steroids should be withdrawn as soon as feasible. Monthly follow-up should be maintained until the target is achieved. In established RA with poor prognostic features, DMARD combinations can be progressively escalated with objective assessments every 2–3 months. Biological DMARDs in combination with methotrexate may be used in resistant cases.

DMARD Combinations

The basic aim of combining DMARDs is to increase the efficacy of treatment. Multiple drugs act synergistically at multiple points in the cascade of inflammation. Since mechanisms of action of most DMARDs remain poorly understood, combinations have been worked out quite empirically. Examples of clinical trials of efficacy of combination DMARD therapy include TICORA, COBRA and FinRaco.[13–15]

DMARDs may be combined in 3 different ways:

Step-up: Start with single agent and then keep on adding more DMARDs if the disease does not improve in 3 months. For example, in TICORA trial, patients were started on sulphasalazine in a stepwise escalating treatment protocol. Assessment was done every month to ensure tight control of disease. Methotrexate and hydroxychloroquine were added if the set goal of keeping DAS in 'low disease activity' range was not achieved. The dose of sulphasalazine and methotrexate was increased to maximum to achieve the goal. In addition, intra-articular steroid injections were used frequently (up to 3 joints per month). If this did not work, cyclosporine and methotrexate combination was substituted. Next step included substitution with leflunomide and injection gold.

Step-down: Start with 3 or 4 DMARDs, achieve remission and later withdraw one drug at a time, leaving one DMARD at the end for maintenance of remission. For example, in COBRA trial, patients were given a combination of sulphasalazine (2 gm/day), methotrexate (7.5 mg/week) and prednisolone (initially 60 mg/day, tapered in 6 weekly steps to 7.5 mg/day). By 28 weeks, steroid had been withdrawn and at 40 weeks, methorexate was discontinued. Sulphasalazine was then continued as a single DMARD maintenance therapy.

Parallel: Here multiple agents are combined from the start and are continued on a long-term basis. An agent is stopped in the event of adverse effect. In case of lack of efficacy of the regimen in about 3 months, all drugs may be stopped. For example, in Fin-RACo trial, combination therapy (sulphasalazine, methotrexate, hydroxychloroquine, and low-dose prednisolone) was continued for 2 years. The adverse effects of combination therapy were comparable with those of monotherapy with sulphasalazine + prednisolone.

When to Stop DMARD Therapy?

Attempts to discontinue DMARDs are usually met with relapse of disease. The current policy is to continue DMARDs indefinitely.

DMARDs during Pregnancy and Lactation

Sulphasalazine and hydroxychloroquine have been found safe during pregnancy as well as lactation.[16]

Complete Remission Rates

Drug-sustained remission rates with conventional DMARDs have been steadily rising in recent years. Tight control strategy as employed in TICORA trial, yielded a remission rate of 65%.[13] However, in routine clinical practice, remission rates are lower because of multiple reasons.

Role of Vaccinations

According to the treatment guideline of the American College of Rheumatology,[11] all patients commencing or while taking conventional DMARD therapy should be vaccinated, when indicated (based on age and risk) against pneumococcus, influenza, herpes zoster, human papillomavirus and hepatitis B.

The same also holds true for killed vaccines in case of biologic drugs. Herpes zoster vaccine is a live-attenuated vaccine and should not be given to patients already on biologic drug/tofacitinib. A gap of 4 weeks is advised between zoster vaccination and commencement of biologic therapy. This vaccine is indicated in all RA patients aged >50 years who are about to start biologic therapy or tofacitinib.

Key Points

1. Remission is the goal of treatment of RA in modern practice.
2. Combinations of DMARDs are most appropriate for RA with poor prognostic factors in early as well as established disease.
3. All patients with RA commencing DMARD therapy should receive vaccination against pneumococcus, influenza, herpes zoster and hepatitis B, where indicated (based on age and risk).

REFERENCES

1. van der Woude D, Young A, Jayakumar K, Mertens BJ, Toes REM, van der Heijde D, et al. Prevalence of and predictive factors for sustained disease-modifying antirheumatic drug-free remission in rheumatoid arthritis. Results from two large early arthritis cohorts. Arthritis Rheum 2009; 60:2262–71.
2. Kirwan J. The effect of glucocorticoids on joint destruction in rheumatoid arthritis. The Arthritis and Rheumatism Council Low-dose Glucocorticoid Study Group. N Engl J Med 1995; 333:142–6.
3. Hamilton RA, Kremer JM. Why intramuscular methotrexate works better than oral drug in patients with rheumatoid arthritis. Br J Rheumatol 1997; 36: 86–90.
4. Visser K, Katchamart W, Loza E, Martinez-Lopez JA, Salliot C, Trudeau J, et al. Multinational evidence-based recommendations for the use of methotrexate in rheumatic disorders with a focus on rheumatoid arthritis: integrating systematic literature research and expert opinion of a broad international panel of rheumatologists in the 3E initiative. Ann Rheum Dis 2009; 68:1086–93.
5. Moder KG, Tefferi A, Cohen MD, Menke DM, Luthra HS. Haematologic malignancies and the use of methotrexate in rheumatoid arthritis: a retrospective study. Am J Med 1995; 99:276–281.
6. Cardiel MH. First Latin American position paper on the pharmacological treatment of rheumatoid arthritis by the Latin American Rheumatology Associations of the Pan-American League of Associations for Rheumatology (PANLAR) and the Grupo Latinoamericano de Estudio de Artritis Reumatoide (GLADAR). Rheumatology 2006; 45: ii7-ii22.
7. Sharma A, Bhilave, N, Sharma K., Varma I. Metabolic syndrome in Indian patients with rheumatoid arthritis and its correlation with disease activity. Arthritis Res Ther 2012; 14(Suppl 1): P66.
8. Pullar T, Bamji A. National guidelines for the monitoring of second line drugs. Produced by the British Society for Rheumatology, July 2000. Website address: www.rheumatology.org.uk
9. Marmor MF, Kellner U, Lai TYY, Lyons JS, Mieler WF. Revised recommendations on screening for chloroquine and hydroxychloroquine retinopathy. Ophthalmology 2011; 118:415–22.
10. Russo PAJ, Wiese MD, Smith MD, Ahern MJ, Barbara JA, Shanahan EM. Leflunomide for Inflammatory Arthritis in End-stage Renal Disease on Peritoneal Dialysis: A Pharmacokinetic and Pharmacogenetic Study. Ann Pharmacother 2013: 47:e15. doi: 10.1345/aph.1R542.
11. Singh JA, Saag KG, Bridges SL, Akl EA, Bannuru RR, Sullivan MC, et al. 2015 American College of Rheumatology guideline for the treatment of rheumatoid arthritis. Arthritis Rheum 2016; 68: 1–26.
12. Smolen JS, Landewé R, Bijlsma J, Burmester G, Chatzidionysiou K, Dougados M, et al. EULAR recommendations for the management of rheumatoid arthritis with synthetic and biological disease-modifying antirheumatic drugs: 2016 update. Ann Rheum Dis 2017; 76:960–77.
13. Grigor C, Capell H, Stirling A, McMahon AD, Lock P, Vallance R, et al. Effect of a treatment strategy of tight control for rheumatoid arthritis (the TICORA study): a single-blind randomised controlled trial. Lancet 2004; 364:263–69.
14. Boers M, Verhoeven AC, Markusse HM, van de Laar MA, Westhovens R, van Denderen JC, et al. Randomised comparison of combined step-down prednisolone, methotrexate and sulphasalazine with sulphasalazine alone in early RA. Lancet 1997; 350: 309–18.
15. Mottonen T, Hannonen P, Leirisalo-Repo M, Nissila M, Kautiainen H, Korpela M, et al. Comparison of combination therapy with single-drug therapy in early rheumatoid arthritis: a randomised trial. FIN-RACo trial group. Lancet 1999; 353:1568–73.
16. Hazes JMW, Coulie PG, Geenen V, Vermeire S, Carbonnel F, Louis E, et al. Rheumatoid arthritis and pregnancy: evolution of disease activity and pathophysiological considerations for drug use. Rheumatology (Oxford) 2011; 50:1955–68.

FURTHER READING

1. Buer JK. A history of the term "DMARD". Inflammopharmacology 2015; 23:163–71.

MCQs

1. **Disease modifying antirheumatic therapy is better than non-steroidal anti-inflammatory drugs because:**
   a. It quickly relieves joint pain and swelling
   b. It eventually reverses the deformities
   c. It alters the course of disease favourably in the long-term
   d. It is less expensive

2. **Which of the following statements about corticosteroids is wrong?**
   a. They are often included in the initial treatment regimen for RA
   b. They are able to reduce the burden of inflammation reliably and quickly
   c. Studies have shown their disease modifying effect in RA
   d. Prednisolone at a dose of 5 to 7.5 mg/day does not cause any adverse effects and should be used in RA on a long-term basis in view of the beneficial effects.

3. **Following are true about methotrexate *except*:**
   a. It is the DMARD with longest 'retention' (continuation) rate
   b. In lower doses, its oral absorption is very good
   c. NSAIDs can increase its toxicity due to a drug interaction
   d. It frequently causes cirrhosis in RA if continued beyond 5 years

4. **Which of the following is not true about hydroxychloroquine in RA?**
   a. Hydroxychloroquine has been shown to be safe in pregnancy by a few studies
   b. Hydroxychloroquine frequently causes blindness in patients with RA.
   c. Its combination with methotrexate is proven to be additive
   d. Its efficacy is no better than that of chloroquine

5. **Identify the correct statement on sulphasalazine**
   a. Sulphasalazine is a sulphonamide with good oral absorption
   b. Its break-down product sulphapyridine is the active ingredient for RA
   c. Its maximum dose for RA is 2 gm daily
   d. It has no effect on male reproductive system

6. **Identify the statement which is not true for leflunomide**
   a. It is highly teratogenic and should be strictly avoided in pregnancy
   b. It stays in the body for 2–3 years
   c. Loading dose of 100 mg/day must be given for the first 3 days
   d. One of the well-recognized side effects is severe weight loss

7. **What is not true about minocycline?**
   a. It is a DMARD suitable for mild RA
   b. It can cause drug-induced lupus
   c. Its mechanism of action is not antibacterial
   d. Its efficacy for RA is less than that of hydroxychloroquine

8. **Which of the following is an example of step-up therapy?**
   a. TICORA trial
   b. COBRA trial
   c. Triple therapy (SSZ + HCQ + MTX)
   d. Fin-RACo trial

9. **Which of the following is not a poor prognostic factor for RA?**
   a. RF level of 320 IU/ml
   b. Positivity for anti-CCP
   c. High socioeconomic status
   d. Extra-articular manifestations

10. **Which statement is true about DMARDs in general?**
    a. A drug-induced remission is achieved in nearly all patients
    b. DMARDs can be stopped after a stable remission of 6 months duration
    c. The concept of window of opportunity is true for DMARD therapy
    d. Methotrexate and cyclosporine should not be combined

Chapter

23

Biologics in Autoimmune and Rheumatic Diseases

Ved Chaturvedi, Gautam Dhar Choudhary, Molly Mary Thabah

INTRODUCTION

In rheumatic diseases, the balance between endogenous anti-inflammatory mediators and pro-inflammatory mediators is disturbed. The production of pro-inflammatory cytokines such as tumour necrosing factor alpha (TNFα), interleukin1 (IL-1), and others in rheumatic disease is overwhelming. New therapeutic strategies in rheumatology include blocking the effects of inflammatory cytokines by monocloncal antibodies directed against TNFα, IL-1 and IL-6 receptor, B cell depletion therapy with anti-CD20 antibody agent amongst others. These agents are known as biologics or biologic response modifiers because they modify the immune response by blocking the effect of pro-inflammatory cytokines or by acting on various immune cells such as the B lymphocyte or the interaction between the T cell and the antigen presenting cell (APC), often resulting in rapid control of clinical disease activity.

Biologics are genetically engineered, produced in the form of monoclonal antibodies or soluble cytokine receptor protein products. Clinical trials show that these biological agents are more effective than traditional agents because they can alter joint remodeling in addition to attenuating symptoms.

It is known that almost 50% of patients with rheumatoid arthritis (RA) are disabled within 10 years of the onset of disease and survival is reduced. However, in the last 2 decades with the availability of biologics the management of RA have changed considerably.

In clinical trials with biologics for diseases like RA and psoriatic arthritis the most commonly used outcome measures for disease activity are American College of Rheumatology (ACR) 20, 50 and 70, which indicate 20%, 50% or 70%, improvement in parameters namely number of tender and swollen joint, C-reactive protein (CRP), pain on visual analogue scale (VAS) and physician or patient global assessment.

Trials in RA also use the disease activity score in 28 joints (DAS28), simplified disease activity index (SDAI), clinical disease activity index (CDAI) to measure disease activity. The extent of joint damage is quantified using the the van der Heijde modified sharp score which quantifies erosions and joint space narrowing. Trials also assess the impact on functional disability and quality of life. The common tools used are the health assessment questionnaire (HAQ) and short form 36 (SF-36). Indices relevant to psoriasis are psoriasis area severity index (PASI) and in ankylosing spondylitis the Bath Ankylosing Spondylitis Disease Activity Index (BASDAI), Bath Ankylosing Spondylitis Functional Activity Index (BASFI) and Bath Ankylosing Spondylitis Metrological Index (BASMI).

Biological agents: The biologics (Table 23.1) which are available and approved for clinical use are:

1. Anti-TNFα, e.g. infliximab, etanercept, adalimumab, golimumab, certolizumab pegol
2. Anti-CD20 antibody agent, e.g. rituximab
3. B cell activating factor (BAFF) inhibitor—belimumab
4. Interleukin (IL)-6 receptor antagonist, e.g. tocilizumab, sarilumab, sirukumab

Table 23.1: Biologics in rheumatology clinical practice

Biologic	*Disease conditions*	*Dosage*	*Side effects*
Infliximab	RA, AS, cutaneous PAN	3–5 mg/kg IV infusion at 0, 2, 6 weeks, and then every 8 weeks	Infections, tuberculosis, flu like symptoms, malignancy
Etanercept	RA, AS, psoriasis, PsA, JIA	25 mg sc biweekly or 50 mg sc weekly	Myalgia, urticaria, tuberculosis, demyelinating disorders, pancytopenia, congestive heart failure, psoriasis
Adalimumab	RA, AS, PsA	40 mg sc every other week	Infection, tuberculosis
Golimumab	RA, AS, PsA	50 mg sc once a month	Dizziness, injection site reactions
Certolizumab pegol	RA	200 mg sc at 0, 2, and 4 weeks, then 200 mg sc every other week	
Abatacept	RA	10 mg/kg IV infusion at 0, 2, 4 weeks, and 4 weekly thereafter	Infusion reactions
Rituximab	RA, SLE, AAV, dermatomyositis	RA—375 mg/m^2 infusion every 4 weeks × 4 doses or 1000 mg IV infusion on day 0 and day 15 **AAV induction therapy (RITUXIVAS and RAVE trial)**—375 mg/m^2 infusion every 4 weeks x 4 doses **AAV maintenance therapy (MAINRITSAN trial)**—500 mg on day 0 and 14, and at months 6, 12, and 18	Headache, infections, malignancy
Tocilizumab	RA, adult onset stills disease, Takayasu arteritis, GCA, PMR	8 mg/kg IV infusion every month for 6 months	Infection, lipid abnormalities
Secukinumab	PsA, AS, RA	10 mg/kg sc at week 0, 2, and 4, and followed by 150 mg sc every 4 weeks	
Belimumab	SLE (excluding renal and CNS lupus)	10 mg/kg IV infusion at week 0, 2, 4, and thereafter every 4 weeks	

Abbreviations: RA, rheumatoid arthritis, SLE, systemic lupus erythematosus, PAN, polyarteritis nodosa; PsA, psoriatic arthritis; JIA, juvenile idiopathic arthritis; AAV, ANCA associated vasculitis; GCA/PMR, giant cell arteritis/polymyalgia rheumatica; sc, subcutaneously.

5. Co-stimulatory receptor blockade—CTLA4-Ig (Abatacept) is a novel fusion protein designed to modulate the T cell co-stimulatory signal mediated through the CD28–CD80/86 pathway
6. IL-1 receptor antagonist, e.g. anakinra, canakinumab, rilonacept
7. IL-17A inhibitors—secukinumab, ixekinumab, brodalumab
8. IL-12/23 inhibitors—ustekinumab
9. Biologics used in osteoporosis—denosumab, romozumab
10. JAK inhibitors (discussed separately in another chapter)

1. ANTI-TNF THERAPY

i. Infliximab

a. Rheumatoid arthritis

Infliximab, a TNFα inhibitor has emerged as a key treatment for RA, and spondyloarthritis. It is a chimeric anti-TNFα monoclonal antibody with a predominance of human amino acid sequences. It has been used successfully in RA, ankylosing spondylitis (AS), giant cell arteritis and polymyalgia rheumatica.

In RA the dose is 3 mg/kg/dose at 0, 2, 6 weeks and thereafter every 8 weeks. Premedication with 100 mg of methylprednisolone is preferred by a few to prevent any anaphylaxis. Mantoux test, chest radiographs are prerequisite prior to infliximab therapy to rule out tuberculosis. Before each infusion, disease activity score by DAS28 and functional activity by health associated questionare (HAQ) are assessed and compared with previous results to determine the efficacy of the drug.

The clinical and radiological efficacy of repeated infusions of infliximab in RA was evaluated in a number of drug trials but the anti-TNF trial in RA with concomitant therapy (ATTRACT)[1,2] is a leading trial which proved efficacy of infliximab in RA. The study enrolled 428 eligible patients with advanced disease, as evidenced by a mean duration of disease ranging from 9 to 12 years. These patients were randomly allocated to placebo or 3 mg/kg or 10 mg/kg of infliximab at weeks 0, 2, and 6 followed by infusions every four or eight weeks. The methotrexate (MTX) dose was kept constant. Infliximab treatment was efficacious at all doses compared with placebo. The ACR20 response rates were achieved in 50–60% by the 30-week end point. Significantly more patients in the infliximab treatment groups also achieved ACR50 (26–31% v 5%) or ACR70 (8–18% v 0%) responses than those in the MTX placebo group. The tender and swollen joint counts, pain scores and serum concentrations of CRP improved by comparable values. The ACR response rates were generally sustained through one year of treatment.

Follow up of the BEST trial which compared the occurrence of drug free remission, functional ability, and radiological damage after 4 years of treatment according to four different treatment strategies for RA show superiority of combination therapy with infliximab for RA. Patient with recent onset, active RA (n = 508) were randomly assigned to four different treatment strategies: (1) sequential monotherapy; (2) step-up combination therapy; (3) initial combination therapy with prednisone and (4) initial combination therapy with infliximab. Treatment was adjusted based on 3 monthly disease activity score (DAS) assessments, aiming at a DAS <2.4. From the third year, patients with a sustained DAS<1.6 discontinued treatment. In total 43% of patients were in remission (DAS<1.6) including 18% in the group 4 and 14%, 12% and 8% in group 1–3 respectively at 4 years.[3]

Another study included patients with active refractory and erosive RA who were treated with intravenous infusions of infliximab in combination with MTX.[4] The objectives of this study were to evaluate the continuation rate of infliximab and its clinical effect over a 7-year period and to document the reasons for discontinuation. After 7 years, 160 of 511 patients (31%) were still on infliximab treatment. The major reasons for infliximab discontinuation included lack of efficacy, adverse events. Mean DAS for patients still on treatment with infliximab decreased from 5.7 at baseline to 3 at year 4 and remained that low until year.[7] Low disease activity (defined as DAS28 <3.2) was present in 60.9% of patients, and 45.5% achieved remission (DAS28 <2.6).[4]

Since treatment with infliximab incurs a huge economic burden, every patient on treatment often asks the question as to when to stop the therapy. Studies show that infliximab may be discontinued with disease remission in a few patients.[5] Serum antibodies to infliximab called human antichimeric antibody (HACA) were detected in the range of 17.4% of all infliximab treated patients. However, higher doses of infliximab and concomitant MTX treatment were associated with lower incidences of serum antibodies to infliximab.[6]

The British society of rheumatology (BSR) recommends use of infliximab when patient of RA has failed to respond to more than one disease modifying anti-rheumatic drugs (DMARDs). In contrast the American College of Rheumatology

(ACR) recommends use of biologics such as infliximab in early RA with high disease activity.

b. Ankylosing Spondylitis (AS)

In AS treatment options was limited to non-steroidal anti-inflammatory drugs (NSAIDs) only, infliximab has been successfully used at a dosage of 5 mg/kg/dosage at 0, 2, 6, and then every 8 weeks. The potential clinical efficacy of infliximab therapy for AS was examined in an open label study of 21 patients with active spondyloarthritis (ankylosing spondylitis n = 10; psoriatic arthritis = 9; and undifferentiated spondyloarthritis n = 2). The study showed that infliximab 5 mg/kg given at weeks 0, 2, and 6 could produce substantial improvements in different measures of disease activity indices, namely BASDAI, BASFI, and BASMI. Moreover, a maintenance regimen of 5 mg/kg every 14 weeks was able to maintain improvement in some cases.[7] One study concluded that infliximab is an effective therapy for early sacroiliitis, providing a reduction in disease activity by week 16. This study was the first to show that infliximab is effective for reducing clinical and imaging evidence of disease activity in patients with MRI-determined early axial spondylarthritis.[8] One study assessed the effect of infliximab on structural changes in AS over 4 years. Conventional radiographs of the cervical and the lumbar spine of 33 AS patients at baseline (BL), after 2 years [FollowUp1 (FU1)] and after 4 years [Follow Up 2 (FU2)] of infliximab therapy were scored by the modified Stokes ankylosing spondylitis spinal score (m SASSS). The mean change over 4 years was 1.6 ± 2.6 m SASSS units (p = 0.001), 0.9 ± 2.3 for BL–FU1, and 0.7 ± 1.6 for FU1–FU2. There was less radiographic progression in comparison with published data from the OASIS cohort.[9]

ii. Etanercept

Etanercept is a genetically engineered fusion protein consisting of two identical chains of the recombinant extracellular human TNF-receptor p75 monomer fused with the Fc domain of human IgG1. It effectively binds TNF and lymphotoxin-α inhibiting their activity. The dose of Etanercept is 25 mg subcutaneous twice a week or 50 mg once a week.

a. Rheumatoid Arthritis

Etanercept is effective and has minimal toxicity in adults with active RA who do not respond to other DMARDs. Efficacy of etanercept in RA has been shown in the ERA (early rheumatoid arthritis) trial and the TEMPO trial with improvements in ACR responses and decrease laboratory parameters.[10,11]

The Enbrel ERA trial[10] evaluated clinical and radiographic outcomes in 632 patients with early and active RA who received monotherapy with either etanercept or MTX for 2 years. These patients were randomized to receive either twice—weekly subcutaneous etanercept (10 mg or 25 mg) or weekly oral MTX (means dosage 19 mg per week) for at least 1 year. Following the blinded phase of the trial, 512 patients continued to receive the therapy to which they had been randomized for up to 1 additional year, in an open label manner. At 24 months, 72% patients receiving 25 mg etanercept achieved ACR 20 response compared to 59% receiving MTX. The mean changes in radiological erosion score in the etanercept group were significantly lower than those in the MTX group.

In another study, RA patients with an average moderate disease activity despite previous MTX monotherapy, who got combination treatment with etanercept and MTX had inhibited radiographic progression and improved radiographic outcomes. The study supports the use of combination therapy in RA patients with moderate disease activity.[12] Another study evaluated safety, efficacy, and radiographic progression in patients with early RA who got long-term treatment with etanercept. The rates of serious adverse events, serious infections did not increase with long-term exposure to etanercept. Efficacy was sustained in patients who completed 5 years of etanercept treatment a (N = 201), even in those who decreased or discontinued use of MTX or corticosteroids. No radiographic progression was seen in 55% of patients with 5-year radiographs.[13]

b. Ankylosing Spondylitis

Evidence from MRI scans collected as part of a 6-month open-label descriptive study showed that etanercept help control entheseal lesions in patients with AS. At the end of the study 86% of the patients treated had resolution or improvement in MRI sites of inflammation.[14] Similarly patients with early axial spondyloarthritis having active inflammatory lesions detected by whole-body MRI improved significantly more with etanercept versus sulfasalazine-treated patients.[15]

The long-term safety and efficacy of etanercept treatment in AS is also shown. A study evaluating the long-term safety and efficacy of etanercept in AS, where 257 of 277 patients (92%) were enrolled in an Open Label Extension (OLE), after 192 weeks of treatment with etanercept, the most common adverse effects were injection site reactions, headache and diarrhoea. Of patients who received etanercept in both the RCT and OLE, and were still in the trial, 71% were ASAS20 responders at week 96, and 81% were responders at week 192. ASAS 5/6 response rates were 61% at week 96 and 60% at week 144, and partial remission response rates were 41% at week 96 and 44% at week 192. Placebo patients who switched to etanercept in the OLE showed similar patterns of efficacy maintenance.[16]

c. Psoriatic Arthritis (PsA) and Juvenile Arthritis (JIA)

Etanercept has been successfully tried in PsA[17] and JIA[18] with significant reduction in disease activity and improvement in quality of life. However, a few case reports have mentioned development of psoriasis in patients after etanercept therapy.

A study conducted on a total of 672 patients randomized 1:1:1:1 to receive placebo or etanercept 25 mg once weekly, 25 mg twice weekly, or 50 mg twice weekly subcutaneous (SC) injection for 12 weeks. Etanercept recipients who completed the initial treatment period continued on their assigned dosage for an additional 12 weeks. Among patients randomized to etanercept 25 or 50 mg twice weekly, 32% and 47% of patients, respectively, achieved Psoriasis Area Severity Index (PASI) 75 at week 12, increasing to 41 and 54% at week 24. In another study, a sustained benefit of etanercept treatment, including inhibition of radiographic progression, in patients with PsA was noticed even after 2 years.[19]

A study evaluated the clinical efficacy and tolerability of etanercept in the treatment of active and progressive PsA in patients who have previously demonstrated an inadequate response to standard DMARDs and NSAIDs.[20] Twenty-seven (27/32) patients (84.3%) completed 3 years (144 weeks) of continuous treatment, while 5/32 (15.6%) patients were withdrawn from the study. At week 144, a significant improvement in DAS28 was registered with a reduction in mean DAS28 from 5.3 at baseline to 1.8, while 25/27 patients (92.5%) achieved PASI 75 with a mean PASI score of 0.7; the mean pain visual analogue scale (pain-VAS) score decreased from 64.2 at baseline to 2 at week 144, corresponding to an improvement of 94.7%.[20]

Currently, etanercept is the best evaluated biologic for the treatment of JIA. The first trial of etanercept in JIA was published in 2000.[18] During the first open part of this study, improvement was seen in 51 of 69 (74%) children with JIA (74%). The responders took part in a placebo controlled trial that confirmed the efficacy of the drug by achieving at least 50% improvement in 70% of children compared with 25% on placebo. This study included children who failed or did not tolerate MTX. Following this trial etanercept was approved in Europe in 2000 for the treatment of children ages 4–17 years with refractory, polyarticular JIA.

Another study was undertaken to evaluate the long-term safety and effectiveness of etanercept alone or in combination with MTX in children with selected categories of JIA.[21] Patients aged 2–18 years received MTX alone ≥10 mg/m^2/week [approximately 0.3 mg/kg/week], etanercept alone (0.8 mg/kg/week, maximum dose 50 mg), or etanercept plus MTX for 3 years in an open-label, non-randomized study. Exposure-adjusted rates of adverse events were similar among the 3 treatment groups. Serious adverse events and medically important infections were also similar among the 3 treatment groups. Scores for physician's global assessment and total active joints improved from baseline, and improvement was maintained for three years in those continuing to receive medication.[22]

iii. Adalimumab

Adalimumab (Humira) is a recombinant, fully human IgG1 monoclonal antibody that binds specifically to TNFα. In the ARMADA trial patients with active RA were randomly assigned to receive 20, 40, or 80 mg of adalimumab or placebo every other week in combination with MTX.[22] An ACR20 response was achieved in 47%, 67%, and 65% patients receiving adalimumab (20 mg, 40 mg,

80 mg) plus MTX at week 24 when compared to 14% of patients who got placebo plus MTX. A larger trial confirmed the efficacy of 40 mg adalimumab given every other week in inhibiting the progression of structural joint damage, reducing signs and symptoms, and improving physical function of patients with active RA.[23]

Similarly, the efficacy of adalimumab for AS was shown in a multicenter, randomized (2:1 ratio), double-blind, placebo-controlled study evaluating subcutaneous injection of adalimumab, 40 mg every other week, compared with placebo for 24 weeks.[24] The primary efficacy end point was the percentage of patients with a ASAS20 response at week 12. This trial showed that 58% of adalimumab patients (121 of 208) achieved an ASAS20 response, compared with 20.6% of placebo-treated patients (22 of 107) at week 12. Adalimumab was safe, well-tolerated and produced significant reduction in symptoms and signs of active AS.[24]

iv. Certolizumab Pegol

Pegylated certolizumab causes significant benefit in reducing the signs and symptoms, reduction in radiological score in RA as seen in the Rheumatoid Arthritis Prevention of Structural Damage (RAPID)I, RAPID II, and efficacy and safety of certolizumab pegol—4 Weekly dosage in rheumatoid arthritis (FAST4WARD) study.[25–27] This agent is used in combination with MTX. The dose is 400 mg given as two subcutaneous injections of 200 mg at 0, 2, and 4 weeks followed by 200 mg subcutaneously every other week.

v. Golimumab

Golimumab is a human immunoglobulin G1 monoclonal antibody that is specific for human TNF.[28] It was created using genetically engineered mice that were immunized with human TNF, resulting in an antibody with human-derived variable and constant regions. Golimumab binds to both the soluble and transmembrane forms of TNF, preventing the binding of TNF to its receptors and thereby inhibiting the biological activity of TNF.

Golimumab in combination with MTX is approved by the FDA for the treatment of moderately to severely active RA. It is also approved for active PsA (alone or in combination with MTX) and for active AS. The efficacy of Golimumab in RA was proven in three studies, denoted RA-1 (also known as GO-AFTER), RA-2 (also known as GO-FORWARD), and RA-3, involving 1,542 adult patients with moderately to severely active RA.[28,29]

2. B CELL DEPLETION THERAPY

B lymphocytes are an essential component of the adaptive (acquired) immune response. They develop initially in the fetal liver and then transferred to the bone marrow at around 12–16 weeks of fetal life where they undergo maturation with the assistance of stromal cells and cytokines. The surface antigen, CD20, acts as a cross membrane calcium ion channel and plays a key role in B cell activation. Targeting CD20 is proven therapeutic in the management of diseases that are largely B cell mediated.

Rituximab is a chimeric mouse-human monoclonal antibody consisting of a human kappa constant region, a human IgG1 Fc portion, and a murine variable region which recognizes human CD20. Rituximab can selectively deplete B cells by antibody-dependent cell mediated cytotoxicity and direct induction of B cell apoptosis.

a. Rheumatoid Arthritis

Rituximab (RTX) is indicated in patients who have not responded or showed resistant to anti-TNFα therapy. REFLEX and DANCER trials have shown the efficacy of RTX in RA.[30,31] The dose regime used in RA is 1000 mg administered on day 1 and day 15 with a premedication of 100 mg of methylprednisolone. Remission of the disease has been found to be as long as 18 months. Rituximab depletes peripheral CD20+ B cells, without decreasing the mean immunoglobulin levels (IgG, IgM, and IgA).[32] Most adverse events occur with the first infusion and are typically mild-to-moderate severity.

Studies to compare the effectiveness of RTX with that of anti-TNF agent in the management RA who had an inadequate response to anti-TNF therapy have shown better outcomes with RTX.[33,34] A prospective cohort study nested within the Swiss clinical quality management RA cohort included all patients who had an inadequate response to at least 1 anti-TNF agent and subsequently received either 1 cycle of RTX or an alternative anti-TNF agent. Total of 116 patients with RA were included; 50 patients received 1 cycle of rituximab, and

66 patients received a second or a third alternative anti-TNF agent. At base line, there were no significant differences between the two groups in age, sex, disease duration, and disease activity. Reduction of disease activity at 6 months measured by DAS28 was more favourable in the group that received RTX. A systematic review also showed that RTX improve symptoms of patients who had inadequate response to or are intolerant to TNF-alpha inhibitors.[35]

A study reported the long-term safety of RTX in RA using data from clinical trials where safety analyses were based on 5013 patient-years of RTX exposure.[36] The most frequent adverse event (AE) was infusion-related reactions seen in 25% of patients during the first infusion. Less than 1% of infusion-related reactions were considered serious. The rates of AE and serious AE were 17.85 events/100 patient-years (95% CI 16.72–19.06) which were stable following each course of infusion. The overall serious infection rate was 4.31/100 patient-years (95% CI 3.77, 4.92). Infections and serious infections over time remained stable across 5 courses at 4–6 events/100 patient-years.[36]

b. Systemic Lupus Erythematosus (SLE)

RTX has been widely used in the treatment of patients with SLE who have failed to respond to conventional immunosuppression.[37] In two major clinical trials the EXPLORER[38] and LUNAR[39] the primary end points was not met, but many smaller open-labeled studies have shown a good response to RTX.[40] A small study has shown that RTX is an effective therapy when used early in the disease and is a useful method of reducing the cumulative steroid burden in patients with predominantly non-renal SLE.[41] In another study, thirty-one sequential patients with relapsing or refractory SLE, 11 of whom had active lupus nephritis, received RTX either 375 mg/m^2/week × 4 (n = 16) or 1000 mg × 2 (n = 15).[42] The median follow-up was 30 months. Thirty of 31 (97%) patients had depleted peripheral B cells. Twenty-seven of 31 (87%) patients achieved remission. Renal response occurred in 10/11 patients (4 complete, 6 partial) with active glomerulonephritis. Clinical improvement was reflected by reductions of disease activity, proteinuria and daily prednisolone dose. Eighteen of 27 (67%) patients relapsed after a median of 11 months. Re-treatment with RTX was effective. Authors concluded that RTX was effective in relapsing or refractory SLE with or without renal involvement. Even though some patients relapsed on RTX, re-treatment was effective.

c. Vasculitis

RTX has a role in the management of ANCA associated vasculitis (AAV), namely granulomatosis with polyangiitis (GPA) and microscopic polyangiitis (MPA). Two randomized controlled trials established the non-inferiority of RTX as initial induction therapy compared to cyclophosphamide (CYC). In the rituximab in vasculitis (RITUXVAS) study, 44 newly diagnosed AAV were randomised to receive either RTX (375 mg/m^2 per week for 4 weeks) or IV CYC.[43] The CYC group were on azathioprine for maintenance therapy. Complete remission was achieved in 82% of patients who received RTX, and 91% who received CYC. It was important to note that the RTX arm also received CYC on weeks 1 and 3. At 2 years the relapse rate was similar in the two groups.

In the rituximab for ANCA-associated vasculitis (RAVE) trial, 197 patients were randomized to receive RTX 375 mg/m^2/week for 4 weeks or to oral CYC 2 mg/kg for 3–6 months.[44] All patients in CYC arm further received azathioprine for 12–15 months. At the end of 6 months, 64% of those who received RTX were in remission and steroid free versus 54% of those who received CYC, showing the non-inferiority of RTX to induce remission in AAV. Patients who had glomerulonephritis and pulmonary hemorrhage also responded to RTX. However, it is important to note that only half of the patients had glomerulonephritis, sick patients such as those with renal failure, i.e. serum creatinine >4 mg/dL were excluded from the RAVE trial. Also almost 50% patients had relapsing disease. Follow-up data suggest the efficacy of RTX compared to CYC. Hence RTX is a good alternative induction agent to IV or oral CYC in the appropriate clinical context.

In the MAINRITSAN trial, 115 patients (mainly with GPA) after induction with CYC and steroids were randomized to receive either 500 mg of RTX on days 0 and 14 and at months 6, 12, and 18 after study entry or daily azathioprine until months 22.[45] At 28 months, 29% of patients in the azathioprine group had major relapses compared with only 5% of the RTX group. The RITAZAREM study (RTX vasculitis maintenance study) is testing the efficacy of RTX at a higher dose (1,000 mg every 4 months) in relapsing patients after remission induction with RTX.[46] In the RITAZAREM study patients with AAV will be recruited at the time of relapse and will receive RTX and glucocorticoid induction therapy. If the disease is controlled by 4 months, patients will be randomized in a 1:1 ratio to receive RTX (1000 mg every 4 months for five doses) or azathioprine (2 mg/kg/day) as maintenance therapy. Patients will be followed for a minimum of 36 months. The primary outcome is the time to disease relapse. Results of this study was awaited.[46] The other important role of RTX is in the management of refractory granulomatous disease, i.e. GPA localized to the orbit, pachymeningits, and airway disease.[47]

In summary, RTX has a role in induction of remission, maintenance of remission in patients at high risk of relapse, in patients who have experienced multiple relapses, and when CYC need to be avoided when there is a high risk of infertility or malignancy.

d. Other Conditions

RTX has also been found to be effective in the treatment of type II mixed cryoglobulinaemia, dermatomyositis.

3. B CELL ACTIVATING FACTOR (BAFF) INHIBITOR- BELIMUMAB

Belimumab is a human monoclonal antibody that inhibits B-cell activating factor (BAFF).[48] It is approved for treatment of autoantibody positive SLE excluding nephritis and CNS lupus. Belimumab binds primarily to circulating soluble BAFF, therefore not inducing antibody-dependent cellular cytotoxicity. Belimumab does reduce the number of circulating B cells, but less durably than anti-CD20 monoclonal antibodies.[49] The BLISS-52 trials and BLISS-76 trials showed that Belimumab plus standard SLE therapy resulted in significant improvement outcomes than did placebo plus standard therapy at 52 weeks.[50,51]

In both these trials SLE patients were randomized to receive either Belimumab at 1mg/kg, 10 mg/kg, or placebo at enrolment, 2 weeks, 4 weeks, and every 4 weeks thereafter, in addition to standard SLE therapy. The inclusion criteria were SLE patients fulfilling the ACR criteria, who were autoantibody positive, and had moderate disease activity (SLEDAI ≥6). While BLISS 52 enrolled 867 patients from Eastern Europe, Asia and Latin America and had a follow-up for 52 weeks, the BLISS 72 trial recruited patients from North America and Europe and were follow-up for 76 weeks. Belimumab reduced number of flares, allowed lower prednisone doses, and decreased serologic activity. Belimumab improved overall SLE activity in musculoskeletal and mucocutaneous domain. Less worsening was seen in hematological, immunological and renal domains.[50] The most common indication for belimumab treatment is arthritis and mucocutaneous disease. In all these trials patients with active severe lupus nephritis were excluded but post-hoc analysis showed that rates of renal flare, renal remission, proteinuria reduction favoured belimimab. Therefore, belimumab may also have renal benefit. Belimumab has also undergone phase II clinical trials for RA and primary Sjögren's syndrome.[52,53]

4. IL-6 RECEPTOR ANTAGONIST

There are three agents in this category. Tocilizumab which was the first IL-6 receptor blocker to be approved for clinical use. The newer agents are sarilumab and sirukumab.

i. Tocilizumab (TCZ)

a. RA

TCZ is an humanised IL-6 receptor antagonist. IL-6 plays a pivotal role in mediating both local and systemic manifestations of RA. The AMBITION trial compared TCZ monotherapy (8 mg/kg every 4 weeks) with MTX monotherapy over 24 weeks has found better efficacy with TCZ.[54] It is the first biologic agent that has been found to be superior to MTX as initial therapy with a rapid onset of effect. The most common adverse effects were increase in lipids, decrease in neutrophils, and skin infections.

The LITHE study assessed radiographic progression, physical function, disease activity, and safety in RA patients who had inadequate response to MTX.[55] At week 104 radiographic progression was significantly lower for patients initially randomized to TCZ-MTX than for patients initially randomized to placebo-MTX. The mean of change of HAQ-DI was also significantly lower in patients initially randomized to TCZ-MTX than in patients randomized to placebo-MTX. Signs and symptoms of RA were maintained or showed improvement. Therefore, compared with placebo-MTX, TCZ-MTX significantly inhibited structural joint damage and improved physical function in patients with RA who previously had inadequate response to MTX.

b. Takayasu Arteritis, Giant Cell Arteritis (GCA) and Polymyalgia Rheumatica (PMR)

A retrospective study from Spain showed that patients with Takayasu arteritis who were refractory to corticosteroids and conventional immunosuppressive agents responded to TCZ. Improvements occurred early in the first 3 months of therapy, and after median follow-up of 15 months, 7 out of 8 patients were asymptomatic.[56] Apart from Takayasu arteritis TCZ therapy also is useful and leads to clinical and serologic improvement in patients with refractory/relapsing giant cell arteritis (GCA) and polymyalgia rheumatica (PMR).[57]

c. Adult Onset Still's Disease, Juvenile Idiopathic Arthritis (JIA)

Patients with AOSD often do not respond to conventional therapy. In this scenario TCZ has been found to be useful to bring relief in arthritis and fever. A multi-center retrospective study of 34 patients who had inadequate response to corticosteroids and immunosuppressive agent (including a biological agent) showed that TCZ caused rapid improvements in both clinical and laboratory parameters.[58] TCZ treatment caused marked reduction in arthritis, fever, lymphadenopathy, and cutaneous manifestations. Inflammatory markers including CRP, ESR, and ferritin all reduced dramatically. The dosage of prednisolone was also reduced from a median of 13.8 mg (baseline) to 2.5 mg at 12 months. Patients with systemic JIA also responds better to TCZ than etanercept.[59]

ii. Sarilumab

It is a new monoclonal antibody (human IgG1) targeting IL-6R. Adults with moderate to severe RA and inadequate response to MTX were randomised in 1:1:1 ratio to receive sarilumab 150 mg or 200 mg or placebo every 2 weeks with MTX for 52 weeks.[60] It was found that sarilumab treated patients responded symptomatically, functionally and radiographically.

Another RCT showed that RA patients with moderate to severe disease activity who had inadequate response to anti-TNF when randomised to receive sarilumab had significant improvements in signs, symptoms and physical function.[61] Sarilumab as monotherapy demonstrated a clear head-to-head superiority over adalimumab in MTX-intolerant subjects.

5. CTLA-4 INHIBITOR (ABATACEPT)

Abatacept is a fusion protein linking the extracellular domain of human cytotoxic T-lymphocyte associated antigen 4 (CTLA-4) to the Fc portion of human IgG1. Abatacept works by competing for the binding between CD28 on the T cell and CD80/86 on the antigen-presenting cell. This is an important co-stimulatory signal that is essential for T cell activation. Blocking T cell activation results in a reduction of the autoimmune process and inflammation and clinical improvement is demonstrated. It is effective in patients who have an inadequate response to either MTX or one or more TNF-inhibitors.

The effect of abatacept in anti-TNF inadequate responders was examined in a phase III trial, abatacept trial in treatment of anti-TNF inadequate (ATTAIN) responders.[62] Abatacept led to improvement in quality of life (QoL) for patients who failed anti-TNF therapy. The AIM study recruited patients who had failed at least 15 mg/week of MTX for at least 3 months and were randomized to either abatacept 10 mg/kg or placebo.[63] These patients who have failed MTX showed good response with abatacept. The abatacept study of safety in use with other rheumatoid (ASSURE trial) arthritis therapies showed that combination of abatacept with synthetic DMARD was safe and improved physical function.[64] But combination of abatacept with biologic therapy was associated with increases in adverse events. Therefore, abatacept should not be combined with biologic therapy.

6. IL-1 INHIBITORS

There are three IL-1 inhibitors currently in clinical use: Anakinra, rilonacept and canakinumab.

i. Anakinra

Anakinra is the earliest (one of the first) approved biologic response modifier. This IL-1 receptor antagonist is useful for the treatment of RA, AOSD, and systemic onset JIA. It is administered by subcutaneous injection at a dosage 100 mg (0.6 ml) once daily and has efficacy in treating RA when used alone or in combination with MTX.[65] However, it is not commonly used for the treatment of RA since the TNF-inhibitors show much greater efficacy in comparison and have a more convenient dosing regimen. It is a valuable drug in the treatment of AOSD and systemic onset JIA where it works much better than the TNF inhibitors. A multicentre, randomised, double-blind, placebo-controlled trial the ANAJIS trial has shown that anakinra is effective in systemic onset JIA and associated with normalisation of blood gene expression profiles in the patients who responded.[66]

ii. Rilonacept

Rilonacept also known as IL-1 Trap (marketed by Regeneron Pharmaceuticals under the trade name ARCALYST) is an IL-1inhibitor. Rilonacept is a dimeric fusion protein consisting of the extracellular domain of human IL-1 receptor and the Fc domain of human IgG1 that binds and neutralizes IL-1.

a. Hereditary Auto-inflammatory Syndromes

Rilonacept is used for the treatment of hereditary autoinflammatory syndromes: Cryopyrin-associated periodic syndromes (CAPS), including familial cold autoinflammatory syndrome, Muckle-Wells syndrome and neonatal onset multisystem inflammatory disease. Patients with CAPS who were in a phase 3 study entered a 72-week long open label extension study, where adults received subcutaneous rilonacept 160 mg, and pediatric patients received 2.2 mg/kg/week.[67] Rilonacept treatment resulted in reduction of symptoms, the number of flare days, and inflammatory markers all normalised. The most common adverse event was injection site reaction and upper respiratory tract infections. Rilonacept is approved by the FDA for treatment of CAPS.

b. Gout

Rilonacept has been tried in treatment of gout based on the role of NLRP3 (cryopyrin) inflammasome IL-1β pathway in the pathogenesis of gout. Rilonacept given 160 mg subcutaneously weekly was able to bring 70% fewer gout flares per patient when compared to placebo.[68] This was shown in the phase 3 PRE-SURGE trial. The PRE-SURGE trial met primary endpoint of reduction in the mean number of gout flares per patient during the 16-week treatment period in patients initiating uric acid-lowering therapy. Rilonacept was generally well tolerated with the incidence of serious adverse events well-balanced across the placebo and rilonacept group. Rilonacept has been tried in the treatment of acute gouty arthritis.[69] It was shown that during acute gouty arthritis, rilonacept plus indomethacin, and rilonacept alone did not provided additional pain relief when compared to indomethacin alone.[69] Hence there may be a role of rilonacept only in prevention of gout flares but not in reducing the pain of acute attacks.

iii. Canakinumab

a. Gout

Canakinumab a fully human monoclonal antibody directed at IL-1β could provide fast and effective relief from acute gout flares in patients whom NSAIDs or other anti-inflammatory therapies are contraindicated.[70] Results from two randomised, multicentre, active-controlled, double-blind trials the β-RELIEVED and β-RELIEVED II trials show that one dose of canakinumab 150 mg was more effective than the triamcinolone acetonide (40 mg) in reducing pain, in delaying time to a first new flare, risk of new flares over a 24 week period.[70] In this trial more than 80% patients had comorbidities, they were enrolled in the trials up to 5 days after the onset of an acute flare, patients were intolerant or unresponsive to NSAIDs and/or colchicine. For patients with acute gout and limited treatment options, canakinumab might be an important alternative to relieving pain, inflammation, and reducing risk of new flare.

7. IL-17A INHIBITORS

There are three IL-17A inhibitors—secukinumab, ixekizumab, brodalumab.

i. Secukinumab

Secukinumab is a humanized monoclonal antibody targeting IL-17A is useful for PsA, AS, and RA. This specific target inhibits the migration of monocytes and neutrophils around the site of inflammation. Secukinumab is a promising drug and found to be effective for treatment of PsA.[71] Apart from its efficacy in PsA, secukinumab has been shown in a 3-year follow-up study to achieve an ASAS 20/40 response of 80%/61% in the IV

150 mg group and 75%/50% in the IV 75 mg group after 156 weeks.[72] There were also sustained improvements in BASDAI, BASFI, and BASMI with no major safety issues. Secukinumab is given 300 mg at 0, 1, 2, 3 and 4 weeks and thereafter after every 4 weeks.

Secukinumab in a dose of 10 mg/kg at baseline and weeks 2 and 4, followed by subcutaneous 150 mg given every 4 weeks is effective in active RA patients who have had an inadequate response to or were intolerant to anti-TNF agents.[73]

ii. Ixekizumab

Ixekizumab is one of the newer monoclonal antibodies directed to IL-17A is shown to be effective and safe in treatment of patients with PsA who have failed or had inadequate response to anti-TNF agents.[74]

iii. Brodalumab

It is an anti-IL-17A approved for treatment of psoriasis.[75]

8. IL-12/23 INHIBITORS

Ustekinumab

Ustekinumab is a human monoclonal antibody that targets the p40 subunit of interleukin-12 (IL-12) and interleukin-23 (IL-23), preventing them from binding to their receptors on T-cells and natural-killer cells. The IL-12/23 proteins are important in regulating the immune system and to play a role in inflammatory diseases such as PsA. Ustekinumab is administered by subcutaneous injection at doses of 45 to 90 mg in weeks 0 and 4, and every 12 weeks thereafter. A double-blind control trial has shown significant reduction in signs and symptoms of PsA and skin lesions compared with placebo, and the drug was well tolerated.[76] However, long-term studies are needed to further characterize ustekinumab efficacy and safety for PsA.

Ustekinumab is licensed in European Union for the treatment of moderate to severe plaque psoriasis after failure of systemic therapies. It is also in phase II/III trial for Crohn's disease.

9. BIOLOGICS FOR OSTEOPOROSIS

i. Denosumab

Denosumab is a fully human monoclonal antibody for the treatment of osteoporosis. Beyond postmenopausal osteoporosis it is also used in cancer induced bone loss, bone metastases, RA, multiple myeloma, and giant cell tumor of bone. Denosumab targets receptor activator of nuclear factor-kappaB (RANK) ligand (RANKL), a protein that acts as the primary signal for bone removal, thereby inhibiting the formation and function of osteoclast.

In a phase III FREEDOM trial involving 7,808 women aged 60 to 90 years, denosumab given subcutaneously twice yearly for 36 months, reduced the risk of new vertebral fracture with a cumulative incidence of 2.3% in the denosumab group versus 7.2% in placebo group (a relative decrease of 68%).[77] Benefit was more in the subset of women with more severe disease, i.e. those who had ≥2 prevalent vertebral fractures and/or ≥1 prevalent vertebral fractures with moderate or severe deformity at the beginning of the study. The DATA study compared to teriparatide to denosumab in the treatment of postmenopausal osteoporosis, where it was found that increase in BMD was greater in combination (teriparatide + denosumab) therapy compared to single therapy.[78]

ii. Romozumab (Sclerostin Inhibitor)

Romozumab is a monoclonal antibody that binds to sclerostin and increases bone formation. In a large RCT women with postmenopausal osteoporosis were randomized to receive either romozumab 210 mg subcutaneously (sc) monthly for 12 months or placebo; thereafter both the groups received 60 mg romozumab sc every 6 months for 12 months.[79] At 12 months there were significantly lesser new vertebral fractures in the romosozumab group compared to placebo, and at 24 months after the transition to denosumab. Occurrence of hyperostosis and other adverse events such as infection was equal in both the groups.

INFECTIONS AND CANCER RISK WITH BIOLOGICAL AGENTS

1. For the anti-TNF agents, one particular infection that has to be looked for is tuberculosis (TB). Screening for TB before starting treatment is prudent since there is a high risk for reactivation of latent TB early in the course of anti-TNF therapy.[80] Any latent TB should be treated preferably for a month before initiating anti-TNF therapy.

2. There are reports of exacerbations of hepatitis in chronic hepatitis B patients and carriers. Prior to initiating biologic all patients should be screened for HBV and HCV infection, especially in endemic regions.[81] If positive, these patients must be followed up with liver function tests, to identify active viral replication that might require anti-viral treatment.
3. Prolonged suppression of the immune response may allow for the re-activation of slow viruses such as the JC polyoma virus, which can cause progressive multifocal leukoencephalopathy (PML). Occurrence of PML is very uncommon and most studies are limited to case reports. In rheumatology RTX is associated with highest risk of PML.[82]
4. Many of the usual signs of sepsis may be suppressed in patients treated with TNF-inhibitors, since their ability to mount anti-inflammatory response may be seriously compromised.
5. Data from the UK show that the risk for lymphoma in RA patients receiving anti-TNF agents is not different from biological naïve RA patients.[83] Similarly, anti-TNF did not increase risk of solid organ malignancy. On the contrary, by reduction of inflammation RA patients who get biologics have a decreased risk of mortality and cardiovascular events.[84]

Conclusion and Impact in Current Clinical Practice

Biological response modifiers represent advancement in the treatment of autoimmune disease. In the past, there were patients in whom all existing therapies were unable to control the inflammation and subsequent destruction of their joints. But currently, disease activity can be well controlled to the extent that function improves almost to normal and radiographic changes of destruction stop progressing. The biologics has an impact not only the disease activity but also the health-related quality of life, patient's functional capacity, and co-morbidities. As an example, Infliximab has been found to correct anemia of chronic disease in RA by preventing bone marrow suppression by TNF. For diseases like Takayasu's arteritis and AOSD where DMARDs was of suboptimal efficacy, tocilizumab has produced long-term remission.

A major problem in the use of these agents is the cost. Nevertheless, the efficacy of these agents prompts their widespread acceptance by physicians and currently 40% of patients with RA in the United States are receiving a biologic agent.

Future Trends

Other targets for therapy that are promising include anti-CD22, anti-CD40-CD40L, etc. It would be ideal if pathogenesis was driven solely by one of these key molecules. In the future, there will be better ways of choosing the best therapy for an individual patient based on predictive factors. The relative advantages and disadvantages of conventional DMARDS versus biologics for RA and AS are summarised in Table 23.2. For diseases such as RA the current best practice is to optimize the use of conventional combination DMARDs, before prescribing a biologic agent.

Table 23.2: Efficacy of biologics compared to standard DMARDs in rheumatoid arthritis (RA) and axial spondyloarthritis (AS)

S No	*Domains*	*Conventional DMARD*	*Biologics*
1.	Disease activity	Have high efficacy in RA but are of doubtful benefit in AS	Have high efficacy in both diseases In RA MTX is to be continued
2.	Radiological damage	Retards radiographic progression in RA	Retards radiographic progression in RA, but in AS data is emerging that biologics can retard radiographic progression.
3.	Extra-articular manifestations	Improves uveitis to some extent	Biologics is effective for most extra-articular manifestations of AS and RA including anaemia
4.	Adverse reactions	Rare but can be corrected by periodic monitoring	Very few. TB and chronic viral hepatitis to be screened before administration of biologics
5.	Cost effectiveness	Cheaper than biologics	Initially was costlier but with advent of biosimilars it has become affordable. Early induction leads to significant benefit

Key Points

- Biologics are genetically engineered agents also known as biologic response modifiers because they modify the immune response by blocking the effect of pro-inflammatory cytokines or by acting on various immune cells.
- The efficacy of the anti-TNF agents in RA, axial SpA (AS), PsA are established.
- Apart from its role in treatment of RA, Tocilizumab is emerging as a useful agent to control disease activity of systemic onset JIA, AOSD, Takayasu arteritis, giant cell arteritis, polymyalgia rheumatica.
- Rituximab has a prominent role in management of difficult RA, severe lupus nephritis, induction and maintenance treatment of ANCA associated vasculitis. Rituximab is better than azathioprine as maintenance therapy by reducing relapses of AAV.
- Belimumab added to standard therapy in SLE reduces number of flares, lower prsednisone doses, and decreased serologic activity. It is useful for SLE activity in musculoskeletal and muco-cutaneous, and results in less worsening of hematological, immunological and renal disease.
- IL-1 antagonist Anakinra is useful for AOSD, systemic onset JIA
- Rilonacept a IL-1 inhibitor antagonist finds a prominent role in treatment of autoinflammatory syndromes: Cryopyrin-associated periodic syndromes.
- In gout rilonacept is effective to reduced gout flares, while for patients with acute gout and limited treatment options, canak numab appears an important alternative to relieving pain, inflammation, and reducing risk of new flare.
- Secukinumab (anti-IL-17A) is emerging as very useful for PsA, and AS.

REFERENCES

1. Maini R, St Clair EW, Breedveld F, Furst D, Kalden J, Wesman M, et al. Infliximab (Chimeric anti-tumor necrosis factor α monoclonal antibody) versus placebo in rheumatoid arthritis patients receiving concomitant methotrexate: a randomized phase III trial. Lancet 1999; 354:1932–39.
2. Lipsky PE, van der Heijde DM, St Clair EW, Furst DE, Breedveld FC, Kalden JR, et al. Infliximab and methotrexate in the treatment of rheumatoid arthritis. N Engl J Med 2000; 343:1594–602.
3. van der Kooij SM, Goekoop-Ruiterman YPM, de Vries-Bouwstra JK, Guler-Yuksel M, Kerstens PJSM, van der Lubbe PAHM, et al. Drug-free remission, functioning and radiographic damage after 4 years of response-driven treatment in patients with recent-onset rheumatoid arthritis Ann Rheum Dis 2009;68: 6 914–921.
4. van der Cruyssen B, Durez P, Westhovens R, De Keyser Fet al. Seven-year follow-up of infliximab therapy in rheumatoid arthritis patients with severe long-standing refractory disease: attrition rate and evolution of disease activity. Arthritis Res Ther 2010;12:R77.
5. van der Bijl AE, Goekoop-Ruiterman YP, de Vries-Bouwstra JK, Ten Wolde S, Han KH, van Krugten MV, et al. Infliximab and methotrexate as induction therapy in patients with early rheumatoid arthritis. Arthritis Rheum. 2007; 56(7): 2129–34.
6. Maini RN, Breedweld FC, Kalden JR, Smolen JS, Davis D, Macfarlane JD, et al. Therapeutic efficacy of multiple intravenous infusions of anti-tumor necrosis factor α monoclonal antibody combined with low dose weekly methotrexate in rheumatoid arthritis. Arthritis Rheum 1998; 41:1552–63.
7. van den Bosch F, Kruithof E, Baeten D, de Kayser F, Mielants H, Veys EM. Effects of a loading dose regimen of three infusions of chimeric monoclonal antibody to tumor necrosis factor α (infliximab) in spondyloarthopathy: an open pilot study. Arthritis Rheum 2000:59:427–33.
8. Barkham N, Keen HI, Coates LC, O'Connor P, Hensor E, Fraser AD, et al. Clinical and imaging efficacy of infliximab in HLA-B27. Positive patients with magnetic resonance imaging-determined early sacroiliitis. Arthritis Rheum 2009; 60(4):946.
9. Baraliakos X, Listing J, Brandt J, Haibel H, Rudwaleit M, Sieper J, et al. Radiographic progression in patients with ankylosing spondylitis after 4 yrs of treatment with the anti-TNF-α antibody infliximab. Rheumatology 2007; 46:1450–1453.
10. Genovese MC, Bathon JM, Martin RW, Fleischmann RM, Tesser JR, Schiff MH, et al. Etanercept versus methotrexate in patients with early rheumatoid arthritis. Two-year radiographic and clinical outcomes. Arthritis Rheum 2002; 46: 1443–50.
11. van der Heijde D, Klareskog L, Rodriguez-Valverde V, Codreanu C, Bolosiu H, Melo-Gomes J, et al. Comparison of Etanercept and Methotrexate, Alone and Combined in the treatment of Rheumatoid Arthritis Two-Year Clinical and Radiographic Results from the TEMPO study, a Double-Blind, Randomized Trial. Arthritis Rheum 2006; 54 : 1063–74.
12. van der Heijde D, Burmester G, Melo-Gomes J, Codreanu C, Martin Mola E, Pedersen R, et al. Inhibition of radiographic progression with combination etanercept and methotrexate in patients with moderately active rheumatoid arthritis previously

treated with monotherapy. Ann Rheum Dis. 2009; 68(7): 1113–8.

13. Genovese MC, Bathon JM, Fleischmann RM, Moreland LW, Martin RW, Whitmore JB, et al. Long-term safety, efficacy, and radiographic outcome with etanercept treatment in patients with early rheumatoid arthritis. J Rheumatol. 2005; 32(7): 1232–42.
14. Marzo-Ortegah, Mcgonagle D, O'connor P, Emery P: Efficacy of etanercept in the treatment of the entheseal pathology in resistant spondylarthropathy: A clinical and magnetic resonance imaging study. Arthritis Rheum 2001; 44: 2112–7.
15. Song IH, Hermann K, Haibel H, Althoff CE, Althoff C, Listing J, et al. Effects of etanercept versus sulfasalazine in early axial spondyloarthritis on active inflammatory lesions as detected by whole-body MRI (ESTHER): a 48-week randomised controlled trial. Ann Rheum Dis. 2011;70(4):590.
16. Davis JC Jr, van der Heijde DM, Braun J, Dougados M, Clegg DO, Kivitz AJ, etal. Efficacy and safety of up to 192 weeks of etanercept therapy in patients with ankylosing spondylitis. Ann Rheum Dis 2008; 67: 346–352.
17. Mease PJ, Goffe BS, Metz J, Vander Stoep A, Fink B, Burge DJ, et al. Etanercept in the treatment of psoriatic arthritis and psoriasis: a randomized trial. Lancet 2000; 356:385–90.
18. Lovell DJ, Giannini EH, Reiff A, Cawkwell GD, Silverman ED, Nocton JJ, et al. Etanercept in children with polyarticular juvenile rheumatoid arthritis. For The pediatric Rheumatology Collaborative Study Group. N Engl J Med 2000; 342:763–69.
19. Mease PJ, Kivitz AJ, Burch FX, Siegel EL, Cohen SB, Ory P, et al. Continued inhibition of radiographic progression in patients with psoriatic arthritis following 2 years of treatment with etanercept. Rheumatol. 2006;33(4):71.
20. Mazoota A, Esposito M, Schipani C, Chimenti S. Long-term experience with etanercept in psoriatic arthritis patients: a 3-year observational study. J Dermatalog Treat 2009;20(6):354–8.
21. Giannini EH, Ilowite NT, Lovell DJ, Wallace CA, Rabinovich CE, Reiff A, et al. Long-term safety and effectiveness of etanercept in children with selected categories of juvenile idiopathic arthritis. Arthritis Rheum. 2009; 60(9): 2794–804.
22. Weinblatt ME, Keystone EC, Furst DE, Moreland LW, Weisman MH, Birbara CA, et al. Adalimumab, a fully human anti-tumor necrosis factor alpha monoclonal antibody, for the treatment of rheumatoid arthritis in patients taking concomitant methotrexate: the ARMADA trial. Arthritis Rheum. 2003; 48(1): 35–45.
23. Keystone EC, Kavanaugh AF, Sharp JT, Tannenbaum H, Hua Y, Teoh LS, et al. Radiographic, clinical and functional outcomes of treatment with adalimumab (a human anti-tumor necrosis factor monoclonal antibody) in patients with active rheumatoid arthritis receiving concomitant methotrexate therapy: a randomized, placebo-controlled, 52-week trial. Arthritis Rheum 2004; 50(5): 1400–11.
24. van der Heijde D, Kivitz A, Schiff MH, Sieper J, Dijkmans BA, Braun J, et al. Efficacy and safety of adalimumab in patients with ankylosing spondylitis: results of a multicenter, randomized, double-blind, placebo-controlled trial. Arthritis Rheum. 2006; 54(7): 2136–46.
25. Smolen J, Landewé RB, Mease P, Brzezicki J, Mason D, Luijtens K, et al. Efficacy and safety of certolizumab pegol plus methotrexate in active rheumatoid arthritis: the RAPID 2 study. A randomised controlled trial. Ann Rheum Dis. 2009; 68(6): 797–804.
26. Fleischmann R, Vencovsky J, van Vollenhoven RF, Borenstein D, Box J, Coteur G, et al. Efficacy and safety of certolizumab pegol monotherapy every 4 weeks in patients with rheumatoid arthritis failing previous disease-modifying antirheumatic therapy: the FAST4WARD study. Ann Rheum Dis. 2009; 68(6): 805–11.
27. Keystone E, van der Heijde D, Mason DH, Landewé R, van Vollenhoven R, Combe B, et al. Certolizumab pegol plus methotrexate is significantly more effective than placebo plus methotrexate in active rheumatoid arthritis. Arthritis Rheum. 2008;58:3319–29.
28. Smolen JS, Kay J, Doyle MK, Landewé R, Matteson EL, Wollenhaupt J, et al. Golimumab in patients with active rheumatoid arthritis after treatment with tumour necrosis factor inhibitors (GO-AFTER study): a multicentre, randomised, double-blind, placebo-controlled, phase III trial. Lancet 2009; 374: 210–21.
29. Keystone EC, Genovese MC, Klareskog L, Hsia EC, Hall ST, Miranda PC, et al. Golimumab, a human antibody to tumour necrosis factor given by monthly subcutaneous injections, in active rheumatoid arthritis despite methotrexate therapy: the GO-FORWARD study. Ann. Rheum. Dis 2009; 68: 789–96.
30. Cohen SB, Emery P, Greenwald MW, Dougados M, Furie RA, Genovese MC, et al. REFLEX trial group. Rituximab for rheumatoid arthritis refractory to anti-tumor necrosis factor therapy: Result of a multicenter, randomized, double-blind, placebo-controlled, phase III trial evaluating primary efficacy and safety at twenty-four weeks. Arthritis Rheum 2006; 54: 2793–806.

31. Emery P, Fleischmann R, Filipcwicz-Sosnowska A, Schechtman J, Szczepanski L, Kavanaugh A, et al. DANCER Study Group. The efficacy and safety of rituximab in patients with active rheumatoid arthritis despite methotrexate treatment: results of a phase IIB randomized, double-blind, placebo-controlled, dose-ranging trial. Arthritis Rheum 2006; 54(5):1390–400.

32. Keystone E, Fleischmann R, Emery P, Furst DE, van Vollenhoven R, Bathon J, et al. Safety and efficacy of additional courses of rituximab in patients with active rheumatoid arthritis: an open-label extension analysis. Arthritis Rheum. 2007; 56(12):3896–908.

33. Roll P, Dörner T, Tony HP. Anti-CD20 therapy in patients with rheumatoid arthritis: predictors of response and B cell subset regeneration after repeated treatment. Arthritis Rheum. 2008; 58(6): 1566–75.

34. Keystone E, Burmester GR, Furie R, Loveless JE, Emery P, Kremer J, et at. Improvement in patients-reported outcomes in a rituximab trial in patients with severe rheumatoid arthritis refractory to anti-tumor necrosis factor therapy. Arthritis Rheum 2008; 59: 785–93.

35. Finckh A, Ciurea A, Bruhart L, Kyburz D, Möller B, Dehler S, et al. B cell depletion may be more effective than switching to an alternative anti-tumor necrosis factor agent in rheumatoid arthritis patients with inadequate response to anti-tumor necrosis factor agents. Arthritis Rheum 2007; 56:1417–23.

36. Hernández-Cruz B, García-Arias M, Ariza Ariza R, Martín Mola E. Rituximab in rheumatoid arthritis: a systematic review of efficacy and safety. Reumatol Clin. 2011; 7(5): 314–22.

37. van Vollenhoven RF, Emery P, Bingham CO 3rd, Keystone EC, Fleischmann R, Furst DE, et al. Long term safety of patients receiving rituximab in rheumatoid arthritis clinical trials. J Rheumatol. 2010; 37(3): 558–67.

38. Merrill JT, Neuwelt CM, Wallace DJ, Shanahan JC, Latinis KM, Oates JC, et al. Efficacy and safety of rituximab in moderately-to-severely active systemic lupus erythematosus: the randomized, double-blind, phase II/III systemic lupus erythematosus evaluation of rituximab trial. Arthritis Rheum 2010; 62:222–33.

39. Rovin BH, Furie R, Latinis K, Looney RJ, Fervenza FC, Sanchez-Guerrero J, et al. LUNAR Investigator Group. Efficacy and safety of rituximab in patients with active proliferative lupus nephritis: the Lupus Nephritis Assessment with Rituximab study. Arthritis Rheum. 2012; 64(4): 1215–26.

40. Terrier B, Amoura Z, Ravaud P, Hachulla E, Jouenne R, Combe B, et al. Safety and efficacy of rituximab in systemic lupus erythematosus: results from 136 patients from the French Autoimmunity and Rituximab registry. Arthritis Rheum 2010; 62:2458–66.

41. Ezeonyeji AN, Isenberg DA. Early treatment with rituximab in newly diagnosed systemic lupus erythematosus patients: a steroid-sparing regimen. Rheumatology (Oxford). 2012; 51(3):476–81.

42. Catapano F, Chaudhry AN, Jones RB, Smith KG, Jayne DW. Long-term efficacy and safety of rituximab in refractory and relapsing systemic lupus erythematosus. Nephrol Dial Transplant. 2010; 25(11): 3586–92.

43. Jones RB, Tervaert JW, Hauser T, Luqmani R, Morgan MD, Peh CA, et al. European Vasculitis Study Group. Rituximab versus cyclophosphamide in ANCA-associated renal vasculitis. N Engl J Med. 2010;363(3):211–20.

44. Stone JH, Merkel PA, Spiera R, Seo P, Langford CA, Hoffman GS, et al; RAVE-ITN Research Group. Rituximab versus cyclophosphamide for ANCA-associated vasculitis. N Engl J Med. 2010 Jul 15;363(3):221–32.

45. Guillevin L, Pagnoux C, Karras A, Khouatra C, Aumaître O, Cohen P, et al. French Vasculitis Study Group. Rituximab versus azathioprine for maintenance in ANCA-associated vasculitis. N Engl J Med. 2014; 371(19):1771–80.

46. Gopaluni S, Smith RM, Lewin M, McAlear CA, Mynard K, Jones RB, et al. RITAZAREM Investigators. Rituximab versus azathioprine as therapy for maintenance of remission for anti-neutrophil cytoplasm antibody-associated vasculitis (RITAZAREM): study protocol for a randomized controlled trial. Trials 2017; 18(1):112.

47. Shimojima Y, Kishida D, Hineno A, Yazaki M, Sekijima Y, Ikeda SI. Hypertrophic pachymeningitis is a characteristic manifestation of granulomatosis with polyangiitis: A retrospective study of anti-neutrophil cytoplasmic antibody-associated vasculitis. Int J Rheum Dis. 2017; 20(4): 489–496.

48. Bossen C, Schneider P. BAFF, APRIL and their receptors: structure, function and signaling. Semin Immunol. 2006; 18(5): 263–75.

49. Cheema GS, Roschke V, Hilbert DM, Stohl W. Elevated serum B lymphocyte stimulator levels in patients with systemic immune-based rheumatic diseases. Arthritis Rheum 2001;44:1313–9.

50. Manzi S, Sánchez-Guerrero J, Merrill JT, Furie R, Gladman D, Navarra SV, et al; BLISS-52 and BLISS-76 Study Groups. Effects of belimumab, a B lymphocyte stimulator-specific inhibitor, on disease activity across multiple organ domains in patients with systemic lupus erythematosus: combined results from two phase III trials. Ann Rheum Dis. 2012; 71(11): 1833–8.

51. Dooley MA, Houssiau F, Aranow C, D'Cruz DP, Askanase A, Roth DA, et al; BLISS-52 and -76 Study

Groups. Effect of belimumab treatment on renal outcomes: results from the phase 3 belimumab clinical trials in patients with SLE. Lupus. 2013; 22(1): 63–72.

52. Stohl W, Merrill JT, McKay JD, Lisse JR, Zhong ZJ, Freimuth WW, et al. Efficacy and safety of belimumab in patients with rheumatoid arthritis: a phase II, randomized, double-blind, placebo-controlled, dose-ranging Study. J Rheumatol. 2013; 40(5): 579–89.
53. Mariette X, Seror R, Quartuccio L, Baron G, Salvin S, Fabris M, et al. Efficacy and safety of belimumab in primary Sjögren's syndrome: results of the BELISS open-label phase II study. Ann Rheum Dis. 2015; 74(3): 526–31.
54. Jones G, Sebba A, Gu J, Lowenstein MB, Calvo A, Gomez-Reino JJ, et al. Comparison of tocilizumab monotherapy versus methotrexate monotherapy in patients with moderate to severe rheumatoid arthritis: the AMBITION study. Ann Rheum Dis. 2010; 69(1): 88–96.
55. Fleischmann RM, Halland AM, Brzosko M, Burgos-Vargas R, Mela C, Vernon E, et al. Tocilizumab inhibits structural joint damage and improves physical function in patients with rheumatoid arthritis and inadequate responses to methotrexate: LITHE study 2-year results. J Rheumatol. 2013; 40(2): 113–26.
56. Loricera J, Blanco R, Hernández JL, Castañeda S, Humbría A, Ortego N, et al. Tocilizumab in patients with Takayasu arteritis: a retrospective study and literature review. Clin Exp Rheumatol. 2016; 34(3 Suppl 97): S44–53.
57. Unizony S, Arias-Urdaneta L, Miloslavsky E, Arvikar S, Khosroshahi A, Keroack B, et al. Tocilizumab for the treatment of large-vessel vasculitis (giant cell arteritis, Takayasu arteritis) and polymyalgia rheumatica. Arthritis Care Res (Hoboken). 2012; 64(11): 1720–9.
58. Ortiz-Sanjuán F, Blanco R, Calvo-Rio V, Narvaez J, Rubio Romero E, Olivé A, et al. Efficacy of tocilizumab in conventional treatment-refractory adult-onset Still's disease: multicenter retrospective open-label study of thirty-four patients. Arthritis Rheumatol. 2014; 66(6): 1659–65.
59. Davies R, Gaynor D, Hyrich KL, Pain CE. Efficacy of biologic therapy across individual juvenile idiopathic arthritis subtypes: A systematic review. Semin Arthritis Rheum. 2017; 46(5): 584–593.
60. Genovese MC, Fleischmann R, Kivitz AJ, Rell-Bakalarska M, Martincova R, Fiore S, et al. Sarilumab plus methotrexate in patients with active rheumatoid arthritis and inadequate response to methotrexate: results of a phase III study. Arthritis Rheumatol. 2015; 67(6): 1424–37.
61. Fleischmann R, van Adelsberg J, Lin Y, Castelar-Pinheiro GD, Brzezicki J, Hrycaj P, et al. Sarilumab and nonbiologic disease-modifying antirheumatic drugs in patients with active rheumatoid arthritis and inadequate response or intolerance to tumor necrosis factor inhibitors. Arthritis Rheumatol. 2017; 69(2): 277–290.
62. Genovese MC, Schiff M, Luggen M, Becker JC, Aranda R, Teng J, et al. Efficacy and safety of the selective co-stimulation modulator abatacept following 2 years of treatment in patients with rheumatoid arthritis and an inadequate response to anti-tumour necrosis factor therapy. Ann Rheum Dis. 2008; 67(4): 547–54.
63. Kremer JM, Genant HK, Moreland LW, Russell AS, Emery P, Abud-Mendoza C, et al. Effects of abatacept in patients with methotrexate-resistant active rheumatoid arthritis: a randomized trial. Ann Intern Med. 2006; 144(12): 865–76.
64. Weinblatt M, Combe B, Covucci A, Aranda R, Becker JC, Keystone E. Safety of the selective co-stimulation modulator abatacept in rheumatoid arthritis patients receiving background biologic and nonbiologic disease-modifying antirheumatic drugs: A one-year randomized, placebo-controlled study. Arthritis Rheum. 2006; 54(9): 2807–16.
65. Fleischmann RM, Tesser J, Schiff MH, Schechtman J, Burmester GR, Bennett R, et al. Safety of extended treatment with anakinra in patients with rheumatoid arthritis. Ann Rheum Dis. 2006; 65(8):1006–12.
66. Quartier P, Allantaz F, Cimaz R, Pillet P, Messiaen C, Bardin C, et al. A multicentre, randomised, double-blind, placebo-controlled trial with the interleukin-1 receptor antagonist anakinra in patients with systemic-onset juvenile idiopathic arthritis (ANAJIS trial). Ann Rheum Dis. 2011; 70(5):747–54.
67. Hoffman HM, Throne ML, Amar NJ, Cartwright RC, Kivitz AJ, Soo Y, et al. Long-term efficacy and safety profile of rilonacept in the treatment of cryopryin-associated periodic syndromes: results of a 72-week open-label extension study. Clin Ther. 2012; 34(10): 2091–103.
68. Sundy JS, Schumacher HR, Kivitz A, Weinstein SP, Wu R, King-Davis S, Evans RR. Rilonacept for gout flare prevention in patients receiving uric acid-lowering therapy: results of RESURGE, a phase III, international safety study. J Rheumatol. 2014; 41(8):1703–11.
69. Terkeltaub RA, Schumacher HR, Carter JD, Baraf HS, Evans RR, Wang J, et al. Rilonacept in the treatment of acute gouty arthritis: a randomized, controlled clinical trial using indomethacin as the active comparator. Arthritis Res Ther. 2013; 15(1): R25. doi: 10.1186/ar4159.

70. Schlesinger N, Alten RE, Bardin T, Schumacher HR, Bloch M, Gimona A, et al. Canakinumab for acute gouty arthritis in patients with limited treatment options: results from two randomised, multicentre, active-controlled, double-blind trials and their initial extensions. Ann Rheum Dis. 2012; 71(11): 1839–48.

71. McInnes IB, Mease PJ, Kirkham B, Kavanaugh A, Ritchlin CT, Rahman P, et al. FUTURE 2 Study Group. Secukinumab, a human anti-interleukin-17A monoclonal antibody, in patients with psoriatic arthritis (FUTURE 2): a randomised, double-blind, placebo-controlled, phase 3 trial. Lancet. 2015; 386 (9999):1137–46.

72. Baraliakos X, Kivitz AJ, Deodhar AA, Braun J, Wei JC, Delicha EM, et al. MEASURE 1 Study Group. Long-term effects of interleukin-17A inhibition with secukinumab in active ankylosing spondylitis: 3-year efficacy and safety results from an extension of the Phase 3 MEASURE 1 trial. Clin Exp Rheumatol. 2017 May 15. [Epub ahead of print]

73. Blanco FJ, Möricke R, Dokoupilova E, Codding C, Neal J, Andersson M, et al. Secukinumab in Active Rheumatoid Arthritis: A Phase III Randomized, Double-Blind, Active Comparator- and Placebo-Controlled Study. Arthritis Rheumatol. 2017; 69(6): 1144–1153.

74. Nash P, Kirkham B, Okada M, Rahman P, Combe B, Burmester GR, et al. SPIRIT-P2 Study Group. Ixekizumab for the treatment of patients with active psoriatic arthritis and an inadequate response to tumour necrosis factor inhibitors: results from the 24-week randomised, double-blind, placebo-controlled period of the SPIRIT-P2 phase 3 trial. Lancet. 2017; 389 (10086):2317–2327.

75. Reichert JM. Antibodies to watch in 2017. MABS 2017; 9(2):167–181.

76. Gottlieb A, Menter A, Mendelsohn A, Shen YK, Li S, Guzzo C, et al. Ustekinumab, a human interleukin 12/23 monoclonal antibody, for psoriatic arthritis: randomised, double-blind, placebo-controlled, crossover trial. Lancet. 2009; 373 (9664): 633–40.

77. Cummings SR, San Martin J, McClung MR, Siris ES, Eastell R, Reid IR, et al. FREEDOM Trial. Denosumab for prevention of fractures in postmenopausal women with osteoporosis. N Engl J Med. 2009; 361(8):756–65.

78. Tsai JN, Uihlein AV, Lee H, Kumbhani R, Siwila-Sackman E, McKay EA, et al. Burnett-Bowie SA, Neer RM, Leder BZ. Teriparatide and denosumab, alone or combined, in women with postmenopausal osteoporosis: the DATA study randomized trial. Lancet. 2013; 382(9886): 50–6.

79. Cosman F, Crittenden DB, Adachi JD, Binkley N, Czerwinski E, Ferrari S, et al. Romosozumab Treatment in Postmenopausal Women with Osteoporosis. N Engl J Med. 2016; 375(16):1532–1543.

80. Malaviya AN, Kapoor S, Garg S, Rawat R, Shankar S, Nagpal S, et al. Preventing tuberculosis flare in patients with inflammatory rheumatic diseases receiving tumor necrosis factor-alpha inhibitors in India—An audit report. J Rheumatol. 2009; 36(7): 1414–20.

81. Temel T, Cansu DÜ, Korkmaz C, Kasifoglu T, Özakyol A. The long-term effects of anti-TNF-α agents on patients with chronic viral hepatitis C and B infections. Int J Rheum Dis. 2015; 18(1):40–5.

82. Clavel G, Moulignier A, Semerano L. Progressive multifocal leukoencephalopathy and rheumatoid arthritis treatments. Joint Bone Spine. 2017 Mar 18.

83. Mercer LK, Galloway JB, Lunt M, Davies R, Low AL, Dixon WG, et al. Risk of lymphoma in patients exposed to antitumour necrosis factor therapy: results from the British Society for Rheumatology Biologics Register for Rheumatoid Arthritis. Ann Rheum Dis. 2017; 76(3):497–503.

84. de La Forest Divonne M, Gottenberg JE, Salliot C. Safety of biologic DMARDs in RA patients in real life: A systematic literature review and meta-analyses of biologic registers. Joint Bone Spine. 2017; 84(2):133–140.

FURTHER READING

1. Braun J, Kiltz U, Heldmann F, Baraliakos X. Emerging drugs for the treatment of axial and peripheral spondyloarthritis. Expert Opin Emerg Drugs. 2015; 20:1–14.

2. Nam JL, Takase-Minegishi K, Ramiro S, Chatzidionysiou K, Smolen JS, van derHeijde D, et al. Efficacy of biological disease-modifying antirheumatic drugs: a systematic literature review informing the 2016 update of the EULARrecommendations for the management of rheumatoid arthritis. Ann Rheum Dis. 2017; 76:1113–36.

3. Smolen JS, Landewé R, Bijlsma J, Burmester G, Chatzidionysiou K, Dougados M, et al. EULAR recommendations for the management of rheumatoidarthritis with synthetic and biological disease-modifying antirheumatic drugs: 2016 update. Ann Rheum Dis. 2017; 76:960–77.

Chapter

24

Small Molecules in Rheumatic Disorders

Sundeep Upadhyay

INTRODUCTION

The Need for Newer Therapies for Inflammatory Joint Disorders

The development and use of biologic monoclonal antibodies for the treatment of inflammatory joint disorders (IJDs) such as rheumatoid arthritis (RA), psoriatic arthritis (PsA), and ankylosing spondylitis (AS) was one of the major advancements of the last three decades in the field of rheumatology-immunology. However, monoclonal antibodies have some disadvantages—firstly these are proteins and have to be administered parenterally and are associated with allergies-anaphylaxis and immunogenicity. Secondly, even though they are dramatically effective for the treatment for IJDs, around 30 to 40% of patients have a sub-optimal response.[1] Finally, since they are mostly targeted to single inflammatory cytokine, and since RA, AS, and PsA are associated with multiple cytokine perturbations, the theoretical implication of incomplete disease control despite the neutralization of one key cytokine seems obvious. While it is true that the antagonism of one cytokine (e.g. IL-6) is associated with several other downstream effects[2] through the inhibition of other cytokines, and inhibition of inflammation by other mechanisms, the efficacy of a given biologic antibody far exceeds the effects achieved solely with single cytokine neutralization. But finally, the redundancy of the immunologic pathways and immune plasticity ensures that biologic agents become less effective over time.[3] Considering the two phenomenon of "Immune escape mechanisms" and "immunogenicity: anti-drug antibodies", both of which reduce the efficacy of biologic agents over time, efforts were made to identify key pathways that could in theory neutralize multiple pathological cytokines.

BEYOND BIOLOGICS

Advances in the Field of the Molecular Biology of Signal Transduction

Advances in the field of cell and molecular biology made possible the understanding of how extracellular signals sent their messages inside the cell all the way to the nucleus. That cytokines, hormones, and other such ligands bind to their receptors was known for long, but beginning in the 90s the details of the mechanisms of how these extracellular signals brought about changes in the cell and influenced the nucleus and its DNA to make changes in the protein produced by that cell, were discovered (*viz.* the transducing of extracellular signals to intracellular changes).[4] These were the signal transduction pathways and these discoveries were made possible by the discovery of the protein kinases. These are enzymes that phosphorylate the serine, threonine and tyrosine residues of the intra-cytoplasmic chains of receptors upon the docking of the ligand onto the receptors. Subsequent to this, a sequence of events that includes second and third messenger structural changes (e.g. dimerization of signal transducers and activator of transcription—STAT) and the translocation of these to the nucleus takes place.[5] The most important of these enzymes are the ninety-plus types of tyrosine kinases which are associated with the T-Cell Receptor (TCR), B-Cell receptor (BCR), the NK cell and the Fc receptor signalling. For RA, the following tyrosine kinases have been targeted: the MAP-kinases (mitogen activated protein kinases for, e.g. the p-38 MAP kinase), spleen tyrosine kinases (STKs), Janus kinases (JAKs) and the Bruton's tyrosine kinase (BTKs) (Table 24.1).[6] Type I and type II cytokines which bind to T cells and NK cells, activate the Janus kinase protein tyrosine kinase enzymes, which then phosphorylate the cytokine receptors. This induces a structural change in the intracytoplasmic portion of the receptor molecules allowing the STATs to attach to the cytoplasmic tail of the receptors and then finally phosphorylate and dimerize. Once dimerized, these STATs move into the nucleus to regulate gene expression (Fig. 24.1).[6] In B cells, antigen binding leads to activation of the following protein tyrosine kinases: BTKs, SRc family of kinases, and the TEC-family of tyrosine kinases. MAP kinases and the nuclear factor of activated T cells (NF-kB) are also intertwined in the complex interplay of these signaling pathways.[7]

The therapeutic potential of blocking these pathways selectively to influence protein production was realized with the development of the tyrosine-kinase inhibitor Imatinib. Imatinib was approved for chronic myeloid leukemia (CML) treatment in oncology and has been extremely successful.[8] This success spurred interest in other oncology indications like non-small cell carcinoma of lung and renal cell carcinoma. Finally, the efficacy and safety profile of these agents in various phase II trials was tested in autoimmune inflammatory rheumatic diseases (AIRDs)/IJDs especially RA, ankylosing spondylitis (AS), psoriatic arthritis (PsA) (Table 24.1).

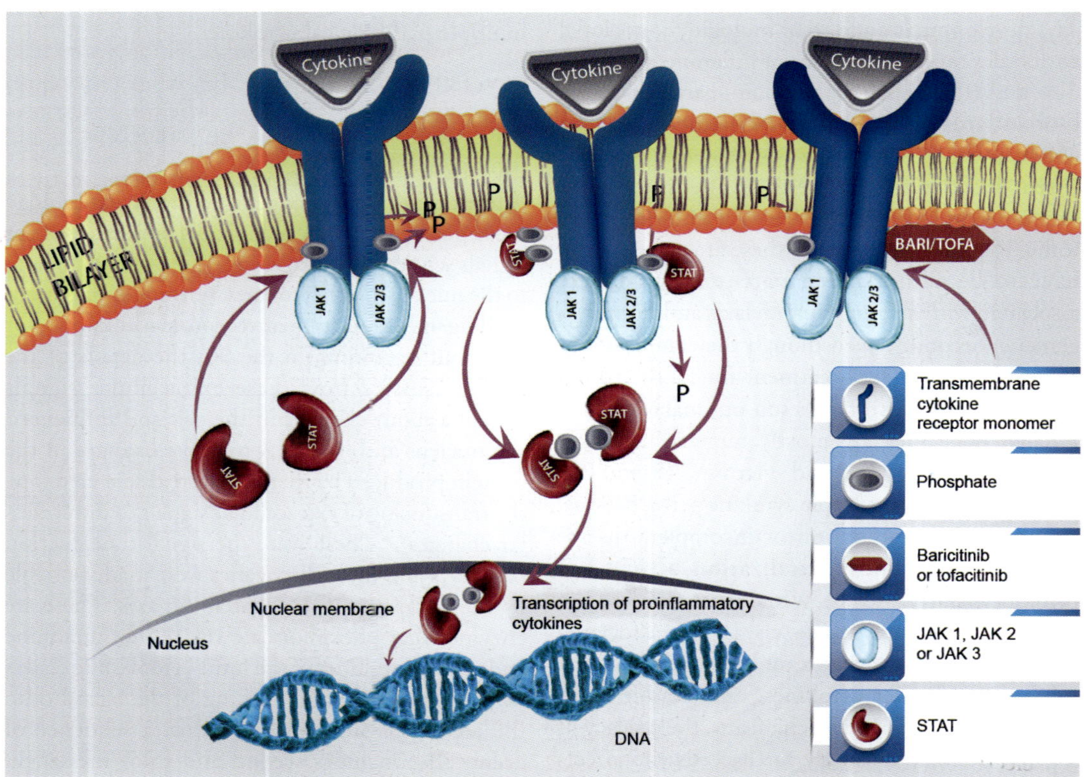

Fig. 24.1: Normally cytokines which bind to T cells and NK cells, activate the Janus kinase (JAKs) protein tyrosine kinase enzymes, which then phosphorylate the cytokine receptors. This induces a structural change in the intra-cytoplasmic portion of the receptor molecules allowing the STATs to attach to the cytoplasmic tail of the receptors and then finally phosphorylate and dimerize. Once dimerized, these STATs move into the nucleus to regulate gene expression. Tofacitinib and baricitinib inhibit JAK- STAT kinases

Table 24.1: Targetted synthetic DMARD (tsDMARD) drug development in rheumatoid arthritis

Pathway	*Mechanism of action*	*Comments*
MAP-kinases: Extracellular cellular signal related kinases (ERKs), p38 MAP kinases, c-JUN N-terminal kinases	Intracellular enzymes-p38 MAP-kinase which is a key regulator is activated by inflammatory cytokines, growth factors. Activates TNF, IL-1, COX-2, MMPs and IL-6	Pamapimod, SCIO-469 and VX 702: all p38 MAP-kinase inhibitors. Doubtful efficacy and side effects like rash, dizziness, headaches and LFT abnormalities; clinical development abandoned.
SYK inhibitors	Binds to ITAMs of Fc gamma receptors of B cells, synoviocytes, macrophages and phagocytes. Downstream MAP-kinases, PLC gamma and PI3K activated. IL-6 and MMPs upregulated.	Fostamatinib phase III: some benefit clinically but radiologically same as placebo. AEs significant: diarrhoea, neutropaenia, LFTs abnormalities, dizziness, etc. Clinical development for RA is abandoned. MK-8457: similar side effects and not much improvement in TNF non-responders; clinical development in RA is abandoned.
BTK inhibitors	TEC family of protein tyrosine kinase enzymes. BCR, TLR and FcR signalling. Upstream activation by LYN, and SYK which leads to downstream activation of NFκ B and NFAT.	Approved for lymphoma and CLL For RA in early stages of development.
MEK inhibitors		ARRY-438162, oral inhibitor of MEK1/MEK2. Clinically in significant change in MTX-IR pts. Abandoned

Abbreviations: MAP Kinases, mitogen-activated protein kinases; SYK, spleen tyrosine kinase; ITAMs, immune-receptor tyrosine-based activation motif; PLC, phospholipase C; PI3K, phosphatidylinositide 3-kinases; IL-6, interleukin-6; MMP, matrix metalloproteinases; TNF, tumour necrosis factor; BTK, Bruton's tyrosine kinase; BCR, B cell receptor; TCR, T cell receptor; CLL, chronic lymphocytic leukemia; MEK inhibitors, inhibit the MAP kinases enzymes MEK1, MEK2; MTX-IR, methotrexate inadequate response.

The JAK-STAT Pathway

The Most Important Signal Transduction Pathway in the Clinical Development of Drugs for RA/and other Rheumatic Disorders[9,10]

Brief summary of the biology and applied aspects of the JAK-STAT pathways

1. These are a family of 4 tyrosine kinase receptor enzymes (JAK-1, JAK-2, JAK-3 and TYK-2) that are important for cytokine signalling.
2. Selective inhibition of these enzymes influences cytokine-receptor signalling pathways via their action on signal transducers and activators (STATs) of protein transcription.
3. The disease modifying effect in RA and other autoimmune inflammatory rheumatic diseases (AIRDs) is through the modulation of cytokine production which in turn is brought about by the engagement of STATs with the DNA.
4. Different JAK-STAT inhibitors have different specificity for the isoforms of the enzymes. Tofacitinib is now thought to be more of a Pan-JAK inhibitor but earlier, the selectivity was believed to be for JAK1/JAK3. Similarly, baricitinib is a selective JAK1/JAK2 inhibitor, filgotinib is a selective JAK-1 inhibitor and Decernotinib is a JAK-3 inhibitor.
5. It is unclear currently as to what profile of selective inhibition by these JAK-STAT inhibitors provides the right mix of efficacy and safety. This is so because many of the JAKs along with the production of inflammatory-cytokines, also

have a housekeeping role and are associated with haematopoiesis and cell growth, etc.

6. Theoretically, JAK1 and JAK3 selectivity is good since JAK2 is involved in erythropoiesis and myelopoiesis.

TOFACITINIB: PAN-JAK INHIBITOR (ESPECIALLY JAK1/JAK3)[7,11]

Tofacitinib is a selective JAK1/JAK3 inhibitor developed by Pfizer for AIRDs. Inhibition of JAK1/3 signalling (Fig. 24.1)will affect the following cytokines: IL-2, IL-7, IL-4, IL-15 and IL-21, etc. Since IL-6 signals through JAK1, there is also a reduction of IL-6 related effects, at least theoretically. The IL-6 inhibitor tocilizumab used for RA is associated with neutropaenia and dyslipidaemia in treated patients, which is also seen in tofacitinib treated patients, and therefore IL-6 also seems to be inhibited by tofacitinib. More recently, it has been shown in a study that the IL-17 production by T-helper cells is also inhibited in patients treated with tofacitinib, but TNF signalling remains unaffected by tofacitinib. It is approved for use in RA in several countries including the US and the EU (European Union). More recently, it is in phase III trials for PsA, psoriasis and IBD (Crohn's disease). It is the first targeted synthetic DMARD (tsDMARD) approved for use in RA and the approval is for is RA patients who have inadequate response to methotrexate (MTX-IR). It appears prominently in the RA management algorithms of the American College of Rheumatology (ACR) and European League Against Rheumatism (EULAR). Given its success both in the clinical trials and in the real-world clinics at 5 mg twice daily, it has also been launched as a single dose extended release 11 mg tablet. At doses used clinically it is a pan-JAK inhibitor (except lower inhibition of TYK2 inhibition). The clinical development after successful phase II studies included 4800 patients over six clinical phase III RCTs. A long-term Extension study is ongoing currently. Most recently, the 8.5 year efficacy and safety data for tofacitinib were released.[12] Table 24.2 shows in brief the Phase-III RCTs involving tofacitinib in RA patients over several indications.[13]

Table 24.2: Published phase III studies of tofacitinib in rheumatoid arthritis (RA)[18]

Study	*Background and Interventions*	*Result change from baseline*
ORAL scan	MTX-IR, background treatment with MTX. Radiological study plus clinical outcomes	ACR 20, DAS-28 ESR: both at 6 months and HAQ-DI at 3 months. 6-month change in radiology
ORAL start	MTX-naïve. Baseline radiology-mTSS	m-TSS at 6 months, ACR 70 at 6 months
ORAL solo	Any one DMARD (biologic/non-biologic) failure. Tofacitinib vs placebo. Background stable doses of antimalarial	3 months- ACR 20, DAS 28 ESR <2.6 and HAQ-DI at 6 months
ORAL sync	One or more DMARD failure- Tofacitinib vs placebo Background DMARD treatment.	ACR 20 at 6 months, DAS-28 ESR <2.4 at 6 months, HAQ-DI at 3 months
ORAL standard	MTX-IR, active control Adalimumab. Background MTX.	ACR 20, DAS 28-ESR <2.6, both at 6 months, HAQ-DI at 3 months
ORAL step	TNF-IR. Background MTX treatment.	ACR 20, DAS 28-ESR <2.6, both at 3 months, HAQ-DI at 3 months
ORAL strategy (results not published)	Comparison of monotherapy Tofacitinib (Tofa) with adalimumab (ADA) plus MTX; also has Tofa + MTX arm	ACR 50 Monotherapy non-inferior to ADA + MTX Tofa + MTX non-inferior to ADA + MTX

Abbreviations: MTX, methotrexate; IR, inadequate response; ACR 20, American College of Rheumatology 20 response; bDMARD, biologic DMARD; DAS 28, disease activity score 28 joints; ESR, erythrocyte sedimentation rate; HAQ-DI, health assessment questionnaire-disability index; mTSS, van der Heijde modified total sharp score.

Tofacitinib is in particular associated with an increased incidence of herpes zoster. The other infections are not unlike any other biologic agent- upper respiratory infections, bronchitis, UTIs and GI side effects (nausea and vomiting), etc. Serious infection incidence-rates are also comparable to rituximab, etanercept, adalimumab, etc.[12] Among malignancies, lung cancer was most common but gastric cancer (especially in Japan), breast cancer, etc. have also been reported.

BARICITINIB: SELECTIVE JAK1/JAK2 INHIBITOR[15]

Baricitinib is the first once daily oral selective JAK1 and JAK-2 inhibitor developed for the treatment of RA. Its potency for JAK-1 and JAK-2 is 100-fold as compared to its potency for JAK-3 inhibition in functional kinase assays. While Eli-Lilly has been developing the drug various clinical phases, it was discovered by Incyte. Currently the drug is in the open label extension phase in many Asian and European countries. The mechanism of action of the drug is through selective inhibition of JAK-1 and JAK-2 enzymes, which primarily influence immune function and regulate cell growth (type I and II cytokines). The various clinical phases of the drug development are shown in Table 24.3.[16]

Baricitinib was submitted for regulatory approvals and review in the first quarter of 2016 in the United States (US), European Union (EU), and Japan. In December 2016 the European Medicines Agency's Committee for Medicinal Products for Human Use (CHMP) recommended approval of baricitinib for the treatment of adults with moderate to severe active RA and in mid-February 2017, the EMA granted full approval. As of late April 2017, there was news of refusal of the US FDA for license-grant to baricitinib till more safety data become available. The safety of baricitinib across all phase I to phase III as well as data from RA-BEYOND have been published recently (Table 24.3).[17] The events were as follows: a higher rate of herpes zoster, anaemia, lymphopaenia, transaminitis and mild azotachmia. Less than 1% discontinued baricitinib due to lab abnormalities.

Table 24.3: Phase III studies of baracitinib in rheumatoid arthritis (RA)[19]

Name of trial	*RA patients*	*Results obtained*	*Results*
RA-BEACON-phase III	Inadequate response to bDMARDs (including anti-TNF)	Nov 2016	HRQoL on SF-36, fatigue and physical function as superior as compared to placebo
RA-BEGIN-phase III	Safety and efficacy in MTX naïve pts	Feb 2015	1. Non-inferiority of baricitinib monotherapy to MTX 2. Baricitinib superior to MTX on ACR 20 responses
RA-BUILD-phase III	csDMARD failed or intolerant RA	Sep 2015	Endpoints for ACR 20, DAS28 and SDAI achieved at 12 weeks. Radiographic at 24 weeks
RA-BEAM-phase III	Safety and efficacy of baricitinib + MTX in RA-compared to a placebo OR adalimumab for 52 weeks	Oct 2015	Baricitinib superior to placebo, and superior to adalimumab
RA-BEYOND	Open label long-term extension phase	2014 0nwards	Ongoing

Abbreviations: HRQoL, health related quality of life; SF 36, short form 36; MTX, methotrexate; ACR 20, American College of Rheumatology 20 responses; bDMARD, biologic DMARD; csDMARD, conventional synthetic DMARD; DAS 28, disease activity score 28 joints; SDAI, simplified disease activity index.

FILGOTINIB: SELECTIVE JAK1 INHIBITOR[16]

Filgotinib is an oral selective JAK-1 inhibitor which has been proven efficacious even as monotherapy in patients with active RA patients in a dose ranging (phase II study (DARWIN 2).[18] Subsequently, filgotinib is in the phase III development for MTX naïve ("FINCH 3") and for MTX-IR ("FINCH1") as ongoing multicentre, double blind clinical trials over several countries. Filgotinib also induced clinical remission in patients with moderate to severe Crohn's disease (phase II FITZROY study).

SMALL MOLECULES IN THE TREATMENT OF PSORIATIC ARTHRITIS, ANKYLOSING SPONDYLITIS AND CROHN'S/IBD

Tofacitinib and baricitinib have been extensively investigated for both PsA and AS (Table 24.4).

Non-nib Molecules

A new class of "non-nib molecules" has been used in rheumatic disorders with success. Phosphodiesterase 4 (PDE-4) inhibition (Apremilast) is one such example (Fig. 24.2) causes the inhibition of cAMP to AMP, which in turn is related to protein kinase A (PKA) activation.[19] PKA activation has

Table 24.4: Small molecule drug development in non-RA rheumatic diseases

Study name	*Description and design*	*Outcome*
Opal BROADEN-tofacitinib in PsA	First JAK inhibitor in PsA. DMARD-IR population and reference arm-adalimumab	Tofacitinib superior to placebo and not inferior to adalimumab
Opal BEYOND-tofacitinib in PsA	First PsA study for TNF-IR	
Opal BALANCE-tofacitinib in PsA	Open label extension of the above two studies	3 years post marketing approval, for safety, efficacy, and tolerability.
OCTAVE induction 1 and 2-tofacitinib in moderate to severe ulcerative colitis	Tofacitinib 10 mg twice daily is successful in inducing remission and mucosal healing	Primary and secondary endpoints met
Ankylosing spondylitis (AS)- Tofacitinib	First study to show efficacy of JAK inhibitor in AS. ASAS rates 23 and 27% for 5 and 10 mg twice daily; better than 5% with placebo, better function and BASDAI	Primary endpoints attained; (but the drug is no longer slated for development in AS)
"PALACE"-Apremilast-a small molecule DMARD.	Apremilast is not strictly a signal transduction molecule (not a "nib")	Apremilast: Approved for use in Ps and PsA
Baricitinib for psoriasis and also for diabetic nephropathy, atopic dermatitis, SLE and very early phase GCA	Ongoing phase II	Ongoing phase II
DEKAVIL-selective JAK1; tissue targeting of IL-10	Ongoing phase II-RA, phase I-IBD	IL-10 key inflammatory cytokine in IBD
IRAK-4 (targets innate immunity): esp IFN-γ signalling	TLR-phase 1 ongoing; also early development phase in SLE, gout, RA, Ps	

Abbreviations: PsA, psoriatic arthritis; DMARD-IR, DMARD inadequate response; TNF-IR, TNF inadequate response; ASAS, Assessment of spondyloarthritis International Society; BASDAI, bath ankylosing spondylitis disease activity index; Ps, psoriasis; IRAK-4, interleukin-1 receptor-associated kinase 4; SLE, systemic lupus erythematosus; GCA, giant cell arteritis; IBD, inflammatory bowel disease, TLR, toll like receptor.

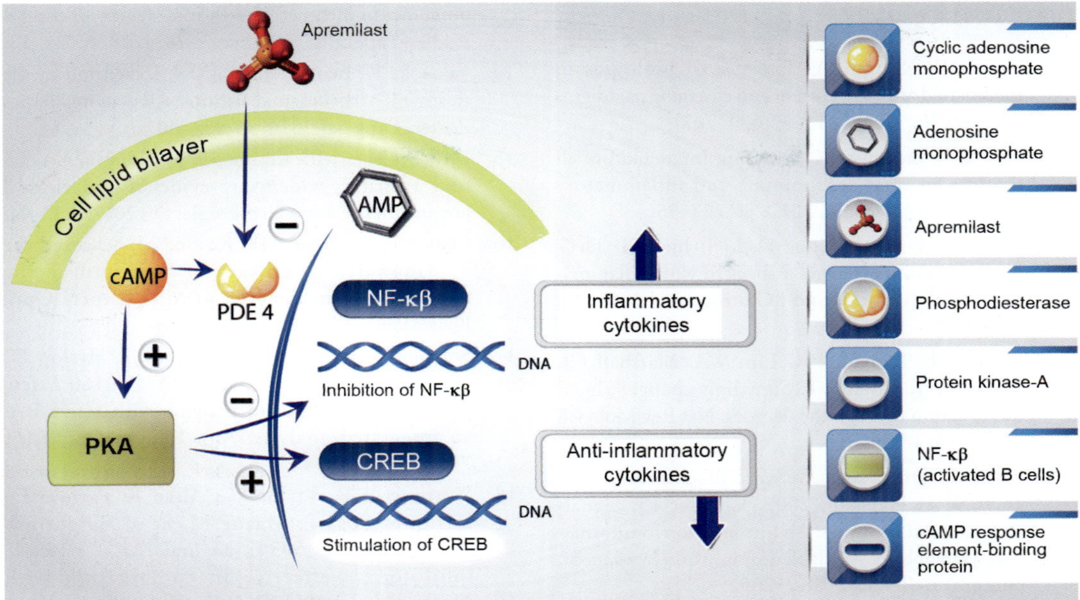

Fig. 24.2: Apremilast is a phosphodiesterase 4 (PDE-4) inhibitor

two effects: through its effect on the cAMP response element binding protein pathway (CREB) to increase anti-inflammatory cytokines and decrease NF-κ B related inflammatory cytokine secretion (Fig. 24.2). Other phosphodiesterase inhibitors in rheumatic disorders are the selective PDE5 inhibitors that mediate their effects in pulmonary artery hypertension through the cGMP pathway (sildenafil, tadalafil and verdanafil). Apremilast is approved for treatment of severe plaque psoriasis and PsA. A phase II study did not find efficacy in active RA.

Conclusion

The era of small molecules (JAK-STAT inhibitors, nibs, etc.) for the treatment of rheumatic disorders has truly arrived. These molecules are also called target synthetic DMARDs (tsDMARD). Tofacitinib and baricitinib offer advantages over biologic DMARD (bDMARD) in that they can be given orally, rapid onset of action, and efficacy as monotherapy.[13,15] Current indication of tofacitinib and baracitinib in RA is inadequate response to MTX or other conventional synthetic DMARD (csDMARD), or failure of bDMARD. Their effect on radiographic progression in RA is also scant. Similarly, in PsA, the tsDMARD are indicated when patient fails csDMARD and/or bDMARD. Long-term safety data are awaited and will influence prescribing decisions in both RA, PsA, and AS.

Key Points

- Monoclonal antibodies have some disadvantages- firstly these are proteins and can be associated with allergies-anaphylaxis and immunogenicity. Secondly, around 30 to 40% of patients have a sub-optimal response, finally since they are mostly targeted to single inflammatory cytokine, there is theoretical implication of incomplete disease control despite the neutralization of one key cytokine.
- Small molecules seem to overcome all the above because they work by inhibition of JAKs, inhibition of resultant gene expression and protein production.
- What the studies on cellular level pathophysiology reveal in the next decade of bench-to-bedside research, is what will finally determine whether small molecules or large-biologic molecules, become the primary modality of choice after csDMARD in AIRDs.
- It is even likely that a combination of these molecules (biologics plus small molecules) may be the next best option after the failure on csDMARDs.
- Two JAK-STAT inhibitors tofacitinib and baricitinib now figure prominently in the treatment guidelines of RA, PsA.

REFERENCES

1. Curtis JR, Singh JA. The use of biologics in rheumatoid arthritis; current and emerging paradigms of care. Clin Ther. 2011; 33:679–707.
2. Tanaka T, Tadamista K. Targeting Interleukin-6: all the way to treat autoimmune and inflammatory diseases. Int J Biol Sci. 2012; 8:1227–36.
3. Ueno A, Ghosh A, Hung D, Li J, Jijon H. Th17 plasticity and its changes associated with inflammatory bowel disease. World J Gastroenterol 2015; 21: 12283–95.
4. Hynes NE, Ingham PW, Lim WA, Marshall CJ, Massague J, Pawson T. Signaling change: signal transduction through the decades. Nat Rev Mol Cell Biol 2013; 14:39–398.
5. Berg JM, Tymoczko JL, Stryer L. Biochemistry. New York: W H Freeman; 2002. Chapter 15, Signal-Transduction Pathways: An Introduction to Information Metabolism. Available from: https://www.ncbi.nlm.nih.gov/books/NBK21205.
6. Yablonski D, Weiss A. Mechanisms of signaling by the hematopoietic-specific adaptor proteins, SLP-76 and LAT and their B cell counterpart, BLNK/SLP-65. Adv Immunol 2001; 79:93–128.
7. Cohen S. Novel intracellular targeting agents in rheumatic disease. In: Firestein GS, Budd RC, Gabriel SE, McInes IB, editors. Kelley and Firestein's Textbook of Rheumatology. Philadelphia: Elsevier Saunders; 2015. P. 1044–60.
8. Druker BJ, Guilhot F, O'Brien SG. Five year follow-up of patients receiving imatinib for chronic myeloid leukemia. N Engl J Med 2006; 355:2408–17.
9. Walker JG, Smith MD. The Jak-STAT pathway in rheumatoid arthritis. J Rheumatol 2005; 32: 1650–53.
10. Murray PJ. The JAK-STAT signaling pathway: Input and Output integration. J Immunol 2007; 178:2623–29.
11. Kaur K, Kalra S, Kaushal S. Systematic review of tofacitinib: a new drug for the management of rheumatoid arthritis. ClinTher. 2014; 37:1074–86.
12. Cohen SB, Tanaka Y, Mariette X, Curtis JR, Lee EB, Nash P, et al. Long-term safety of rheumatoid arthritis up to 8.5 years: integrated analysis of data from the global clinical trials. Ann Rheum Dis 2017; 76:1253–62.
13. Boyce EG, Vyas D, Rogan EL, Valle-Oseguera CS, O'Dell KM. Impact of tofacitinib on patient outcomes in rheumatoid arthritis – review of clinical studies. Patient Relat Outcome Meas 2016; 7:1–12.
14. Yamaoka K. Benefit and Risk of Tofacitinib in the treatment of rheumatoid arthritis: AfFocus on Herpes Zoster. Drug Saf 2016; 39:823–40.
15. Kriya B, Cohen MD, Keystone E. Baricitinib in rheumatoid arthritis: evidence-to-evidence and clinical potential. Ther Adv Musculoskelet Dis 2017; 9:37–44.
16. Richez C, Truchetet ME, Kostine M, Schaeverbeke T, Bannwarth B. Efficacy of baricitinib in the treatment of rheumatoid arthritis. Expert Opin Pharmacother 2017; 18:1399–1407.
17. Smolen J, Genovese M, Takeuchi T, Hyslop D, Macias WL, Rooney TP, et al. THU0166 Safety profile of baricitinib in patients with active RA: an integrated analysis. Ann Rheum Dis 2016; 75(suppl.2):243–44.
18. Westhovens R, Taylor PC, Alten R, Pavlova D, Enriquez-Sosa F, Mazur M, et al. Filgotinib (GLPG0634/GS-6034), an oral JAK-1 selective inhibitor, is effective in combination with Methotrexate (MTX) in patients with active rheumatoid arthritis and insufficient response to MTX: results from a randomized, dose finding study (DARWIN1). Ann Rheum Dis 2017; 76:998–1008.
19. Keating GM. Apremilast: A Review in Psoriasis and Psoriatic Arthritis. Drugs 2017; 77:459–72.

FURTHER READING

1. Gossec L, Smolen JS, Ramiro S, de Wit M, Cutolo M, Dougados M, et al. European League Against Rheumatism (EULAR) recommendations for the management of psoriatic arthritis with pharmacological therapies: 2015 update. Ann Rheum Dis. 2016; 75:499–510.
2. Singh JA, Hossain A, Tanjong Ghogomu E, Kotb A, Christensen R, Mudano AS, et al. Biologics or tofacitinib for rheumatoid arthritis in incomplete responders to methotrexate or other traditional disease-modifying anti-rheumatic drugs: a systematic review and networkmeta-analysis. Cochrane Database Syst Rev. 2016; (5):CD012183.
3. Singh JA, Hossain A, Tanjong Ghogomu E, Mudano AS, Maxwell LJ, Buchbinder R, et al. Biologics or tofacitinib for people with rheumatoid arthritis unsuccessfully treated with biologics: asystematic review and network meta-analysis. Cochrane Database Syst Rev. 2017; 3:CD012591.

Chapter

25

Stem Cell Transplantation in Autoimmune Connective Tissue Disorders

Velu Nair, Subramanian Shankar, Y Uday

INTRODUCTION

Autoimmune diseases (AIDS) result from failure of immune tolerance resulting in tissue damage by immune complexes, auto-antibodies or cytotoxic T-cells. An inflammatory cascade is set up that results in further tissue destruction. The contribution of genetic factors, infections and hormonal influences to the development of AIDS is well known.[1] Immunosuppressive and immunomodulatory agents used for treatment of these disorders are unable to induce clinically significant durable remissions.

The possibility of haematopoietic stem cell transplantation (HSCT) being a therapeutic modality for AIDS was entertained when it was noted that patients with a combination of AIDS and a severe haematological disease (leukaemia, aplastic anaemia) were cured of both diseases following allogeneic HSCT (allo-HSCT).[2] Therapeutic potential of allo-HSCT has been demonstrated in animal models of fully developed AIDS.[3] A 'graft versus autoimmunity' effect in AIDS similar to graft *vs* leukemia (GVL) effect was postulated. Interestingly, it was found that experimental adjuvant arthritis and experimental allergic encephalomyelitis could be cured by means of total body irradiation (TBI) followed by autologous HSCT (ASCT). This led to the concept of using HSCT as a form of immunotherapy while simultaneously deescalating the conditioning regimen to non-myeloablative levels. It was therefore hypothesised that the newly developing T cells might be tolerant to self-antigens.

Favourable results have been reported by European Group for Blood and Marrow Transplantation/European League Against Rheumatism (EBMT/EULAR) registry in systemic lupus erythematosus (SLE), systemic sclerosis (SSc) and multiple sclerosis (MS) with varied responses in rheumatoid arthritis (RA).[2] This article reviews the current status of HSCT in AIDS.

RATIONALE FOR USE OF HSCT IN AUTOIMMUNE DISORDERS

Curative therapy of AIDS requires resetting of immune tolerance. This would necessitate depleting the expanded pool of autoreactive T lymphocytes and B lymphocytes, retarding the process of immune senescence in the residual lymphocyte populations, restoring the integrity of regulatory networks, and, at the same time, preserving a pool of memory cells capable of responding to environmental pathogens.

HSCT is a medical treatment by which non-functioning, dysfunctional or malignant cells are eliminated using high-dose chemotherapy with or without radiation therapy. This is followed by stem cell rescue sourced from bone marrow (BM) or peripheral blood stem cell (PBSC). In AIDS, the main target is to achieve lymphoablation. The newly generated lymphoid cells will have a new immunological repertoire which hopefully is not auto-reactive. HSCT aims to 'reset' the dysregulated immune system of AID patients.

HOW DOES ASCT WORK: PROLONGED IMMUNOSUPPRESSION OR RESETTING IMMUNE POINT?

Despite full immune reconstitution, remission was sustained in only some patients of AIDS after the ASCT. In patients with SLE, humoral responses to vaccination were lost after ASCT as expected, but it also led to the loss of anti-dsDNA in these patients.[4] The query previously was that if the 'auto-aggressive' immune system was reinstated with the ASCT, how is the disease coming under control. Now it is realised that the 'auto-aggressive' immune system is debulked and not fully ablated, allowing the re-establishment of the normal immune regulation, in part due to the increased T-reg numbers and activity.[5] Several groups have described reduced deposition of collagen, increased microvasculature and decreased skin thickening in patients with SSc following ASCT, none of these easily explained by the either sustained immunosuppression or direct impact on the endothelial cells or fibroblasts.[6] These suggest influence on a vitiated inflammatory niche which is yet to be fully described.[7]

SUCCESS OF HSCT IN AUTOIMMUNE DISEASES

HSCT can be either autologous (donor is self) or syngeneic (identical monozygotic twins) or allogeneic (donor and recipient are different but human leucocyte antigen (HLA) matched, related or unrelated). Presently the trend is to mostly offer ASCT for AIDS in view of its safety (Fig. 25.1) (low transplant related mortality or TRM).

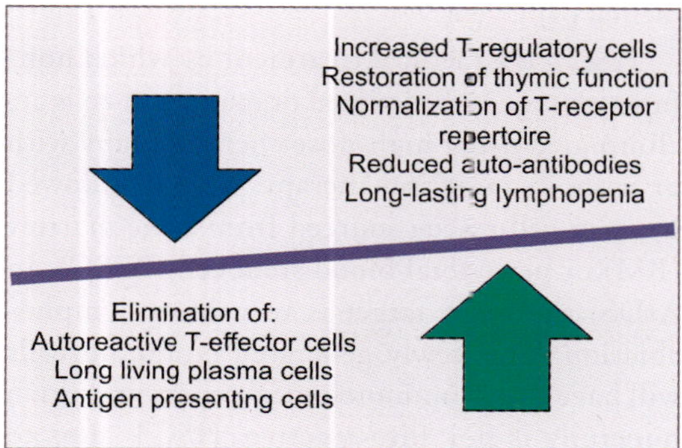

Fig. 25.1: Mechanism of development of immune tolerance secondary to ASCT in AIDS

The first clinical use of this concept was done in the year 1996 when immune ablation followed by ASCT was done in systemic sclerosis. This was followed by a series of case reports and case series reporting variable success in AIDS with ASCT, such that by 2011, EBMT/EULAR reported ASCT in 469 patients with multiple sclerosis (MS), 266 with systemic sclerosis (SSc), 95 with systemic lupus erythematosus (SLE) and 59 with inflammatory bowel disease (IBD).[8] In north America AIBMTR reported a similar experience of 399 patients.[9] The increased TRM with ASCT was offset by the durable drug free remissions, but this was seen only in one-third of the registered cases. This increased experience led to generation of the guidelines and recommendations for selecting ideal patients for ASCT.[8] The major conclusion from these guidelines suggest that patients with severe, progressive AIDS refractory to conventional therapy are the ideal candidates for the ASCT. But it was soon realised that the answers to these questions can only be gathered through prospective long-term studies (Table 25.1). Role of ASCT in SSc, SLE and RA is being reviwed here.

Systemic Sclerosis

SSc is an immunologic disorder and possibly most difficult to treat amongst various AIDS with conventional immunosuppression. Hence patients with SSc were considered most appropriate candidates for investigation of ASCT.[2,10] The EBMT/EULAR registry reported the efficacy of ASCT in SSc patients in phase I/ II studies from 1996 to 2002 with minimum 6-month follow-up. In a cohort of 57 patients, results demonstrated a marked and lasting impact on skin involvement allowing a fall in skin score of > 25% of initial values in almost 80% patients followed for 2 years. The TRM was 8%. Two-thirds of patients showed durable complete and partial remission up to 3 years post ASCT. Based on this, a phase III study the Autologous Stem Cell Transplantation International Scleroderma (ASTIS) trial was designed under the auspices of EBMT/EULAR.[11] A total of 156 patients were randomly assigned to receive HSCT (n = 79) or cyclophosphamide (n = 77). During a median follow-up of 5.8 years, 53 events occurred: 22 in the HSCT group (19 deaths and 3 irreversible organ failures) and 31 in the control group (23 deaths and 8 irreversible organ failures). Though

Table 25.1: Prospective trials on role of HSCT in autoimmune diseases

Study	*Disease*	*Country*	*Conditioning regimen*	*Treatment*	*Outcomes*
ASTIC www.astic.eu	Crohn's disease	UK n-45	Cy:200 mg/kg + ATG:7.5 mg/kg + Unselected graft	Early *vs.* late HSCT	Early transplant effective 1 TRM in early transplant arm
KISS www.clinicaltrials.gov	Crohn's disease	USA		HSCT *vs.* standard therapy	
ASSIST www.clinicaltrials.gov	Systemic sclerosis	USA/Brazil n-19	Cy: 200 mg/kg ATG:6.5 mg/kg Unselected graft	HSCT *vs.* cyclophosphamide	Achieved end point 2 relapses
ASTIS www.astistrial.com	Systemic sclerosis	UK n-156	Cy: 200 mg/kg + ATG (rabbit), 7.5 mg/kg with CD 34 selection	HSCT *vs.* monthly cyclophosphamide 750 mg/m^2	Improved event free survival 10% TRM
SCOT www.clinicaltrials.gov	Systemic sclerosis	USA n-76	Cy 120 mg/kg + ATG (equine): 90 mg/kg + TBI 800 cGY + CD34 selection	HSCT *vs.* monthly cyclophosphamide 750 mg/m^2	Recruitment completed 2011
HSCT for MS failing Interferon www.clinicaltrials.gov	Multiple sclerosis	USA/ Canada/ Brazil		HSCT *vs.* standard of care	
ASTIMS www.astims.org	Multiple sclerosis	Europe		HSCT *vs.* mitoxantrone	
ASTIL www.ebmt.org	Systemic lupus erythematosus	Europe, pending		HSCT phase II	

ASCT was associated with increased treatment-related mortality in the first year after treatment, it conferred a significant long-term event-free survival benefit. In a first ever case, the efficacy of ASCT was shown in a case of a young girl with aggressive form of local scleroderma.[12]

Systemic Lupus Erythematosus

In the last few decades, mortality in SLE has reduced because of aggressive immunosuppressive drugs and improved supportive care. However, patients with visceral organ involvement showing sub-optimal response to immunosuppressive therapy may benefit from HSCT. There is some evidence that patients with SLE have bone marrow dysfunction, with a marked reduction in CD34+ cells (elevated levels of apoptosis) and possibly reduced stem cell proliferation capability. Following ASCT, the number of apoptotic CD34+ cells is greatly reduced compared to the pre-treatment period.[13] The EBMT/EULAR registered 53 cases of SLE who underwent ASCT between 1995 and 2002 while 50 cases of SLE were treated with ASCT were reported from USA between 1997 and 2005. In the US study, the overall survival (OS) at 5 years was 80% while the probability of disease free survival (DFS) was 50% with 4% TRM. Secondary analysis demonstrated stabilization of renal function and significant improvement in SLEDAI score and carbon monoxide diffusion lung capacity. There was also improvement in serological markers in form of lower ANA/anti dsDNA titres and improvement in complement levels. A similar report on benefits of ASCT in 23 patients from 2001 to 2008 was reported.[13] These results show that in refractory SLE, ASCT controls disease activity, improves serological markers and stabilizes or reverses organ dysfunction. Definitive randomized trials such as ASTIL are likely to establish the utility of ASCT in SLE.[13]

Rheumatoid Arthritis and Juvenile Idiopathic Arthritis

Following multiple studies of stem cells in the treatment of RA (autologous, syngenic and allogeneic) in experimental animal models in the 80s, numerous case reports and case series in humans were published of the effect of ASCTs in RA. EBMT and autologous bone marrow transplant registry (ABMTR) retrospectively analysed data of most of them that comprised 73 patients from 15 centres.[14] The patients, belonging to prebiologic era, had been extensively treated with a mean of 5 DMARDs. Most of the patients received conditioning with high dose cyclophosphamide (mainly 200 mg/kg) with or without anti-thymocyte globulin (ATG) and were rescued with unmanipulated autologous peripheral blood stem cells. There was no documented transplant related mortality.[14] Relapse was experienced in most patients within 6 months of transplant and they were started back on DMARD therapy, which provided control in approximately 50% of the patients. Patients with seronegative RA had a significantly better response than seropositive disease. Phase III studies were attempted, but recruitment was compromised by the increasingly widespread use of biological anti-rheumatic agents.

Retrospective analysis of follow-up data on 34 children with refractory JIA who were treated with ASCT at nine different European transplant centres showed that 53% patients achieved complete drug-free remission and 18% achieved partial response at 12 to 60 months follow-up. However, 21% of these patients were resistant to ASCT. TRM was 9% and disease related mortality was 6%. These results show that ASCT in severely ill patients with JIA induces a drug-free remission of the disease in a substantial proportion of patients with a significant risk of mortality.

With the widespread use of biological therapies, and their relatively good safety profile, the use of ASCT in RA and JIA is relatively rare now. ASCT is now only reasonably considered in relatively rare patients whose disease has resisted conventional and biological treatments, and small numbers of cases continue to be registered with the EBMT.[15]

CHALLENGES OF HSCT IN AUTOIMMUNE DISEASES

Transplant Related Mortality

In AIDS, immediate death due to disease is a rare phenomenon. A major challenge remains death due to premature atherosclerosis and cardiac events secondary to systemic inflammation. So, a transplant related mortality (TRM) of 5–10%, though much lower than the HSCT in haematological conditions, is still unacceptable. It is hypothesised that eventually the number of deaths due to premature atherosclerosis will surpass the TRM, but the same needs to be proven in prospective trials with adequate follow-up over long-term. The various causes and preventive measures for transplant related complications are enumerated in Table 25.2 and Fig. 25.2.

Relapse/Non-response

Two-thirds of the patients transplanted for AIDS do not respond to ASCT or relapse if responded. The predictive factors for response to ASCT are

Table 25.2: Complications secondary to ASCT in AID

Complications	*Early/Late cause*	*Preventive measures*
Infection	Early: Aplastic period Late: T cell reconstitution phase	Prophylactic antibiotics Vaccination
Lung toxicity	Early: Secondary to TBI	Lung shield/volume mediated TBI
Scleroderma renal crisis (in patients with SSc)	Early: Conditioning phase causing rapid fluid/ electrolyte shifts and high-dose glucocorticoid given as prophylaxis for ATG-induced cytokine storm	ACE inhibitors
Macrophage activation syndrome (in patients with SSc)	Early: Probably infection triggered and secondary to the profound immunosuppression resulting from TBI and CD34 purging	Reduced intensity regimen rather than myeloablative or non-myeloablative regimen
Second autoimmunity	Late: To platelets, erythrocytes, thyroid; rarely fatal (acquired haemophilia-A in MS)	Resolves T-reg network is reconstituted

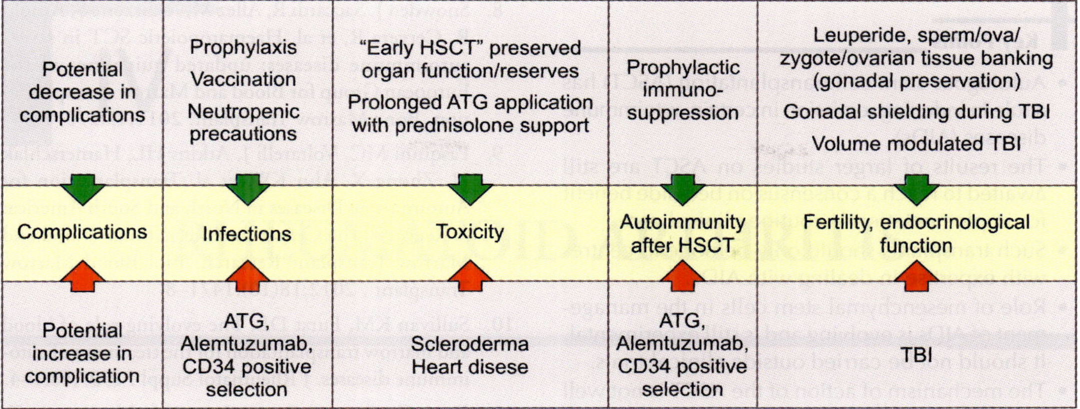

Fig. 25.2: Factors affecting complications after HSCT

not well studied. It is hypothesised that the T cell regulation plays a major role in determining success of HSCT. [16] Those patients with high number of CD3, CD4, CD27, CD45RA, CD45RB, and CD45RO on synovium have statistically increased chances of sustained clinical remission. Other factors such as increased CRP and higher activity on immunoglobulin scan are associated with clinical remission post HSCT but were not statistically significant.[16]

Centre Effect

It has been seen in all trials that the TRM following HSCT depends on the expertise of the transplant team and measures taken to prevent infections. To this end, it has been suggested that the procedure should be performed in accredited stem cell transplant units with experience in performing HSCT. Auto and allo-HSCT have been performed worldwide in both high-efficiency particulate air (HEPA) filtered units and non-HEPA filtered air-conditioned rooms. Clearly, the former, if available, should be preferred.

RESEARCH AREAS

The issues of patient selection, timing of HSCT, type of HSCT (Auto *vs* Allo), source of stem cells (Haematopoietic, Mesenchymal or Embryonic) and need for post-HSCT immunosuppression especially in disorders such as RA need to be addressed by multi-centric randomized trials. HSCT should ideally be carried out prior to irreversible organ dysfunction and preferably when disease activity is under control. Because transplant physicians do not have expertise in AIDS it is vital to actively engage rheumatologists in protocol conception for conditioning and post-transplant maintenance therapy. National and international co-ordination would be helpful in developing guidelines regarding patient selection and defining clinical and scientific endpoints. Hence, all treating teams should comprise both disease and transplant experts working towards a common goal achieving maximal benefit with minimal TRM using HSCT in the treatment of AIDS.

Conclusion

Most clinical studies reported for HSCT in AIDS have involved ASCT and have been carried out in patients who have failed to respond to conventional therapies. Many of these patients had advanced active disease, often with organ dysfunction which explains the wide range in TRM from 1 to 12%. All clinical trials have established that ASCT can achieve partial or complete remission in various AIDS. However, the durability of this response has been variable ranging from months to years. It has been noted that the beneficial effects of ASCT are accrued due to its immunosuppression and immunomodulatory effects. ASCT holds the maximum promise for SSc and SLE, while the success of biological therapy has made ASCT less attractive in RA.

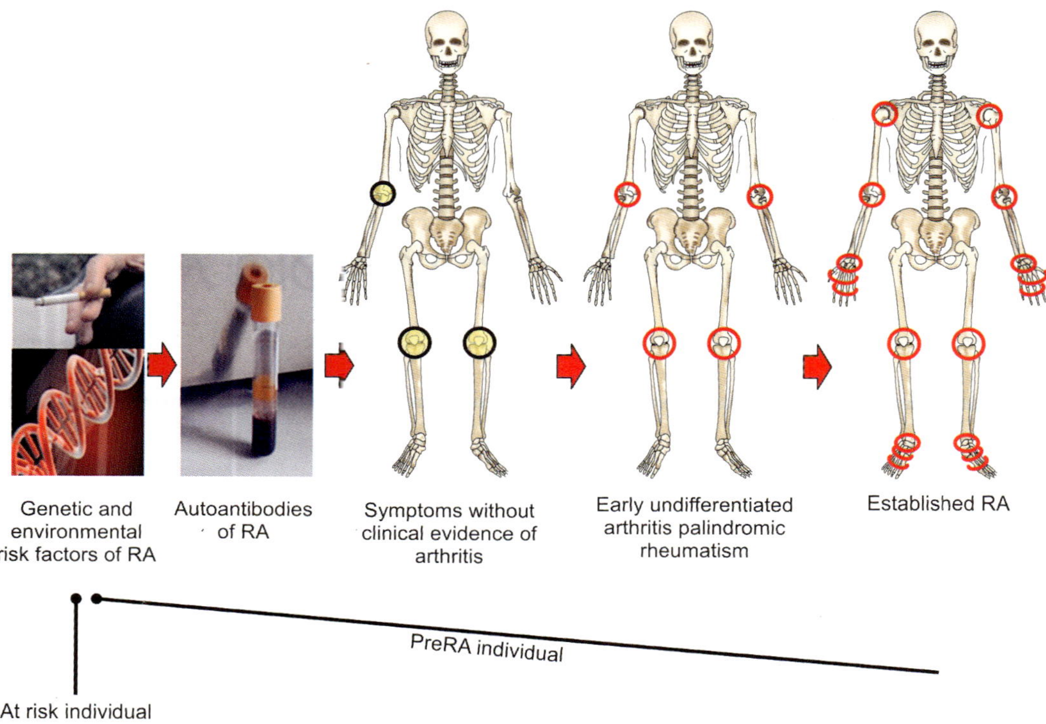

Fig. 26.1: Phases of pre-rheumatoid arthritis (pre-RA)

Table 26.1: Pre-RA

a. Genetic risk factors of RA b. Environmental risk factors for RA c. Systemic autoimmunity associated with RA	Asymptomatic
d. Symptoms without clinical evidence of arthritis e. Early undifferentiated arthritis.	Symptomatic

a. Genetic Risk Factors of RA[3]

Almost 70% of RA patients has a genetic link with HLADR4 compared to 30% in controls giving a relative risk of developing RA with this gene of 4 to 5 fold. The susceptibility to RA is associated with the third hypervariable region of DRβ1 allele which is termed the susceptibility epitope (SE). Other genes such as protein tyrosine phosphatase 22 (PTPN22) and peptidyl arginase deiminase (PADI-4) also caries approximately two fold risk of developing RA. Many other genes which carry a lower risk are also described.

Twin studies show that 12–15% of identical twins develop RA compared to 4% in non-identical twins. The disease rate of first degree family members is 0.8% compared to 0.5% in the general population. It is found that the known genetic risk alleles and heredity can explain only 16% of the overall disease burden.[3] Many other factors like epigenetics can further alter the gene expression in favour of developing RA. It is now clear that genes only modestly increase the risk of RA while the environment is likely to play a stronger role.

b. Environmental Risk Factors of RA[4]

A number of environmental risk factors clearly contribute to RA susceptibility like smoking, infections, periodontitis, obesity, hormonal and dietary factors (Fig. 26.2).

Smoking: Smoking is the best defined risk factor for seropositive RA. Smoke is a strong stimulus for protein citrillination and generation of ACPA antibodies.[5] HLA susceptibility epitope (HLA -SE) has increased ability to bind the citrullinated protein. A smoker with two copies of HLA SE has the chance of developing RA is 40 fold, indicating

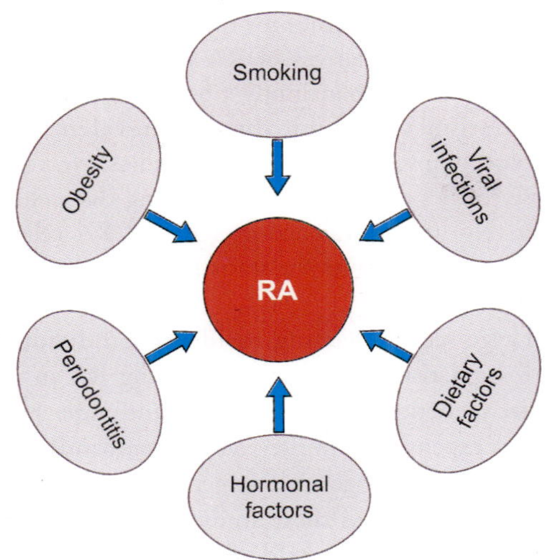

Fig. 26.2: Environmental risk factors for rheumatoid arthritis (RA)

the interaction between genetic and environmental risk factors in the development of RA.[6] It is also found that there is a long latency (up to 20 years) after cessation of smoking to return to the risk level of a non-smoker.

Infections: It is found that infective agents could contribute to the initiation or perpetuation of RA by a variety of mechanisms like molecular mimicry, toll-like receptor activation or direct invasion of the synovium, etc. Parvovirus 19, mycoplasma, enteric bacteria, Epstein-Barr virus are the common implicated infection agents. In India, epidemic of chikungunya virus infection also contributed to increase the prevalence of RA in susceptible population.[7]

Periodontitis: Periodontitis is a chronic inflammatory disease of gingiva. It is a disease with similar pathogenesis, similar risk factors and epidemiological association as RA. Various studies from India and abroad show that the occurrence and severity of periodontitis was found to be higher in RA subjects suggesting a positive relationship between these two chronic inflammatory diseases.[8] RA patients with more periodontal symptoms have higher levels of RA disease activity. In RA periodontitis is a strong predictor of ACPA. The bacteria causing periodontitis, namely porphyromonas gingivalis, causes citrullination through its own PADI enzyme.[9] Hence controlling periodontitis may reduce the risk of developing RA.

Dietary risk factors: Many dietary factors are found to modify the risk of RA in susceptible population. Studies show that omega 3 fatty acids ameliorate clinical symptoms of RA and fish oil consumption has a protective effect. The antioxidants in fruits and vegetables (vitamin C, vitamin E, carotenoids, lycopene) also show a protective effect. Red meat intake and vitamin D deficiency increases the risk of RA. Excessive coffee consumption is also a risk factor for ACPA positive RA. High salt intake is found to be a risk factor in recent studies.

Hormonal factors: Even though RA is three times more common in women than men, oestrogen containing contraceptives and other hormone replacement therapies are not reducing the risk of RA. Pregnancy ameliorate RA in majority of women. RA risk increases with breast feeding due to surge of proinflammatory prolactin. It is also found that nulliparity increases the risk of RA. Obesity is also an important risk factor for RA in the Western world, as overweight increases the levels of circulatory leptins and stimulate proinflammatory cytokines like TNF-α and interleukin-1.[10]

c. Systemic Autoimmunity Associated with RA

People who are detected to have positive RF and/or ACPA without any symptoms of RA are classified under this group. Multiple studies showed that RA-related antibodies are present years prior to the diagnosis of RA. Studies from stored prediagnosis blood samples from biobanks of Finland and Northern Sweden demonstrated that RF and ACPA were present years prior to the onset of clinically apparent RA. In prospective studies from the UK, initially healthy family members of patients with RA showed the presence of RF preceding the onset of clinically apparent RA. It is now clear that in comparison to controls, a combination of both ACPA and any RF isotype is highly specific for future development of RA.

d. Symptoms without Clinical Evidence of Arthritis

It is not uncommon to see patients presenting with arthralgia or joint stiffness with positive RF and/or ACPA, without any demonstrable arthritis on physical examination. Some of these patients later progress to classical RA after a variable period of time and hence it is considered as a 'pre-RA' phase.

e. Unclassified Arthritis (Undifferentiated Arthritis)

It is a situation where there is arthritis of more than one joint without fulfillment of criteria for RA or any other connective tissue disease. It is a diagnosis of exclusion based on the failure to satisfy classification criteria for a well-recognized rheumatic condition. However, these patients have a great potential for development of persistent inflammatory arthritis (Fig. 26.3). It is found that 30–40% of patients with undifferentiated arthritis progress to classical RA on long-term follow up.[11]

Leiden Prediction Rule[12] (Fig. 26.4) and other scoring systems are useful to identify people who will later progress to RA. A Leiden prediction rule of 6 or below, 91% will not progress to RA. If the prediction score is 8 or more, 84% will progress to RA.

Prevention of RA

Screening and follow-up of people at the risk of developing RA is appropriate for developing prevention programmes for the disease. Screening of first degree relatives of RA patients, twins of RA patients, autoantibody positive individuals, population with high disease prevalence are employed in various prospective studies. However, these studies are expensive comparing the yield as relatively low prevalence of RA and RA related autoantibodies limits the statistical power of these studies.[13]

Large-scale screening to identify individuals with high risk features of developing RA in the future (genetic factors, RA related antibodies) will be very expensive. Low prevalence of autoantibodies in RA requires large-scale screening to identify at risk individuals. As these studies are not suitable for the developing world, prevention of RA by modifying the environmental factors is evolving as a cheap and effective strategy to prevent RA (Table 26.2).[14]

Patients who are symptomatic definitely need treatment. Many studies show that patients with undifferentiated arthritis benefit from methotrexate. Combining multiple DMARDs or DMARDs with corticosteroids and biological agents may be even more beneficial.[15] However, more randomized controlled trials with larger follow-up time are needed to prove which treatment provide best results or alter the disease course.

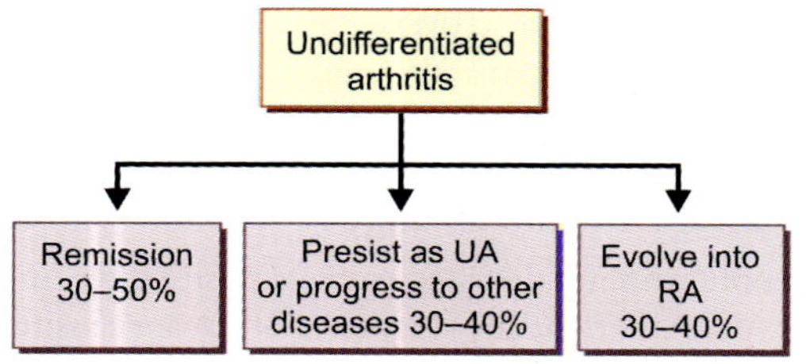

Fig. 26.3: Outcome of undifferentiated arthritis (UA)

Variables		Score
Age (years)		X 0.02
Female		1
Joint distribution	Small joints of hand/feet	0.5
	Symmetry	0.5
	Upper limb	1.0
	Upper and lower limb	1.5
Morning stiffness	26–90 mm	1.0
(on 100 mm VAS)	>90 mm	2.0
Tender joints (n)	4–10	0.5
	>10	1.0
Swollen joints	4–10	0.5
	>10	1.0
CRP (mg/L)	5–50	0.5
	>50	1.5
RF positivity		1
ACPA positivity		2

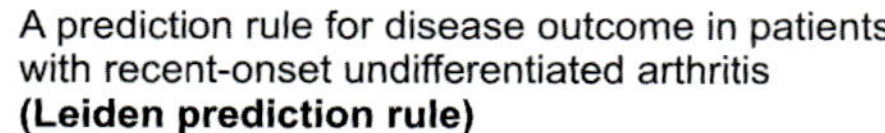

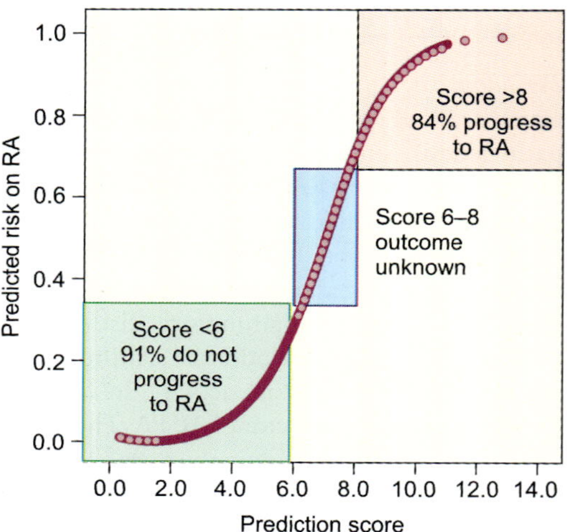

Fig. 26.4: Leiden prediction rule

Table 26.2: Modifying the risk factors to prevent RA

- Avoid/quit smoking
- Good dental hygiene: Treat periodontitis early
- Balanced diet containing fish oil, antioxidants, vitamin D
- Avoid excess coffee and foods with high salt content
- Optimizing body weight
- Prevention of infection

Key Points

- Pre-rheumatoid arthritis (pre-RA) is the preclinical period of the disease that precedes the onset of clinically apparent RA.
- It includes the interaction between genetic and environmental risk factors, development of disease related auto-antibodies and joint symptoms and signs which may be considered nonspecific or unclassified for rheumatoid arthritis.
- Better understanding of the pre-RA stage will be useful to develop screening programs and preventive strategies for RA.

REFERENCES

1. Raza K, Klareskog L, Holers VM. Predicting and preventing the development of rheumatoid arthritis Rheumatology(Oxford) 2016; 55:1–3.
2. Gerlag DM, Raza K, van Baarsen LG, Brouwer E, Buckley CD, Burmester GR, et al. EULAR recommendations for terminology and research in individuals at risk of rheumatoid arthritis: report from the Study Group for Risk Factors for Rheumatoid Arthritis. Ann Rheum Dis. 2012; 71:638–41.
3. Turk SA, van Beers-Tas MH, van Schaardenburg D. Prediction of future rheumatoid arthritis. Rheum Dis Clin North Am 2014; 40:753–70.
4. van Steenbergen HW, Huizinga TW, van der Helm-van Mil AH. The preclinical phase of rheumatoid arthritis: What is acknowledged and what needs to be assessed? Arthritis Rheum 2013; 65:2219–32.
5. Sugiyama D, Nishimura K, Tamaki K, Tsuji G, Nakazawa T, Morinobu A, et al. Impact of smoking as a risk factor for developing RA: A meta-analysis of observational studies. Ann Rheum Dis 2010; 69:70–81.
6. Karlson EW, Deane K. Environmental and gene-environment interactions and risk of rheumatoid arthritis. Rheum Dis Clin North Am 2012; 38: 405–26.
7. Binoy JP, Geetha P, Shanu PM. Clinical profile and long-term sequel of chikungunya fever. Indian J Rheumatol 2011; 6(1):12–19.
8. Joseph R, Rajappan S, Nath SG, Paul BJ. Association between chronic periodontitis and rheumatoid arthritis: a hospital-based case-control study. Rheumatol Int. 2013; 33:103–9.
9. Wolff B, Berger T, Frese C, Max R, Blank N, Lorenz HM, et al. Oral status in patients with early RA; a prospective case control study. Rheumatology (Oxford) 2014; 53:526–31.
10. de Hair MJ, Landewé RB, van de Sande MG, van Schaardenburg D, van Baarsen LG, Gerlag DM, et al. Smoking and overweight determine the likelihood of developing rheumatoid arthritis. Ann Rheum Dis 2013; 72:1654–8.
11. van der Helm-van Mil AH, le Cessie S, van Dongen H, Breedveld FC, Toes RE, Huizinga TW. A prediction rule for disease outcome in patients with recent-onset undifferentiated arthritis: how to guide individual treatment decisions. Arthritis Rheum 2007; 56:433–40.
12. van der Helm-van Mil AH, le Cessie S, van Dongen H, Breedveld FC, Toes RE, Huizinga TW. A prediction rule for disease outcome in patients with recent-onset undifferentiated arthritis. Arthritis Rheum. 2007; 56:433–40.
13. Klareskog L, Gregersen PK, Huizinga TW. Prevention of autoimmune rheumatic disease: state of the art and future perspectives. Ann Rheum Dis 2010; 69: 2062–66.
14. Lahiri M, Morgan C, Symmons DP, Bruce IN. Modifiable risk factors for RA: prevention, better than cure? Rheumatology (Oxford). 2012; 51:499–512.
15. Wevers-de Boer KV, Heimans L, Huizinga TW, Allaart CF. Drug therapy inundifferentiated arthritis: a systematic literature review. Ann Rheum Dis. 2013; 72:1436–44.

FURTHER READING

1. Bykerk VP, Costenbader KH, Deane K. Preclinical rheumatic disease. Rheum Dis Clin North Am. 2014; 40(4):xv–xviii.
2. Deane KD, El-Gabalawy H. Pathogenesis and prevention of rheumatic disease: focus on preclinical RA and SLE. Nat Rev Rheumatol. 2014; 10:212–28.
3. Karlson,EW, van Schaardenburg, D, van der Helm-van Mil AH. Strategies to predict rheumatoid arthritis development in at-risk populations. Rheumatology (Oxford) 2016; 55:6–15.

Chapter

27

Classification Criteria for Rheumatoid Arthritis

Banwari Sharma

INTRODUCTION

Most systemic rheumatic diseases lack a single pathognomonic or diagnostic feature. Features like joint pain and swelling, morning stiffness, fever, fatigue, skin rash are common to several diseases, also many of these features do not occur concurrently but sequentially. When these features present in a particular combination along with certain laboratory markers than that helps in identifying a specific disease.

Classification criteria help differentiate patients with a particular disease from patients with a potentially similar condition as well as from the normal population. Not all patients with rheumatoid arthritis (RA) are alike. Patients with RA are a heterogeneous group of people, each with their unique challenges. Classification criteria are created in an attempt to produce a homogeneous group of patients that can then be used for clinical and basic research. Rheumatologists in clinical practice are not limited only to classification criteria for diagnosis of RA. Their clinical acumen and sophisticated imaging modalities guide their final judgement. While using classification criteria, the clinician should never lose sight of the fact that therapeutic decisions in an individual patient should not be governed only by fulfillment or lack of fulfillment of criteria.

Evolution of RA Classification Criteria

Classification Criteria for RA Before 1987

Classification criteria for RA were first proposed, in 1956 and published by the American Rheumatism Association (ARA) in 1958.[1,2] These criteria incorporated 11 clinical, serological, radiological and histological features with 19 exclusions. The 1958 criteria sets classified disease as 'definite', 'probable' and 'possible' RA (Table 27.1). The 'definite' group required 5 criteria with 6 weeks of symptoms while the 'probable' RA required 3 criteria with 4 weeks of joint symptoms. The 1958 revision added the category of 'classic' RA for patients exhibiting 7 out of 11 criteria.

Within short period of time clinicians realized two important drawbacks in these criteria, first was the inability to pick-up disease during remission and the other drawback was inclusion of synovial fluid analysis and biopsy (synovial or rheumatoid nodule), that is never done in majority of RA patients.

The revision of the same criteria, known as the New York Criteria, was introduced in 1967, but the criteria were poorly accepted by rheumatology fraternity due to lack of a defined cut-point for definite cases of RA.[3]

1987 American Rheumatism Association (ARA: ACR) Classification Criteria

The criteria for RA were revised in 1987 after a long gap of 30 years.[4] The 1987 ARA criteria were developed using RA cases and controls attending tertiary care referrals. The patients included had established longstanding disease, e.g. mean disease duration 7.7 years. These criteria have kept five of the original 1958 ARA criteria. The classification subcategories were dropped and definite and classic RA were relabelled simply as RA. The term probable

Table 27.1: Comparison of 1987 and 1958 classification criteria for RA[4]

1987	*1958*
Morning stiffness in and around joints for at least 1 h	Morning stiffness
Soft tissue joint swelling observed by physician in at least 3/14 joint groups (Right or left: MCP, PIP, wrist, elbow, knee ankle, MTP)	Swelling of a joint
Symmetric swelling of one joint area	Swelling of another joint
Rheumatoid nodule	Rheumatoid nodule
Rheumatoid factor by method positive in <5% normal population	Rheumatoid factor
Radiographic changes in wrist or hands: erosions or juxta-articular osteoporosis	Radiographic changes
—	Pain on movement or tenderness in a joint
—	Symmetric swelling
—	Mucin clot
—	Synovial biopsy
—	Nodule biopsy
RA: 4/7	Classical RA: 7/11 Definite RA: 5/11, Probable RA: 3/11

Abbreviations: MCP, metacarpophalangeal; MTP, metatarsophlangeal; PIP, proximal interphalangeal.
Source: Arnett FC, Edworthy SM, Bloch DA, et al. The American Rheumatism Association 1987 revised criteria for the classification of rheumatoid arthritis. Arthritis Rheum 1988; 31: 315–24.

RA was replaced by terms such as undifferentiated polyarthritis, undifferentiated oligoarthritis, or undifferentiated monoarthritis, and the lengthy list of exclusions, previously part of the 1958 criteria, were dropped.

According to the 1987 criteria a patient was classified as having RA if at least four of these seven criteria were satisfied, four of the criteria must have been present for at least six weeks: morning stiffness, arthritis of three or more joint areas, arthritis of the hands, and symmetric arthritis (Table 27.1). Rheumatoid factor (RF) was included as a criterion, but anti-citrullinated peptide antibody (ACPA) testing was not available at that time. The other two criteria were rheumatoid nodules and radiographic erosive changes typical of RA, but these are generally not present in the early stages of disease.

These criteria had a sensitivity of 91–94% and specificity of 89% when comparing RA with non-RA. Over a period of time a few shortcomings became apparent. The first was the poor performance characteristics of 1987 criteria in early RA. When used for early inflammatory polyarthritis, the 1987 ARA criteria for RA had a low ability to discriminate between patients who developed progressive erosive disease and those who did not.[5] In various studies it has been shown that the 1987 ACR criteria, when applied to early RA, have a sensitivity ranging from 40 to 90% and specificity from 50 to 90%.[5–9] The another drawback of 1987 criteria was the inclusion of radiographic features. The radiologic criterion of erosions is encountered in a very small proportion (~13%) of patients in the first 3 months of disease onset limiting its utility.[10] However, as many as 50–70% patients may have erosive disease by 2 years thereby underscoring the importance of early treatment.[11]

Need for New Criteria

At the time of 1987 RA criteria development, the accepted paradigm for treatment of RA was the "pyramid approach" or "start low and go slow," and therapeutic options were limited. For standards of practice at that time, they were appropriate. Over the past two decades, the therapeutic "pyramid" for RA has been inverted. Early aggressive therapy resulted in improvement in clinical outcome, as well as disease-associated morbidity and, mortality in

Key Points

1. Classification criteria for RA were first published in 1958 and revised in 1967.
2. The 1987 RA classification criteria were good for established disease but had low sensitivity to identify early RA.
3. The 2010 ACR/EULAR criteria for RA aimed for early identification of poor prognosis arthritis patients from recent onset undifferentiated arthritis patient group.
4. Validation studies of the 2010 RA criteria have shown that 12–42% of the patients who fulfill the 2010 RA criteria at presentation don't develop RA, but rather develop another chronic disease or have a self-limiting condition.

REFERENCES

1. Bennett GA, Cobb S, Jacox R, Jessar RA, Ropes MW. Proposed diagnostic criteria for rheumatoid arthritis. Bull Rheum Dis. 1956; 7:121–4.
2. Ropes MW, Bennett GA, Cobb S, Jacox R, Jessar RA. 1958 Revision of diagnostic criteria for rheumatoid arthritis. Bull Rheum Dis.1958; 9 175–6.
3. Bennett PH, Burch TA. New York Symposium on population studies in the rheumatic diseases: new diagnostic criteria. Bull Rheum Dis 1967; 17:453–8.
4. Arnett FC, Edworthy SM, Bloch DA, McShane DJ, Fries JF, Cooper NS, et al. The American Rheumatism Association 1987 revised criteria for the classification of rheumatoid arthritis. Arthritis Rheum. 1988; 31:315–24.
5. Harrison BJ, Symmons DP, Barrett EM, Silman AJ. The performance of the 1987 ARA classification criteria for rheumatoid arthritis in a population based cohort of patients with early inflammatory polyarthritis. American Rheumatism Association. J Rheumatol 1998; 25:2324–30.
6. Aletaha D, Breedveld FC, Smolen JS. The need for new classification criteria for rheumatoid arthritis. Arthritis Rheum 2005; 52:3333–6.
7. Saraux A, Berthelot JM, Chalès G, Le Henaff C, Thorel JB, Hoang S, et al. Ability of the American College of Rheumatology 1987 criteria to predict rheumatoid arthritis in patients with early arthritis and classification of these patients two years later. Arthritis Rheum 2001; 44:2485–91.
8. Hülsemann JL, Zeidler H. Diagnostic evaluation of classification criteria for rheumatoid arthritis and reactive arthritis in an early synovitis outpatient clinic. Ann Rheum Dis 1999; 58:278–80.
9. Visser H, le Cessie S, Vos K, Breedveld FC, Hazes JM. How to diagnose rheumatoid arthritis early: a prediction model for persistent (erosive) arthritis. Arthritis Rheum 2002; 46:357–65.
10. Machold KP, Stamm TA, Eberl GJ, Nell VK, Dunky A, Uffmann M, et al. Very recent onset arthritis—clinical, laboratory and radiological findings during the first year of disease. J Rheumatol 2002; 29:2278–87.
11. Plant MJ, Jones PW, Saklatvala J, Ollier WE, Dawes PT. Patterns of radiological progression in rheumatoid arthritis: results of an 8 year prospective study. J Rheumatol 1998; 25:417–26.
12. Aletaha D, Neogi T, Silman AJ, Funovits J, Felson DT, Bingham CO 3rd, et al. 2010 Rheumatoid arthritis classification criteria: an American College of Rheumatology/European League Against Rheumatism collaborative initiative. Arthritis Rheum 2010; 62:2569–81.
13. Aletaha D, Neogi T, Silman AJ, Funovits J, Felson DT, Bingham CO 3rd, et al. 2010 rheumatoid arthritis classification criteria: an American College of Rheumatology/European League Against Rheumatism collaborative initiative. Ann Rheum Dis 2010; 69:1580–8.
14. Radner H, Neogi T, Smolen JS, Aletaha D. Performance of the 2010 ACR/EULAR classification criteria for rheumatoid arthritis: a systematic literature review. Ann Rheum Dis 2014; 73:114–23.

FURTHER READING

1. Ortiz EC, Shinada S Evolution of classification criteria for rheumatoid arthritis: how do the 2010 criteria perform? Rheum Dis Clin North Am. 2012; 38:345–53.
2. Sokolove J, Strand V. Rheumatoid arthritis classification criteria—It's finally time to move on! Bull NYU Hosp Jt Dis 2010; 68:232–8.

Chapter

28

Articular and Extra-articular Manifestations of Rheumatoid Arthritis

S Chandrashekara

INTRODUCTION

Rheumatoid arthritis (RA) is a disease with predominant manifestations involving joints. Extra-articular manifestations appear during the later stages of disease. The onset of manifestations may be acute/intermediate, chronic (insidious) and rarely palindromic. The involvement of joint/s may also vary from single to multiple joints. Monoarticular or oligoarticular, involving one to three joints for substantial period, is less common (1 to 15%) in RA patients. The usual presentation is polyarticular, symmetrical and often having the initial involvement of small joints of hands/feet. The polyarticular presentation accounts for more than 80% of the manifestations. The remaining 10–15% may have non-classical presentations like oligoarticular, predominantly having arthritis of large joints, shoulders and hip. Rarely, the patients may present with dominant extra-articular features like fever, scleritis, vasculitis and carpel tunnel syndrome. Sometimes, the initial presentation may be bursitis and tendinitis. Subsequently, the disease may progress to the remaining joints. RA can present with diverse clinical features but the most common presentation is polyarthritis of small joints.

ARTICULAR FEATURES

General symptoms: Small joints of the hands are initially involved. Typically, early morning stiffness may last for more than 1 hour and may gradually increase to persist for the entire day. The resting stiffness—the stiffness of the joint following prolonged rest—is a typical feature of inflammatory arthritis. This may be accompanied by swelling and limitation in joint movement (Fig. 28.1). The range of movement may be reduced from minimal to complete loss of mobility. The pain, ranging from mild discomfort to severe incapacitating pain, varies from patient to patient. The squeeze sign, tenderness on squeezing the hand, is considered as one of the earliest indicator of RA. In late RA, the joint deformity may be associated with or without pain. The specific joint areas affected are discussed below.

Hands: In RA, metacarpophalangeal joints and wrist are involved, sparing the distal interphalangeal joints (DIP). This combination helps us to differentiate RA from psoriatic and osteoarthritis

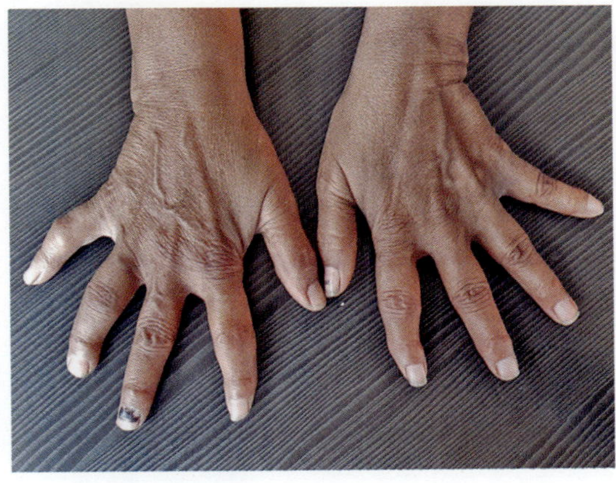

Fig. 28.1: Presentation of subtly swollen proximal interphalangeal joints and mild deformities in a patient with rheumatoid arthritis

(where DIP is involved). Wrist involvement may be accompanied by median nerve compression, and restriction of flexion and extension movements at the wrists. The classic hand deformities, which are commonly seen in RA, are not specific to the disease. The ulnar drifting of fingers at metacarpophalangeal (MCP) joints, i.e the drifting of the fingers away from the thumb is the classic and unique sign of early deformities of rheumatoid hands. The other finger deformities include Boutonnière deformity (flexion at PIP, hyperextension at DIP and MCP hyperextension) swan neck deformity (hyperextension at PIP and flexion at DIP) (Fig. 28.2), and rarely mallet finger (only flexion at DIP). In addition, multiple combinations of dislocation and deformities are possible.[1, 2]

Foot: This is the second common site involved in RA. The initial symptoms could be swelling and stiffness including ankles. Sometimes the swelling may be diffuse and may be confused with pedal oedema. DIP joints are not affected in foot also. Lateral deviation of the greater toe, which may sometimes override the second toe, is the most common deformity noted. Reduction in ball of toes, and flattening and collapse of arch of foot with unstable ankles are the common deformities. Hammer toe, a flexion deformity at PIP, is also frequently seen.[2]

Larger joints: Knees, hip, elbows, shoulders are involved.[2] Swelling and limitation of movement are the early features. Deformities with dislocation of articular surface, ultimately resulting in fibrous or bony ankyloses, are commonly noted. The elbow joints, which include proximal annular joint of radioulnar joints, often get easily dislocated resulting in early restriction of pronation and supination.

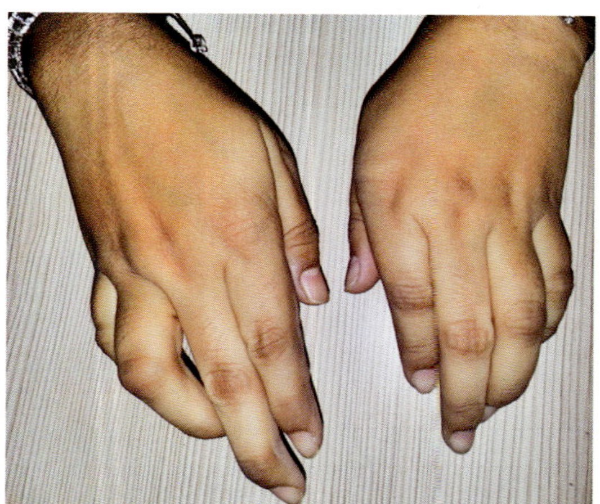

Fig. 28.2: Swan neck deformity of the index and middle fingers, and Boutonnière deformity of the ring finger of the right hand

Back and spine: Early involvement of axial joints of spine is rare in RA. The back pain in RA is not significantly different from the general population.[3] The involvement of neck, especially atlanto-axial joint, is seen in later stages of RA. The initial symptoms may be neck pain, especially at the back of the neck. The subtle symptoms include feeling of being pushed while walking, and tingling and pain sensation across the neck sometimes radiating down the neck. If not identified early, the symptoms of cord compression may gradually appear. In recent years, the incidence of cord compression has significantly reduced.

Other musculoskeletal involvement: Initial presentation could be bursitis and rarely tendinitis. The initial complaints may be vague pain in muscles and tendon insertion. The bursitis of prepatellar bursa, wrist and elbow is seen in significant percentage of patients.

EXTRA-ARTICULAR PRESENTATIONS

The possible extra-articular features are detailed in Table 28.1.

Table 28.1: Common extra-articular manifestations in rheumatoid arthritis

Sicca symptoms
Ocular: Iridocyclitis, uveitis, episcleritis, scleritis, retinal vasculitis
Psychiatric: Anxiety, depression
Neurological: Carpel tunnel syndrome, compressive neuropathy, peripheral neuropathy
Pulmonary: Bronchiolitis obliterans, pleuritis, interstitial lung disease
Cardiac: Pericarditis
Hematological: Anemia, lymphadenopathy, splenomegaly, hepatomegaly
Bone: Atlanto-axial dislocation, enthesitis
Osteoporosis
Rheumatoid nodules
Small vessel vasculitis, purpura, gangrene
Amyloidosis
Atherosclerosis

Muscle: Degeneration of muscle fibres are found in all RA patients. Muscle disease in RA can occur in one or more of five forms, namely reduction by atrophy of type II fibres, neuropathy and degeneration, steroid myopathy, rare myositis overlap, and chronic myopathy.

Skin: The most frequent skin lesion in RA is rheumatoid nodules (Fig. 28.3) and occur in 20% to 30% of RA patients. The skin atrophy, palmar erythema, and Raynaud's phenomenon (rare) are seen in feet and hands. Vasculitis lesions with nail fold infarcts, pyoderma gangrenosum, palpable purpura and livedo reticularis are seen.[4]

Fistula development: Rarely cutaneous sinuses may develop near joints and they are often sterile or septic. The fistula either connects the skin to a joint or bursa.

Eye: The ophthalmic manifestations include keratoconjunctivitis sicca (generally associated with Sjögren's syndrome), scleritis, episcleritis, and secondary cataracts or keratitis leading to vision loss.[5] Scleritis causes severe ocular pain and dark red discolouration. It can be localized, generalized or necrotizing. Granulomatous resorption can cause scleromalacia perforans, which can be vision threatening. Perilimbal ischaemic ulcer can be caused by cryoproteins.

Hematological abnormalities: Majority of patients with RA have mild normocytic normochromic anemia.Anemia of chronic disease is noted in 25–35% of the subjects.[6,7] The response of patients with these manifestations to iron therapy, vitamin B12 or folate therapy is around 25%. Incidence of eosinophilia (5% of TC) and thrombocytosis are reported in some patients.

Felty's syndrome: It is characterized by the clinical triad of neutropenia, splenomegaly, and nodular RA, and is seen in less than 1% of patients.[8] Although its incidence is declining, this condition typically occurs at late stages of severe RA.

Vasculitis: Clinical vasculitis may manifest as distal artheritis ranging from splinter haemorrhage to gangrene, cutaneous ulceration, peripheral neuropathy, palpable purpura and rarely as arteritis of viscera (heart, lungs, kidneys, liver, spleen pancreas, and testes)[9] (Fig. 28.4).

Neurovascular diseases: Mild to severe distal sensory[10] or motor neuropathy is one of the common clinical manifestations. Rheumatoid pachymeningitis is a rare complication of RA. Autonomic nervous disturbances and compression neuropathy can also occur. Carpal tunnel syndrome is one of the common forms of entrapment neuropathy.

Renal diseases: Renal involvement is rare, but drugs used in RA like salicylates and phenacetin can cause renal papillary necrosis. Membranous nephropathy is associated with gold salts and penicillamine.

Pulmonary diseases: The pulmonary manifestations of RA include pleural disease, interstitial fibrosis,

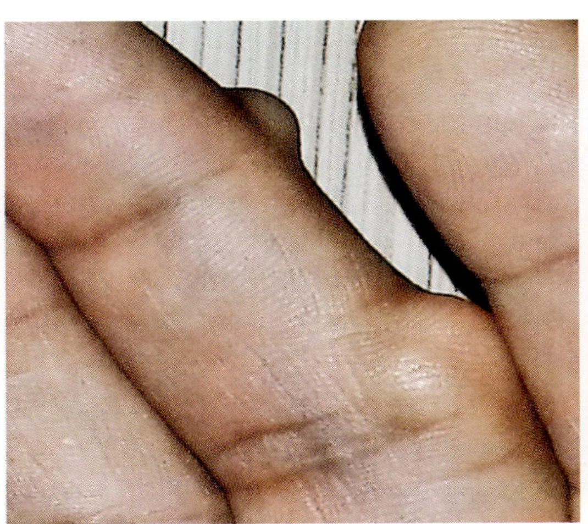

Fig. 28.3: Rheumatoid nodules

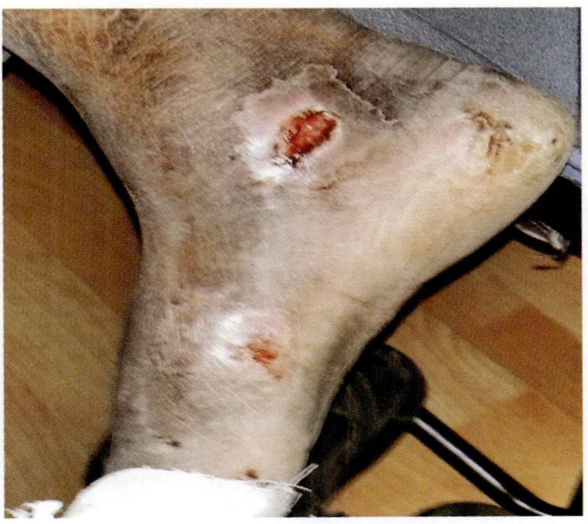

Fig. 28.4: Vasculitic ulcer

nodular lung disease, bronchiolitis, pulmonary hypertension and small airway disease.[11] Pleural disease could be pleurisy or pleural effusion. Plural fibrosis may appear as a sequela of both the conditions. Interstitial fibrosis and pneumonitis are seen at later stages of RA. The presence of fine, diffuse, dry rales with cough, and breathlessness are the common symptoms. The patient may not complain of breathlessness due to limited mobility. X-rays indicate diffuse reticular or reticulonodular pattern in both the lungs. High-resolution computed tomography (HRCT) and pulmonary function tests (PFTs) may assist in confirming the diagnosis.

Nodules associated with nodular lung disease may be single or multiple, and may appear as coin lesions. Nodules may cavitate creating bronchopleural fistula. Malignant transformation is rare.

Bronchiolitis is a rare complication in which proteinaceous exudates in bronchioles causes respiratory insufficiency and death. Pulmonary hypertension occurs in 30% of RA patients and are often asymptomatic. Small airway diseases occur in 50% of RA patients as a part of generalized exocrinopathic process.

Cardiac complications: Pericardial involvement is noted in 50% of patients only at autopsy.[12] Echocardiographic features of pericardial effusion are found in 31% of RA patients. Cardiac tamponade and constrictive pericarditis are rare. Myocarditis, which may be often granulomatous or interstitial, are reported. Endocardial involvement like mitral regurgitation and aortic regurgitation is also noted. Conduction defects like AV block occur due to granulomatous involvement of AV node bundle of HISS. Rarely, amyloidosis can lead to heart block. Coronary vasculitis is also seen and may lead to myocardial infarction.[12]

To conclude RA is a systemic disease with predominant involvement of the joints. The implementation of treatment strategy focussed on active disease control has reduced the incidence of extra-articular manifestations as well the deformities and disabilities secondary to the disease. The targeted treatment strategy has improved the therapeutic outcome. Generally, the disease is noted to become milder with time.

Key Points

1. The predominant manifestations of RA involve joints.
2. Polyarthritis of small joints is one of the common clinical presentations in RA.
3. The extra-articular manifestations of RA include ophthalmologic manifestations, cardiopulmonary disease, vasculitis, neuropathy, glomerulonephritis, and Felty's syndrome.
4. The implementation of targeted treatment approach focussed on active disease control has contributed to reduce the incidence of extra-articular manifestations significantly.

REFERENCES

1. Grassi W, De Angelis R, Lamanna G, Cervini C. The clinical features of rheumatoid arthritis. Eur J Radiol. 1998;27 Suppl1:S18–24.
2. Apfelberg DB, Maser MR, Lash H, Kaye RL, Britton MC, Bobrove A. Rheumatoid hand deformities: pathophysiology and treatment. West J Med. 1978; 129:267–72.
3. Neva MH, Häkkinen A, Isomäki P, Sokka T. Chronic back pain in patients with rheumatoid arthritis and in a control population: prevalence and disability-a 5-year follow-up. Rheumatol 2011; 50:1635–9.
4. Chua-Aguilera CJ, Möller B, Yawalkar N. Skin manifestations of rheumatoid arthritis, juvenile idiopathic arthritis, and spondyloarthritides. Clin Rev Allergy Immunol. 2017 Jul 27. doi: 10.1007/s12016-017-8632–5.
5. Murray PI, Rauz S. The eye and inflammatory rheumatic diseases: The eye and rheumatoid arthritis, ankylosing spondylitis, psoriatic arthritis. Best Pract Res Clin Rheumatol 2016; 30:802–825.
6. Agrawal S, Misra R, Aggarwal A. Anemia in rheumatoid arthritis: high prevalence of iron-deficiency anemia in Indian patients. Rheumatol Int. 2006; 26:1091–5.
7. Ravindran V, Jain S, Mathur DS. The differentiation of anaemia in rheumatoid arthritis: parameters of iron-deficiency in an Indian rheumatoid arthritis population. Rheumatol Int. 2008; 28:507–11.
8. Owlia MB, Newman K, Akhtari M. Felty's Syndrome, Insights and Updates. Open Rheumatol J. 2014; 8:129–36.

9. Kishore S, Maher L, Majithia V. Rheumatoid Vasculitis: a diminishing yet devastating menace. Curr Rheumatol Rep. 2017;19:39.
10. Biswas M, Chatterjee A, Ghosh SK, Dasgupta S, Ghosh K, Ganguly PK. Prevalence, types, clinical associations, and determinants of peripheral neuropathy in rheumatoid patients. Ann Indian Acad Neurol. 2011; 14:194–7.
11. Shaw M, Collins BF, Ho LA, Raghu G. Rheumatoid arthritis-associated lung disease. Eur Respir Rev 2015; 24(135):1–1.
12. Voskuyl AE. The heart and cardiovascular manifestations in rheumatoid arthritis. Rheumatology (Oxford). 2006; 45 Suppl4: iv 4–7.

FURTHER READING

1. Chandrashekara S, Shobha V, Dharmanand BG, Jois R, Kumar S, Mahendranath KM, et al. Factors influencing remission in rheumatoid arthritis patients: results from Karnataka rheumatoid arthritis comorbidity (KRAC) study. Int J Rheum Dis. 2016 Jul 25. doi: 10.1111/1756-185X.12908. [Epub ahead of print].
2. Handa R, Rao UR, Lewis JF, Rambhad G, Shiff S, Ghia CJ. Literature review of rheumatoid arthritis in India. Int J Rheum Dis. 2016; 19:440–51.
3. Pathan E, Joshi VR. Rheumatoid arthritis and the kidney. J Assoc Physicians India. 2004; 52:488–94.

Chapter

29

Psychiatric Manifestations of Rheumatoid Arthritis

Sonal Mehra

INTRODUCTION

Rheumatoid arthritis (RA) is a chronic disease characterized by persistent synovitis, systemic inflammation, and autoantibodies like rheumatoid factor (RF) and anti-citrullinated peptide antibodies (ACPA). Extra-articular manifestations are observed in up to 40% of patients. The neurological complications in RA effect both the central nervous system (CNS) and peripheral nervous system. In the CNS they are divided in brain (neurological and psychiatric manifestations) and spinal cord involvement, while PNS involvement include compressive and non-compressive neuropathies. In this article the psychiatric manifestations of RA are discussed.

Individuals with RA have an increased prevalence of both physical disability and psychiatric comorbidity. They have worse outcomes in several health-related quality of life domains, including increased psychological distress, decreased quality of sleep, and increased use of passive pain-coping strategies. Although prevalence rates for all psychiatric disorders are not consistently elevated in individuals with RA, affective (mood) disorders seem to be considerably more common in individuals with RA than in the general population.[1–3]

AFFECTIVE DISORDERS

Depression

The psychiatric condition most commonly associated with RA is depression. A meta-analysis of 72 studies on depression in individuals with RA found a prevalence of 16.8% for major depressive disorder.[3] Likely causes for depression in RA include more advanced age, severe forms of disease, pain, work disability. Depression seems to decrease adherence to treatment[4] and is associated with increased disability,[5] impaired quality of life[6,7] and reduced rates of clinical remission of RA. Even in other medical conditions patients with moderate or severe depression are less likely than those with mild or no depression to adhere to medical treatment.[4] A bad consequence of depression in RA is a higher risk of suicide and mortality.

Anxiety

One study found a 16% lifetime prevalence of anxiety disorders in individuals with RA.[8] As with depression, symptoms of anxiety also correlate with impaired health-related quality of life and poor responses to medical treatment in patients with RA.[9]

Regular screening for depression and anxiety is recommended in early during the course of treatment. Good psychological support improves patient outcome and quality of life.[10,11] Other cause of mood disorders such as hyperthyroidism, Cushing's syndrome, and steroid induced psychosis must be excluded.[12]

Obsessive Compulsive Disorder (Other Personality Traits)

Researchers from Italy studying the relationship of stress and RA, found that stressful life events preceded onset of RA in 86% cases. In a small preliminary study of 15 patients, 60% patients have a correlation of disease flare ups with micro-events. Interestingly 40% patients were found to have obsessive compulsive personality disorder, 40% showed borderline personality, and 7% showed schizoid and dependent disorder.[13] Authors felt that personality disorders could pathogenically be due to altered stress response system. People with obsessive compulsive personality have obsessive thoughts like fear of contamination, religious obsessions, and sexual obsessions, etc. Sexual obsessions (includes sexual thoughts with family or strangers) are fairly common but underreported. All these thoughts cause distress and anxiety. Compulsive behavior includes excessive hand washing, cleaning, etc. In the context of a patient suffering with rheumatic disease a person with obsessive compulsive personality may have obsessive thoughts that the disease can never be effectively treated, which in the long run can prove detrimental to disease control. Another study evaluating psychiatric manifestations, and personality traits in early RA with disease duration less than 1 year showed that 7 of 22 (32%) early RA patients had obsessive compulsive symptoms and depression compared to 6 of 34 (17%) controls. These early RA patients also had prominent social dysfunction stress and somatization. They tended to adopt a less adaptive life style when compared to controls. Psychological stress occurred independent of disease activity. Hence showing that even in early RA, patients manifest psychological distress in the form of obsessive compulsive symptoms, depression, and have less adaptive defence style, thereby emphasizing the importance of assessing psychiatric co-morbidity soon after diagnosis.[14]

SCHIZOPHRENIA

In contrast to the increased rates of mood disorders, rates of schizophrenia seem to be decreased in individuals with RA compared with the general population. One explanation for is the inflammatory cytokine profile of individuals with schizophrenia which features increased levels of markers of immune cell activation, including soluble IL2 receptor and increased circulating levels of proinflammatory cytokines, such as IL6.[15–17]

FATIGUE AND SLEEP DISTURBANCES

Studies have shown close relationships between fatigue, psychological distress and depression.[18] Almost 80% of individuals with RA experience clinically significant fatigue, which impairs both their quality of lifeand their physical functioning.[19–22] Fatigue is related to pain intensity,[19] RA disease activity, inflammatory processes (which could result in enhanced pain processing), and presence of concurrent affective mood disorders such as depression.[23–25]

COGNITIVE DYSFUNCTION

Cognitive dysfunctions are more frequently observed in RA when compared to controls especially in visual–spatial and planning functions.[26,27] The risk factors identified for cognitive dysfunction in RA are low education level, low income, use of oral glucocorticoids and presence of cardiovascular disease risk factors.[28] Patients with cognitive impairment have increased functional difficulties,less adherence and poorer quality of life.[26–28]

When cognitive complains are present the patient must be evaluated with neuropsychiatric testing. Mood disorders must be screened and MRI brain is indicated to identify structural cause of decline in cognition such as multi-infarct state. Drugs might be responsible too. Hence it is important to look for temporal relation to drug initiation and discontinuation. Methotrexate can be associated with persistent headaches, especially in high doses.[29] Rituximab, gold salts and anti-TNF drugs can in rare instances cause demyelination.[30,31] Some biological agents are associated with leukoencephalopathy that may present as cognitive impairment, such as Tocilizumab.[32]

MECHANISM

RA being a chronic inflammatory state contributes to inflammation by dysregulating the hypothalamic-pituitary-adrenal (HPA) axis. This is manifested as impairment of normal anti-inflammatory responses to stress. There is increased sympathetic tone at rest but decreased sympathetic response during stress.[33] Chronic inflammatory states in general (one of them being RA) alters dopaminergic transmission in the brain resulting in lower basal levels of presynaptic dopamine, decreased dopamine release from basal ganglia in response to pain states, and decreased D_2 and D_3 receptor activation in the nucleus accumbens. These changes enhance central pain processing and decreases positive affective responses, resulting in deranged coping strategies manifesting as depression, decreased motivation, poor treatment adherence, and social withdrawal.[33] These factors in turn contribute to poor disease control.

Another mechanism of psychiatric disturbance in RA is the emotional distress that occur with decline in physical health and function contributing to chronic inflammatory state as mentioned above thus creating a vicious cycle.[33]

Social factors interact with psychological manifestations of RA. As it is known that cognitive, affective, and behavioural factors are closely related and interdependent. Individuals who maintain optimistic perspectives or high levels of self-efficacy are more likely to have healthy repertoires for coping with RA when compared to individuals with pessimistic perspectives or low self-esteem.

Social environment can impact a patient's psychological reactions to RA; patients with supportive family and friends are better in adjusting with RA related declines in physical and emotional psychological health, whereas those in stressful environments have worse symptoms, reduced physical function, and poor psychological adjustment. Individuals with RA can also adversely influence their immediate social environments through self-imposed isolation or initiating conflict with others when pain or frustration levels are high, which could then have consequences for future inter-personal emotional interaction.

INTERVENTIONS FOR IMPROVING PSYCHOLOGICAL HEALTH IN RA PATIENTS

Cognitive-Behavioural Therapy (CBT)

CBT is a form of psychotherapy to treat anxiety, depression, and obsessive compulsive disorder. It is a goal oriented practical approach to problem solving. The patient is encouraged to talk, identify, and reframe distorted patterns of thinking into more-adaptive and less-distressing thoughts. CBT approach in RA improves depression, fatigue, medication adherence, and improves the patients perceptions of social support.[34] But a patient with advanced RA might not benefit to the same extent as with early disease.

Mindfulness-based Stress Reduction

Mindfulness-based stress reduction (MBSR) is effective in improving general well-being of the patient. MBSR is a program that includes "mindfulness" to help people living with chronic pain. A key component of MBSR is mindful meditation. Meditation helps by reducing reactivity to emotions, allow the patient to accept his/her physical disease by engaging in mindful and purposeful action. Apart from meditation MBSR involves focusing attention process on breathing and bodily sensations (body awareness), and being mindful of activities in daily life such as eating or walking, and yoga. Apart from general well-being MBSR can reduce psychological distress and decrease pain intensity.[35,36] However, MBSR do not affect RA disease activity. MBSR is taught to patients by certified trainers over a period of 8 to 10 weeks.

Other principles included in CBT are relaxation techniques, effective problem-solving, and goal setting, as well as education about the nature of the disease.

Effective Control of RA Disease Activity

A recent study showed that both methotrexate and etanercept improved disease activity, quality of life, health associated questionnaire disability index, with faster more pronounced changes in DAS28-CRP in the etanercept group. But the total sleep time, sleep efficiency, and improvements in stage II sleep was much better in the group that received etanercept.[37]

Conclusion

In RA proinflammatory cytokines (increases in IL-6 and TNFα) play a role in initiating "sickness behavior" which includes symptoms such as social withdrawal, social disconnection, mood disorders. It is found that controlling for increases in social disconnection eliminated the relationship between exposure to inflammatory challenge and depressed mood. Thus, inflammation can have social psychological consequences, which may play a role in cytokine-related depressive symptoms.[38]

TNFα, IL-1β, and IL-6 are three main cytokines that have been implicated in communication between the periphery and the brain by neural and humoral pathways. However, in recent years communication via peripheral immune-cell-to-brain and the gut-microbiota-to-brain routes have received increasing attention for their ability to modulate brain function.[39]

Key Points

- Apart from the physical examination findings and laboratory results,emphasis on the psychosocial factors unique to each patient, could yield improvements in the efficacy of current treatments, slowing disease activity and promoting long-term psychological health and physical function.
- Encouraging patients for cognitive behavioral therapy and yoga enhance the ability to cope with the stress of disease and reducing the fatigue and pain intensity.
- Counselling about the disease aspects is essential for long term treatment adherence and better disease activity control.

REFERENCES

1. Lee YC, Lu B, Edwards RR, Wasan AD, Nassikas NJ, Clauw DJ, et al.The role of sleep problems in central pain processing in rheumatoid arthritis. Arthritis Rheum 2013; 65:59–68.
2. Covic T, Cumming SR, Pallant JF, Manolios N, Emery P, Conaghan PG, et al. Depression and anxiety in patients with rheumatoid arthritis: prevalence rates based on a comparison of the Depression, Anxiety and Stress Scale (DASS) and the hospital, Anxiety and Depression Scale (HADS). BMC Psychiatry 2012; 12:6.
3. Matcham F, Rayner L, Steer S, Hotopf M.The prevalence of depression in rheumatoid arthritis: a systematic review and meta-analysis. Rheumatology (Oxford) 2013; 52:2136–48.
4. DiMatteo MR, Lepper HS, Croghan TW. Depression is a risk factor for noncompliance with medical treatment: meta-analysis of the effects of anxiety and depression on patient adherence. Arch Intern Med 2000; 160:2101–7.
5. Margaretten M, Barton J, Julian L, Katz P, Trupin L, Tonner C, et al.Socioeconomic determinants of disability and depression in patients with rheumatoid arthritis. Arthritis Care Res (Hoboken) 2011; 63:240–6.
6. Hyphantis T, Kotsis K, Tsifetaki N, Creed F, Drosos AA, Carvalho AF, et al. The relationship between depressive symptoms, illness perceptions and quality of life in ankylosing spondylitis in comparison to rheumatoid arthritis. Clin Rheumatol 2013; 32: 635–44.
7. Piccinni A, Maser JD, Bazzichi L, Rucci P, Vivarelli L, Del Debbio A, et al. Clinical significance of lifetime mood and panic-agoraphobic spectrum symptoms on quality of life of patients with rheumatoid arthritis. Compr Psychiatry 2006; 47:201–8.
8. Lok EY, Mok CC, Cheng CW, Cheung EF. Prevalence and determinants of psychiatric disorders in patients with rheumatoid arthritis. Psychosomatics 2010; 51:338–8.
9. Kekow J, Moots R, Khandker R, Melin J, Freundlich B, Singh A. Improvements in patient-reported outcomes, symptoms of depression and anxiety, and their association with clinical remission among patients with moderate-to-severe active early rheumatoid arthritis. Rheumatology (Oxford) 2011; 50:401–9.
10. Mok CC, Lok EY, Cheung EF.Concurrent psychiatric disorders are associated with significantly poorer quality of life in patients with rheumatoid arthritis. Scand J Rheumatol 2012; 41:253–9.
11. Lopresti AL, Maker GL, Hood SD, Drummond PD. A review of peripheral biomarkers in major depression: the potential of inflammatory and oxidative stress biomarkers. Prog Neuropsychopharmacol Biol Psychiatry 2014; 48:102–11.
12. Cabrera-Marroquín R, Contreras-Yáñez I, Alcocer-Castillejos N, Pascual-Ramos V. Major depressive episodes are associated with poor concordance with therapy in rheumatoid arthritis patients: the impact on disease outcomes.Clin Exp Rheumatol 2014; 32:904–13.
13. Marcenaro M, Prete C, Badini A, Sulli A, Magi E, Cutolo M. Rheumatoid arthritis, personality, stress response style, and coping with illness. A preliminary survey. Ann N Y Acad Sci 1999; 876:419–25.
14. Hyphantis TN, Bai M, Siafaka V, Georgiadis AN, Voulgari PV, Mavreas V, et al. Psychological distress and personality traits in early rheumatoid arthritis: A preliminary survey. Rheumatol Int 2006; 26:828–36.
15. Jeste DV, Gladsjo JA, Lindamer LA, Lacro JP. Medical comorbidity in schizophrenia. Schizophr Bull 1996; 22:413–30
16. Potvin S, Stip E, Sepehry AA, Gendron A, Bah R, Kouassi E. Inflammatory cytokine alterations in schizophrenia: a systematic quantitative review. Biol Psychiatry 2008; 63:801–8.
17. Gorwood P, Pouchot J, Vinceneux P, Puéchal X, Flipo RM, De Bandt M, et al. Rheumatoid arthritis and schizophrenia: a negative association at a dimensional level. Club Rhumatisme et Inflammation. Schizophr Res 2004; 66:21–9.
18. Smedstad LM, Moum T, Vaglum P, Kvien TK.The impact of early rheumatoid arthritis on psychological distress. A comparison between 238 patients with RA and 116 matched controls. Scand J Rheumatol 1996; 25:377–82.
19. Pollard LC, Choy EH, Gonzalez J, Khoshaba B, Scott DL. Fatigue in rheumatoid arthritis reflects pain, not disease activity. Rheumatology (Oxford) 2006; 45: 885–9.

20. Oncü J, Baþoðlu F, Kuran B. A comparison of impact of fatigue on cognitive, physical, and psychosocial status in patientswith fibromyalgia and rheumatoid arthritis. Rheumatol Int 2013; 33:3031–7.
21. Rupp I, Boshuizen HC, Jacobi CE, Dinant HJ, van den Bos GA.Impact of fatigue on health-related quality of life in rheumatoid arthritis. Arthritis Rheum 2004; 51:578–85.
22. Nikolaus S, Bode C, Taal E, van de Laar MA. Fatigue and factors related to fatigue in rheumatoid arthritis: a systematic review.Arthritis Care Res (Hoboken). 2013; 65:1128–46.
23. Joharatnam N, McWilliams DF, Wilson D, Wheeler M, Pande I, Walsh DA. A cross-sectional study of pain sensitivity, disease-activity assessment, mental health, and fibromyalgia status in rheumatoid arthritis. Arthritis Res Ther 2015; 17:11.
24. Fifield J, McQuillan J, Tennen H, Sheehan TJ, Reisine S, Hesselbrock V, et al. History of affective disorder and the temporal trajectory of fatigue in rheumatoid arthritis. Ann Behav Med 2001; 23: 34–41.
25. Belt NK, Kronholm E, Kauppi MJ. Sleep problems in fibromyalgia and rheumatoid arthritis compared with the general population. Clin Exp Rheumatol 2009; 27:35–41.
26. Bartolini M, Candela M, BrugniM, Catena L, Mari F, Pomponio G, et al. Are behaviour and motor performances of rheumatoid arthritis patients influenced by subclinicalcognitive impairments? A clinical and neuroimaging study. Clin Exp Rheumatol 2002; 20:491–7.
27. Appenzeller S, Bertolo MB, Costallat LT. Cognitive impairment in rheumatoid arthritis. Methods Find Exp Clin Pharmacol 2004; 26:339–43.
28. Shin SY, Katz P, Wallhagen M, Julian L.Cognitive impairment in persons with rheumatoid arthritis. Arthritis Care Res (Hoboken). 2012; 64:1144–50.
29. Quinn CT, Kamen BA. A biochemical perspective of methotrexate neurotoxicity with insight on non-folate rescuemodalities. J Investig Med 1996; 44:522–30.
30. Kur-Zalewska J, Swarowska-Knap J, T³ustochowicz W. Neurological disorders with demyelinating brain white matter lesions in a patient with rheumatoid arthritis treated with etanercept. Pol Arch Med Wewn 2008; 118:234–7.
31. Mok CC. Rituximab for the treatment of rheumatoid arthritis: an update. Drug Des Devel Ther. 2013; 8:87–100.
32. Sato H, Kobayashi D, Abe A, Ito S, Ishikawa H, Nakazono K, et al. Tocilizumab treatment safety in rheumatoid arthritis in a patient with multiple sclerosis: a case report. BMC Res Notes 2014; 7:641.
33. Sturgeon JA, Finan PH, Zautra AJ. Affective disturbance in rheumatoid arthritis: psychological and disease-related pathways. Nat Rev Rheumatol. 2016; 12:532–42.
34. Ferwerda M, van Beugen S, van Middendorp H, Spillekom-vanKoulil S, Donders ART, Visser H, et al. A tailored-guided internet-based cognitive behavioral intervention for patients with rheumatoid arthritis as an adjunct to standard rheumatological care: results of a randomized controlled trial. Pain. 2017; 158:868–78.
35. Pradhan EK, Baumgarten M, Langenberg P, Handwerger B, Gilpin AK, Magyari T, et al. Effect of Mindfulness-Based Stress Reduction in rheumatoid arthritis patients. Arthritis Rheum 2007; 57:1134–42
36. Zautra AJ, Davis MC, Reich JW, Nicassario P, Tennen H, Finan P, et al. Comparison of cognitive behavioral and mindfulness meditation interventions on adaptation to rheumatoid arthritis for patients with and without history of recurrent depression. J Consult Clin Psychol 2008; 76:408–21.
37. Detert J, Dziurla R, Hoff P, Gaber T, Klaus P, Bastian H, et al. Effects of treatment with etanercept versus methotrexate on sleep quality, fatigue and selected immune parameters in patients with active rheumatoid arthritis. Clin Exp Rheumatol 2016; 34:848–56.
38. Eisenberger NI, Inagaki TK, Mashal NM, Irwin MR. Inflammation and social experience: an inflammatory challenge induces feelings of social disconnection in addition to depressed mood. Brain Behav Immun 2010; 24:558–63.
39. D'Mello C, Swain MG. Immune-to-brain communication pathways in inflammation-associated sickness and depression. Curr Top Behav Neurosci 2017; 31:73–94.

Chapter

30

Principles of Management of Rheumatoid Arthritis

VR Joshi, Taral Parikh, C Balakrishnan

The principles of management of rheumatoid arthritis (RA) are based on its natural history, impact on the patient, the society, and the therapeutic measures available. It is a chronic inflammatory systemic disease, mainly affecting synovial joints. The disease evolves over years from a preclinical state to early undifferentiated arthritis, early RA, established RA, and finally to an advanced, destructive, disabling disease.[1] Persistent joint inflammation leads to progressive joint damage, physical and functional disability, impaired quality of life, local and systemic complications, and shortens life expectancy. There are three over-riding principles of management, namely

- Early diagnosis,
- Early treatment with disease modifying anti-rheumatic drugs (DMARDs), and
- Targeted treatment, the target being remission or low disease activity (LDA)

The clinical manifestations of RA (excluding extra-articular complications) that need treatment are listed in Box 30.1.

Box 30.1 Manifestations of RA requiring treatment

Articular	*General*
Synovitis	Fatigue
Pain and stiffness	Sleep disturbances
Damage	Mood disorders
Deformities	Weight loss
	Deconditioning
	Co-morbidities

A. EARLY DIAGNOSIS

Early diagnosis and treatment are critical as joint damage can develop within a few weeks of disease onset and early treatment with DMARDs can control the disease, prevent/or retard joint damage, and improve disease outcomes.[2]

Until 2010, 1987 ACR criteria (though classification criteria) were used to diagnose RA.[3] However these were useful to diagnose established RA, when in all probability, joint damage would have developed, defeating the very idea of preventing or retarding joint damage with early treatment. The 2010 ACR/EULAR criteria were developed to overcome this shortcoming and are now used to diagnose early RA.[4] Over the years the concept of early RA has evolved from a period of 3 years to that of 3–6 months. Pushing the concept of early diagnosis further carries the inherent risk of treating patients of undifferentiated arthritis who would otherwise not develop RA.

B. WINDOW OF OPPORTUNITY

The concept of a window of opportunity suggests the period when it is possible to alter the disease process, i.e. reverse it or even induce remission. Treatment during this period is more likely to achieve this than treatment initiated in the later stages of disease.[5]

C. TREAT TO TARGET (T2T) STRATEGY[6]

This is the presently accepted concept to treat RA. It implies treating patients with an aim to achieve the target decided. Accordingly during the active stage of disease the patient is followed upfrequently, (at a predefined interval, usually of 1–3 months) to modify the therapy (dose or medication) till the target is achieved; ideally within 6 months of treatment initiation (the so-called induction phase). The concept further stipulates a maintenance phase during which the treatment is scaled down to a level that continues to keep the disease under control.

D. REMISSION AND LOW DISEASE ACTIVITY AS THE THERAPEUTIC TARGET

Inducing remission or low disease activity is the main target of treatment (Treat to Target—T2T).[6] Many composite disease activity measures (which are a single composite of pooled core elements) are available to assess disease activity; of these DAS-28 (DAS-28 ESR or CRP), CDAI (clinical disease activity index), SDAI (simplified disease activity index), RAPID-3 (routine assessment patient index data), and PASI/ PAS II (patient assessment scale) are commonly used (Box 30.2). Anyone of these can be used in day-to-day clinical practice.

The ACR/EULAR definition of remission (Boolean based and index based definitions) requires more stringent criteria (Box 30.3).

Low disease activity (LDA) is accepted as the target when remission is not possible, especially so in patients with long-standing disease or with associated co-morbidities. LDA is characterized by minimal joint damage, and disability, with functional capacity preserved.[8]

It is important to realize that, remission as defined presently, does not mean a complete total remission.[8] MRI and US can detect persistent disease activity. Definition of remission based on imaging (USG or MRI) is at present not available. Further, the definition of remission does not include joint damage and functional capacity.[8] These parameters are assessed with imaging and health assessment questionnaire (Indian HAQ).[9] Modified sharp score, van der Heijde modification of sharp score and Larsen score are commonly employed to assess joint damage.

Seronegative RA

The initial impression that patients with seronegative RA have a milder disease course is not borne out by recent studies. Anti-carbamylated protein (anti-Carb p) antibodies are linked to more severe outcome in patients with ACPA negative RA. Present recommendation is to treat seronegative RA patients on the same lines as that of seropositive patients.[8]

Box 30.2 Instruments used to measure rheumatoid arthritis disease activity[7]

	Thresholds for disease activity			
Instrument	*Remission*	*Low*	*Moderate*	*High*
Disease activity score in 28 joints (DAS28)	<2.6	≥2.6 to <3.2	≥3.2 to 5.1	>5.1
Simplified disease activity index (SDAI)	≤3.3	>3.3 to 11	> 11.0 to ≤26	>26
Clinical disease activity index (CDAI)	≤2.8	>2.8 to 10	> 10.0 to 22.0	> 22.0
Patient activity scale (PAS or PAS II)	<0.25	0.26 to 3.70	≥3.71 to 8.0	> 8.0
Routine assessment patient index data (RAPID)	≤1.0	>1.0 to 2.0	>2.0 to 4.0	>4.0

Box 30.3 ACR/EULAR 2011 provisional definition of RA remission[7]

For clinical trials:

Boolean-based definition

At any time point, patient must satisfy all of the following:

- Tender joint count ≤1
- Swollen joint count ≤1
- Patient global assessment ≤1 (0–10 scale)
- C-reactive protein ≤1 mg/dl

Index-based definition

Simplified disease activity index score ≤3.3

For clinical practice:

Boolean-based definition

At any time point, patient must satisfy all of the following:

- Tender joint count ≤1
- Swollen joint count ≤1
- Patient global assessment ≤1 (0–10 scale)

Index-based definition

Clinical disease activity index score ≤2.8

E. PRE-THERAPY EVALUATION

- Pre-therapy evaluation is necessary to plan treatment strategy and decide therapeutic target. The evaluation includes.
  - Staging of disease—early, established, or long standing
  - Assessing disease activity, functional and physical capacity, quality of life, and radiological damage
  - Defining therapeutic target

F. PHARMACOTHERAPY

DMARDs, corticosteroids, and NSAIDs form the cornerstone of pharmacotherapy.

DMARDs have been categorized by the European Medicine Agency (EMA) as[10]

- *cs DMARDs* (Conventional synthetic DMARDs): Methotrexate, leflunomide, sulfasalazine, hydroxychloroquine, gold salts, and others.
- *bo DMARDS* (Biological originator DMARDs): The 5 TNF inhibitors (TNFi), infliximab, etanercept, golimumab, adalimumab, and cetrolizumab pegol; anti-CD20 antibody—rituximab; anti-IL-6 receptor antibody—tocilizumab, anti-IL-1 antibody—anakinra, and co-stimulation blocker—abatacept.
- *ts DMARDs* (Targeted synthetic DMARDs): Janus kinase inhibitors—tofacitinib and others.
- *bs DMARDs* (Biosimilar DMARDs): Rituximab-Mabtas, Reditux; Adalimumab-AdaliRel; Etanercept-Intacept; Infliximab-Infimab.

Presently many targeted therapies are undergoing development and evaluation.

DMARDs

All DMARDs share the potential to modify disease course and disease outcomes. They are defined as agents that reduce or reverse signs, symptoms, and disability, improve quality of life, and work capacity, and retard progression of joint damage.

Corticosteroids

Steroids act rapidly and have disease modifying potential. Liberal and prolonged steroid therapy is to be avoided due to steroid side-effects. The present indications for steroids are—low dose (<10 mg/d) bridging therapy to tide over the time required (usually <6 months) for DMARDs to act, higher doses including pulse therapy to treat flares, and systemic complications, and intra-articularly to treat persistent synovitis of a few joints. For some patients with chronic long standing RA, prolonged low dose steroid treatment may be required.

NSAIDs

NSAIDs are effective anti-inflammatory, analgesic drugs. They are often the initial choice of therapy, mainly to provide relief from pain, inflammation, stiffness, and joint swelling. Their main drawbacks are: (i) Do not possess disease modifying potential and hence do not alter the disease course, (ii) can cause serious gastrointestinal, renal, and cardiovascular side-effects. In the elderly, NSAIDs should be used only if absolutely necessary. Gastro-protective Cox-2 inhibitors are preferred but the other side-effects of NSAIDs remain. The pain relieving efficacy of NSAIDs is not dose dependent. There is a sealing effect. For persistent pain non-opioid or opioid analgesics like paracetamol, dextropropoxyphene, and tramadol may be used.

Biologics

Biologics have many advantages. They act on specific pathologic processes, effect faster disease control, are effective even when conventional DMARDs fail, and prevent or retard radiological worsening of joints. The main disadvantage is high cost, more so in the Indian context. Further, in early RA biologics as mono-therapy are possibly no more effective than methotrexate monotherapy. Other issues are parenteral administration, risk of infections, and unknown long-term adverse effects. There is no difference in the response amongst the biologics.

Biosimilar is a therapeutic product which is similar in terms of quality, safety and efficacy to an already licensed reference biotherapeutic product (WHO definition).

Guidelines

EULAR and ACR guidelines have been published recently[11, 12] The Indian guidelines to treat RA were published in 2008.[13]

The principles of treatment include early diagnosis and institution of treatment, using T2T strategy to achieve remission or at least low disease activity (LDA). It does not matter which therapy is

Joint Protection

Joint protection involves avoidance of excessive stress on the joints, and lifestyle modification. During acute inflammation splints provide rest to swollen, painful joints.

Prevention and Correction of Deformities

Deformities limit activities and function. Hands, knees, feet, and shoulders are commonly affected. Custom made orthoses are preferred. Serial casting is sometimes applied to correct the deformities. Splints can be either rest splints or working splints, the latter permits joint movement and prevents tendency to develop deformities. Assistive devices and gait aids are required for patients with significant affection of lower limb joints.

Nutrition and Diet

Diet is often discussed and prescribed. RA results in loss of fat-free body mass (rheumatoid cachexia), reduction in muscle strength, osteoporosis, and anaemia (due to inflammation and GI loss). A balanced diet with adequate nutrients is necessary. Supplements of iron, with folic acid, Vitamin B_{12}, calcium and vitamin D are advised. Anti-oxidant omega-3 fatty acids are not of any proven efficacy.

Vegetarian, vegan, and mediterranean diets have been advised. Some studies report improvement in the disease process. There is, however, no consensus on diets. Patients should be allowed to consume diet that suits them best, so long as it is balanced and provides the necessary nutrients.

Fatigue

Fatigue is an important symptom of RA, and also an indirect marker of disease activity. It affects patient's quality of life, functional status, and rehabilitation. Productivity is diminished. Therapy involves treatment of anaemia, hypothyroidism, and depression along with control of inflammation, energy conservation, adequate sleep, rest, and physical activity. Moderate intensity aerobic exercises are beneficial.

Cognitive Behavioural Therapy

Cognitive behavioural therapy includes explaining the therapy, its necessity, how it helps, and how to cope with the illness. Cognitive behaviour therapy is expected to encourage patient's involvement in disease management. Family's understanding of the disease and its impact on patient plays an important role. It is advisable to get the family involved in therapeutic program.

Mood Disorders

Depression is common. It aggravates pain and impairs outcome. Treatment includes psychotherapy and antidepressants. Cognitive behavioural therapy helps.

Sleep Disturbances

Sleep fragmentation (easily disrupted light sleep) and multiple mid-sleep awakenings are reported by the patients. Patients are advised to maintain sleep diary noting sleep details and the factors affecting sleep. Management is medical, psychological, environmental corrections, sleep hygiene and behavioural treatment. Tricyclic anti-depressants are indicated with associated fibromyalgia.

Deconditioning

Deconditioning is due to the disease and diminished physical activity. There is decreased cardiorespiratory efficiency, decreased muscle function, fatigue, osteoporosis, atrophy of type II (fast twitch) muscle fibers, loss of muscle mass, impaired function, incoordination, and an increased tendency to falls. Energy cost of movements is increased. Tolerance to pain, and sleep is affected. Deconditioning possibly impairs the immune function also. Treatment of RA, especially with steroids aggravates the situation. Prevention and treatment includes regular daily physical activity of adequate intensity, duration, and frequency. Attitudes and beliefs can affect drive to perform exercises.

Adherence

Adherence to pharmacological and non-pharmacological therapeutic regimens is important but is prone to default. Factors affecting adherence are prolonged treatment, complicated treatment regimens, lack or slowness of response, depression, family factors, finances, etc. Therapeutic strategies can be educational, organizational, and behavioural.

Surgery

The most common surgical intervention is joint replacement. The only indications for joint replacement are intractable joint pain at rest or with minimal activity and severely declined functional capacity, affecting the quality of life. Synovectomy, once a popular intervention is virtually out of vogue. Other surgical interventions are for correction of hand deformities, tendon attrition, atlanto-axial instability, cervical spine involvement with neurological symptoms and/or deficit, and limb deformities.

FINAL RECOMMENDATION

Therapy of RA is rapidly evolving. It is necessary to remain updated.

Key Points

- Perform initial assessment, and stage the disease
- Aim early diagnosis and DMARD (mono or combined) ± low dose steroid therapy
- Use boDMARDs and tsDMARDs in patients with inadequate response to csDMARDs
- Perform 1–3 monthly assessment of response
- Aim remission/low disease state within 3–6 months
- Maintain remission/low disease state
- Treat flares with short courses of steroids
- Use intra-articular steroid injections for persistent synovitis of a few joints
- Simultaneously institute supportive measures including physiotherapy
- Treat systemic complications, with high dose steroids/immunosuppression

Acknowledgement

We thank Mrs Madhuri Kelkar for the secretarial help.

REFERENCES

1. Thomas R, Cope AP. Pathogenesis of rheumatoid arthritis. In: Watts RA, Conaghan PG, Denton CP, Foster H, Issacs J, Müller U, Editors. Oxford Text Book of Rheumatology. Oxford: Oxford University Press; 2013. p. 840–48.
2. Nell VP, Machold KP, Eberl G, Stamm TA, Uffmann M, Smolen JS. Benefit of very early referral and very early therapy with disease modifying anti-rheumatic drugs in patients with early rheumatoid arthritis. Rheumatology (Oxford) 2004; 43:906–14.
3. Arnett FC, Edworthy SM, Bloch DA, McShane DJ, Fries JF, Cooper NS, et al. The American Rheumatism Association 1987 criteria for the classification of rheumatoid arthritis. Arthritis Rheum 1988, 31:315–324.
4. Aletaha D, Neogi T, Silman AJ, Funovits J, Felson DT, Bingham CO 3rd, et al. 2010 rheumatoid arthritis classification criteria: An American College of Rheumatology/European League against Rheumatism collaborative initiative. Ann Rheum Dis. 2010; 69:1580–88.
5. Van Nies JA, Krabben A, Schoones JW, Huizinga TW, Kloppenburg M, van der Helm-Van Mil AH. What is the evidence for the presence of a therapeutic window of opportunity in rheumatoid arthritis? A systematic literature review. Ann Rheum Dis. 2014; 73:861–70.
6. Smolen JS, Aletaha D, Bijlsma JW, Breedveld FC, Boumpas D, Burmester G, et al. Treating rheumatoid arthritis to target: recommendations of an international task force. Ann Rheum Dis. 2010; 69: 831–37.
7. Hissaria P. Review of rheumatoid arthritis disease outcome measures: Recommendations and its relevance in private practice. Indian J Rheumatol 2015; 10:133–39.
8. Smolen JS, Aletaha D. Rheumatoid arthritis therapy reappraisal: strategies, opportunities, and challenges. NatRev Rheumatol 2015; 11:276–89.
9. Kumar A, Malaviya AN, Pandhi A, Singh R. Validation of an Indian version of health assessment questionnaire in patients with rheumatoid arthritis. Rheumatology 2002; 41:1457–59.
10. Smolen JS, van der Heijde D, Machold KP, AletahaD, Landewé R. Proposal for a new nomenclature of disease-modifying antirheumatic drugs. Ann Rheum Dis 2014; 73:3–5.
11. Smolen JS, Landewé R, Breedveld FC, Buch M, Burmester G, Dougados M, et al. EULAR recommendations for the management of rheumatoid arthritis, with synthetic and biologic disease-modifying anti rheumatic drugs: update. Ann Rheum Dis. 2013; 73:492–509.
12. Singh JA, Saag KG, Bridges SL Jr, Akl EA, Bannuru RR, Sullivan MC, et al. 2015 American College of Rheumatology Guideline for the treatment of rheumatoid arthritis. Arthritis Rheum. 2016; 68:1–26.
13. Misra R, Sharma BL, Gupta R, Agarwal S, Agarwal P, Grover S, et al. Indian Rheumatology Association consensus statement on the management of adults with rheumatoid arthritis. Indian J Rheumatol 2008; 3(5):1–16.
14. Haschka J, Englbrecht M, Hueber AJ, Manger B, Kleyer A, Reiser M, et al. Relapse rates in patients with rheumatoid arthritis in stable remission tapering or stopping antirheumatic therapy; interim results from the prospective randomized controlled retro study. Ann Rheum Dis 2016; 75:45–51.

Pregnancy and Rheumatoid Arthritis Activity

Disease activity of RA often ameliorates during pregnancy resulting in partial improvement in clinical symptoms or even complete remission. The improvement, though, is often short-lived with the disease often relapsing after parturition. Improvement of RA during pregnancy was first reported by Hench in 1938, and subsequently confirmed in numerous retrospective and prospective studies. Among the studies which have examined RA activity during pregnancy, the PARA study was a prospective one and notable for having relatively more strict criteria for measuring disease activity. In this study, 48% patients among those with an initial DAS28-CRP >3.2 showed moderate improvement during gestation, whereas patients with DAS28-CRP ≤3.2 (low disease activity) remained stable throughout the pregnancy.[18] Thus, pregnancy influenced the disease course more in patients who had moderate to high disease activity at the time of conception.[18] A negative or low-titre positivity of anti-citrullinated peptide antibodies and therapy with pregnancy safe DMARD or TNF-inhibitors prior to conception are associated with amelioration of disease activity during pregnancy. In a recent prospective study studying the disease course in female patients with RA in subsequent pregnancies, while treatment in both pregnancies remained similar, only 14.8% of the patients had a comparable disease activity course.[19] 63% patients in this study had a postpartum flare following both pregnancies. The occurrence of a flare postpartum was associated with a flare after the subsequent pregnancies ($p = 0.003$), despite the differences in treatment between both pregnancies.[19]

Management of Rheumatoid Arthritis and Pregnancy

Preconceptional Assessment and Planning

Obtaining a reproductive history is vital in all women diagnosed with RA. For patients in reproductive age group, the desire to initiate or extend family should be determined. Women not interested in conceiving at present but with desire to do so in the future should receive education about the risks and possibility of adverse outcomes associated with medications prescribed for treatment in context of pregnancy (Table 31.1). Contraception counselling while on DMARDs is imperative and should be based on the patient's social, cultural, and educational background.

Male Patients with RA on DMARDs Planning Conception

A paternal diagnosis of RA has not been associated with adverse fetal outcomes. The data on safety of methotrexate on male fertility is variable. The prescription insert for methotrexate states caution advising waiting for at least 3 months after cessation of the drug before attempting conception, the period correlating to the period required for spermatogenesis (74 days). Recently, a prospective trial found no significant increase in the incidence of major congenital anomalies or spontaneous abortions in male patients exposed preconceptionally to methotrexate in dosages used for therapy of RA.[20]

Before planning conception, the patient should preferably have low disease activity or be in remission.

Post-conceptional Management

A significant number of pregnancies are unplanned. Hence, when a woman with RA on therapy is detected to be pregnant, a detailed discussion for possible risks of each individual medication on both mother and the fetus should be done. This should also include discussion about possible but uncertain improved disease activity during pregnancy and postpartum flare. Certain medications, e.g. methotrexate and leflunomide, would require immediate discontinuation. Decisions to maintain pregnancies should be made after a detailed counselling on possible adverse fetal outcomes due to medication exposure which also informs the patient about the possibility of a normal fetus despite having been exposed to medications.

In case of women receiving leflunomide and conceiving unintentionally, the patient should undergo washout procedure with cholestyramine in order to facilitate drug excretion. Cholestyramine 8 g three times a day for 11 days should be administered, with post-washout checking of the plasma leflunomide level with a target of <0.02 mg/L measured 2 weeks apart. Additional cholestyramine should be considered for patients in whom leflunomide levels remain elevated.

Table 31.1: Recommendations for antirheumatic medications in RA patients planning conception[22]

Drugs	*Recommendations*
Methotrexate	Discontinue 1–3 months before pregnancy, continue folic acid until the end of the first trimester
Leflunomide	Discontinue 2 years before or conduct the washout procedure before pregnancy
Tocilizumab	Discontinue 3 months before pregnancy
Rituximab	Discontinue before pregnancy, in exceptional cases, treatment possible before first trimester
Abatacept	Discontinue 10 weeks before pregnancy
Tofacitinib	Discontinue 2 months before pregnancy
Glucocorticoids	Use lowest dose possible, short-term pulses can be considered, long-term treatment requires peripartal stress doses of glucocorticoids
NSAID	COX-II inhibitors should be stopped before pregnancy; non-selective COX-inhibitors can be used until gestational week 32, use lowest effective dose
Sulfasalazine	Continue during pregnancy with folic acid supplementation
Antimalarials	Continue during pregnancy;hydroxychloroquine should be preferred in pregnancy; combine with sulfasalazine to increase the effect
TNF inhibitors	Can be continued in pregnancy; complete monoclonal antibodies can be considered until gestational week 20, etanercept and certolizumab may be considered for use throughout pregnancy

Abbreviations used: RA—rheumatoid arthritis; TNF—tumor necrosis factor; NSAIDs—non-steroidal anti-inflammatory drugs; COX-II - Cyclo-oxygenase II

Assessment of Disease Activity during Pregnancy

Pregnancy influences ESR considerably more than it does CRP, and it also affects the patients assessment of global health (GH) based on a visual analogue scale (VAS).Hence, RA disease activity during pregnancy should be monitored preferably using DAS28-CRP without GH.

Health of the Infant after Prenatal TNFi Exposure

Data on the neonatal well-being after antenatal exposure to TNFi are scarce. A recent systematic review in women with inflammatory bowel disease did not find a higher incidence of infection in infants up to a year after maternal gestational TNFi exposure.[23] Recently, a study compared the outcome in 55 offsprings of women with inflammatory bowel disease who received TNFi therapy during gestation up to week 25 (stop group) with those having received TNFi beyond 30th week of gestation (continue group).[24] At one year age, growth, rate of infections, allergies, atopic dermatitis, and adverse events with administered vaccines were similar in the two groups, no difference being observed between infants exposed to TNFis and 459 infants with no such exposure.[24]

Administration of any live vaccine should be delayed till 6 months after birth in infants with exposure to TNFis during the second half of pregnancy.[25]

Key Points

1. Pregnancy should be planned once the patient has stable or inactive disease.
2. The optimal therapeutic strategy should be chosen keeping in mind the balance between efficacy and potential adverse effects of medications on the conceptus.
3. Detailed patient counselling prior to therapy initiation and the planning of conception should be part and parcel of patient management.

(Contd.)

(Contd.)

4. DMARDs which are not safe in pregnancy (especially methotrexate) should be discontinued before conception.
5. A good control of RA disease activity during antenatal period is vital for positive maternal and fetal outcomes.

REFERENCES

1. Wallenius M, Salvesen K, Daltveit A, Skomsvoll J. Rheumatoid arthritis and outcomes in first and subsequent births based on data from a national birth registry. Acta ObstetGynecolScand 2014; 93:302–7.
2. Clowse M, Chakravarty E, Costenbader K, Chambers C, Michaud K. Effects of infertility, pregnancy loss, and patient concerns on family size of women with rheumatoid arthritis and systemic lupus erythematosus. Arthritis Care Res 2012; 64:668–74.
3. Reed S, Vollan T, Svec M. Pregnancy Outcomes in Women with Rheumatoid Arthritis in Washington State. Matern Child Health 2006; 10:361–66.
4. Jawaheer D, Zhu J, Nohr E, Olsen J. Time to pregnancy among women with rheumatoid arthritis. Arthritis Rheum 2011; 63:1517–21.
5. de Man Y, Hazes J, van der Heide H, Willemsen S, de Groot C, Steegers E, et al. Association of higher rheumatoid arthritis disease activity during pregnancy with lower birth weight: Results of a national prospective study. Arthritis Rheum 2009; 60:3196–3206.
6. Hargreaves E. A survey of rheumatoid arthritis in West Cornwall: A report to the Empire Rheumatism Council. Ann Rheum Dis 1958; 17:61–75
7. Kay A, Bach F. Subfertility before and after the Development of Rheumatoid Arthritis in Women. Ann Rheum Dis 1965; 24:169–73.
8. Junco D, Annegers J, Coulam C, Luthra H. The Relationship between Rheumatoid Arthritis and Reproductive Function. Rheumatology. 1989; XXVIII(suppl 1):33–33
9. Wallenius M, Skomsvoll J, Irgens L, Salvesen K, Nordvag B, Koldingsnes W, et al. Fertility in women with chronic inflammatory arthritides. Rheumatology 2011; 50:1162–67
10. Wallenius M, Skomsvoll J, Irgens L, Salvesen K, Nordvåg B, Koldingsnes W, et al. Parity in patients with chronic inflammatory arthritides childless at time of diagnosis. Scand J Rheumatol 2012; 41:202–7.
11. Tibbetts T, DeMayo F, Rich S, Conneely O, O'Malley B. Progesterone receptors in the thymus are required for thymic involution during pregnancy and for normal fertility. Proc Nat Acad Sci 1999; 96:12021–12026.
12. Forger F, Marcoli N, Gadola S, Moller B, Villiger P, Ostensen M. Pregnancy induces numerical and functional changes of CD4+CD25high regulatory T cells in patients with rheumatoid arthritis. Ann Rheum Dis 2007; 67:984–90.
13. Elenkov I, Wilder R, Bakalov V, Link A, Dimitrov M, Fisher S. IL-12, TNF-α, and Hormonal Changes during Late Pregnancy and Early Postpartum: Implications for Autoimmune Disease Activity during These Times. J Clin Endocrinol Metabol 2001; 86: 4933–38.
14. Munoz-Suano A, Kallikourdis M, Sarris M, Betz A. Regulatory T cells protect from autoimmune arthritis during pregnancy. J Autoimmunity 2012; 38:J103–J108.
15. Tham M, Schlör G, Yerly D, Mueller C, Surbek D, Villiger P. Reduced pro-inflammatory profile of γδT cells in pregnant patients with rheumatoid arthritis. Arthritis Res The 2016; 18(1).
16. Crocker I. Neutrophil function in pregnancy and rheumatoid arthritis. Ann Rheum Dis 2000; 59: 555–64.
17. Förger F, Østensen M. Is IgG galactosylation the relevant factor for pregnancy-induced remission of rheumatoid arthritis?. Arthritis Res Ther 2010; 12:108.
18. de Man Y, Dolhain R, van de Geijn F, Willemsen S, Hazes J. Disease activity of rheumatoid arthritis during pregnancy: Results from a nationwide prospective study. Arthritis Rheum 2008; 59: 1241–48.
19. Ince-Askan H, Hazes J, Dolhain R. Is Disease Activity in Rheumatoid Arthritis during Pregnancy and after Delivery Predictive for Disease Activity in a Subsequent Pregnancy?. J Rheumatol 2015; 43:22–25.
20. Weber-Schoendorfer C, Hoeltzenbein M, Wacker E, Meister R, Schaefer C. No evidence for an increased risk of adverse pregnancy outcome after paternal low-dose methotrexate: an observational cohort study. Rheumatology 2013; 53:757–63.
21. Krause M L, Makol A. Management of rheumatoid arthritis during pregnancy: challenges and solutions. Open Access Rheumatol 2016; 8:23–36.
22. Förger F, Villiger PM. Treatment of rheumatoid arthritis during pregnancy: present and future. Expert Rev Clin Immunol 2016; 12:937–44

23. Nielsen OH, Loftus EV, Jess T. Safety of TNF-alpha inhibitors during IBD pregnancy: a systematic review. BMC Med 2013; 11:174.
24. de Lima A, Zelinkova Z, van der Ent C, Steegers EA, van der WoudeCJ .Tailored anti-TNF therapy during pregnancy in patients with IBD: maternal and fetal safety. Gut 2016; 65:1261–8
25. GötestamSkorpen C, Hoeltzenbein M, Tincani A, Fischer-Betz R, Elefant E, Chambers C, et al. The EULAR points to consider for use of antirheumatic drugs before pregnancy, and during pregnancy and lactation. Ann Rheum Dis. 2016; 75:795–810.

FURTHER READING

1. Flint J, Panchal S, Hurrell A, van de Venne M, Gayed M, Schreiber K, et al. BSR and BHPR Standards, Guidelines and Audit Working Group. BSR and BHPR guideline on prescribing drugs in pregnancy and breastfeeding-Part I: standard and biologic disease modifying anti-rheumatic drugs and corticosteroids. Rheumatology (Oxford) 2016; 55:1693–97.
2. Østensen M, Wallenius M. Fertility and pregnancy in rheumatoid arthritis. Indian J Rheumatol 2016;11 (Suppl S2):122–7.

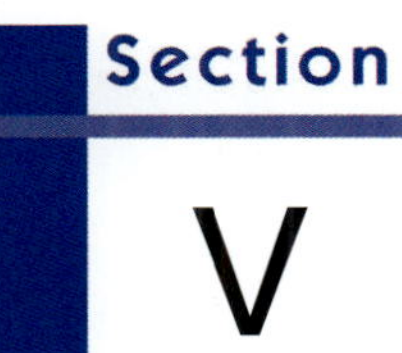

Section V

SPONDYLOARTHRITIS

Chapter

32

Classification of Spondyloarthropathies

Sanjiv Kapoor

INTRODUCTION

Spondyloarthropathies or spondyloarthritides are constellation of inflammatory arthritides including ankylosing spondylitis (AS), non-radiographic axial SpA (nr-axSpA), peripheral SpA, reactive arthritis (Reiter's syndrome), arthritis/spondylitis associated with psoriasis, arthritis/spondylitis associated with inflammatory bowel diseases (IBD) and juvenile-onset SpA. There is an overlap of symptoms and complications within this constellation of disorders. Inflammatory back pain, peripheral arthritis, anterior uveitis and enthesitis are the prime clinical features.[1] Patients with psoriasis or IBD may present with skin psoriasis or colitis as presenting symptoms.[2] Manifestations involving other organs are very rare. Most often, there is a lag in time between the onset of symptoms and diagnosis of spondyloarthropathies. Early diagnosis is of importance but distinguishing the subtypes are challenging and that is where the classification criteria play an important role.

Cardinal Features Considered in Classifications

Since the onset of symptoms is deceptive and radiographic changes occur later, the diagnosis is delayed by many years.[3] Lack of awareness among the physicians and ambiguity in assigning the clinical features to spondyloarthropathies also contributes to delay in diagnosis. Therefore, it is necessary to identify the prime clinical features and employ sensitive diagnostic tools for detecting the disease in its early stage.[1] Classification criteria are set to facilitate the diagnosis of the subtypes of spondyloarthropathies, subsequent initiation of targeted treatment and measure the prognosis.

Inflammatory Back Pain

Inflammatory back pain is a common complaint and it reflects inflammation of sacroiliac joints, spine and spinal entheses. Clinically it is difficult to differentiate chronic mechanical back pain from that related to spondyloarthropathies. A subjective examination to assess patient history of back pain is more crucial in evaluating the cause of back pain.

Calin Criteria

According to Calin et al., inflammatory back pain is considered if any four of the following five criteria is fulfilled.[4]

a. Insidious onset of back pain
b. Patient aged <40 years
c. Pain last for at least 3 months
d. Pain associated with morning stiffness
e. Alleviation of pain with exercise or activity

Originally, 17-point questionnaire was used and only five of these criteria emerged as a strong indicator of inflammatory back pain with a sensitivity and specificity 60% and 97%, respectively.[5] Calin criteria does not define duration of morning stiffness and insidious onset of back pain.

Rudwaleit Criteria[6]

Rudwaleit et al. compared the various combinations of the features of back pain and proposed different classification as well as diagnostic criteria, which was based on the following features.

a. Morning stiffness of >30 minutes duration
b. Improvement in back pain following exercises but not with rest
c. Sleep disruption, especially during the second half of the night because of back pain
d. Alternating buttock pain

When any two of the above criteria were met, then the sensitivity was 70.3% and specificity was 81.2%.

ASAS Assessment Criteria for Inflammatory Back Pain

The Assessment of SpondyloArthritis International Society (ASAS) classification criteria for inflammatory back pain were developed based on the clinical and diagnostic expertise of experts from nine countries. During the workshop, 13 experts examined 20 patients with chronic back pain and suspected axial spondyloarthropathy. Each of the patients was subjected to detailed evaluation of clinical history, physical examination by the 13 experts. Each expert was required to give 4–7 judgments for each patient and each inflammatory back pain parameter. Based on the outcome of the analysis of judgment of experts, the ASAS inflammatory back pain criteria were developed. Fulfilling four out of the following five criteria (iPAIN) is confirmative of inflammatory back pain.

a. Insidious onset
b. PAIN at night
c. Age at onset <40 years
d. Improvement with exercise
e. No improvement with rest

All these criteria were part of the criteria defined by Calin et al. and Rudwaleit et al., however, these criteria were confirmed based on the judgment of experts rather than comparing patients with spondyloarthropathy to those with other types of chronic back pain. Fulfilling four out of five parameters resulted in a sensitivity and specificity of 77.0% and 91.7%, respectively. Subsequently, the new set of inflammatory back pain criteria was validated in a separate cohort of patients (*n* = 648) and it resulted in a sensitivity of 79.6% and specificity of 72.4%.[5,7]

Imaging

Modified New York (mNY) criteria for ankylosing spondylitis regards unilateral sacroiliitis grade 3–4, and bilateral sacroiliitis grade 2–4 as important diagnostic criteria.[8] The severity of sacroiliac joint changes is usually classified according to the New York criteria in 5 stages.

- Grade 0: Normal findings.
- Grade 1: Suspicious changes.
- Grade 2: Minimal abnormality with small areas of erosions or sclerosis without alteration in the joint width.
- Grade 3: Unequivocal abnormality—moderate or advanced sacroiliitis consisting of erosion, sclerosis, widening, narrowing, and/or partial joint fusion (ankylosis).
- Grade 4: Severe abnormality in the form of total ankylosis.

Outcome of radioimaging of the sacroiliac joints and the spine is crucial in the diagnosis and classification of spondyloarthropathies; however, early changes are hardly perceivable on X-ray in a subset of patients and thus the specificity is low in the early stages of the disease. On the contrary, magnetic resonance imaging (MRI) captures active inflammation at sacroiliac joints and spine in its early stages. Besides the mNY, European Spondyloarthropathy Study Group (ESSG) criteria and the Amor criteria have not included MRI in their classification criteria.[8–10] The ASAS classification criteria for axial spondyloarthropathy has included MRI in its imaging criteria.[11] ASAS criteria requires the presence of subchondral or periarticular bone marrow edema in sacroiliac joints (in plane resolution of 0.4 to 0.6 mm, with slice thickness of 3 to 4 mm) on fat-saturated, T2-weighted or short-tau inversion recovery (STIR) sequences, with two or more lesions visible on one slice or a single lesion visible on two or more consecutive slices.[12] Axial SpA with definite radiographic sacroiliitis on X-rays can be classified as AS, while patients showing MRI features of sacroiliitis are classified as nonradiographic axial spondyloarthritis (nr-axSpA).

HLA B-27

Genetic factors play an important role in the pathogenesis of spondyloarthropathies. Human leukocyte antigen (HLA) B-27 is one of the common predisposing genetic factors for spondyloarthropathies. However, its association varies among different ethnic groups.[13]

Advancement in Classification Criteria

Diagnosing spondyloarthropathies is always a challenge. In 1963, Rome clinical criteria for ankylosing spondylitis were proposed. Three years later in 1966, the Rome clinical criteria were modified and the New York criteria were introduced. In 1984, the New York criteria were modified to more sensitive classification criteria, the modified New York criteria (mNY criteria) for ankylosing spondylitis.[8] The Amor criteria was proposed in 1990–1991 and the ESSG criteria was proposed in 1991.[9,10] A combination of clinical symptoms and a definite radiographic finding of sacroiliitis is included in all the three criteria—mNY criteria, Amor criteria and ESSG criteria. The latest ASAS classification criteria of axial spondyloarthritis has included MRI in the imaging arm.[11]

Different Classification Criteria for Ankylosing Spondylitis and Spondyloarthropathies

The initial clinical presentation of spondyloarthropathies can help in defining the disease characteristic and initiate treatment (Table 32.1).[14] Diagnosis is complicated as it is not possible to assign a single feature to the constellation of spondyloarthropathies. Hence, diagnosis of spondyloarthropathies is based on a combination of symptoms, the findings of physical examination, imaging and laboratory investigations.

Classification criteria encompass different features of spondyloarthropathies such as symptoms, signs, laboratory findings, imaging, genetic factors and etiological agents. The criteria defined for research purpose are also employed in a clinical setting. However, there is a distinction between diagnostic and classification criteria; diagnostic criteria are set to help clinicians in diagnosing and must be sensitive enough to recognize and diagnose spondyloarthropathies in its early stages as well as be applicable to individual patients. On the contrary, classification criteria as such enable the categorization of homogenous patients within clinical studies. In practice, owing to lack of appropriate diagnostic tool or criteria, physicians use the classification criteria originally developed for clinical research.[10,14]

Modified New York (mNY) Criteria for Classification

The mNY classification criterion is the updated version of the New York criteria, wherein the pain criterion was replaced with the Rome pain criterion. The diagnosis comprised of the clinical and imaging arms (Table 32.2).[8]

European Spondyloarthropathy Study Group Criteria

European Spondyloarthropathy Study Group criteria were proposed to include the classification criteria for the entire constellation of spondyloarthropathies; however, it was specifically developed to include patients with undifferentiated spondylarthropathy (Table 32.3a).[9]

Amor criteria: A score of 6 or more classifies a patient as having spondyloarthropathy (Table 32.3b).[10]

Table 32.1: Generic classification with characteristic disease features

Clinical presentation	*Diagnostic implication*
Patients with axial spondyloarthropathies (predominant axial involvement)	Inflammation of the sacroiliac joints, the spine, and thoracic cage
Patients with peripheral spondyloarthropathies (predominant peripheral joint manifestations)	Peripheral arthritis, enthesitis and dactylitis

Table 32.2: Modified New York criteria for ankylosing spondylitis

1. Low back pain for at least 3 months duration improved by exercise and not relieved by rest
2. Limitation of lumbar spine motion in sagittal and frontal planes
3. Chest expansion decreased relative to normal values for age and sex
4a. Unilateral sacroiliitis grade 3–4
4b. Bilateral sacroiliitis grade 2–4
Definite ankylosing spondylitis if (4a or 4b) and any clinical criterion (1–3)

Table 32.3a: European spondyloarthropathy study group classification criteria
These criteria resulted in a sensitivity of 87% and a specificity of 87%.
Inflammatory spinal pain or synovitis (Asymmetric, predominantly in lower extremities)
Plus, one of the following
• Family history: First- or second-degree relatives with ankylosing spondylitis,
• Psoriasis, acute iritis, reactive arthritis, or inflammatory bowel disease
• Past or present psoriasis, diagnosed by a physician
• Past or present ulcerative colitis or Crohn's disease, diagnosed by a physician and confirmed by radiography or endoscopy
• Past or present pain alternating between the two buttocks
• Past or present spontaneous pain or tenderness at examination of the site of the insertion the Achilles tendon or plantar fascia (enthesitis)
• Episode of diarrhea occurring within 1 month before onset of arthritis
• Nongonococcal urethritis or cervicitis occurring within 1 month before onset of arthritis
• Bilateral grade II–IV sacroiliitis or unilateral grade III or IV sacroiliitis
• Grades are 0: Normal; 1: Possible; 2: Minimal; 3: Moderate; 4: Completely fused (ankylosed)

Table 32.3b: Amor criteria for the classification of spondyloarthropathies	
Amor criteria	
• Clinical symptoms or history of scoring	*Points*
• Lumbar or dorsal pain at night or morning stiffness of lumbar or dorsal pain	1
• Asymmetrical oligoarthritis	2
• Buttock pain	1
• If alternate buttock pain	2
• Sausage like toe or digit	2
• Heel pain or other well-defined enthesopathy	2
• Iritis	1
• Nongonococcal urethritis or cervicitis within 1 month before the onset of arthritis	1
• Acute diarrhea within one month before the onset of arthritis	1
• Psoriasis, balanitis, or inflammatory bowel disease (ulcerative colitis or Crohn's disease)	2
Radiological findings	
• Sacroiliitis (bilateral grade 2 or unilateral grade 3)	3
Genetic background	
• Presence of HLA-B27 and/or family history of ankylosing spondylitis, reactive arthritis, uveitis, psoriasis, or inflammatory bowel disease	2
Response to treatment	
• Clear-cut improvement within 48 h after NSAIDs intake or rapid relapse of the pain after their discontinuation	2
A patient is considered as suffering from a spondyloarthropathy if the sum is ≥ 6	

NSAID: Nonsteroidal anti-inflammatory drug; HLA: Human leukocyte antigen.

ASAS Classification Criteria of Axial Spondyloarthritis

Since radiographic changes of sacroiliitis could not be detected in the earlier stage of spondyloarthropathies, the ASAS wanted to accommodate such patients in the classification criteria. In 2004, ASAS suggested the inclusion of such patients as those with pre-radiographic phase of ankylosing spondylosis. In 2009 the ASAS proposed and validated the new candidate criteria for axial spondyloarthropathies that included patients with and without definite radiographic sacroiliitis.[11]

All ASAS members across the world (25 centers in 16 countries) enrolled 661 patients and of them 649 patients were analyzed. Experts were provided with two sets of candidate criteria for the classification of axial spondyloarthritis, which differed only in the clinical arms. They also re-evaluated the patients with certain modification in the criteria. The specificity and sensitivity and both the original and refined data sets were evaluated. After the study, the members convened a meeting and arrived at final classification criteria (Tables 32.4a and 32.4b).[11]

Peripheral Spondyloarthritis

Peripheral spondyloarthritis can include patients with peripheral arthritis, dactylitis and enthesitis without back pain. Compared to the ESSG and Amor classification, the new ASAS classification criteria for peripheral arthritis seems to perform better (Tables 32.5a and 32.5b).[15]

The criteria are applicable to patients with peripheral arthritis (usually predominantly of lower limbs and/or enthesitis, and/or dactylitis. Sensitivity of the criteria was 77.8%, the specificity 82.2%.

Array of Spondyloarthropathies

The current and classic classification of spondyloarthropathies are summarized in Table 32.6.

Ankylosing Spondylitis/Axial Radiographic SpA

In India, prevalence of ankylosing spondylitis ranges between 7 and 9.8 per 10,000.[17] Axial involvement is one of the characteristics of the disease. Radiographic sacroiliitis is a definite finding during the course of the disease in majority of the patients. The diagnosis of ankylosing spondylitis can be based on the mNY (Table 32.2) and axial radiographic SpA can be diagnosed by ASAS criteria (Table 32.4a).

Nonradiographic Axial Spondyloarthritis (nr-axSpA)

SpA with predominantly axial involvement in absence of plain radiographic changes of sacroiliitis

Table 32.4a: Assessment in SpondyloArthritis International Society (ASAS) classification criteria for axial spondyloarthropathies

In patients with ≥3 months back pain (with/ without peripheral manifestations) and age at onset <45 years:

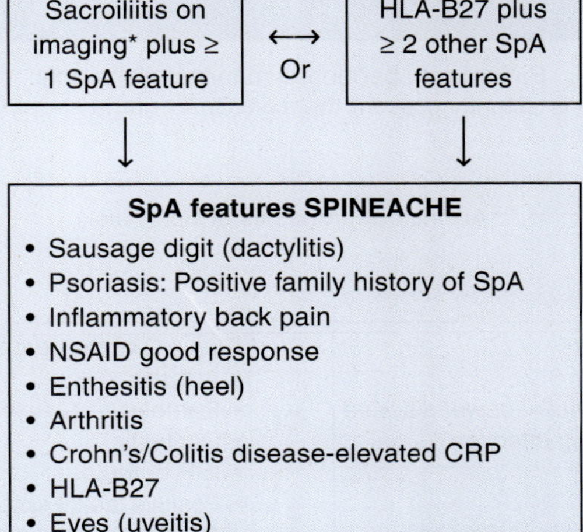

**Sacroiliitis on imaging*

X-ray: Bilateral grades II–IV or unilateral grades III–IV, according to the modified New York criteria.

MRI: Active (acute) inflammation on MRI

Table 32.4b: Specification of the variables used for ASAS criteria for classification of axial SPA

Clinical criteria	*Definition*
Inflammatory back pain	Inflammatory back pain according to expert (*see* inflammatory back pain): Four out of five of the following parameters present: (1) age at onset < 40 years, (2) insidious onset, (3) improvement with exercise, (4) no improvement with rest, (5) pain at night (with improvement upon getting up).
Arthritis	Past or present active synovitis diagnosed by a doctor.
Family history	Presence in first-degree or second-degree relatives of any of the following: (a) ankylosing spondylitis, (b) psoriasis, (c) uveitis, (d) reactive arthritis, (e) inflammatory bowel disease.
Psoriasis	Past or present psoriasis diagnosed by a doctor.
Inflammatory bowel disease	Past or present, Crohn's disease or ulcerative colitis diagnosed by a doctor.
Dactylitis	Past or present dactylitis diagnosed by a doctor.
Enthesitis	Heel enthesitis: Past or present spontaneous pain or tenderness at examination at the site of the insertion of the Achilles tendon or plantar fascia at the calcaneous.
Uveitis anterior	Past or present uveitis anterior, confirmed by an ophthalmologist.
Good response to NSAIDs	At 24–48 h after a full dose of NSAIDs the back pain is not present anymore or much better.
HLA-B27	Positive testing according to standard laboratory techniques.
Elevated CRP	CRP above upper normal limit in the presence of back pain, after exclusion of other causes for elevated CRP concentration.
Sacroiliitis by X-rays	Bilateral grades II–IV or unilateral grades III–IV, according to the modified New York criteria.
Sacroiliitis by MRI	Active inflammatory lesions of sacroiliitis joints with definite bone marrow oedema/osteitis suggestive of sacroiliitis associated with spondyloarthritis.

Table 32.5a: Assessment in SpondyloArthritis International Society (ASAS) classification criteria for peripheral spondyloarthritis

Patient with peripheral manifestations only
(if back pain is actually present the axial SpA criteria should be applied)

↓

Arthritis or enthesitis or dactylitis

↓

Plus ≥ 1 of		Plus ≥ 2 of the remaining
• Psoriasis • Inflammatory bowel disease • Preceding infection • HLA-B27 • Uveitis • Sacroiliitis on imaging	Or	• Arthritis • Enthesitis • Dactylitis • IBP in the past • Positive family history for SpA

*Sacroiliitis on imaging
X-ray: Bilateral grade II–IV or unilateral grade III–IV, according to the modified New York criteria.
MRI: Active (acute) inflammation on MRI

Table 32.5b: Definition of SpA features for use of ASAS classification criteria for peripheral SpA

SpA feature	*Definition*
Entry criteria arthritis	Current peripheral arthritis compatible with SpA (usually asymmetric and/or predominant involvement of the lower limb), diagnosed clinically by a doctor.
Enthesitis	Current enthesitis, diagnosed clinically by a doctor.
Dactylitis	Current dactylitis, diagnosed clinically by a doctor.
Additional SpA features	
Inflammatory back pain in the past	Inflammatory back pain in the past according to the rheumatologist's judgement.
Arthritis	Past or present peripheral arthritis compatible with SpA (usually asymmetric and/or predominantly involvement of the lower limb), diagnosed clinically by a doctor.
Enthesitis	Enthesitis: Past or present spontaneous pain or tenderness.
Uveitis	Past or present anterior, confirmed by an ophthalmologist.
Dactylitis	Past or present dactylitis, diagnosed by a doctor.
Psoriasis	Past or present psoriasis, diagnosed by a doctor.
Inflammatory bowel disease	Past or present Crohn's disease or ulcerative colitis diagnosed by a doctor.
Preceding infection	Urethritis/cervicitis or diarrhea within 1 month before the onset of arthritis/enthesitis/dactylitis.
Family history for SpA	Presence in first-degree (mother, father, sister, brother, children) or second-degree (maternal and paternal grandparents, aunts, uncles, nephews) relatives of any of the following; (1) ankylosing spondylitis, (2) psoriasis, (3) acute uveitis, (4) reactive arthritis, (5) inflammatory back pain.
HLA-B27	Positive testing according to standard laboratory techniques.
Sacroiliitis by imaging	Bilateral grades II–IV or unilateral grades III–IV sacroiliitis on plain radiograph, according to the modified New York criteria, or active sacroiliitis on MRI according to the ASAS consensus definition.

Table 32.6: Current and classic classification of spondyloarthritis[16]

Current classifications

Axial spondyloarthritis

- With radiographic sacroiliitis
- Without radiographic sacroiliitis
  - Sacroiliitis on MRI
  - HLA-B27 positivity plus two clinical or lab criteria

Peripheral spondyloarthritis

- With psoriasis
- With inflammatory bowel disease (Crohn's disease or ulcerative colitis)
- With preceding infection
- Without psoriasis or inflammatory bowel disease or preceding infection

Classic classifications

- Ankylosing spondylitis
- Reactive arthritis (infection-associated arthritis)
- Psoriatic spondyloarthritis
  - Predominantly peripheral
  - Predominantly axial
- Enteropathic spondyloarthritis (associated with inflammatory bowel disease)
  - Predominantly peripheral
  - Predominantly axial
- Juvenile-onset spondyloarthritis (enthesitis-related juvenile idiopathic arthritis)
- Undifferentiated spondyloarthritis

is known as nr-axial SpA. In axial SpA, sacroiliitis is not detectable on X-ray in the early stage of the disease and in this context, MRI is a better imaging option. Now the ASAS classification criteria is better for confirming nr-axial spondyloarthritis. HLA-B27 positive patients with two clinical or laboratory features of SpA can also be classified as nr-ax SpA (Table 32.4a).

Peripheral SpA

ASAS peripheral criteria is used for diagnosis in patients suffering from peripheral arthritis, enthesitis (at any site), and or dactylitis. The patient should not have concurrent inflammatory back pain at the time of presentation. If patient has both manifestations, they should be classified as axial SpA with peripheral arthritis. The ASAS criteria are inclusive of disorders such as psoriasis, IBD and reactive arthritis.

Psoriatic Arthritis (Psoriatic Arthritis Study Group Criteria)

Most of the patients with psoriasis can be classified as SpA. They are labelled as axial or peripheral SpA with psoriasis. The classification of psoriatic arthritis (CASPAR) study group is an international group of researchers in psoriatic arthritis. They developed the CASPAR criteria which are simple and user-friendly (Table 32.7).[18] Based on its potential usefulness and performance, it is more likely to be introduced as the universal classification criteria for psoriatic arthritis. The CASPAR criteria have specificity of 98.7% and sensitivity of 91.4%. Patients with rheumatoid type (polyarticular) or with DIP involvement can be classified by using CASPAR criteria.

Summary

As a group of overlapping diseases and disorders, the diagnosis of spondyloarthropathies is challenging.

Table 32.7: Classification of psoriatic arthritis study group criteria for the classification of psoriatic arthritis (PsA)[18]

The CASPAR classification criteria for PsA
To be classified as having PsA, a patient must have inflammatory articular disease (joint, spine, entheseal) with ≥ 3 of the following 5 points.

Criterion	*Description*
1. Evidence of psoriasis (one of a, b, c):	
a. Current psoriasis	Psoriasis skin or scalp disease currently present, as judged by a rheumatologist or a dermatologist.
b. Personal history of psoriasis	A history of psoriasis obtained from patient or family physician, dermatologist, rheumatologist, or other qualified health care professional.
c. Family history of psoriasis	A history of psoriasis in a first-or second-degree relative by patient report.
2. Psoriatic nail dystrophy	Typical psoriatic nail dystrophy, including onycholysis, pitting and hyperkeratosis observed on current physical examination.
3. Negative test result for RF	By any method except latex but preferably by ELISA or nephelometry, according to the local laboratory reference range.
4. Dactylitis (one of a, b):	
a. Current	Swelling of an entire digit
b. History	A history of dactylitis recorded by a rheumatologist
5. Radiological evidence of juxta-articular new bone formation	Ill-defined ossification near joint margins (excluding osteophytes formation) on plain X-ray films of hand or foot.

CASPAR, classification criteria for psoriasis arthritis: PsA, psoriatic arthritis: RF, rheumatoid factor: ELISA, Current psoriasis score 2; all other items score 1

Early diagnosis is critical for initiating prompt treatment; however, there is a considerable delay in the diagnosis of spondyloarthropathies. An array of classification criteria was developed and validated to identify patients with spondyloarthropathies from the general population. The basic features of all these classification criteria included inflammatory back pain, imaging and detection of HLA B-27. A combination of clinical symptoms and a definite radiographic finding of sacroiliitis are included in all the three criteria—mNY criteria, Amor criteria and ESSG criteria. MRI in the imaging arm was included only in the ASAS. Owing to lack of proper diagnostic tool, physicians most often use the classification criteria for diagnosing spondyloarthropathies. However, the diagnostic classification criteria has to be sensitive enough to recognize spondyloarthropathies in its early stages as well as be applicable to individual patients.

Key Points

1. Low back pain in a patient who is less than 45 years age group, spondyloarthritis should be considered.
2. Assessment of Spondyloarthritis International Society (ASAS) classification criteria of spondyloarthritis is used. Patients can be classified as:
   a. Axial spondyloarthritis
      i. With radiographic sacroiliitis
      ii. Without radiographic sacroiliitis (NR SPA) Sacroiliitis on MRI
   b. Peripheral SPA
3. X-ray Pelvis AP view and/or MRI SI joint STIR images showing active inflammatory lesions of sacroiliitis joints with definite bone marrow oedema/osteitis is suggestive of sacroiliitis associated with spondyloarthritis.
4. HLA B27 (PCR method) is positive in majority of patients.

REFERENCES

1. Braun J, Sieper J. Early diagnosis of spondyloarthritis. Nat Clin Pract Rheumatol. 2006; 2:536–45.
2. Zochling J, Brandt J, Braun J. The current concept of spondyloarthritis with special emphasis on undifferentiated spondyloarthritis. Rheumatology 2005; 44:1483–91.
3. Harper BE, Reveille JD. Spondyloarthritis: Clinical Suspicion, Diagnosis, and Sports. Curr Sports Med Rep. 2009; 8:29–34.
4. Calin A, Porta J, Fries JF, Schurman DJ. Clinical history as a screening test for ankylosing spondylitis. JAMA. 1977; 237:2613–4.
5. Weisman MH. Inflammatory back pain. Rheum Dis Clin North Am. 2012; 38:501–12.
6. Rudwaleit M, Metter A, Listing J, Sieper J, Braun J. Inflammatory back pain in ankylosing spondylitis: a reassessment of the clinical history for application as classification and diagnostic criteria. Arthritis Rheum. 2006; 54:569–78.
7. Sieper J, van der Heijde D, Landewé R, Brandt J, Burgos-Vagas R, Collantes-Estevez E, et al. New criteria for inflammatory back pain in patients with chronic back pain: a real patient exercise by experts from the Assessment of Spondyloarthritis International Society (ASAS). Ann Rheum Dis. 2009; 68: 784–8.
8. van der Linden S, Valkenburg HA, Cats A. Evaluation of diagnostic criteria for ankylosing spondylitis. A proposal for modification of the New York criteria. Arthritis Rheum.1984; 27:361–8.
9. Dougados M, van der Linden S, Juhlin R, Huitfeldt B, Amor B, Calin A, et al. The European Spondylarthropathy Study Group preliminary criteria for the classification of spondylarthropathy. Arthritis Rheum. 1991; 34:1218–27.
10. Akgul O, Ozgocmen S. Classification criteria for spondyloarthropathies. World J Orthop. 2011; 2:107–15.
11. Rudwaleit M, van der Heijde D, Landewé R, Listing J, Akkoc N, Brandt J, et al. The development of Assessment of Spondyloarthritis International Society classification criteria for axial spondyloarthritis (part II): validation and final selection. Ann Rheum Dis 2009; 68:777–83.
12. Rudwaleit M, Jurik AG, Hermann KG, Landewé R, van der Heijde D, Baraliakos X, et al. Defining active sacroiliitis on magnetic resonance imaging (MRI) for classification of axial spondyloarthritis: a consensual approach by the ASAS/OMERACT MRI group. Ann Rheum Dis 2009; 68:1520–7.
13. Londono J, Santos AM, Peña P, Calvo E, Espinosa LR, Reveille JD, et al. Analysis of HLA-B15 and HLA-B27 in spondyloarthritis with peripheral and axial clinical patterns. BMJ Open. 2015; 5(11):e009092.
14. van Tubergen A, Weber U. Diagnosis and classification in spondyloarthritis: Identifying a chameleon. Nat Rev Rheumatol. 2012; 8:253–61.
15. Rudwaleit M, van der Heijde D, Landewé R, Akkoc N, Brandt J, Chou CT, et al. The Assessment of Spondyloarthritis International Society classification criteria for peripheral spondyloarthritis and for spondyloarthritis in general. Ann Rheum Dis. 2011; 70:25–31.

16. Taurog JD, Chhabra A, Colbert RA. Ankylosing Spondylitis and Axial Spondyloarthritis. N Engl J Med 2016; 374:2563–74.

17. Dean LE, Jones GT, MacDonald AG, Downham C, Sturrock RD, Macfarlane GJ. Global prevalence of ankylosing spondylitis Rheumatology (Oxford). 2014; 53(4):650–7.

18. Taylor W, Gladman D, Helliwell P, Marchesoni A, Mease P, Mielants H. Classification criteria for psoriatic arthritis: development of new criteria from a large international study. Arthritis Rheum 2006; 54:2665–73.

FURTHER READING

1. Taurog JD, Chhabra A, Colbert RA. Ankylosing Spondylitis and Axial Spondyloarthritis. N Engl J Med 2016; 374:2563–74.

Chapter

33

Ankylosing Spondylitis: Clinical Features and Management

Alakendu Ghosh, Subhankar Haldar

INTRODUCTION

Ankylosing spondylitis (AS) is the prototype of seronegative spondyloarthritis primarily affecting the axial skeleton (sacroiliac joints and spine) but also affects the peripheral joints, entheses and many extra-articular sites such as eyes, heart, lungs, ultimately leading to functional disability and premature death.[1] Overtime patients develop ankylosis of the spine. Inflammatory back pain and radiographic sacroiliitis is the hallmark of this disease.

The age of onset is the second or third decade of life and males are affected two to three times more than females.[1,2] Female have less severe form of disease, and the cervical spine and peripheral joints are more involved in female than male.[3] If the age at onset is younger than sixteen years a patient is classified as juvenile ankylosing spondylitis or enthesitis related arthritis.[4] The prevalence of the disease is varied in different geographic areas. The course of disease progression is fluctuating and variable. The first ten years of the disease pattern often determines subsequent outcomes.[5]

The disease has strong association with HLA-B27 haplotype. But only 1–6% of HLA-B27 positive adults suffer from AS. Positive family history of AS in the first-degree relatives confers 10–30% greater risk of developing AS.[6] In Caucasians 90% of patients with AS are HLA-B27 positive, while in Afro-Caribbeans HLAB27 is positive in only 50%. HLA-B27 negative AS patients have significantly older age at disease onset and a less frequent prevalence of acute anterior uveitis when compared to B27 positive AS. Diagnostic delay is also longer as compared to B27 positive AS.[7]

The 1984 modified New York classification criteria is an established diagnostic criteria for AS.[8] These criteria have radiographic sacroiliitis as a mandatory feature, which is usually a late manifestation of AS (Table 33.1). Also the limitation of lumbar spine mobility and limitation of chest

Table 33.1: 1984 Modified New York diagnostic criteria for ankylosing spondylitis[8]

Clinical criteria	i. Low back pain and stiffness for more than three months that improves with exercise but is not relieved by rest ii. Limitation of motion of the lumbar spine in both the sagittal and frontal planes iii. Limitation of chest expansion relative to normal values correlated for age and sex
Radiological criteria	i. Bilateral sacroiliitis ≥ grade 2 ii. Unilateral sacroiliitis grade 3 or 4

Definite AS—patient with one radiological criteria plus at least one clinical criteria
Probable AS—patients with all three clinical criteria or radiological criteria
Reference: van der Linden S, et al. Arthritis Rheum. 1984; 27:361–8

expansion may not be the first presenting complaints. Application of these criteria for diagnosis may cause delay of 5–10 years before a diagnosis of AS can be made among other factors.[9] Current emphasis nowadays is on early diagnosis, and early initiation of treatment in order to decrease the inflammatory burden and improve outcome.

PATHOLOGY AND PATHOGENESIS

Enthesitis is the hallmark pathological feature of AS. John Ball first described enthesopathy and he reported that inflammation, bone erosion and new bone formation in AS starts at the entheses. Entheses are the sites of attachment of ligaments, tendons and joint capsules to bone and they are present in vertebrae, sacroiliac joints and as well as in peripheral joints. The Achilles enthesis is the most characterized entheses in the peripheral skeleton and has been described as an organ, termed as achilles enthesis organ.[10] This organ consists of ligaments, tendons and capsules attached to the cancellous bone of the calcaneum through the fibrocartilaginous connections and the synovium of the retrocalcaneal bursa is also the part of this organ. New bone formation at entheses of vertebrae is called syndesmophytes which bridge the vertebrae and fuse them across to form so called 'bamboo spine' in AS.[11]

Immunologically there is interaction between class I MHC molecule HLA-B27 and T lymphocytes, with tumor necrosis factor (TNF-α) identified as a key regulatory cytokine in the inflammatory cascade but TNF-α is not closely involved in syndesmophyte formation or bone erosion.

T cells, B cells, bone marrow derived macrophages and cells involved in neoangiogenesis are confined at the sites of enthesitis.[12–14] The bone destructive collagen-degrading matrix metalloproteinase 1 (MMP1) and collagenolytic proteinase cathepsin-K in the osteoclasts are also present at the sites of enthesitis of the vertebrae.[15] In AS, bone erosion is caused by osteoclastic activity and which is different from bone erosion in RA, where bone destructive MMP3 and RANK (receptor activator of nuclear factor kB) ligand play the major role.[16] Probably for this reason, anti-TNF arrests the bone erosion less frequently in AS than RA.

Several clinical and animal studies have shown that inflammation and bone destruction are coupled in RA but uncoupled in AS (Fig. 33.1). Anti-TNF therapy controls the inflammation and can prevent the bone erosion in RA, but not in AS. In AS, there is marked improvement so far the inflammation is concerned but the entheseal bone erosion and new bone formation (mostly syndesmophytes in vertebrae) continue.

The process of new bone/syndesmophytes formation is presented in Fig. 33.2. The roles of the bone morphogenic proteins (BMPs), Wnt (wingless) pathways and the role of TNF appear conflicting.[17] The investigators have reported that there is increased BMPs activation at the achilles enthesis.[18,19] Marked decrease of sclerostin (an endogenous inhibitor of bone formation) in osteocytes even in patients with AS who have already developed sydesmophytes[20] and lower serum concentration of DKK1 (Dickkopf-related protein1) in patients with AS[21] leads to continuous new bone formation despite recent development of therapy like anti-TNF.

Synovitis is rare in AS and when present can be difficult to distinguish from rheumatoid arthritis (RA). Immunohistochemistry can differentiate synovitis of AS from RA. M1 proinflammatory macrophages are present in RA synovia whereas M2 regulatory macrophages consist in high number in synovia of AS[22, 23].

Genetics of AS

Risk factors for AS include HLA B27 positivity, positive family history, male gender, and frequent GI infections.[24] Among these HLA B27 is most important, as it is observed that there is 16 times higher chance of developing the disease among HLA B27 positive relatives.[6] The relative risk of developing AS in a sibling or first-degree relative of a patient with AS is 75% (study from Iceland).[25] The recurrence ratio in AS drops rapidly with

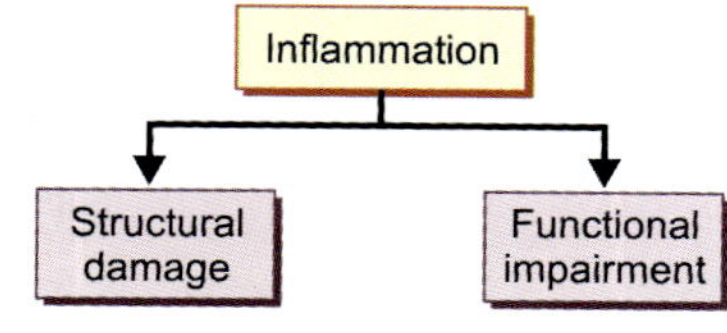

Fig. 33.1: The inflammation is coupled to structural damage and both features contribute to loss of function

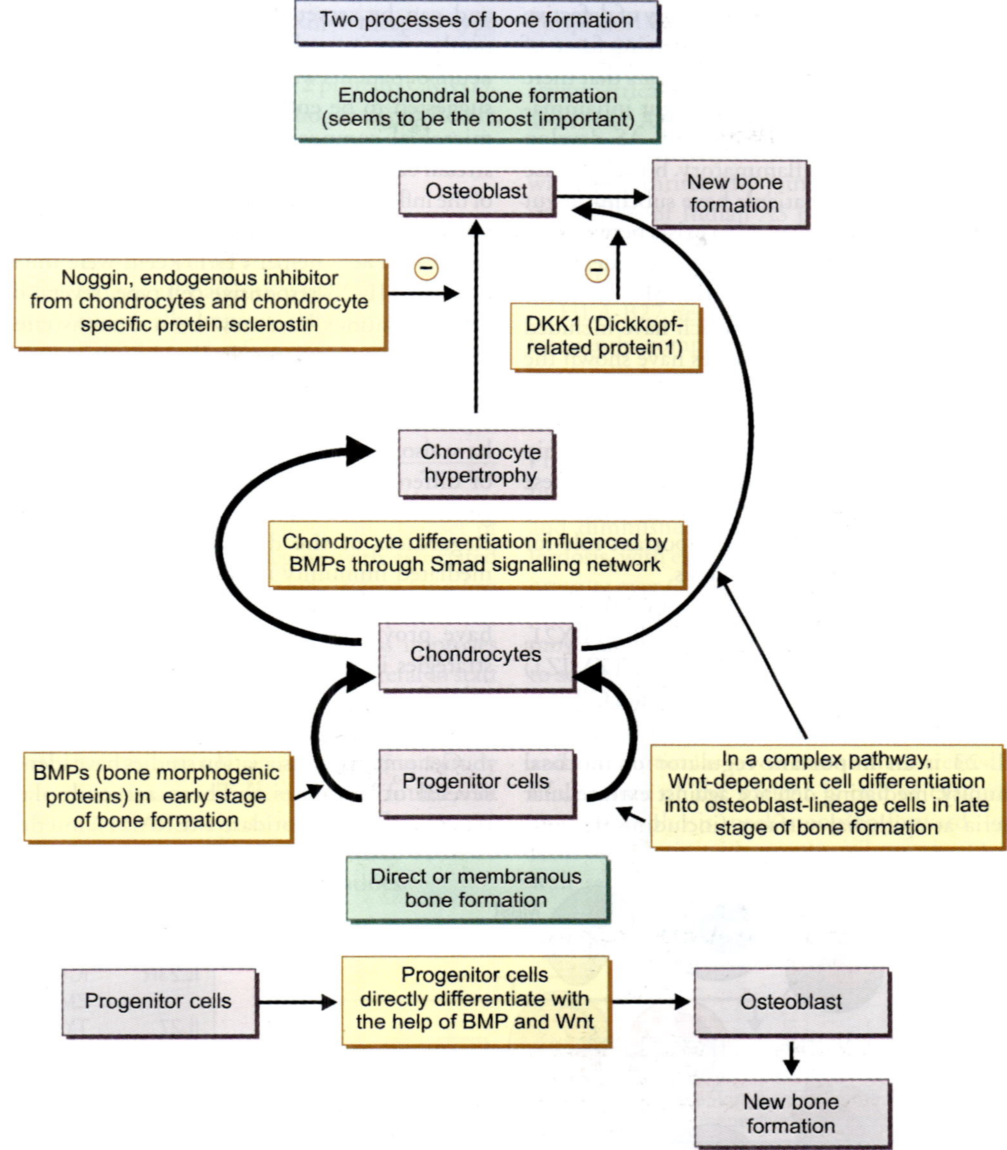

Fig. 33.2: Process of new bone/syndesmophytes formation

increasing distance of relationship to the proband, suggesting the presence of substantial non-MHC genetic risk factors for the disease.[26] Apart from HLAB27 there are must be other genetic or environmental risk factors for AS, as studies from African population (Zambia) show that AS can develop in the absence of HLAB27.[27] To date more than 32 loci have been associated with AS. These include additional MHC loci, such as HLA-B alleles other than HLA-B27 as well as non-MHC loci.[28]

Of the different subtypes of HLA-B27, *2705 is the predominant in most populations, while *2708 is common in Asians. There are a number of hypotheses regarding how the HLA genes cause or predispose to the illness. One hypothesis is by molecular mimicry where HLA-B27 binds to self or microbial peptides and presents them to CD8 positive Tcells, second is aberrant folding or processing of the heavy chain of HLA-B27, and the third is free heavy chain homodimers hypothesis.

53. van Denderen JC, van der Paardt M, Nurmohamed MT, de Ryck YM, Dijkmans BA, van der Horst-Bruinsma IE. Double blind, randomised, placebo controlled study of leflunomide in the treatment of active ankylosing spondylitis. Ann Rheum Dis. 2005; 64:1761–4.

54. Braun J, Sieper J. Therapy of ankylosing spondylitis and other spondyloarthritides: established medical treatment, anti-TNF-alpha therapy and other novel approaches. Arthritis Res 2002; 4:307.

55. Braun J, Bollow M, Seyrekbasan F, Häberle HJ, Eggens U, Mertz A, et al. Computed tomography guided corticosteroid injection of the sacroiliac joint in patients with spondyloarthropathy with sacroiliitis: clinical outcome and followup by dynamic magnetic resonance imaging. J Rheumatol 1996; 23:659–64.

56. Maksymowych WP, Jhangri GS, Fitzgerald AA, LeClercq S, Chiu P, Yan A, et al. A six-month randomized, controlled, double-blind, dose-response comparison of intravenous pamidronate (60 mg versus 10 mg) in the treatment of nonsteroidal anti-inflammatory drug-refractory ankylosing spondylitis. Arthritis Rheum. 2002; 46:766–73.

57. Breban M, Gombert B, Amor B, Dougados M. Efficacy of thalidomide in the treatment of refractory ankylosing spondylitis. Arthritis Rheum. 1999; 42: 580–1.

58. Braun J, Bollow M, Neure L, Seipelt E, Seyrekbasan F, Herbst H, et al. Use of immunohistologic and in situ hybridization techniques in the examination of sacroiliac joint biopsy specimens from patients with ankylosing spondylitis. Arthritis Rheum. 1995; 38: 499–505.

59. Ghosh A. Treatment of ankylosing spondylitis with special reference to biologics: Single Centre Experience. J Indian Rheumatol Assoc 2004; 12: 54–57.

60. Baraliakos X, Haibel H, Listing J, Sieper J, Braun J. Continuous long-term anti-TNF therapy does not lead to an increase in the rate of new bone formation over 8 years in patients with Ankylosing spondylitis. Ann Rheum Dis 2014; 73:710–15.

61. Arends S, Brouwer E, Efde M, van der Veer E, Bootsma H, Wink F, et al. Long-term drug survival and clinical effectiveness of etanercept treatment in patients with ankylosing spondylitis in daily clinical practice. Clin Exp Rheumatol. 2017; 35:61–68.

62. Baraliakos X, Listing J, Fritz C, Haibel H, Alten R, Burmester GR, et al. Persistent clinical efficacy and safety of infliximab in ankylosing spondylitis after 8 years—early clinical response predicts long-term outcome. Rheumatology (Oxford). 2011; 50:1690–9.

63. Sieper J, van der Heijde D, Dougados M, e Brown LS, Lavie F, Pangan AL. Early response to adalimumab predicts long-term remission through 5 years of treatment in patients with ankylosing spondylitis. Ann Rheum Dis. 2012; 71:700–6.

64. McInnes IB, Mease PJ, Kirkham B, Kavanaugh A, Ritchlin CT, Rahman P, et al; FUTURE 2 Study Group. Secukinumab, a human anti-interleukin-17A monoclonal antibody, in patients with psoriatic arthritis (FUTURE 2): a randomised, double-blind, placebo-controlled, phase 3 trial. Lancet. 2015; 386 (9999):1137–46.

65. Malaviya AN, Kapoor S, Garg S, Rawat R, Shankar S, Nagpal S, et al. Preventing tuberculosis flare in patients with inflammatory rheumatic diseases receiving tumour necrosis factor-alpha inhibitors in India—An audit report. J Rheumatol. 2009; 36: 1414–2010.

FURTHER READING

1. Baraliakos X, Braun J. Non-radiographic axial spondyloarthritis and ankylosing spondylitis: what are the similarities and differences? RMD Open. 2015; 1(Suppl 1):e000053.

2. Malaviya AN, Rawat R, Agrawal N, Patil NS. The Non-radiographic Axial Spondyloarthritis, the Radiographic Axial Spondyloarthritis, and Ankylosing Spondylitis: The Tangled Skein of Rheumatology. Int J Rheumatol. 2017; 2017:1824794. doi: 10.1155/2017/1824794.

Chapter

34

Psoriatic Arthritis: Clinical Features and Management

V Krishnamurthy

INTRODUCTION

Psoriatic arthritis (PsA) is an inflammatory arthritis associated with psoriasis and is a member of spondyloarthritis (SpA) family. It is heterogenous in nature by its musculoskeletal manifestations like axial arthritis, peripheral arthritis, enthesitis, dactylitis and extra-articular manifestations affecting the skin, nails, eyes and organ systems. The rheumatoid factor and anti-cyclic citrullinated peptide are generally negative and HLA B27 can be positive in those with spinal involvement.

Epidemiology

Among patients with psoriasis 7 to 42% develop arthritis and in 85% of them skin psoriasis precedes arthritis, whereas in 15% psoriasis develops later after the onset of arthritis and rarely never at all (psoriatic arthritis sine psoriasis). The most common skin lesions are plaque psoriasis and psoriasis vulgaris but other patterns can be seen. The exact prevalence and incidence of psoriatic arthritis varies in different populations. The prevalence is 0.04 to 0.1% and the incidence is about 3.6 to 6.6 per 100,000 per annum. The sex distribution is equal and the mean age of onset is 30 to 55 years of age. The incidence of developing arthritis in a patient with psoriasis is constant (74 per 1000 person-years) while the prevalence increases over time once psoriasis is diagnosed and is 20.5% over 30 years.

In a study from southern India it was found that males were more affected than females in a ratio of 2:1, the peak incidence was more common in the fourth and fifth decade (69%) and arthritis followed the skin lesion in 50.8% of the patients.[1]

First-degree relatives and monozygotic twins have a higher risk for psoriasis and PsA. Psoriasis may be an etiologic and trigger factor for inflammatory arthritis. HIV infection is found to be associated with higher incidence of psoriasis.[2]

ETIOLOGY

The definite etiology of PsA is not clear. Several factors may play a role.[2] It is suspected that entheseally derived antigens may trigger immune response in the adjacent synovium causing psoriatic arthritis.

Genetics

Familial aggregation and paternal transmission are noted with chromosome 16q and indicate a role for genes. HLA class I molecules are associated with psoriasis and PsA. HLA-C*06 is associated with early onset and severe disease more in psoriasis than PsA. HLA-B27, B08, B38 and B39 are associated with PsA, more so with HLA-B27*05:02 and B39*01:01 subsets. HLA-B27 is associated with symmetrical sacroiliitis, enthesitis and dactylitis, whereas B08 is associated with asymmetrical sacroilitis, joint ankylosis, deformity and dactylitis.

A study of the class I major histocompatibility complex (MHC) chain related gene A (MICA) showed MICA-A9 polymorphism was associated with polyarthritis. The corresponding allele MICA-002, was increased in PsA, whereas the Cw*0602 allele was significantly increased in both psoriasis and PsA. Thus MICA-002 may be a possible candidate gene for the development of PsA.[3] There are differences in haplotypes between psoriasis vulgaris and PsA.[4]

The Collaborative Association Study of Psoriasis showed that HLA-C and IL-23R SNP's are more strongly associated with psoriasis and IL-12B with PsA. Other genes that are associated with psoriatic arthritis are TNF 308 allele and non-MHC regions—PSORS region on chromosome 4, 6 and 17.[4]

A new susceptibilty gene region TRAF 31P2 on chromosome 6p, involved in IL-17 signalling and interacting with Rel/nuclear factor-kappa B is associated with PsA.[5]

None of the studies have shown the usefulness of these genetic markers in identifying PsA in psoriasis and till date no candidate gene has been identified.

Infection

Adequate proof for the role of infection is lacking. PsA in HIV infection could be due to increase in CD8 T cells rather than infection. Guttate psoriasis and preceeding streptococcal infection has been found in children.[6]

Trauma and Stress

Trauma and stress can act by a mechanism of up regulation of proinflammatory cytokines and release of neuropeptides in psoriasis and PsA and do not have a direct action. PsA can develop following trauma. Koebner phenomenon where psoriatic skin lesions develop at sites of skin trauma occurs in 52% of patients with psoriasis.[6]

PATHOGENESIS

T cells

Studies on active skin lesions, synovial fluid and response to biological agents have implicated the role of T helper cells Th1 and Th17 cells. Cytokines of these T helper cells play an active role in inflammation.[2] There is an increase in CD8+ T cells in the epidermis and entheseal sites, whereas CD8+ and CD4+ are present in the dermis. The synovium is infiltrated with CD163+ macrophages and CD3+ Tcells, which could be a future biomarker.[2]

Angiogenesis

Increased vascularity is a hallmark of psoriasis and arthritis.[2] Thickening and perivascular infiltration of capillaries and small arteries are seen in psoriatic skin lesions and synovium in contrast to straight and branching vessels seen in rheumatoid synovium. Increase in concentration of vascular growth factors and angiogenesis play a role in the pathogenesis.[2]

Cytokines

TNF-α, nuclear factor κB (NFκB), IL-1β, IL-10, IL-12 and IL-17 are expressed both in skin and in synovial membrane of patients with PsA in greater concentrations. There is an increase in Th 17 cells in PsA with increase in IL-17, IL-12, and IL-23.[2,6,7]

These cytokines sustain the inflammation and hence the favourable response to TNF blockade. IL-17 and IL-23 potentiate osteoclastogenesis and bone erosion in PsA. Inhibition of common p40 subunit of IL-12/IL-23 leads to clinical improvement. There is a central involvement of Th17-Th22-IL-23 axis in both psoriasis and PsA with upregulation of TNF production.

Matrix Metalloproteinases (MMP)

Matrix metalloproteinases and their tissue inhibitors (TIMPs) are present in the synovium of PsA and are upregulated. These cause cartilage destruction in PsA. Serum levels of MMP3 that decrease after anti-TNF therapy could be used as a biomarker for response to treatment.[2,6,7]

Bone Remodelling

There is bone destruction in the form of osteolysis, asymmetric eccentric erosions and pencil in cup deformities in PsA as well bone formation in the form of periosteitis, spur formation, enthesophytes and bony ankylosis. This is marked by increase in serum bone alkaline phosphatase. TNF plays a role in osteoclast activation. In PsA there is upregulation of RANKL protein and low expression of osteoprotogerin (OPG) resulting in bone destruction.[7]

CLINICAL FEATURES AND CLASSIFICATION

PsA usually develops in the setting of an established diagnosis of skin psoriasis but some patients might be unaware that they have skin psoriasis.[8] Physicians should carefully look for psoriatic skin lesions in the scalp, intergluteal/perianal areas and nails since they are at higher risk of developing PsA. In general there is no clinical correlation between severity of skin lesion and arthritis. Patients can manifest fatigue, stress, disability and comorbid conditions apart from arthritis.

In 1971 Moll and Wright classified PsA in to 5 clinical types. This classification is still being followed (Table 34.1).[9] The most common pattern of arthritis seen among Indian PsA is polyarthritis followed by oligoarthritis.[1] Oligoarthritis when it occurs is asymmetrical, whereas the polyarticular type is predominantly symmetrical in distribution.

The Classification criteria for Psoriatic Arthritis (CASPAR) have high specificity but less sensitivity and is often used in clinical trials.[10] To be eligible for classification as PsA the patient must have inflammation of joint or spine or entheses with 3 or more of the following: (1) Psoriasis, (2) psoriatic nail disease, (3) negative rheumatoid factor (RF), (4) dactylitis, (5) radiographic evidence of new bone formation (Table 34.2).[10]

Table 34.1: The Moll and Wright classification criteria for psoriatic arthritis[9]

According to Moll and Wright there are five clinical patterns
1. Distal interphalangeal (DIP) joint arthritis
2. Arthritis mutilans
3. Symmetrical polyarthritis
4. Asymmetrical oligoarthritis
5. Spondylitis

Table 34.2: Classification criteria for psoriatic arthritis (CASPAR) criteria[10]

Inflammatory arthritis (joint or spine or enthesitis) with more than 3 points from any of the following category
1. Psoriasis: Current (2), history of (1), family history of (1)
2. Nail Dystrophy (1): Onycholysis, pitting, hyperkeratosis
3. Negative for rheumatoid factor (1)
4. Dactylitis: Current (1), history of (1)
5. Radiographs of hands or feet (1): Juxta-articular new bone formation

However, the pattern of joint involvement in PsA is not fixed; it may fluctuate and overlap and can be influenced by treatment.

DIP Arthritis

DIP arthritis is characteristic of PsA and occurs early in the disease (Figs 34.1 and 34.2). It is often associated with dactylitis and nail abnormalities (Table 34.3). The prevalence is about 3.9% to 10% of all PsA. Disproportionate swelling of the IP joint of thumb is found to be associated with 28.7% of PsA and may have a positive predictive value of 84 percent.

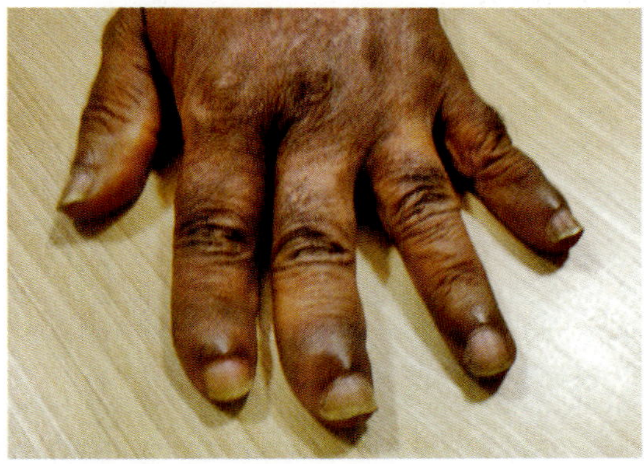

Fig. 34.1: DIP arthritis with nail changes

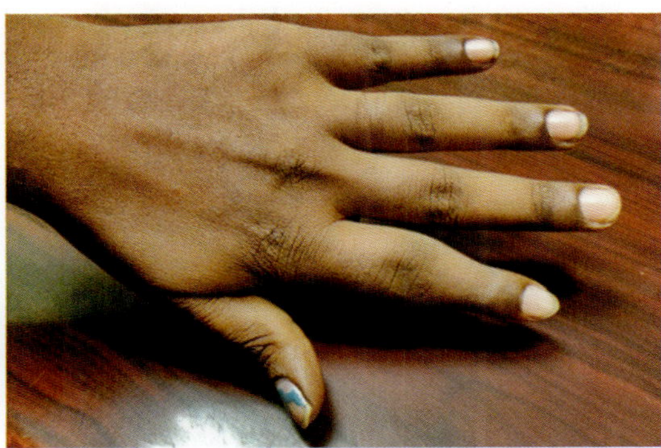

Fig. 34.2: Dactylitis or sausaging of digits

Table 34.3: Nail changes in psoriatic arthritis

1. Pitting
2. Ridging
3. Onycholysis
4. Yellowish nail margins
5. Dystrophic hyperkeratosis
6. Combination of the above

Mono and Oligoarticular Arthritis

It is the most common presenting pattern and the prevalence ranges from 11 to 70%. They progress to involve additional joints and have a male preponderance.

Symmetric Polyarthritis

Clinically indistinguishable from rheumatoid arthritis (RA), it involves the small joints of hands and feet with a female preponderance. They often have longer disease duration and are associated with erosive arthritis simulating RA.

Arthritis Mutilans

This is characterized by erosion and destruction of the joint. The clinical features are shortening of digits, flail joints caused by subluxation and telescoping of digits ('doigt en lorgnette') opera glass finger (Fig. 34.3). It is rare and occurs with delay in treatment. The Group for Research and Assessment of Psoriasis and Psoriatic Arthritis (GRAPPA) has defined arthritis mutilans as the involvement of joints of hand and feet but not axial joints, with radiographic changes of erosion involving the entire articular surface of both sides of a joint and/or pencil-in-cup change and osteolysis. Even if one joint is affected it is considered for the diagnosis. It has female preponderance and the prevalence is less than 5%. It is often associated with sacroiliitis.[6, 8]

Spondyloarthritis

Predominant spinal manifestation is uncommon. However spinal involvement can be made out clinically and by radiological methods in almost 40 to 70% of patients and its incidence can increase with prolonged disease without treatment. Severe peripheral arthritis and the presence of HLA-B27 are risk factor for spondyloarthritis.[11] Spinal involvement closely resembles AS in the presence of symmetrical or asymmetrical sacroiliitis and ankylosis of spinal joints but differs by the lower incidence of zygapophyseal joint involvement (Table 34.4). Atlanto axial subluxation of cervical spine similar to RA can occur and be clinically silent. Hence special attention should be paid to detect this joint involvement prior to anaesthesia.

Dactylitis

Swelling of whole digit (sausaging) is called dactylitis (Fig. 34.4) and it is due to joint inflammation, soft tissue oedema, bone oedema and tenosynovitis. Fingers and toes are affected and occur in 29 to 33.5% of patients.

Table 34.4: Spine in psoriatic arthritis

- Unilateral sacroiliitis is common
- Zygapophyseal fusion is less common
- Skip lesions of syndesmophytes
- Atlanto-axial subluxation can occur

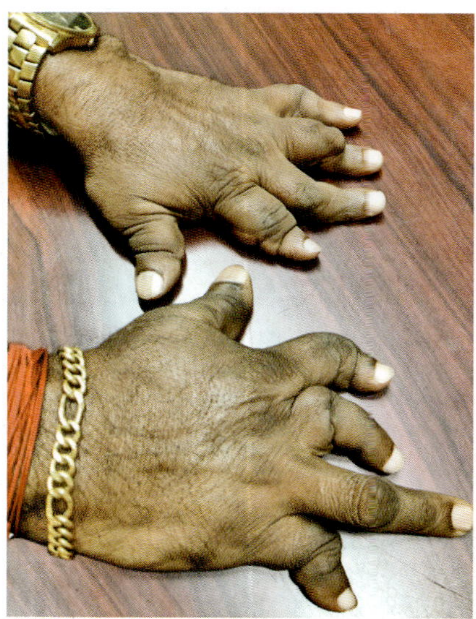

Fig. 34.3: Arthritis mutilans

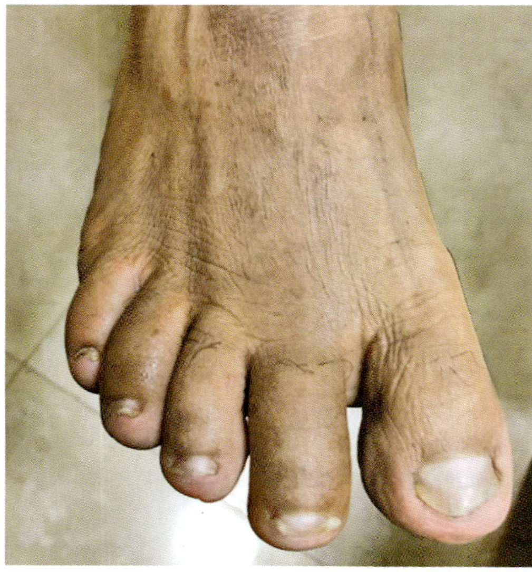

Fig. 34.4: Sausaging of toes (dactylitis)

Enthesitis

Inflammation at sites of tendon and ligament insertion (enthesitis) occurs in 20 to 40% of cases and can be a presenting feature in 4% of PsA. Enthesitis is a characteristic feature of PsA.[12] Achilles tendonitis and plantar fascitis are the most common sites, while iliac crest, insertions of quadriceps and patellar tendons, rotator cuff and epicondyles of elbow are the other common sites affected.

Peripheral Oedema

Peripheral oedema of feet or lymphoedema can occur due to extensor tenosynovitis and enthesitis.

Other Associated Features

The syndrome of synovitis, acne, pustulosis, hyperostosis and osteitis (SAPHO) can occur as a subgroup of PsA in 3% of patients and is associated with psoriasis vulgaris or palmoplantar pustulosis.[13] Anterior chest wall involvement is revealed by scintigraphy in these patients.

Psoriatic onycho pachydermo periostitis is a rare manifestation of PsA and is characterized by severe nail dystrophy, soft tissue swelling, DIP involvement and periostitis of terminal phalanx.[14] Periosteal reaction of the terminal phalanx is radiologically described as ivory phalanx.

Usually PsA patients have mild to moderate skin lesions. There is no correlation between joint and severity of skin lesions, whereas there is a close correlation between nail changes and arthritis. About 60 to 80% of PsA patients have nail changes. Nail changes include pitting, ridging (Fig. 34.5), hyperkeratosis and onycholysis. Arthritis occurs within 2 years of onset of nail changes and DIP arthritis is more commonly associated with nail changes. Patients with polyarticular disease and elevated inflammatory markers carry poor prognosis with 47% risk of erosions in first 2 years. There is a significant number of undiagnosed PsA in patients with psoriasis.[6,8]

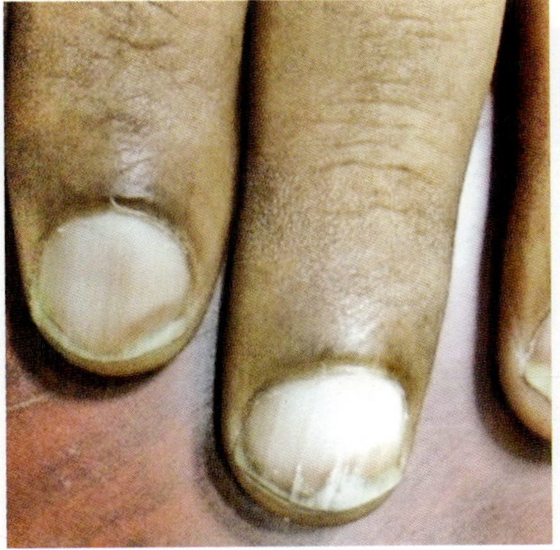

Fig. 34.5: Longitudinal ridging of nails with dystrophy

Extra-articular Manifestations

Extra-articular disease is less in PsA compared to RA. Uveitis occurs in 7 to 18%, more bilateral in patients with spinal involvement. Higher prevalence of inflammatory bowel disease has been noted in some studies.

Co-morbidities

A number of studies had highlighted cardiovascular risk factors and metabolic syndrome in psoriatic arthritis.

JUVENILE PSORIATIC ARTHRITIS (JPsA)[15]

Psoriatic arthritis occurring before the age of 16 years is labeled as Juvenile psoriatic arthritis (JPsA), which accounts for 2 to 5% of all Juvenile Idiopathic arthritis (JIA) cases. The criterion based by Southwood *et al* is used for diagnosis (Table 34.5). In majority of children the skin changes occur after the onset of arthritis. Eliciting family history of psoriasis is important. There is a female preponderance with a bimodal onset of 2–4 years and 10–12 years. Oligoarthritis at presentation may progress to polyarthritis. Looking for nail changes is important in clinching the diagnosis. Roughly 5% of adults with PsA have the onset in childhood.

Table 34.5: Criteria for the diagnosis of Juvenile psoriatic arthritis

Diagnosis	*Criteria*
Definite psoriatic arthritis	Arthritis onset before age 16 years of age and either typical psoriasis or Three minor criteria from: • Dactylitis • Nail pitting • Psoriasis like rash • Family history of psoriasis (first or second degree relative)
Probably psoriatic arthritis	Arthritis onset before 16 years of age and two minor criteria

OUTCOME MEASURES[16]

- PASI (psoriasis assessement severity index) score is used for assessing the extent of skin involvement.
- MDA (minimal disease activity) criteria is an easy method to assess disease activity criteria). Five out of seven criteria are taken as the cut off.
- DAPSA (Disease Activity in PSoriatic Arthritis) is a measure of disease activity index for PsA. It incorporates tender and swollen joints, CRP, patient and physician assessments.
- PsAJAI (Psoriatic Arthritis Joint Activity Index) used in radiological improvement.
- CPDAI (Composite Psoriatic Disease Activity Index) has 5 domains consisting of peripheral joints, skin, entheses, dactylitis and spinal manifestations. Disease activity, joint function and quality of life of patients are assessed in outcome measures.
- Other measures are PASDAS (Psoriatic Arthritis Disease Activity Score) and the GRACE measure (Grappa Composite Exercise).
- ACR 20, 50, 70 response criteria, as used in RA are still substituted for assessment.

Remission is difficult to achieve in PsA. Younger age, lower functional impairment and high CRP have been associated with higher remission rates.

INVESTIGATIONS

Laboratory

There are no specific laboratory markers for psoriatic arthritis. Acute phase reactants are raised and in particular ESR and CRP correlate well with inflammation. The other markers are anaemia of chronic disease, elevated serum amyloid A, hypoalbuminemia, hyper-gammaglobulinemia and fibrinogenemia. In 5 to 16% of patients low levels of RFand in 5% ACPA can be positive. A positive test for rheumatoid FA does not rule out PsA. Hyperuricemia can occur and denotes renal impairment and metabolic abnormalities. It is not due to increase in cell turnover in skin as was thought earlier.

Radiography

Peripheral Joints

Sites of enthesitis show new bone formation. Erosive changes are more common at DIP joints and there can be associated soft tissue swelling. Subchondral osteoporosis is uncommon. Proliferative changes are due to periostitis and new bone formation along the shafts of metacarpals and metatarsal described as whiskering.[17] Destructive changes, resorption or osteolysis result in pencilling or whittling of phalanx causing the pencil-in-cup deformity (Figs 34.6 and 34.7).

Spinal Joints

Asymmetrical sacroiliitis is more common than symmetrical lesion. Syndesmophytes are thick and non-marginal, differing from AS.[17] They occur as skip lesions than as flowing syndesmophytes. In general the spinal lesions are less and not as severe as seen in ankylosing spondylitis but cervical involvement is more common.

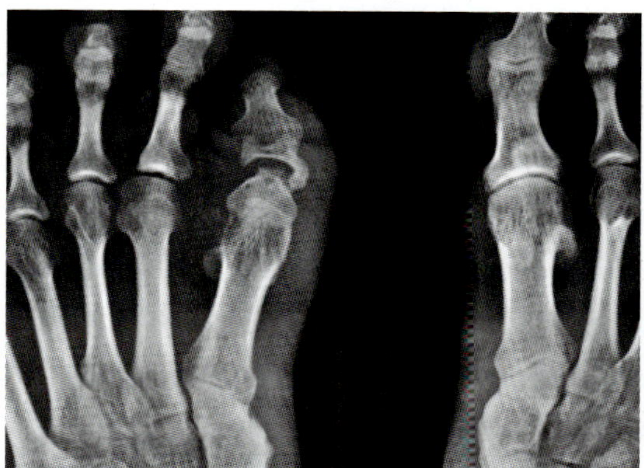

Fig. 34.6: Plain X-rays of hands showing pencil-in-cup appearance

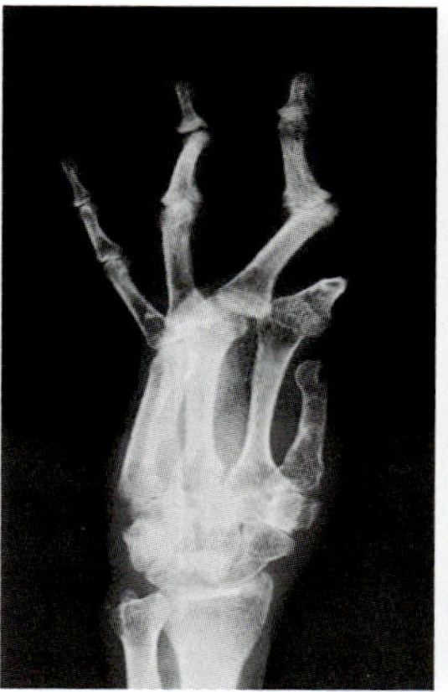

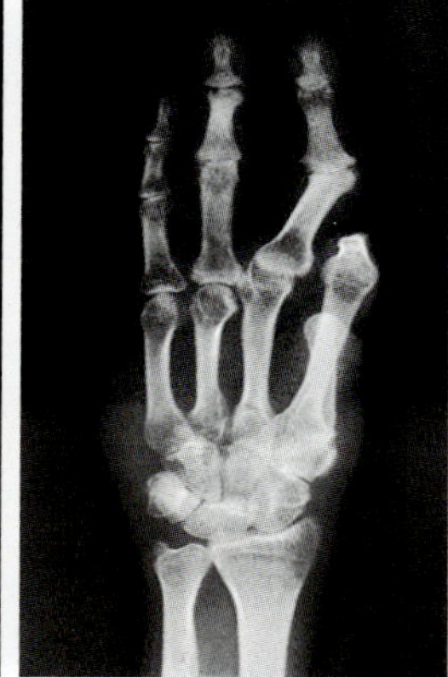

Fig. 34.7: Plain X-ray of hands showing osteolysis

Musculoskeletal Ultrasound[17]

It is useful in detecting early PsA. It detects at entheseal sites, entheseal thickening/formation, hypoechoic changes, increased vascularity in power Doppler, tenosynovitis and bony erosions.

Magnetic Resonance Imaging

MRI is not routinely used. It can detect bone marrow oedema at enthesitis and early sacroilitis. It is also useful as an outcome measure in response to biological agents.[18] A new MRI scoring system called PsAMRIS has been proposed for this.

Other Imaging Techniques

High resolution CT is used where MRI is contraindicated. MRI has superseded other imaging modalities like scintigraphy and PET CT.

DIAGNOSIS

Diagnosis is clinical and a high index of suspicion in a case of inflammatory arthritis is needed to clinch the diagnosis. Detailed history of psoriasis including family history and looking for skin lesions at the scalp, behind the ears (Fig. 34.8), umbilicus, gluteal folds and nail changes are essential. Axial signs, enthesitis, dactylitis, and radiographic changes help in diagnosing the condition.

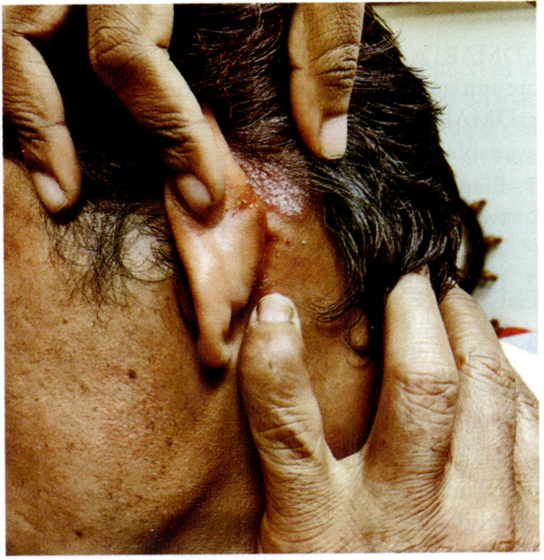

Fig. 34.8: Skin lesion behind the ear

Differential Diagnosis

It can be differentiated from RA in that it is asymmetrical with DIP involvement and has skin and nail changes. Asymmetrical sacroiliitis with non-marginal syndesmophytes differentiates it from AS. It is sometimes difficult to differentiate PSA from reactive arthritis (ReA). However ReA predominantly causes lower limb joints arthritis and is associated with eye involvement and occurs in young males. DIP arthritis has to be differentiated from Herberden's nodes of osteoarthritis, gouty tophi in gouty arthritis and multicentric reticulo histiocytosis where periungual nodules are present.

Prognosis and Mortality

Prolonged disease with polyarthritis, radiological damage, elevated acute phase reactants and extra-articular manifestations are associated with poor prognosis and increase in mortality. Association with HLA antigens like HLA-B27 and HLA-B39 are indicators for poor prognosis in psoriatic arthritis. Radiologic damage, increased ESR, prior use of DMARDs, extensive and progressive disease is associated with increased mortality.

MANAGEMENT

Management of PsA is multidisciplinary and consists of non-pharmacological, pharmacological, rehabilitative and surgical therapies along with skin management with the help of the dermatologist. Treatment decisions are made between the patient and the rheumatologist considering the efficacy, safety and costs.[19,20] Early treatment is important for good prognosis and should be started within 3 months of onset.

The main aim of treatment is reaching the target of remission, or minimal/low disease activity, by regular monitoring and appropriate adjustment of therapy. The management includes treat to target, treating extra-articular manifestations, metabolic syndrome, cardiovascular diseases and comorbid conditions.

Patients with peripheral arthritis respond well to selective and non-selective COX-2 inhibitors. It would be better if dermatologists and rheumatologists can manage concurrent problems with the use of a single drug, which is however difficult. Disease modifying drugs methotrexate, cyclosporine and leflunomide alone or in combination are effective.

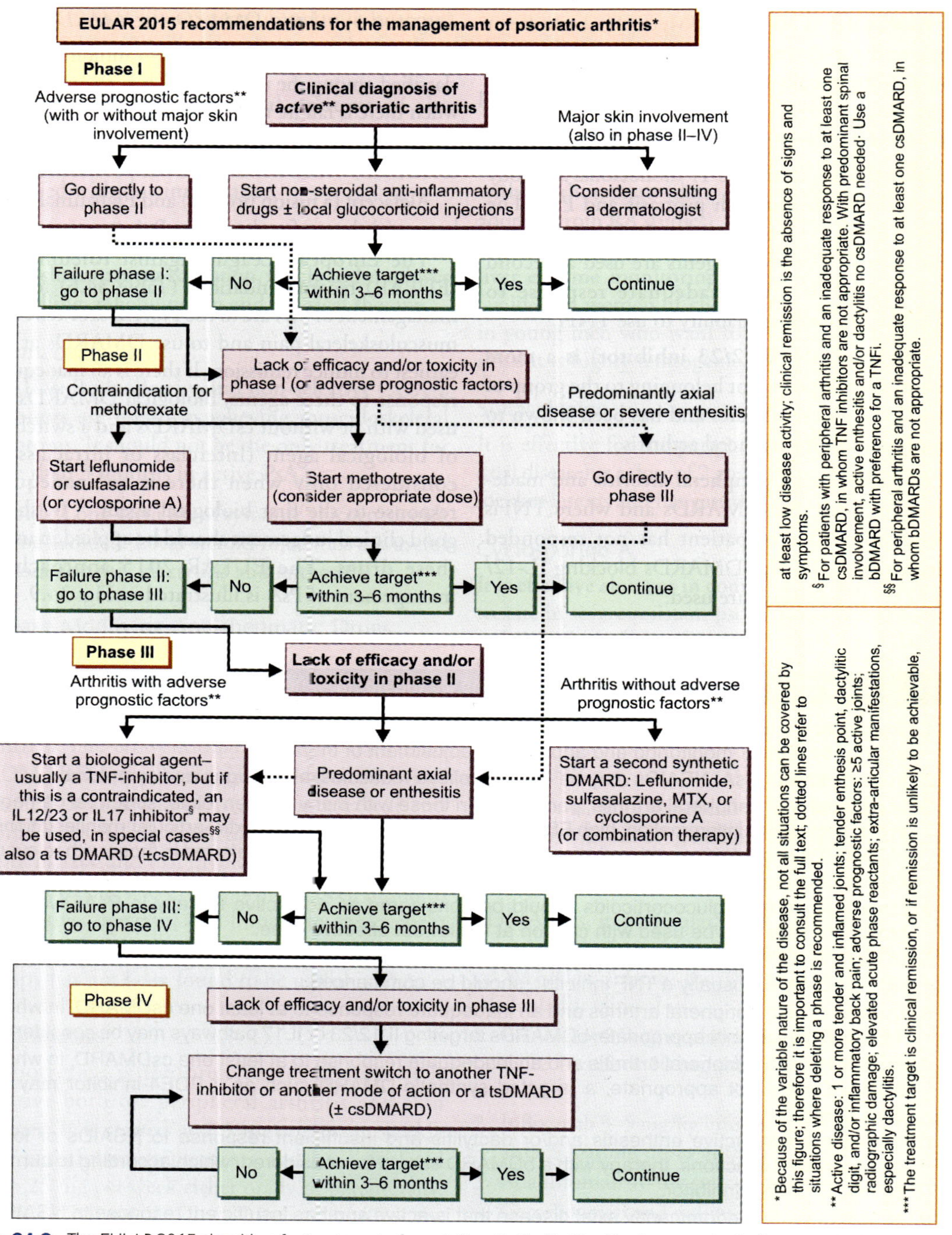

Fig. 34.9: The EULAR 2015 algorithm for treatment of psoriatic arthritis (PsA) with pharmacological non-topical treatments

Abbreviations: bDMARD, biological DMARD; csDMARDs, conventional synthetic DMARD; DMARD, disease-modifying antirheumatic drug; EULAR, European League Against Rheumatism; IL, interleukin; MTX, methotrexate; PsA, psoriatic arthritis; TNFi, tumour necrosis factor inhibitors; DMARD, targeted synthetic DMARD.

(Reproduced from Gossec L, et al. European League Against Rheumatism (EULAR) recommendations for the management of psoriatic arthritis with pharmacological therapies: 2015 update. Ann Rheum Dis 2016;75:499–510 with permission from BMJ Publishing Group Ltd.)

Conclusion

Diagnosing and treating PsA can be challenging. Immunosuppressive therapy including biological agents, methotrexate and cyclosporin are contra-indicated in psoriatic arthritis secondary to HIV infection. Targeted therapies with biologicals and small molecules are a new hope in the treatment of psoriasis and psoriatic arthritis.

Key Points

- PsA is an inflammatory arthritis.
- Elicit history for psoriasis in a patient presenting with arthritis.
- Examine for psoriatic skin lesions in hidden areas.
- In PsA the sex distribution is equal.
- HLA B27 is associated with bilateral sacroiliitis.
- There is no specific laboratory marker for diagnosis.
- Characteristic radiological changes are present.
- Can mimic several inflammatory arthritic conditions.
- NSAIDs and methotrexate play a major role in the management.
- Biological agents slow radiological progression.

REFERENCES

1. Rajendran CP, Ledge SC, Rani KP, Madhavan R. Psoriatic arthritis. J Assoc Physicians India 2003; 51:1065–68.
2. Richlin C, McGonagle D. Etiology and pathogenesis of Psoriatic arthritis; In Hochberg CM, Silman JA, Smolen SJ, Weinblatt EM, Weisman HM, Editors. Rheumatology. Philadelphia: Mosby; 2011. pp.1195–1204.
3. Gonzalez S, Martinez-Borra J, Torre-Alonso JC, Gonzalez-Roces S, Sanchez del Río J, Rodriguez Pérez A, et al. The MICA-A9 triplet repeat polymorphism in the transmembrane region confers additional susceptibility to the development of psoriatic arthritis and is independent of the association of Cw*0602 in psoriasis. Arthritis Rheum. 1999; 42:1010–6.
4. Chandran V. The genetics of psoriasis and psoriatic arthritis. Clin Rev Allergy Immunol. 2013; 44: 149–56.
5. Elder JT. Genome-wide association scan yields new insights into the immunopathogenesis of psoriasis. Genes Immun. 2009; 10:201–9.
6. Oliver F, Elmamoun. Psoriatic Arthritis. In: Firestein GS, Budd RC, Gabriel SE, McInnes IB, O'dell JR, Editors. Kelely and Firestein's Text book of Rheumatology. Philadelphia, PA: Elsevier Saunders; 2017. pp.1285–1308.
7. Anthony M. Turkiewicz and Moreland LW. Psoriatic Arthritis- Current Concepts on Pathogenesis-Oriented Therapeutic Options. Arthritis Rheum 2007; 56:1051–66.
8. Bruce IN. Clinical features of Psoriatic arthritis. In Hochberg CM, Silman JA, Smolen SJ, Weinblatt EM, Weisman HM, Editors. Rheumatology. Philadelphia: Mosby; 2011; pp.1183–1193.
9. Wright V, Moll JM. Psoriatic arthritis. Bull Rheum Dis. 1971; 21:627–32.
10. Taylor W, Gladman D, Helliwell P, Marchesoni A, Mease P, Mielants H; CASPAR Study Group. Classification criteria for psoriatic arthritis: development of new criteria from a large international study. Arthritis Rheum. 2006; 54:2665–73.
11. Marsal S, Armadans-Gil L, Martínez M, Gallardo D, Ribera A, Lience E. Clinical, radiographic and HLA associations as markers for different patterns of psoriatic arthritis. Rheumatology (Oxford). 1999; 38: 332–7.
12. Scarpa R. Peripheral enthesopathies in psoriatic arthritis; J Rheumatol 1998; 25:2259–90.
13. Winchester R. Psoriatic arthritis and the spectrum of syndromes related to the SAPHO (synovitis, acne, pustulosis, hyperostosis, and osteitis) syndrome. Curr Opin Rheumatol. 1999; 11:251–6.
14. Boisseau-Garsaud AM, Beylot-Barry M, Doutre MS, Beylot C, Baran R. Psoriatic onycho-pachydermo-periostitis. A variant of psoriatic distal interphalangeal arthritis? Arch Dermatol. 1996; 132:176–80.
15. Southwood TR, Petty RE, Malleson PN, Delgado EA, Hunt DW, Wood B, et al. Psoriatic arthritis in children. Arthritis Rheum. 1989; 32:1007–13.
16. Her M, Kavanaugh A. A review of disease activity measures for psoriatic arthritis: what is the best approach? Expert Rev Clin Immunol. 2014; 10: 1241–54.
17. Sudol-Szopinska I, Matuszewska G, Kwiatkowska B, Pracol G. Diagnostic imaging of psoriatic arthritis. Part I: etiopathogenesis, classifications and radiographic features. J Ultrason. 2016; 16 (64):65–77.
18. Sudol-Szopinska I, Pracol G. Diagnostic imaging of psoriatic arthritis. Part II: magnetic resonance imaging and ultrasonography. J Ultrason. 2016; 16 (65):163–74.
19. Mease JP. Management of Psoriatic arthritis. In: Hochberg CM, Silman JA, Smolen SJ, Weinblatt EM, Weisman HM, Editors. Rheumatology. Philadelphia: Mosby; 2011; 120:1205–1210.

20. Gossec L, Smolen JS, Ramiro S, de Wit M, Cutolo M, Dougados M, et al. European League Against Rheumatism (EULAR) recommendations for the management of psoriatic arthritis with pharmacological therapies: 2015 update. Ann Rheum Dis. 2016; 75:499–510.

21. Mease JP, McInnes BI, Kirkham B, Kavannagh A, et al. Secukinumab inhibition of IL-17A in patients with psoriatic arthritis. N Engl J Med 2015;373; 14:1329–39.

FURTHER READING

1. Fitzgerald O, Winchester R. Psoriatic arthritis: from pathogenesis to therapy. Arthritis Res Ther. 2009; 11(1):214.
2. Smolen JS, Schöls M, Braun J, Dougados M, Fitz Gerald O, Gladman DD, et al. Treating axial spondyloarthritis and peripheral spondyloarthritis, especially psoriatic arthritis, to target: 2017 update of recommendations by an international task force. Ann Rheum Dis. 2017 Jul 6.

Chapter

35

Enteropathic Arthritis

Rajiva Gupta, Natasha Negalur

INTRODUCTION

Enteropathic arthritis belongs to the group of seronegative spondyloarthropathies (SpA), which includes reactive arthritis, psoriatic arthritis, idiopathic ankylosing spondylitis (AS) and unclassifiable SpA. Enteropathic arthritis is a form of SpA associated with inflammatory bowel disease (IBD) namely ulcerative colitis (UC) and Crohn's disease (CD). Joint involvement is also known to occur in other gastrointestinal diseases like Whipple's disease, intestinal bypass surgery and in celiac disease. Some rheumatologists also include post-dysentery reactive arthritis under this terminology, but in this book it will be classified under reactive arthritis.

EPIDEMIOLOGY

About 3–4 million people worldwide get affected with IBD. Arthropathy is the most common extra-intestinal manifestation seen in IBD. The prevalence widely varies between studies due to differences in diagnostic criteria and may be underestimated due to transient nature of oligoarticular arthritis or the response of arthritis to corticosteroids given for IBD flares. Most of the studies on prevalence of arthritis in IBD are cross-sectional. As extra-intestinal manifestations develop over time, the true incidence or prevalence cannot be estimated from cross sectional studies. A systemic review and meta-analysis found that in patients of IBD, 10% have sacroiliitis, 3% have AS, 13% have peripheral arthritis, 1–54% have enthesitis and 0–6% have dactylitis.[1] A recently published Indian study showed that in CD ($n = 62$) and UC ($n = 58$), 23% have peripheral arthritis, 18% have AS, and 50% have osteopaenia/osteoporosis.[2] In other population based studies, cumulative incidence of SpA in CD ($n = 311$) and UC ($n = 365$) was found to be 0.5% and 22% respectively.[3–4]

Risk factors for arthritis include active disease, family history of IBD, erythema nodosum or pyoderma gangrenosum.[5,6] Women may be at a greater risk for peripheral arthritis, whereas men tend to have more frequent axial involvement.

PATHOGENESIS

Although the association between SpA and IBD is largely established, the exact mechanism by which joint disease occurs in bowel inflammation is not known. Interestingly, this association is both ways—approximately 10–15% of IBD cases are complicated by SpA, while ileal inflammation resembling IBD occurs in two thirds of cases of SpA. There is a common genetic background between AS and IBD as seen by the fact that both occur simultaneously in a patient with a higher prevalence in families and increased risk and cross risk ratios among relatives. It has been postulated that it is the recirculation of antigen- specific memory T cells from the gut to the joints that is responsible for initiating the joint inflammation. A likely chain of events would involve gastrointestinal infection with a bacterial micro-organism, followed by gut inflammation and inflammation in the joints. This especially applies to the development of peripheral arthritis where arthritis activity runs in parallel to gut activity, but it does not explain the development of axial involvement, which is independent of gut pathology. Role of IL-23 has also been postulated in these patients.[7–8]

GUT PATHOLOGY IN SPONDYLOARTHROPATHY

Because the pattern of joint disease seen in patients with IBD and asymptomatic gut inflammation in SpA is similar, it has been postulated that the intestinal inflammation may play a pathogenic role in the arthritis. Mielants *et al* from Belgium have documented that subclinical gut inflammation is prevalent in more than 50% of patients with idiopathic AS.[9] There are two major types of intestinal inflammation in SpA. An acute inflammation resembling infectious enterocolitis is seen with largely intact architecture and neutrophilic infiltration in the lamina propria. The chronic inflammation is more suggestive of early Crohn's disease, with distortion of villi and crypts, apthoid ulceration and mononuclear cell infiltration in the lamina propria.[10] Mielants et al. in their follow-up study of 217 patients with spondyloarthropathies who had an initial ileocolonoscopy found that of the 123 patients regularly followed up, 40 (32%) had normal gut histology, 28 (23%) had acute and 55 (45%) chronic inflammatory gut lesions. Eleven (5%) of these 217 patients went on to develop frank IBD, all these patients had AS, and the majority had chronic lesions in the initial ileocolonoscopy and were HLA-B27 negative.[11] In an Indian study, 39 patients of suspected enteropathic arthritis were studied. Patients were grouped into 3 categories, namely those with normal bowel histology, those with mild non-specific chronic changes, and those with histology suggestive of IBD. Patients with non-specific chronic gut inflammation had higher occurrence of axial involvement (with or without peripheral articular involvement), as compared to those with normal gut histology (8/9 *vs* 10/21, P = 0.049), and this pattern was similar to that in patients with IBD.[12] Although gut inflammation in SpA has been well-documented, it is not established yet whether these lesions contribute to the pathogenesis of SpA, or whether patients with inflammatory gut lesions have a different disease course than those without.

CLINICAL FEATURES OF ENTEROPATHIC ARTHRITIS

Arthritis is the most common extra-intestinal manifestation of inflammatory bowel disease (IBD). Classically, joint involvement in enteropathic arthritis has two main patterns:

1. Axial involvement resembling idiopathic AS or sacroiliitis.
2. Peripheral asymmetric arthritis.

These are not mutually exclusive. Other rheumatologic manifestations which are known to occur in IBD are metastatic granulomas of bone and joint, clubbing, periostitis, amyloidosis, psoas abscess leading to septic arthritis of the hip, osteoporosis, osteomalacia, granulomatous vasculitis and complications of steroid therapy.[13] In a recent study of IBD, of the 155 patients with joint/back pain, 13 had chronic back pain, 80 peripheral joint complaints, and 62 axial and peripheral joint complaints. The Assessment in Spondyloarthritis International Society (ASIS) criteria for axial and peripheral SpA were fulfilled in 12.3% of patients, with 9.7% (n = 15) receiving a rheumatological diagnosis of arthritis.[14]

SPONDYLOARTHROPATHIES

The European Spondyloarthropathy Study Group (ESSG) classification criteria for SpA enable the inclusion of previously neglected cases of undifferentiated SpA.[15] Axial involvement in IBD mostly manifests as either isolated radiographic sacroiliitis, which is often asymptomatic, or ankylosing spondylitis (AS), which is almost indistinguishable from idiopathic ankylosing spondylitis.

Previously, patients with inflammatory back pain (IBP) without radiographic abnormalities were not considered as part of enteropathic arthritis. But more recent studies have also looked into the prevalence of IBP in IBD.[6] In a recent prospective study of 103 patients with IBD, 30% patients had IBP, 18% patients had asymptomatic sacroiliitis, and AS was found in about 10% patients.[16] These figures contrasted from those seen in a recent population based study of 654 newly diagnosed patients with IBD and followed up for 6 years where the occurrence of AS was 3.7%, IBP without AS was found in 18% patients, and radiographic sacroiliitis (RSI) was estimated to be present in 2%

patients.[17] There was a much lower frequency of RSI in this study compared to hospital based studies probably reflecting the shorter duration of disease of patients and differences in radiological scoring technique and a referral bias in the hospital studies. In another study, 81 patients (55 CD and 26 UC) with remittent and low active IBD without joint symptoms were studied with 27.1% showing RSI at baseline. All patients were HLA-B27 negative. At 3 years, 18.1% presented chronic IBP symptoms with bone oedema at MRI.[18]

The clinical features of AS in IBD are similar to idiopathic AS. The low back pain is characteristically insidious in onset and is associated with morning stiffness. Pain is better with exercise, worse with rest and is diffuse, involving the entire low back and buttocks. Differences between idiopathic AS and IBD related SpA are listed in Table 35.1.[17, 22–27] Peripheral joints may also be involved along with axial disease mainly the hip and shoulders.

Isolated sacroiliitis is often asymptomatic and most patients are HLA-B27 negative and often do not progress to AS. Magnetic resonance imaging (MRI) is more sensitive to detect early sacroiliac changes. In one study, MRI picked up inflammatory sacroiliitis in 16.7% (n = 31).[19] Symptoms of sacroiliitis when they appear do not correlate with the activity of the underlying bowel disease. Sacroiliitis is often symmetrical in the IBD related disease and in idiopathic AS while it is typically asymmetrical or unilateral in reactive arthritis and in psoriatic arthropathy.

PERIPHERAL JOINT INVOLVEMENT

In the past, the prevalence of peripheral synovitis in CD has been underestimated, probably strict criteria were not employed in distinguishing UC from CD. Peripheral arthritis occurs slightly more in CD than UC, especially with colonic involvement.

i. Pattern of Peripheral Arthritis

The main pattern of joint involvement is a pauciarticular, episodic, non-destructive synovitis involving large more than small, and lower limb more than upper limb joints. The classification has never been uniform until Orchard and co-workers from Oxford described the natural history and articular distribution of peripheral arthropathy in 637 patients of UC and 483 patients with CD.[6] In this study peripheral joint involvement was divided into two groups according to the number of joints involved (Table 35.2). The most commonly affected joint in type 1 arthropathy (pauciarticular) is the knee joint; the small joints of the hands are very rarely involved. In type 2 arthropathy (polyarticular) the metacarpophalangeal joints are the most commonly involved. The peripheral arthritis associated with IBD is typically non-erosive and non-deforming in the long-term, although erosive disease has also been reported. In a series by McEwen et al, 52 patients were studied, out of which persistent joint damage was seen in 5 patients.[20] In CD septic arthritis of the hip has been reported which leads to rapid destruction of the joint and requires aggressive therapy.

ii. Relation of Peripheral Arthritis with Intestinal Symptoms

Typically, joint involvement occurs concurrently or after IBD has been diagnosed, although synovitis may predate the diagnosis of IBD in some patients. In the study by Orchard et al it was found that 31% of the type 1 arthropathy in UC and 24% in CD presented at or 3 years before the diagnosis. In contrast type 2 arthropathy was rarely apparent before diagnosis in both UC and CD.[6] Peripheral arthritis is most commonly seen in extensive disease in UC, but in CD the arthritis is more common in patients who have colonic involvement and less in small intestine lesion. In patients with ulcerative

Table 35.1: Clinical and radiological difference between idiopathic AS and IBD-related AS[17, 22–27]

Variable	*Idiopathic AS*	*IBD-related AS*
Age of onset	Usually in 3rd decade	Any age
Gender	More common in males	Both males and females
Axial disease	Severe	Less severe
Radiological changes	Squaring of vertebrae and Romanus lesions	Milder
HLA-B27	80–90%	50–70%

Table 35.2: Classification of peripheral arthropathy in enteropathic arthritis[6]

Clinical feature	*Type 1 (pauciarticular)*	*Type 2 (polyarticular)*
Number of joints	Less than 5 joints	Five or more joints
Onset and duration	Acute, self-limiting attacks (<10 weeks)	Symptoms usually persist for months to years
Type of joints	Most common knee	Most common metacarpophalangeal
Association with course of IBD	Often coincides with relapses of IBD	Runs a course independent of IBD
Extra-intestinal manifestation of IBD	Strongly associated with extra-intestinal manifestations of IBD	Associated with uveitis but not with other extra-intestinal manifestations
Association with HLA	Association with HLA-B27, B35 and DR 103	Association with HLA-B44
Antedates diagnosis of IBD	Can occur prior to diagnosis of IBD	Usually occurs along with or after IBD diagnosis

colitis, total colectomy may prevent further attacks of peripheral arthritis as it provides removal of all diseased bowel, therefore eliminating the arthritogenic stimulus. However, in Crohn's disease surgical removal of all diseased bowels is unusual.

OTHER MANIFESTATIONS

Clubbing of fingers, uveitis, and skin manifestations are seen in IBD, with higher frequency in Chron's disease (Table 35.3).[21] Uveitis in IBD is more often bilateral with a tendency for chronicity and poor response to topical steroid therapy.

MANAGEMENT

Management of peripheral arthritis generally depends upon controlling the underlying bowel disease; particularly since the natural course of type 1 arthropathy parallels that of the underlying bowel activity. Axial arthropathy, which is characterized by flares and remissions, the persisting disability of type 2 arthropathy and the failure of medical or surgical therapy of the underlying disease in altering the progressive nature of the axial arthropathy, make them more challenging to manage. Treatment of isolated CD, UC, axial SpA and peripheral SpA

Table 35.3: Extra intestinal manifestations of inflammatory bowel disease (IBD)[21]

Musculoskeletal manifestations	Pauciarticular arthritis Axial arthropathies and ankylosing spondylitis
Dermatological manifestations	Erythema nodosum Pyoderma gangrenosum Aphthous stomatitis/oral ulcerations Sweet syndrome
Ocular manifestations	Episcleritis Scleritis Uveitis
Hepatobiliary manifestations	Primary sclerosing cholangitis Gall stones Granulomatous hepatitis Portal vein thrombosis
Pulmonary manifestations	Chronic bronchitis Bronchiectasis
Metabolic manifestations	Osteopenia/Osteoporosis Osteomalacia

are defined. However, treatment of arthritis in IBD is patient oriented.

1. Non-steroidal Anti-inflammatory Drugs (NSAIDs)

Most patients with mild spondyloarthropathy will respond to NSAIDS. Although these drugs are known to increase gut permeability and cause inflammation of the small intestine (NSAIDs enteropathy),[28] it has been observed that gut inflammation in spondyloarthropathies resolve in some patients who continue to take NSAIDs. NSAIDs enteropathy is mainly asymptomatic, localized to the mid small intestine rather than the ileum and rarely involve colon.

2. DMARDs

Sulfasalazine (SSZ)

This drug inhibits the function of nuclear factor kappa B (NF-κB) that influences the production of pro-inflammatory cytokines. Sulfasalazine is of proven value in the induction and maintenance of remission of UC in particular. Its role in remission maintenance in CD is less well-defined. The beneficial effect of SSZ on peripheral arthritis in IBD is probably because of the effect of the 5-aminosalicylate (5-ASA) moiety of the drug on reducing gut inflammation. SSZ also reduces peripheral joint disease activity in idiopathic AS, but it has only a modest effect, if any, on the axial disease.[29] Recently a multicentric randomized controlled study on the role of SSZ in the treatment of undifferentiated SpA and early AS showed that SSZ was no better than placebo for the treatment of the signs and symptoms of SpA however, SSZ was more effective than placebo in the subgroup of patients with IBP and no peripheral arthritis.[30]

Methotrexate (MTX)

MTX has shown utility in CD but efficacy in UC for gut inflammation is not established; further evaluation is ongoing in two randomized multicenter trials (Meteor in Europe, and Merit in US) for the efficacy of MTX in UC.[31] MTX may prove beneficial for arthropathy in IBD as well as for IBD itself. Hydroxycholorquine and azathioprine are not effective for arthritis in IBD. Other peripheral musculoskeletal manifestations, such as enthesitis or dactylitis, are also indications for DMARD therapy.

3. Anti-tumour Necrosis Factor (Anti-TNF) Therapy

TNF inhibitors are highly effective for IBD patients who are steroid dependent or refractory to conventional treatment and have been shown to produce good response in musculoskeletal manifestations as well.[32] They have revolutionized the treatment of arthropathy in IBD.

Infliximab is highly effective in moderate to severe CD and UC, promoting fistula closure, mucosal healing, and sparing use of steroids. Initially given for refractory CD, it was also found to have significant improvement in joint symptoms.[33] Prolonged response of axial and peripheral joint symptoms as well as remission maintenance in inactive CD was observed.[34] Infliximab is effective for axial and peripheral joint manifestations of IBD.[35]

Adalimumab is also effective for AS and CD. Results from the CARE study demonstrated the efficacy of adalimumab in treatment of the extra intestinal manifestations (EIMs) of CD.[36] Care was a large multi-centre phase IIIb open-label clinical trial of 945 CD patients conducted in Europe. There was a benefit for arthropathy where incidence of arthritis reduced from 8.7% at baseline (82 patients) to 2.1% (20 patients) at week 20. A reduction in arthralgia and sacroiliitis was also seen. There was no effect seen for AS in this study, but the incidence of AS in this population of patients with IBD was very low.[36] In another study adalimumab was again found to be effective in reducing EIM of CD.[37]

Etanercept is very effective in idiopathic AS, it is also effective for arthritis and axial disease of Crohn's related SpA but it has no effect on the colitis.[38,39] Moreover recent data show development of new IBD among patients treated with etanercept.[40] Thus, use of etarnercept is not encouraged in IBD and enteropathic arthritis.

Ustekinumab a fully human monoclonal immunoglobulin (IgG1) against the interleukin (IL)-12/23 shared P40 subunit has been shown in phase 2 studies to have clinical effectiveness in CD gut inflammation, and may be particularly beneficial in patients refractory to infliximab.[41] However, their utility in IBD associated arthropathy is yet to be determined.

WHIPPLE'S DISEASE

This is a rare multisystem disease caused by the bacteria *Trophyrema whipplei*. In its classic form Whipple's disease presents with weight loss, diarrhoea, fever, joint pain, neurological disease, adenopathy and uveitis.[42] It may also present as a culture negative endocarditis. Joint involvement occurs in up to 90% patients with Whipple's disease.[43] Joint involvement occurs in the form of intermittent migratory polyarthralgia of the large joints and it is characteristically non-erosive. The articular symptoms are episodic and often followed by periods of remission. Oligoarticular involvement may also occur. The joint involvement is often accompanied by intestinal symptoms and isolated presentation is uncommon. Whipple's disease should be considered in the differential diagnosis of seronegative polyarthralgia in a male patient. Axial involvement with sacroiliitis and lumbar pain may also be seen.[44] Earlier methods of diagnosis relied solely on the detection of diastase resistant, PAS positive, foamy macrophages in biopsy tissue. Diagnosis is now being complemented by the use of PCR to identify the organism in involved tissue like duodenal biopsy.[45] Immunohistological methods to detect *T. whipplei* in involved tissue are currently being developed and will probably have wide applications.

The disease being rare (only 1000 odd cases being reported so far) there are no randomized trials of antibiotic therapy. At presentation patients should be treated with intravenous antibiotics, which cross the blood-brain barrier like ceftriaxone. Cotrimoxazole is the oral antibiotic of choice, because it crosses the uninflammed blood-brain barrier and its administration is observed to be associated with less likelihood of neurological relapse. Patients are usually treated for 12 to 24 months with close clinical, histological and laboratory monitoring by PCR. For patients who do not have neurologic involvement the combination of doxycycline 200 mg per day and hydroxychloroquine 200 mg three times a day works as bactericidal agents. Interferon gamma can be offered to patients with resistant neurological disease.

CELIAC DISEASE

Classically celiac disease is a gluten sensitive enteropathy that occurs in genetically predisposed individuals and responds to the withdrawal of gluten from the diet. The human major histocompatibility molecules DQ2 and DQ8 are essential genetic factors for the development of celiac disease with majority of the patients carrying DQ2 alleles. The triggering antigen is wheat gliadin, which gets deaminidated by tissue transglutaminase. The deaminidated gliadin then binds to the DQ2 molecule of the antigen-presenting cell causing an immune reaction in the lamina propria. The clinical manifestations of celiac disease are protean and it has been called "the great modern-day imposter". Adults present with gastrointestinal symptoms including diarrhoea, weight loss, and abdominal discomfort. The extra-intestinal features results from nutrient malabsorption and include anaemia, hypocalcaemia, osteopaenia, infertility, and peripheral neuropathy. Inflammatory arthritis is a well described. The major pattern of arthritis described is large joint, seronegative, non-erosive, non-deforming oligo or polyarticular type. In some patients arthritis may be the first presenting complaint and may occur in the absence of bowel disease.[46–48] Axial involvement in the form of sacroiliitis has been found in 14 out of 22 adult celiac disease patients studied by bone scintigraphy. Diagnosis is suggested by typical history supported by the presence of IgA anti-endomysial or IgA anti-tissue transglutaminase antibodies along with small intestinal histology, which shows villous atrophy and intraepithelial lymphocytosis. Gluten free diet is helpful in relieving the arthritis.[49]

COLLAGENOUS COLITIS

Collagenous colitis (CC) is a condition characterized histologically by a diffuse thickening (>10 mm) of the subepithelial collagen layer of the colon, along with intraepithelial lymphocytosis and infiltration of the lamina propria with inflammatory cells. It is an inflammatory condition affecting mainly middle aged or elderly females and presents with watery diarrhoea, crampy abdominal pain, weight loss and a normal colon by endoscopy.[50] Another related condition with a similar mode of presentation is lymphocytic colitis characterized by diffuse inflammation of the lamina propria and with intraepithelial lymphocytic inflammation. The name microscopic colitis has been given to include both collagenous colitis and lymphocytic colitis.[51] Among the various rheumatological conditions (autoimmune thyroid disease, idiopathic uveitis, polymyalgia rheumatica, giant cell arteritis, myasthenia gravis, scleroderma, Sjögren's syndrome, idiopathic pulmonary fibrosis and lupus), collagenous colitis is most commonly associated with rheumatoid arthritis.[52, 53] Most patients are seropositive and have an erosive disease. Rheumatoid arthritis usually precedes the intestinal symptoms by many years and occurs independent of the exacerbation of the arthritis. Therefore, collagenous colitis should be considered in the differential diagnosis of watery diarrhoea in a patient with rheumatoid arthritis. Collagenous colitis has also been reported to occur in patients with SpA with axial involvement.[54] Peripheral joint involvement is either oligoarthritis or polyarthritis of both large and small joints and not necessarily with lower limb predilection. Joint involvement may antedate or coincide with the onset of bowel disease in half of the patients. Treatment consists of oral glucocorticoids, sulfasalazine, NSAIDs and anti-diarrhoeals.

ARTHROPATHY ASSOCIATED WITH INTESTINAL BYPASS SURGERY

The surgical management of morbid obesity by the induction of an intestinal bypass (to reduce the intestinal mucosal surface) is associated with a large number of metabolic and inflammatory complications. Bypass enteropathy and arthropathy are the major inflammatory complications. Arthropathy occurs in 35 to 52% of patients post-operatively and seems to occur more in jejuno-colic bypass than in jejunoileal bypass. Delamere *et al* found that 5 out of 9 subjects in their study who underwent jejuno-colic bypass had arthropathic symptoms.[55] This could be due to bacterial contamination and overgrowth that occurs in the blind loop.[56] Increased levels of complement-containing cryoprecipitates and circulating complexes have been demonstrated in patients who develop joint symptoms after bypass procedures.[56,57] The arthropathy is episodic, migratory non-erosive, involving both large and small joints and reversal of the bypass is associated with complete and permanent remission of the arthritis.

Conclusion

Arthritis, being the most common extra-intestinal manifestation of IBD, needs co-operation between gastroenterologist and rheumatologist in the management of disease as a whole. Corticosteroids, DMARDs and TNF inhibitors have shown efficacy in the treatment of enteropathic arthritis.

Key Points

1. Axial involvement in IBD manifests as either isolated radiographic sacroiliitis, or ankylosing spondylitis (AS). Six to 20% of patients with IBD have axial disease. Symptoms of IBP are indistinguishable from that of idiopathic AS, however it occurs at any age and there is no significant gender difference. Axial disease occurs independent of bowel disease.
2. Peripheral arthritis occurs in 5–20% of IBD. Peripheral joint involvement is divided into type 1 and 2 arthropathy according to number of joints involved, duration of symptoms, relation with gut symptoms and genetic predisposition.
3. Corticosteroids are the mainstay of treatment. NSAIDs, though effective for enteropathic arthritis can worsen IBD.
4. SSZ is widely used and is effective for both the gut and joints.
5. Infliximab and adalimumab are effective in causing resolution of enteropathic arthritis (both axial and peripheral arthritis).
6. Etanercept though effective in idiopathic AS disease is not effective for IBD. New data show that treatment with etanercept (particularly in spondyloarthritis) is associated with the development of IBD- "paradoxical IBD".
7. Newer biologics targeting IL-12/23 Ustekinumab is showing promising results for enteropathic arthritis.

REFERENCES

1. Karreman M, Luime J, Hazes J, Weel A. THU0404 The Prevalence of Axial and Peripheral Spondyloarthritis in Inflammatory Bowel Disease: A Systematic Review & Meta-Analysis. Annals Rheumatic Dis. 2016; 75(Suppl 2): 334.
2. Bandyopadhyay D, Bandyopadhyay S, Ghosh P, De A, Bhattacharya A, Dhali G et al. Extra-intestinal manifestations in inflammatory bowel disease: Prevalence and predictors in Indian patients. Indian J Gastroenterol. 2015; 34:387–94.
3. Shivashankar R, Loftus E, Tremaine W, Bongartz T, Harmsen W, Zinsmeister A et al. Incidence of Spondyloarthropathy in Patients with Crohn's Disease: A Population-based Study. J Rheumatol. 2012; 39:2148–52.

53. Bennuci M, Bardazzi G, Magaro L, Li Gobbi F, Mannoni A, Serni U. A case report of a man with rheumatoid factor positive rheumatoid arthritis associated with collagenous colitis [letter]. Clin Exp Rheumatol 2001; 19:475.

54. Narvaez J, Montala N, Busquets-Perez N, Nolla JM, Valverde J. Collagenous colitis and spondyloarthropathy. Arthritis Rheum 2006; 55: 507–12.

55. Delamere JP, Baddeley RM, Walton KW. Jejuno-ileal bypass arthropathy: its clinical features and associations. Ann Rheum Dis 1983; 42:553–57.

56. Fisch C, Schiller P, Harr T, Maclachlan D. First presentation of intestinal bypass syndrome 18 yr after initial surgery. Rheumatology (Oxford) 2001; 40:351.

57. Rose E, Espinoza LR, Osterland CK. Intestinal bypass arthritis with circulating immune complexes and HLA-B27. J Rheumatol 1977; 4:129–34.

Chapter

36

Reactive Arthritis

Madhuri HR, Damodaram P, Narsimulu G

INTRODUCTION

Reactive arthritis (ReA) is defined as a sterile arthritis following a genitourinary or gastro-intestinal infection. It is to be differentiated from post-infectious arthritis by the features it shares with other spondyloarthropathies (SpA) and its extra articular manifestations.[1] It is now classified along with SpA.

HISTORY

A case series of 5 patients with the triad of arthritis, urethritis and conjunctivitis was reported in the 19th century. Although Reiter's syndrome was a popular eponym, the description was actually initially published by Fiessinger and Edgar LeRoy. Reiter had also mistakenly attributed the cause of urethritis and conjunctivitis to a spirochaetal organism.[2] Due to evidence of Reiter's participation in various Nazi activities in the Second World War, the term Reiter's syndrome was later removed from the medical literature and it is now renamed reactive arthritis (ReA).

EPIDEMIOLOGY

It is common in young adults between the ages of 20 to 40 years. Both sexes are equally affected when ReA follows gastrointestinal infections. But there is a male predominance following Chlamydial infections. The incidence varies between 1.3 to 30/100,000 population depending on the geographic area, triggering infections and diagnostic methods used.[3]

PATHOGENESIS

ReA develops as a result of genetic and infectious agent interactions. While other types of arthritis are due to overactive immune response, in ReA there is a downregulation of cytokines like TNF-α and Interferon γ (IFNγ). This cytokine profile has been proven in experimentally induced ReA.

In chlamydial infections, a type I immune response leads to increase in IFN-γ levels and help in killing the phagocytosed bacteria. A dominant Type II immune response predisposes to susceptibility to infection and persistence of chronic infection.

Macrophages are antigen presenting cells responsible for the innate immune response. Depending on the differentiation of T cells profile they cause, macrophages are divided into M1 (causing Th1) or M2 macrophages (Th2 profile). M1 macrophages are required for control of intracellular pathogens.[4] In SpA dominance of M2 macrophages has been described.[5]

The classical pathogens which have been described to be trigger ReA include gastrointestinal infections with *Shigella flexneri*, *Campylobacter jejuni*, *Campylobacter coli*, *Yersinia enterocolitica* and *Y. pseudotuberculosis*; genitourinary infections with Chlamydia trachomatis, and respiratory infections with *Chlamydia pneumoniae*. Other organisms which have also been reported to cause ReA include *Ureaplasma urealyticum*, *Clostridium difficile*, *Salmonella typhimurium*, *Salmonella enteriditis*, *Mycobacterium bovis* (BCG), *Neisseria gonorrhoea*, enterotoxigenic *E. coli* , *Mycoplasma genitalium* and many others.[6]

Organisms implicated in ReA are usually gram-negative and intracellular organisms and contain lipopolysaccharide (LPS) in the cell membranes. In genetically predisposed individuals, there is persistence of the organism in dormant state in the synovial cavity. These persistent antigens lead to an immune response in the individual in the form of a sterile arthritis. This is well-described in chlamydial infection, where the organism persists in a dormant state in the synovial cavity. The organism can be identified accurately by PCR. Some characteristics of various serovars of Chlamydia such as ability to survive in hypoxic environments, variations in the outer membrane proteins also dictate the duration of infection. Other factors are heavy metal sensitisation, role of repeated infections, and host genetic variability like polymorphisms in TLR2, deficiency of IL-12 or IFN-γ, or deficiency of type 2 cytokines, such as IL-10, influences susceptibility or resistance to infection.[7] High synovial IFN-γ and TNF-α levels correlate with resistance. HLA-B27 is a risk factor for the occurrence of severe arthritis, prolonged course and extra-articular manifestations.

CLINICAL FEATURES

After 1 to 4 weeks of a gastrointestinal or genito-urinary infection clinical symptoms appear in around 7 to 12% of affected individuals.

Articular Manifestations

The most common manifestation is an acute mono or oligoarthritis, predominantly affecting the lower limbs although polyarthritis and upper limb involvement can also occur. Enthesitis and dactylitis may also occur. Inflammatory back pain is found in around 30% of the individuals.

Extra-articular Manifestations

Ocular manifestations are more common with chlamydial ReA than post-dysentery ReA. Ocular manifestations include conjunctivitis, acute anterior uveitis. Conjunctivitis seen in 30 to 60% of patients is usually bilateral, occurs early in the course of the disease, but typically after urethritis. Conjunctivitis is more common with sexually acquired and after shigella infection than with other post-dysentery ReA.[8] There is male to female ratio of 1.3:1. Uveitis is typically characterized by acute unilateral attacks of inflammation of anterior chamber (seen in 5% of patients) presenting as ocular pain, erythema, and photophobia. Chronic inflammation of the eye can lead to progressive intraocular damage and visual loss.

Cutaneous Manifestations

Cutaneous manifestations in ReA occur in up to 50% of patients. Keratoderma blenorrhagicum is a hyperkeratotic lesion usually on the soles or palms, seen in about 25% of affected men. It begins as a clear vesicle on an erythematous base, developing into nodules. It eventually crusts over, and adjacent lesions coalesce.[9] These cutaneous lesions are difficult to distinguish from pustular psoriasis, both clinically and histologically. Nail changes similar to those of psoriasis are seen in up to 15% of the patients with ReA.

Circinate balanitis is a painless shallow erythematous ulcer of the glans penis, seen in up to 25% of men. In circumcised men it is dry, plaque like, and hyperkeratotic, resembling psoriasis or keratoderma. In uncircumcised men it is typically moist, shallow ulcer surrounding the meatus. Superficial, usually painless oral ulcer occurs in approximately 5–15% of the patients. They are typically located in the hard palate or tongue but may also be found on the soft palate, gingival and cheeks.[10]

Cardiac involvement in the form of aortic regurgitation, AV nodal block and pericarditis can occur in up to 10% of affected individuals.[11]

INVESTIGATIONS

Investigations show an elevated ESR and CRP, thrombocytosis, normal or elevated WBC count. There may be mild transaminitis. There are no diagnostic tests for ReA. Synovial fluid may be examined for crystal in cases of suspicion of gout and for microbiological culture in case of suspicion of septic arthritis. The culture is typically sterile in cases of ReA.

In every case an effort to identify the causative organism must be made. The organism can be isolated in only about 60% of the cases. Stool or urine culture may be required. For chlamydial infection nucleic acid amplification technique (PCR) of early morning urine sample is the preferred method. Other sample may be urogenital swab testing but urine sample is the preferred and easier method. Serological testing by enzyme immunoassay, complement fixation and micro-immunofluorescence is useful only in cases of upper genital tract infections by *Chlamydia trachomatis*. PCR is also useful in cases of respiratory infections by Chlamydia.

In enteric infections, stool culture is usually unyielding as the patient has already recovered from the infection and the organism is no longer identifiable from the stool sample. However, Salmonella and Yersinia may persist in the gut for weeks after the symptoms have resolved. Special techniques like preserving the sample at low temperatures of 4–8°C before culture help improve the yield of Yersinia infections. Serological tests to identify Salmonella (Widal test), Campylobacter and Yersinia infections are available, but only rising titres can be considered significant as there is substantial variation in the titres in normal population based on the geographic area and methods used. Many infections are asymptomatic leading to difficulty in diagnosis.

ROLE OF HLA B27 TESTING

Testing for HLA-B27 is not required for diagnosis but its presence along with other features like elevated inflammatory markers, metatarso-phalangeal joint involvement and genitourinary involvement could predict ReA with a sensitivity of 70% and specificity of 93%.[6] HLA-B27 positivity is seen in 60–80% cases in hospital based surveys, though the incidence in the community and in milder cases may not be as high. HLA B27 positivity predicts severe, prolonged course and higher incidence of extra-articular symptoms.

IMAGING

Radiographs are not useful in the diagnosis of ReA. Newer imaging modalities like ultrasonography and Power Doppler detect enthesitis, bursitis and subclinical synovitis. MRI of the sacroiliac joints may show inflammation or spinal imaging may show axial involvement as in SpA.[12]

DIAGNOSIS AND CLASSIFICATION CRITERIA

The diagnosis is mainly clinical. A history of a preceding gastrointestinal or genitourinary infection within the last 1–4 weeks, followed by an acute inflammatory mono or oligoarthritis with or without other extra-articular manifestations is the common clinical scenario of ReA.

In 1995 at the Third International Workshop on reactive arthritis in Berlin diagnostic criteria for ReA (Table 36.1) were developed.[13] The criteria include typical peripheral arthritis, which is predominantly lower, asymmetric oligoarthritis plus evidence of preceding infection. Where there is clear urethritis or diarrhoea in the preceding 4 weeks laboratory confirmation was not essential, but when clear infection is not there then there having to be laboratory evidence of prior infection such as a positive stool culture, urethral culture or by PCR or serology. HLA-B27 was not essential for diagnosis. Other conditions causing asymmetric arthritis such as gout should be ruled out. A history of preceding infection, in the last 1 day to a maximum of 4 weeks, is considered to be most relevant for diagnosis. A diagnosis of ReA does not require the presence of HLA-B27 or extra-articular features.

The 1995 criteria were updated in 1999 at the Fourth International Workshop on Reactive Arthritis (Table 36.2).[14] It was agreed that a history of a preceding symptomatic infection is thought to be most relevant for a diagnosis of ReA. The minimal interval between preceding symptoms and arthritis is proposed to be 1–7 days, maximally 4 weeks. To diagnose definite ReA there has to be asymmetric lower limb oligoarthritis or monoarthritis, and preceding symptomatic infection, and

Table 36.1: Diagnostic criteria for reactive arthritis proposed at the Third International Workshop on reactive arthritis at Berlin, 1995

Peripheral arthritis
Predominantly lower limb, asymmetric oligoarthritis
Plus

Evidence of preceding infection
Where there is clear diarrhoea or urethritis in the preceding 4 weeks, no laboratory confirmation is required

Where there is no clear evidence of infection in prior four weeks laboratory confirmation is desirable

Exclusion criteria
Other causes of monoarthritis or oligoarthritis should be excluded.

Table 36.2: Diagnostic criteria for Reactive Arthritis proposed at the Fourth International Workshop on Reactive Arthritis in Berlin, 1999

Major criteria
1. Arthritis, with two of three of the following findings: • Asymmetric • Monoarthritis or oligoarthritis • Affecting predominantly lower limbs
2. Preceding symptomatic infection, with one or two of the following findings: • Enteritis (diarrhoea for at least 1 day, 3 days to 6 weeks before the onset of arthritis) • Urethritis (dysuria or discharge for at least 1 day, 3 days to 6 weeks before the onset of arthritis)
Minor criteria (at least one of the following)
1. Evidence of triggering infection Positive nucleic acid amplification test in the morning urine or urethral or cervical swab for *Chlamydia trachomatis* Positive stool culture for enteric pathogens associated with reactive arthritis
2. Evidence of persistent synovial infection (positive result on immunohistologic analysis or polymerase chain reaction assay for Chlamydia)
Definition of reactive arthritis *Definite reactive arthritis: Both major criteria and one relevant minor criterion* *Probable reactive arthritis: Both major criteria but no relevant minor criteria or* *Major criteria and one or more minor criteria* *Exclusion criteria: Other causes for acute arthritis*

there should be a lab evidence of triggering infection or evidence of synovial persistence of the microbe.

A search for Chlamydia in urine/urethra/cervix is recommended, while in the case of diarrhoea enterobacteria should be searched for in stool and antibodies against them in serum. A differentiation between acute and chronic ReA with a cut off of 6 months was recommended.[14]

In this meeting the term "reactive arthritis" is used only if the clinical picture and the microbes involved are HLA-B27 and spondyloarthropathy (SpA) associated, whereas the term "infection related arthritis" is used for all other arthritides related to or associated with infections. Presence of SpA related symptoms may contribute to the diagnosis.

NATURAL HISTORY

The average duration of ReA is 3 months. It has a self-limiting course. Roughly 15% patients will progress to chronic arthritis or develop into SpA.[16,17] Factors which determine the progression to chronic arthritis or SpA are presence of HLA B27.[17] Patients who are B27 positive also have much more severe acute disease.[16] The prognosis of ReA triggered by Yersinia, Salmonella, and Shigella are well-studied.[17,18] A 10-year follow-up study of 85 patients with Yersinia arthritis showed that one third patients had radiologic evidence of sacroiliitis.[17] In a study from Finland, 8 of 50 (16%) patients with Salmonella arthritis developed SpA. The follow-up period was 5–22 years with mean of 11 years.[18]

TREATMENT

Most of the cases of ReA are self-limited. Symptomatic relief with NSAIDs is sufficient in most of the cases. In case of active disease in spite of NSAIDs intra-articular glucocorticoids can be given in case of mono or oligoarthritis.

Disease modifying antirheumatic drugs (DMARDs) are useful in patients with chronic arthritis, HLA-B27 positivity, severe arthritis, and polyarthritis. Sulfasalazine is effective in inducing remission in peripheral arthritis.[19]

Acute anterior uveitis requires treatment with topical glucocorticoids and mydriatics. Some cases may progress to chronic uveitis. Keratoderma and pustular lesions are treated with topical glucocorticoids. In severe cases, methotrexate maybe used and may have some benefit for more prolonged arthritis. Circinate balanitis and oral mucosal lesions usually respond to mild topical glucocorticoids.

Role of Biologics

As most of the patients respond to NSAIDs, intra-articular steroids or sulfasalazine, the evidence regarding use of biologics is limited to case reports. The TNF inhibitors infliximab, etanercept and adalimumab have been used in the treatment of refractory ReA.[20] Response has been noted in the tender joint count, swollen joint count as well as extra-articular manifestations. Synovial biopsy specimens have shown partial improvement after treatment.

Role of Antibiotics

As ReA is by definition a sterile arthritis, there appears to be no role for antibiotics in the management. But since Chlamydia infection is known to persist in the synovium, this has led to trials of prolonged courses of antibiotics for treatment of Chlamydia induced ReA. A double blind RCT showed that 17 of 27 patients (63%) patients with Chlamydia induced ReA receiving combination antibiotics improved compared to 6/15 patients (20%) receiving placebo.[21] However, a meta-analysis concluded that the role of antibiotics in ReA is uncertain.[22]

Conclusion

ReA is a sterile arthritis occurring 1–3 weeks after an infection, common in young adults, presents as asymmetric oligoarthritis, mostly of lower limbs. Extra-articular manifestations like uveitis, cutaneous lesions and cardiac involvement may occur. Most cases are self-limited though some may become chronic. Treatment is by NSAIDs, intra-articular corticosteroids and sulfasalazine in chronic cases. Biologicals have been used in chronic arthritis or extra-articular manifestations.

Key Points

- Reactive arthritis (ReA) is a sterile joint inflammation that develops after a distant infection.
- ReA should be a part of the differential diagnosis of undifferentiated oligoarthritis or monoarthritis
- Musculoskeletal symptoms usually begin 2 to 4 weeks following a gastrointestinal or genitourinary infection.
- Chronic inflammation of the eye can lead to progressive intraocular damage and visual loss.
- Urine analysis should be carried out at diagnosis and should be repeated during the follow-up, in order to detect possible aseptic pyuria resulting from urethritis.
- Joint fluid should always be aspirated when possible. Gram-stain and bacterial culture should be performed to exclude septic arthritis.
- Hip arthritis, HLA-B27, and high ESR are associated with a more chronic course.
- Sulphasalazine can be given in chronic ReA in patients with peripheral joint involvement.

REFERENCES

1. Colmegna I, Cuchacovich R, Espinoza L. HLA-B27-Associated Reactive Arthritis: Pathogenetic and Clinical Considerations. Clin Microbiology Rev 2004; 17:348–69.
2. Keynan Y, Rimar D. Reactive arthriti—the appropriate name. Isr Med Assoc J 2008; 10: 256–58.
3. Kvien TK, Glennas A, Melby K, Granfors K, Andrup O, Karstensen B, et al. Reactive arthritis: incidence, triggering agents and clinical presentation. J Rheumatol 1994; 21:115–22.
4. Morrison SG, Su H, Caldwell HD, Morrison RP. Immunity to murine *Chlamydia trachomatis* genital tract reinfection involves B cells and CD4+ T cells but not CD8+ T cells. Infect Immun 2000; 68: 6979–87.
5. Baeten D, Kruithof E, De Rycke L, Boots AM, Mielants H, Veys EM, et al. Infiltration of the synovial membrane with macrophage subsets and polymorphonuclear cells reflects global disease activity in spondyloarthropathy. Arthritis Res Ther 2005; 7: 359–69.
6. Petersel DL, Sigel LH. Reactive arthritis. Infect Dis Clin North Am 2005; 19:863–89.
7. Rottenberg ME, Gigliotti-Rothfuchs A, Wigzell H. The role of IFNγ in the outcome of chlamydial infection. Curr Opin Immunol 2002; 14:444–51.
8. Monnet D, Braban M, Hudry C, Dougados M, Brézin AP. Ophthalmic findings and frequency of extra-articular manifestations in patients with HLA-B27 uveitis, a study of 175 cases. Ophthalmology 2004; 111:802–9.
9. Hill Gastion JS, Lillicarp MS. Arthritis and Enteric infection. Best Pract Res Clin Rheumatol 2003; 17: 219–39.
10. Arnette FC. Seronegative Spondyloarthropathies. Bull Rheum Dis. 1987; 37:1–12.

11. Selmi C, Gershwin M. Diagnosis and classification of reactive arthritis. Autoimmun Rev. 2014; 13: 546–49.
12. Carter JD, Hudson AP. Reactive arthritis: clinical aspects and medical management. Rheum Dis Clin North Am. 2009; 35:21–44.
13. Kingsley G, Sieper J. Third International Workshop on Reactive Arthritis. 23–26 September 1995, Berlin, Germany. Report and abstracts. Ann Rheum Dis. 1996; 55:564–84.
14. Braun J, Kingsley G, van der Heijde D, Sieper J. On the difficulties of establishing a consensus on the definition of and diagnostic investigations for reactive arthritis. Results and discussion of a questionnaire prepared for the 4th International Workshop on Reactive Arthritis, Berlin, Germany, July 3–6, 1999. J Rheumatol. 2000; 27:2185–92.
15. Kvien T, Glenna S A, Melby K. Prediction of diagnosis in acute and subacute oligoarthritis of unknown origin. Rheumatology. 1996; 35:359–63.
16. Leirisalo M, Skylv G, Kousa M, Voipio-Pulkki L, Suoranta H, Nissilä M, et al. Follow-up study on patients with Reiter's disease and reactive arthritis, with special reference to HLA-B27. Arthritis Rheum. 1982; 25:249–59.
17. Leirisalo-Repo M, Suoranta H. Ten-year follow-up study of patients with Yersinia arthritis. Arthritis Rheum. 1988; 31:533–7.
18. Leirisalo-Repo M, Helenius P, Hannu T, Lehtinen A, Kreula J, Taavitsainen M, et al. Long-term prognosis of reactive salmonella arthritis. Ann Rheum Dis. 1997; 56:516–20.
19. Clegg D, Reda D, Abdellatif M. Comparison of sulfasalazine and placebo for the treatment of axial and peripheral articular manifestations of the seronegative spondyloarthropathies: A Department of Veterans Affairs co-operative study. Arthritis Rheum 1999; 42:2325–29.
20. Meyer A, Chatelus E, Wendling D, Berthelot JM, Dernis E, Houvenagel E, et al: Club Rhumatisme et inflammation. Safety and efficacy of anti-tumour necrosis factor? therapy in ten patients with recent-onset refractory reactive arthritis. Arthritis Rheum 2011; 63:1274–80.
21. Carter JD, Espinoza LR, Inman RD, Sneed KB, Ricca LR, Vasey FB, et al. Combination antibiotics as a treatment for chronic Chlamydia-induced reactive arthritis: a double-blind, placebo-controlled, prospective trial. Arthritis Rheum. 2010; 62:1298–307.
22. Barber C, Kim J, Inman RD, Esdaile J, James MT. Antibiotics for treatment of reactive arthritis: a systematic review and metaanalysis. J Rheumatol 2013; 40:916–28.

FURTHER READING

1. Colmegna I, Cuchacovich R, Espinoza LR. HLA-B27-associated reactive arthritis: pathogenetic and clinical considerations. Clin Microbiol Rev. 2004; 17:348–69.
2. Morris D, Inman RD. Reactive arthritis: developments and challenges in diagnosis and treatment. Curr Rheumatol Rep. 2012; 14:390–4.
3. Misra R, Gupta L. Epidemiology: Time to revisit the concept of reactive arthritis. Nat Rev Rheumatol. 2017; 13:327–28.

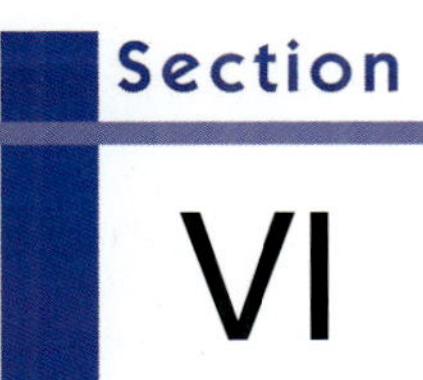

CONNECTIVE TISSUE DISEASES

Chapter

37

Clinical Manifestations of Systemic Lupus Erythematosus

Sukumar Mukherjee, Somnath Bhar

INTRODUCTION

Systemic Lupus Erythematosus (SLE) is a multisystem prototype autoimmune heterogenous illness characterized by myriad of systemic features primarily due to immune dysregulation at multiple levels of the immune cascade with hyperactivation of B cell activity and diminished T cell suppressor activity leading to increased tissue specific and nontissue specific circulatory antibodies. Having highly variable features like constitutional symptoms, glomerulonephritis, neuropsychiatric disease and cutaneous manifestation SLE is hardly curable, but most of the patients experience remission and their survival has improved over the years.[1]

EPIDEMIOLOGY

The annual incidence of SLE in US and Europe has been estimated to be 1 to 23 per 100,000 per year. In Europe, the prevalence of the disease generally ranges between 20 and 50 cases per 100,000 people.[2] Prevalence rate is higher in Afro-Carribean population. However, low incidence of SLE is observed in West African countries. In a study near Delhi, the prevalence of SLE was found to be 3.2 per 100,000 population.[3] However, with a large population in India the total burden of SLE is considerable and worldwide more than 85% of SLE patients are female. The female to male ratio as studied by Pande et al. in 1993 in India is around 9:1.[4] The highest incidence of SLE is often seen in the premenopausal age. Sixty per cent of SLE patients have disease onset between 16 and 55 years of age, 20% present before the age of 16 and 15% after the age of 55.[2] Male lupus patients have more severe disease with higher mortality, older age of diagnosis, less photosensitivity and more serositis than females.

Clinical Features

The onset of SLE may be acute or insidious and any organ system can be involved. Fever, malar rash, arthralgias are common presenting features of SLE.[5] Fatigue is a clinical marker of active disease, the mechanism of which is unclear. The major organ involvement like renal and CNS tend to occur early in the natural course of disease and can be life-threatening. The natural history is characterized by relapse and remission; however, permanent complete remission is rare. The comparative data of frequency of occurrence of some common features of SLE is shown in Table 37.1.[6]

Mucocutaneous Involvement

The skin lesions in SLE are classified as specific and nonspecific (Table 37.2). Lupus specific skin lesions show histopathologic features of interface dermatitis. The butterfly facial erythematosus rash (confluent macular or papular erythema) (Fig. 37.1) sparing nasolabial furrow is characteristic. Moreover, the rash is also seen in the preauricular, auricular and retroauricular areas. The discoid lesion (Fig. 37.2), bullous lesions (Fig. 37.3), urticarial and vasculitic lesions have also been described in SLE. Discoid lupus are circular with erythematosus margin with surface scaring, follicular plugging. In

Table 37.1: Frequency of clinical features of SLE

Features	*Indian data (%)*	*Western data (%)*
Arthritis	72–92	86–94
Alopecia	52–80	50
Skin rash	74–90	60
Photosensitivity	10–62	33–62
Malar rash	37–76	72–90
Oral ulcers	41–61	30
Fever	74–91	80
Lymphadenopathy	26–47	50
Neuro-psychiatric	19–63	20–45
Renal	35–73	29–73
Cardiac	10–29	20–30
Pleuropulmonary	9–54	36–57

Table 37.2: Gilliam classification of skin lesions associated with lupus (partial representation)

A. Lupus erythematosus (LE) specific skin lesions	*B. LE-nonspecific skin lesions*
1. Acute cutaneous LE (ACLE)	1. Cutaneous vascular disease (vasculitis)
Localised ACLE	2. Nonscarring alopecia
Generalized ACLE	3. Sclerodactyly
2. Subacute cutaneous LE (SCLE)	4. Rheumatoid nodules
Annular SCLE	5. Calcinosis cutis
Papulosquamous SCLE	6. LE-nonspecific bullous lesions
3. Chronic cutaneous LE (CCLE)	7. Urticaria
Classical discoid LE (DLE)	8. Papulonodular mucinosis
Lichenoid DLE	9. Cutis laxa/anetoderma
	10. Acanthosis nigricans
	11. Erythema multiforme
	12. Leg ulcers
	13. Lichen planus

Source: Gilliam JN, Sontheimer RD. Distinctive cutaneous subsets in the spectrum of lupus erythematosus. J Am Acad Dermatol. 1981; 4:471–5

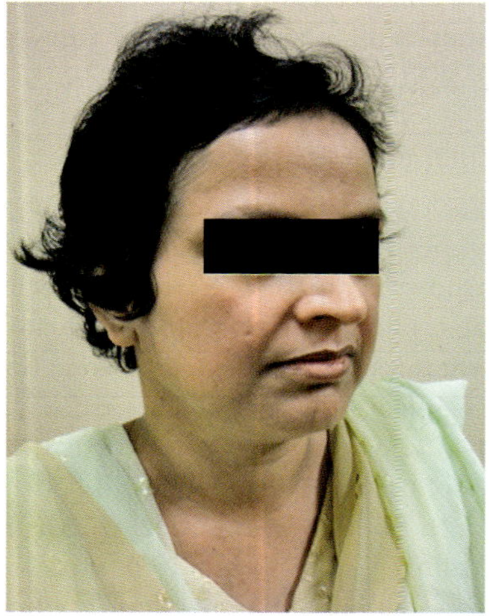

Fig. 37.1: Widespread facial rash

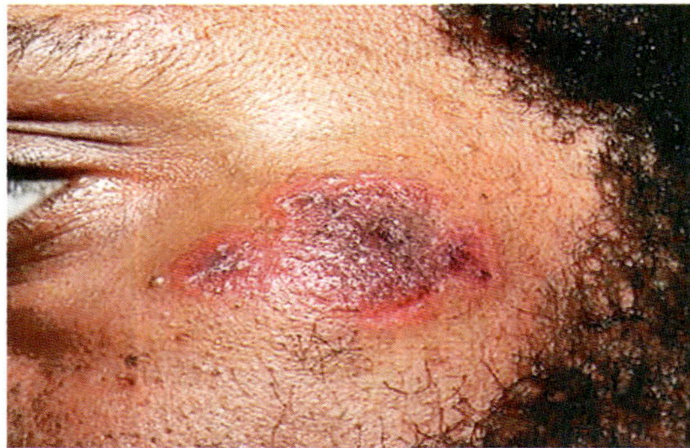

Fig. 37.2: Discoid lupus

antiphospholipid syndrome secondary to SLE, livedo reticularis is typical. Multiple small painless oral ulcers (Fig. 37.4) and diffuse alopecia

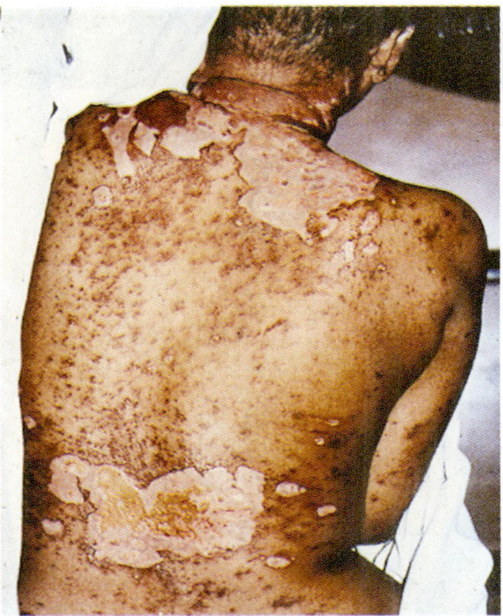

Fig. 37.3: Bullous eruptions in SLE

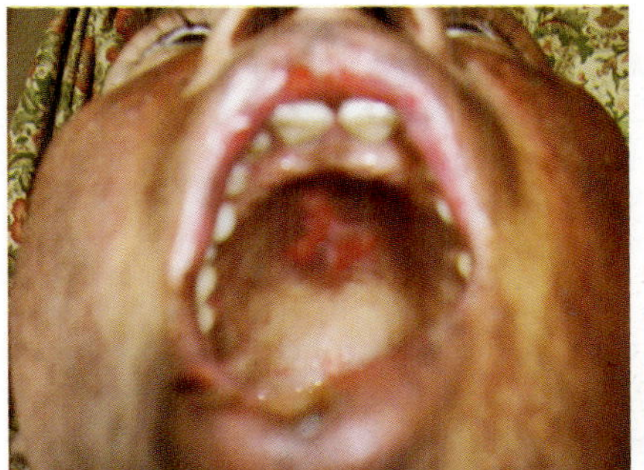

Fig. 37.4: Oral ulcer

are not uncommon. Involvement of the mucous membrane occur in 25–45% of SLE patients.[8] In the scalp scarring alopecia is irreversible and non-scarring alopecia hair can regrow. 'Lupus hair' is characterized by short irregular sized hair at the frontal hairline that easily fractures.[7] Subacute cutaneous lupus erythematosus (Fig. 37.5) is characterized by the presence of photosensitive, superficial non-indurated non-scaring, annular or papulosquamous psoriasis-like lesions. Photosensitivity occurs in 60–100% of SLE patients. Raynaud's phenomenon is mild and uncommon. In southern India Raynaud's phenomenon is rare, whereas lymphadenopathy is commonly encountered.[9] Sometimes diffuse hyperpigmentation, urticaria or diffuse maculopapular rash are also observed. Diffuse hand erythema is not uncommon (Fig. 37.6). Digital gangrene due

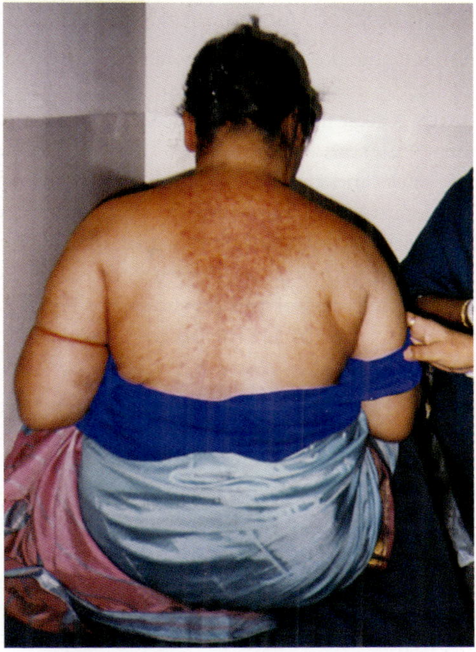

Fig. 37.5: Sub-acute cutaneous lupus (anti-Ro-positive)

to progressive vasculitis is also observed in a few patients with active SLE (Fig. 37.7).

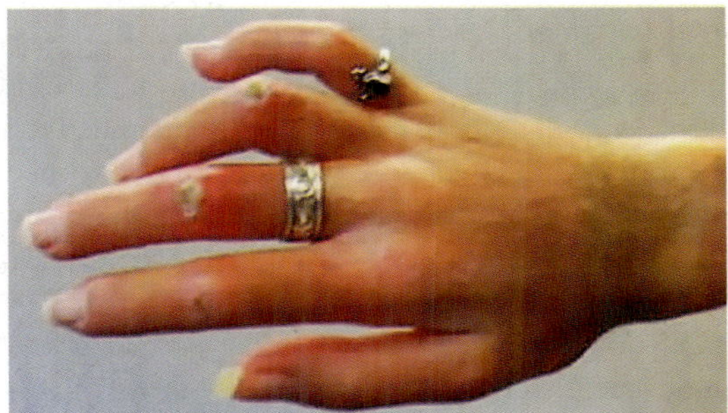

Fig. 37.6: Raynaud's phenomenon in SLE

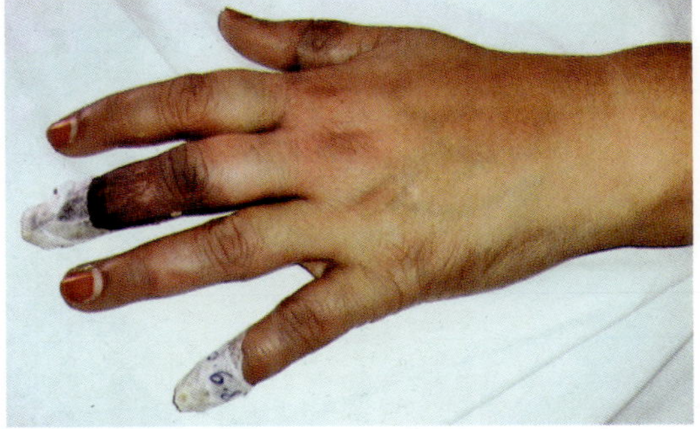

Fig. 37.7: Digital gangrene

Musculoskeletal Manifestation

Arthralgias are pretty common in 90% patients and may be the initial presentation; but non-erosive arthritis is seen in about 10% patients. Arthritis is polyarticular and symmetrical involving hands, wrist, knees and ankles. The synovitis is often mild to moderate with less joint effusion.[10] A few patients with erosive arthritis variety indistinguishable from rheumatoid arthritis are designated as Rhupus. Avascular necrosis of hip or shoulder (Fig. 37.8) and septic arthritis are suspected when there is persistent pain and swelling in a joint. The avascular necrosis is mostly attributed to antiphospholipid syndrome (APS), prolonged steroid therapy or immobilization. Osteoporosis can also occur in the long run and may either be attributed to disease activity or offending drugs like steroid. Tenosynovitis and tendon rupture can also occur. Myalgia, muscle weakness or tenderness are found in 60% patients, whereas inflammatory myositis complicates the course in 5–10% of patients. Muscle weakness, specifically proximal myopathy, is caused by steroid or chloroquine therapy or rarely due to myasthenia like syndrome. The muscle enzymes are rarely elevated in such situations.

Neurological Manifestations

The neurological involvement is widespread remains a challenge in terms of pathogenesis, assessment and treatment. In 1999, a consensus document defined the neurological and psychiatric manifestation in 19 different syndromes as given in Table 37.3.[7]

The most common manifestations of diffuse CNS lupus are cognitive dysfunction (20–30%) headache (20–40%) and seizures (7–10%).[11] There is a clear distinction between CNS manifestations due to active lupus, vasculitis and those due to APS. However, opportunistic CNS infections may complicate the clinical picture in an immunocompromised host like SLE. Vascular occlusion may present like a stroke in young with cerebral infarction (Fig. 37.9).

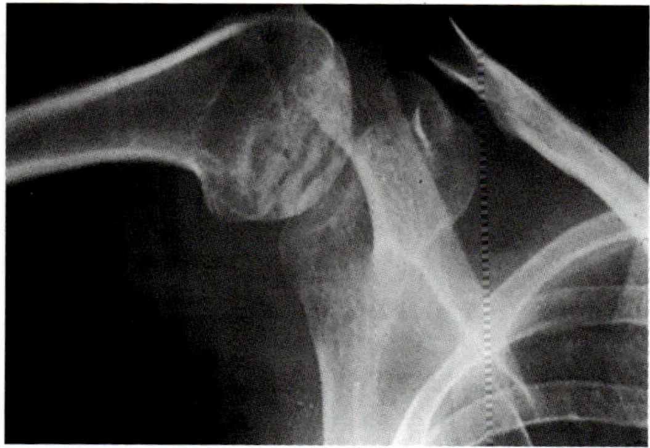

Fig. 37.8: Avascular necrosis in right shoulder

Table 37.3: Neuropsychiatric manifestation of SLE[11]

Central nervous system
- Aseptic meningitis
- Cerebrovascular disease
- Demyelinating syndrome
- Headache
- Movement disorder
- Myelopathy
- Seizure disorder
- Acute confusional state
- Anxiety disorder
- Cognitive dysfunction
- Mood disorder
- Psychosis

Peripheral nervous system
- Acute inflammatory demyelinating polyradiculopathy
- Autonomic disorder
- Mononeuropathy, single or multiplex
- Cranial neuropathy
- Myasthenia gravis
- Plexopathy
- Polyneuropathy

Overall, CNS disease is associated with significant morbidity and mortality in lupus. Primary neuropsychiatric disease, CNS infection and steroid psychosis are the differential diagnosis to be entertained in patients with cerebral lupus. Multiple peripheral nerves may simultaneously be involved in mononeuritis multiplex (Fig. 37.10).

Renal Manifestation

Nephritis is the most important complication of SLE and occurs mostly in the first 36 months of the disease.[7] The renal disease differs from patient to patient in clinical pattern, severity and prognosis. Ninety per cent of the lupus patients may have evidence of nephritis on renal biopsy but around 50% of the patients show clinical manifestation. The classification of lupus nephritis is essentially histology-based on light and electron microscopy

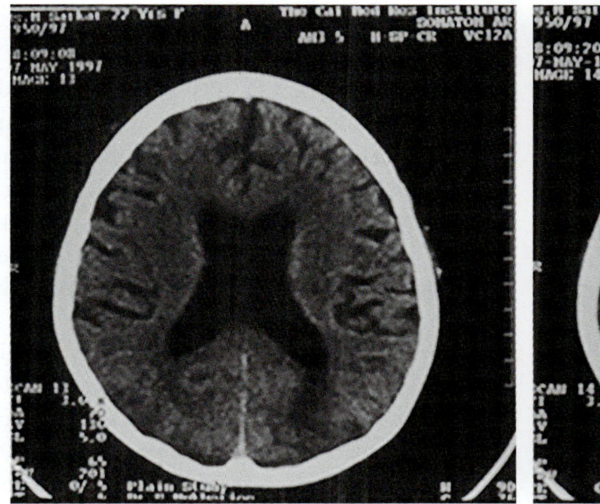

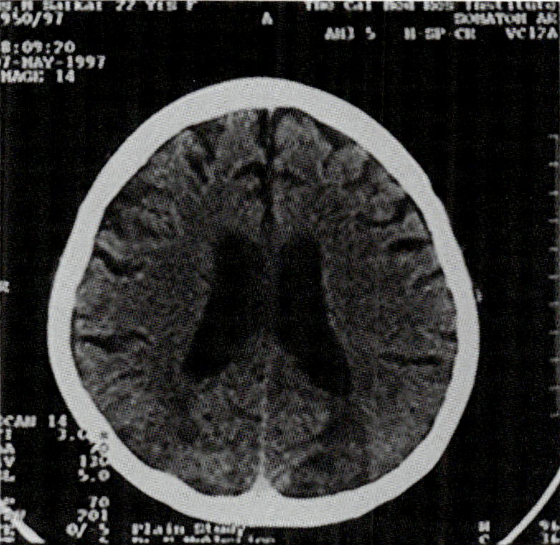

Fig. 37.9: Infarct in left occipital lobe in SLE

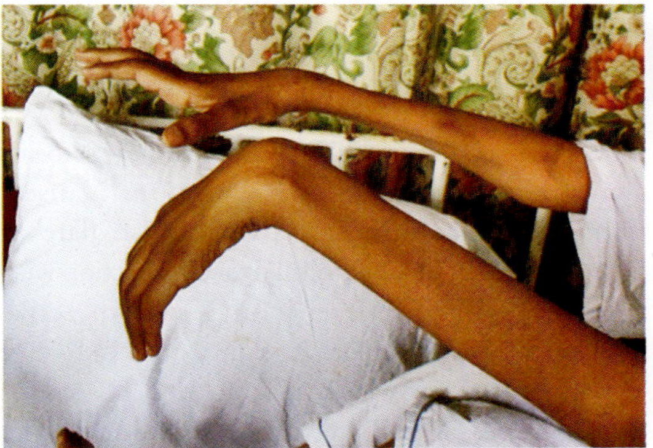

Fig. 37.10: Left wrist drop and right claw hand (mononeuritis multiplex)

as well as immunofluorescence staining. Lupus nephritis is thus classified according to World Health Organization (WHO) as in Table 37.4. As per this classification, classes 1 and 2 refer to either normal glomeruli or minimal mesangial proliferation respectively. Classes 3 and 4 refer to focal and diffuse proliferative glomerulonephritis respectively, class 5 to membranous glomerulonephritis and class 6 to advanced sclerotic glomerular lesions with chronic kidney disease. Details are provided in Table 37.4.[7]

This histologic classification is not exclusive and there is likelihood of transformation from one class to other. Therapy is also variable according to class of the disease. Thus renal biopsy is indicated for case identification, treatment and disease activity monitoring of lupus (Table 37.5).

Table 37.4: The revised 1995 World Health Organization (WHO) classification of lupus nephritis

1. Normal glomeruli (less than 5% patients)
   a. Normal by all techniques
   b. Normal on light microscopy but deposits on immunohistology and/or electron microscopy
2. Pure mesangial alterations (15% patients)
   a. Mesangial widening and/or mild hypercellularity
   b. Mesangial cell proliferation
3. Focal segmental glomerulonephritis (associated with mild/moderate mesangial alterations, and/or segmental epimembranous deposits) (20% patients)
   a. Active necrotizing lesions
   b. Active and sclerosing lesions
   c. Sclerosing lesions
4. Diffuse glomerulonephritis (severe mesangial/mesangiocapillary with extensive subendothelial deposits. Mesangial deposits always present and frequently subepithelial deposits) (50% patients)
   a. With segmental lesions
   b. With active necrotizing lesions
   c. With active and sclerosing lesions
   d. With sclerosing lesions
5. Diffuse membranous glomerulonephritis (15% patients)
   a. Pure membranous glomerulonephritis
   b. Associated with lesions of category II (A or B)
6. Advanced sclerosing glomerulonephritis

Table 37.5: Indications of renal biopsy (any of the following)

1. Increasing serum creatinine without compelling alternative causes
2. Confirmed proteinuria of ≥1 gram/24 hours
3. Proteinuria >0.5 gram/24 hours plus hematuria
4. Proteinuria >0.5 gram/24 hours plus cellular casts

Patients with classes 1 and 2 have better prognosis than other classes. Class 4 glomerulonephritis usually have microscopic hematuria and proteinuria (>500 mg per 24 hrs). If the disease is untreated at this stage, virtually all patients develop ESRD within 2 years of diagnosis. A small group of SLE patients with nephrotic range proteinuria may have membranous glomerulonephritis on renal biopsy and they may have thromboembolic manifestations and hyperlipidemia. In some cases digital gangrene due to small vessel vasculitis is reported.[1]

Pulmonary Manifestations

The most common pulmonary feature of SLE is pleurisy with or without effusion. Pulmonary infiltrates also occur as a manifestation of active SLE and it may be confused with pulmonary infection. The life threatening changes in lungs include interstitial pneumonia leading to fibrosis and intra-alveolar haemorrhage. In this country, pulmonary TB, whether miliary, infiltrative or cavitary form, are common as a result of immunosuppression and need to be distinguished from other lung shadows. Rarely, shrinking lung syndrome presenting with dyspnoea on exertion without any ausculatory signs is encountered.[2] Pulmonary hypertension is a rare but potentially life-threatening complication.

Cardiac Manifestations

Pericarditis is the most frequent form of cardiac involvement and on rare occasion it may lead to pericardial effusion or cardiac tamponade. The other features are myocarditis and Libman-Sacks endocarditis mostly involving mitral or aortic valves. The consequences of these are mostly heart failure, arrhythmias or embolic events. The accelerated atherosclerosis in SLE are related to continuing active disease with coronary vasculitis and prolonged high dose of steroids in combination with other traditional risk factors.

Haematological Features

The normocytic normochromic anemia due to active disease is found in 70% patients.[2] Coombs' positive autoimmune haemolytic anaemia and particularly lymphopenia are significantly associated with disease activity. Thrombocytopenia leading to purpura (Fig. 37.11) can be due to autoimmune destruction of platelets or because of co-existent APS. In childhood and adolescence idiopathic thrombocytopenic purpura can antedate SLE by years. ESR is elevated while CRP mostly remains within normal limits. A rise in CRP denotes either a significant inflammation or associated infection. Lymphadenopathy occurs in approximately 40% of patients and these patients likely to have constitutional symptoms. Splenomegaly occurs in 10–45% of patients particularly during active disease and is not necessarily associated with cytopenias.

Gastrointestinal Manifestation

The diffuse abdominal pain of autoimmune serositis may be an uncommon presentation. Nausea, vomiting and diarrhea can be due to SLE 'flare' or drug induced. The subclinical liver dysfunction with raised transaminase level is suggestive of active disease. Acute pancreatitis is observed in 2–8% patients presenting with 'acute abdomen'. Vasculitis of intestine in SLE may pose like one of the emergencies with ischaemia, perforation, bleeding and sepsis.[2]

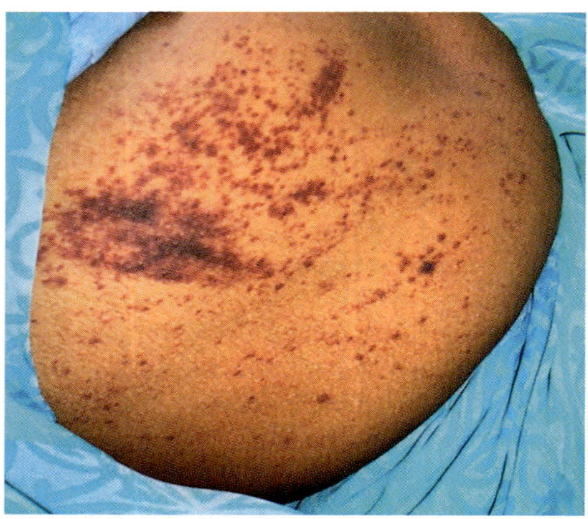

Fig. 37.11: Purpuric spots

Ocular Manifestations

Non-specific conjunctivitis, dry eye (sicca) syndrome are common in SLE and does not threaten vision. In contrast retinal vasculitis and optic neuritis are vision threatening manifestations of SLE. In patients on long-term steroid therapy, cataract and glaucoma need to be screened periodically.

Antiphospholipid Syndrome (APS)

Around 20–25% SLE patients are positive for anti-phosholipid antibodies but only 10% have APS characterized by increased risk of systemic vascular thrombosis (venous and arterial), recurrent abortion, livedo reticularis, thrombocytopenia, ischemic CNS disease and seizures.[8] Premature and accelerated coronary atherosclerosis is often worse SLE patients with APS.

SPECIAL SITUATIONS

1. **Pregnancy and SLE:** Fertility is unaffected in SLE except during acute phase of illness. During pregnancy SLE patients may have 'flare'. It is advisable to plan pregnancy while the disease is in remission. Active SLE is associated with increased risk of miscarriage, stillbirth and low birth weight babies. The SLE patients with APS carry an inordinate risk of foetal wastage. Low dose steroids, azathioprine and hydroxychloroquine are safe in pregnancy.[8]

   Newborn may have neonatal lupus from SLE mother with anti-Ro antibodies. The disease is self-limited or the child can have congenital heart block.

   Oestrogen containing hormones and intra-uterine contraceptive devices are best avoided in lupus patients. Hormone therapy with progesterone alone is permitted. Barrier methods can be safe.
2. **Paediatric SLE and neonatal lupus syndrome:** The prevalence of paediatric SLE constitutes 20–25% of total SLE population. This special group has higher frequency of hepatosplenomegaly, chorea, renal disease and avascular necrosis.[12] They should be treated as in adult SLE. Most of these patients survive longer with current treatment.

   The neonatal lupus syndrome is characterized by photosensitive skin rash, hepatosplenomegaly, cytopenias and carditis. The serious consequences are endomyocardial fibroelastosis, and congenital complete heart block in infants when born out of anti-Ro positive SLE mother. Treatment is supportive when features are transient. Permanent pacing is indicated in symptomatic persistent congenital heart block.
3. **Drug induced SLE:** This is a small subgroup of patients who develop milder lupus on exposure to certain drugs like procainanide, hydralazine, isoniazid, omeprazole, methyldopa, chlorpromazine and anti-TNF drugs. This is characterized by positive antinuclear antibodies and anti-histone antibodies. Anti-ds-DNA is absent except in those cases associated with use of IFNα (interferon α) and anti-TNF (anti-tumour necrosis factor) therapies. In most patients serological features are accompanied by mild clinical disease including fever, myalgia, rash, arthritis, and serositis. Renal and neuro-psychiatric manifestations are seen in anti-TNF induced lupus as opposed to the rest. The patient may remain well on withdrawal of culprit agents.
4. **Overlap syndrome:** This syndrome is said to exist when there is overlap between clinical and serological criteria for two or more connective tissue disorders, namely SLE with scleroderma or SLE with rheumatoid arthritis or dermatomyositis. Mixed connective tissue disease (MCTD) is specifically characterized with Raynaud's phenomenon, swollen dorsum of hands, dermatomyositis and mild polyarthritis and positive anti-U1NP.

DIAGNOSIS

The diagnosis of SLE is based on characteristic clinical features and autoantibodies. The clinical classification criteria has been laid down in 1982 and further updated by Hochberg in 1997 (Table 37.6).[13]

A person shall be said to have SLE if any 4 or more of the 11 criteria are present, serially or simultaneously during any interval of observation (95% specificity and 85% sensitivity). These classification criteria were laid down for purpose of research. In daily practice a patient can still have a clinical diagnosis of SLE without fulfilling the classification criteria.

Table 37.6: Revised American College of Rheumatology (ACR) classification criteria for SLE: 1997 update[13]

Item	*Definition*
Malar rash	Fixed erythema, flat or raised, over the malar eminences, sparing nasolabial folds
Discoid rash	Erythematous, raised patches with adherent keratotic scaling and follicular plugging; atrophic scarring may occur in older lesions
Photosensitivity	Skin rash as a result of unusual reaction to sunlight by history or on physical examination
Oral ulcers	Oral or nasopharyngeal ulceration, usually painless, observed by physician
Non-erosive arthritis	Involving 2 or more peripheral joints, characterized by tenderness, swelling or effusion.
Pleuritis/pericarditis	a. Pleuritis—convincing h/o pleuritic chest pain or rub or pleural effusion on physical examination, *Or* b. Pericarditis—documented by ECG, rub or pericardial effusion
Renal disorder	a. Persistent proteinuria > 0.5 gm/ day or >3+ by dipstick OR b) Cellular casts – may be red cell, Hb, granular, tubular or mixed
Neurological disorder	a. Seizures—in the absence of offending drugs, or known metabolic derangement, e.g. uraemia, ketoacidosis or electrolyte imbalance, *Or* a. Psychosis—in the absence of offending drugs, or known metabolic derangement, e.g. uraemia, ketoacidosis or electrolyte imbalance
Immunological disorder	a. Anti-DNA: Antibody to native DNA in abnormal titre, *Or* b. Anti-Sm: Presence of antibody to Sm nuclear antigen, *Or* c. Positive finding of aPL antibodies based on: 1. ↑ serum level of IgG or IgM aCL or 2. a positive test result for lupus anticoagulant, using a standard method or 3. a false positive test for syphilis for at least 6 months and confirmed by TPI or FTA abs test
Haematological disorder	a. Haemolytic anaemia with reticulocytosis, *Or* b. Leukopenia < 4000/cu mm on 2 or more occasions, *Or* c. Lymphocytopenia < 1500 on 2 or more occasions, *Or* d. Thrombocytopenia < 100, 000/ cu mm in the absence of offending drugs.
Positive ANA	An abnormal titre of ANA by immunoflurescence or an equivalent assay at any point in time in the absence of drug

The fluorescent antinuclear antibody (FANA) is positive in greater than 95% of patients. The various staining patterns (i.e. homogenous, speckled, rim, nucleoler) can be demonstrated depending on the content of different autoantibodies in the serum. This test has high sensitivity but low specificity. Anti-dsDNA are highly specific for SLE but present only in about 50–60% of patients. It is tested by ELISA, *crithidia lucillae assay*, immunofluroscence assay or Farr assay. The anti-dsDNA positivity has got good correlation with progression of lupus nephritis. The reduced value of C3 is closely associated with renal damage. The antibodies to Smith (anti-Sm) has high specificity for SLE but present only in about 10% cases. Recently anti-nucleosome antibody is reported to be associated with lupus nephritis.

ACR 1997 criteria have certain limitations. It is validated on established diseases and using this criteria one may miss the early or limited disease. Some systems like mucocutaneous system have been over-represented. Moreover, all systems have equal contribution without any weight.[2]

Recent revision of ACR 2012 has been made by the Systemic Lupus International Collaborating Clinics (SLICC) after 8 years of study in order to improve clinical relevance and incorporate new knowledge in the field of lupus (Table 37.7).[14] However, this classification is not meant to be used for the diagnosis of lupus.

Table 37.7: 2012 ACR/SLICC* classification criteria for systemic lupus erythematosus[14]

Clinical criteria	*Immunologic criteria*
1. Acute cutaneous lupus	1. ANA
2. Chronic cutaneous lupus	2. Anti-dsDNA
3. Oral or nasal ulcers	3. Anti-Sm
4. Non-scarring alopecia	4. Antiphospholipid antibodies
5. Arthritis	5. Low complement (C3, C4, CH50)
6. Serositis	6. Direct Coombs' test (do not count in the presence of hemolytic anaemia)
7. Renal	
8. Neurologic	
9. Hemolytic anaemia	
10. Leukopenia	
11. Thrombocytopenia (<100,000/mm^3)	

*SLICC: Systemic Lupus International Collaborating Clinics

Requirements: ≥4 criteria; at least 1 clinical and 1 laboratory are required for diagnosis. Or biopsy proven lupus nephritis with positive ANA and anti-dsDNA implies a diagnosis of SLE.

[*Source*: Petri M, et al. Arthritis Rheum 2012; 64:2677–86]

PROGNOSTIC MARKERS

Male sex, persistently high SLEDAI score, complicated pregnancy and status of autoantibodies are strongly related to prognosis. Serum anti-dsDNA antibody titre is strongly associated with lupus nephritis, progression to end stage renal disease and poor survival. Anti-phospholipid antibodies are strongly associated with APS, cerebrovascular disease and contributes significantly to mortality. Anti-Ro (SSA) is associated with neonatal lupus and congenital heart block in babies born to Ro positive mothers. Antibodies to other extractable nuclear antigens (anti-Ro, La, Sm, RNP) are found to be associated with dermatological manifestations and less severe renal involvement.[2]

COMORBID CONDITIONS

Lupus patients are particularly susceptible to infection due to immune dysregulation and immunosuppressive treatment. Infections are responsible for 20–55% of all deaths in SLE.[2] In India, tuberculosis is a major threat; though any infection like bacterial, viral or fungal can complicate the scenario. Premature atherosclerosis and coronary artery disease can occur in long run and this is related to disease activity as well as cumulative dose of steroids. Osteoporosis and increased fracture risk is another concern mainly for the women suffering from lupus. A few malignancies, namely non-Hodgkin's lymphoma, lung cancer and cervical cancer have been reported to be associated with SLE.

EMERGENCIES IN LUPUS

Emergencies can arise in lupus patients due to active disease with vital organ involvement, comorbidities or drug toxicity. Infections particularly lung infections can be fatal. Premature atherosclerosis can lead to acute coronary syndrome or cerebrovascular events. Acute abdomen due to mesenteric ischaemia or acute pancreatitis warrants prompt investigation and management. Malignancies may also give rise to life-threatening situation. Antiphospholipid syndrome may present with acute thrombotic events in brain, limbs and acute coronary events. Crescentic glomerulonephritis with rapidly progressive renal failure may pose acute life-threatening emergency. Diffuse alveolar hemorrhage is another devastating life-threatening complication of lupus requiring aggressive immunosuppression.

Key Points

- SLE is an autoimmune inflammatory disease with gender bias and heterogenous manifestations. Remission and recurrences are common in the natural history of the disease.
- Severity of the disease is expressed by major organ involvement. Comorbid disorders are common cause of morbidity and mortality.

Antimalarial Drugs

Antimalarial drugs, namely chloroquine, hydroxychloroquine are prescribed for patients with cutaneous and joint manifestations. These drugs may be associated with retinal toxicity, and in such patients, quinacrine is used. Antimalarial drugs usage is associated with significant reduction in disease activity and flare rate.[8] Hydroxychloroquine has been shown to have beneficial effects on lipid profile, subclinical atherosclerosis and also improves survival in lupus patients. Hydroxychloroquine reduces lupus activity and is safe for use in pregnant women.

DMARDs

Methotrexate (MTX) has been used as steroid-sparing treatment for articular and cutaneous manifestations of lupus. Leflunomide can be used for treating arthritis and also has been tried in refractory lupus nephritis.[9]

CYTOTOXIC DRUGS

Therapy with cyclophosphamide (CYC) is of indisputable efficacy in lupus nephritis. CYC is used also for treatment of CNS lupus and other severe manifestations of the disease. CYC retains renal function, and lessen the risk of evolution to end-stage renal disease. Following induction therapy, a maintenance regimen is necessary to decrease the risk of flares.

For patients with moderate to severe disease, monthly pulses of intravenous methyl prednisolone with CYC or mycophenolate are given during the induction period and for patients with milder forms of disease, less intensive regimens of CYC followed by azathioprine or mycophenolate is recommended.[10] CYC has demonstrated potency in life-threatening extra-renal lupus manifestations such as severe thrombocytopenia, neurologic disease, abdominal vasculitis, acute pneumonitis/alveolar haemorrhage, and extensive skin disease. Chlorambucil, an aromatic alkylating agent, has comparable effects on immune functions as those described for CYC.

ANTIMETABOLITES

Azathioprine (AZA): AZA, a purine synthesis inhibitor, is used as maintenance drug for major organ involvement particularly for nephritis. It also demonstrated efficacy in non-renal conditions like arthritis, severe cutaneous disease, pneumonitis, hepatitis, protein-losing enteropathy, haemolytic-anaemia, and thrombocytopenia.[11,12]

Mycophenolate mofetil (MMF): MMF is a prodrug of mycophenolic acid. It is an inhibitor of inosine monophosphate dehydrogenase. It is used as induction agent in management of moderately severe nephritis and also as maintenance regimen.[13,14] It is also used in extra renal manifestations like refractory skin and haematological manifestations.

BIOLOGIC THERAPIES

B cell-targeted therapies namely Rituximab and Belimumab are used in certain scenarios. Rituximab is tried in refractory lupus, including refractory lupus nephritis and based on data derived from observational studies rituximab is able to induce induce remission of refractory renal disease. Belimumab a monoclonal antibody directed against BAFF is effective in autoimmune cytopenias and other non-cerebral, non-renal manifestations namely mucocutaneous and musculoskeletal disease. Co-stimulation blockers with abatacept had mixed results in lupus management. Tumour necrosis factor inhibitors are associated with drug induced lupus and presently have no role in lupus management.

MANAGEMENT OF SPECIFIC DISEASE MANIFESTATIONS

Mucocutaneous and Joint Disease

Mild malar rash respond to sun avoidance. Moderate to severe rash require topical corticosteroids, tacrolimus with moderate to high dose corticosteroids. Antimalarial drugs may take to three months to work.

Hematologic Disease

Peripheral cytopenias are common and usually mild in SLE. They need thorough clinical and laboratory evaluation to exclude secondary causes. Mild cytopenias require no specific therapy. In more severe cases high dose glucocorticoids are the mainstay of treatment. Steroid-sparing agents can be added during steroid tapering. Rituximab may be considered in patients with refractory cytopenias.

Lupus Nephritis

Involvement of kidney is the major cause of morbidity and mortality in lupus. It requires aggressive immunosuppressive therapy.

Cyclophosphamide (CYC) forms a cornerstone drug in the management of lupus nephritis. The efficacy of CYC given as "pulse" intravenous (IV) monthly dose was first demonstrated beginning in the early 1980s by a series of trials conducted at the National Institutes of Health (NIH).[15] The dose of IV CYC used was 0.75 to 1 gm/m^2 every month for 6 months, then quarterly for the next 18 months. IV CYC was given along with oral prednisolone. The initial pivotal trial demonstrated clear benefit of IV CYC + prednisolone group over prednisolone alone in maintaining renal function at 5 years. Subsequently by the year 2005-2006 the Euro Lupus Nephritis Trial (ELNT) demonstrated that IV CYC is equally effective when given in much lower doses of 500 mg pulses every fortnightly for 6 doses.[16]

Proliferative Nephritis

For all patients with lupus nephritis renoprotective measures should be instiuted namely strict BP control, reduction of preoteinuria with ACE inhibitors, avoidance of nephrotoxins, and cessation of smoking. Patients with class 1 and class II lupus nephritis have good prognosis and no specific therapy is required.

For class III and class IV lupus nephritis induction of remission is initiated with pulse IV CYC and oral prednisolone. One can choose between the NIH regime or the ELNT regime. There is presently no published RCT from India comparing the NIH regime versus the ELNT regime. Most centers use the NIH regime which is "modified" in that IV CYC pulse is given for 6 months along with tapering doses of prednisolone, this is followed by maintenance with oral azathioprine and low dose prednisolone. The duration of maintenance therapy is for at least 24 months. With CYC 80% achieve renal remission with initial therapy, 30% experience renal flares, 5–20% progress to ESRD.

The need for a newer immunosuppressive therapy was realized because of the toxicity of CYC. CYC induced premature ovarian failure is a major concern more so because SLE predominantly affects young women (and men) who are in the reproductive age group. The risk of premature ovarian failure increases with age of the patient.[17] The risk is 12% in patients aged ≤25 years; 27% in 26–30 years; and 62% in women aged ≥31 years.[17] The risk further increases with cumulative dose of cyclophosphamide; 12% with 7 monthly doses vs 39% with 15 or more doses. Cumulative CYC dose is also associated with malignancy and infections.

Because of all these concerns MMF was explored and data from two large trials by Ginzler *et al* and Appel *et al* (the ALMS trail) showed equal efficacy of MMF and IV CYC for induction of remission of lupus nephritis. MMF is effective in induction of remission and as maintenance therapy of lupus nephritis.[18,19] Recently Rathi *et al* from India showed that MMF is not inferior to IV CYC (ELNT regime) for induction of remission in less severe proliferative lupus nephritis.[20]

The American College of Rheumatology (ACR) guidelines suggest that either IV CYC or MMF can be used as induction regimen.[21] For maintenance therapy either azathioprine or MMF is used. All these is given against a background of prednisolone.

Membranous Nephropathy

For non-nephrotic-range proteinuria with preserved renal function, high-dose glucocorticoids with or without AZA/mycophenolate as induction regimen and maintenance regimen with low-dose glucocorticoids with or without AZA is used.

For moderate nephrotic-range proteinuria with preserved renal function and severe nephrotic-range proteinuria with impaired renal function at presentation, high-dose glucocorticoids with or without AZA/mycophenolate is recommended.

Renal flares management: Mild to moderate flares treated with glucocorticoids in combination with AZA or MMF. For severe nephritic flares, cytotoxic therapy with monthly pulses of CYC and pulse steroids. Rituximab (RTX) can also be considered.

Patients with end stage renal disease require renal replacement therapy. Comorbidities like hypertension and dyslipidaemias are independent risk factors for renal failure and cardiovascular morbidity. Angiotensin converting enzyme inhibitors, angiotensin receptor blockers have antiproteinuric effects and are initial choice of treatment for controlling blood pressure. The other drugs which can be used for controlling blood pressure are calcium channel blockers and centrally acting drugs. Aggressive management of dyslipidemias with statins prevents thromboembolic events.

Central Nervous System Disease

Non-lupus related causes like infections, metabolic disturbances, and drug effects should be excluded prior to initiation of specific treatment strategy. Timely initiation of immunosuppression results in improved long-term outcomes. Corticosteroids, immunosuppressants, antiplatelet/anticoagulant treatment and symptomatic drugs are used depending on the presumptive pathogenic mechanism.[22] Acute confusional state, aseptic meningitis, myelitis, optic neuritis, and severe psychosis are indications for high dose glucocorticoids with IV CYC followed by maintenance with less intensive immunosuppressive therapy (predisolone and AZA). Antiphospholipid antibody-associated NPSLE, antiphospholipid syndrome-associated thrombotic events are indications for antithrombotic or antiplatelet therapy. Intravenous immunoglobulins (IVIG), plasmapheresis is indicated for refractory disease not caused by anti-phospholipid syndrome. The mangement of other manifestations such as cerebrovascular accidents (strokes), control of seizures, depression is same as for the general population.

Antiphospholipid Syndrome (APS)

APS associated thrombosis requires anticoagulation therapy with warfarin with a target INR 2 to 3 to prevent recurrent events. For patients with recurrent thrombotic events, the target INR should be 3 to 4. Hydroxychloroquine is used along with anticoagulation as it also has antithrombotic properties through inhibiting platelet aggregation.

For the pregnant SLE-APS patient and a history of pregnancy complications or thrombosis or both unfractionated heparin/low molecular weight heparin and aspirin is indicated throughout pregnancy as it significantly increases live birth rates. In the postpartum period anti-thrombotic coverage with low molecular weight heparin or unfractionated heparin for 4–6 weeks is required for all APS patients irrespective of their thrombotic history. Long-term anticoagulation is recommended in women with thrombosis in the past. Warfarin can also be considered in place of heparin in postpartum period. Both heparin and warfarin are safe during breastfeeding.

For the pregnant SLE patient with positive antiphospholipid antibodies but has neither thrombotic event nor pregnancy morbidity in the past, there is no hard evidence to support the use of heparin or aspirin throughout the pregnancy.

Key Points

- The management of patients with SLE should be individualized.
- Management of major organ involvement involves induction with the combination of high-dose glucocorticoids with pulses of CYC or MMF followed by maintenance therapy with less intensive immunosuppression to maintain low disease activity or to secure remission.
- Hydroxychloroquine is cornerstone drug in the management of lupus. It is proven to improve long-term outcomes and prevents flares.

REFERENCES

1. Doria A, Gatto M, Iaccarino L, Punzi L. Value and goals of treat-to-target in systemic lupus erythematosus: knowledge and foresight. Lupus. 2015; 24:507–15.
2. Ting WW, Sontheimer RD. Local therapy for cutaneous and systemic lupus erythematosus: practical and theoretical considerations. Lupus. 2001; 10:71–84.
3. Kamen DL. Management of SLE: adjunctive measures. In: Wallace DJ, Hahn BH, editors. Dubois' lupus erythematosus and related syndromes. 8th ed. Philadelphia: Saunders; 2013. P. 633–9.
4. Panopalis P, Yazdany J. Bone health in systemic lupus erythematosus. Curr Rheumatol Rep. 2009; 11:177–84.
5. Koneru S, Shishov M, Ware A, Farhey Y, Mongey AB, Graham TB, etal. Effectively measuring adherence to medications for systemic lupus erythematosus in a clinical setting. Arthritis Rheum. 2007; 57:1000–6.
6. Kiros KA, Boumbas DT. Systemic glucocorticoid therapy in systemic lupus erythematosus. In: Wallace DJ, Hahn BH, editors. Dubois' lupus erythematosus and related syndromes. 8th ed. Philadelphia: Saunders; 2013. p. 591–600.
7. Albert DA, Hadler NM, Ropes MW. Does corticosteroid therapy affect the survival of patients with systemic lupus erythematosus? Arthritis Rheum. 1979; 22:945–53.

8. Ruiz-Irastorza G, Ramos-Casals M, Brito-Zeron P, Khamashta MA. Clinical efficacy and side effects of antimalarials in systemic lupus erythematosus: asystematic review. Ann Rheum Dis. 2010; 69:20–8.
9. Remer CF, Weisman MH, Wallace DJ. Benefits of leflunomide in systemic lupus erythematosus: a pilot observational study. Lupus. 2001; 10:480–3.
10. Houssiau FA, Vasconcelos C, D'Cruz D, Sebastiani GD, de Ramon Garrido E, Danieli MG, et al: The 10-year follow-up data of the Euro-Lupus Nephritis Trial comparing low-dose and high-dose intravenous cyclophosphamide. Ann Rheum Dis. 2010; 69:61–4.
11. Abu-Shakra M, Shoenfeld Y. Azathioprine therapy for patients with systemic lupus erythematosus. Lupus. 2001; 10:152–3.
12. Callen JP, Spencer LV, Burruss JB, Holtman J. Azathioprine. An effective, corticosteroid-sparing therapy for patients with recalcitrant cutaneous lupus erythematosus or with recalcitrant cutaneous leukocytoclastic vasculitis. Arch Dermatol. 1991; 127:515–22.
13. Contreras G, Pardo V, Leclercq B, Lenz O, Tozman E, O'Nan P, et al. Sequential therapies for proliferative lupus nephritis. N Engl J Med. 2004; 350:971–80.
14. Chan TM, Li FK, Tang CS, Wong RW, Fang GX, Ji YL, Lau CS, et al. Efficacy of mycophenolate mofetil in patients with diffuse proliferative lupus nephritis. Hong Kong-Guangzhou Nephrology Study Group. N Engl J Med 2000; 343:1156–62.
15. Steinberg AD, Steinberg SC. Long-term preservation of renal function in patients with lupus nephritis receiving treatment that includes cyclophosphamide versus those treated with prednisone only. Arthritis Rheum. 1991; 34:945–50.
16. Houssiau FA, Vasconcelos C, D'Cruz D, Sebastiani GD, Garrido Ed Ede R, Danieli MG, et al. Immunosuppressive therapy in lupus nephritis: the Euro-Lupus Nephritis Trial, a randomized trial of low-dose versus high-dose intravenous cyclophosphamide. Arthritis Rheum. 2002; 46:2121–31.
17. Boumpas DT, Austin HA 3rd, Vaughan EM, Yarboro CH, Klippel JH, Balow JE. Risk for sustained amenorrhea in patients with systemic lupus erythematosus receiving intermittent pulse cyclophosphamide therapy. Ann Intern Med. 1993; 119:366–9.
18. Ginzler EM, Dooley MA, Aranow C, Kim MY, Buyon J, Merrill JT, et al. Mycophenolate mofetil or intravenous cyclophosphamide for lupus nephritis. N Engl J Med. 2005; 353:2219–28.
19. Appel GB, Contreras G, Dooley MA, Ginzler EM, Isenberg D, Jayne D, et al; Aspreva Lupus Management Study Group. Mycophenolate mofetil versus cyclophosphamide for induction treatment of lupus nephritis. J Am Soc Nephrol. 2009; 20:1103–12.
20. Rathi M, Goyal A, Jaryal A, Sharma A, Gupta PK, Ramachandran R, et al. Comparison of low-dose intravenous cyclophosphamide with oral mycophenolate mofetil in the treatment of lupus nephritis. Kidney Int. 2016; 89:235–42.
21. Hahn BH, McMahon MA, Wilkinson A, Wallace WD, Daikh DI, Fitzgerald JD, et al; American College of Rheumatology. American College of Rheumatology guidelines for screening, treatment, and management of lupus nephritis. Arthritis Care Res (Hoboken). 2012; 64:797–808.
22. Bertsias GK, Ioannidis JP, Aringer M, Bollen E, Bombardieri S, Bruce IN,et al. EULAR recommendations for the management of systemic lupus erythematosus with neuropsychiatric manifestations: report of a task force of the EULAR standing committee for clinical affairs. Ann Rheum Dis. 2010; 69:2074–82.

FURTHER READING

1. Lazzaroni MG, Dall'Ara F, Fredi M, Nalli C, Reggia R, Lojacono A, et al. A comprehensive review of the clinical approach to pregnancy and systemic lupus erythematosus. J Autoimmun. 2016; 74:106–117.
2. Thakral A, Klein-Gitelman MS. An Update on Treatment and Management of Pediatric Systemic Lupus Erythematosus. Rheumatol Ther. 2016; 3:209–19.
3. Thong B, Olsen NJ. Systemic lupus erythematosus diagnosis and management. Rheumatology (Oxford). 2016 Dec 24. pii: kew401. doi:10.1093/rheumatology/kew401.

Chapter

39

Pregnancy in Systemic Lupus Erythematosus

S Rajeswari

Catching a tiger by its tail is far easier and less risky than tackling lupus at its peak especially, pregnancy flares. What have studies and experience taught us? It is universal knowledge that lupus in remission for 6 months clinically and by laboratory parameters can go ahead. What happens during pregnancy in aggressive lupus? What is the effect on the mother and fetus? How to tackle infertility? These issues will be addressed under the following headings:

- Preconception counselling
- Preservation of oocytes
- Identifying the right individual
- Organ flares in pregnancy—renal, hematological, etc.

  Complications—eclampsia, preeclampsia, hemolysis elevated

  Liver enzymes and low platelets (HELLP)
- How to monitor and treat mother during antenatal and peri-partum
- Complications in fetus
- Monitoring fetus and treatment of foetal complications
- Pregnancy in SLE patients with antiphospholipid syndrome (APS)

Pregnancy by itself is an altered immune state. The 'T' regulatory cells ($CD4^+/CD25^+$) which are immunosuppressive and produce foetal tolerance are increased. There is a shift from T helper cell 1(Th1) (cell mediated) to T helper cell 2 (Th2) (antibody mediated) immune response, i.e. Th2 polarisation which is responsible for the maternal immune tolerance of the fetus.[1] Migration of foetal cells and foetal DNA into maternal circulation promotes tolerance to paternal antigens.[2] IL-4, IL-10, transforming growth factor-β (TGF-β) and interferon γ (IFN-γ) are secreted under the influence of oestrogens and gestagens. Tumour necrosis factor (TNF) is decreased.

Both in pre-eclampsia and SLE, T regulatory cells are decreased. SLE which is a Th2 mediated disease worsens in pregnancy. Decreased oestrogen, progesterone, and Th2 cytokines are observed in the third trimester of pregnancy.[2]

PRECONCEPTION COUNSELLING

Cryopreservation and reimplantation are techniques in vogue, the most beneficial being pre-pubertal patients.

PROVISIONAL CRITERIA FOR OVARIAN CRYOPRESERVATION[3]

- Not more than 30 years
- No existing children
- Reasonable chance of surviving five years
- More than 50% ovarian dysfunction by therapy
  - Age more than 15 years with no previous chemo or radiotherapy.
  - Age less than 15 years eligible if previous mild chemotherapy.

CHECKLIST FOR COUNSELLING AND PREGNANCY PLANNING FOR PATIENTS WITH SLE

I. *Risk assessment*
   a. Age
   b. Previous pregnancies
   c. Disease activity assessment
   d. Irreversible damage

II. *Autoantibodies*
   a. Anti phospholipid antibodies
   b. Anti Ro/La

III. *Current treatment adjustments*

IV. *Pregnancy*–Contraindicated if
   a. SLE Disease Activity Index (SLEDAI >8)
   b. High irreversible damage

V. Drugs which are contraindicated to be replaced by safe ones–wait for remission for at least 2 to 3 months

VI. Treat active disease

VII. *Prophylaxis:*
   a. Low dose Aspirin: 1st trimester onwards till delivery to reduce pre-eclampsia especially in patients with lupus nephritis.
   b. Prophylaxis: Pre-eclampsia and thrombosis in patients with positive antiphospholipid antibodies.[2]
   c. Pregnancy: Planned one only when the disease is in remission both clinically and by investigations for at least 6 months.

HOW TO IDENTIFY THE RIGHT INDIVIDUAL?

- A planned pregnancy has successful rates.
- Contraception mandatory till disease is quiescent.
- Intrauterine device (IUD), hormonal and barrier methods are commonly used.
- Hormonal methods are either combined pills or progestogen only, subcutaneous implants, injectables, skin patches and vaginal rings. Combined is relatively safe for stable and low active SLE. Combined oral contraceptives are contraindicated in APS, hypertension, obesity, severe disease and history of thrombosis. Progestogen only preparations are safe. IUD are safe except for pelvic infection.

Situations where Pregnancy is not Advisable

- Severe pulmonary hypertension (systemic pressure > 50 mm Hg)
- Severe restrictive lung disease (Forced Vital Capacity < 1L)
- Advanced renal insufficiency (serum creatinine > 2.8 mg%)
- 24 hours urine protein > 0.5 gm
- Advanced heart failure
- Previous severe pre-eclampsia or HELLP despite treatment.

Situations where Pregnancy is Deferred

- Severe disease flare and stroke within 6 months.
- Active, lupus nephritis.
- Recent major thrombosis within 2 years.

COMPARISON OF LUPUS PREGNANCY AND NORMAL PREGNANCY[2,4]

There are various similarities and differences in the hormonal and immune responses between a pregnant lupus patient and woman with normal pregnancy as illustrated in Table 39.1.

Normal pregnancy is associated with hemodilution. There is progesterone induced smooth muscle relaxation and compression of the ureters by the gravid uterus, the end result being pyelonephritis and symptomatic urolithiasis. The other manifestations mimicking lupus include arthralgia, myalgia, facial and palmar rash, oedema of face, hands and legs, hearing loss, shortness of breath and carpal tunnel syndrome.

DIFFERENCES BETWEEN PRE-ECLAMPSIA AND LUPUS NEPHRITIS IN PREGNANCY[5,6]

It is important to differentiate a pre-eclampsia from lupus nephritis in pregnancy as depicted in Table 39.2.

Definitions

It is worthwhile remembering certain definitions to understand the issues related with lupus pregnancy.

Maternal flare: Any clinical event attributable to disease activity that required a change in therapy.

Table 39.1: Comparison of immunological and hormonal responses between lupus pregnancy and normal pregnancy

S.No.	*Parameters*	*Pregnancy*		*Outcomes*
		Normal	*Lupus*	
1.	TH17-IL17	↑	↑	Pre-eclampsia, fetal loss
2.	Estradiol + Progesterone (2nd and 3rd trimesters)	↑	↓	Impaired placental function and fetal loss
3.	IL-10	1st trimester - ↓ 3rd trimester - ↑	↑ in all trimesters	B cell stimulation
4.	Treg cells	↑	↓	Disease activity
5.	Chemokines CXCL-8, IL-8 CXCL-9, MIG CXCL-10, IP-10	↓	↑	↑ Flares and complications
6.	Ficolin 3	↓	↑	Haemolysis
7.	IFN-α	↓	↑	Pre-eclampsia
8.	C4d	↓	↑	↓ Placental weight ↓ Birth weight
9.	Prolactin	↓	↑	
10.	IL-6	↓	↓	
11.	sTNFαR		↑	

Abbreviations: IL, interleukin; CXCL, C-X-C motif chemokine ligand; IP-10, Interferon gamma-induced protein 10; IFNα, interferon alpha; sTNFαR, soluble tumour necrosis factor alpha receptor.

Table 39.2: Differences between pre-eclampsia and lupus nephritis in pregnancy

S.No.	*Parameters*	*Normal*	*Preclampsia*	*Lupus nephritis*
1.	Hypertension	—	After 20 weeks	Anytime
2.	Haemolysis	—	Severe	Present
3.	Platelets	↓	Normal or ↓	Normal or ↓
4.	GFR	↑		
5.	Creatinine clearance	↑ by 30%		
6.	Serum creatinine	<0.9 (↓)	Normal or ↑	Normal or ↑
7.	LFT	Normal except ↑ ALP	↑	↑
8.	Uric acid	↓	↑ >5.5 mg%	Normal
9.	Anti-dsDNA		Absent	↑
10.	24 hours urine a. Calcium b. Protein		 <195 mg% —	 >195 mg% Doubling of previous value or >3 g/day
11.	Urine sediments		Inactive	Active-cellular RBC casts, dysmorphic RBC's and cylinders
12.	Other organs		Occasionally-CNS HELLP	Active Non renal +
13.	Steroid response		No	Yes
14.	Serum sFlt-1		↑	—
15.	Placental growth factor		+	—

Abbreviations: GFR, glomerular filtration rate; HELLP, hemolysis Elevated Liver enzymes, Low Platelet; sFlt-1, Soluble Fms-like tyrosine kinase 1 or Soluble vascular endothelial growth factor (SVEGF) receptor.

Pre-existing lupus nephritis: Confirmed by renal biopsy, documented proteinuria and on high dose steroid before pregnancy, on greater than 15 mg prednisolone within 4 months of conception.[7]

Renal flare: New onset proteinuria greater than 0.5 g/day, urinary cellular casts or by renal biopsy after delivery (from conception to one month post-partum).

Acute kidney injury: 1.5 fold increase in serum creatinine compared to baseline and serum creatinine more than 0.9 mg%.[7]

Miscarriage: Fetal loss at <24 weeks of gestation.

Still birth: Intrauterine fetal death (IUFD) after 24 weeks of gestation.

Neonatal death: Death of live infant within 28 days after delivery.

Intrauterine growth retardation (IUGR): Growth index less than 10th percentile of any population according to gestational weeks using obstetric ultrasound.

Low APGAR score: APGAR Score <7 at one and 5 minutes of delivery.

Pre-eclampsia (Toxemia):[6] A pregnancy complication occurring in the mother after 20 weeks of pregnancy, that manifests with new onset high blood pressure (BP) >140/90 mmHg and proteinuria >0.3 g/day, without hypertension and <0.3 g/day at baseline.

Fetal loss: Death of a fetus after 10 weeks of gestation.

Preterm delivery: Delivery before 37 weeks of pregnancy.[6]

ORGAN FLARES

Multiorgan flares are common in a pregnant lupus patient especially when there is active disease. About 50% experience a flare during pregnancy.[6] However, renal flares are very difficult to differentiate from pre-eclampsia, HELLP and pregnancy induced hypertension (PIH). Lupus flares range from 7% to 33%. In active disease, at conception it increases to 60%.[8,9] Post-partum flares also occur. Many studies have revealed the increased incidence of mucocutaneous, musculoskeletal and most significantly renal and haematological flares.[4] Active SLE with decreased platelet count in 1st trimester produces 44% foetal losses.[6] Hypertension and pre-eclampsia occur in 35% of patients with lupus nephritis. HELLP syndrome occurs in 30–50% of lupus pregnancies as early as 15–20 weeks of gestation.[10] Disease in remission 6 to 12 months before pregnancy produces less flares.

Poor Prognostic Markers

1. Active disease 6 months prior to pregnancy
2. Maternal hypertension
3. Previous foetal loss
4. Active renal lupus
5. Serum creatinine >2.8 mg%
6. SLE onset during pregnancy
7. Presence of APS
8. High anti-dsDNA titres
9. Low complement
10. Proteinuria
11. Thrombocytopaenia
12. Comorbid states
    a. Diabetes mellitus
    b. Hypertension
    c. Pulmonary hypertension
    d. Older age at conception
    e. Renal failure
    f. Pre-eclampsia (30%)[1], eclampsia, HELLP,
    g. Thrombophilia[2]
    h. Sepsis, pneumonia, anaemia
    i. Ante-partum and post-partum haemorrhage[6]
    j. Deep vein thrombosis (0.4%) and stroke (0.32%)[6]
    k. Pulmonary thromboembolism

Lupus Nephritis

A renal flare is associated with proteinuria, hypertension, haematuria, low complement and anti-dsDNA antibodies. Duration of a renal flare is an independent predictor of chronic kidney disease. Pre-existing lupus nephritis is associated with more renal flares.

Superimposed pre-eclampsia is associated with worsening hypertension and 100% increase in proteinuria in patients with baseline hypertension and proteinuria greater than 0.3 g/day respectively.[7] 50% decrease in proteinuria by 6 months is associated with four times likelihood of achieving complete remission. Successful pregnancies are seen in 65 to 92% of lupus nephritis patients.

Renal Transplantation

Successful pregnancies are possible in renal transplant recipients. But there is an increased frequency of pre-eclampsia, low birth weight and premature births. Pregnancy outcomes are better in patients on hydroxychloroquine, low dose aspirin in the first trimester (primary prophylaxis for pre-eclampsia), clinically inactive SLE, serum creatinine <1.5 mg%, non significant proteinuria <500 mg/day and well-controlled hypertension.[11]

Other Organ Flares

Multiple studies have shown that whatever flare occurred in pregnancy had been present in the 6 months prior to conception. These were nephritis, serositis followed by hematological, cutaneous and musculoskeletal flares.[12]

Drugs in Pregnancy and Lupus

Treatment of lupus in pregnancy should be aggressive and cautious. Many drugs can be teratogenic. Table 39.3 gives a list of drugs used in pregnancy and lactation. Rituximab can cause B cell depletion in neonate and hence is contraindicated during pregnancy. A patient who has received Rituximab should be advised against pregnancy for 6–12 months after receiving the last dose.

Foetal Complications

30% of SLE patients have anti-Ro and La (SS A and SS B) antibodies which cross the placenta by active transport between 16th and 30th weeks of gestation and cause neonatal lupus syndrome (NNLS), the most common manifestation. Corticosteroid use carries a worse prognosis. Incidence of preterm delivery is 25% and small for growth is 6–35%. Intrauterine growth retardation (IUGR) occurs due to placental insufficiency. Other risk factors are hypertension and hypothyroidism. 1 in 5 pregnancy losses occur due to a serum creatinine of >2.5–2.8 mg%, proteinuria >500 mg/day, GFR <60 ml/min and low platelet count.[10]

Organs Involved in NNLS

1. *Cutaneous neonatal lupus (CNL):* It is the most common (5%) manifestation occurring within the first 2 weeks of life and disappears within 3 to 6 months. Common lesions are geographical, erythematous photosensitive lesions over sun exposed areas, resembling subacute cutaneous lesions but leave no sequelae.

Table 39.3: Drugs for lupus in pregnancy and lactation

S.No.	*Stages*	*Names*	*FDA*	*Safe in Lactation*
	Antenatal			
1.	Without APS	HCQ 200–400 mg	C	Yes
	With APS	HCQ Aspirin—Low dose Heparin, dalteparin		Yes
2.	a. Flares	Prednisolone 10 mg/day		Yes
		Prednisolone <20 mg/day		
		Avoid NSAIDs after 28 weeks[14]	C	Yes
		Acetaminophen	C	Yes
	b. Severe flares	Pulse steroid	C	Yes
3.	Breastfeeding	Oral corticosteroids (wait for 4 hours after taking pill)	C	
4.	Severe flares immunosuppresion			
	a. Azathioprine	Safe 2 mg/kg/day	D	No
	b. Cyclosporine	Lowest effective dose	C	No
	c. IVIG		C	Yes
	d. Cyclophosphamide	Last resort	D	No

2. *Cardiovascular*

a. Complete heart block (CHB)—occurs in 2% and is the most severe. The risk increases to 18% if a previous child had CHB and up to 50% if there are two affected children. Along with CNL the risk increases to 6 to 10 fold. Anti-Ro and La positivity is increased in mothers of 80% of children with CHB. Salient features include:
   - Inflammation and fibrosis involving AV node and myocardium
   - Develops between 16 to 24 weeks of gestation
   - Fetal heart rate <60/min
   - Rarely cardiomyopathy
   - Requires permanent pacemaker
   - Perinatal death in 20%
   - 10 year mortality 20–35%.

b. 1st and 2nd degree heart block are also found and progress to CHB in childhood.

c. Endocardial fibroelastosis–less frequently.

   Early diagnosis can be made by:
   - ECG
   - Ultrasonogram
   - Fetal ECHO (18–28 weeks)
   - Trans abdominal ECG
   - Fetal kinetocardiogram.[2]

   Treatment of CHB:
   - Betamethasone and dexamethasone (4–8 mg/day) till end of pregnancy
   - IVIG 1 g/kg on 14th and 18th week of gestation–passive transfer of acquired auto-immunity
   - Plasmapheresis
   - Hydroxychloroquine

3. *Preterm birth*

4. *Low birth weight*

MATERNAL AND FETAL MONITORING[1,2]

Regular maternal and fetal monitoring improves the pregnancy outcomes as shown in Table 39.4.

Contraception

- Till lupus is under control
- Oral–depot progesterone
- IUCD

Antiphospholipid Syndrome

Untreated APS results in 45% to 90% fetal losses.

Management includes

1. Low dose aspirin (81 mg/day)–pre-conception
2. Unfractionated heparin 5000 units subcutaneous (sc) 12th hourly or
3. Dalteparin 5000 units sc/day plus
4. Aspirin
5. Hydroxychloroquine.

Previous venous or arterial thrombosis

- Low molecular weight heparin (LMWH): Enoxaparin—0.6–0.7mg/kg sc twice a day (for venous events); 1 mg/kg sc twice a day (for arterial events).

Peri-partum management

- Therapeutic doses of LMWH are decreased to prophylactic doses 24 hours before labour and restarted after 2–4 hours after epidural catheter is removed.
- Epidural anaesthesia–safe 12 to 24 hours after last dose of LMWH.
- Switched to warfarin 5–7 days post-partum or enoxaparin 1.5 mg/kg/day sc.
- Warfarin may be used in selected patients in 2nd and 3rd trimester with INR monitoring. It is safe in lactation.

Refractory APS

Hydrocychloroquine, reduces aPL binding to syncytiotrophoblasts and restores placental expression of Annexin A5, down regulates TLR4 expression, aPL induced inhibition of promigratory IL-6 and enhances trophoblast migration.[13]

Special Situations

1. **Pulmonary hypertension:**
   a. Phosphodiesterase 5 inhibitors
   b. Prostaglandin derivative—epoprostenol
2. Systemic hypertension—labetolol
3. Statins (StAmP trial)—to ameliorate early onset pre-eclampsia
4. Folic acid – 5 mg/day, 12 weeks before and after conception
5. Stop smoking and alcohol
6. Prophylactic calcium 1g and vitamin D to prevent osteoporosis. Vitamin D either alone or when combined with LMWH decreases aPL induced trophoblastic cellular immune responses and inhibits anti-angiogenic factor soluble fms-related tyrosine kinase release induced by LMWH.[13]

Table 39.4: Maternal and fetal monitoring in pregnancy with lupus

S.No.	*Visits*	*Laboratory*	*Maternal*	*Ultrasound*
1.	Obstetrics a. Monthly till 20 weeks b. Every 2 weeks till 28 weeks c. Every week till delivery	1. 1st visit • CBC • PT, aPTT • LAC • aCL–IgG, IgM • β_2 GPI–IgG, IgM Repeat 12 weeks later • Anti Ro, La, Sm, RNP • Blood sugar	Hypertension ECHO Lupus Activity Index	1st trimester Routine 16 weeks • 1st fetal echo • Echo every 2 weeks • Biophysical profile
2.	Rheumatology	• Serum creatinine • Uric acid • AST, ALT • Anti dsDNA • C3, C4, CH50 • Urine-protein, active sediments, protein/creatinine ratio, dysmorphic RBCs • Creatinine clearance • Urine culture	Disease activity SLEDAI etc.	24 weeks • Fetal growth • Amniotic fluid • Umbilical artery (fetal placental blood flow) • Uterine artery for pre-eclampsia and IUGR
3.	Quarterly	• CBC • Platelet count • Anti-dsDNA, • C3, C4, CH50 • Uric acid • AST, ALT • 24 hours urine protein • Protein/creatinine ratio, dysmorphic RBCs • Abnormal renal tests–consider renal biopsy before 32 weeks of gestation		• Umbilical and uterine artery Doppler ↑ Resistance • Absent or reverse diastolic flow

Key Points

1. Planned pregnancies are successful with lupus in remission.
2. Meticulous and methodical screening for disease activity and neonatal lupus is mandatory.
3. Co-morbid states should be kept under control.

REFERENCES

1. Stanhope TJ, White WM, Moder KG, Smyth A, Garovic VD. Obstetric nephrology: lupus and lupus nephritis in pregnancy. Clin J Am Soc Nephrol. 2012; 7:2089–99.
2. de Jesus GR, Mendoza-Pinto C, de Jesus NR, Dos Santos FC, Klumb EM, Carrasco MG, et al. Understanding and managing pregnancy in patients with lupus. Autoimmune Dis. 2015; 2015:943490.
3. Anderson RA, Wallace WH, Baird DT. Ovarian cryopreservation for fertility preservation: indications and outcomes. Reproduction. 2008; 136:681–9.
4. Yang H, Liu H, Xu D, Zhao L, Wang Q, Leng X, et al. Pregnancy-Related Systemic Lupus Erythematosus: Clinical Features, Outcome and Risk Factors of Disease Flares — A Case Control Study. PLoS ONE 2014; 9(8): e104375.
5. Lateef A, Petri M. Managing lupus patients during pregnancy. Best Pract Res Clin Rheumatol. 2013; 27: 435–47.
6. Wallace D, Hahn BH. Dubois' Lupus Erythematosus and Related Syndromes, 8th Edition, 2013, Philadelphia, Elsevier Saunders.

7. Koh JH, Ko HS, Lee J, Jung SM, Kwok SK, Ju JH, Park SH. Pregnancy and patients with preexisting lupus nephritis: 15 years of experience at a single center in Korea. Lupus. 2015; 24:764–72.
8. Guballa N, Sammaritano L, Schwartzman S, Buyon J, Lockshin MD. Ovulation induction and *in vitro* fertilization in systemic lupus erythematosus and antiphospholipid syndrome. Arthritis Rheum. 2000; 43:550–6.
9. Erkan D, Sammaritano L. New insights into pregnancy-related complications in systemic lupus erythematosus. Curr Rheumatol Rep 2003; 5:357–63.
10. Doria A, Tincani A, Lockshin M. Challenges of lupus pregnancies. Rheumatology (Oxford). 2008; 47 (Suppl 3):iii9–12.
11. Campise M, Giglio E, Trespidi L, Messa P, Moroni G. Pregnancies in women receiving renal transplant for lupus nephritis: description of nine pregnancies and review of the literature. Lupus. 2015; 24:1210–3.
12. Tedeschi SK, Massarotti E, Guan H, Fine A, Bermas BL, Costenbader KH. Specific systemic lupus erythematosus disease manifestations in the six months prior to conception are associated with similar disease manifestations during pregnancy. Lupus. 2015; 24:1283–92.
13. Chighizola CB, Raschi E, Borghi MO, Meroni PL. Update on the pathogenesis and treatment of the antiphospholipid syndrome. Curr Opin Rheumatol. 2015; 27:476–82.

FURTHER READING

1. Jones A, Giles I. Fertility and pregnancy in systemic lupus erythematosus. Indian J Rheumatol 2016; 11: 128–34.

Chapter

40

Update on Antiphospholipid Syndrome

Arvind Ganapati, Vishad Vishwanath, Debashish Danda

HISTORY

In 1952, a circulating substance leading to prolongation of coagulation was first described in SLE patients.[1] It was observed that there were increased rates of pregnancy loss in these patients. In 1963 Bowie et al, observed that the circulating anticoagulant in blood caused paradoxical increase in occurrence of thrombosis.[2] It was soon realised that the patients who had the circulating anticoagulant also had biological false positivity for syphilis (BFP). The circulating anticoagulant was later termed lupus anticoagulant (LAC). Subsequent research showed that lupus anticoagulant was nothing but an antiphospholipd antibody (aPL ab).[3] In 1985, Hughes formally described the syndrome of arterial and venous thromboses and recurrent pregnancy loss associated with BFP, which is now known as the antiphospholipid antibody syndrome (APS).[4] Sapporo criteria, the first classification criteria, for this syndrome was described in 1999.[5] The criterion was modified during the 11th International Congress on Antiphospholipid Antibodies held at Sydney in 2006.[6]

The last 3 decades have witnessed remarkable advances in the understanding of manifestations, pathogenesis, diagnosis and treatment of antiphospholipid syndrome (APS). In this chapter, we will endeavour to highlight these information on the pathogenesis, diagnosis and management of APS.

ANTIPHOSPHOLIPID ANTIBODIES

Anionic phospholipids were initially described as the target antigens for antiphospholipid antibodies (aPL abs). These phospholipids included cardiolipin (diphosphatidyl glycerol), phosphatidyl serine, phosphatidyl inositol and phosphatidyl ethanolamine. But subsequent studies showed that the major antigenic targets are in fact certain proteins which complex with these phospholipids. The most important of these proteins was identified as the Beta 2 glycoprotein 1 (β2GP1). Later antibodies to about fifty other antigens were identified which include those reacting with prothrombin, thrombin, phosphatidyl serine prothrombin complex, annexin V, high and low molecular weight Kininogen, oxidized LDL, tissue plasminogen activator and coagulation factors XII and VII.[7,8] A subset of these antibodies, however, do bind purely to phospholipids. These phospholipid binding antibodies belong to the category of natural antibodies; and probably have a role in homeostasis and have been found to be elevated in infections.[9,10]

Epidemiology

Some studies found up to 10% of healthy blood bank donors positive for antiphospholipid antibodies, however, only 2% of them remain persistently positive.[11,12] Studies also found LA in 0.3% and aCL in 2–9% of normal pregnancies and these figures were comparable to those in healthy non-pregnant women.[13]

Etiology of Antiphospholipid Syndrome (APS)

APS may occur as a manifestation of another disease (secondary) or idiopathic (primary). Primary APS constitutes more than 50% of all cases of APS and is by far the most common acquired thrombophilia.[14]

The most common cause of secondary APS is SLE.[15] The prevalence of antiphospholipid antibodies in lupus patients range from 12 to 44% for anti-cardiolipin (aCL) antibody, 15–34% for LAC, and 10–19% for anti-β2GP1 antibodies.[19] The prevalence of antibodies appears to increase with duration of disease. The Hopkins Lupus cohort, a 20 years prospective follow-up cohort, reported a prevalence of 47%, 26% and 32.5% respectively for aCL, LAC and anti-β2GP1 in patients with SLE.[19,20] It is also known that aPL abs may be one of the first antibodies to appear in SLE and may predate the diagnosis of SLE by up to 7 years.[16] The presence of these antibodies, especially LAC, is a predictor for thrombosis as, up to 50% of patients with LAC develop deep vein thrombosis.[17]

Higher than normal prevalence of aPL ab has been described in other autoimmune diseases as well, though their relationship to disease manifestations are less well established. Prevalence rates have been between 7 and 37% in RA and along with antibodies to oxidized LDL, antiphospholipid antibodies may have a role in cardiovascular morbidity in RA too.[18,19] However, direct link between aPL ab and thrombosis have not been established well in RA. Use of infliximab in RA has also been associated with new occurrence of aPL ab.[20]

In scleroderma, though their prevalence may not be higher than general population, presence of aCL has been reported to be associated with more vascular manifestations as digital pitting scars, pulmonary arterial hypertension and capillary changes.[21] Presence of LAC in Sjögren's syndrome is associated with increased risk of thrombosis.[22] Presence of the rarer antiphospholipid antibodies like antiphosphatidyl serine has been linked to cutaneous PAN.[23] There are reports of higher prevalence of antiphospholipid antibodies in various systemic vasculitic conditions such as granulomatosis with polyangitis, giant cell arteritis (GCA), Behçet's syndrome and polyarteritis nodosa (PAN), but these are not pathogenic and are likely to be an epiphenomenon to systemic inflammation.

Malignancies, both solid and hematological, have been associated with higher prevalence of aPLs[24,25] which may contribute to increased risk of thrombosis in these patients.[26]

Infections and APL Positivity

Infections are associated with aPL antibody positivity, but these are mostly transient, IgM variety, of low affinity and these antibodies seldom cross react to β2GP1. Approximately 20% of HCV positive patients have associated aPL, but only a fraction of them have anti-β2GP1 antibody.[27,28] HCV associated aPL ab positivity, however, is not associated with excess of thrombosis. aPL antibodies are also seen in up to 50% of patients infected with HIV and a fraction of them are even β2GP1 positive.[10] However, no increased rates of thrombosis have been confirmed in these patients. One study had suggested that aPL ab in HIV may be associated with an increase risk of cutaneous necrosis and avascular necrosis.[28] High prevalence of aPL ab has been demonstrated in HBV infected individuals, but this is not found to be associated with thrombosis. Other viral infections as EBV, CMV, varicella and Parvovirus have also been associated with aPL ab positivity.[27]

aPL ab positivity occurs in many bacterial infection including leprosy, tuberculosis, malaria, rickettsial infection, syphilis and leptospirosis. But significantly increased risk of thromboses has not been consistently demonstrated in these infections. Amongst them, leprosy is reported to have high prevalence of anti-β2GP1 antibodies.[16,28]

Drugs and APS

aPL abs have been known to occur following exposure to drugs which include quinidine, procainamide, phenothiazines, phenytoin, hydralazine, propranolol, interferon and infliximab.[29] These antibodies are usually transient in nature, mostly of IgM isotype and do not cross react with β2GP1. Thrombosis, however, has been reported with quinidine, procainamide and phenytoin. Although formation of aPL abs have been documented with infliximab, etanercept and adalimumab, there is no report of increased risk for thrombosis associated with their use.[30]

Pathogenesis of Antiphospholipid Syndrome

aPL, though previously thought to directly recognize anionic phospholipids, are actually directed against phospholipid-binding protein and their main antigenic target is β2GP1, which can induce humoral and cellular immune responses.

β2GP1 Structure and Function

β2GP1 is a protein of approximately 50-kD, composed of five "sushi" domains, of which domain V mediates the binding of the molecule to anionic phospholipids, while domain I seems to be the main target of antibodies associated with an increased risk of thrombosis.[31] Besides, β2GP1 exists at least in two different conformations: A circular plasma conformation in which domain I interacts with domain V and an "activated" open conformation.[32,33] After the positively charged patch of domain V binds to anionic surfaces, the open conformation is obtained. Thus, β2GP1 exposes the hidden epitopes, especially the cryptic epitope on the domain I, which is recognized by anti-β2GP1 antibodies in the APS.[34]

Increasing evidence suggests that misfolded β2GPI proteins are rescued from degradation and transported to the cell surface without processing to peptides, when they associate with the peptide binding groove of HLA class II molecules in the endoplasmic reticulum (ER).[35] These misfolded β2GP1 proteins associated with MHC class II molecules are transported intact to the cell surface without processing to peptides. Furthermore, these complexes efficiently stimulate β2GP1-specific B cells.[36]

Role of Toll-like Receptors (TLR) in APS

TLR4 provides a third signal for B cell activation which enables B cells respond better to signals from both B cell antigen receptor (BCR) and CD40, thus upregulating B cell activation, surface molecules expression, anti-β2GP1 ab production, and cytokines secretion as well as facilitating B cells to function like an antigen presenting cell (APC). At the same time, TLR4 also enhances production of antibodies by B cells via upregulation of expression of B cell activating factor (BAFF).[37]

Functionally the anti-β2GP1 antibodies can bind to more than one β2GP1 molecule. This binding to anti-β2GP1 on cell membrane induces clustering of β2GP1 on the cell surface. This binding also enhances the affinity of β2GP1 for anionic phospholipids on cell surface. These clusters, and not the monomeric forms are in turn capable of activating a myriad of receptors present on the membranes of endothelial cells, monocytes, platelets, trophoblasts and decidual cells. The receptors include ApoER2, glycoprotein 1 b alpha, heparin sulphate, annexin II, TLR 2 and TLR 4.[38,39] The binding to some of these receptors leads to cell activation through trasnsciption pathways as the NFK β and P38 MAPKs. This in turn triggers procoagulant or inflammatory pathways depending on the type of cell activated.

The prevailing view is that pathogenic aPL bind anionic phospholipids such as cardiolipin only in the presence of a protein cofactor or bind directly to the protein cofactors, of which β2GP1 is considered the most important. Cofactor dependence or direct cofactor binding is believed to distinguish pathogenic from non-pathogenic aPL. Cofactor-independent aPL are generally regarded as irrelevant for the APS. However, this central dogma of pathogenesis, has been challenged in a recent review, as these observations has never been substantiated by appropriate clinical studies.[40]

Etiopathogenesis of Thromboses and Pregnancy Loss

Thrombosis

Passive infusion of aPL antibodies in mice have been associated with thrombosis through increased expression of endothelial adhesion molecules, tissue factor and nitric oxide.[56] The antibody form complexes chiefly with β2GP1 bound on the cell surface leading to activation of downstream molecules finally leading to endothelial cell activation and thromboses. The role of the recently described anti-prothrombin antibodies in thrombosis have also been studied and it appears that the binding of the antibody leads to activation of prothrombinase and a 'gain of function of prothrombin' which aids coagulation.[41] The outcome of aPL ab binding to cell will depend on the function of cells. For example, anti-β2GP1 antibody binding with β2GP1 present on platelet surface can activate APO E2 receptor on platelet surface. This can lead to downstream activation of P38/MAPK pathway ultimately leading to increased thromboxane release.[42] It has also been suggested that the anti-phospholipid antibodies are capable of disrupting the activated protein C complex, a powerful anti-coagulant and this process may contribute to the prothrombotic effect.[43] Other potential causes for thrombosis in APS includes disruption of the anticoagulant annexin A5 coating on the phospholipid surface by aPL antibodies and interference with fibrinolytic pathways.[41]

Neutrophil extracellular traps (NETs) have recently been recognized as an important activator of the coagulation cascade, as well as an integral component of arterial and venous thrombi. In a study, sera and plasma from patients with primary APS were shown to have elevated cell-free DNA and NETs, as compared to healthy volunteers. Freshly isolated neutrophils from patients with APS also showed high levels of spontaneous NET release. Exposure of sera of healthy controls to monoclonal aPL antibodies from APS patients, especially those targeting β2GP1, also enhanced NET release.[44]

Pregnancy Loss

Several different mechanisms have been suggested for pregnancy loss:

1. Placental infarction: Given the prothrombotic effects of anti-phopsholipid antibodies, it is possible that aPL abs lead to formation of thrombus and hence insufficiency in the materno-fetal circulation.[45,46] Supporting this theory is the finding that annxein A5, an anti-coagulant protein has high expression on placenta and it binds with phospholids on cell membrane.[47] This coating denies the clotting pathway of the much needed phospholipids for initiating thrombosis. It has been postulated that binding of anti-β2GP1 antibodies destabilse the annexin A5 coating leading to thromboses. However, histopathological studies have failed to demonstrate thrombus in placental samples from many of the miscarried fetuses from mothers with APS raising questions on this theory for fetal loss.

2. Proinflammatory mileu and complement pathway: In a pregnant murine model based on BALB/c mice and FcRy mice, infusion with aPL had caused increased fetal losses and exhibited fetal growth retardation. Mice with complement deficiency or complement blockage showed protection against aPL-induced pregnancy complications, as did treatment with heparin (which has an anti-complement effect) as opposed to fondaparinux with no anti-complement effect.[48,49] Increased complement deposition products were observed in the decidua of pregnant mice infused with aPL, whereas mice depleted of neutrophils or mice treated with heparin did not show pregnancy loss or growth restriction as well as there was a lack of complement deposition in their decidual tissue.[48] The findings from these animal studies were supported in human studies by Shamonki et al, who showed that placentae of women with aPL have increased complement deposition.[50]

3. Anti-trophoblastic effects of aPL abs: It was proposed that the pathogenesis of aPL related recurrent pre-embryonic loss might differ from the pathogenesis of morbidity occurring in late pregnancy.[51] aPL have direct pathogenic effects on placentation and apoptosis of trophoblast cells, which may play an important role, particularly in early recurrent miscarriages.[52]

Though there is debate on the exact mechanism of pregnancy loss in APS, it is well accepted that the complement pathway plays a major role in its pathogenesis.

The Two Hit Hypothesis

A large number of asymptomatic individuals are seropositive for aPL abs. But only a fraction of them actually develop symptoms. Thus two hit hypotheses has thus been invoked to explain this observation in pathogenesis of APS. Accordingly, an aPL ab positive patient may not develop thrombosis unless there is a second insult. Most common insult appears to be infections.[53] Other second hits may include use of OCPs, other prothrombotic risk factors, etc.

Clinical Features of APS

APS can cause thromboembolism in any part of the body and the manifestations depend on the organ involved.

Systemic Vascular Thrombosis

Deep vein thrombosis (DVT) in the lower extremities is the commonest presentation and up to half of these patients develop pulmonary embolism. In one prospective study, aPL abs were elevated in up to 20% of patients with DVT or pulmonary embolus before the event.[54] Cervera et al reported that venous thrombosis is the presenting feature in up to 30% of patients with APS.[15] Superficial vein thrombosis can exist concomitantly or as an isolated feature. Spectrum of venous thromboses can range from upper extremity thrombosis and cerebral venous sinus thrombosis (Fig. 40.1) to thrombosis involving superior and inferior vena cava, hepatic veins (Fig. 40.2) (Budd-Chiari syndrome), portal vein, renal veins and retinal veins.

Arterial thromboses may manifest with ischaemia of digits (Fig. 40.3) or infarction in central nervous. system, myocardium, kidney, gall bladder, mesentery, etc.

Central Nervous System (CNS)

Ischemic stroke is the first presentation of APS in up to 30% of patients.[15] They are the leading causes of mortality in APS. APS is one of the most common causes of young stroke. In the RATIO study, the odds ratio for ischemic stroke among patients with lupus anti-coagulant positivity was 43.1.[55] In one Indian study, the prevalence of aPL abs was 29.4% in patients with young stroke.[56] Most often these infarcts involve the middle cerebral arteries. Co-existing risk factors such as age, smoking, OCP use and high CRP levels are associated with an increased risk for stroke.

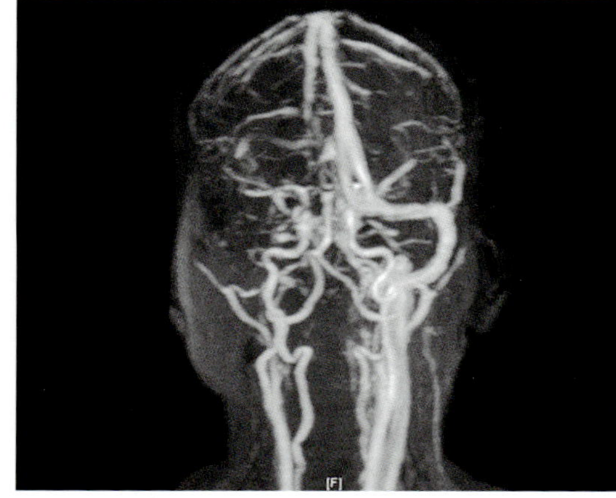

Fig. 40.1: MR venogram of a patient with APS causing thrombosis of left transverse and sigmoid dural venous sinuses. Partial thrombus is also seen in the right transverse sinus

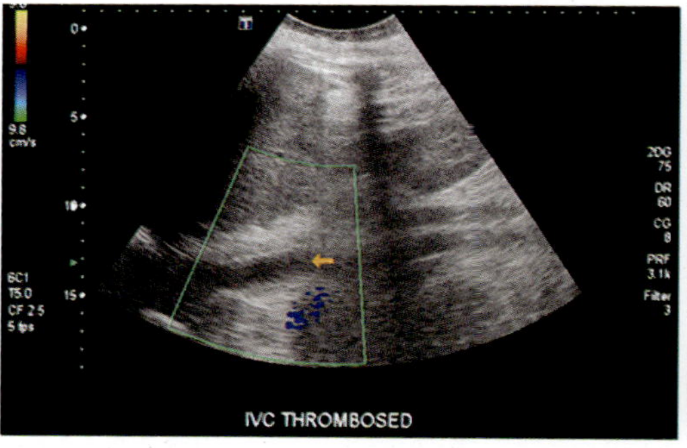

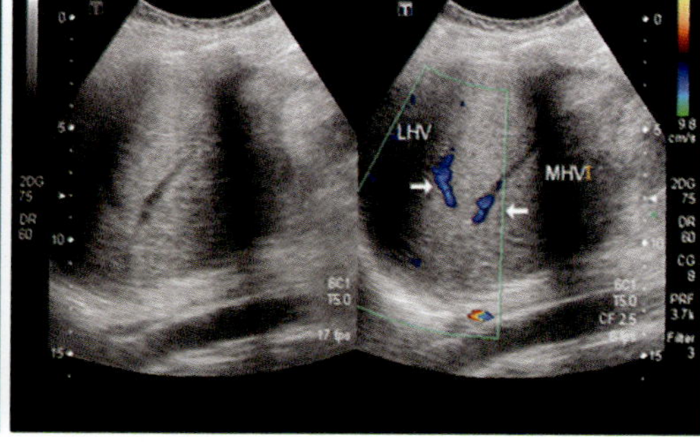

Fig. 40.2: 2D color Doppler image of a patient with APS with thrombosed inferior vena cava (left) and partial occlusion of left hepatic and main hepatic vein in the same patient (right)

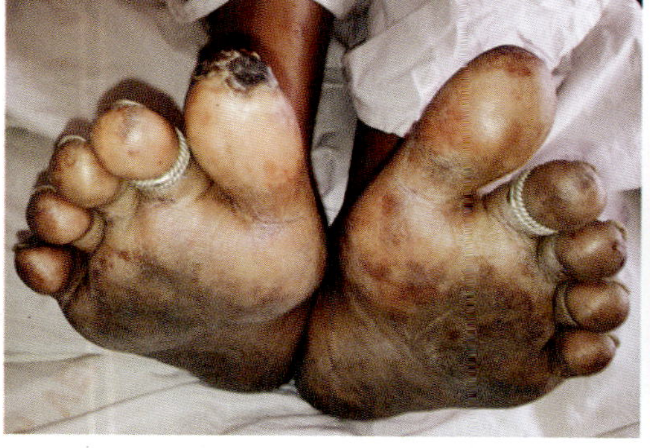

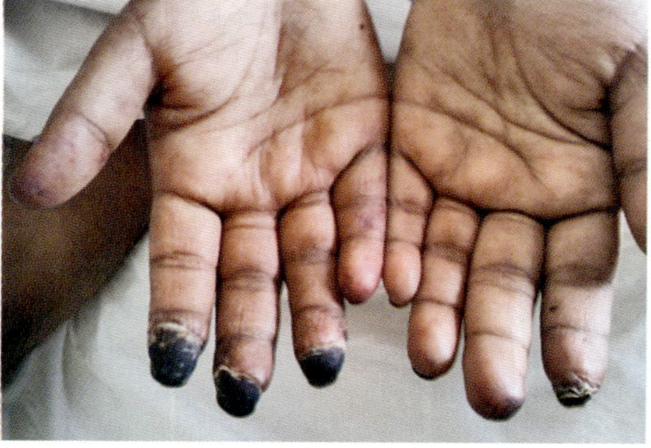

Fig. 40.3: Gangrene of toes and fingers in an APS patient

APS screening hence should be done in all cases of young stroke. Sneddon's syndrome, a triad of livido reticularis, hypertension and stroke is one of the manifestations of APS. Rarely APS can present with dissociated sensory loss and weakness due to anterior spinal artery involvement. Multi-infarct dementia is another rare but devastating manifestation of APS.[57]

Equally important, but less frequently recognized features of neurological APS include the non thrombotic neurological manifestations in APS.[58] The cause of these may be small vessel ischaemia, local inflammation or direct neurotoxicity. These manifestations range from cognitive dysfunction, migrainous headache, seizures, chorea, transverse myelitis and optic neuritis. In fact, presence of antiphospholipid antibodies in SLE patients have been found to be the major risk factors for neuropsychiatric manifestations independent of the presence of thromboses. Asymptomatic LA positivity also has been identified as a predictor of cognitive dysfunction in SLE.

Renal

All renal vasculature can be affected in APS.

Both renal artery and renal vein thrombosis are known to occur in APS.[59] Renal artery thrombosis, if unilateral can cause infarction and scarring and if bilateral cause acute renal shutdown.

Renal artery stenosis without thrombosis is an increasingly recognized manifestation APS. It appears that APS induced endothelial dysfunction and proliferation, rather than frank thrombosis is the cause for this. Renal vein thrombosis can present with massive hematuria and nephrotic range proteinuria.

APS nephropathy is a well-recognized manifestation and can closely resemble thrombotic thrombocytopenic purpura (TTP) or lupus nephritis, when it occurs in SLE patients. Histology would reveal microvascular thromboses. Alternatively TTP itself can be a manifestation of APS. Proliferative glomerulonephritis has also been associated with APS. In renal transplant recipients, antiphospholipids antibodies are a major risk factor for renal allograft rejection. The vascular endothelium of proliferating intrarenal vessels from patients with APS nephropathy showed indications of activation of the mTORC pathway. Patients with APS nephropathy who required transplantation and were receiving sirolimus had no recurrence of vascular lesions and had decreased vascular proliferation on biopsy, as compared to patients with antiphospholipid antibodies who were not receiving sirolimus.[60]

Cardiac

Valve thickening especially of the mitral valve can be seen in up to a third of APS patients. This can be associated with regurgitation, stenoses and vegetations.[61] Antiphospholipid antibodies also cause accelerated atherosclerosis and endothelial dysfunction. Myocardial infarction can occur due to premature atherosclerosis or acute thrombosis. Recurrent in-stent restenosis is a feature in those patients undergoing coronary stenting. APS can cause intracardiac emboli, dilated cardiomyopathy and both thromboembolic and non-thromboembolic pulmonary arterial hypertension.

Pulmonary

Antiphospholipid lung syndrome is a rare and life threatening manifestation of APS, often an warning manifestation of catastrophic APS (CAPS) and is characterized by pulmonary microthrombosis, thromboembolism of lung arteries, pulmonary hypertension, adult respiratory distress syndrome and diffuse alveolar haemorrhage.[62]

Dermatological Manifestation

There are several non criteria manifestations in skin. This includes livido reticularis (Fig. 40.4), ulcerations, digital gangrene, subungal splinter haemorrhages, superficial vein thrombosis, purpura, atrophic blanche, extensive cutaneous necrosis and primary anetoderma, a localized flaccid sac-like skin lesion.[63]

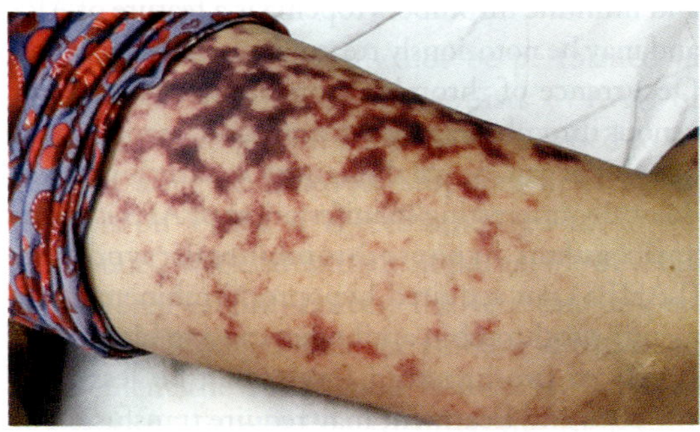

Fig. 40.4: Livedo reticularis pattern in an APS patient

In spite of these guidelines there is still lack of uniformity in reporting. In one study, the reproducibility of LAC was as low as 33% between reference laboratories.

Recently a new assay has been developed that measures the anti-β2GP1 dependent LAC, which is likely to be more sensitive in detecting pathogenic LAC than the functional assay.

Anti-cardiolipin and Anti-β2GP1 Antibodies

The detection of these antibodies are based on solid phase ELISA.[71] It is recommended to test for both IgG and IgM isotypes. The reporting is done as MPL or GPL units for aCL. There are no universal standards units for anti-β2GP1. Like LAC, these assays are also plagued with problems of wide inter-assay variability. Differences in the purity of the β2GP1 antigen used, differences in oxidation status of the antigen used, variations in calcium ion concentration in different sera, lack of standard calibrators for the assay and insufficient definition of positive cut-off values are some of the causes for this variability.[72]

Laboratory Markers as Predictors of Clinical Events

The exact predictive values of the APLs are not known, but all data suggest that LAC positivity has the highest risk for clinical events.[73] However, this test is positive in only 20–40% of APS patients. Studies have also shown that LAC positivity correlates with the titer of anti-β2GP1 antibodies.

LAC positivity has been associated with increased occurrence of arterial as well as venous thromboses and pregnancy loss . Further it has been shown that a test which specifically detects the β2GP1 dependent lupus anticoagulant is more strongly associated with thromboses (OR 10.2 *vs* 40.3) as compared to standard LA assay.[74] However, it should be remembered that some asymptomatic individuals can have LAC positivity especially in low titers. The predictive value of other antibodies for clinical events is actually less evident. Higher concentration of both aCL and anti-β2GP1 also correlate well with clinical events, albeit to a lesser extent as compared to LAC.[75]

It is also been increasingly realized that the positivity of more than one of the three standard lab tests has a higher predictive value for events. This is a strong argument in favour of performing all three tests in any given patient. Based on this understanding, Miyakis et al have suggested to classify aPL ab positivity into 2:[6]

Class 1. Two or more laboratory criteria present

Class 2a. Isolated LAC positive

Class 2b. Isolated aCl positive

Class 2c. Isolated anti-β2GP1 positivity

In a study on the aPL profile of 175 APS patients by Tincani et al, it was found that 35%, 28% and 37% of patients respectively had all three, two out of 3 and single test positivity. Of the patients with only 1 positive test, 75% had β2GP1 antibody, 19% had LAC and 6% had a Cl positivity.[25]

Seronegative APS (SNAPS)

A subset of patients present with classical clinical manifestations of APS, are persistently negative for any of the anti-phospholipid antibodies tested routinely. These patients can be described to have sero negative APS (SNAPS). It has been argued that the absence of antibodies represent technical issues rather than true absence. Different causes have been suggested. These include:

a. Transient fall in aPL ab levels following thrombosis due to consumption; therefore, if tested immediately following an acute episode of thrombosis, the levels may be falsely low.[76]

b. Conventionally only IgM and IgG subtypes of the aCL and anti-β2GP1 antibodies are measured. However, some of the patients with APS have been found to have only IgA subtype and hence their presence is unrecognized.[77]

c. Wide intra and inter observer variability has been observed in testing for antiphospholipid antibodies and hence it is possible that the assay errors contribute to a substantial number of SNAPS cases.

d. Antibodies against antigens other than β2GP1 have been detected in subsets of patients with APS. Notable among these are antibodies directed against prothrombin and vimentin. The presence of these antibodies have been detected in sera of patients in absence of the classical antibodies.[78] Newer autoantibodies in APS are summarized in Table 40.1.

Table 40.1: Newer antibodies in antiphospholipid syndrome (APS)

Antibody	*Salient features and associations*
Antibodies to phosphatidyl ethanolamine	Significantly associated with fetal loss and/or thrombosis, present in the absence of the laboratory criteria of APS[79,80]
aPLs to negatively charged phospholipids other than cardiolipin	Antibodies against phosphatidic acid (PA), phosphatidylserine (PS) and phosphatidylinositol (PI) Significant correlation with recurrent pregnancy loss in the absence of other APLs[81]
Anti-domain I antibodies of β2GP1	Associated with thrombosis (predominantly venous) more than their counterparts targeted to other domains of β2GP1 and with pregnancy morbidity[34,82]
Anti-domain 4/5 antibodies	IgG anti-β2GP1-Dm 4/5 are present in individuals with single positivity for IgG β2GP1 and not associated with thrombo-embolic events, as opposed to IgG anti-β2GP1-Dm 1 which is present in triple positive patients and associated with thromboembolic events The ratio of anti-Dm1 to anti-Dm 4/5 may be a useful tool for the diagnosis of antiphospholipid syndrome[83,84]
Antibodies to vimentin/cardiolipin complex	Sensitive markers for APS as there is a persistent presence of IgG and IgM anti-vimentin/cardiolipin complex antibodies in almost all patients with APS and a large portion of patients with SNAPS Overlapping presence of anti-vimentin/cardiolipin antibodies in SLE and RA patients, however, weakens its specificity[85]
Antibodies to prothrombin	Predicts first or recurrent risk of thrombosis in patients with APS[86]
Antibodies to prothrombin/ phosphatidylserine complex	Positive association between aPS/PT (IgG and/or IgM isotype) and arterial and/or venous thrombosis Higher sensitivity and specificity than aCL Solid correlation and surrogate marker of LA and it can be used as a confirmatory test for APS[87]
Annexin A5 antibody (aAnxA5)	Clinical correlation is inconsistently reported in relation to pregnancy-related morbidity No association with thrombosis[88–90]
IgA aCLs	Prevalence in SLE in African American, Afro-Caribbean and Hispanics to be 16%, 21% and 14%, respectively[91]
IgA anti-β2GPI antibodies	More associated with deep venous thrombosis than the IgM isotype[92]

APS in Special Scenario

Catastrophic APS (CAPS)

Catastrophic APS is a serious form of this illness presenting with/or leading to multi-organ failure. It is a life-threatening medical emergency with a mortality of about 50%. It may be the first presentation of APS in about 50% of this complication and half of the episodes of CAPS are precipitated by infections. It involves at least 3 organs over a period of days or weeks with histological evidence of multiple occlusions of large or small vessels.[93,94]

The kidney is the organ most commonly involved followed by the lungs, CNS, heart, adrenal and skin. Acute respiratory distress syndrome due to non-thrombotic microangiopathy is the most common pulmonary manifestation, followed by pulmonary embolism, and alveolar haemorrhage. Cerebral involvement may be in the form of stroke, cortical venous thrombosis or diffuse encephalopathy. The gastrointestinal system could also be involved resulting in gut ischemia and abdominal pain. Patients often have thrombocytopenia and micoangiopathic hemolytic anemia. Peripheral smear examination may reveal schistocytes, but this finding is rarer in comparison to that in TTP which is a close differential diagnosis. Disseminated intravascular coagulation (DIC), which is extremely rare in primary or secondary APS, is noted in about 25% of patients with catastrophic antiphospholipid syndrome.

Microvascular manifestations of catastrophic APS include renal thrombotic microangiopathy, adult respiratory distress syndrome, cerebral microthrombi with microinfarctions and myocardial microthrombi.

The pathophysiology of this syndrome is still poorly understood; however, initial thrombosis in a patient with APS may alter the balance of hemostasis and cause a "thrombotic storm", leading to multiple microthrombi in the body.

The precipitating factors include:

1. Infections of upper respiratory tract, intestine, urinary tract and leg ulcers.
2. Surgical procedures including angioplasty, dental extractions, endoscopic retrograde cholangiography, dilatation and curettage, caesarean section, hysterectomy and lung biopsy. But, even trivial events like needle stick injury can be a trigger.
3. Drugs: Withdrawal of anticoagulant or introduction of oral contraceptive agents, thiazide diuretics and captopril can also precipitate a CAPS event.

Management of Anti-phospholipid Syndrome (Fig. 40.5)

The presentation of APS can be varied from calf vein thrombosis to life threatening CAPS and the management strategy will differ based on the severity and spectrum of the organ involvement.

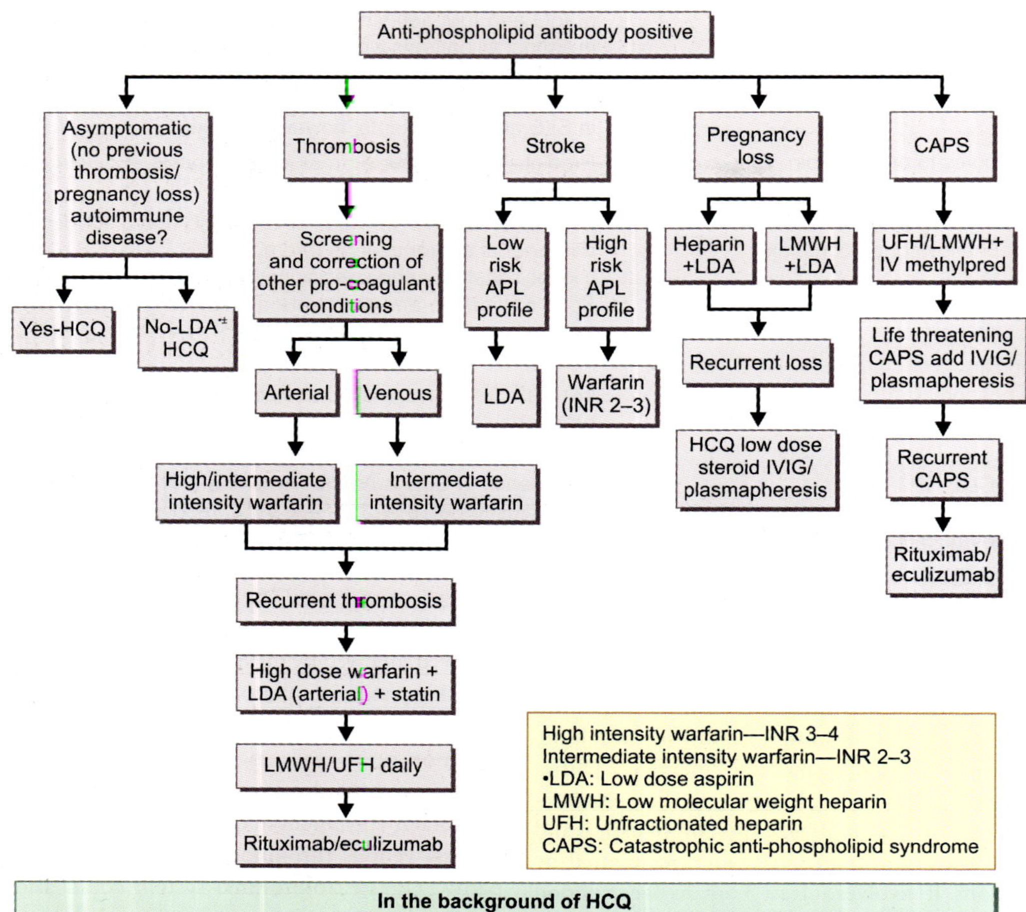

Fig. 40.5: Flowchart for management of APS

Primary Prevention

The risk of thrombosis in otherwise healthy individuals with anti-phospholipid antibodies is considered to be small (<1%/year).[120] But in aPL ab positive patients with SLE, without prior thrombotic event, the risk of thrombosis is in the range of 0–3.8% per annum.[121] Of these, 46–76% of the events are associated with a concurrent non-aPL thrombotic risk factors which suggests that a double hit is needed to cause a clinical thrombotic event in this scenario. Highlights of 2 studies with best methodological quality in primary prophylaxis of APL positive asymptomatic patients are presented in Table 40.2.

It appears that premenopausal women with obstetric APS also have an increased risk for subsequent thromboses. One study showed an incidence rate for thrombosis of 7.2 per 100 patient years in obstetric APS as against an expected background rate of 1 per 10,000 patient years in normal premenopausal women.[97] Hence it may be advisable to consider them for long-term antiplatelet therapy.

Secondary Prevention

Patients with stroke and APL positivity are at increased risk of recurrence; Table 40.3 shows 2 RCTs on secondary prevention.

Although currently anti-coagulation is the therapeutic option of choice in APS, newer management strategies also have been evolving.

Oral Anticoagulation

Oral anticoagulation is the treatment of choice for long-term secondary thrombo-prophylaxis. However, there is much debate on the dose and

Table 40.2: Highlighting 2 studies on primary thromboprophylaxis in APL positive asymptomatic subjects

Study	*Type of study*	*Treatment*	*Number of patients*	*Duration*	*Event rate*	*Main findings*
ALPASA study Erkan et al 2007[95]	RCT	Low dose Aspirin *vs* none	98	Mean 2.4 years	2.75/100 patient years *vs* 0/100 patient years	Low dose aspirin not beneficial for primary prophylaxis in APL positive subjects
ALIWAPAS study Cuadrado et al 2014[96]	RCT	Low dose aspirin (LDA) *vs* low dose aspirin + low intensity anticoagulation	166	5 years	4.9% *vs* 4.8%	LDASA+ low intensity anticoagulation offers same benefit as LDASA alone for primary prophylaxis in APL positive subjects

Table 40.3: Highlighting 2 studies on secondary prevention of stroke patients with APL positivity

Study	*Study design*	*Number of patients*	*Treatment*	*Observation duration*	*Outcome*	*Main findings*
APASS study Levine et al 2004[98]	RCT subgroup analysis	720	ASA 325 mg/day *vs* warfarin target INR 1.4–2.8	2 years	Recurrence rate 22.18% *vs* 26.15%	LDA, low intensity anticoagulation equally effective in secondary stroke prevention
Okuma et al 2010[99]	RCT	20	LDA *vs* LDA + warfarin target INR 2–3	3.9 years	Stroke-free survival of 25% *vs* 75%	Warfarin + LDA was more effective than LDA alone in secondary prevention of stroke

duration of oral anticoagulant therapy. Although some experts suggest that the duration of anti-coagulation should be decided on a case to case basis, with a shorter course of anti-coagulation for low-risk APS (e.g. intermediate titre positivity of aCL/β2GP1 or definite correctable precipitating factor for thrombosis), there is now good agreement that it is best to treat all APS patients with past thrombotic event with life long anticoagulation.

Venous Thrombotic Events

Anticoagulation with a target INR just below or around 3.0 has been shown to confer effective protection against venous recurrences.[100] The two randomized clinical studies comparing moderate- and high-intensity anticoagulation in patients with a definite diagnosis of APS failed to report any difference between the two regimens.[101,102] Both studies were specifically designed to demonstrate superiority of high-intensity warfarin in terms of better prevention of recurrent thrombosis as compared to moderate-intensity anticoagulation. In the study by Crowther on 114 APS patients, the incidence of recurrent thrombosis was even higher among patients receiving high-intensity warfarin (10.7%) as compared to those in the moderate-intensity arm (3.4%), although this difference did not achieve statistical significance. Conversely, in the 2005 WASP trial, the incidences of recurrence were 11.1% and 5.5% among patients receiving moderate-intensity and high-intensity warfarin respectively. However, both the studies had limited statistical power, because of inadequate sample size.

Arterial Thrombotic Events

The randomized trials by Crowther and Finazzi recruited patients with a history of arterial events, even though the latter were under represented as compared to subjects experiencing venous thrombosis (24% in the study by Crowther and 32% in the trial by Finazzi). These two studies were not in support of a superiority of high intensity as compared to moderate-intensity anticoagulation. As a consequence, a 2006 systematic review which included only these randomized controlled trials recommended only moderate anticoagulation.[103] However, a standard target INR range of 2.0 to 3.0 was not sufficient in preventing recurrences among patients presenting with arterial events in many studies on oral anticoagulation. These observations suggest and emphasise the requirement of high-intensity anticoagulation in arterial thrombosis.[15,104–106]

Direct Oral Anti-coagulants

Recently, Cohen et al. reported the impact of rivaroxaban on thrombin generation in APS patients.[107] This controlled, randomized, open label, non-inferiority trial (RAPS) compared 54 APS patients treated with rivaroxaban versus 56 APS patients treated with warfarin. The primary endpoint was the mean percentage change in endogenous thrombin potential (ETP) determined by thrombin generation before and after the introduction of rivaroxaban. Patients with at least one venous thrombosis during no or sub-therapeutic anticoagulant therapy were included, and they were considered to be low-risk APS patients. High-risk APS patients, namely those with previous arterial manifestations or with recurrent VTE while on therapeutic dose of warfarin, were not included. Moreover, only 25% of patients with triple positivity were included. Overall, patients treated with rivaroxaban had a significant twofold-increase in thrombin potential, suggesting a higher thrombotic risk, in comparison with warfarin users. However, authors stated that no increased thrombotic risk was noticed in the rivaroxaban arm as compared to standard-intensity warfarin users, because no clinical event occurred during the short follow-up (210 days).

A systematic review analysed 122 APS patients treated with direct-acting oral anticoagulants (DOACs); among them, 19 experienced recurrent thrombosis while on DOACs. Of note, triple positivity (positivity of all three laboratory criteria for APS) was associated with a 3.5-fold increased risk for recurrent thrombosis.[108]

Several trials in thrombotic APS using strong clinical endpoints (TRAPS trial evaluating rivaroxaban and ASTRO-APS trial evaluating edoxaban) are ongoing.[109,110] Results of these trials are expected to address these issues regarding efficacy and safety of DOACs in APS.

Hydroxychloroquine (HCQ)

HCQ is an antimalarial drug with anti-inflammatory and antithrombotic properties. In addition, HCQ has been shown to exert immunomodulatory effects: It prevents activation of TLR3, TLR7, and TLR9, inhibits antigen processing and presentation, and reduces circulating immune complexes.[111] *In vitro* models of thrombotic APS have demonstrated HCQ to inhibit GPIIb/IIIa expression on aPL-activated platelets,[112] to reverse the formation of aPL-β2GP1 PL bilayer complexes[113] and to prevent the aPL-induced disruption of the Annexin A5 shield.[114] Patients receiving a combination of HCQ and oral anticoagulation experienced less recurrences as compared to those on anticoagulants only. However, the extrapolation of data is affected by the limitations of small sample size of 40 patients and the follow-up lasted only 36 months.[115] It is generally recommended that all patients with thrombotic APS as well as those asymptomatic aPL carriers must take HCQ indefinitely.

Statins

Statins inhibit cholesterol synthesis in the mevalonate pathway by blocking the 3-hydroxy-3-methyl-glutarylcoenzyme A (HMG-CoA) reductase. It has a wide array of additional pleiotropic anti-thrombotic and anti-inflammatory effects in APS, apart from its usual dyslipidemia benefits.

Fluvastatin and simvastatin were both reported to suppress anti-β2GP1 antibody-induced endothelial adhesiveness and to reduce monocyte adhesion to the endothelium, while rosuvastatin inhibited the upregulation of VCAM induced by aPL.[116] Erkan et al observed a significant reduction in half of the evaluated proinflammatory and procoagulant parameters (IL15, VEGF, TNFα, IP10, CD40L and TF) in a cohort of 41 asymptomatic aPL carriers after three months of treatment with fluvastatin.[117]

Corticosteroids

Contrary to expectations, corticosteroids have not been beneficial in the management of thrombotic manifestation or pregnancy morbidity in APS. However, use of corticosteroid is beneficial in some of the non-criteria manifestations of APS as Coombs' positive hemolytic anaemia, immune thrombocytopenia, demyelinating syndrome, and CAPS. Benefit in obstetrical scenario is discussed below.

Management of Obstetric APS

Unfractionated Heparin (UFH) is the treatment of choice for prevention of obstetric APS.[118,119] The mechanism of action of UFH in obstetric APS is multifaceted. UFH not only inhibits thrombus formation, it also prevents complement activation, inhibits binding of aPL abs to the trophoblastic surface, promotes trophoblast invasiveness and inhibits apoptosis of trophoblasts. UFH is used in doses of 5000–10000U twice daily to reduce pregnancy loss.

The other treatment modalities include low dose aspirin (LDA) and low molecular weight heparin (LMWH). Although a few studies had addressed the issue, the effectiveness of LDA alone has not been conclusively proven in APS. Studies comparing LDA alone and LDA with UFH have shown that the combination is better than LDA alone.[118,119] Though evidence is scarce, most experts recommend to use LDA in combination with UFH. A recent Cochrane based systematic review determined that combination of UFH with LDA reduces risk of pregnancy loss in comparison with LDA alone (RR of 0.46, CI 0.29-0.71).[120] It is recommended to start LDA preconception and heparin at the first confirmation of pregnancy. It is recommended to stop heparin 24 hours prior to the expected delivery and restart 6 hours later to be continued till 6–8 weeks. Warfarin can replace heparin if there was prior thromboembolism in addition to obstertrical events. LDA also should be stopped from the last month of pregnancy till childbirth to prevent premature closure of ductus arteriosus in the baby.

Recent trials using enoxaparin and nadroparin in RPL due to APS have found effects comparable to UFH and may evolve as the drug of choice, because of better side effect profile including lesser osteoporosis, lower incidence of heparin induced thrombocytopenia (HIT) and the ease of monitoring these agents.[121,122] It is a good practice to supplement the mothers with vitamin D and calcium, to avoid heparin induced osteoporosis.

Management of Refractory APS

In all cases of refractory APS, a meticulous search for non-aPL ab risk factors should be done and addressed accordingly. The compliance with prescribed medications should also be ensured.

Refractory Thrombotic APS

Recurrent thrombosis have been known to occur despite attainment of target INR in APS and management of this is highly challenging and empirical. A high intensity warfarin therapy with target INR between 3 and 4 may be helpful in individuals with no other bleeding risk factors. Addition of low dose aspirin and or statin can be considered at this stage, while HCQ should continue right from the stage of initial diagnosis. In patients not responding or intolerant of high intensity Warfarin, long term anticoagulation with LMWH or heparin has been found to be of use. In one retrospective study, enoxaparin used over a median duration of 36 months was found to be safe and effective in refractory thrombosis.[123]

Rituximab has been shown to be effective non-thrombotic manifestations of APS in a pilot open label phase II study,[124] however, results of the first randomized controlled trial using rituximab in anticoagulation resistant APS (RITAPS) are yet to be known. Newer targets like complement pathway inhibitors, inhibitors of NF-κB pathway and MAP kinase pathways offer promising therapeutic avenues.[125]

Management of Refractory Obstetric APS

Initial studies on use of high dose prednisolone showed non-favourable obstetric outcome as well as higher rate of infections, preterm deliveries, gestational diabetes and hypertension. Bramham et al studied 20 females with history of 89 pregnancies and only 4 of them had favourable outcome in terms of live births. These patients were given prednisolone at 10 mg/day from 14 weeks of gestation. Twenty-five pregnancies occurred in them, of which 16 (64%) resulted in successful outcome. This dose of prednisolone used in the study was not associated with any significant side effect.[126]

The other potential treatment modality is intravenous immunoglobulins (IVIG). But IVIG is prohibitively expensive and small studies have not shown significant benefits. Again HCQ must continue as a sheet anchor agent in all cases.

Management of CAPS

Current best therapeutic approach for CAPS is the so-called triple therapy, which includes the administration of anticoagulation, steroids with IVIG, and/or plasma exchange. Rituximab can be considered in patients with severe CAPS, especially in those with prominent haematologic and/or microthrombotic manifestations. On the basis of a limited number of case reports, eculizumab has been used with favourable outcomes in patients with CAPS to prevent new events following kidney transplant and in patients with CAPS refractory to standard therapy.[127]

Mortality in APS

The Euro phospholipid cohort is the largest prospective cohort of APS with a total of 1000 APS patients. In the 5-year follow-up of this study, the mortality was 5.3%. The cause of mortality in decreasing order of frequency was bacterial infections (20.8%), myocardial infarction (18.9%), stroke (13.2%), haemorrhage (11.3%), malignancy (11.3%), catastrophic APS (9.4%) and pulmonary embolism (9.4%).

The mortality is nevertheless higher in patients presenting with CAPS and can be as high as 50%. Cerebral involvement, followed by cardiac dysfunction and infections are the major causes of mortality in CAPS.[94]

Controversies in APS

There is much debate on characterization of mechanism of thrombosis and pregnancy loss, choice of ideal diagnostic assay and their standardization, existence of seronegative APS and issues in management of APS. A major limitation in the quality of data in APS is the poor standardization and reproducibility of assays used for detection of aPL abs. Many of the so-called seronegative APS may actually be due to poor standardization of assay. There is conflicting data on the predictive ability of the three major antibody assays in APS. However, it appears that LAC is the most specific for clinical event in overall scenario, though the sensitivity is low. Another dilemma to the clinician is in the management of thrombosis especially arterial thrombosis. The weight of evidence appears slightly in favour of high intensity as against low intensity anticoagulation for arterial thromboses. Similarly the treatment of APS associated stroke also is a controversial issue, with larger studies showing no advantage of oral anticoagulation over aspirin in this scenario. However, the quality of this data has been severely criticised. Controversies also abound in the management of asymptomatic carriers of aPL ab. Recent recommendations, however, have been in favour of treating high-risk aPL ab carriers with low dose aspirin. DOACs should be used with caution in APS patients as results of randomized control trials with clinical primary endpoints assessing clinical efficacy and safety are awaited, which may establish whether the prescription of DOACs could be a safe alternative to warfarin.

Key Points

- Presence of 1 clinical and 1 laboratory criteria are essential to fulfil revised Sapporo classification criteria for the diagnosis of APS.
- In patients satisfying only clinical arm of revised Sapporo criteria, probability of seronegative APS exists—anti-PSPT and anti-domain antibodies are useful in seronegative APS.
- HCQ should be considered in management of all APS cases as sheet anchor drug.
- Lifelong anticoagulation is recommended in all thrombotic APS cases (except in low-risk stroke patients).
- Newer DOAC agents still have to prove their clinical efficacy in large scale controlled trials.
- 5-year mortality in Euro phospholipid cohort is approximately 5%; high mortality (≥50%) in CAPS.

REFERENCES

1. Conley CL HR. A hemorrhagic disorder caused by circulating anticoagulant in patients with disseminated lupus erythematosus. 1952; 31:621–2.
2. Bowie EJ, Thompson JH, Pascuzzi CA, Owen CA. Thrombosis in systemic lupus erythematosus despite circulating anticoagulants. J Lab Clin Med. 1963; 62:416–30.
3. Thiagarajan P, Shapiro SS, De Marco L. Monoclonal immunoglobulin M lambda coagulation inhibitor with phospholipid specificity. Mechanism of a lupus anticoagulant. J Clin Invest. 1980; 66:397–405.
4. Hughes GRV. The anticardiolipin syndrome. 1985; 3:285–6.
5. Wilson WA, Gharavi AE, Koike T, Lockshin MD, Branch DW, Piette JC, et al. International consensus statement on preliminary classification criteria for definite antiphospholipid syndrome: report of an international workshop. Arthritis Rheum. 1999; 42:1309–11.
6. Miyakis S, Lockshin MD, Atsumi T, Branch DW, Brey RL, Cervera R, et al. International consensus statement on an update of the classification criteria for definite antiphospholipid syndrome (APS). J Thromb Haemost 2006; 4:295–306.
7. Roubey RA. Antiphospholipid syndrome: antibodies and antigens. Curr Opin Hematol. 2000; 7:316–20.
8. Alessandri C, Conti F, Pendolino M, Mancini R, Valesini G. New autoantigens in the antiphospholipid syndrome. Autoimmun Rev. 2011; 10:609–16.
9. Horstman LL, Jy W, Bidot CJ, Ahn YS, Kelley RE, Zivadinov R, et al. Antiphospholipid antibodies: paradigm in transition. J Neuroinflammation. 2009; 6:3.
10. Petrovas C, Vlachoyiannopoulos PG, Kordossis T, Moutsopoulos HM. Anti-phospholipid antibodies in HIV infection and SLE with or without anti-phospholipid syndrome: comparisons of phospholipid specificity, avidity and reactivity with beta2-GPI. J Autoimmun. 1999; 13:347–55.
11. Shi W, Krilis SA, Chong BH, Gordon S, Chesterman CN. Prevalence of lupus anticoagulant and anticardiolipin antibodies in a healthy population. Aust N Z J Med. 1990; 20:231–6.
12. Vila P, Hernández MC, López-Fernández MF, Batlle J. Prevalence, follow-up and clinical significance of the anticardiolipin antibodies in normal subjects. Thromb Haemost. 1994; 72:209–13.
13. Lockwood CJ, Romero R, Feinberg RF, Clyne LP, Coster B, Hobbins JC. The prevalence and biologic significance of lupus anticoagulant and anticardiolipin antibodies in a general obstetric population. Am J Obstet Gynecol. 1989; 161:369–73.
14. Roldan V, Lecumberri R, Muñoz-Torrero JFS, Vicente V, Rocha E, Brenner B, et al. Thrombophilia testing in patients with venous thromboembolism. Findings from the RIETE registry. Thromb Res. 2009; 124:174–7.
15. Cervera R, Piette J-C, Font J, Khamashta MA, Shoenfeld Y, Camps MT, et al. Antiphospholipid syndrome: clinical and immunologic manifestations and patterns of disease expression in a cohort of 1,000 patients. Arthritis Rheum. 2002; 46:1019–27.
16. Biggioggero M, Meroni PL. The geoepidemiology of the antiphospholipid antibody syndrome. Autoimmun Rev. 2010; 9:A299–304.
17. Petri M. Update on anti-phospholipid antibodies in SLE: the Hopkins' Lupus Cohort. Lupus. 2010; 19:419–23.
18. Ostrowski RA, Robinson JA. Antiphospholipid antibody syndrome and autoimmune diseases. Hematol Oncol Clin North Am. 2008; 22:53–65.

19. Sherer Y, Gerli R, Vaudo G, Schillaci G, Gilburd B, Giordano A, et al. Prevalence of antiphospholipid and antioxidized low-density lipoprotein antibodies in rheumatoid arthritis. Ann N Y Acad Sci. 2005; 1051: 299–303.
20. Jonsdottir T, Forslid J, van Vollenhoven A, Harju A, Brannemark S, Klareskog L, et al. Treatment with tumour necrosis factor alpha antagonists in patients with rheumatoid arthritis induces anticardiolipin antibodies. Ann Rheum Dis. 2004; 63:1075–8.
21. Marie I, Jouen F, Hellot M-F, Levesque H. Anticardiolipin and anti-beta2 glycoprotein I antibodies and lupus-like anticoagulant: prevalence and significance in systemic sclerosis. Br J Dermatol. 2008; 158:141–4.
22. Pasoto SG, Chakkour HP, Natalino RR, Viana VST, Bueno C, Lianza AC, et al. Lupus anticoagulant: a marker for stroke and venous thrombosis in primary Sjögren's syndrome. Clin Rheumatol. 2012; 31:1331–8.
23. Kawakami T, Yamazaki M, Mizoguchi M, Soma Y. High titer of anti-phosphatidylserine-prothrombin complex antibodies in patients with cutaneous polyarteritis nodosa. Arthritis Rheum. 2007; 57: 1507–13.
24. Reinstein E, Shoenfeld Y. Antiphospholipid syndrome and cancer. Clin Rev Allergy Immunol. 2007; 32:184–7.
25. Tincani A, Taraborelli M, Cattaneo R. Antiphospholipid antibodies and malignancies. Autoimmun Rev. 2010; 9:200–2.
26. Font C, Vidal L, Espinosa G, Tàssies D, Monteagudo J, Farrús B, et al. Solid cancer, antiphospholipid antibodies, and venous thromboembolism. Autoimmun Rev. 2011; 10:222–7.
27. Sène D, Piette J-C, Cacoub P. Antiphospholipid antibodies, antiphospholipid syndrome and infections. Autoimmun Rev. 2008; 7:272–7.
28. Ramos-Casals M, Cervera R, Lagrutta M, Medina F, García-Carrasco M, de la Red G, et al. Clinical features related to antiphospholipid syndrome in patients with chronic viral infections (hepatitis C virus/HIV infection): description of 82 cases. Clin Infect Dis 2004; 38:1009–16.
29. Dlott JS, Roubey RAS. Drug-induced lupus anticoagulants and antiphospholipid antibodies. Curr Rheumatol Rep. 2012; 14:71–8.
30. Visvanathan S, Wagner C, Smolen J, St Clair EW, Hegedus R, Baker D, et al. IgG and IgM anticardiolipin antibodies following treatment with infliximab plus methotrexate in patients with early rheumatoid arthritis. Arthritis Rheum. 2006; 54: 2840–4.
31. Meroni PL. Pathogenesis of the antiphospholipid syndrome: an additional example of the mosaic of autoimmunity. J Autoimmun. 2008; 30:99–103.
32. Xie H, Sheng L, Zhou H, Yan J. The role of TLR4 in pathophysiology of antiphospholipid syndrome-associated thrombosis and pregnancy morbidity. Br J Haematol. 2014; 164:165–76.
33. Xie H, Kong X, Zhou H, Xie Y, Sheng L, Wang T, et al. TLR4 is involved in the pathogenic effects observed in a murine model of antiphospholipid syndrome. Clin Immunol 2015; 160:198–210.
34. de Laat B, Pengo V, Pabinger I, Musial J, Voskuyl AE, Bultink IEM, et al. The association between circulating antibodies against domain I of beta2-glycoprotein I and thrombosis: an international multicenter study. J Thromb Haemost JTH. 2009; 7:1767–73.
35. Tanimura K, Jin H, Suenaga T, Morikami S, Arase N, Kishida K, et al. β2-Glycoprotein I/HLA class II complexes are novel autoantigens in antiphospholipid syndrome. Blood. 2015; 125:2835–44.
36. Jiang Y, Arase N, Kohyama M, Hirayasu K, Suenaga T, Jin H, et al. Transport of misfolded endoplasmic reticulum proteins to the cell surface by MHC class II molecules. Int Immunol. 2013; 25:235–46.
37. Cheng S, Wang H, Zhou H. The Role of TLR4 on B Cell Activation and Anti-β2GPI Antibody Production in the Antiphospholipid Syndrome. J Immunol Res. 2016; 2016:1719720.
38. Lutters BCH, Derksen RHWM, Tekelenburg WL, Lenting PJ, Arnout J, de Groot PG. Dimers of beta 2-glycoprotein I increase platelet deposition to collagen via interaction with phospholipids and the apolipoprotein E receptor 2'. J Biol Chem. 2003; 278:33831–8.
39. Romay-Penabad Z, Aguilar-Valenzuela R, Urbanus RT, Derksen RHWM, Pennings MTT, Papalardo E, et al. Apolipoprotein E receptor 2 is involved in the thrombotic complications in a murine model of the antiphospholipid syndrome. Blood. 2011; 117: 1408–14.
40. Lackner KJ, Müller-Calleja N. Pathogenesis of the antiphospholipid syndrome revisited: time to challenge the dogma. J Thromb Haemost JTH. 2016; 14:1117–20.
41. Ramesh S, Morrell CN, Tarango C, Thomas GD, Yuhanna IS, Girardi G, et al. Antiphospholipid antibodies promote leukocyte-endothelial cell adhesion and thrombosis in mice by antagonizing eNOS via β2GPI and apoER2. J Clin Invest. 2011; 121:120–31.
42. Vega-Ostertag ME, Ferrara DE, Romay-Penabad Z, Liu X, Taylor WR, Colden-Stanfield M, et al. Role

of p38 mitogen-activated protein kinase in antiphospholipid antibody-mediated thrombosis and endothelial cell activation. J Thromb Haemost. 2007; 5:1828–34.

43. Zuily S, Ait Aissa K, Membre A, Regnault V, Lecompte T, Wahl D. Thrombin generation in antiphospholipid syndrome. Lupus. 2012 Jun; 21(7):758–60.

44. Yalavarthi S, Gould TJ, Rao AN, Mazza LF, Morris AE, Núñez-Álvarez C, et al. Release of neutrophil extracellular traps by neutrophils stimulated with antiphospholipid antibodies: a newly identified mechanism of thrombosis in the antiphospholipid syndrome. Arthritis Rheumatol 2015; 67:2990–3003.

45. Ogishima D, Matsumoto T, Nakamura Y, Yoshida K, Kuwabara Y. Placental pathology in systemic lupus erythematosus with antiphospholipid antibodies. Pathol Int. 2000; 50:224–9.

46. Magid MS, Kaplan C, Sammaritano LR, Peterson M, Druzin ML, Lockshin MD. Placental pathology in systemic lupus erythematosus: a prospective study. Am J Obstet Gynecol. 1998; 179:226–34.

47. Rand JH, Wu X-X, Quinn AS, Taatjes DJ. The annexin A5-mediated pathogenic mechanism in the antiphospholipid syndrome: role in pregnancy losses and thrombosis. Lupus. 2010; 19:460–9.

48. Girardi G, Berman J, Redecha P, Spruce L, Thurman JM, Kraus D, et al. Complement C5a receptors and neutrophils mediate fetal injury in the antiphospholipid syndrome. J Clin Invest. 2003; 112:1644–54.

49. Girardi G, Redecha P, Salmon JE. Heparin prevents antiphospholipid antibody-induced fetal loss by inhibiting complement activation. Nat Med. 2004; 10:1222–6.

50. Shamonki JM, Salmon JE, Hyjek E, Baergen RN. Excessive complement activation is associated with placental injury in patients with antiphospholipid antibodies. Am J Obstet Gynecol. 2007; 196:167.e1–5.

51. Derksen RHWM, de Groot PG. The obstetric antiphospholipid syndrome. J Reprod Immunol. 2008; 77:41–50.

52. Di Simone N, Meroni PL, de Papa N, Raschi E, Caliandro D, De Carolis CS, et al. Antiphospholipid antibodies affect trophoblast gonadotropin secretion and invasiveness by binding directly and through adhered beta2-glycoprotein I. Arthritis Rheum. 2000; 43:140–50.

53. Shoenfeld Y, Blank M, Cervera R, Font J, Raschi E, Meroni P-L. Infectious origin of the antiphospholipid syndrome. Ann Rheum Dis. 2006; 65:2–6.

54. Ginsburg KS, Liang MH, Newcomer L, Goldhaber SZ, Schur PH, Hennekens CH, et al. Anticardiolipin antibodies and the risk for ischemic stroke and venous thrombosis. Ann Intern Med. 1992; 117:997–1002.

55. Urbanus RT, Siegerink B, Roest M, Rosendaal FR, de Groot PG, Algra A. Antiphospholipid antibodies and risk of myocardial infarction and ischaemic stroke in young women in the RATIO study: a case-control study. Lancet Neurol. 2009; 8:998–1005.

56. Mishra MN, Rohatgi S. Antiphospholipid antibodies in young Indian patients with stroke. J Postgrad Med. 2009; 55:161–4.

57. Sanna G, Bertolaccini ML, Cuadrado MJ, Khamashta MA, Hughes GRV. Central nervous system involvement in the antiphospholipid (Hughes) syndrome. Rheumatology (Oxford) 2003; 42:200–13.

58. Muscal E, Brey RL. Antiphospholipid syndrome and the brain in pediatric and adult patients. Lupus. 2010; 19:406–11.

59. Tektonidou MG. Renal involvement in the antiphospholipid syndrome (APS)-APS nephropathy. Clin Rev Allergy Immunol. 2009; 36:131–40.

60. Canaud G, Bienaimé F, Tabarin F, Bataillon G, Seilhean D, Noël L-H, et al. Inhibition of the mTORC pathway in the antiphospholipid syndrome. N Engl J Med. 2014; 371:303–12.

61. Tenedios F, Erkan D, Lockshin MD. Cardiac manifestations in the antiphospholipid syndrome. Rheum Dis Clin North Am. 2006; 32:491–507.

62. Scheiman Elazary A, Cohen MJ, Aamar S, Dranitzki Z, Tayer-Shifman O, Mevorach D, et al. Pulmonary hemorrhage in antiphospholipid antibody syndrome. J Rheumatol. 2012; 39:1628–31.

63. Weinstein S, Piette W. Cutaneous manifestations of antiphospholipid antibody syndrome. Hematol Oncol Clin North Am. 2008; 22:67–77.

64. Schreiber K, Hunt BJ. Pregnancy and Antiphospholipid Syndrome. Semin Thromb Hemost. 2016; 42: 780–8.

65. Wang J-G, Xie Q-B, Yang N-P, Yin G. Primary antiphospholipid antibody syndrome: a case with bilateral sudden sensorineural hearing loss. Rheumatol Int. 2009; 29:467–8.

66. Uthman I, Godeau B, Taher A, Khamashta M. The hematologic manifestations of the antiphospholipid syndrome. Blood Rev. 2008; 22:187–94.

67. Mazodier K, Arnaud L, Mathian A, Costedoat-Chalumeau N, Haroche J, Frances C, et al. Lupus anticoagulant-hypoprothrombinemia syndrome: report of 8 cases and review of the literature. Medicine (Baltimore). 2012; 91:251–60.

68. Mehdi AA, Salti I, Uthman I. Antiphospholipid syndrome: endocrinologic manifestations and organ involvement. Semin Thromb Hemost. 2011; 37:49–57.

69. Yang P, Kruh JN, Foster CS. Antiphospholipid antibody syndrome. Curr Opin Ophthalmol. 2012; 23(6):528–32.

70. Pengo V, Tripodi A, Reber G, Rand JH, Ortel TL, Galli M, et al. Update of the guidelines for lupus anticoagulant detection. Subcommittee on Lupus Anticoagulant/Antiphospholipid Antibody of the Scientific and Standardisation Committee of the International Society on Thrombosis and Haemostasis. J Thromb Haemost JTH. 2009; 7 1737–40.

71. Lakos G, Favaloro EJ, Harris EN, Meroni PL, Tincani A, Wong RC, et al. International consensus guidelines on anticardiolipin and anti-β2-glycoprotein I testing: report from the 13th International Congress on Antiphospholipid Antibodies. Arthritis Rheum. 2012; 64:1–10.

72. de Groot PG, Urbanus RT. Screening: Guidelines for antiphospholipid antibody detection. Nat Rev Rheumatol. 2011; 8:125–6.

73. Galli M, Luciani D, Bertolini G, Barbui T. Lupus anticoagulants are stronger risk factors for thrombosis than anticardiolipin antibodies in the antiphospholipid syndrome: a systematic review of the literature. Blood. 2003; 101:1827–32.

74. de Laat B, Mertens K, de Groot PG Mechanisms of disease: antiphospholipid antibodies from clinical association to pathologic mechanism. Nat Clin Pract Rheumatol. 2008; 4:192–9.

75. Danowski A, de Azevedo MNL, de Souza Papi JA, Petri M. Determinants of risk for venous and arterial thrombosis in primary antiphospholipid syndrome and in antiphospholipid syndrome with systemic lupus erythematosus. J Rheumatol. 2009; 36:1195–9.

76. Drenkard C, Sánchez-Guerrero J, Alarcón-Segovia D. Fall in antiphospholipid antibody at time of thromboocclusive episodes in systemic lupus erythematosus. J Rheumatol. 1989; 16:614–7.

77. Samarkos M, Asherson RA, Loizou S. The clinical significance of IgA antiphospholipid antibodies. J Rheumatol. 2001; 28:694–7.

78. Cervera R, Conti F, Doria A, Iaccarino L, Valesini G. Does seronegative antiphospholipid syndrome really exist? Autoimmun Rev. 2012; 11:581–4.

79. Sugi T, Matsubayashi H, Inomo A, Dan L, Makino T. Antiphosphatidylethanolamine antibodies in recurrent early pregnancy loss and mid-to-late pregnancy loss. J Obstet Gynaecol Res. 2004; 30:326–32.

80. Sanmarco M, Gayet S, Alessi M-C, Audrain M, de Maistre E, Gris J-C, et al. Antiphosphatidylethanolamine antibodies are associated with an increased odds ratio for thrombosis. A multicenter study with the participation of the European Forum on antiphospholipid antibodies. Thromb Haemost. 2007; 97:949–54.

81. Yetman DL, Kutteh WH. Antiphospholipid antibody panels and recurrent pregnancy loss: prevalence of anticardiolipin antibodies compared with other antiphospholipid antibodies. Fertil Steril. 1996; 66: 540–6.

82. de Laat B, Derksen RHWM, Urbanus RT, de Groot PG. IgG antibodies that recognize epitope Gly40-Arg43 in domain I of beta 2-glycoprotein I cause LAC, and their presence correlates strongly with thrombosis. Blood. 2005; 105:1540–5.

83. Pengo V, Ruffatti A, Tonello M, Hoxha A, Bison E, Denas G, et al. Antibodies to Domain 4/5 (Dm4/5) of β2-Glycoprotein 1 (β2GP1) in different antiphospholipid (aPL) antibody profiles. Thromb Res. 2015; 136:161–3.

84. Andreoli L, Chighizola CB, Nalli C, Gerosa M, Borghi MO, Pregnolato F, et al. Clinical characterization of antiphospholipid syndrome by detection of IgG antibodies against β2 -glycoprotein i domain 1 and domain 4/5: ratio of anti-domain 1 to anti-domain 4/5 as a useful new biomarker for antiphospholipid syndrome. Arthritis Rheumatol 2015; 67:2196–204.

85. Ortona E, Capozzi A, Colasanti T, Conti F, Alessandri C, Longo A, et al. Vimentin/cardiolipin complex as a new antigenic target of the antiphospholipid syndrome. Blood. 2010; 116:2960–7.

86. Bizzaro N, Ghirardello A, Zampieri S, Iaccarino L, Tozzoli R, Ruffatti A, et al. Anti-prothrombin antibodies predict thrombosis in patients with systemic lupus erythematosus: a 15-year longitudinal study. J Thromb Haemost JTH. 2007; 5:1158–64.

87. Atsumi T, Ieko M, Bertolaccini ML, Ichikawa K, Tsutsumi A, Matsuura E, et al. Association of autoantibodies against the phosphatidylserine-prothrombin complex with manifestations of the antiphospholipid syndrome and with the presence of lupus anticoagulant. Arthritis Rheum. 2000; 43:1982–93.

88. Arnold J, Holmes Z, Pickering W, Farmer C, Regan L, Cohen H. Anti-beta 2 glycoprotein 1 and anti-annexin V antibodies in women with recurrent miscarriage. Br J Haematol. 2001; 113:911–4.

89. Nojima J, Kuratsune H, Suehisa E, Futsukaichi Y, Yamanishi H, Machii T, et al. Association between the prevalence of antibodies to beta(2)-glycoprotein I, prothrombin, protein C, protein S, and annexin V in patients with systemic lupus erythematosus and thrombotic and thrombocytopenic complications. Clin Chem. 2001; 47:1008–15.

90. de Laat B, Derksen RHWM, Mackie IJ, Roest M, Schoormans S, Woodhams BJ, et al. Annexin A5 polymorphism (-1C→T) and the presence of anti-annexin A5 antibodies in the antiphospholipid syndrome. Ann Rheum Dis. 2006; 65:1468–72.

91. Molina JF, Gutierrez-Ureña S, Molina J, Uribe O, Richards S, De Ceulaer C, et al. Variability of anticardiolipin antibody isotype distribution in 3 geographic populations of patients with systemic lupus erythematosus. J Rheumatol. 1997; 24:291–6.

92. Mehrani T, Petri M. Association of IgA Anti-beta2 glycoprotein I with clinical and laboratory manifestations of systemic lupus erythematosus. J Rheumatol. 2011; 38:64–8.

93. Cervera R, Espinosa G. Update on the catastrophic antiphospholipid syndrome and the "CAPS Registry." Semin Thromb Hemost. 2012; 38:333–8.

94. Cervera R, Asherson RA. Catastrophic antiphospholipid syndrome: therapeutic developments. Expert Rev Clin Immunol. 2007; 3:277–85.

95. Erkan D, Harrison MJ, Levy R, Peterson M, Petri M, Sammaritano L, et al. Aspirin for primary thrombosis prevention in the antiphospholipid syndrome: a randomized, double-blind, placebo-controlled trial in asymptomatic antiphospholipid antibody-positive individuals. Arthritis Rheum. 2007; 56:2382–91.

96. Cuadrado MJ, Bertolaccini ML, Seed PT, Tektonidou MG, Aguirre A, Mico L, et al. Low-dose aspirin vs low-dose aspirin plus low-intensity warfarin in thromboprophylaxis: a prospective, multicentre, randomized, open, controlled trial in patients positive for antiphospholipid antibodies (ALIWAPAS). Rheumatology (Oxford) 2014; 53:275–84.

97. Erkan D, Merrill JT, Yazici Y, Sammaritano L, Buyon JP, Lockshin MD. High thrombosis rate after fetal loss in antiphospholipid syndrome: effective prophylaxis with aspirin. Arthritis Rheum. 2001; 44: 1466–7.

98. Levine SR, Brey RL, Tilley BC, Thompson JLP, Sacco RL, Sciacca RR, et al. Antiphospholipid antibodies and subsequent thrombo-occlusive events in patients with ischemic stroke. JAMA. 2004; 291:576–84.

99. Okuma H, Kitagawa Y, Yasuda T, Tokuoka K, Takagi S. Comparison between single antiplatelet therapy and combination of antiplatelet and anticoagulation therapy for secondary prevention in ischemic stroke patients with antiphospholipid syndrome. Int J Med Sci. 2009; 7:15–8.

100. Prandoni P, Simioni P, Girolami A. Antiphospholipid antibodies, recurrent thromboembolism, and intensity of warfarin anticoagulation. Thromb Haemost. 1996; 75:859.

101. Crowther MA, Ginsberg JS, Julian J, Denburg J, Hirsh J, Douketis J, et al. A comparison of two intensities of warfarin for the prevention of recurrent thrombosis in patients with the antiphospholipid antibody syndrome. N Engl J Med. 2003; 349:1133–8.

102. Finazzi G, Marchioli R, Brancaccio V, Schinco P, Wisloff F, Musial J, et al. A randomized clinical trial of high-intensity warfarin vs. conventional antithrombotic therapy for the prevention of recurrent thrombosis in patients with the antiphospholipid syndrome (WAPS). J Thromb Haemost 2005; 3:848–53.

103. Lim W, Crowther MA, Eikelboom JW. Management of antiphospholipid antibody syndrome: a systematic review. JAMA. 2006; 295:1050–7.

104. Wittkowsky AK, Downing J, Blackburn J, Nutescu E. Warfarin-related outcomes in patients with antiphospholipid antibody syndrome managed in an anticoagulation clinic. Thromb Haemost. 2006; 96: 137–41.

105. Khamashta MA, Cuadrado MJ, Mujic F, Taub NA, Hunt BJ, Hughes GR. The management of thrombosis in the antiphospholipid-antibody syndrome. N Engl J Med. 1995; 332:993–7.

106. Ruiz-Irastorza G, Khamashta MA, Hunt BJ, Escudero A, Cuadrado MJ, Hughes GRV. Bleeding and recurrent thrombosis in definite antiphospholipid syndrome: analysis of a series of 66 patients treated with oral anticoagulation to a target international normalized ratio of 3.5. Arch Intern Med. 2002; 162: 1164–9.

107. Cohen H, Hunt BJ, Efthymiou M, Arachchillage DRJ, Mackie IJ, Clawson S, et al. Rivaroxaban versus warfarin to treat patients with thrombotic antiphospholipid syndrome, with or without systemic lupus erythematosus (RAPS): a randomised, controlled, open-label, phase 2/3, non-inferiority trial. Lancet Haematol. 2016; 3:e426–436.

108. Dufrost V, Risse J, Zuily S, Wahl D. Direct Oral Anticoagulants Use in Antiphospholipid Syndrome: Are These Drugs an Effective and Safe Alternative to Warfarin? A Systematic Review of the Literature. Curr Rheumatol Rep. 2016; 18:74.

109. Woller SC, Stevens SM, Kaplan DA, Branch DW, Aston VT, Wilson EL, et al. Apixaban for the Secondary Prevention of Thrombosis Among Patients With Antiphospholipid Syndrome: Study Rationale and Design (ASTRO-APS). Clin Appl Thromb 2016; 22:239–47.

110. Pengo V, Banzato A, Bison E, Zoppellaro G, Padayattil Jose S, Denas G. Efficacy and safety of rivaroxaban vs warfarin in high-risk patients with antiphospholipid syndrome: Rationale and design of the Trial on Rivaroxaban in AntiPhospholipid Syndrome (TRAPS) trial. Lupus. 2016; 25:301–6.

111. Kuznik A, Bencina M, Svajger U, Jeras M, Rozman B, Jerala R. Mechanism of endosomal TLR inhibition by antimalarial drugs and imidazoquinolines. J Immunol Baltim Md 1950. 2011; 186(8):4794–804.

112. Espinola RG, Pierangeli SS, Gharavi AE, Harris EN, Ghara AE. Hydroxychloroquine reverses platelet activation induced by human IgG antiphospholipid antibodies. Thromb Haemost. 2002; 87(3):518–22.

113. Rand JH, Wu X-X, Quinn AS, Chen PP, Hathcock JJ, Taatjes DJ. Hydroxychloroquine directly reduces the binding of antiphospholipid antibody-beta2-glycoprotein I complexes to phospholipid bilayers. Blood. 2008 ; 112:1687–95.

114. Rand JH, Wu X-X, Quinn AS, Ashton AW, Chen PP, Hathcock JJ, et al. Hydroxychloroquine protects the annexin A5 anticoagulant shield from disruption by antiphospholipid antibodies: evidence for a novel effect for an old antimalarial drug. Blood. 2010; 115: 2292–9.

115. Schmidt-Tanguy A, Voswinkel J, Henrion D, Subra JF, Loufrani L, Rohmer V, et al. Antithrombotic effects of hydroxychloroquine in primary antiphospholipid syndrome patients. J Thromb Haemost 2013; 11:1927–9.

116. Meroni PL, Raschi E, Testoni C, Tincani A, Balestrieri G, Molteni R, et al. Statins prevent endothelial cell activation induced by antiphospholipid (anti-beta2-glycoprotein I) antibodies: effect on the proadhesive and proinflammatory phenotype. Arthritis Rheum. 2001; 44:2870–8.

117. Erkan D, Willis R, Murthy VL, Basra G, Vega J, Ruiz-Limón P, et al. A prospective open-label pilot study of fluvastatin on proinflammatory and prothrombotic biomarkers in antiphospholipid antibody positive patients. Ann Rheum Dis. 2014; 73:1176–80.

118. Kutteh WH. Antiphospholipid antibody-associated recurrent pregnancy loss: treatment with heparin and low-dose aspirin is superior to low-dose aspirin alone. Am J Obstet Gynecol. 1996; 174:1584–9.

119. Rai R, Cohen H, Dave M, Regan L. Randomised controlled trial of aspirin and aspirin plus heparin in pregnant women with recurrent miscarriage associated with phospholipid antibodies (or antiphospholipid antibodies). BMJ. 1997; 314:253–7.

120. Empson M, Lassere M, Craig J, Scott J. Prevention of recurrent miscarriage for women with antiphospholipid antibody or lupus anticoagulant. Cochrane Database Syst Rev. 2005; (2):CD002859.

121. Fouda UM, Sayed AM, Ramadan DI, Fouda IM. Efficacy and safety of two doses of low molecular weight heparin (enoxaparin) in pregnant women with a history of recurrent abortion secondary to antiphospholipid syndrome. J Obstet Gynaecol 2010; 30:842–6.

122. Ruffatti A, Favaro M, Tonello M, De Silvestro G, Pengo V, Fais G, et al. Efficacy and safety of nadroparin in the treatment of pregnant women with antiphospholipid syndrome: a prospective cohort study. Lupus. 2005;14:120–8.

123. Vargas-Hitos JA, Ateka-Barrutia O, Sangle S, Khamashta MA. Efficacy and safety of long-term low molecular weight heparin in patients with antiphospholipid syndrome. Ann Rheum Dis. 2011; 70:1652–4.

124. Erkan D, Vega J, Ramón G, Kozora E, Lockshin MD. A pilot open-label phase II trial of rituximab for non-criteria manifestations of antiphospholipid syndrome. Arthritis Rheum. 2013; 65:464–71.

125. Scoble T, Wijetilleka S, Khamashta MA. Management of refractory anti-phospholipid syndrome. Autoimmun Rev. 2011; 10:669–73.

126. Bramham K, Thomas M, Nelson-Piercy C, Khamashta M, Hunt BJ. First-trimester low-dose prednisolone in refractory antiphospholipid antibody-related pregnancy loss. Blood. 2011; 117:6948–51.

127. Rodriguez-Pintó I, Espinosa G, Cervera R. Catastrophic antiphospholipid syndrome: The current management approach. Best Pract Res Clin Rheumatol. 2016; 30:239–49.

Chapter

41

Sjögren's Syndrome

R Porkodi

INTRODUCTION

Sjögren's syndrome (SS) is an autoimmune epithelitis affecting primarily the exocrine glands and epithelium, characterized by dry eyes and dry mouth. SS can be seen alone (primary SS) or in association with another autoimmune disease like rheumatoid arthritis (RA) or systemic lupus erythematosus (SLE) when it is called secondary SS. Prevalence varies from 0.5 to 5%. This is a disease predominantly affecting middle-aged females, though children and elderly are also affected.

Etiology and Pathogenesis

Autoimmunity may be triggered by viruses, hormones in a genetically susceptible individual, becomes chronic through immune regulatory mechanisms and result in tissue lesions due to inflammatory process.

Immunogenetics

SS is associated with increased frequencies of HLA B8, HLA DW3 and HLA DR3.[1] In Caucasian individuals the HLA DR B1* 0301/DQA1* 0501/ DQB1* 0201 haplotype has the strongest association with the production of SS-A and SS-B antibodies.[1] IRF 5 and STAT 4 gene polymorphism are involved in the activation of type-1 interferon pathway.

Environmental Triggers

Epstein-Barr virus (EBV) infection of the epithelial cells may lead to neo-antigen expression and increased rate of apoptosis. The apoptosis is mediated through the Fas-Fas ligand interaction and the perforin/granzyme B pathway. Lymphoid cell activation by EBV results in cleavage of alpha Fodrin to 120 KD fragments which occur concomitantly with cellular apoptosis and expression of ZEBRA protein which is a marker for the activation of the life cycle of EBV.[2] Thus EBV induces the formation of autoantigens linked to SS. Other viruses like cytomegalovirus, HIV, HTLV-1 and HCV are found in the epithelial cells of the salivary glands and create a chronic local immune response similar to the autoimmune sialadenitis in primary SS.

Immunopathogenesis

Ductal and acinar epithelial cells express HLA-class II molecules and act as antigen presenting cells. CD4 cells are attracted and become activated. Pro inflammatory cytokines IL-1, IL-6 and TNF-alpha are produced and are responsible for inflammatory perpetuation and stimulation of B cells. B cells contain intra-cytoplasmic immunoglobulins with anti-Ro (SS-A) and or anti-La(-B) reactivity. The most common auto-antibodies in SS are directed against these ribonucleic protein antigens Ro and La. Anti-La antibodies are more specific for SS while anti-Ro antibodies might be present in SLE and RA. The Ro antibodies are directed against 52 KD protein in SS and 60 KD protein in SLE.[3] B cell activation is initially polyclonal, later may evolve to oligoclonal or monoclonal activation and may end as malignant monoclonal proliferation. Activation of IFN-alpha, B cell activating factor

(BAFF) and antibodies to muscarinic receptor-3 are new steps in the road to develop SS.

CLINICAL MANIFESTATIONS

Ocular

Symptoms: Patients usually present with red itchy painful eyes, complain of a gritty feeling or irritation from a 'grain of sand'. Symptoms include blurring of vision, photophobia, redness, ocular fatigue or inability to tolerate contact lenses.

Physical signs: Dilatation of bulbar conjunctival vessels, pericorneal injection and lacrimal gland enlargement. Thick mucus secretion may cause blurring of vision. Xerophthalmia may lead to corneal abrasions and infection with gram-positive organisms.

Tests

a. Schirmer's test is considered positive when there is less than 5 mm of wetting in 5 minutes when a filter paper strip is inserted in the lower eye lid.
b. Slit lamp examination after instillation of dyes over the cornea will show epithelial defects. Fluorescein stains epithelial defects. Rose Bengal binds devitalized cells. Lissamine green is less irritating and equally sensitive.
c. Tear break up time is another clinical test. A drop of fluorescein is applied to the eye and the time between the last blink and appearance of dark non-fluorescent areas in the tear film is measured. A rapid tear break up time indicates an abnormality.

Oral

Symptoms: Patients complain of dry mouth and need a constant supply of water to be comfortable. Eating is difficult without supplemental liquids. Some may have dysgeusia. Dry burning throat leads to dry cough, difficulty in speaking continuously or hoarseness of voice. Dental caries, periodental disease may be the first symptom of dry mouth due to loss of antibacterial action of saliva. *Candida albicans* infection may occur.

Signs: Enlargement of parotid is frequent and may precede xerostomia by several years. Sublingual and submandibular glands may also be affected. Absence of pooling of saliva under the tongue is diagnostic of xerostomia. Loss of filiform papillae from the dorsal surface of the tongue, cracked or sore tongue or tongue sticking to the roof, extensive oral erythema, angular chelitis are often found.

Tests

a. Sialometry measures salivary flow rates with or without stimulation for the various salivary glands.
b. Sialography shows sialectasis and is a sensitive test for SS.
c. Scintigraphy measures the 99m TC pertechnate uptake by the salivary gland. The uptake is reduced and secretion of labeled saliva is delayed or absent.
d. Salivary gland MRI shows an increase in the fat areas and a decrease in intact lobule area. MR sialography provides information on ductal changes; dynamic contrast enhanced MRI quantifies micro vascular function.
e. Salivary gland ultrasonography: Reduced volume of the submandibular glands, parenchymal inhomogeneity are highly specific markers of SS.[4]
f. Minor salivary gland biopsy: Histopathological features include focal aggregation of at least 50 lymphocytes, plasma cells, macrophages adjacent to and replacing normal acini and the consistent presence of these foci in all or most of the glands in the specimen.

In a study from Chennai[5] dry mouth and dry eyes were the presenting symptoms in 55.6% of patients and these were also the commonest finding in 89% and 86% respectively, similar to the study from Lucknow (86.4%).[6]

Other Xerosis

Dryness of all mucus membranes occur. Dry nose leads to congestion, crusting and epistaxis. Xerotrachea can lead to dry cough and hoarseness. Vaginal dryness leads to pruritus, irritation and dyspareunia. Sweat secretion is reduced leading to pruritus and excoriation. Loss of exocrine function may lead to pancreatic hypofunction and hypochlorhydria.

EXTRAGLANDULAR MANIFESTATIONS

Systemic Symptoms

Easy fatiguability, low grade fever, myalgia, arthralgia may occur.

Musculoskeletal Symptoms

Arthritis occurs in 54–84% of SS patients. Arthritis may precede overt sicca manifestations and be associated with morning stiffness. Chronic polyarthritis may lead to Jaccoud's arthropathy. Erosions are rare. Arthritis is the most common extra-articular symptom and was seen in 77.8% of patients in the study from Chennai.[5]

Skin

Skin manifestations include purpura (10–15%), annular erythema (5–10%) pernio like lesions and urticarial lesions. Flat purpura is seen in patients with hypergammaglobulinemia. Palpable purpura is a feature of vasculitis. Raynaud's phenomenon is seen in 35% and usually precedes the sicca symptoms by many years.

Pulmonary Manifestations

Cough occurs in 40–50% secondary to xerotrachea or to bronchial hyperresponsiveness. Bronchiolitis, lymphoid interstitial pneumonia, bronchiectasis and fibrosis can occur.

Cardiovascular Manifestations

Pericardial effusion (usually mild and asymptomatic) is the commonest feature. Autonomic disturbances like orthostatic intolerance, secretomotor dysfunction, male sexual dysfunction, urinary dysfunction, gastroparesis have been reported. Primary pulmonary hypertension has also been reported.

Gastro-intestinal Manifestations

Dysphagia is common either due to hyposalivation or due to oesophageal dysmotility. Chronic atrophic gastritis leads to epigastric pain and nausea. Autoimmune hepatitis has been reported in 25% with smooth muscle antibodies being present in 7–33%. Antimitochondrial antibodies occur in 7–13% of patients and suggests a potential association with primary biliary cirrhosis.[7] Acute or chronic pancreatitis has been reported.

Renal Manifestations

Lymphocytic infiltration of the tubulointerstitium results in distal renal tubular acidosis (RTA). The first presenting symptom of SS has been hypokalemic paralysis. In such patients, sicca symptoms are mild and previously unrecognized. In a study from Chennai the presenting symptom of hypokalemic paralysis was seen in 5.5% of SS patients.[5] RTA has also been reported from Mumbai.[8]

Glomerular disease is rare. Membranous or membranoproliferative glomerulonephritis has been described. Type 2 mixed cryoglobulinemia and low C4 levels seen in those patients. Interstitial cystitis may present as nocturia, suprapubic or perineal pain. Recurrent renal colic may be due to renal stones.

Neurological Manifestations

Cranial neuropathy involving trigeminal, facial or optic nerve has been reported. Sensorineural hearing loss is noted in one-half of SS patients on audiometric testing. Peripheral sensory motor neuropathy occurs as a consequence of small vessel vasculitis. Pure sensory neuropathy is a characteristic neurological complication of primary SS, caused by damage to the sensory neurons of the dorsal root and gasserian ganglion. Various CNS manifestations have been observed in Sjögren's patients including hemiparesis and hemisensory defects, seizure, movement disorders, transverse myelopathy and dementia. Patients with SS may present with multiple sclerosis like disease or with asymptomatic white matter lesions in MRI.

Vasculitis

Vasculitis of small and medium-sized vessels is found in 5% of SS patients manifesting as purpura, urticaria, skin ulceration or mononeuritis multiplex.

Systemic vasculitis can involve the kidneys, lungs, gastrointestinal tract, spleen, breast and reproductive tract. Large vessel vasculitis with absent peripheral pulses has been reported from Chennai.[9]

Autoimmune Thyroiditis

Thyroid disease has been seen in more than one-third of patients with SS as evidenced by the presence of antithyroid antibodies and elevated levels of thyroid stimulating hormones.

Lymphoproliferative Disease

Risk factors for the development of lymphoma in patients with SS are parotid enlargement, splenomegaly, lymphadenopathy, low C4 complement level and type-II mixed cryoglobulinemia. Skin vasculitis, peripheral nerve involvement, fever, anemia, lymphopenia were observed more frequently in SS patients with lymphoma than in those without lymphoma. Lymphomas are of B cell origin expressing IgM immunoglobulins and producing rheumatoid factors. Low grade lymphomas remit without therapy, while intermediate and high grade lymphomas have bad prognosis. Rare occurrence of both non-Hodgkin's lymphoma and multiple myeloma in a patient with primary SS has been reported.[10]

SECONDARY SJÖGREN'S SYNDROME

Sjögren's syndrome can be associated with other auto-immune rheumatic diseases (Table 41.1).

Prognostic Factors

The favourable prognostic factors include:

- Arthritis
- Raynaud's phenomenon
- Interstitial nephritis
- Lung and liver involvement

The adverse prognostic factors are presence of:

- Purpura
- Glomerulonephritis
- Low C4 levels
- Mixed monoclonal cryoglobulinemia

Table 41.1: Causes of secondary Sjögren's syndrome

1. Rheumatoid arthritis
2. Systemic lupus erythematosus
3. Systemic sclerosis
4. Mixed connective tissue disease
5. Primary biliary cirrhosis
6. Myositis
7. Vasculitis
8. Thyroiditis
9. Chronic active hepatitis
10. Mixed cryoglobulinemia

Diagnosis

Work up for patients with suspected Sjögren's syndrome is based on stepwise approach given by the 2002 American European consensus group modification of the European community criteria.[11]

The six criteria are:

1. *Ocular symptoms*: A positive response to at least one of the following questions:
   a. Have you had daily persistent troublesome dry eyes for more than 3 months?
   b. Do you have a recurrent sensation of sand or gravel in the eyes?
   c. Do you use tear substitutes more than three times a day?
2. *Oral symptoms*: A positive response to at least one of the following questions.
   a. Have you had a daily feeling of dry mouth for more than 3 months?
   b. Have you had recurrently or persistently swollen salivary glands as an adult?
   c. Do you often drink liquids to help in swallowing dry food?
3. *Ocular signs*: Objective evidence of ocular involvement defined as a positive result for at least one of the following two tests:
   a. Schirmer's test performed without anaesthesia (5 mm in 5 min.)
   b. Rose Bengal score or other ocular dry score 4 according to the (Bijsterveld scoring system)
4. *Histopathology*: In minor salivary glands (obtained through normal appearing mucosa) focal lymphocytic sialadenitis evaluated by an expert histopathologist with a focus score of 1, defined as a number of lymphocytic foci (which are adjacent to normal appearing mucus acini and contain more than 50 lymphocytes) per 4 mm^2 of glandular tissue.
5. *Salivary gland involvement*: Objective evidence of salivary gland involvement defined by a positive result for at least one of the following diagnostic tests:
   a. Unstimulated whole salivary flow (15 ml in 15 min.)
   b. Parotid sialography showing the presence of diffuse sialectasis (punctate, cavitatory or destructive pattern), without evidence of obstruction in the major duct.

c. Salivary scintigraphy showing delayed uptake, reduced concentration and or delayed excretion of tracer.

6. Antibodies to Ro/SSA or La/SSB antigens or both.

Patients are classified as having primary SS when they fulfil four or more of the six classification criteria; either criterion 4 or criterion 6 is mandatory.

Another recent set of criteria proposed by the Sjögren's International Collaborative Clinical Alliance Cohort (SICCAC) have been accepted by the American College of Rheumatology (ACR) and require the presence of two of the following three criteria:

1. Positivity for anti-Ro/SS-A antibody or anti-La/SS-B antibody or rheumatoid factor (RF) together with antinuclear antibodies (ANA) at a dilution of 1:320.
2. Minor salivary gland biopsy showing focal lymphocytic sialadenitis with a focus score of at least one.
3. Ocular sicca with ocular staining score of at least 3.

These criteria have a sensitivity >92% and specificity >95%.[12]

Investigations findings in a patient with Sjögren's syndrome are given in Table 41.2.

Differential diagnosis of Sjögren syndrome is summarized in Table 41.3.

TREATMENT

Ocular Disease

Xerophthalmia can be controlled with artificial tears made of polyvinyl alcohol or methyl cellulose. Lubricating ointments and hydroxypropyl cellulose inserts can be tried. Ciclosporine A 0.05% drops twice daily has been approved by FDA. Existing tears may be retained in the eyes by blocking their drainage or by inhibiting their evaporation. Punctate occlusion by inserting collagen or silicon plugs or by electrocautery retains the tears in the

Table 41.2: Findings on investigations in Sjögren's syndrome

Routine investigations	*Autoantibodies*	*Special investigations*
Cytopenia	Antibodies to Ro (SS-A) 66%	Cryoglobulins
Raised erythrocyte sedimentation rate (ESR)	Antibodies to La (SS-B) 62%	Low levels of C3, C4, CH50
Hypergammaglobulinemia	Rheumatoid factor	
	Antinuclear antibody	
	Anti-gastric parietal cell antibody	
	Anti-thyroglobulin antibody	
	Anti-thyroid microsomal antibody	
	Anti-smooth muscle antibody	
	Anti-salivary duct antibody	

Table 41.3: Differential diagnosis of Sjögren's syndrome

Conditions mimicking Sjögren's syndrome
Diffuse infiltrative lymphocytosis in HIV, HTL-V1
Hepatitis C infection
Sarcoidosis
Tuberculosis
Amyloidosis
Lymphoma
Ig G4 related disease
Hyperlipidemia
Dryness due to ageing
Previous radiation to head and neck

eyes. Tear evaporation can be prevented by wearing goggles with side chambers. Lateral tarsography is indicated in severe dryness to reduce the ocular surface. Corneal transplantation is advised for corneal perforation. Drug with anticholinergic side effects like phenothiazine, tricyclic antidepressants, antispasmodics, antiparkinsonian drugs and cigarette smoking should be avoided.

Oral Disease

Secretagogues stimulate muscarinic M1 and M3 receptors in salivary glands leading to increased salivation. Pilocarpine is administered as 5 mg QID, cevemeline is administered as 30 mg TID. Methotrexate and ciclosporine A improve the subjective symptoms of dryness.

Other Xerosis

Vaginal dryness requires lubricant jellies, while moisturizing lotions can be applied for dry skin.

Systemic Disease

Hydroxychloroquine has been used to treat arthralgia, myalgia and fatigue. Systemic corticosteroids, azathioprine, mycophenolate, cyclophosphamide are used for vasculitis, peripheral neuropathy, diffuse interstitial pneumonitis, glomerulonephritis. Renal tubular acidosis can be treated with potassium chloride and potassium citrate. Frank lymphomas require chemotherapy and radiotherapy.

Biologics

Biological agent infliximab has been found to be useful for treating ocular, oral and systemic manifestations of SS. Treatment with Rituximab an anti-CD-20 chimeric humanized monoclonal antibody specific for the B cell surface molecule CD20 has resulted in significant improvement of subjective symptoms, an increase in salivary gland function and improvement in systemic features. There was rapid depletion of peripheral B cells within a few weeks and a decrease in IgM RF levels.[13] Dose is 375 mg/m^2 once weekly for 4 weeks. Epratuzumab, fully humanized monoclonal antibody specific for B cell surface molecule CD 22 given in 4 doses of 360 mg/mm^2 has shown improvement.[14]

SS patients with elevated BAFF levels, hypergammaglobulinemia elevated levels of autoantibodies might be candidates for anti-BAFF therapy like Belimumab 10 mg/kg at week 0, 2, 4 and every 4 weeks.[15] T cell targeted therapy with abatacept has also been used in SS.

Follow Up

Patients with stable disease limited to mucosal surfaces may require only annual evaluation, while those with extra-glandular manifestations should be evaluated every six months and those with end organ damage every three months. Fertile anti-Ro/La positive women should be advised about the risk of fetal congenital heart block.

Key Points

- Sjögren's syndrome is an autoimmune disease affecting exocrine glands.
- Characterized by dry eyes, dry mouth, salivary gland and lacrimal gland enlargement.
- Elevated ESR, hypergammaglobulinemia, presence of Ro or La-antibodies and positive minor salivary gland biopsy showing lymphocytic infiltration is diagnostic.
- Dry eyes treated with artificial tears, dry mouth with secretagogues.
- Systemic symptoms require corticosteroids, hydroxychloroquine, cyclophosphamide and biologics.

REFERENCES

1. Mann D. Immunogenetics of Sjögren's in: Talal N, Moutsopoulas HM, Kassan SS eds. Sjögren's Syndrome. Clinical and immunological aspects. Berlin: Springer–Verlag, 1987. P 235–43.
2. Inoue H, Tsubota K, Ono M, Kizu Y, Mizuno F, Takada K, et al. Possible involvement of EBV-mediated alpha-fodrin cleavage for organ-specific autoantigen in Sjögren's syndrome. J Immunol. 2001; 166:5801–9.
3. Ben-Chetrit E, Fox RI, Tan EM. Dissociation of immune responses to the SS-A(Ro) 52-kd and 60-kd polypeptides in systemic lupus erythematosus and Sjögren's syndrome. Arthritis Rheum. 1990; 33:349–55.
4. Tzioufas AG, Moutsopoulos HM. Ultrasonography of salivary glands: an evolving approach for the diagnosis of Sjögren's syndrome. Nat Clin Pract Rheumatol. 2008; 4:454–5.
5. Porkodi R, Rukmangatharajan S, Kanakarani P, PArthiban M, Vasanthy N, Madhavan R, et al.

Primary Sjögren's syndrome—clinical and immunological features. J Indian Rheumatol Assoc 2003; 11:63–65.

6. Hissaria P, Aggarwal A, Dabadghao S et al. Rarity of primary Sjögren's syndrome in India. JIRA 2000; 8(suppl)1:S.36.
7. Lindgren S, Manthorpe R, Eriksson S. Autoimmune liver disease in patients with primary Sjögren's syndrome. J Hepatol 1994, 20:354.
8. EM, Pathan JL Oak, CG Jyotish: Extra glandular manifestations in secondary Sjögren's syndrome—JIRA 1999, 4:116–17.
9. Porkodi R, Ramakrishnan S, Krishnamurthy V, Rajendran CP, Chandrsekaran AN et al. Primary Sjögren's syndrome with vasculitis. J Indian Rheumatol Assoc 1993; 1:42–43.
10. N Kumar, CA Kelly. Primary Sjögren's syndrome presenting with non-Hodgkin's lymphoma and terminating with myeloma. JIRA 1999; 7:4:114–15.
11. Vitali C, Bombardieri S, Jonsson R, Moutsopoulos HM, Alexander EL, Carsons SE, et al; European Study Group on Classification Criteria for Sjögren's syndrome. Classification criteria for Sjögren's syndrome: a revised version of the European criteria proposed by the American-European Consensus Group. Ann Rheum Dis. 2002; 61:554–8.
12. Shiboski SC, Shiboski CH, Criswell L, Baer A, Challacombe S, Lanfranchi H, et al. Sjögren's International Collaborative Clinical Alliance (SICCA) Research Groups. American College of Rheumatology classification criteria for Sjögren's syndrome: a data-driven, expert consensus approach in the Sjögren's International Collaborative Clinical Alliance cohort. Arthritis CareRes (Hoboken). 2012; 64:475–87.
13. Pijpe J, van Imhoff GW, Spijkervet FK, Roodenburg JL, Wolbink GJ, Mansour K, et al. Rituximab treatment in patients with primary Sjögren's syndrome: an open-label phase II study. Arthritis Rheum. 2005; 52:2740–50.
14. Steinfeld SD, Tant L, Burmester GR, Teoh NK, Wegener WA, Goldenberg DM, et al. Epratuzumab (humanised anti-CD22 antibody) in primary Sjögren's syndrome: an open-label phase I/II study. Arthritis Res Ther. 2006; 8(4):R129.
15. Szodoray P, Jonsson R. The BAFF/APRIL system in systemic autoimmune disease with special emphasis on Sjögren's syndrome. Scan J Immunol 2005; 62:421–28.

FURTHER READING

1. Jonsson R, Theander E, Sjöström B, Brokstad K, Henriksson G. Autoantibodies present before symptom onset in primary Sjögren syndrome. JAMA. 2013; 310:1854–5.
2. Ng WF, Bownan SJ. Primary Sjögren's syndrome too dry and too tired. Rheumatology (Oxford) 2010; 49:844–53.
3. Ramos-Casals M, Tzioufas AG, Stone JH, Sisó A, Bosch X. Treatment of primary Sjögren syndrome: a systematic review. JAMA. 2010; 304:452–60.

Chapter

42

Mixed Connective Tissue Disease

Rajkiran Dudam

INTRODUCTION

Mixed connective tissue disease (MCTD) was first described by Gordon C Sharp and his colleagues in 1972.[1] It is characterized by overlapping clinical features of systemic lupus erythematosus (SLE), scleroderma and polymyositis (PM)/dermatomyositis (DM) and high serum titres of anti-U1RNP antibodies. On the other hand, overlap syndrome is a condition where patient has clinical features which fulfil the classification criteria for more than one recognized rheumatic disease. Undifferentiated connective tissue disease (UCTD) is a condition where patient do not fulfil the classification criteria of any defined rheumatic disease. On follow-up these patients might develop new symptoms which might fulfil the criteria or they might continue to be the same and not progressing to any specific rheumatic disease. Such patients usually have positive ANA test results by indirect immunofluorescence assay (IIFA).

Frequency

MCTD is rare in India, and there are no estimates of the prevalence of the disease in India.[2] A nationwide survey in Norway found the point prevalence of MCTD in Norway is 3.8/100,000 adults, with an incidence of 2.1/100,000/year[3] with male to female ratio 3.3. The mean age at diagnosis was 37 years, with an onset usually between 15 and 25 years of age.

Classification Criteria

There are many classification criteria for MCTD, but the Alarcon-Segovia criteria (Table 42.1) are most sensitive (62.5%) and specific (86.2%) for identification of MCTD.[4] The other criteria used to diagnose MCTD are the modified Sharps's criteria, Kasukawa criteria, and Kahn.[5]

Table 42.1: Alarcon-Segovia criteria for diagnosis of mixed connective tissue disease (MCTD)[4]

A. Serologic criteria

Anti-RNP at hemagglutination titre of >1:1600

B. Clinical criteria

1. Swollen hands
2. Synovitis
3. Myositis
4. Raynaud's phenomenon
5. Acrosclerosis

MCTD is present if criteria A is accompanied by three or more clinical criteria—one of which must include synovitis or myositis.

CLINICAL FEATURES

The most frequent clinical features of MCTD includes Raynaud's phenomenon, arthralgia and/or arthritis, sausage-shaped swollen fingers, gastro-oesophageal reflux, myositis, rash (Table 42.2).[5,6] Pulmonary hypertension (PH) is the most serious complication and common disease related cause of death.[7] Apart from high titer speckled pattern fluorescent ANA and elevated anti-U1 RNP antibodies the other laboratory features include anaemia of chronic disease, leucopenia predominantly lymphopenia, hypergammaglobulenemia, and elevated creatinine phosphokinase (CPK) levels.

Table 42.2: Common clinical and laboratory features of mixed connective tissue disease (MCTD)[4–6]

1. Raynaud's phenomena
2. Swollen fingers
3. Arthralgia/arthritis
4. Sclerodactyly
5. Acrosclerosis
6. Skin rash
7. Oesophageal reflux
8. Myalgias/myositis
9. Pulmonary hypertension
10. Pleuritis/pericarditis
11. Interstitial lung disease
12. High titers of anti-U1 RNP
13. Antibodies against U1-70 kD small nuclear ribonucleoprotein (snRNP)

Skin

Raynaud's phenomenon is usually the initial manifestation[6] and it is characterized by initial pallor (white colour) on exposure to cold followed by cyanosis (blue) due to depleted oxygen supply and later rubor (red) once the blood supply returns when area is warmed and then back to normal. This phenomenon is often accompanied with numb, tingling and painful pins and needles sensation. Nailfold capillarioscopy (NFC) can be helpful in distinguishing between benign primary Raynaud's and secondary Raynaud's due to MCTD and other autoimmune connective tissue diseases. NFC changes observed in MCTD are similar to those found in systemic sclerosis. In a typical patient of MCTD Raynaud's phenomenon is accompanied by puffy fingers and sometimes hand oedema (Fig. 42.1).

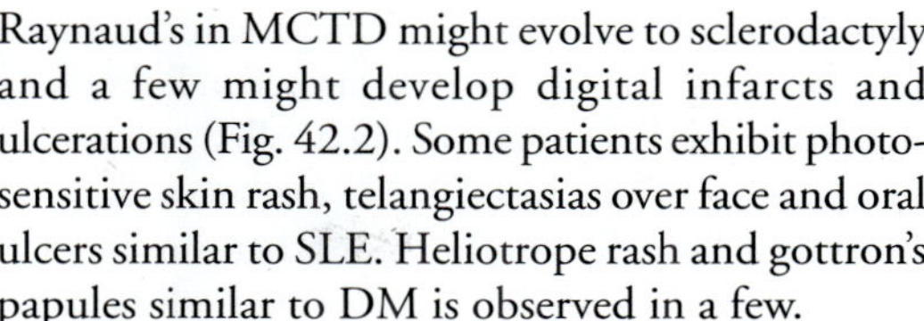

Raynaud's in MCTD might evolve to sclerodactyly and a few might develop digital infarcts and ulcerations (Fig. 42.2). Some patients exhibit photosensitive skin rash, telangiectasias over face and oral ulcers similar to SLE. Heliotrope rash and gottron's papules similar to DM is observed in a few.

Fever

MCTD can sometimes present as pyrexia of unknown origin.

Joints

Up to 90% patients have arthralgias and arthritis.[6] It can be in the form of symmetric arthritis involving hands and wrists. Rheumatoid factor can be positive in 25 to 50% cases. Arthritis is usually non-erosive and jaccouds kind of arthropathy has also been observed. Presence of anti-citrullinated peptide antibodies (ACPA) is associated with erosive disease.

Muscle

Myalgias and myositis is observed in 25 to 75% patients.[6,7] Myositis is infrequent at disease onset but can be observed in majority during follow up. Patients usually have elevated CPK levels with mild muscle discomfort rather than significant weakness. MCTD myositis is less severe than PM/DM myositis and usually steroid responsive. Histologically MCTD myositis resembles pattern in DM.

Heart and Pulmonary Hypertension

Cardiac involvement can occur in 13 to 65% patients.[8] Pericarditis is the most common cardiac

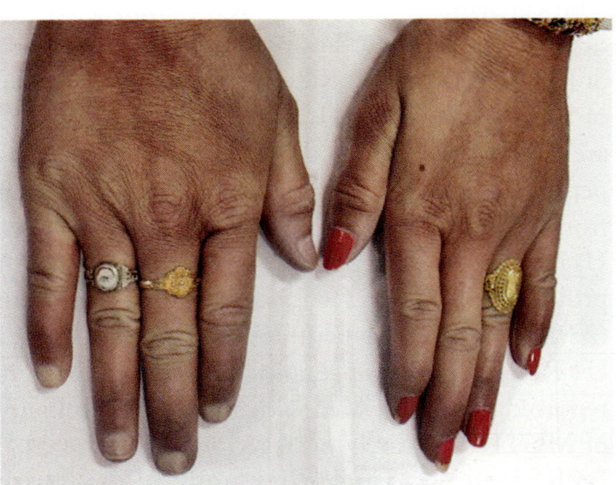

Fig. 42.1: Puffy fingers

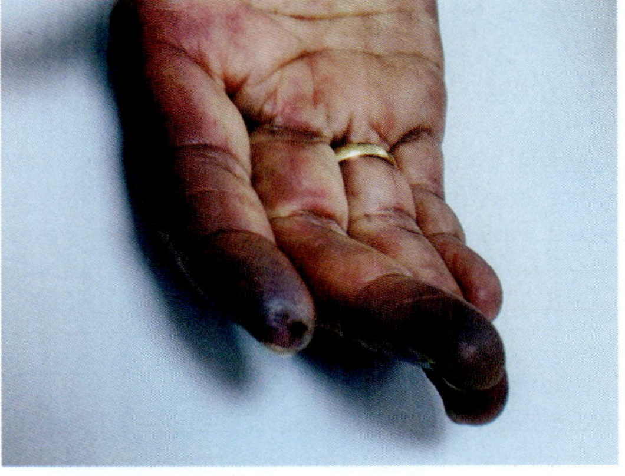

Fig. 42.2: Raynaud's phenomenon with digital tip infarct

manifestation, seen in 30 to 43% patients.[8] Cardiac tamponade is rare. Myocarditis can occur. Pulmonary hypertension (PH) can be isolated called pulmonary arterial hypertension (PAH) or secondary to interstitial lung disease (ILD) or other causes. PAH may be undiagnosed in early stage while some patients present with isolated PAH. PAH should be suspected in patients with exertional dyspnea and a screening 2D echocardiography is to be done to estimate right ventricular systolic pressure (RVSP). A mean resting pulmonary artery pressure of more than 25 mm measured by cardiac catheterization is definitive for diagnosis of PH.

Presence of PAH determines the prognosis of MCTD.[7] MCTD is an important cause of CTD-PAH. Data from the REVEAL registry show that 52 of 641 (8.1%) patients with CTD-PAH was due to MCTD.[9] The causes of CTD-PAH in the REVEAL registry was systemic sclerosis (SSc) $n = 399$, followed by SLE $n = 110$ then MCTD $n = 52$, and RA $n = 28$. A nationwide survey from Norway found that the prevalence of PAH in MCTD is 3.7%.[10] Patients with PAH have elevated levels of N-terminal pro-brain natriuretic peptide (NT-ProBNP).[9] In scleroderma PAH is usually due to interstitial pulmonary fibrosis, whereas in MCTD it is due to bland intimal proliferation and medial hypertrophy of pulmonary arterioles.

Lungs

Interstitial lung disease may be observed in up to 50% cases. ILD is demonstrated by HRCT chest, majority of patients have septal thickening with ground glass opacities. Pleuritis and rarely pulmonary haemorrhage are also described.

Gut

Oesophageal reflux can present as heart burn, dysphagia, odynophagia and food regurgitation. Bacterial overgrowth, malabsorption syndrome, and other intestine manifestations similar to SSc can occur. Patients with anti-SSA (Ro) antibody positivity can develop sicca symptoms, submandibular gland enlargement and photosensitive malar rash.[11]

Nervous System

Neurological manifestations include trigeminal neuralgia which sometimes can be initial presentation, sensorineural deafness, headache, aseptic meningitis, psychosis, stroke, seizures, transverse myelitis, cauda equina, and cerebral haemorrhage can be rarely seen.

Kidney

MCTD patients with anti-dsDNA antibodies can have membranoproliferative glomerulonephritis similar to SLE. Membranous nephropathy has been reported from Japan.[12] Interstitial nephritis in patients with concomitant Sjögren's syndrome, and rarely scleroderma renal crisis kind of presentation have been reported.[13]

Protean Presentation

Recently it is described that MCTD presents can have three distinct clinical phenotype.[14]

- One group with predominant vascular manifestations, i.e. Raynaud's, PAH, livedo reticularis, antiphospholipid antibodies, and thrombosis. Bland intimal proliferation with medial hypertrophy that affects small and medium size vessels is characteristic of vascular lesion of MCTD. Vascular involvement differs from scleroderma as it is less associated with fibrosis and presents immunoglobulin and complement deposits in vessel wall.
- Another group presents with pulmonary fibrosis, ILD, myositis, and esophageal dysmotility.
- Third group are patients with ACPA positivity, erosive arthritis, musculoskeletal damage, gastrointestinal symptoms, and osteoporotic fractures.[14]

Etiology and Pathogenesis

Clinical symptoms can manifest within a year of development of auto-antibodies for U1RNP. The SnRNP (small nuclear Ribonuclear protein) is made of 5 snRNA molecules (U1, U2, U4, U6, U7). They are uridine rich hence the U, and have 11 polypeptides with a molecular weight of each ranging from 11 to 70 kD. The antigenic epitopes for anti-U1RNP reside on p70 protein (70 kD), the A protein (33 kD) and occasionally the C protein (22 kD). These polypeptides are complexed with U1 RNA. Antibodies reacting to p70 protein were reported to be associated with anti-RNP specificity of MCTD and rarely occurred in SLE patients.[15] Both adaptive and innate immunity are thought to play an important role in pathogenesis. Adaptive immune response involving B cells and production of anti-U1RNP and innate immunity through TLR 7 and TLR3. Genetic association with major histocompatibility genes human leukocyte antigen (HLA)–DRB1*04/*15 have been identified.[16] Disappearance of anti-snRNP autoantibodies occurs during prolonged remission.[7]

MANAGEMENT

Management of MCTD requires an assessment of disease severity, and the major manifestation (Table 42.3). Most treatment strategies are largely based on experiences of management of SLE, RA, and SSc.

Management of PAH

The AMBITION trial showed that upfront combination therapy with oral ambrisentan and tadalafil results in 50% reduction in the rate of clinical worsening in PAH with NYHA functional II or III.[17] Most of pulmonary vasodilator trials in PAH have included both idiopathic PAH and CTD-PAH patients, and as expected the majority of CTD-PAH is contributed by SSc. However, these effects may be valid in MCTD as well.

Immunosuppression along with pulmonary vasodilator therapy may produce improvement in functional class and could stabilize pulmonary hypertension in patients with SLE or MCTD.[18,19]

Table 42.3: Treatment of mixed connective tissue disease (MCTD) according to manifestation and severity of manifestations

Clinical feature	*Mild*	*Moderate to severe*
Arthralgia/arthritis	NSAIDs Hydroxychloroquine	Systemic corticosteroids Methotrexate B cell depleting agents (Rituximab)
Raynaud's phenomenon	Protective measures such as • minimizing exposure to cold • wearing protective gloves • avoid trauma • avoid betablockers • stop smoking	Calcium channel blockers topical nitrates PDE 5 inhibitors Prostaglandin analogues (Iloprost) Anticoagulation for digital gangrene
Rash	Avoid sun exposure Hydroxychloroquine Topical steroids	Systemic steroids Azathioprine Methotrexate
Oesophageal reflux	Raise head end of bed while sleeping avoid smoking and caffeine	Proton pump inhibitors H_2 antagonists
Myositis	Steroids, hydroxychloroquine	Steroids Methotrexate Azathioprine IVIG B cell depleting agents (rituximab)
Pulmonary hypertension	PDE5 inhibitors	Calcium channel blockers PDE5 inhibitors ET receptor antagonist (Bosentan, Ambrisentan) Cyclophosphamide Anticoagulation
Pericarditis/pleuritis	NSAIDs Steroids	NSAIDs Steroids Azathioprine

In a report on the role of intensive immunosuppression of CTD-PAH from a French referral center, of the 28 patients (SLE, n = 13; MCTD, n = 8; limited cutaneous SSc, n = 5; diffuse SSc, n = 1 and rheumatoid arthritis, n = 1) who received monthly IV bolus of cyclophosphamide 600 mg/m^2 for at least 3 months, plus oral prednisone 0.5 to 1 mg/day, only 8 of 28 patients responded. The 8 responders were SLE (n = 5), and MCTD (n = 3), none of the patients with SSc responded.[18]

The same group reported their experience with intensive immunosuppressive therapy (consisting of monthly IV pulse of cyclophosphamide 600 mg/m^2 for 6 months and oral glucocorticoid therapy) in 16 patients with SLE or MCTD associated PAH.[19] There were 8/16 patients (50%) who responded. Six of the 8 non-responders subsequently responded to pulmonary vasodilators.

Although strong evidence in the form of RCTs to favour intensive immunosuppressive therapy is limited, but considering the overall grim prognosis of the PAH it may be worthwhile to try cyclophosphamide and prednisolone in PAH, especially in early PAH.[18,19] For patients with NYHA class II and above, oral ambrisentan (10 mg per day) and tadalafil (up to 40 mg per day) can be tried.

Prognosis

Classically MCTD is described as a benign and steroid responsive condition with less renal and neurological involvement. Patients may evolve to SSc or SLE or any other rheumatic disease.[20] A long-term follow up study of 47 patients showed that about 62% have a favourable outcome, while the remaining have continued active disease or died, but in this series of 47 patients it was shown that they rarely evolved into SLE or SSc.[7]

Pulmonary hypertension was the most frequent cause of death. Data from a Hungarian series showed the 5, 10, and 15-year survival rates after the diagnosis were 98%, 96%, and 88% respectively.[21] In this series, the causes of death in declining order of frequency was PAH, infections, thrombotic thrombocytopenic purpura, and cardiovascular events. Mortality was associated with the presence of antiphospholipid and anti-endothelial cell antibodies.[20]

Patient Education

Education about the disease is very important while managing any rheumatic disease. Many patients and their family would have never heard of the disease, and the treating physician may be the first to introduce this term to them. Hence a detailed counseling about the disease, symptoms, and its natural course at initial meeting would make management easier for the patients and physician.

Key Points

- MCTD has overlap clinical features of SLE, SSc, polymyositis /dermatomyositis which may not be evident at initial presentation but may develop over a period of time.
- Raynaud's phenomenon is seen in almost all patients.
- High titers of anti-U1RNP antibodies are mandatory for diagnosis.
- Pulmonary arterial hypertension is the leading cause of death.

REFERENCES

1. Sharp GC, Irvin WS, Tan EM, Gould RG, Holman HR. Mixed connective tissue disease—an apparently distinct rheumatic disease syndrome associated with a specific antibody to an extractable nuclear antigen (ENA). Am J Med. 1972; 52:148–59.
2. Sood A, Kumar A, Pande I, Malaviya AN. Does mixed connective tissue disease exist in India? Br J Rheumatol. 1995; 34:539–41.
3. Gunnarsson R, Molberg O, Gilboe IM, Gran JT; PAHNOR1 Study Group. The prevalence and incidence of mixed connective tissue disease: a national multicentre survey of Norwegian patients. Ann Rheum Dis. 2011; 70:1047–51.
4. Alarcón-Segovia D, Cardiel MH. Comparison between 3 diagnostic criteria for mixed connective tissue disease. Study of 593 patients. J Rheumatol. 1989; 16:328–34.
5. Tani C, Carli L, Vagnani S, Talarico R, Baldini C, Mosca M, Bombardieri S. The diagnosis and classification of mixed connective tissue disease. J Autoimmun. 2014; 48–49:46–9.
6. Ungprasert P, Crowson CS, Chowdhary VR, Ernste FC, Moder KG, Matteson EL. Epidemiology of Mixed Connective Tissue Disease, 1985-2014: A Population-Based Study. Arthritis Care Res (Hoboken). 2016; 68:1843–48.

7. Burdt MA, Hoffman RW, Deutscher SL, Wang GS, Johnson JC, Sharp GC. Long-term outcome in mixed connective tissue disease: longitudinal clinical and serologic findings. Arthritis Rheum. 1999; 42:899–909.

8. Ungprasert P, Wannarong T, Panichsillapakit T, Cheungpasitporn W, Thongprayoon C, Ahmed S, et al. Cardiac involvement in mixed connective tissue disease: a systematic review. Int J Cardiol. 2014; 171: 326–30.

9. Chung L, Liu J, Parsons L, Hassoun PM, McGoon M, Badesch DB, et al. Characterization of connective tissue disease-associated pulmonary arterial hypertension from REVEAL: identifying systemic sclerosis as a unique phenotype. Chest. 2010; 138: 1383–94.

10. Gunnarsson R, Andreassen AK, Molberg Ø, Lexberg ÅS, Time K, Dhainaut AS, et al. Prevalence of pulmonary hypertension in an unselected, mixed connective tissue disease cohort: results of a nationwide, Norwegian cross-sectional multicentre study and reviewof current literature. Rheumatology (Oxford). 2013; 52:1208–13.

11. Setty YN, Pittman CB, Mahale AS, Greidinger EL, Hoffman RW. Sicca symptoms and anti-SSA/Ro antibodies are common in mixed connective tissue disease. J Rheumatol. 2002; 29:487–9.

12. Ichikawa K, Konta T, Sato H, Ueda Y, Yokoyama H. The clinical and pathological characteristics of nephropathies in connective tissue diseases in the Japan Renal Biopsy Registry (J-RBR). Clin Exp Nephrol. 2017 Mar 2.

13. Vij M, Agrawal V, Jain M. Scleroderma renal crisis in a case of mixed connective tissue disease. Saudi J Kidney Dis Transpl. 2014; 25:844–8.

14. Szodoray P, Hajas A, Kardos L, Dezso B, Soos G, Zold E, et al. Distinct phenotypes in mixed connective tissue disease: subgroups and survival. Lupus. 2012; 21:1412–22.

15. Paradowska-Gorycka A. U1-RNP and TLR receptors in the pathogenesis of mixed connective tissue disease Part I. The U1-RNP complex and its biological significance in the pathogenesis of mixed connective tissue disease. Reumatologia. 2015; 53:94–100.

16. Hoffman RW, Maldonado ME. Immune pathogenesis of Mixed Connective Tissue Disease: a short analytical review. Clin Immunol. 2008; 128:8–17.

17. Galiè N, Barberà JA, Frost AE, Ghofrani HA, Hoeper MM, McLaughlin VV, et al. Initial use of ambrisentan plus tadalafil in pulmonary arterial hypertension. N Engl J Med 2015; 373:834–44.

18. Sanchez O, Sitbon O, Jaïs X, Simonneau G, Humbert M. Immunosuppressive therapy in connective tissue diseases-associated pulmonary arterial hypertension. Chest 2006; 130:182–9.

19. Jais X, Launay D, Yaici A, Le Pavec J, Tchérakian C, Sitbon O, et al. Immunosuppressive therapy in lupus- and mixed connective tissue disease-associated pulmonary arterial hypertension: a retrospective analysis of twenty-three cases. Arthritis Rheum. 2008; 58(2):521–31.

20. Cappelli S, Bellando Randone S, Martinoviæ D, Tamas MM, Pasaliæ K, et al. “To be or not to be,” ten years after: evidence for mixed connective tissue disease as a distinct entity. Semin Arthritis Rheum. 2012; 41:589–98.

21. Hajas A, Szodoray P, Nakken B, Gaal J, Zöld E, Laczik R, et al. Clinical course, prognosis, and causes of death in mixed connective tissue disease. J Rheumatol. 2013; 40:1134–42.

FURTHER READING

1. Hasegawa EM, Caleiro MT, Fuller R, Carvalho JF. The frequency of anti-beta2-glycoprotein I antibodies is low and these antibodies are associated with pulmonary hypertension in mixed connective tissue disease. Lupus. 2009; 18:618–21.

2. Rajagopala S, Thabah MM. Pulmonary hypertension associated with connective tissue disease. Indian J Rheumatol 2017; 12:38–47.

Chapter

43

Scleroderma

Jyotsna Oak, Rupali Mathur

INTRODUCTION

The word "Scleroderma" is derived from two Greek words, "sclero" meaning hard and "derma" meaning skin. Scleroderma or systemic sclerosis (SSc) is an autoimmune disorder of unknown etiology characterized by fibrosis and tightening of skin as well as microvascular injury in affected organs. SSc may affect multiple organs simultaneously or at different times and thus has a wide spectrum of clinical manifestations and severity. The involved target organs include skin, muscles, joints, nerves, vasculature, kidney, heart, lung and gastrointestinal tract. The disease can have limited cutaneous involvement or diffuse cutaneous involvement. Diffuse cutaneous SSc is defined as skin thickening proximal to the elbows and knees (upper arm, thighs, anterior chest and abdomen) that is documented at any time during the illness. Limited cutaneous SSc is defined as skin thickening limited to distal extremities and face throughout the illness (Table 43.1).

EPIDEMIOLOGY

SSc has a worldwide distribution. The disease is rare in childhood. Women are affected more frequently than men. The highest incidence is in the age group of 30–35 years. The peak age of onset ranges from 25 to 54 years in Caucasian women and from 34 to 44 years in African women. Racial factors seem to play a role in disease susceptibility as well as disease expression.

There is no data on the prevalence of SSc in India. In USA the prevalence in the general population has been estimated in the range of 24.2 per 100,000, whereas in Europe it is 7.1 to 15.8 per 100,000.

PATHOGENESIS

The pathological hallmark of SSc is non-inflammatory proliferative obliterative vasculopathy affecting small arteries and arterioles in multiple vascular beds, loss of capillaries and fibrosis that is most prominent in skin, lungs and heart. The characteristic feature is overproduction and accumulation of collagen and extracellular matrix protein like fibronectin, tenascin, fibrillin-1 and glycosaminoglycans in skin and other organs. The disease process involves immunological mechanisms and vascular endothelial cell activation and/or injury. It also involves fibroblasts activation resulting in production of excessive collagen.

Table 43.1: Classification of systemic sclerosis

Presenting feature	*Limited*	*Diffuse*
Raynaud's phenomenon	Long duration	Short duration
Skin involvement	Distal limbs and face	Proximal part of limbs and trunk
Tendon friction rub	Absent	Present +/–
Nailfold capillaries	Dilatation	Dilatation and loss
Auto-antigen	Centromere	Topoisomerase 1

An early event that precedes fibrosis is vascular injury. Small vessel vasculopathy and fibrosis characterize pathology of SSc. The vessels involved are not only small arteries, arterioles and capillaries in the skin but also organs like GI tract, kidney, heart and lungs. Microvascular injury leads to proliferation of intima and smooth muscles with deposition of matrix and perivascular fibrosis, leading to obliteration vasculopathy. Endothelial injury plays a major role.

Antiendothelial antibodies (AECA) seen in 25% of the SSc patients have been shown to mediate antibody dependent cytotoxicity against endothelial cells. This also leads to microvascular damage. Pathogenesis involves a complex interplay between endothelial cells, fibroblasts and the immune systems triggered by environmental factors. Injury to endothelial cells and basal lamina lead to proliferation of intima and smooth muscle cells with deposition of matrix and perivascular fibrosis leading to narrowing of lumen and ischemia. Figure 43.1 shows the pathogenesis of SSc.

The etiological agents implicated include viruses, silica exposure, vinyl chloride, organic solvents, bleomycin and even smoking.

CLINICAL FEATURES

Raynaud's Phenomenon

Over 95% of the patients will experience Raynaud's phenomenon as an initial symptom.[1] It is defined as episodic vasoconstriction of small arteries and arterioles of fingers, toes and sometimes the tip of nose and earlobes.

Patients experience pallor and/or cyanosis followed by rubor on rewarming. Pallor and cyanosis are usually associated with coldness and numbness of fingers and toes.

Episodes are brought by cold exposure, vibration and emotional stress. The pathogenic mechanism of Raynaud's phenomenon is uncertain, but probably represents an imbalance between vasoconstrictor and vasodialator mediator in response to physiological level of stimulation by cold or emotions. In diffuse SSc skin changes are seen typically within a year of Raynaud's phenomenon. In case of limited cutaneous SSc it will precede the disease by several months or even years. Telangiectasias are local dilated loops of small blood vessels in the skin which are prone to hemorrhage on trauma (Fig. 43.2).

The hallmark of SSc is taut, hidebound skin. This may take several years to develop. In clinical practice the following subsets are easily identified.

Localized Scleroderma

This is usually observed in dermatology practice in the form of morphea and linear scleroderma. Morphea presents as plaques of fibrotic skin and

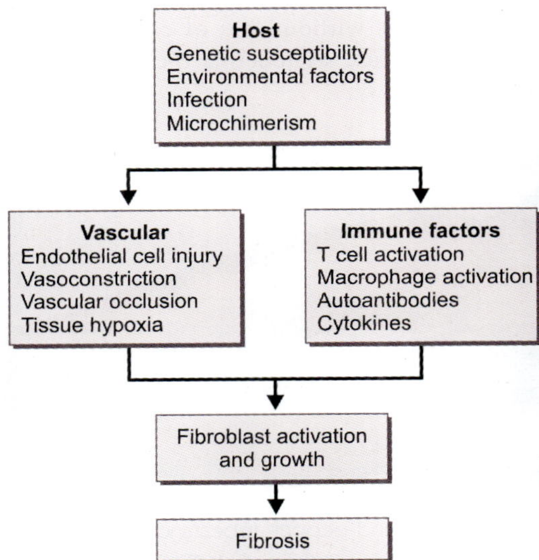

Fig. 43.1: Pathogenesis of systemic sclerosis

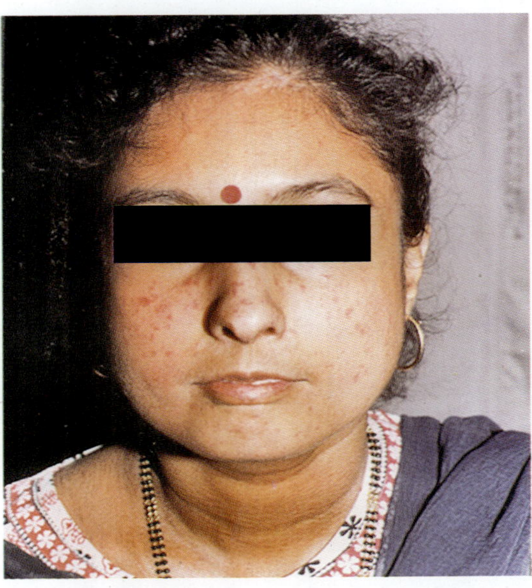

Fig. 43.2: Telangiectasia in a patient with systemic sclerosis

subcutaneous tissue without systemic disease, while linear scleroderma is seen as fibrotic bands that occur on the extremities involving skin and deeper tissue.

Limited Cutaneous SSc

In limited cutaneous skin changes affect face, neck, and extremities distal to the elbows and knees. Involvement of skin is the hallmark of SSc and there are usually three phases, edematous, indurative, and atrophic phase. The edematous phase is characterized by the complaints of stiff hands with associated puffiness. Subsequently, the skin appears thick and tight, tendon fibrosis results into contracture of joints. The thumb is usually spared. Perioral fibrosis leads to 'pursed lip' and puckered mouth. This along with pinched nose and masked-like face constitutes the typical scleroderma facies (Fig. 43.3).

Digital scar/pseudo-clubbing; finger contractures, ischaemic ulcers, calcinosis and telangiectasia are common findings. Calcinosis is uncommon in Indian patients.[1,2]

An occasional patient may present with 'CREST' (Calcinosis, Raynaud's phenomenon, Esophageal involvement, Sclerodactyly and Telangiectasia) syndrome. Late in the disease course pulmonary hypertension may develop.

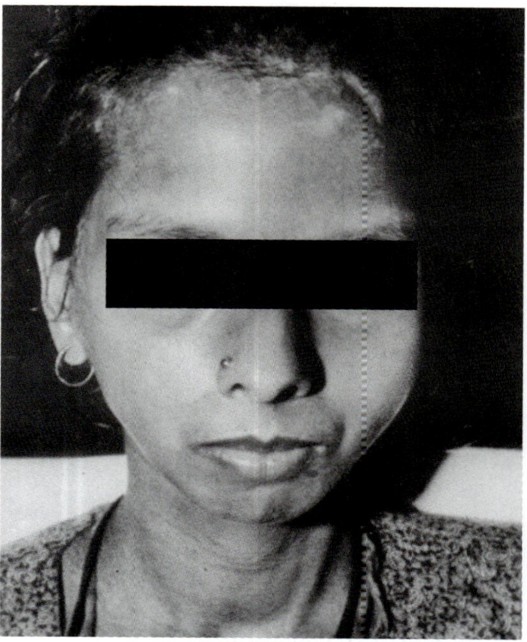

Fig. 43.3: Scleroderma facies

Diffuse Cutaneous SSc

This is characterized by extensive skin involvement extending proximal to the knees and elbows and usually affecting the trunk. The edematous phase may begin distally in the extremities and advances proximally. The skin becomes firm, thickened and eventually tightly bound to underlying subcutaneous tissue (indutrative phase). In diffuse SSc skin changes will become generalized involving extremities followed by trunk and face over a period of time varying from months to years. Rapid progression of these changes over 1–3 years is associated with greater risk of visceral disease particularly heart, lungs and kidneys. Also in diffuse scleroderma, skin changes usually peak for 3–5 years and then slowly improve. On the other hand, the skin of limited cutaneous SSc will usually have a more gradual progression.

In the extremities the taut skin over fingers gradually limits full extension and flexion contractures develop. Ulcers may appear on volar pads of fingertips and over elbows, and malleoli. The volar pads of fingertips develop pitting scars. There is resorption of terminal phalynx. Skin may show diffuse hypopigmentation with sparing of pigment around hair follicles giving a typical 'salt and paper' appearance (Fig. 43.4). Skin loses hair, oil and sweat glands and appears dry and coarse.

Wide angle microscopy may be helpful in distinguishing patients of limited cutaneous scleroderma who demonstrate enlargement of capillary loops without loss of capillary beds, whereas in diffuse scleroderma there is a loss of capillary loops.

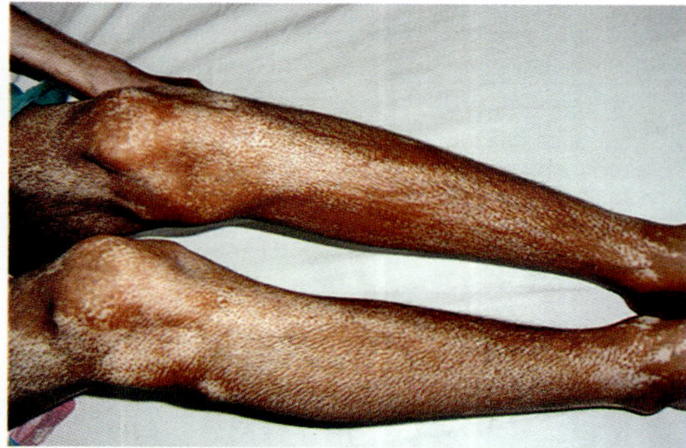

Fig. 43.4: Salt and pepper appearance of the skin in scleroderma (*Figure Courtesy*: Dr Vikas Agarwal, Professor of Clinical Immunology, SGPGIMS, Lucknow)

Musculoskeletal Features

More than half of the patients with SSc complain of pain, swelling and stiffness of fingers and knees. A symmetrical polyarthritis resembling RA is also observed. In advanced cases leathery crepitations can be palpated over joints. Muscle weakness is observed in patients with severe diffuse disease and a few patients develop myositis which is identical to polymyositis.

Pulmonary Involvement

Lung disease is a frequent manifestation and has replaced renal disease as a leading cause of mortality.

Interstitial lung disease is present in almost all patients of diffuse SSc and half of these patients may be asymptomatic. Patients present with dry cough and progressive dyspnoea on exertion. Bibasilar velcro rales are typically present on auscultation.

Pulmonary function tests show reduction in vital capacity, low diffusing capacity and low PO_2 with exercise. Chest film shows reticular appearance, honey combing or mottling especially of lower two-thirds of lung (Fig. 43.5). HRCT can detect early pulmonary changes showing 'ground glass' appearance due to alveolitis (Fig. 43.6). Bronchoalveolar lavage (BAL) may detect increased number of alveolar macrophages or neutrophils as evidence of alveolitis.

Patients with diffuse scleroderma who have anti-topoisomerse 1 antibody are particularly at risk of developing pulmonary fibrosis. In a subset of patients of limited cutaneous SSc after many years of disease, pulmonary hypertension may develop.

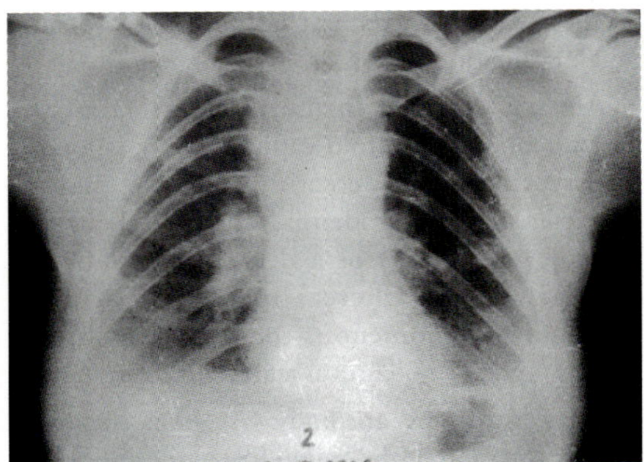

Fig. 43.5: Chest X-ray in scleroderma showing bilateral basal haziness due to interstitial lung disease

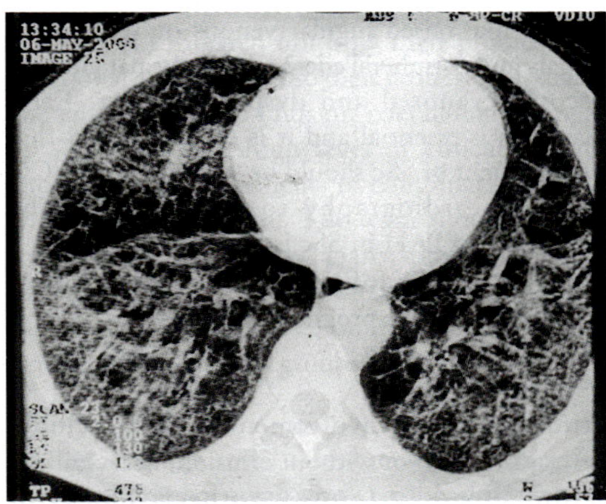

Fig. 43.6: HRCT chest showing areas of ground glass opacities and honey combing

This is due to narrowing and obliteration of pulmonary arteries and arterioles by intimal fibrosis and medial hypertrophy. It is manifested initially by exertional dyspnoea and eventually by right-sided heart failure. The prognosis is poor.[3] With development of pulmonary hypertension, the mean duration of survival is less than 2 years. Other pulmonary complications of SSc include aspiration pneumonia, pulmonary haemorrhage, pneumothorax, respiratory failure and increased risk of lung cancer.

The factors signifying poor prognosis of lung disease are male sex, low DLCO, presence of early lung involvement, severe Raynaud's phenomenon, and cigarette smoking.

Pulmonary Arterial Hypertension

When SSc has associated pulmonary arterial hypertension (PAH), it can lead to life-threatening consequences. Pulmonary hypertension is defined as resting mean pulmonary artery pressure of greater than 25 mm of Hg with normal pulmonary capillary wedge pressure. The gold standard for diagnosis is by right heart catheterization. But for all practical purposes pulmonary hypertension is measured by echocardiography.

PAH is associated with limited cutaneous SSc, late age of onset SSc, low DLCO and numerous telangiectasia. The clinical features are dyspnea on exertion, fatigue and chest pain. Physical examination may reveal systolic murmur of tricuspid regurgitation, loud pulmonary components of S2, and S3 gallop. Signs of right heart failure may be

Kidney

- Baseline renal function tests, i.e. GFR, 24 hrs urinary protein, serum creatinine, urine analysis, creatinine clearance.

GI Tract

- Upper GI contrast radiography
- Oesophageal manometry
- Assessment of malabsorption (stool examination)

Lungs/Heart

- Plain chest radiograph, HRCT, pulmonary function test including DLCO (diffusing capacity for carbon monoxide), arterial blood gas analysis, and echocardiography are recommended tests.
- In selected patients bronchoalveolar lavage, lung biopsy, right heart catheterization, serum CK, troponin, ECG monitoring is necessary.
- Assessment of pulmonary hypertension (PAH) by 2-D echo or right heart catheterization.

Musculoskeletal

- Muscle enzymes (CPK; aldolases), electromyography (EMG), muscle biopsy

The initial assessment, described above, is essential for the treatment plan for individual patient.

Organ-based Treatment

Skin Involvement

Lesions of localized scleroderma, including morphea, appear to soften with ultraviolet-A (UVA) light therapy, other options include potent topical corticosteroids, calsipotriol and methotrexate.

Many drugs have been used in the treatment of SSc without any consistent or prolonged benefit. In uncontrolled studies, penicillamine has been reported to reduce skin thickening and prevent development of significant organ involvement.[10] This drug interferes with inter and intramolecular cross-linking of collagen and is also immunosuppressive which reduces collagen production. Starting dose of D-penicillamine is 125 mg/day and increased to 1.5 gm/day within 1–3 months. There is no major difference between low dose and higher dose of D-penicillamine.[11] It is a toxic drug which can cause glomerulonephritis with nephrotic syndrome, aplastic anaemia, leucopenia, thrombocytopenia and myasthenia gravis. Other side effects include fever, rash, anorexia, loss of taste. Patient should have monthly blood count and urine analysis. D-penicillamine is hardly used these days.

Other drugs used in SSc are cyclophosphamide, methotrexate, azathioprine and other immunosuppressive drugs like chlorambucil, flurouracil. These are reserved for those patients with rapidly progressive disease. Because of the poor prognosis of scleroderma patients who have rapid progressive disease with organ involvement, high dose immunosuppressive therapy with cyclophosphamide followed by autologous stem cell transplantation has been tried.[12,13] Though stem cell transplant has long-term benefits, it is associated with post-transplant mortality of about 10%.[12]

Immunosuppressive therapy when given at the early inflammatory phase could potentially alter the natural course of the disease. Mycophenolate mofetil (MMF) is a promising drug. Improvement in mRSS after treatment with MMF was seen as early as 3 months and continued through 12 months follow-up.[14]

Intense pruritis in early stages of scleroderma is managed with antihistaminics, skin lubricating creams, low dose oral steroids are rarely helpful. Telangiectasia mainly on face can be treated with laser therapy. Calcinosis causes great distress. Pharmacological agents like probenecid, colchicine and warfarin have failed to dissolve or prevent new calcinosis, diltiazem, calcium channel blocker may be useful treatment for calcinosis as it also helps in pulmonary hypertension.[15] Calcinosis can be removed surgically, if suitable.

Many other treatment options like high doses of 1,25-dihydroxy vitamin D_3, retinoid, interferon, cycloserine, photochemotherapy, plasmapheresis are studied in scleroderma with disappointing results.[16] Other therapies investigated for treatment of skin thickening include the use of thalidomide, humanized anti-transforming growth factor (TGF) beta-1-monoclonal antibody and relaxin hormone.[17,18]

Raynaud's Phenomenon

The management of Raynaud's phenomenon is directed at control of vasospasm. Patient is advised to:

- Avoid cold and stress, dress warmly, and wear mittens and socks

- Avoid smoking and drugs like ergotamine/amphetamine/β-blockers

Drug therapy for Raynaud's phenomenon includes:

- Calcium channel blockers—nifedipine
- Vasodilators such as prazosin, methyl dopa
- Ketanserin, losartan, bosentan
- Prostacyclin analogues—epoprostenal, iloprost
- Pentoxyfylline
- Antiplatelet drugs—aspirin, dipyridamole

Surgical therapy

- Surgical sympathectomy
- Digital sympathectomy (radical microarteriolysis)

Treatment of digital ulcers

- Soak with warm water
- Topical and oral antibiotics
- Debridement and amputation

In above mentioned treatment, calcium channel blockers nifedepine, diltiazem and amlodipine can be effective but side effects like palpitation, light headedness limits their use. Usual dose of nifedipine is 60–90 mg per day.

Nitroglycerine paste applied locally and oral losartan (angiotensin II receptor antagonist) reduces severity and frequency of Raynaud's phenomenon. Ketanserine, fluoxetine are oral serotonin antagonist decreases 5-hydroxytryptamine which plays a major role in Raynaud's phenomenon.

Prostacyclin analogs like iloprost given intravenously 0.5–3.0 ng/kg/min for 3 to 5 days or orally (50–150 microgram twice a day) is a potent vasodilator and also inhibits platelet adhesion and aggregation, alters neutrophil function and repairs damaged endothelium. It reduces severity and frequency of Raynaud's phenomenon. Other prostacyclin analogs are epoprostenal and beraprost.

Pentoxifyllin and antiplatelets like aspirin and dipyridamole improves the blood flow by reducing platelet adhesion and increasing deformability of red blood cells.

If patient dose not respond to medical treatment, surgery may be necessary including surgical sympathectomy (cervical or lumbar), radical microarteriolysis (digital sympthectomy), deb ridement and amputation of digits.

Lung Involvement

ILD is the leading cause of death in patients of SSc. Early diagnosis is critical for which lung function test and HRCT imaging are used for screening and management. There are two main types of parenchymal lung disease in scleroderma.[19]

- Fibrosing alveolitis leading to interstitial lung disease
- Pulmonary arterial hypertension

Inflammatory Fibrosing Alveolitis

At initial assessment patient is assessed for having inflammatory alveolitis (active disease) and interstitial fibrosis (chronic state) by HRCT and bronchoalveolar lavage. In active disease drug therapy is helpful. Goh and his colleagues have proposed a simple staging system whereby extensive disease (more than 20% HRCT involvement) would indicate inmunosuppressive treatment, whereas limited disease may not.[20] Once extensive lung fibrosis is present, the best response to treatment expected is stabilization of the disease. It should not be regarded as failed treatment.

The major drugs that are currently used to treat ILD are corticosteroids, cyclophosphamide, azathioprine and MMF.[21–25]

Immunosuppressants

There are different regimens for active lung disease. One approach is to initiate low dose prednisolone 30–40 mg/day plus cyclophosphamide 600–800 mg/monthly intravenously or 2 mg/kg orally daily. Prednisolone is tapered slowly over 3 months. After 6 months the lung condition is reassessed and dose of cyclophosphamide is individualized. The optimal duration of therapy is unclear. Other approach is to replace cyclophosphamide by azathioprine (1–2 mg/kg/day) to avoid adverse effects.

Cyclophosphamide (CYC) is fast emerging as very important treatment modality of SSc-ILD. The scleroderma lung study (SLS) (n = 158) demonstrated that CYC given orally at a dose of 1–2 mg/kg per day improved lung function test, dyspnoea score and quality of life over 12 months compared with placebo.[23]

CYC given intravenously at a dose of 600 mg/m^2 was compared with placebo in 45 SSc patients with Scl-ILD. The treatment included 6 infusions of CYC given at 4 weekly interval followed by oral azathioprine 2.5 mg/kg per day or placebo for 6 months. Prednisolone was co-administered in active treatment group. Patients who received IV CYC had stable lung functions over 2 years.[24] Azathioprine is used as maintenance therapy after monthly pulse CYC.[24]

Mycophenolate moefetil (MMF) is a safe, less toxic alternative to CYC. Several uncontrolled prospective or retrospective case studies suggest that MMF is as effective as IV CYC in stabilizing or improving lung functions in SSc-ILD.[25] MMF is given up to 1.5 gm twice a day for two years.

Rituximab is a monoclonal antibody directed against the B cells CD 20 antigen and is a potential therapy for patients with scleroderma ILD. In one study rituximab therapy was associated with significant improvement in FVC and DLCo and stablization of HRCT findings.[26]

Autologous haematopoietic stem cell transplant (HSCT) remains an experimental method for the treatment of scleroderma lung disease due to its treatment-related mortality, infections and other complications.[13]

It is a policy to continue treatment for two years before making decision about efficacy. Once disease becomes stable, consideration is given to withdrawing therapy gradually. But once there is lung fibrosis, treatment may not be helpful. Patient should receive polyvalent pneumococcal and influenza vaccination.

Pulmonary Hypertension

The therapeutic options for treatment of pulmonary hypertension with scleroderma include nifedipine, anticoagulation, prostacyclin analogs, endothelin receptor antagonists (Bosentan).

Prognosis of pulmonary hypertension is very poor. If it is refractory to drug, we can consider heart–lung or single lung transplantation if there is no other major organ involvement.

Nifedipine

It improves pulmonary vascular resistance but there are no definitive studies that shows an improvement in survival. The daily dose is 60–90 mg/day.

Anticoagulation

Anticoagulation with warfarin can improve the survival. Adjust the dose of warfarin to maintain INR of 2.5.

Prostacyclin Analogs

Epoprostenol by continuous IV infusion via indwelling catheter improves exercise capacity, reduces dyspnea, reduces pulmonary pressures and resistance, and improves cardiac output. The invasiveness of delivery system, adverse effects and high cost limit the use of epoprostenol.[27,28]

Iloprost given as 5-day infusion (6 hours per day) every 3 months is found to improve general health, improve the skin, and increase the DLCO.[29] Iloprost as inhalation therapy is also effective in severe pulmonary hypertension. Flushing and jaw pain are more common with iloprost.[30]

Endothelin Receptor Antagonists

Bosentan is an oral nonselective endothelin receptor antagonist. It is safe and effective orally. The dose of bosentan is 125–250 mg twice daily. Bosentan has been proved to improve exercise capacity, functional class and haemodynamic measures in PAH. It can produce increase in serum amino-transferase enzyme levels.

Sitaxentan is selective endothelin receptor antagonist given at a dose of 100 mg/day. It significantly improved exercise capacity and hemodynamics.

Ambrisentan has longer duration of action and is given at a dose of 5 or 10 mg once a day.

Phosphodiesterase 5 Inhibitors

Sildenafil is as efficacious as epoprostenal for reducing pulmonary vascular resistance. It improves ventilation—perfusion matching and oxygenation. Other drugs in this group are tadalafil and verdenafil. Tadalafil once daily dosing is preferable to sildenafil which has to be given 8 to 6 hourly. Ambrisantan in combination with tadalafil (AMBITION trial) is effective to reduce symptoms, reduce dyspnea, improve NYHA class and the six-minute walk distance.[31,32] This combination can be given upfront in patients with scleroderma PAH in order to bring about symptomatic relief, and to increase exercise capacity. This combination also has short-term survival benefit and reduces number of hospital admissions.[31,32]

Soluble guanylate cyclase stimulator riociguat is also used in PAH.

Lung and heart–lung transplantation for PAH remains the treatment of last resort.

Renal Involvement

Renal hypertension crisis in scleroderma is more common in first five years. High blood pressure and proteinuria may be initial presentation. Control of blood pressure is the mainstay of treatment of scleroderma renal crisis. ACE inhibitors are first line antihypertensive drugs.[33] Other drugs include calcium channel blockers, beta blockers, clonidine, etc. Dialysis may be required in patients with progressive renal failure. Patients are usually not candidates for kidney transplantation because of other system involvement. Steroids are associated with a higher risk of SRC and patients with steroids should be carefully monitored for blood pressure and renal function.

Gastrointestinal Involvements

Since 40% of patients with scleroderma related oesophageal reflux are asymptomatic, the emperic use of acid reducing agents is generally recommended. Reflux oesophagitis is treated with frequent meals, antacids, elevation of head of bed, avoiding tea, coffee, alcohol and chocolates which reduces lower oesophageal sphincter tone. H_2 blocker (ranitidine, cimetidine), proton pump inhibitor (omeprazole, pantoprazole, etc), metoclopramide and itopride which increases GI motility are useful drugs.

Malabsorption and bacterial overgrowth can be treated with antibiotics like ciprofloxacin, metronidazole and doxycyclin. Parental nutrition may be useful in some patients. Octreotide may sometimes be useful in pseudo-obstruction. Surgical treatment may be necessary in some patients.

Musculoskeletal Involvement

Baseline assessment of the patients for joint involvement includes a survey for swollen, tender or deformed joints. Arthritis symptoms require NSAIDs or low dose of steroids to control inflammation and pain. Physiotherapy is important in limiting contractures. It should include both active and passive exercises. Encouragement from the therapist and physician is also critical if a regular programme is to be performed. Surgery has an uncertain role in realignment of contracted fingers, but it does not greatly improve function.

A mild myopathy with a little biochemical or histological change is a common feature of scleroderma. Treatment is similar to idiopathic polymyositis. Glucocorticosteroids alone or in combination with methotrexate, azathioprine or other immunosuppressive agents are used. Steroid increases risk of renal crisis.

Cardiac Involvement

The direct cardiac manifestations of SSc (those not induced by systemic or pulmonary hypertension) include pericarditis, pericardial effusion, myocarditis, myocardial fibrosis, arrhythmias and coronary artery disease.

Pericarditis without tamponade is treated with NSAID and corticosteroids. Cardiac tamponade typically requires pericardiocentesis. If effusion is loculated or there is special need for biopsy or patient had coagulopathy then direct surgical drainage is required.

Myocardial fibrosis is present in up to 90% of autopsies. It produces cardiomyopathies (dilated or restrictive) and arrhythmias. Restrictive cardiomyopathy does not respond to steroids. Myocarditis requires combination of glucocorticoids and CYC. Digoxin, diuretics, antiarrhythmatics are required for heart failure.

Patient with dry mouth and dry eyes are treated with pilocarpine hydrochloride tablets and artificial tears respectively.

Pregnancy in SSc

The direct effect of pregnancy on SSc is greater possibility of renal crisis when SSc is of recent onset. There is a risk of prematurity and small full-term infant because of SSc induced placental vascular abnormalities.

PROGNOSIS

Systemic sclerosis of either the limited or diffuse cutaneous subset confers substantial increase in the risk of premature death. As the pathophysiology of of the disease becomes better understood, more specific therapies may be developed in future.

Key Points

1. Scleroderma or systemic sclerosis (SSc) is an autoimmune disorder of unknown etiology characterized by fibrosis and tightening of skin as well as microvascular injury in affected organs.
2. Involvement of the lungs by interstitial lung disease (ILD) and PAH is life threatening.
3. Deterioration of lung functions is maximum in the first 2 years of disease. Hence, once SSc is diagnosed, pulmonary function tests (PFTs) should be formed at regular intervals at least every 6 months in the first 2 years of diagnosis in order to detect fall in the FVC. HRCT chest is used to confirm ILD.
4. The management of active alveolitis of ILD consists of pulse IV CYC monthly pulse with low dose prednisolone for 6 months followed by maintenance with azathioprine.
5. MMF up to 2 gm per daily is as effective as IV CYC pulse therapy.
6. Ambrisentan and tadalafil combination is effective to reduce symptoms, improve exercise capacity in PAH.
7. Rituximab is a potential therapy for scleroderma lung not responding to CYC or MMF.

REFERENCES

1. Kumar A, Malviya AN, Tiwari SC, Singh RR, Kumar A, Pande JN. Clinical and laboratory profile of systemic sclerosis in northern India. J Assoc Physicians India 1990, 38:765–8
2. Krishnamurthy V, Porkodi R, Ramakrishnan S, Rajendran CP, Madhavan R, Achuthan K, et al. Progressive systemic sclerosis in south India. J Assoc Physicians India. 1991; 39:254–7.
3. Steen VD, Conte C, Owens GR, Medsger TA, Severe restrictive lung disease in systemic sclerosis. Arthritis Rheum 1994; 37:1283–9.
4. White B, Moore WC, Wigley FM, Xiao H Q, Wise RD. Cyclophosphamide'is associated with pulmonary function and survival benefit in patients with scleroderma and alveoitis. Ann of Intern Med 2000; 132:947–54.
5. Garg A, Prakash K, Gopinath PG, Malviya AN. Radionuclide esophageal transit time in progressive systemic sclerosis. Ind J Med Res 1984; 79:110–3.
6. Steen VD. Renal involvement in systemic sclerosis. Clin Dermatol 1994; 12:253–8.
7. Desai V, Ghanekar MA, Sequeira RD, Joshi VR. Renal involvement in scleroderma. J Assoc Physicians India. 1990; 38:768–70.
8. Preliminary criteria for the classification of systemic sclerosis (scleroderma). Subcommittee for scleroderma criteria of the American Rheumatism Association Diagnostic and Therapeutic Criteria Committee. Arthritis Rheum. 1980; 23:581–90.
9. van den Hoogen F, Khanna D, Fransen J, Johnson SR, Baron M, Tyndall A, et al. 2013 classification criteria for systemic sclerosis: an American college of rheumatology/European league against rheumatism collaborative initiative. Ann Rheum Dis. 2013; 72:1747–55.
10. Steen VD, Medsger TA Jr, Rodnan GP. D-Penicillamine therapy in progressive systemic sclerosis (scleroderma): a retrospective analysis. Ann Intern Med. 1982 ; 97:652–9.
11. Clements PJ, Furst DE, Wong WK, Mayes M, White B, Wigley F, et al. High-dose versus low-dose D-penicillamine in early diffuse systemic sclerosis: analysis of a two-year, double-blind, randomized, controlled clinical trial. Arthritis Rheum. 1999; 42:1194–203.
12. Naraghi K, van Laar JM. Update on stem cell transplantation for systemic sclerosis: recent trial results. Curr Rheumatol Rep. 2013; 15:326.
13. Sullivan KM, Shah A, Sarantopoulos S, Furst DE. Review: Hematopoietic stem cell transplantation for scleroderma: effective immunomodulatory therapy for patients with pulmonary involvement. Arthritis Rheumatol. 2016; 68:2361–71.
14. Mendoza FA, Nagle SJ, Lee JB, Jimenez SA. A prospective observational study of mycophenolate mofetil treatment in progressive diffuse cutaneous systemic sclerosis of recent onset. J Rheumatol. 2012; 39:1241–7.
15. Dolan AL, Kassimos D, Gibson T, Kingsley GH. Diltiazem induces remission of calcinosis in scleroderma. Br J Rheumatol. 1995; 34:576–8.
16. Badea I, Taylor M, Rosenberg A, Foldvari M. Pathogenesis and therapeutic approaches for improved topical treatment in localized scleroderma and systemic sclerosis. Rheumatology (Oxford). 2009; 48:213–21.
17. Denton CP, Merkel PA, Furst DE, Khanna D, Emery P, Hsu VM, et al. Cat-192 Study Group; Scleroderma Clinical Trials Consortium. Recombinant human anti-transforming growth factor beta1 antibody therapy in systemic sclerosis: a multicenter, randomized, placebo-controlled phase I/II trial of CAT-192. Arthritis Rheum. 2007; 56:323–33.
18. Seibold JR, Korn JH, Simms R, Clements PJ, Moreland LW, Mayes MD, et al. Recombinant human relaxin in the treatment of scleroderma. A randomized, double-blind, placebo-controlled trial. Ann Intern Med. 2000; 132:871–9.

19. Bolster MB, Silver RM. Assessment and management of scleroderma lung disease. Curr Opin Rheumatol. 1999; 11:508–13.
20. Goh NS, Desai SR, Veeraraghavan S, Hansell DM, Copley SJ, Maher TM, et al. Interstitial lung disease in systemic sclerosis: a simple staging system. Am J Respir Crit Care Med. 2008; 177:1248–54.
21. Silver RM, Warrick JH, Kinsella MB, Staudt LS, Baumann MH, Strange C. Cyclophosphamide and low-dose prednisone therapy in patients with systemic sclerosis (scleroderma) with interstitial lung disease. J Rheumatol. 1993; 20:838–44.
22. Steen VD, Lanz JK Jr, Conte C, Owens GR, Medsger TA Jr. Therapy for severe interstitial lung disease in systemic sclerosis. A retrospective study. Arthritis Rheum. 1994; 37:1290–6.
23. Tashkin DP, Elashoff R, Clements PJ, Goldin J, Roth MD, Furst DE, et al; Scleroderma Lung Study Research Group. Cyclophosphamide versus placebo in scleroderma lung disease. N Engl J Med. 2006; 354:2655–66.
24. Hoyles RK, Ellis RW, Wellsbury J, Lees B, Newlands P, Goh NS, et al. A multicenter, prospective, randomized, double-blind, placebo-controlled trial of corticosteroids and intravenous cyclophosphamide followed by oral azathioprine for the treatment of pulmonary fibrosis in scleroderma. Arthritis Rheum. 2006; 54:3962–70.
25. Shenoy PD, Bavaliya M, Sashidharan S, Nalianda K, Sreenath S. Cyclophosphamide versus mycophenolate mofetil in scleroderma interstitial lung disease (SSc-ILD) as induction therapy: a single-centre, retrospective analysis. Arthritis Res Ther. 2016; 18:123.
26. Daoussis D, Liossis SN, Tsamandas AC, Kalogeropoulou C, Paliogianni F, Sirinian C, et al. Effect of long-term treatment with rituximab on pulmonary function and skin fibrosis in patients with diffuse systemic sclerosis. Clin Exp Rheumatol. 2012; 30 (2 Suppl 71):S17–22.
27. Klings ES, Hill NS, Ieong MH, Simms RW, Korn JH, Farber HW. Systemic sclerosis-associated pulmonary hypertension: short- and long-term effects of epoprostenol (prostacyclin). Arthritis Rheum. 1999; 42:2638–45.
28. Badesch DB, Tapson VF, McGoon MD, Brundage BH, Rubin LJ, Wigley FM, et al. Continuous intravenous epoprostenol for pulmonary hypertension due to the scleroderma spectrum of disease. A randomized, controlled trial. Ann Intern Med. 2000; 132:425–34.
29. Biasi D, Carletto A, Caramaschi P, Zeminian S, Pacor ML, Corrocher R, et al. Iloprost as cyclic five-day infusions in the treatment of scleroderma. An open pilot study in 20 patients treated for one year. Rev Rhum Engl Ed. 1998; 65:745–50.
30. Olschewski H, Simonneau G, Galiè N, Higenbottam T, Naeije R, Rubin LJ, et al; Aerosolized Iloprost Randomized Study Group. Inhaled iloprost for severe pulmonary hypertension. N Engl J Med. 2002; 347: 322–9.
31. Coghlan JG, Galiè N, Barberà JA, Frost AE, Ghofrani HA, Hoeper MM, et al; AMBITION investigators. Initial combination therapy with ambrisentan and tadalafil in connective tissue disease-associated pulmonary arterial hypertension (CTD-PAH): subgroup analysis from the AMBITION trial. Ann Rheum Dis. 2017; 76:1219–27.
32. Galiè N, Barberà JA, Frost AE, Ghofrani HA, Hoeper MM, McLaughlin VV, et al; AMBITION Investigators. Initial Use of Ambrisentan plus Tadalafil in Pulmonary Arterial Hypertension. N Engl J Med. 2015; 373:834–44.
33. Zawada ET Jr, Clements PJ, Furst DA, Bloomer HA, Paulus HE, Maxwell MH. Clinical course of patients with scleroderma renal crisis treated with captopril. Nephron. 1981; 27:74–8.

FURTHER READING

1. Kowal-Bielecka O, Fransen J, Avouac J, Becker M, Kulak A, Allanore Y, et al; EUSTAR Coauthors. Update of EULAR recommendations for the treatment of systemic sclerosis. Ann Rheum Dis. 2017; 76:1327–1339.

Chapter

44

Lung Involvement and Pulmonary Arterial Hypertension in Connective Tissue Diseases

Siddharth Jain, Uma Kumar

INTRODUCTION

Connective tissue diseases (CTDs) consist of a spectrum of multisystem autoimmune disorders characterized by immune dysregulation and immune mediated organ dysfunction. They include rheumatoid arthritis (RA), systemic lupus erythematosus (SLE), idiopathic inflammatory myopathies (IIM), Sjögren's syndrome (SS), systemic sclerosis (SSc), mixed connective tissue disease (MCTD). Lung involvement in CTDs is very diverse and consists of lung parenchymal involvement that is interstitial lung disease (ILD), airway involvement (bronchiolitis and bronchiectasis), vascular involvement (pulmonary hypertension), pleural involvement (pleuritis, pleural effusion) and respiratory muscle involvement. Among these various types of pulmonary involvement, ILD is the most common and the most serious form. The different patterns of lung involvement in individual CTDs has been summarised in Table 44.1.

IDIOPATHIC INTERSTITIAL PNEUMONIA vs LUNG DOMINANT-CTD vs AUTOIMMUNE-FEATURED ILD

Idiopathic interstitial pneumonias (IIPs) are a group of diffuse inflammatory and/or fibrotic lung disorders clustered together based on similar clinical, radiologic and histopathologic features.[1,2] The diagnosis of IIP requires the exclusion of known causes of interstitial pneumonia such as toxins, drugs or CTD.[1] Determining that an ILD is CTD-associated is important both therapeutically and prognostically. Identifying underlying CTD in patients presenting with a presumable IIP can be challenging as the boundary between IIP and CTD-ILD is not clearly defined. In fact, many patients with IIP may have clinical features that suggest an underlying autoimmune etiology but do not meet the established criteria for a definite CTD. Differing, but overlapping, terms have been used to describe these patients in the past, including "undifferentiated

Table 44.1: Patterns of lung involvement in individual CTDs

	Airways	*Pleura*	*Vascular*	*Interstitial lung disease*
Systemic sclerosis	–	–	++++	++++
Rheumatoid arthritis	++	++	+	+++
Polymyositis/dermatomyositis	–	–	++	++++
Sjögren's syndrome	+++	+	+	+++
Systemic lupus erythematosus	+	++++	+	+

CTD associated ILD" (UCTD-ILD), "lung-dominant CTD (LD-CTD)" or "autoimmune-featured ILD".[3–5] These terms are controversial, overlapping and not universally accepted. The ERS/ATS task force thus has recently proposed the term "interstitial pneumonia with autoimmune features" (IPAF) to identify individuals with IIP and features suggestive of, but not definitive for, a CTD.[6] The task force statement also offers classification criteria based on a combination of features from three domains (i) a clinical domain comprising of specific extra-thoracic clinical features, (ii) a serologic domain comprising of specific autoantibodies, and (iii) a morphologic domain comprising of specific chest imaging, histopathologic or pulmonary physiologic features.[6] Development of a unique IPAF cohort should now provide the desired platform for prospective studies to provide a better understanding of this cohort, its management, and its similarity/differences compared to the well-characterized, classifiable forms of CTD-ILD.

CTD-ILD

ILD is a heterogeneous group of diffuse lung parenchymal disorders with variable clinical, radiographic, and histopathological features. CTD-ILD is the third most common cause of ILD overall. The relationship of ILD with CTD is intricate, wherein, ILD is associated with a definite well-established CTD in 60% cases, with an "undifferentiated CTD" in 25% cases[7] and it predates CTD in remaining 15% cases, as the first, and possibly the sole, manifestation of an otherwise occult CTD. Almost all patients with CTD are at risk of ILD; some are more likely to be associated with it than others. The prevalence of CTD-ILD is not known. In an Indian study, the most common pulmonary involvement observed on chest radiography and HRCT of thorax in CTD patients was ILD (38.5%) while pulmonary arterial hypertension (PAH) was observed in 14.9% patients.[8] It has been seen that a particular CTD is more often associated with certain type of ILD as discussed later.

How does CTD-ILD differ from IIP?

CTD-ILD differs from IIP in the fact that it usually shows a combination of histopathologic patterns and its response to immunosuppressive therapy is favourable. CTD-ILD is associated with a more favourable prognosis than idiopathic ILD of equivalent severity (except in patients with rheumatoid arthritis having usual interstitial pneumonia).[9–11] This could be due to early detection and relatively slow progression of parenchymal disease in CTD-ILD. It could also be due to the fact that CTD patients are often on immunosuppressive therapy. Therefore, it is often argued that presence of a mixed histologic pattern must raise the possibility of a forme fruste of CTD even if patient has no evidence of CTD at that point of time.[12]

Problems Encountered in the Diagnosis of CTD-ILD

Many patients with CTD have subclinical ILD (abnormalities on high resolution CT in asymptomatic individuals). Most of the drugs commonly used in the treatment of CTD have known pulmonary toxicities. The drugs may even mask or unmask underlying respiratory diseases. ILD can also occur from infections that develop as a consequence of immune suppression. To complicate the picture further, clinical presentation of CTD-ILD is non-specific. Hence, a very high index of clinical suspicion is needed for the diagnosis of CTD-ILD. It is extremely important to rule out all other causes of parenchymal lung disease, e.g. pulmonary infection, drug toxicity, environment exposure, etc., before diagnosing CTD-ILD.

Clinical Features

History of dyspnoea, cough, chest pain, hemoptysis, progression of symptoms, occupational history, family history, past history, drug history are important. Detailed cardiovascular and respiratory system examination must be done. Clubbing, increased respiratory rate, crackles, rhonchi/wheeze, decreased breath sounds , pleural rub, loud P2, right ventricular heave or evidence of right heart failure may be seen on examination.

Investigations

The armamentarium of investigations available for evaluation of CTD-ILD is summarised in Table 44.2.

Treatment

There are no guidelines *per se* regarding the treatment of CTD-ILD. Apart from a few RCTs in scleroderma-ILD, there are very few studies to

Table 44.2: Investigations for CTD-ILD

Investigation	*Role in CTD-ILD*
Pulmonary function testing (PFT)	Routine spirometry, TLC, DLCO and blood gases must be done in all patients; important for monitoring disease progression and treatment response. May be normal in early stages.
Cardiopulmonary exercise testing (CPET)	Can differentiate between cardiac and pulmonary origin of dyspnoea even in early stages of disease.
Chest X-ray	May be normal (early ILD) or show diffuse reticular shadows, airspace opacities, nodules, pleural effusion/ thickening, mediastinal lymphadenopathy, prominent pulmonary artery conus, etc.
HRCT chest	Investigation of choice for diagnosis of ILD
Bronchoscopy	FOB helps visualise proximal airway; BAL helps diagnose small airway involvement/infection/haemorrhage. BAL fluid cytokine and chemokine analysis is under research as markers of disease severity and predictors of outcome.
Lung biopsy (open/VATS assisted)	Rarely needed; when there is diagnostic dilemma/suspicion of malignancy

Abbreviations: TLC—total lung capacity; DLCO—diffusion capacity of lung for carbon monoxide; FOB—fibre optic bronchoscopy; BAL—bronchoalveolar lavage; VATS—video assisted thoracoscopy; HRCT—high resolution computerised tomography

devise an evidence-based approach to managing CTD-ILD. This often leads to a personal, experience based treatment and/or extrapolation of data available for scleroderma to other CTDs. It is important to identify which patients to treat (by assessing the risks versus the benefits) and management is currently limited only to cases with progressive or clinically significant disease. Immunosuppression with corticosteroids and cytotoxic medications forms the mainstay of pharmacologic treatment. Non-pharmacological approaches to treatment must also be considered in all CTD-ILD patients. Therapeutic considerations include: (i) underlying CTD, (ii) extent and rate of progression of ILD, (iii) likelihood of response based on radiological patterns, (iv) age of the patient, (v) ability of the patient to comply with treatment, (vi) current therapies the patient is on.

Additional supportive treatment for CTD-ILD patients includes: (i) drug therapy reconsideration (withdrawl of unnecessary drugs), (ii) smoking cessation, (iii) pulmonary rehabilitation, (iv) assessment of the need for supplementary oxygen, (v) management of gastroesophageal reflux, (vi) nutritional building, (vii) vaccination against pneumococcus, influenza, (viii) consideration of lung transplant in appropriate patients.

The treatment of individual CTD-ILDs has been summarized in Table 44.3 for convenience.

Recent Updates in the Management of CTD-ILD

Scleroderma Lung Study 2 published recently has shown a comparable efficacy of CYC and MMF at 2 years with a better safety profile and less treatment withdrawal in patients receiving MMF.[13] Evidence is building in favour of a possible role of rituximab as a disease modifying agent in systemic sclerosis, and as an alternative drug for scleroderma ILD with a favourable side effect profile.[14] Pirfenidone was found to be safe and well-tolerated in SSc-ILD patients in a 16-week, open-label trial recently, supporting the need for its further investigation in SSc-ILD.[15] Hematopoietic stem cell transplantation may have a role to play in systemic sclerosis, as shown in the ASSIST trial.[16] A multinational trial of nindetanib for SSc-ILD is currently ongoing.

PULMONARY ARTERIAL HYPERTENSION

What is Pulmonary Arterial Hypertension?

Pulmonary arterial hypertension (PAH) is defined by an elevated mean pulmonary artery pressure ≥25 mm Hg at rest with a normal pulmonary capillary wedge pressure of ≤15 mm Hg and

Table 44.3: Frequency, histopathologic pattern and treatment of ILD in specific CTDs

CTD	*Frequency of ILD (%)*	*Characteristic histopathologic pattern*	*Treatment*
Systemic sclerosis	45 (clinically significant)	NSIP UIP	Low dose steroids to be used to avoid scleroderma renal crisis. CYC and MMF equally efficacious with a better safety profile in MMF (SLS 2). Role of rituximab in refractory cases. Trials of pirfenidone and nindetanib ongoing.
Rheumatoid arthritis	20 to 30	UIP NSIP OP DAD	RA with UIP: Behaves similar to IPF prognostically RA with NSIP: Good response to steroids and immunosuppressants
Polymyositis/dermato-myositis	20 to 50	NSIP UIP COP DAD	High dose steroids (tapered over 3–6 months) + immunosuppressive therapy. Tacrolimus and rituximab in refractory cases.
Sjogren's syndrome	Up to 25	NSIP (most common) LIP OP UIP DAD	Good response to steroids. Some patients may require immunosuppressants (Aza/CYC)
Systemic lupus erythematosus	2 to 8	DAD AIP NSIP LIP OP "Shrinking lung syndrome"	Oral prednisolone/IV methylprednisolone/CYC/plasma pheresis depending on disease severity. Rituximab in refractory cases.

Abbreviations: NSIP: Non-specific interstitial pneumonia, UIP: Usual interstitial pneumonia, OP: Organizing pneumonia, DAD: Diffuse alveolar damage, LIP: Lymphocytic interstitial pneumonia, CYC: Cyclophosphamide, MMF: Mycophenolate Mofetil, Aza: Azathioprine, SLS 2: Scleroderma lung study 2

pulmonary vascular resistance >3 Wood units on right heart catheterization.[17]

Association of PAH with CTD

PAH is a potentially life threatening complication of many CTDs, more frequently SSc, SLE and MCTD, and to a lesser extent RA, dermatomyositis and Sjögren's syndrome.

Classification of PAH and Assessment of its Severity

CTD associated PAH (CTD-PAH) is classified in Category 1 as per the WHO updated clinical classification (Dana Point, 2008).[18] The detailed classification has been provided in Table 44.4. Severity of PAH is assessed by NYHA/WHO functional classification of pulmonary hypertension.[19]

Epidemiology

There is considerable variability in the epidemiological data for PAH depending on the variability in the patient population, definition and the method used for assessing PAH. In a population based study in Japan, 2.6 % of over 3500 patients with CTD had PAH.[20] Prevalence of PAH in SSc patients has been widely reported to vary from 5 to 35%. A study from India showed a prevalence of 32%.[21] PAH has been described in 0.5–14% patients with SLE.[22]

CTD-PAH vs Idiopathic PAH

Females are more commonly affected in CTD-PAH. Compared to idiopathic PAH, patients with CTD-PAH are older, have lower cardiac output and tend to have poorer survival. In a US registry of

myopathic changes on electromyography, and a muscle biopsy showing myofiber degeneration and regeneration with chronic inflammatory infiltrates; in the case of DM, the presence of the heliotrope rash or Gottron's papules. The diagnosis of inclusion body myositis (IBM) was accepted by IMACS as that defined by Greggs et al.[3]

Another approach to classifying the IIMs utilizes the immune responses in these patients. The autoantibodies associated with myositis have an important role in identifying additional groups of patients who share common features and may eventually assist in defining their pathogenesis. Overall, these myositis autoantibodies have been helpful in assisting in the diagnosis of certain patients with confusing presentations and in predicting clinical courses and responses to therapy. However, their significance is being suggested that the Bohan and Peter criteria be modified to add MSA as a criterion.[4,5] However, this inclusion has some limitations, these antibodies are not present in all patients, the immunoprecipitation techniques that are the "gold standard" for identifying these antibodies are available in only a few commercial laboratories. There are important phenotypical differences (muscular or extra muscular manifestations) among the inflammatory myopathies. Recently, over thirty different myositis specific antibodies (Table 45.1) and associated auto-antibodies (Table 45.2) have been identified and are being used as a diagnostic tool. These autoantibodies categorize different patient groups more accurately than the classical international classification criteria for myositis.

PATHOGENESIS

In idiopathic inflammatory myopathies (IIMs) multiple pathogenic pathways cause muscle damage and weakness. The muscle fibre damage and weakness in this group of disorders is due to the autoimmune response to skeletal muscle-derived antigens. There is formation of autoantibodies, autoreactive lymphocytes, with unusual overexpression of major histocompatibility complex (MHC) class I molecules on the surface of the affected myofibers. The innate and adaptive immune system pathways are activated and metabolic defects occur in the skeletal muscle.[6] Innate immune pathways are a link between the adaptive and metabolic pathways. The intrinsic defects in skeletal muscle also contribute to muscle weakness and damage in myositis. There are active interactions between innate, adaptive, metabolic and homeostatic pathways in muscle in these diseases.

CLINICAL PRESENTATION

The presentation of inflammatory myositis is often insidious and the presentation can be very subtle with symptoms such as fatigue and tiredness which can often be the presenting feature and can lead to misdiagnosis. Other systemic symptoms such as weight loss may occur but is usually mild. If the weight loss is persistent and severe in a patient with myositis, then associated malignancy should be considered.

Fever is more common in childhood dermatomyositis, but adults with anti-synthetase antibody syndrome may have fever accompanying or heralding active disease.

The most frequent presentation however, is a gradual onset symmetrical painless and progressive proximal muscle weakness over the course of a few months (Table 45.3). Some patients especially children and young adults have a more acute onset of disease with muscle pain and weakness developing rapidly over the course of several weeks.

Only very small subsets of patients have a very slowly evolving weakness over the course of many years before diagnosis. They are typically older males in their late 60s with weakness of distal extremities who often have pathological features of inclusion body myositis on biopsy. These symptoms are often so slow that many people debate whether this is truly an inflammatory or degenerative disease.

Other features such as pitting edema of the extremities, hoarseness or dysphagia as a result of bulbar palsy nasal regurgitation of liquids due to pharyngeal palsy or breathlessness due to interstitial lung disease may rarely be a presenting feature.

CLINICAL FEATURES

Skeletal Muscle

Patients often complain of difficulty in performing activities requiring proximal muscle strength such as climbing stairs, getting up from squatting position and combing hair.

Table 45.1: Myositis specific antibodies[36]

Name of the antibody	*Target of the antibody*	*Frequency among autoimmune myopathies*	*Acute and severe proximal weakness*	*Muscle pathology*	*Skin lesion*	*Life-threatening complication*
Anti-Jo-1	Histidyl-t-RNA-synthetase	15–20%	+++/++	IMPP/NM	MH	ILD+/cardiac+
Anti-PL-7	Threonyl-t-RNA-synthetase	5%	++/+	IMPP/NM	MH	ILD++
Anti-PL-12	Alanine-t-RNA-synthetase	<5%	+/±	IMPP/NM	MH	ILD++
Anti-EJ	Glycyl-t-RNA-synthetase	<5%	+/±	Not characterised	MH	ILD++
Anti-Zo	Phenylalanyl-RNA-synthetase	<5%	+/±	Not characterised	MH	ILD++
Anti-Ha	Anti-tyrosyl-RNA-synthetase	<5%	+/±	Not characterised	MH	ILD++
Anti-KS	Asparaginyl-RNA-synthetase	<5%	+/±	Not characterised	MH	ILD++
Anti-MI-2	Nucleosome remodelling and deacetylase (NuRD) complex	6%	++	IMPP	DM rash	None
Anti-TIF1	Transcriptional intermediary	6%	++	IMPP	DM rash	None
Anti-MDA5	Melanoma differentiation associated gene	6%	±	Not characterised	DM rash, ulcers MH, palmar papules	ILD+++ (rapidly progressive ILD, especially in Asians)
Anti-SAE	Small ubiquitin-like modifier activating enzyme	<5%	+	Not characterised	DM rash	None
Anti-NXP2	Nuclear matrix protein 2	<5%	++	Not characterised	DM rash calcinosis	Cancer
Anti SRP	Signal recognition particle	5%	+++/±	NM	None	Cardiac++
Anti-HMGCR	3-hydroxy-3 methyl-glutarylcoenzyme A reductase	6%	+++/±	NM (statin associated)	None	None
Anti-cN1A	Cytosolic 5′-nucleotidase1A	10%	–	sIBM	None	None

IMPP: Immune myopathies with perimysial pathology was described by Pestronk and Mozaffar in patients with anti-synthetase syndrome and dermatomyositis[36] NM: Necrotizing myopathy. DM rash: Dermatomyositis rash. MH: Mechanic's hand. ILD: Interstitial lung disease.

Table 45.2: Myositis associated antibodies[37,38]

Name of the antibody	*Target of the antibody*	*Frequency among autoimmune myopathies*	*Muscle pathology*	*Life-threatening complication*	*Associated auto-immune diseases*
Anti-Ro52	TRIM21	30%	IMPP/PM	ILD	SSc, SS, SLE
Anti-Ro-60/SSA	60kDa Ro	10%	IMPP/PM	ILD	SSc, SS, SLE, RA
Anti-La/SSB	La	15%	IMPP/PM	ILD	SSc, SS, SLE, RA
Anti-Ku	Ku	20%	PM	ILD	SSc, SS, SLE
Anti-URNP	Ribonucleoprotein	15%	PM	none	MCTD
Anti-mitochondrial	Pyruvate deshydro-genase-E2	10%	Granulomatous	Cardiac	PBC
Anti-PM/scl	Nucleolar PM/Sc1 macromolecular complex	10%	PM	ILD cardiac	SSc, SLE
Anti-CENP	Centromer	<5%	PM/NM	ILD cardiac	SSc
Anti-topo	Topo-isomerases	<5%	PM/NM	ILD cardiac	SSc
Anti-RNA Pol III	RNA polymerase III	<5%	PM/NM	ILD cardiac	SSc
Anti-fibrillarin	Fibrillarin	15%	PM/NM	ILD cardiac	SSc
Anti-CCP	Cyclic citrullinated peptide	10%	IMPP/PM	ILD	RA
Rheumatoid factor	IgG	20%	IMPP/PM	ILD	SSc, SS, SLE, RA

The reported frequency of antibodies are mainly derived from Kenigeg et al.[37] and Labrador-Horrillo et al.[38] IMPP: Immune myopathies with perimysial pathology,[36] MCTD: Mixed connective tissue disease. NM: Necrotizing myopathy. DM rash: Dermatomyositis rash. ILD: Interstitial lung disease. PM: Polymyositis. SSc Systemic sclerosis. SS: Sjögren syndrome, SLE: Systemic upus erythematosus. PBC: Primary biliary cirrhosis. RA: Rheumatoid arthritis.

Table 45.3: Presenting features of idiopathic inflammatory myositis

Syndrome	*Estimated frequency*
Painless proximal muscle weakness over months	55%
Acute or subacute proximal pain and weakness over weeks	30%
Insidious proximal and distal weakness over the years	10%
Proximal myalgia alone	5%
Dermatomyositis rash alone; extremity oedema	Less than 1%

In polymyositis and dermatomyositis, the lower extremity is usually affected initially in a symmetrical manner. With progression of weakness, patient develops waddling gait and may be unable to rise without assistance. Gradually, the upper extremity symptoms follow and the patient experiences difficulty in lifting the arms overhead doing activity such as combing hair and taking out things from cupboard. The neck flexors may become affected leading to a dropped head. Muscle pains are rare and more commonly occurring in dermatomyositis particularly with exercise. Dysphagia and nasal regurgitation of liquids reflect pharyngeal striated muscle involvement and is a poor prognostic sign. Pharyngeal weakness manifest with hoarseness or change in voice. Oculofacial weakness almost never occurs and its presence should prompt the consideration of another diagnosis. Physical examination using manual muscle testing is necessary to grade muscle weakness.

Skin

The presence of typical skin rash indicates the clinical subset of dermatomyositis. The rash may proceed, developed simultaneously or follow muscle symptoms. Figure 45.1 shows a child with the facial rash of dermatomyositis. Gottron's papules are characteristic scaly and erythematous violaceous plaques located over bony prominences particularly the MCPs and PIPs (Fig. 45.2). Gottron sign, on the other hand, is a macular erythema and generally occurs in the same distribution, or over the knees or elbows. The presence of periorbital rash with oedema called heliotrope rash (Fig. 45.3) is considered a specific sign of dermatomyositis. The less specific signs include photosensitive scalp involvement (Fig. 45.4), shawl sign and V sign, mechanic's hands which refers to cracking and fissuring involving radial and palmar aspects of fingers (Fig. 45.5), calcinosis, panniculitis, acanthosis nigricans, multifocal lipoatrophy among children and poikiloderma.

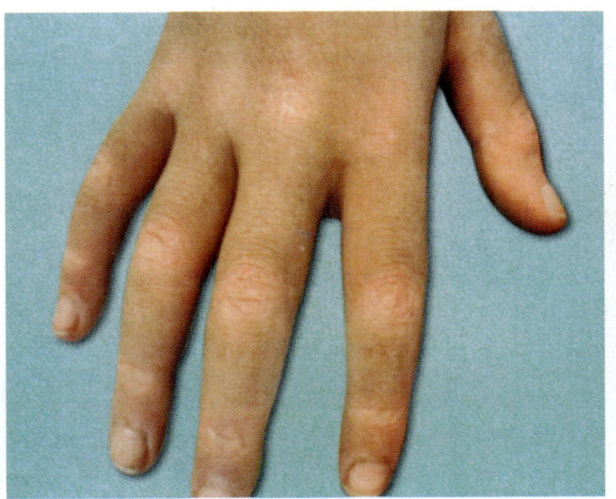

Fig. 45.2: Scaly, erythematous plaques over the PIP and MCP joints also known as Gottron's papules

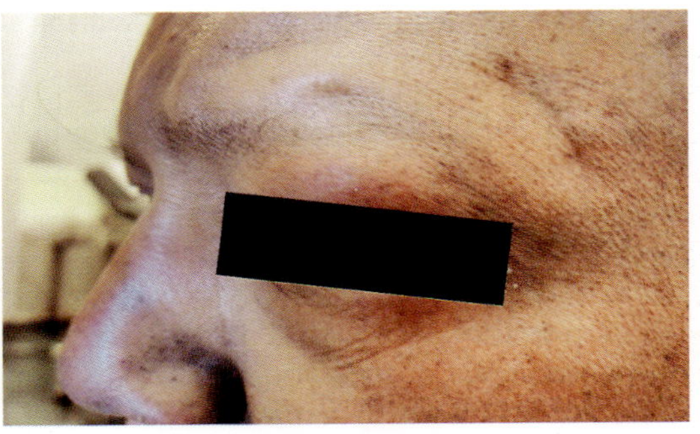

Fig. 45.3: Periorbital oedema

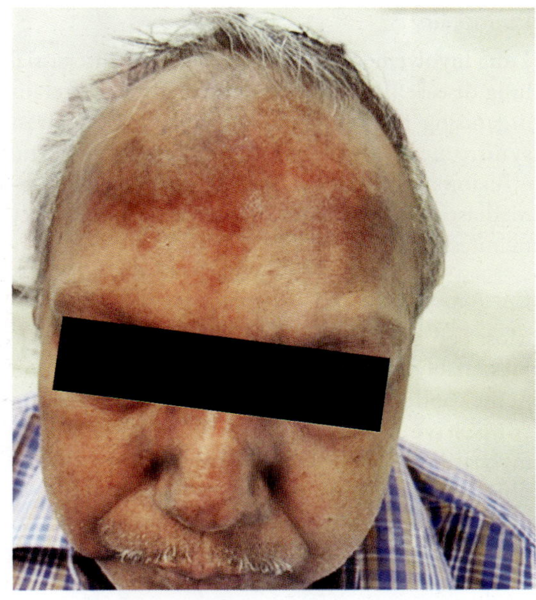

Fig. 45.4: Photosensitive rash over the scalp in a patient with dermatomyositis

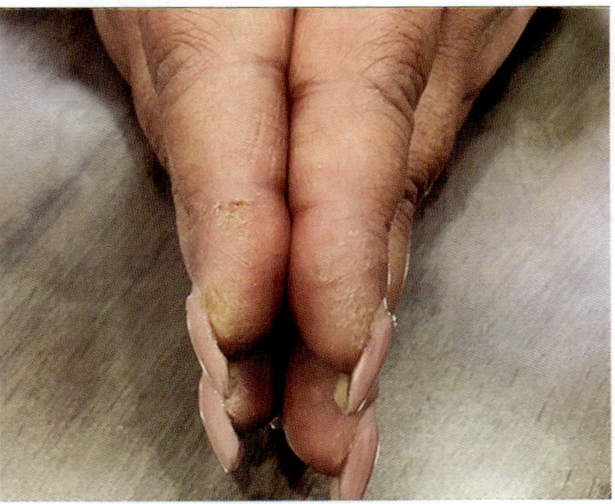

Fig. 45.5: Mechanic hands

Some patients may present with classic biopsy-confirmed rash of dermatomyositis and no muscle weakness, normal enzymes or even a normally electromyography (EMG). These patients are said to have 'amyopathic dermatomyositis' if these findings have been present for 2 years or longer.

Articular

Polyarthritis occurs in a rheumatoid like distribution and is relatively mild but sometimes can be severe in syndromes such as antisynthetase antibody syndrome.

Pulmonary

Lung involvement occurs in the form of interstitial lung disease (UIP, NSIP). Bronchiolitis obliterans organising pneumonia, acute respiratory distress syndrome, pulmonary hypertension, pleural effusions, alveolar haemorrhage and pneumomediastinum are some of the other pulmonary manifestations.

Cardiac

Most common finding is the rhythm disturbance presumably from inflammatory or fibrotic alteration of the conducting system. Pericardial effusions and congestive heart failure typically caused by myocarditis or cardiomyopathy and myocardial fibrosis is often seen.

Gastrointestinal Tract

Dysphagia is commonly due to pharyngeal muscle involvement, dysphonia and nasal regurgitation of fluids also occurs. Rarely oesophageal strictures or megaoesophagus may develop. Post-prandial bloating is seen in distension, obstruction and malabsorption due to small-bowel involvement. Pneumatosis cystoid intestinalis has been rarely reported.

Peripheral Vascular System

Raynaud's phenomena, digital ulcerations and pitting and periungual infarcts are some of the peripheral vascular manifestation.

Renal

Acute renal failure, generally non-oliguric, due to rhabdomyolysis has been seen. Various renal lesion such mesangioproliferative glomerulonephritis (GN), crescentic GN with FSGS membranous nephropathy, minimal-change disease, FSGS, etc. have also been described.

Malignancy in Myositis

It is well known that there is increased risk of malignancy associated with inflammatory myositis. IIMs especially dermatomyositis is associated with malignancies namely lung, pancreas, stomach, colorectal, and non-Hodgkin's lymphoma. Certain clinical features such as epidermal necrosis, cutaneous leukocytoclastic vasculitis, significant weight loss and the presence of amyopathic dermatomyositis strongly suggest the presence of an underlying malignancy. On the other hand, the presence of pulmonary fibrosis, myositis associated or specific serum antibodies or clinically confirmed associated connective tissue disease decreases the likelihood of cancer.

DIFFERENTIAL DIAGNOSIS

The differential diagnosis of muscle weakness is very vast and encompasses the various types of myopathy (Table 45.4). However, certain clues helps in narrowing the differential diagnosis. For example, acute onset of muscle weakness following a febrile episode would strongly suggest a viral cause. Family history would suggest certain heritable myopathies, the presence of pain would suggest a metabolic cause such as osteomalacia or polymyalgia rheumatica. The involvement of facial and ocular muscles is a strong pointer against the diagnosis of inflammatory myositis, and usually indicates the presence of a primary muscle disease such as facioscapulohumeral myopathy. The presence of fasciculations suggests an anterior horn cell disease. Episodic muscle weakness suggests a channelopathy, and precipitating factors such as exercise suggests a metabolic myopathy. Presence of muscle cramps or an asymmetrical weakness suggests an underlying neuropathy.

Muscle Biopsy in IIM

Muscle biopsy is the definite test to establish diagnosis of IIM.[7] Following are the fundamental features seen in muscle biopsy.

Table 45.4: Differential diagnosis of myopathy

Drug induced	Statins, colchicine, hydroxychloroquine, phenytoin, etc.
Metabolic myopathy	Glycogen-storage disorders, lipid storage disorders, etc.
Endocrine causes	Hypothyroidism, hyperparathyroidism, and Addison's disease, etc.
Neuromuscular disorders	Myasthenia gravis, Eaton-Lambert syndrome, amyotropic lateral sclerosis, etc.
Infectious cause	Pyomyositis, parasitic myopathies, viral myopathies, etc.
Miscellaneous	Sarcoid myopathy, mitochondrial myopathies, myotonic dystrophy, channelopathies, etc.

1. Presence of chronic inflammatory infiltrates between individual muscle fibres (endomysial), and around muscle fascicles (perimysial) and around blood vessels (Fig. 45.6)
2. Presence of myofibre necrosis and regeneration.

In addition, the other features are:

A. Autoagressive T cells invading intact muscle fibres.

B. Connective tissue labeling with alkaline phosphatase histochemical stain.

C. Rimmed vacuoles, amyloid deposits, and fibrillary inclusion bodies are seen in inclusion body myositis.

D. Diffuse *de novo* sarcolemmal expression of MHC-1 molecule, perifascicular atrophy/ necrosis or perifascicular MHC-1 expression, thickened capillaries and capillary depletion and tubuloreticular inclusions in endothelial cells are characteristic of dermatomyositis.

Electromyography (EMG)

EMG testing is a sensitive but non-specific method of evaluating inflammatory myopathy. Findings include irritability of myofibrils on needle insertion and at rest (fibrillation potentials, complex repetitive discharges, positive sharp waves) and short duration, low-amplitude, complex (polyphasic) potentials on contraction with early and full recruitment. Specialized techniques including single fibre EMG, macro EMG and quantitative motor unit action potential analysis, increase precision.[8]

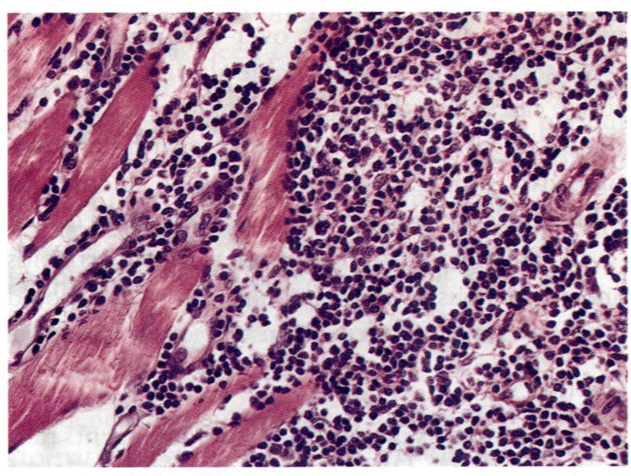

Fig. 45.6: Light microscopic 100X (H and E) examination of muscle in inflammatory myositis showing lymphomononuclear cell infiltrate in the endomysium, between and around individual myofibres. Intrusion into myofibres and myofibre destruction is noted. (*Courtesy*: Dr Andleeb Abrari, MD; DNB; FRC Path, Consultant Pathologist, Max Hospital, New Delhi)

Magnetic Resonance Imaging of IIM

MRI is a useful tool in evaluating anatomy, morphology, and potential functionality of muscles. It is sensitive to identify areas of muscle edema, areas of active involvement, detecting potentially subclinical disease, and providing guidance for muscle biopsy. MRI can be useful in longitudinal course and disease progression or regression and thus prognostic information and evaluating response to treatment.

While the findings of MRI are nonspecific the pattern of involvement can suggest a specific disease entity especially when the contralateral extremity is imaged for comparison and symmetry.[9]

Ultrasound in IIM

Ultrasonography is an alternative approach to image the muscular system.[10] Ultrasound is well suited to evaluate a wide variety of abnormalities involving skeletal muscle. The unique ability of ultrasound to assess the dynamic function of muscles on the real-time or perform Doppler evaluation of vascularity provides information that is not readily obtained with other imaging modalities. With ultrasound, image-guided biopsy can be performed in conjunction with a diagnostic evaluation.[11]

Normally the muscle fascicles appear dark or hypoecoic and perimysium appear echogenic. In IIM, there is increased echogenecity of muscle due to muscle oedema with preservation of underlying fascicular and multipennate morphology. It can also be seen in muscle atrophy which also has diminished muscle volume. Consequently, no single pattern is unique to this disorder but rather an admixture of muscle oedema and atrophy patterns exists.

Capillaroscopy in IIM

Capillaroscopic changes in patients of dermatomyositis known as 'Scleroderma spectrum diseases' are associated with Raynaud's phenomenon.[12] Microhaemorrages and giant capillaries are frequent (Fig. 45.7). Dilated capillaries can sometimes be appreciated with the naked eye (Fig. 45.8). These findings are consistent with microvasculopathy and capillary involvement observed in muscle biopsies. There is architectural derangement of nail fold

important to understand that an echocardiographic study done earlier in the course of the illness may miss these findings. Development of aneurysms more than 6 weeks after the onset of illness is also uncommon. With appropriate treatment, these aneurysms tend to regress over the next few months. Patients who do not have complete resolution of aneurysms may show either a decrease in the size of aneurysm or go on to develop coronary stenosis. This coronary stenosis may be complicated by premature atherosclerosis and lead to significant coronary obstruction and myocardial ischaemia later in life. Myocardial ischaemia that occurs under these circumstances may be indistinguishable from that seen in association with primary atherosclerosis. It is also said that larger aneurysms (i.e. more than 8 mm), and especially the ones in the more proximal regions of the coronaries, are less likely to regress than the others. In such cases thrombosis and rupture can also occur. In patients with persistent aneurysms, coronary artery stenosis can develop several years after the acute episode of KD.

KD is said to be the most common cause of pediatric myocardial infarction. Myocardial infarction in children presents with atypical symptoms as compared to adults. The main presenting complaints are uneasiness, vomiting, shock and abdominal pain. It is important to note that chest pain may not be a significant feature in young children. In almost one-third of children, the infarcts may be asymptomatic. However, the ECG and cardiac enzyme changes are like those in adults.

Treatment

Intravenous immunoglobulin (IVIG) is the drug of choice and is very effective when given in the first 10 days of illness. It reduces the chances of development of coronary abnormalities from 15–25% to less than 3%. IVIG results in a rapid defervescence of fever and normalization of the acute phase reactants. IVIG also improves myocardial function in KD patients having myocarditis. It is given as a single dose intravenous infusion of 2 gm/kg administered over 8–12 hours. While the mechanism of action of IVIG in patients with KD is largely unknown, it is believed that it results in down-regulation of the cytokine cascade. Children with severe forms of KD may require additional doses of IVIG. In 10–15% patients, fever may persist even after 36–48 hours of IVIG administration, or may recur after initial subsidence. Such patients are said to have IVIG resistant KD. Treatment of these patients warrants a second dose of IVIG, or administration of additional therapy in the form of methylprednisolone (30 mg/kg IV followed by a gradual taper over 2–3 weeks), infliximab (5–6 mg/kg IV single dose) or cyclosporine (5 mg/kg/day).

Aspirin is administered for its anti-inflammatory and antithrombotic effects. During the acute phase of illness, aspirin is administered at 30–50 mg/kg/day given every 6 hours. Around the 14th day of illness, when fever has resolved, aspirin is reduced to antithrombotic doses of 3–5 mg/kg/day as a single daily dose. Children with KD who develop coronary artery aneurysms are kept on long-term aspirin. It may be noted that KD is the only vasculitic disorder in which corticosteroids are relatively contraindicated.

A repeat echocardiogram is obtained at 2–3 weeks and again at 6–8 weeks following the onset of illness. Aspirin can be discontinued after the sedimentation rate and platelet counts have normalized (this usually takes 6–8 weeks) and the echocardiograms are reported to be normal. If echocardiography done at 6–8 weeks shows evidence of coronary dilatation and/or aneurysms, low dose aspirin needs to be continued indefinitely and follow-up investigations (coronary angiography/ dobutamine stress echocardiography) are then mandatory. Further management of KD patients with aneurysms is dependent on the severity of coronary disease.

Patients with a single small aneurysm should receive long-term aspirin and avoid physical sports. Patients with giant/multiple aneurysms should receive therapy that is more aggressive and are usually put on warfarin. Such patients often need bypass grafting or balloon angioplasty.

MORTALITY

KD was associated with a mortality rate of 1–2% in the pre-IVIG era. In developed countries this has dropped to 0.08% with improved recognition and appropriate therapy of the disease in the acute phase. At Chandigarh, mortality rate in KD is 0.8%. Deaths are most common 2–12 weeks after the onset of the illness and are usually secondary to the coronary aneurysms and complications thereof. Delays in diagnosis and institution of appropriate therapy are responsible for majority of deaths.

SUMMARY

Kawasaki disease is a medium vessel vasculitis and presents as an acute febrile illness of young children.

KD is, by no means, a rare disease. It has been reported from all parts of the world and is being increasingly recognized in India. KD is believed to be the most common vasculitic disorder of children and is now recognized as the most common cause of acquired heart disease in children in Europe, North America, Japan and several countries in Asia (e.g. Japan, Korea and Taiwan) from where accurate nationwide data are available. Clinical and epidemiological features of KD support an infectious cause, but the aetiology remains elusive. KD causes significant coronary artery disease that may lead to myocardial infarction and sudden death. For the pediatrician, it is important to diagnose KD as early as possible because a significant proportion of untreated patients develop coronary artery abnormalities. These coronary abnormalities can have lifelong sequelae. Intravenous immunoglobulin is the treatment of choice and is highly effective in preventing the coronary complications associated with the disease.

It is important to remember that KD can affect the cardiovascular system in several ways and these sequelae may manifest in young adults. Hence long-term follow-up is mandatory in all children diagnosed as KD, irrespective of their cardiac status in the initial attack.

Key Points

1. Kawasaki disease should be considered in differential diagnosis of illnesses where fever persists for more than 5 days.
2. Diagnosis is clinical, though laboratory parameters can help in arriving at a diagnosis.
3. Long-term morbidity is associated with coronary artery abnormalities.
4. Risk of coronary abnormalities can be decreased if intravenous immunoglobulin is given early, preferably within 10 days of onset of illness.
5. Kawasaki disease, being a systemic vasculitis, can involve other systems as well (e.g. hepatic, pulmonary, gastrointestional and central nervous system).
6. Early recognition can prevent delays in diagnosis and allow institution of appropriate therapy.

REFERENCES

1. Son MB, Sundel RP. In: Petty RE, Laxer RM, Lindsley CB, Wedderburn LR, Eds. Textbook of Pediatric Rheumatology. 7th ed. Philadelphia: Elsevier; 2016. p. 467–83.
2. Singh S. Kawasaki Disease: A clinical dilemma. Indian Pediatr 1999; 36:871–75.
3. Newburger JW, Takahasi M. Gerber MA, et al, Diagnosis, Treatment and Long-Term Management of Kawasaki Disease: A Statement for Health Professionals from the Committee on Rheumatic Fever, Endocardidits and Kawasaki Disease, Council on Cardiovascular Disease in the Young. American Heart Association. Pediatrics 2004; 114:1708–33.
4. Singh S, Aulakh R, Bhalla AK, Suri D, Manoj Kumar R, Narula N, Burns JC. Is Kawasaki disease incidence rising in Chandigarh, North India? Arch Dis Child 2011; 96(2):137–40.
5. Sundel R, Szer I. Vasculitis in childhood. Rheum Dis Clin N Amer 2002; 28:625–54.
6. Singh S, Bansal A, Gupta A, Kumar RM, Mittal BR. Kawasaki disease—a decade of experience from Northern India. Int Heart J 2005; 46:679–89.
7. Muta H, Ishii M, Sakaue T. et al. Older Age is a Risk Factor for the Development of Cardiovascular Sequelae in Kawasaki Disease. Pediatrics 2004; 44:751–54.
8. Mitra A, Singh S, Devidayal, Khullar M. Serum lipids in North Indian children treated for Kawasaki disease. Int Heart J 2005; 46:811–17.
9. Oechslin EC, Arbenz U, Mayer K. Giant and fusiform aneurysms of coronary arteries following early and adequate treatment of suspected Kawasaki disease. Heart 2004; 90:1437.
10. Kato H, Ichinose E, Kawasaki T: Myocardial infarction in Kawasaki disease. Clinical analyses in 195 cases. J Pediatr1986; 108:923–927.
11. Singh S, Kumar L. Kawasaki disease: Treatment with intravenous immunoglobulin during the acute stage. Indian Pediatr 1996; 33:689–92.
12. Kawasaki T. Acute febrile mucocutaneous syndrome with lymphoid involvement with specific desquamation of the fingers and toes in children. Arerugi 1967; 16:178–222.
13. Singh S, Gupta A. Kawasaki Disease. Indian Heart J 2004; 56: 261–62.
14. Narayanan SN, Krishna Veni, Sabiranathan K. Kawasaki disease. Indian Pediatr 1997; 34:139–43.
15. Singh S, Kumar L, Trehan A, Marwaha RK. Kawasaki Disease at Chandigarh. Indian Pediatr 1997; 34: 822–25.

16. Kato H, Ichinose E, Yoshioka F, Takechi T, Matsunaga S, Suzuki K, et al. Fate of coronary aneurysms in Kawasaki disease: serial coronary angiography and long-term follow-up study. Am J Cardiol. 1982; 49: 1758–66.
17. Mitra S, Singh S, Grover A, Kumar L. A child with prolonged pyrexia and peripheral desquamation: Is it Kawasaki Disease? Indian Pediatr 2000; 37:786–89.
18. Singh S, Kansra S. Kawasaki disease. Natl Med J India 2005; 18:20–24.
19. Kushner HI, Macnee R, Burns JC. Impressions of Kawasaki syndrome in India. Indian Pediatr 2006; 43:939–42.
20. Singh S, Gupta MK, Bansal A, et al. A comparison of theclinical profile of Kawasaki disease in children from Northern India above and below 5 years of age. Clin Exp Rheumatol 2007; 25(4):654–7.
21. Singh S, Newburger J, Kuijpers T, Burgner D. Management of Kawasaki Disease in resource limited setting. Pediatr Infect Dis J 2015; 34(1):94–6.
22. Singh S, Vignesh P, Burgner D. The epidemiology of Kawasaki Disease—a global update. Arch Dis Child 2015; 100(11):1084–8.
23. Singh S, Bhattad S, Gupta A, Suri D, Rawat A. Mortality in children with Kawasaki disease: 20 years of experience from a tertiary care centre in North India. Clin Exp Rheumatol 2016; 34 (3 Suppl 97):S129–33.
24. Narsaria P, Singh S, Gupta A, Khullar M, Bhalla A. Lipid profile and fat patterning in children at a mean of 8.8 years after Kawasaki disease: a study from Northern India. Clin Exp Rheumatol 2015; 33(2 Suppl 89):S-171–5.
25. Rawat A, Singh S. Biomarkers for Diagnosis of Kawasaki Disease. Indian Pediatr 2015; 52:473–474.
26. Singh S, Bhattad S, Sharma D, Phillip S. Recent Advances in Kawasaki Disease-Proceedings of the 3rd Kawasaki Disease Summit, Chandigarh, 2014. Indian J Pediatr 2016; 83:47–52.
27. Reddy M, Singh S, Rawat A, Sharma A, Suri D, Rohit MK. Pro-brain natriuretic peptide (ProBNP) levels in North Indian children with Kawasaki disease. Rheumatol Int 2016; 36(4):551–9.
28. Singh S, Sharma A, Jiao FY. Kawasaki disease: issues in diagnosis and treatment. A developing country perspective. Indian J Pediatr 2016; 83(2):140–5.
29. Singh S, Aulakh R, Kawasaki T. Kawasaki disease and the emerging coronary artery disease epidemic in India: is there a correlation? Indian J Pediatr 2014; 81(4):328–32.
30. Prakash J, Singh S, Gupta A, Bharti B, Bhalla AK. Sociodemographic profile of children with Kawasaki disease in North India. Clin Rheumatol 2016; 35:709–13.
31. Burns JC, Herzog L, Fabri O, Tremoulet AH, Rodó X, Uehara R, et al. Kawasaki Disease Global Climate Consortium). Seasonality of Kawasaki disease: a global perspective. PLoS One 2013; 8(9):e74529.
32. Chaudhuri K, Ahluwalia TS, Singh S, Binepal G, Khullar M. Polymorphism in the promoter of the CCL5 gene (CCL5G-403A) in a cohort of North Indian children with Kawasaki disease. A preliminary study. Clin Exp Rheumatol 2011; 29(1 Suppl 64): S126–30.
33. Kashyap R, Mittal BR, Bhattacharya A, Manoj Kumar R, Singh S. Exercise myocardial perfusion imaging to evaluate inducible ischaemia in children with Kawasaki disease. Nuc Med Commun 2011; 32(2):137–41.
34. Gupta A, Singh S, Gupta A, Suri D, Rohit M. Aortic stiffness studies in children with Kawasaki disease: preliminary results from a follow-up study from North India.Rheumatol Int 2014; 34(10):1427–32.
35. Suthar R, Singh S, Bhalla AK, Attri SV. Pattern of subcutaneous fat during follow-up of a cohort of North Indian children with Kawasaki disease: a preliminary study.Int J Rheum Dis 2014; 17(3):304–12.
36. Meena RS, Kumar RM, Gupta A, Singh S. Carotid intima-media thickness in children with Kawasaki disease. Rheumatol Int 2014; 34(8):1117–21.
37. George M, Ahluwalia J, Gupta A, Singh S. Anti-phospholipid antibodies in children with Kawasaki Disease—a preliminary study from North India. Rheumatol Int 2014; 34(6):849–50.
38. Gupta A, Singh S, Rohit M, Suri D, Rawat A, Narula N. Giant coronary aneurysms in Kawasaki Disease. Indian J Pediatr 2014; 81(4):401–2.
39. Madhusudan S, Singh S, Manojkumar R, Gupta A, Suri D, Rawat M. Late symptomatic myocarditis in Kawasaki disease: an unusual manifestation. Indian J Pediatr 2014; 81(4):404–5.]
40. Dogra S, Gehlot A, Suri D, Rawat A, Kumar RM, Singh S.Incomplete Kawasaki Disease followed by Systemic Onset Juvenile Idiopathic Arthritis—The Diagnostic Dilemma. Indian J Pediatr 2013; 80: 783–785.
41. Aggarwal P, Singh S, Suri D, Narula N, Manoj Kumar R. Symptomatic myocarditis in Kawasaki disease. Indian J Pediatr 2012;79(6):813–4.
42. Ghelani SJ, Singh S, Manoj Kumar R. QT interval dispersion in North Indian children with Kawasaki disease without overt coronary artery abnormalities. Rheumatol Int 2011; 31:301–305.
43. Nakano H, Ueda K, Saito A, Nojima K. Repeated quantitative angiograms in coronary arterial aneurysm in Kawasaki disease. Am J Cardiol. 1985; 56:846–51.

Chapter

47

Takayasu Arteritis

MR Sivakumar

HISTORY

The first case of Takayasu arteritis (TA) was described in 1908 by Dr Mikito Takayasu at the Annual Meeting of the Japan Ophthalmology Society as a "wreathlike" appearance of blood vessels in the retina.[1] Since then, this uncommon disease has been described from various parts of the world.

BACKGROUND

TA is a rare form of primary systemic vasculitis that appears to be commoner in Asia than Europe or North America. Cardiovascular system is the major organ involved. Despite the term "pulseless disease", the predominant findings in TA are asymmetric pulse. Absent peripheral pulses occur late in the course of the disease. While 5-year survival rates exceed 90%, the disease has a high incidence of residual morbidity.

PATHOPHYSIOLOGY

TA is characterized by granulomatous inflammation of the aorta and its major branches, leading to stenosis, thrombosis, and aneurysm formation.[2] The lesions of TA are segmental and patchy. Mononuclear infiltration of the adventitia occurs early in the course of the disease, with cuffing of the vasa vasorum. Granulomatous changes may be observed in the tunica media with Langerhans cells and central necrosis of elastic fibres and smooth muscle cells. A panarteritis with infiltrates of lymphocytes, plasma cells, histocytes, and giant cells is present. Later, fibrosis of the media and acellular thickening of the intima compromise the vessel lumen. Stenosis is common with a wide range of symptoms. Endothelial activation leads to a hypercoagulable state predisposing to thrombosis. Congestive heart failure occurs as a result of hypertension, aortic root dilation, or myocarditis. Transient ischaemic attacks, cerebrovascular accidents, mesenteric ischaemia, carotidynia and claudications may occur. One hypothesis for granulomatous vasculitis development is that antigens deposited in vascular walls activate CD4+ T cells, followed by release of cytokines chemotactic for monocytes. These monocytes are transformed into macrophages that mediate endothelial damage and granuloma formation in the vessel wall. Human studies have demonstrated increased expression of intercellular adhesion molecule-1 (ICAM-1) and vascular cell adhesion molecule-1 (VCAM-1) in TA. Anti-aorta antibodies, anti-endothelial cell antibodies, immunoglobulin G, M and properdin are found in pathologic specimens. There is an expansion of both T helper 1 (Th1) and Th17 cells in peripheral blood of patients with TA. Cytokine abnormalities include elevated tumour necrosis factor, interleukin 6 (IL-6) and IL-17. B cells may also play a role, with plasmablast expansion in patients with active TA. IL-6 and B-cell activating factor also increase total numbers of circulating B cells; reduction in disease activity has been anecdotally observed with successful depletion of B cells with anti-CD20.

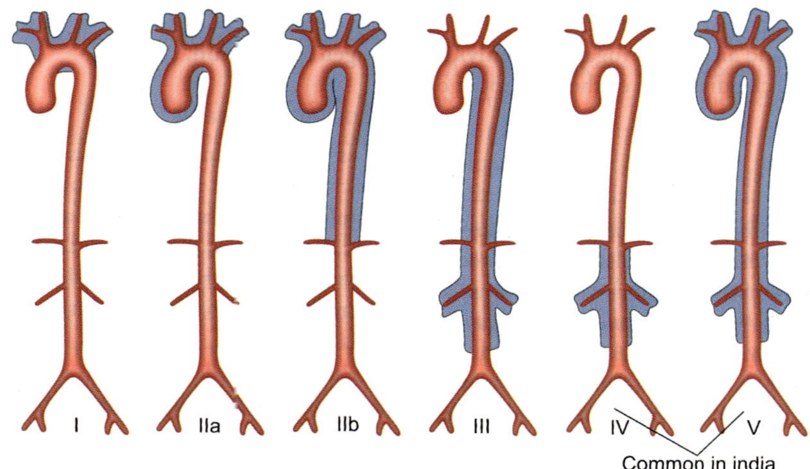

Fig. 47.2: Numano's classification by angiographic criteria[7]

Disease Monitoring

Monitoring of acute phase reactants as a measure of disease activity has limited value. Performing regular imaging of affected vasculature and surveillance imaging for new lesions are essential. The presence of active disease requires aggressive treatment with corticosteroids and immunosuppressants. However, current markers of disease activity are inadequate to identify all patients with disease flare. Serial MRI may reveal vessel wall edema, but whether this measures actual inflammation is unclear. Clinical analysis of TA disease activity, response to treatment and detection of relapse remain suboptimal. Kerr *et al* defined active TA as the new onset or worsening of two or more of the following features: (1) Fever or arthralgia, (2) Raised ESR (>20 mm/hr), (3) Features of vascular ischemia or inflammation such as claudication, diminished or absent pulse, bruit, vascular pain, asymmetric blood pressure in either upper or lower limbs, (4) Typical angiographic features.

Development of Disease Extent Index for Takayasu Arteritis (DEI.Tak), Indian Takayasu Arteritis Activity Score (ITAS) and Takayasu Arteritis Score (TADS) to Assess TA.[8,9]

The Indian Rheumatology Association Vasculitis core group has validated instruments, like DEI.Tak, ITAS and TADS. The Physicians Global Assessment (PGA) is still the basis for treatment decisions in TA. The acute phase response is neither sensitive nor specific enough to be helpful as ESR/CRP is not necessarily elevated, even in acute vascular occlusions. Standard vascular imaging shows damage as well as "activity" and may not reverse with medical therapy.

LABORATORY STUDIES

TA has no specific markers. Complete blood count (CBC) reveals a normochromic, normocytic anaemia in 50% of patients. Acute phase reactants are elevated, with leukocytosis and thrombocytosis. Erythrocyte sedimentation rate, transaminase and hypoalbuminemia are noted. The von Willebrand factor-related antigen (factor VIII-related antigen) may be elevated. Antiendothelial antibodies are present. Antinuclear antibody results are usually negative and rheumatoid factor is elevated in 15% of TA patients. Increased levels of immunoglobulins G, M, and A are present. Serum levels of matrix metalloproteinase (MMP) 2, 3, 9 were elevated in TA patients, but was not related to disease activity. In contrast, MMP-3 and MMP-9 were only elevated in patients with active disease.

IMAGING STUDIES

Invasive digital subtraction angiography is the gold standard for making the diagnosis of TA, while CT angiography or MR angiography (MRA) may also be useful. Drawbacks to arteriography, including morbidity from use of contrast dye can be avoided by using MRA. Arteriography often demonstrates long, smooth, tapered narrowing or occlusions

(Fig. 47.3). Stenoses occur in 90–100% of patients with TA and aneurysm formation in only 27%. Gadolinium-enhanced cardiovascular MRI has also been used to demonstrate myocardial perfusion defects, which may be important in long-term prognosis. Non-contrast, T2-weighted, short-T1 inversion recovery (STIR) images may be used to monitor edema in the aortic wall, which may be a surrogate for inflammation (Fig. 47.4); edema can be found in 94% of patients with clinically active disease.

Three-dimensional MRA imaging provide exciting new data that may improve the understanding of the disease. MRI, MRA and CT scan are useful for serial examinations and diagnosis in

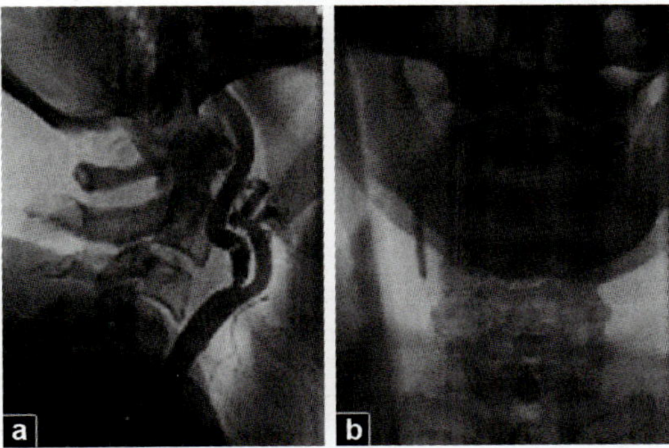

Fig. 47.3: (a) Carotid angiogram showing stenosis of the internal carotid artery at the origin, (b) carotid angioplasty and stenting of the right internal carotid artery

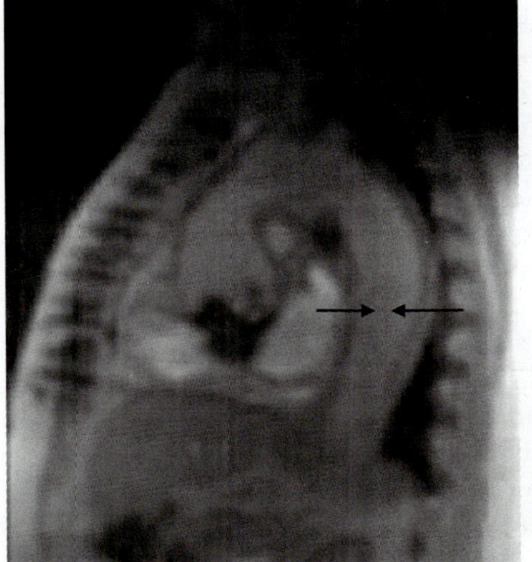

Fig. 47.4: MRI of thoracic aorta showing thickened wall (black arrows)

the early phase of TA. Aortic lesions including stenosis, dilatation, wall thickening and mural thrombi are well visualized on MRI.

^{18}F-Fluorodeoxyglucose PET: ^{18}F-Fluorodeoxyglucose (^{18}F-FDG)-PET metabolic imaging is increasingly used.[18] F-FDG-PET makes use of the fact that deoxyglucose is taken up by metabolically active cells and is not highly metabolized. The accumulation of radiolabeled deoxyglucose delineates arterial wall inflammation in giant cell arteritis (GCA) and TA reveals the extent of disease. The similarity in the pattern of arterial disease in TA and GCA is apparent, although involvement of common carotid and pulmonary arteries is more common in patients with TA. The principal advantage of ^{18}F-FDG-PET-CT is in the diagnosis of early pre-stenotic disease, an event that can be missed by intra-arterial angiography. The utility of ^{18}F-FDG-PET-CT in the follow-up of patients remains unclear. The limitations of ^{18}F-FDG-PET have led to exploration of novel PET ligands for imaging vascular inflammation and in particular [^{11}C]-PK11195, a ligand targeting the peripheral benzodiazepine receptor, which is highly expressed on activated monocytes/macrophages. [^{11}C]-PK11195 PET-CT allows for sensitive and specific detection of active Takayasu arteritis and GCA, with a high target-to-background ratio. Although this investigation is useful for diagnosing TA in the pre-pulseless stage, its value as a tool for monitoring of patients on immunosupressive therapy is yet to be accurately delineated.

Gallium-67 radionuclide scan: This scan may demonstrate increased uptake in the aorta and branches.

Duplex ultrasound: Duplex Doppler of common carotids and subclavian arteries, with measurement of intima media thickness (IMT) may be used to evaluate and monitor disease after treatment (Fig. 47.5). Carotid evaluation often reveals a

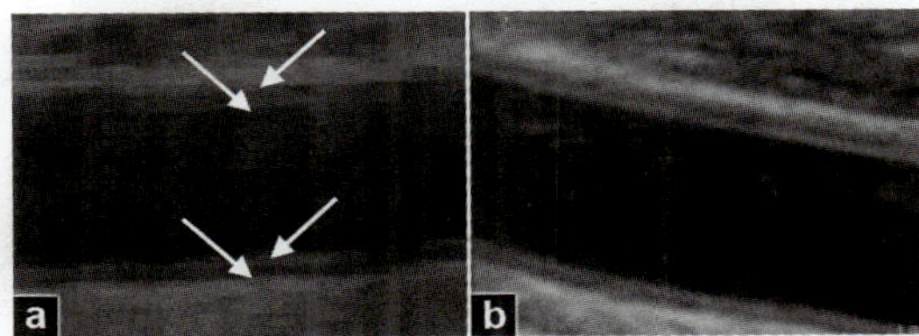

Fig. 47.5: (a) Ultrasound image of the carotid artery showing normal intima media thickness (IMT) in control and (b) increased IMT in a patient with Takayasu

homogenous, circumferential thickening of the vessel wall that is distinguishable from atherosclerotic thickening. Sivakumar *et al* demonstrated that the mean IMT in TA (0.84 ± 0.16 mm) was significantly higher compared to controls (0.57 ± 0.15 mm) $p < 0.0001$. Increased IMT was seen across all types, as classified by pattern of arteriographic involvement. The risk for a vascular event showed a continuous rise as the IMT increased across the range. The relative risk was 1.52 per increase of IMT by 1 SD (0.16 mm). Increased common carotid intima media thickness in all 5 subtypes of TA was observed (Fig. 47.6).[10]

Transcranial Doppler can demonstrate intracranial arterial occlusions and emboli in patients with acute stroke.

Chest radiography: May reveal widening of the ascending aorta, irregular descending aorta, aortic calcifications and rib notchings.

Echocardiography: Usually always performed at baseline and follow-up echocardiography is repeated as indicated by the clinical condition.

HISTOLOGIC FINDINGS FROM ARTERIAL BIOPSIES

Mononuclear infiltration of the adventitia with perivascular cuffing of the vasa vasorum occurs early in the disease (Fig. 47.7). Granulomatous changes may be observed in the tunica media with

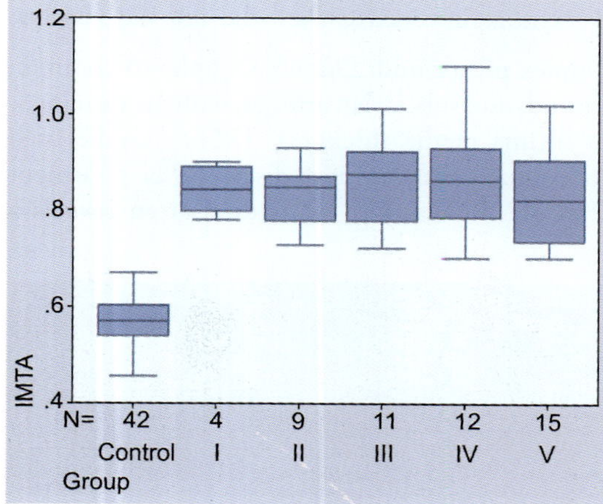

Fig. 47.6: Increased intima media thickness (IMT) in all 5 types of Takayasu arteritis patient compared to controls

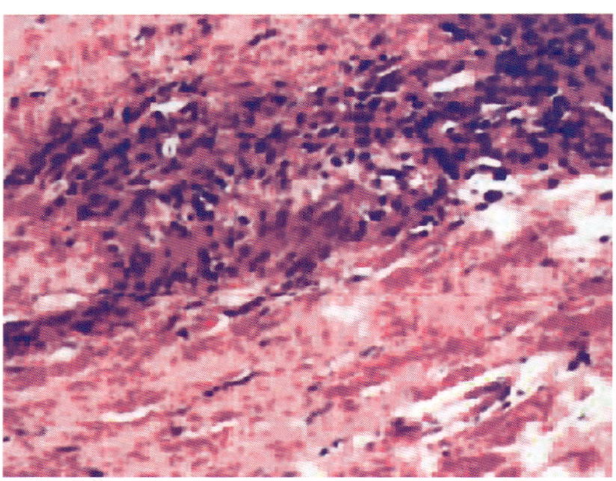

Fig. 47.7: Histopathological picture of arterial wall showing mononuclear cell infiltrate (H&E stain)

Langerhans cells and central necrosis of elastic fibres and smooth muscle cells. Later, fibrosis of the media and acellular thickening of the intima may compromise the vessel lumen. Grossly, wrinkling of the intima is found. Histological specimens seldom are available due to the large vessels affected, with the exceptions of specimens obtained during autopsy and bypass surgery.

Prognosis

The relatively poor outcome for patients with this disease reflects a variety of clinical shortcomings, including a lack of awareness of TA per se. The rarity of the disorder and the heterogeneous nature of its clinical presentation predisposes to late diagnosis and delayed treatment. There is a survival rate of 85–95% at 15 years. Japanese studies support 90–95% 5-year survival rates. In contrast, in a series involving 26 Mexican children aged 3–15 years, the 5-year survival rate was only 35%.

TREATMENT

Medical

Goals of medical therapy are to control active inflammation and to normalize clinical and laboratory parameters while preventing further vascular damage. The EULAR 2009 guidelines for large vessel vasculitis recommend an initial prednisolone dose of 1 mg/kg/day (total maximum dose of 60 mg/day).[11] It is suggested to maintain high dose prednisolone treatment until all evidence of active disease has resolved, then taper prednisolone dosage gradually. About 40% patients relapse on steroid taper. Patients not responding to corticosteroids or those who relapse during corticosteroid taper require an additional agent. Some may be controlled with weekly infusions of methylprednisolone (15 mg/kg, not to exceed 1 g/wk). Extensive use of these infusions is associated with significant steroid induced toxicity.

Immunosuppressive regimens include weekly Methotrexate (7.5–25 mg) or daily or monthly intravenous cyclophosphamide, cyclosporine which offers lower ovarian toxicity, mycophenylate mofetil, which is useful to treat glucocorticoid resistant disease, leflunomide, azathioprine, anti-tumour necrosis factor agents, etanercept and infliximab. The cardiovascular risk associated with TA results in most patients receiving low-dose aspirin and a statin. Matrix metalloproteinase inhibition using minocycline may be a useful adjunctive therapy.

Rituximab, monoclonal anti-CD20 antibody, has been used for disease control.[12] Treatment regimens have been the 4-week lymphoma protocol (375 mg/m^2 weekly for 4 wk) or the 2-week fixed-dose rheumatoid arthritis protocol (1000 mg repeated once 14 day later).

Tocilizumab (monoclonal anti-IL-6 receptor antibody) has induced remission in TA resistant to other treatments. The mammalian target of rapamycin (mTOR) antagonists, sirolimus and everolimus, are known to inhibit myofibroblast proliferation and their use in TA should be explored.

T lymphocyte infiltration is seen in affected arteries and hence Abatacept, a selective modulator of CD80/CD86:CD28 co-stimulatory signal required for full T-cell activation holds considerable promise. The chemokines CCL 19 and CCL 21, and the pro-inflammatory cytokine interleukin-18 all have important roles in the development of granulomatous large vessel vasculitis and hence represent potential therapeutic targets.

Potential novel therapeutic targets include IL-1, IL-6 and IL-18, IFN-γ, activated dendritic cells, Toll-like receptors and myofibroblast proliferation. Drug-eluting stents, such as those containing the antibiotic rapamycin or the antineoplastic drug paclitaxel, might also have a role. There exists only one randomized controlled trial of an immunosupressive agent in TA (abatacept) which did not show an add-on benefit to glucocorticoid in reducing relapses. Rest of the evidences regarding management of TA are drawn from case series and case reports.

Surgery

Patients with fibrotic changes require surgical treatment of symptomatic stenotic or occlusive disease. This can be achieved by percutaneous angioplasty or stenting or by resection and grafting. These procedures should always be preceded and followed by effective immunosuppressive therapy.[10] Percutaneous balloon angioplasty of the aorta normalizes systolic and diastolic blood pressures within 24 hours, with improvement of exercise tolerance and restoration of peripheral pulses.

Endovascular stenting is used in patients with severe stenoses, hypertension, or ischemia during the fibrotic phase of the disease. Multiple stents have been used in children to relieve long-segment renal artery stenosis and attendant renovascular hypertension. Careful pre-procedure analysis of individual lesions is required and trials of stents incorporating antiproliferative drugs such as Sirolimus are indicated.

Follow-up

Measurement of ESR and C-reactive protein may not prove a useful aid. Regular imaging of affected vasculature as well as surveillance imaging for new lesions is essential. Treatment may be monitored with MRI and/or MRA or CT scan. Mural thickening is observed to decrease with corticosteroid treatment. Periodic imaging may identify an active disease by the appearance of new areas of stenosis, despite normal erythrocyte sedimentation rate and the absence of clinical features.

Conclusion

TA is a rare granulomatous large vessel vasculitis more commonly seen in India, affecting most often young females. Assessment of disease activity vis-à-vis damage is difficult. Corticosteroids form the mainstay of treatment; other immmunosuppressives, although commonly used for maintenance, have a little evidence base from randomized controlled trials.

Key Points

1. TA is more commonly seen in women than men in India and Asia, but it also afflicts persons in North America, Europe, Middle East and Africa.
2. The diagnosis of TA is difficult and requires a high index of suspicion. Clinical presentation depends on the location and severity of the aortic branch lesions, ascending aortitis and abdominal aorta and branches are frequently affected in Indian patients.
3. The principal therapeutic intervention for TA is corticosteroids. About 50–60% respond favourably.
4. Therapeutic interventions like percutaneous transluminal angioplasty (PTA) and stenting, bypass surgeries or surgical reconstruction should be performed when the disease is made inactive by the use of effective immunosuppressive therapy.
5. The Disease Extent Index for Takayasu Arteritis (DEI.Tak), Indian Takayasu Arteritis Score (ITAS) and the Takayasu Arteritis Damage Score (TADS) are valuable tools in documenting the spectrum of TA.

REFERENCES

1. Numano F. The story of Takayasu. Arthritis Rheum 2002; 41:103–06.
2. Hotchi M. Pathological studies on Takayasu's arteritis. Heart Vessels 1992; suppl 7:11–17.
3. Maksimowicz-McKinnon K, Clark TM, Hoffman GS. Limitations of therapy and a guarded prognosis in an American cohort of Takayasu arteritis patients. Arthritis Rheum 2007; 56:1000–09.
4. Sivakumar MR, Sanjeev S. Cerebrovascular manifestations and carotid artery intima medial thickness in Takayasu's arteritis evaluated by using the Disease Extent Index forTA (DEI.Tak). Clin Exp Rheumatol 2007; 25(1):S–120.
5. Peter J, David S, Danda D, Peter JV, Horo S, Joseph G. Ocular manifestations of Takayasu arteritis: a cross-sectional study. Retina. 2011; 31:1170–8.
6. Arend WP, Michel BA, Bloch DA, Hunder GG, Calabrese LH, Edworthy SM. The American College of Rheumatology 1990 criteria for the classification of Takayasu arteritis. Arthritis Rheum 1990; 33:1129–34.
7. Ishikawa K. Diagnostic approach and proposed criteria for the clinical diagnosis of Takayasu's arteriopathy. J Am CollCardiol. 1988; 12:964–72.
8. Aydin SZ, Yilmaz N, Akar S, Aksu K, Kamali S, Yucel E, et al. Assessment of disease activity and progression in Takayasu's arteritis with Disease Extent Index-Takayasu. Rheumatology (Oxford). 2010; 49:1889–93.
9. Misra R, Danda D, Rajappa SM, Ghosh A, Gupta R, Mahendranath KM, et al; on behalf of the Indian Rheumatology Vasculitis (IRAVAS) group. Development and initial validation of the Indian Takayasu Clinical Activity Score (ITAS2010). Rheumatology 2013; 52:1795–801.
10. Sivakumar MR. Outcome of vascular interventions in Takayasu arteritis using the Takayasu arteritis damage score. Arthritis Rheum. 2011; 63:S58.
11. Caltran E, Di Colo G, Ghigliotti G, Capecchi R, Catarsi E, Puxeddu I, et al. Two Takayasu arteritis patients successfully treated with rituximab. Clin Rheumatol. 2014; 33:1183–4.
12. Mukhtyar C, Guillevin L, Cid MC, Dasgupta B, de Groot K, Gross W, et al; European Vasculitis Study Group. EULAR recommendations for the management of large vessel vasculitis. Ann Rheum Dis. 2009; 68:318–23.

FURTHER READING

1. Misra DP, Sharma A, Kadhiravan T, Negi VS. A scoping review of the use of non-biologic disease modifying anti-rheumatic drugs in the management of large vessel vasculitis. Autoimmun Rev. 2017; 16:179–191.
2. Nakagomi D, Jayne D. Outcome assessment in Takayasu arteritis. Rheumatology (Oxford). 2016; 55:1159–71.

Chapter

48

Polymyalgia Rheumatica and Giant Cell Arteritis

Mohit Goyal, Bimlesh Dhar Pandey

INTRODUCTION

Polymyalgia rheumatica (PMR) and giant cell arteritis (GCA) are closely associated inflammatory conditions that frequently occur together. PMR is among the most common inflammatory rheumatologic diseases in the elderly while GCA is the most common idiopathic inflammatory disease of the vessels in this age group. Immunogenetics of the two show some overlap but also some clear differences. There are immunogenetic similarities between the two but some very clear differences too have been demonstrated.

Clinical features of PMR are pain and stiffness, in the shoulders, neck, and in pelvic girdles specifically in the mornings. It may be followed by a systemic inflammatory syndrome and a rise in the levels of acute phase reactants. There is a rapid response to glucocorticoids, which has historically been considered a criteria for diagnosis.

GCA also known as Horton's disease or temporal arteritis is a vasculitis of medium and large-sized arteries, is the most common vasculitis in the elderly. Presenting symptomatology includes a systemic inflammatory syndrome, manifestations secondary to vascular involvement, and sometimes cranial ischaemic complications in form of loss of vision and stroke.

HISTORY

The first reports of these conditions dates back to 1888 when Bruce described a case of polymyalgia rheumatica which he called senile rheumatic gout[1] and to 1890 when Hutchinson described "red streaks on the forehead" of an elderly man,[2] later recognized as giant cell arteritis.

Awareness about polymyalgia rheumatica increased when independent reports were published in 1940s and 1950s that described the condition using various names, viz. secondary fibrositis, periarthrosis humeroscapularis, peri-extraarticular rheumatism, myalgic syndrome of the aged, pseudo-polyarthrite rhizomelique, and anarthritic rheumatoid disease.[3] Barber coined the term "polymyalgia rheumatica" in 1957.

Giant cell arteritis came to be recognized as a specific disease only after the article by Horton et al in 1932[4] and other reports during the 1930s and 1940s. In 1937, Horton et al described seven cases with systemic symptoms, fever, headache, jaw claudication and anaemia. Temporal artery biopsies in these cases revealed granulomatous inflammation with giant cells.[5] These lesions were regarded as those of vasculitis and the condition was called "temporal arteritis". In 1941 Gilmour demonstrated similar biopsy findings in patient with inflammation of large vessels and proposed the term "giant cell arteritis".[6]

The two conditions continued to be described as separate entities until 1964 when a definite association between the two was accepted.[7]

EPIDEMIOLOGY

Both these conditions are common inflammatory diseases of older individuals that occur after the age of 50 years and their incidence peaks around the age of 73 years.[8,9] Figure 48.1 shows the how PMR

Table 48.4: Clinical features of polymyalgia rheumatica (PMR) and giant cell arteritis (GCA)

Common features	*Specific to PMR*	*Specific to GCA*
Fever	Low grade fever	High spiking fever
Malaise	Stiffness	Headache
Fatigue	Shoulder, hip and neck pain	Scalp tenderness
Depression	Edema	Jaw claudication
Weight loss	Synovitis and bursitis	Diplopia
	Subjective weakness	Transient visual loss
	Muscle tenderness	Permanent loss of vision

Investigations

The diagnosis of PMR is basically clinical and is supported by elevated acute phase reactants. A normocytic normochromic anaemia is commonly seen. Hemogram may also show thrombocytosis, indicating ongoing systemic inflammation. An erythrocyte sedimentation rate (ESR) higher than 100 mm is common in PMR and C-reactive protein (CRP) is also significantly elevated. But lower ESR and CRP in low disease activity can make diagnosis tricky.[27] Some studies have found that IL-6 levels are sensitive indicators of disease activity.[28] Normal creatine kinase and aldolase levels help to differentiate the condition from polymyositis.

Liver and renal function tests and urine analysis are essentially normal. Muscle biopsy and electromyography are normal. Most patients are negative for rheumatoid factor and anti-citrullinated peptide antibodies.[29] Synovial fluid analysis is suggestive of mild inflammation with predominant polymorphs. There are no specific antibody tests available for PMR and GCA. Relevant tests may be ordered to rule out other causes of vasculitis.

ESR and CRP are elevated in GCA as well. CT angiography (CTA) or magnetic resonance angiography (MRA) of the involved vessels reveal stenoses and/or aneurysms in branches of the aorta and other arteries.[30] The most commonly affected arteries are subclavian, carotid and ascending segment of the aorta. Temporal artery histopathology in GCA shows a mixed-cell infiltration of the vessel wall with lymphomononuclear and inflammatory cells. A mixed-cell, predominantly lymphomononuclear, and[31] granulomata are seen in about 50 percent of the temporal artery biopsies (TABs). TABs have sensitivity of 70 to 90 percent.[32] Guidance by colour duplex sonography (CDUS) increases the chances of biopsy yield. TAB is not useful in cases with isolated PMR.

Apart from CDUS and MRA, 18F-fluorodeoxyglucose positron emission tomography (18F FDG-PET) has emerged as a useful imaging modality. FDG-PET has become the modality of choice. It shows vascular enhancement with involvement of vertebral bursae and a resolution of these findings after successful therapy. FDG-PET gives vascular imaging in cases with isolated PMR as well suggesting an ongoing subclinical vasculitis.[33] These modalities can be used to look bursitis and tenosynovitis around the shoulders and the hips.

Malignancies, chronic infections and drug reactions are important differentials of PMR. Metabolic abnormalities like hypothyroidism, rheumatic conditions like seronegative rheumatoid arthritis, calcium pyrophosphate disease and polymyositis may also mimic PMR.[34]

Diagnostic criteria for PMR[35] (Table 48.5) and GCA[36] (Table 48.6) are composite indices with clinical as well as laboratory variables.

Treatment

PMR has a self-limiting course ranging from months to years with relapses in about 25% of treated patients.[37] Glucocorticoids are the mainstay of treatment of PMR and they lead to complete clinical remission with normalization of ESR in majority of patients. Joint guidelines of ACR and EULAR recommend starting with a dose of 12.5 mg to 25 mg of prednisolone or equivalent with a slow tapering.[38] In case of non-response to steroids in 3 days, alternative diagnoses should be considered. Dose is tailored according to the patient's weight and the severity of symptoms. Tapering is guided by improvement in the clinical symptoms. The dose of steroid and duration of treatment remain a matter of debate.

Table 48.5: The European League Against Rheumatism/American College of Rheumatology(EULAR/ACR) 2012 classification criteria for polymyalgia rheumatica (PMR)[35]

Criterion	*Points without US (0–6)*	*Points with US (0–8)*
Morning stiffness > 45 minutes	2	2
Hip pain or limitation of range of motion	1	1
Absence of rheumatoid factor or anti-CCP	2	2
Absence of other joint involvement	1	1
At least one shoulder with subdeltoid bursitis and/or biceps tenosynovitis and/or glenohumeral synovitis (either posterior or axillary) and at least one hip with synovitis and/or trochanteric bursitis	NA	1
Both shoulders with subdeltoid bursitis, biceps tenosynovitis or glenohumeral synovitis	NA	1
Score required for diagnosis	4 or more	5 or more
Sensitivity	68%	66%
Specificity	78%	81%

2012 provisional classification criteria for polymyalgia rheumatica: a European League Against Rheumatism/American College of Rheumatology collaborative initiative.[35] Required criteria: Age 50 years or older; bilateral shoulder ache; abnormal CRP and/or ESR. US, ultrasound.

Table 48.6: American College of Rheumatology 1990 classification criteria for giant cell arteritis (GCA)

Criterion	*Definition*
Age at disease onset 50 years or more	Development of symptoms or findings beginning at age 50 or older.
New headache	New onset of or new type of localized pain in the head.
Temporal artery abnormality	Temporal artery tenderness to palpation or decreased pulsation unrelated to arteriosclerosis of cervical arteries.
Elevated erythrocyte sedimentation rate (ESR)	ESR of 50 mm/hour or more by Westergren method
Abnormal artery biopsy	Biopsy specimen with the artery showing vasculitis characterized by prominence of mononuclear cell infiltration or granulocyte inflammation, usually with multinucleated giant cells.

The American College of Rheumatology 1990 criteria for the classification of giant cell arteritis.[36] A patient with vasculitis shall be said to have giant cell arteritis if three of the following five criteria are present. The criteria have a sensitivity of 93.5% and a specificity of 91.2%.

Although thromboembolic events have not been implicated in the ischaemic damage in GCA, retrospective analyses have showed that patients put on low-dose aspirin present less often with cranial ischaemic complications on follow-up.[39] A report found reduced relapses with longer duration between relapses and some steroid-sparing effect in patients receiving angiotensin II receptor blockers.[40] In another study, Statins helped glucocorticoid tapering by reducing expression of major histocompatibility complex class II antigen and inhibiting activation and proliferation of T cell.[41]

Steroid sparing agents such as methotrexate may be used. Studies have shown beneficial effect of methotrexate when given at a weekly dose of 10 mg per week which could not be replicated at doses of 7.5 mg per week.[42] In another study, the arm with azathioprine at a daily dose of 150 mg showed a lower mean requirement of prednisolone compared to the placebo arm at 52 weeks.[43] The study

Table 48.7: Agents used in the treatment of polymyalgia rheumatica and giant cell arteritis

Mainstay of treatment	*Steroid sparing agents*	*Agents with additional benefits*
Glucocorticoids	Methotrexate	Low dose aspirin
Tocilizumab	Azathioprine	Angiotensin II receptor blockers
	Mycophenolate	
	Leflunomide	

included patients with PMR as well as GCA. Another series revealed good response to monthly pulses of cyclophosphamide and also a significant steroid-sparing effect in 15 patients with steroid-dependent giant cell arteritis.[44] Two case series showed steroid sparing effect of leflunomide.[45,46] Mycophenolate mofetil has also been shown to have significant reduction in the corticosteroid requirement in data that is limited to case reports.

Hydroxychloroquine was found no better than placebo in a 96-week trial.[47] A randomized trial demonstrated the efficacy of dapsone in reducing risk of relapses[46] but its side effect profile prevented further research. If not contraindicated, NSAIDs may be used for symptomatic relief but they do not provide any additional benefit. They must be used with caution in view of their adverse effect profile.[48] There are individual reports of efficacy of clarithromycin in PMR.

RCTs with infliximab,[49] adalimumab[50] and etanercept[51] have found no significant role for TNF alpha inhibitors. One case report noted clinical improvement with reduction in blood levels of inflammatory markers with rituximab in a patient with GCA refractory to other conventional therapies.[52] A Swiss study on patients with GCA suggested response to Tocilizumab but the blunting of acute phase response by IL-6 inhibition was not taken into account thus raising possibility of a bias in the clinical evaluations.[53] A summary of the important agents used in treatment of PMR and GCA is shown in Table 48.7.

Statement of Unmet Need

There is insufficient evidence for an effective glucocorticoid sparing agent. Prolonged usage of steroid and its reinstitution in relapses leads to significant comorbidities as a result of the adverse effects. There is dearth of treatment options in patients who fail to respond adequately to these agents.

Recent Advances

GiACTA trial (Giant cell arteritis Actemra trial) is an ongoing trial studying the efficacy of Tocilizumab in GCA. The results of the study which show enhanced disease free remission period are encouraging. Phase III results have revealed that 56% of patients who received tocilizumab were able to achieve steroid-free remission of their disease at one year, versus 14% in those who were on a tapering regimen of only steroids for a period of six months.[54] The results have raised hopes for a breakthrough in the management of the condition.

Conclusion

PMR and GCA are diseases chiefly of the elderly, so although still rare, their incidence has increased with increased in the life expectancy. The lack of high quality studies has been a limiting factor in the search for new avenues in diagnosis and management of these conditions. In GCA, while temporal artery biopsy has high specificity, the condition cannot be ruled out in biopsy negative patients. The classification criteria are highly sensitive but not highly specific. For more than half a century, glucocorticoids have been the mainstay of the treatment but they significantly add to the already existing comorbidities in the age group in which these conditions occur. Newer advances in treatment with not only glucocorticoid sparing potential but also with a more favourable efficacy and adverse effect profile are the need of the hour.

Key Points

- PMR and GCA are inflammatory diseases that occur after the age of 50 years, either separately or together.
- Etiology is thought to be an interplay between age, environment and genetic susceptibility.
- Fever, weight loss, arthralgias and myalgias are clinical features common to PMR and GCA.

- Shoulder, hip and neck pain with significant morning stiffness are important features of polymyalgia rheumatica.
- Giant cell arteritis is characterized by headache, scalp tenderness, jaw and upper limb claudication and visual disturbances.
- Characteristic clinical features with elevated acute phase reactants are usually sufficient for diagnosing these conditions.
- CT angiography, MR angiography, colour duplex ultrasound, temporal artery biopsy and FDG-PET are other modalities that can help.
- Glucocorticoids are the mainstay of the treatment. IL-6 inhibition with tocilizumab has shown promise.

REFERENCES

1. Mowat A. Strathpeffer Spa: Dr William Bruce and polymyalgia rheumatica. Ann Rheum Dis 1981; 40(5):503–506.
2. Portioli I. The history of polymyalgia rheumatica/giant cell arteritis. ClinExp Rheumatol. 2000; 18(4 Suppl 20):S1–3.
3. Hunder G. OP1. History of giant cell arteritis (GCA) and polymyalgia rheumatica (PMR). Rheumatology. 2005; 44(suppl_3):iii1.
4. Borchers A, Gershwin M. Giant cell arteritis: A review of classification, pathophysiology, geoepidemiology and treatment. Autoimmun Rev 2012;11(6–7): A544–A554
5. Roberts WC, Zafar S, Ko JM. Morphological features of temporal arteritis. Proceedings (Baylor University Medical Center). 2013; 26(2):109–115.
6. Bengtsson B. Epidemiology of giant cell arteritis. Baillière's Clinical Rheumatology. 1991; 5(3): 379–385.
7. Hamrin B, Jonsson N, Landberg T. Arteritis in "polymyalgia rheumatica." Lancet 1964; 1:397.
8. Kermani TA, Warrington KJ. Polymyalgia rheumatica. Lancet. 2013; 381(9860):63–72.
9. Weyand CM, Goronzy JJ. Medium- and large-vessel vasculitis. N Engl J Med 2003; 349(2):160–169.
10. Salvarani C, Cantini F, Boiardi L, Hunder GG. Polymyalgia rheumatica and giant-cell arteritis. N Engl J Med 2002; 347(4):261–271.
11. Salvarani C, Gabriel SE, O'Fallon WM, Hunder GG. The Incidence of Giant Cell Arteritis in Olmsted County, Minnesota: Apparent Fluctuations in a Cyclic Pattern. Ann Intern Med 1995; 123(3):192.
12. Smeeth L, Cook C, Hall A. Incidence of diagnosed polymyalgia rheumatica and temporal arteritis in the United Kingdom, 1990–2001. Ann Rheum Dis 2006; 65(8):1093–1098.
13. Okumura T, Tanno S, Ohhira M, Nozu T. The rate of polymyalgia rheumatica (PMR) and remitting seronegative symmetrical synovitis with pitting edema (RS3PE) syndrome in a clinic where primary care physicians are working in Japan. Rheumatol Int 2011; 32(6):1695–1699.
14. Carmona FD, Mackie SL, Martin JE, Taylor JC, Valio A, Eyre S et al. A large-scale genetic analysis reveals a strong contribution of the HLA class II region to giant cell arteritis susceptibility. Am J Hum Genet. 2015; 96(4):565–580.
15. Ghosh P, Borg F, Dasgupta B. Current understanding and management of giant cell arteritis and polymyalgia rheumatica. Expert Rev Clin Immunol 2010; 6(6):913–928.
16. Martinez-Taboada V, Alvarez L, RuizSoto M, Marin-Vidalled M, Lopez-Hoyos M. Giant cell arteritis and polymyalgia rheumatica: Role of cytokines in the pathogenesis and implications for treatment. Cytokine. 2008; 44(2):207–220.
17. Alvarez-Rodriguez L, Lopez-Hoyos M, Mata C, Marin M, Calvo-Alen J, Blanco R et al. Circulating cytokines in active polymyalgia rheumatica. Ann Rheum Dis 2009; 69(01):263–269.
18. Caylor TL, Perkins A. Recognition and Management of Polymyalgia Rheumatica and Giant Cell Arteritis. Am Fam Physician. 2013; 88:676–684.
19. Yamashita H, Kubota K, Mimori A. Clinical value of whole-body PET/CT in patients with active rheumatic diseases. Arthritis Res Ther 2014; 16(4).
20. Gonzalez-Gay M, Garcia-Porrua C, Amor-Dorado J, Llorca J. Fever in biopsy-proven giant cell arteritis: Clinical implications in a defined population. Arthritis Care Res2004; 51(4):652–655.
21. Mandell BF. Polymyalgia rheumatica: clinical presentation is key to diagnosis and treatment. Cleveland Clin J Med 2004; 71(6):489–495.
22. Salvarani C, Gabriel S, Hunder G. Distal extremity swelling with pitting edema in polymyalgia rheumatica. Report of nineteen cases. Arthritis Rheum 1996; 39(1):73–80.
23. Gonzalez-Gay M, Barros S, Lopez-Diaz M, Garcia-Porrua C, Sanchez-Andrade A, Llorca J. Giant cell arteritis: disease patterns of clinical presentation in a series of 240 patients. Am J Ophthalmol 2006; 141(6):1173.
24. Aiello P, Trautmann J, McPhee T, Kunselman A, Hunder G. Visual Prognosis in Giant Cell Arteritis. Ophthalmology. 1993; 100(4):550–555.
25. Kermani TA, Warrington KJ. Polymyalgia rheumatica. Lancet. 2013; 381(9860):63–72.
26. Salvarani C, Pipitone N, Versari A, Hunder GG. Clinical features of polymyalgia rheumatica and giant cell arteritis. Nat Rev Rheumatol. 2012; 8(9):509–521.

27. Cantini F, Salvarani C, Olivieri I, Macchioni L, Ranzi A, Niccoli L et al. Erythrocyte sedimentation rate and C-reactive protein in the evaluation of disease activity and severity in polymyalgia rheumatica: A prospective follow-up study. Sem Arthritis Rheum 2000; 30(1):17–24.
28. Weyand C, Fulbright J, Evans J, Hunder G, Goronzy J. Corticosteroid Requirements in Polymyalgia Rheumatica. Arch Intern Med1999; 159(6):577.
29. Lopez-Hoyos M. Clinical utility of anti-CCP antibodies in the differential diagnosis of elderly-onset rheumatoid arthritis and polymyalgia rheumatica. Rheumatology 2004; 43(5):655–657.
30. Prieto–González S, Arguis P, García-Martínez A, Espígol-Frigolé G, Tavera-Bahillo I, Butjosa M et al. Large vessel involvement in biopsy-proven giant cell arteritis: prospective study in 40 newly diagnosed patients using CT angiography. Ann Rheum Dis 2012; 71(7):1170–1176.
31. Lie J. Illustrated histopathologic classification criteria for selected vasculitis syndromes Arthritis Rheum 2010; 33(8):1074–1087.
32. Borchers A, Gershwin M. Giant cell arteritis: A review of classification, pathophysiology geoepidemiology and treatment. Autoimmunity Rev 2012; 11(6–7): A544–A554.
33. Blockmans D, De Ceuninck L, Vanderschueren S, Knockaert D, Mortelmans L, Bobbaers H. Repetitive 18-fluorodeoxyglucose positron emission tomography in isolated polymyalgia rheumatica: a prospective study in 35 patients. Rheumatology 2006; 46(4): 672–677.
34. Gonzalez-Gay MA, Garcia-Porrua C, Salvarani C, Olivieri I, Hunder G. The spectrum of conditions mimicking polymyalgia rheumatica in northwestern Spain. J Rheumatol 2000; 27:2179–2184.
35. Dasgupta B, Cimmino M, Kremers H, Schmidt W, Schirmer M, Salvarani C et al. 2012 Provisional classification criteria for polymyalgia rheumatica: A European League Against Rheumatism/American College of Rheumatology collaborative initiative. Arthritis Rheum 2012; 64(4):943–954.
36. Hunder GG, Bloch DA, Michel BA, Stevens MB, Arend WP, Calabrese LH et al. The American College of Rheumatology 1990 criteria for the classification of giant cell arteritis. Arthritis Rheum. 1990; 33(8): 1122–1128.
37. Pountain G, Hazleman B. ABC of Rheumatology: polymyalgia rheumatica and giant cell arteritis. BMJ. 1995; 310(6986):1057–1059.
38. Dejaco C, Singh Y, Perel P, Hutchings A, Camellino D, Mackie S et al. 2015 Recommendations for the management of polymyalgia rheumatica: a European League Against Rheumatism/American College of Rheumatology collaborative initiative. Ann Rheum Dis. 2015; 74(10):1799–1807.
39. Nesher G, Berkun Y, Mates M, Baras M, Rubinow A, Sonnenblick M. Low-dose aspirin and prevention of cranial ischemic complications in giant cell arteritis. Arthritis Rheum 2004; 50(4):1332–1337.
40. Alba M, García-Martínez A, Prieto-González S, Espígol-Frigolé G, Butjosa M, Tavera-Bahillo I et al. Treatment with angiotensin II receptor blockers is associated with prolonged relapse-free survival, lower relapse rate, and corticosteroid-sparing effect in patients with giant cell arteritis. Sem Arthritis Rheum 2014; 43(6):772–777.
41. Meroni P, Luzzana C, Ventura D. Anti-Inflammatory and Immunomodulating Properties of Statins: An Additional Tool for the Therapeutic Approach of Systemic Autoimmune Diseases?. Clin RevAllergy Immunol 2002; 23(3):263–278.
42. Caporali R, Cimmino MA, Ferraccioli G, Gerli R, Klersy C, Salvarani C, et al. Prednisone plus methotrexate for polymyalgia rheumatica: a randomized, double-blind, placebo-controlled trial. Ann Intern Med. 2004; 141(7):493–500.
43. De Silva M, Hazleman B. Azathioprine in giant cell arteritis/polymyalgia rheumatica: a double-blind study. Ann Rheum Dis 1986; 45(2):136–138.
44. De Boysson H, Boutemy J, Creveuil C, Ollivier Y, Letellier P, Pagnoux C et al. Is there a place for cyclophosphamide in the treatment of giant-cell arteritis? A case series and systematic review. Sem Arthritis Rheum2013; 43(1):105–112.
45. Adizie T, Christidis D, Dharmapaliah C, Borg F, Dasgupta B. Efficacy and tolerability of leflunomide in difficult-to-treat polymyalgia rheumatica and giant cell arteritis: a case series. Int J Clin Prac 2012; 66(9): 906–909.
46. Diamantopoulos A, Hetland H, Myklebust G. Leflunomide as a Corticosteroid-Sparing Agent in Giant Cell Arteritis and Polymyalgia Rheumatica: A Case Series. Bio Med Res Int 2013; 2013:1–3.
47. Bienvenu B, Ly K, Lambert M, Agard C, André M, Benhamou Y et al. Management of giant cell arteritis: Recommendations of the French Study Group for Large Vessel Vasculitis (GEFA). La Revue de Médecine Interne. 2016; 37(3):154–165.
48. Gabriel S, Sunku J, Salvarani C, O'Fallon W, Hunder G. Adverse outcomes of antiinflammatory therapy among patients with polymyalgia rheumatica. Arthritis Rheum 1997; 40(10):1873–1878.
49. Hoffman GS, Cid MC, Rendt-Zagar KE, Merkel PA, Weyand CM, Stone JH, et al. Infliximab for Maintenance of Glucocorticosteroid-Induced Remission of Giant Cell Arteritis. Ann Intern Med 2007; 146(9):621.

50. Seror R, Baron G, Hachulla E, Debandt M, Larroche C, Puéchal X et al. Adalimumab for steroid sparing in patients with giant-cell arteritis: results of a multicentre randomised controlled trial. Ann Rheum Dis 2014; 73(12):2074–2081.
51. Martinez-Taboada V, Rodriguez-Valverde V, Carreno L, Lopez-Longo J, Figueroa M, Belzunegui J et al. A double-blind placebo controlled trial of etanercept in patients with giant cell arteritis and corticosteroid side effects. Ann Rheum Dis 2007; 67(5):625–630.
52. Bhatia A, Ell PJ, Edwards JC. Anti-CD20 monoclonal antibody (rituximab) as an adjunct in the treatment of giant cell arteritis. Ann Rheum Dis 2005; 64(7): 1099–1100.
53. Villiger P, Adler S, Kuchen S, Wermelinger F, Dan D, Fiege V, et al. Tocilizumab for induction and maintenance of remission in giant cell arteritis: a phase 2, randomised, double-blind, placebo-controlled trial. Lancet. 2016; 387(10031):1921–1927.
54. Stone JH, Tuckwell K, Dimonaco S, Klearman M, Aringer M, Blockmans D et al. Efficacy and Safety of Tocilizumab in Patients with Giant Cell Arteritis: Primary and Secondary Outcomes from a Phase 3, Randomized, Double-Blind, Placebo-Controlled Trial. Arthritis Rheumatol [Internet]. 2016;68 Suppl 10. Available from: http://acrabstracts.org/abstract/efficacy-and-safety-of-tocilizumab-in-patients-with-giant-cell-arteritis-primary-and-secondary-outcomes-from-a-phase-3-randomized-double-blind-placebo-controlled-trial/ (accessed 21 January 2017).

FURTHER READING

1. Salvarani C, Cantini F, Boiardi L, Hunder GG. Polymyalgia rheumatica and giant-cell arteritis. N Engl J Med 2002; 347:261–71.

Chapter

49

Polyarteritis Nodosa and Microscopic Polyangiitis

Ramesh Jois, Vijay KR Rao

POLYARTERITIS NODOSA

According to the 2012 revised Chapel Hill nomenclature polyarteritis nodosa (PAN) is defined as a necrotizing arteritis of medium or small arteries without glomerulonephritis or vasculitis in arterioles, capillaries or venules and not associated with antineutrophil cytoplasmic antibodies.[1]

PAN is a rare disease, with an incidence of about 3–4.5 cases per 100,000 population annually. Older estimates placed the prevalence as high as 7.7 cases per 100,000 population in Alaskan Eskimos hyper-endemic for hepatitis-B virus (HBV) infection.

PAN can solely involve a single organ or present systemically. It may affect any organ, but for unknown reasons it spares the pulmonary and glomerular arteries. Medium-sized and small muscular arteries, preferentially at vessel bifurcations are affected resulting in micro-aneurysm formation, thrombosis, aneurysmal rupture with haemorrhage, and consequently organ ischaemia or infarction. Rarity of PAN, diverse clinical presentation and absence of definitive diagnostic criteria contribute to a delayed diagnosis.

PAN has been widely described in the context of HBV while idiopathic or classical PAN (cPAN) is not so common. HBV-PAN may occur at any time during the course of acute or chronic hepatitis B infection, although it typically occurs within six months of infection. The activity of HBV-PAN does not parallel with that of hepatitis. With the development HBV vaccine, the percentage of patients with HBV-PAN has decreased from 36% to less than 5%.[2, 3]

Clinical Features

The common clinical features are highlighted in the American College of Rheumatology (ACR) classification criteria for PAN (Table 49.1).[4] Glomerular disease is not a feature and distinguishes PAN from small vessel vasculitides. Lungs are also typically spared. Other clinical presentations (some rare) are listed in Table 49.2.

The French vasculitis study group (FVSG) reported clinical outcomes in 123 patients with HBV-PAN.[2] These patients when compared to cPAN had more severe disease (higher FFS), higher neuropathy, gastrointestinal manifestations requiring surgery, weight loss, cardiomyopathy,

Table 49.1: American College of Rheumatology (ACR) (1990) classification criteria for PAN[4]

- Weight loss of 4 kg or more
- Livedo reticularis
- Testicular pain/tenderness
- Myalgia or leg weakness/tenderness
- Mononeuropathy or polyneuropathy
- Diastolic blood pressure >90 mmHg
- Elevated blood urea nitrogen >40 mg/dl or creatinine >1.5 mg/dl
- Presence of hepatitis B surface antigen or antibody in serum
- Arteriogram demonstrating aneurysms or occlusions of the visceral arteries
- Biopsy of small-or medium-sized artery showing granulocytes and mononuclear leukocytes in the arterial wall

3 of 10 criteria to be present to classify as PAN
(*Source*: Reference no. 4)

Table 49.2: Other manifestations of PAN

Organ	*Manifestation*
Constitutional symptoms	Fever, weight loss, fatigue
Musculoskeletal	Arthralgia/arthritis
Skin	Subcutaneous nodules, palpable/necrotic purpurae, ulcerations, digital infarction, gangrene.
Eyes	Choroiditis, iritis, retinal vasculitis
Renal	Hypertension (can be malignant), renal impairment (due to ischaemic nephropathy), subcapsular/perirenal haemorrhage, microaneurysm rupture and intraperitoneal haemorrhage.
Cardiac	Coronary arteritis, dissection of aorta and other large arteries.
Gastointestinal	Ischaemic vasculitis causing haemorrhage and small bowel perforation, acute necrotizing pancreatitis.
	Vasculitis of gallbladder or appendix (more often in HBV-PAN)
Central nervous system	Encephalopathy, seizures, stroke, subarachnoid haemorrhage and cranial nerve palsies (Rare)

hypertension, orchitis and elevated liver enzymes. Microaneurysms were more common in the mesenteric than renal artery.

Rare Varities of PAN

Cutaneous PAN: Limited to the skin and do not have systemic features. They usually appear on the legs (sometimes arm and trunk) as livido reticularis, nodular or ulcerative lesions. Usually, benign, they have a relapse and remitting course and are steroid-responsive.

Single-organ PAN: Limited disease has been occasionally described from surgical specimens of appendix, abdominal organs or testis. A careful follow-up is required for systemic involvement.

ADA2 Deficiency: Recessive loss-of-function mutations of CECR1, the gene encoding adenosine deaminase 2 (ADA2) has recently been described as a possible genetic cause of a rare form of PAN.[5] Familial clusters, very early age of onset, early-onset stroke, necrotic cutaneous ulcers, renal and gastrointestinal ischaemia are the predominant features.

Investigations

Systemic inflammatory response (raised ESR, CRP, leucocytosis) is frequently observed. Meticulous screening for HBV is mandatory. ANCA is negative. Angiography remains the gold standard for diagnosis of PAN. MRI/CT angiography reveals multiple stenoses and micro-aneurysm (1–5 mm) at branches of renal, celiac and mesenteric arteries (Fig. 49.1).

Tissue biopsy (e.g. muscle biopsy), where possible, may confirm the diagnosis. Nerve biopsy confirms vasculitis but does not help to differentiate PAN from small vessel vasculitides. Renal biopsy may not be useful since glomerular disease is not seen in PAN and is contraindicated in patients with micro-aneurysm due to high-risk of rupture after biopsy. Inflammatory cell infiltrate in all layers of the vessel wall is typically seen with polymorphonuclear neutrophils in the acute and mononuclear

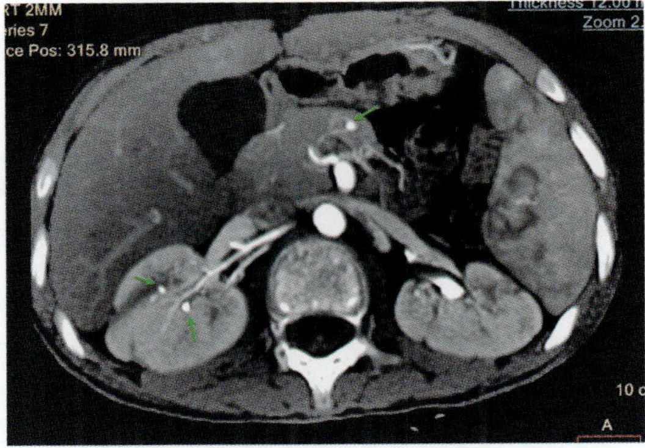

Fig. 49.1: CT Angiography in arterial phase showing micro-aneurysms in right kidney and pancreas (arrows) and multiple renal cortical scars suggestive of renal vascular injury

cells in subacute stage. Fibrinoid necrosis of the vessels causing thrombosis and tissue infarction is seen in the chronic stage.

Outcome and Prognosis

Effective treatment has improved the overall outcome of PAN. A retrospective review of 348 patients with cPAN from the FVSG showed a relapse rate of 28% after a mean follow-up of 68 months in cPAN and 10.6% in HBV-PAN.[2, 3] The mortality was 19.6% and 38.6% in cPAN and HBV-PAN respectively. Persistent active viral replication leads to higher relapses in HBV-PAN.

Untreated, the 5-year survival rate of PAN is 13% and nearly 50% die within the first three months of onset. Steroids have improved the 5-year survival rate to 50–60% and when combined with other immunosuppressants the 5-year survival rates are 76–89% for cPAN and 64–70% for HBV-PAN.[2, 3]

Death associated with PAN occurs as a result of uncontrolled vasculitis, infection related to immunosuppression and vascular complications of the disease (e.g. myocardial infarction and stroke). Mortality is higher in patients with acute abdominal syndromes and malignant hypertension.

In a prospective study of 342 patients with PAN, Guillevin *et al* found five factors to be associated with a poor prognosis (Table 49.3).[6] The five factor score (FFS) developed in 1996 can be used to predict survival and help guide treatment decisions. The revised FFS (2009) has not been used to guide treatment decisions and is only a tool for survival prognostication The parameters included in the 2009 FFS include age >65 yrs, renal insufficiency, cardiac involvement, and gut involvement.

Treatment

Since prognosis and survival depends on the FFS, the choice of therapy is also based on FFS as summarized in Table 49.3. Limited data is available from randomized trials and majority of published data from cohort studies have included PAN along with ANCA-associated vasculitis. Intravenous methylprednisolone is used in life-threatening or severe organ manifestations. Oral corticosteroids (CS) are given with prednisolone at 1 mg/kg/day initially and tapered with an aim to stop by 12–18 months. Cyclophosphamide (CYC) intravenous pulses are preferred over oral (6 pulses at 600 mg/m², 3 infusions fortnightly and 3 infusions at 3-weekly intervals). Maintenance therapy with methotrexate or azathioprine is recommended for at least 18 months. Relapsing disease requires longer duration of treatment.[7]

HBV-PAN is treated with anti-viral therapy (ribavirin, interferon-α, lamivudine, entecavir or tenofovir), plasma exchange and lower dose of steroids for about two weeks. Plasma exchange helps in removing circulating immune complexes and is useful until sero-conversion is achieved, while steroids help in the life-threatening inflammatory process and anti-viral drugs for reduction of viral load.[7]

Key Points

- PAN is a rare medium-vessel vasculitis sometimes associated with hepatitis B virus infection.
- Vascular obstruction and aneurysm formation is the hallmark.
- Early recognition and treatment reduces mortality and morbidity.

Table 49.3: Prognostic five factor score (FFS 1996) and treatment guidance[6]

FFS Parameters	*5-year mortality rate*	*Treatment guidance*
Proteinuria >1 gm/24 h S Creatinine >1.58 mg/dl Gastrointestinal involvement Cardiomyopathy CNS involvement	FFS 0: 12% FFS 1: 25.9% FFS ≥ 2: 45.9%	FFS 0: Steroids alone* FFS ≥1: Induction: Steroids + Cyclophosphamide. Maintenance: Steroids with azathioprine/ methotrexate†

[*Cyclophosphamide is added if the disease is progressive/not adequately controlled by steroids alone. † Mycophenolate, leflunomide, rituximab only sparse data available]
Source: Reference no. 6

MICROSCOPIC POLYANGIITIS

Microscopic polyangiitis (MPA) is an idiopathic autoimmune necrotizing vasculitis of small vessels with anti-neutrophil cytoplasmic antibody (ANCA) positivity in more than 75% of cases. Because it can lead to both pulmonary capilliaritis and glomerulonephritis, MPA is a primary cause of pulmonary-renal syndrome. Previously classified under PAN, MPA is now defined as a distinct clinical entity and grouped along with granulomatosis with polyangiitis(GPA) and eosinophilic granulomatosis with polyangiitis under ANCA-associated vasculitis (1994 Chapel Hill classification criteria).[1]

The average age at onset is about 50 years. The annual incidence is estimated to be 3–24/million and prevalence 25–94/million.[8]

Neutrophil activation and chemotaxis due to the fixation of ANCA to membranous myeloperoxidase (MPO) leads to the release of reactive oxidative species, proteinases and cytokines resulting in endothelial and tissue injury (vasculitis).

Clinical features: The clinical features of MPA are summarized in Table 49.4.[9]

Investigations

Raised ESR and CRP, thrombocytosis, leucocytosis and normocytic anaemia reflect the systemic inflammatory nature of MPA.

Microscopic haematuria and proteinuria (15% have nephrotic range) is found in more than 90% of patients with renal involvement. Raised creatinine at presentation is not uncommon. P-ANCA positivity with raised anti-MPO is see in 75% of patients with MPA, however serial measurements should not be performed to monitor disease activity or guide treatment.

Renal histology in patients with kidney disease is characterized by the presence of focal segmental pauci-immune glomerulonephritis. Extra-capillary crescents are present in nearly all renal biopsies and often involve more than 60% of the glomeruli (Fig. 49.2). The severity of renal impairment and the prognosis for renal function correlates with the presence of glomerular sclerosis, tubular damage and active glomerular disease with crescents. Nerve and skin biopsy reveal necrotizing vasculitis in medium and and small vessels. Granulomas are not a feature of MPA unlike GPA.

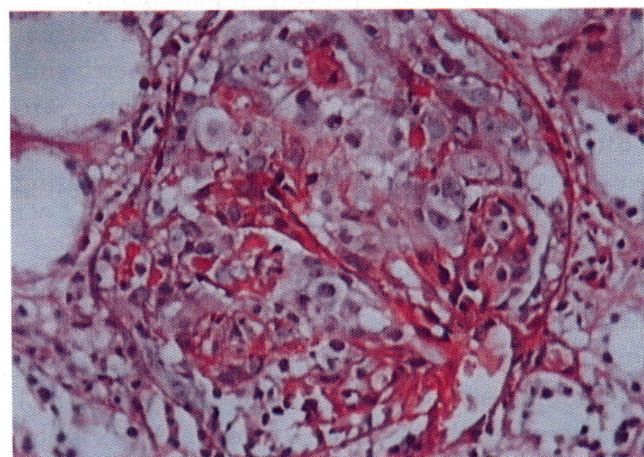

Fig. 49.2: Renal biopsy showing crescentic glomerulonephritis

Table 49.4: Clinical features of microscopic polyangiitis (MPA)[9]

Organ involved	*Frequency*	*Clinical features*
Constitutional	56–76%	Fever, myalgias, arthralgias, weight loss
Renal	79%	Microscopic haematuria, proteinuria and RPGN
Pulmonary	25%	Pulmonary alveolar haemorrhage, ARDS and interstitial lung fibrosis
Neurology	57%	Mononeuritis multiplex, symmetric polyneuropathy, cranial neuropathies (rare)
Cutaneous	62%	Purpura, ulcers, splinter haemorrhages, livedo reticularis, mouth ulcers, digital gangrene
Ocular	1%	Retinal vasculitis, scleritis, episcleritis, blepharitis and choroiditis
Gastrointestinal	30%	Abdominal pain, small/large bowel ischaemia, ulcerations and/or perforations
Cardiovascular	17%	Severe acute congestive heart failure (without MI) and pericarditis

Chapter

50

Granulomatosis with Polyangiitis and Eosinophilic Granulomatosis with Polyangiitis

Shankar Naidu, Manish Rathi, Aman Sharma

Anti-neutrophil cytoplasmic antibody (ANCA) associated vasculitides (AAV) are a group of primary systemic vasculitis which include granulomatosis with polyangiitis (GPA), microscopic polyangiitis (MPA) and eosinophilic granulomatosis with polyangiitis (EGPA). They are classified as small vessel vasculitis according to the Chapel Hill consensus definitions.[1] AAV are characterized by presence of ANCA in majority of the patients and pauci-immune reaction in the tissues. GPA is characterized by granulomatous inflammation, tissue necrosis and vasculitis involving predominantly upper respiratory tract (URT), lungs and kidneys. EGPA is characterized by asthma, peripheral eosinophilia and systemic vasculitis. Lungs and skin are the most common organs involved in EGPA.

EPIDEMIOLOGY

AAV can affect patients of any age group but is more common in middle aged to elderly population and has no gender predilection. GPA is a rare disorder, with annual incidence varying from 0.5 per million in Peru to 10.8 per million in United Kingdom.[2] The prevalence of GPA also depends on the geographical location, with studies from UK showing a point prevalence of 130 per million, whereas studies from USA showed a low point prevalence of 26 per million.[2] The annual incidence of EGPA was reported to be 1 to 3 per million and the prevalence varied from 14 per million in Sweden to 45.7 per million in UK.[2]

PATHOGENESIS

The exact etiology and pathophysiology of AAV is largely unknown, but various studies have indicated that certain infections, environmental factors, epigenetic and genetic changes may predispose patients to develop systemic vasculitis. Ethnicity plays an important role, as indicated by more prevalence of GPA in Caucasians as compared to Asians. Among the genetic factors, HLA DPB1*0401 was found to be associated with GPA and HLA DRB1*07 and HLA DRB4 are more prevalent in EGPA patients.[3,4] Polymorphisms in genes encoding CTLA4, PTPN22 are also identified in patients with GPA, while EGPA patients were found to have polymorphisms in IL-10 gene.[5,6] A few cases of familial inheritance were reported in GPA, with an estimated relative risk of 1.56.[7] Infections can act as a trigger for onset of vasculitic process and also precipitate relapses. *Staphylococcus aureus* nasal carriage has been associated with higher relapses in patients of GPA. Use of drugs like leukotriene modifying agents, like montelukast and zafirlukast, has led to development of EGPA in patients of chronic asthma. Cocaine abuse has been shown to induce PR3 ANCA positive vasculitis.

Classification Criteria

The American College of Rheumatology (ACR) proposed classification criteria for various vasculitides including GPA and EGPA. The 1990 ACR criteria for GPA and EGPA are shown in Tables 50.1 and 50.2 respectively.[8,9] A hierarchical approach was used for classifying the systemic vasculitis according to the European Medical Agency (EMEA) algorithm.[10]

Table 50.1: 1990 ACR classification criteria for GPA

Criterion	*Definition*
Nasal or oral inflammation	Development of painful or painless oral ulcers or purulent or bloody nasal discharge
Abnormal chest radiography	Presence of nodules, fixed infiltrates or cavities
Urinary sediment	Microscopic hematuria (> 5RBCs/hpf) or RBC casts in urine sediment
Granulomatous inflammation on biopsy	Histological changes showing granulomatous inflammation within the wall of an artery or in the perivascular or extravascular area

Two criteria classify GPA with a sensitivity of 88.2% and specificity of 92%

Table 50.2: 1990 ACR classification criteria for EGPA

Criterion	*Definition*
Asthma	History of wheezing or diffuse high pitched rales on expiration
Eosinophilia	Eosinophils >10% on white blood cell differential count
Neuropathy	Development of mononeuropathy, multiple mononeuropathies or polyneuropathy attributable to systemic vasculitis
Pulmonary infiltrates (non-fixed)	Migratory or transitory pulmonary infiltrates on radiographs (not including fixed infiltrates) attributable to systemic vasculitis
Paranasal sinus abnormality	History of acute or chronic paranasal sinus pain or tenderness or radiographic opacification of paranasal sinuses
Extravascular eosinophils	Biopsy including artery, arteriole or venule showing accumulation of eosinophils in extravascular areas

Four criteria classify EGPA with a sensitivity of 85% and specificity of 99.7%.

Clinical Features

Granulomatosis with Polyangiitis

The disease may present with very diverse manifestations, ranging from non-specific constitutional symptoms like fever, malaise, anorexia and weight loss to severe multisystem involvement leading to multiple organ failures or death. The characteristic clinical manifestations include upper respiratory tract involvement, lung infiltrates or cavities and pauci-immune glomerulonephritis in kidneys.

Upper respiratory tract (URT) involvement is seen in 80–90% of patients in various studies.[11,12] URT involvement is in the form of nasal crusting, nasal obstruction (Fig. 50.1), persistent rhinorrhea, bloody nasal discharge, chronic sinusitis. Some patients may have nasal ulcers, nasal septal perforation, collapsed nasal bridge and subglottic stenosis. Ear involvement may present as ear ache,

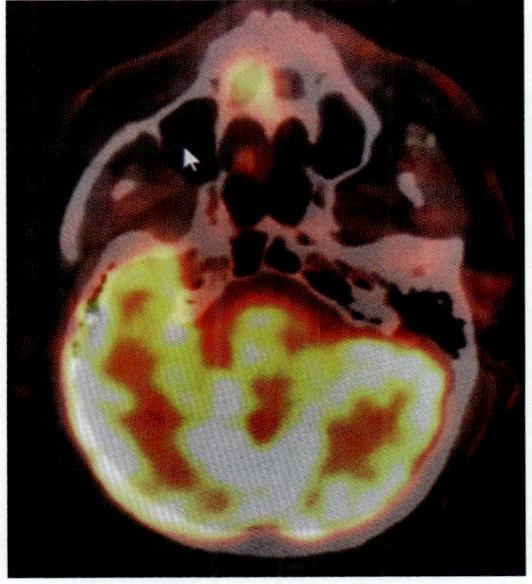

Fig. 50.1: PET scan showing nasal inflammation in GPA

fullness, purulent or bloody ear discharge and decreased hearing. Both conductive and sensorineural hearing loss may be seen in the patients. There can be oral ulceration and strawberry gingivitis in the oral cavity. Sometimes there may be oro-antral fistula also (Fig. 50.2). Pulmonary involvement is seen in 70–80% of patients.[11, 12] Lung involvement is commonly seen in the form of nodules (Fig. 50.3), infiltrates, single of multiple cavities. Reticulonodular opacities, consolidation, pleural effusions and lymphadenopathy are rare manifestations. Diffuse alveolar haemorrhage is a rare but organ and life threatening presentation, requiring urgent immunosuppressive therapy.

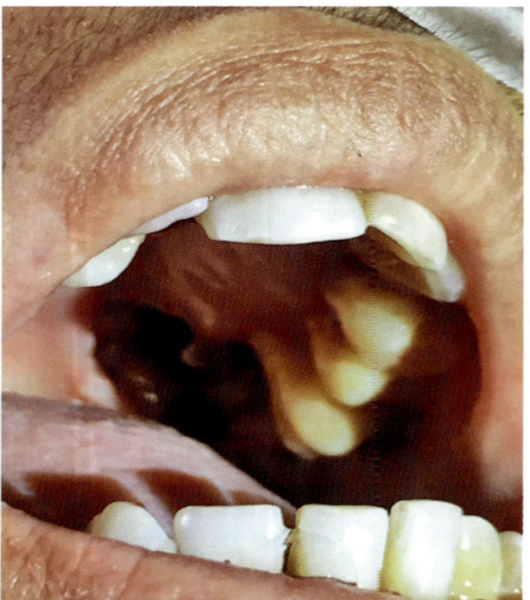

Fig. 50.2: Oroantral fistula in a patient with GPA

Renal involvement may vary from asymptomatic proteinuria to organ threatening rapidly progressive renal failure requiring renal replacement therapy. Renal involvement is seen in about 77% of patients.[11,12] Proteinuria more than 500 mg/day, microscopic hematuria, RBC casts in urine microscopy, elevated serum creatinine indicate the presence of renal involvement. Kidney biopsy shows a pauci-immune glomerulonephritis and helps in prognostication depending on the extent and type of involvement. Skin involvement is seen in 20–30% of patients, with the most common manifestation being purpura.[11,12] Other cutaneous manifestations include urticaria, livedo reticularis, nodules, erythema nodosum and pyoderma gangrenosum. Histopathology from the skin lesions commonly shows leukocytoclastic vasculitis.

Eye involvement may be seen in the form of uveitis, retinal vasculitis, scleritis (Fig. 50.4), episcleritis, conjunctivitis, corneal ulceration, retro orbital mass and optic atrophy. Ocular involvement is seen in 39.4% of patients.[11] Peripheral nervous system involvement is seen in about 25% of patients and presents as peripheral symmetric polyneuropathy or mononeuritis multiplex.[12] Central nervous system involvement is less common than the PNS and is seen in 7% of patients.[12] CNS involvement presents as cranial neuropathies, pachymeningitis or mass lesions. Gastrointestinal involvement is very

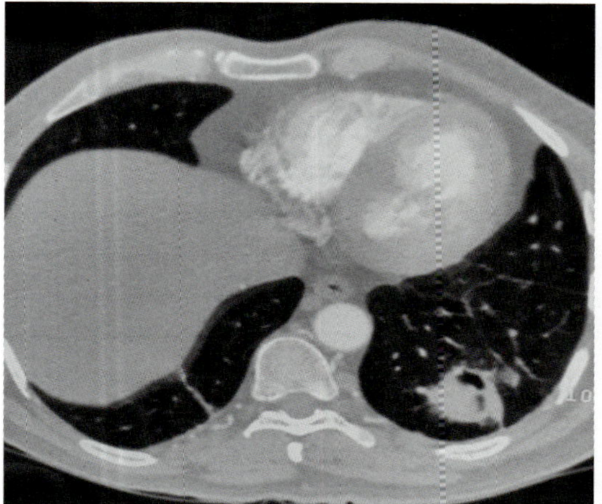

Fig. 50.3: Cavitating lung nodule in GPA

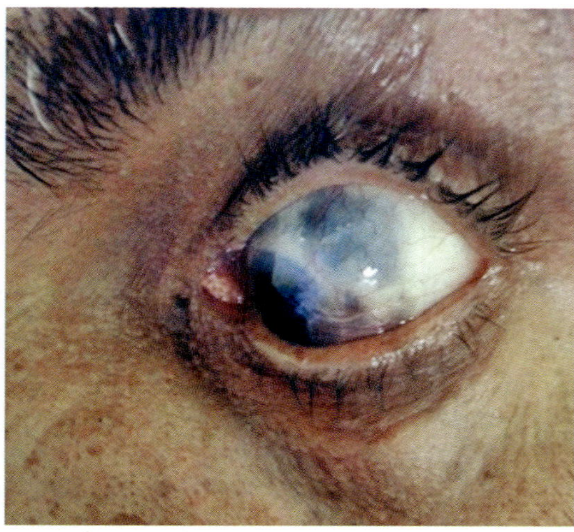

Fig. 50.4: Scleritis in a patient with GPA

rare, seen in less than 5% of patients and may present with GI bleed, mesenteric ischemia or bowel perforation.[12] Cardiac involvement may be seen rarely and may present as myocarditis, pericarditis or conduction abnormalities. Very rarely, genito-urinary, thyroid, breast, liver, parotids and pituitary may be involved.

Eosinophilic Granulomatosis with Polyangiitis

The clinical features of EGPA progress through three different phases, which might not be distinguishable all the time due to overlap in time period. The first phase is the prodrome phase, characterized by atopy, allergic rhinitis and new onset asthma. The second phase is eosinophilic phase, characterized by blood eosinophilia (>10%) and tissue eosinophilia predominantly in lungs. Tissue eosinophilia in lungs is manifested as lung infiltrates. The third phase in the vasculitic phase, in which two or more extrapulmonary organs are involved with small to medium vessel vasculitis, vascular and extravascular eosinophilic granulomatosis. Cutaneous involvement in form of purpura and peripheral nerve involvement are the most common manifestations in vasculitic phase.

Asthma is the most important clinical manifestation, seen in more than 90% of patients of EGPA.[13] It precedes vasculitic phase by 8 to 10 years and is poorly controlled in most patients and requires long-term corticosteroids.[14] Lung infiltrates are present in 38–74% of patients.[13,15] Other less common lung manifestations include lung nodules, pleural effusion, alveolar hemorrhages. Ear, nose and throat involvement is seen in 48–69% of patients and the involvement includes allergic rhinitis, sinusitis and nasal polyposis.[13,15] Cutaneous involvement is seen in 39% of patients and manifests as pupura (Fig. 50.5), subcutaneous nodules, urticarial rash, livedo reticularis and cutaneous necrosis.[13] Peripheral nervous system involvement in the form of peripheral neuropathy and mononeuritis multiplex in 51–64% of patients.[13,15] CNS involvement is rare (5.2%) and includes ischaemic stroke, haemorrhagic stroke and cranial neuropathies.[13]

Renal involvement is less common than in GPA and is seen in about 20% of patients.[13] Proteinuria, elevated serum creatinine, microscopic hematuria indicate renal involvement. Cardiovascular manifestations are present in a quarter of patients, in the form of cardiomyopathy, pericarditis, deep vein thrombosis or pulmonary thromboembolism.[13] Gastrointestinal manifestations, seen in 23–62% of patients, include abdominal pain, vomiting, diarrhoea, eosinophilic colitis, mesenteric vasculitis, bowel infarction and perforation.[13,15] Ocular involvement is present in 6.5% of patients. Constitutional symptoms include fever, myalgia, arthralgia and weight loss. A comparison of various clinical manifestations of GPA and EGPA are shown in Table 50.3.

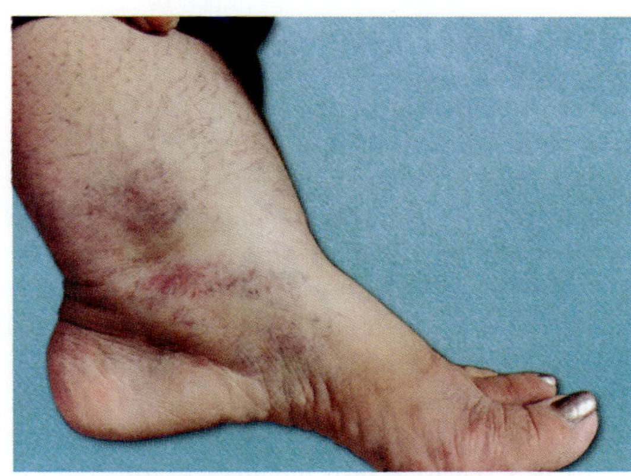

Fig. 50.5: Erythematous non-blanchable skin rash in a patient with EGPA

Investigations

Blood Investigations

Several relevant investigations are required to be done for establishing diagnosis, to determine the

Table 50.3: Comparison of clinical features of GPA and EGPA

Clinical feature	*GPA [8]*	*EGPA [10]*
Kidney	76.5%	21.7%
Lung	66.7%	91.4%
Asthma	NA	91.1%
ENT	84.3%	48%
Eye	39.4%	6.5%
Skin	30.1%	39.7%
Neurologic symptoms	28.3%	55.1%
Cardiovascular	10.4%	27.4%
Gastrointestinal	6.1%	23.2%
ANCA positivity	92.7%	31%
c-ANCA/ PR3 positivity	78.5%	5.2%
p-ANCA/ MPO positivity	10.6%	20.7%
Deaths	13.9%	11.7%
Relapse	46.5%	25.3%

disease severity and to assess for complications arising out of treatment. These patients may have leukocytosis and thrombocytosis. EGPA patients are characterized by peripheral blood hyper eosinophilia. C-reactive protein (CRP), erythrocyte sedimentation rate (ESR) are generally elevated with active vasculitis. Renal function tests, urine routine examination and microscopy, 24-hour urinary protein estimation should be done to look for renal involvement. Hypoalbuminemia is common but deranged liver function tests can rarely be seen. ANCA testing by indirect immunofluorescence (IIF) is a sensitive test and detects the staining pattern, p-ANCA or c-ANCA. Enzyme linked immunosorbent assay (ELISA) for proteinase 3 (PR3) and myeloperoxidase (MPO) identifies the antigen. Routinely, ANCA testing should be done by both IIF and ELISA to label a patient as ANCA negative. Among GPA patients, 10–20% can be ANCA negative while in EGPA 50–60% of patients are ANCA negative.[11–13] Anti-nuclear antibodies (ANA) is usually negative while rheumatoid factor (RF) can be seen in 50% of patients with GPA.

Radiology

Plain X-ray chest should be done to look for infiltrates, cavities, effusion and whereever necessary, high resolution computer tomography (HRCT) chest and/or contrast enhanced CT chest should be done to look for cavities, infiltrates, diffuse alveolar haemorrhage. CT paranasal sinuses may show mucosal thickening, sinusitis and bony destruction. CECT abdomen should be performed to look for evidence of mesenteric ischaemia. CT findings suggesting GI involvement include dilated bowel loops, target sign indicating bowel wall oedema, comb sign indicating increased mesenteric vascularity, pneumatosis intestinalis, gangrene of the bowel and perforation. Neuroimaging includes CECT head and contrast enhanced magnetic resonance imaging (CEMRI) brain to look for CNS involvement. Fibreoptic bronchoscopy and nasal endoscopy may be required in some patients for obtaining tissue samples for histopathology. Echocardiography should be done when cardiac involvement is suspected. Nerve conduction study (NCS) is required when there is peripheral nerve involvement.

Histopathology

The classical triad of GPA consists of granulomas, vasculitis and necrosis. The characteristic renal biopsy finding is the pauci-immune glomerulonephritis. Four categories of glomerulonephritis, focal, crescentic, sclerotic and mixed, have been proposed by the international working group of renal pathologists.[16] Focal class has the best prognosis followed by crescentic group, while sclerotic group has a poor prognosis. Pulmonary tissue can show geographical necrosis with palisading histiocytes, granulomatous vasculitis, necrotising vasculitis or capillaritis.

EGPA is characterized by: (1) Vessel wall infiltrates with eosinophils, (2) Extravascular eosinophilic infiltrates and (3) Extravascular necrotizing eosinophilic granulomas. The tissue for histopathology can be obtained from lungs, skin or nerves. However, the three characteristic lesions are found together in very few patients.

Measurement of Disease Activity

Assessment of disease activity and damage is important for determining the severity of disease and prognosis of the patient. Brimingham Vasculitis Activity Score (BVAS-v3) is a validated score to measure the disease activity.[17] The damage can be measured by Vasculitis Damage Index (VDI).[18] The European Vasculitis Study Group (EUVAS) recommended categorizing patients of ANCA associated vasculitis into five groups based on the severity of disease to assist treatment decisions.[19] The five categories and their definitions are shown in Table 50.4.

Table 50.4: EUVAS categories of AAV

Category	*Definition*
Localized	Upper or lower airway disease without systemic involvement or constitutional symptoms
Early systemic	Systemic disease without organ threatening or life threatening disease
Generalized	Renal or other organ threatening disease, serum creatinine <5.6 mg/dl
Severe	Renal or other organ failure, serum creatinine >5.6 mg/dl
Refractory	Progressive disease unresponsive to glucocorticoids and cyclophosphamide

Treatment

Use of corticosteroids and cyclophosphamide (CYC) has dramatically improved the survival rates of patients with AAV. The treatment can be divided into two phases, remission induction and remission maintenance.

Remission Induction

For remission induction in generalized and severe disease, corticosteroids plus intravenous CYC or Rituximab are the drugs of choice. Pulse CYC is given at a dose of 15 mg/kg every two weeks for 3 doses followed by every three to four weeks for 3–6 months. Corticosteroids are given as IV pulse (methylprednisolone 1 gm) for 3 days followed by oral prednisolone at 1 mg/kg/day. Once remission is achieved, steroids can be tapered to a target dose of 5–7.5 mg/day prednisolone at the end of 6 months. Rituximab, a chimeric anti-CD20 monoclonal antibody, has been shown to be non-inferior to CYC in remission induction. Plasma exchange should be considered in patients with severe renal involvement (serum creatinine >5.8 mg/dl) or alveolar haemorrhage. For patients with concomitant infection and sepsis, intravenous immunoglobulins (IVIg) can be considered. For localized disease, combination of methotrexate (15–25 mg/week) and prednisolone may be used, however, many patients may require CYC during relapses.

Remission Maintenance

Once remission is achieved, immunosuppressants need to be given for a prolonged time to prevent relapses. Drugs commonly used for maintenance are azathioprine (AZA), mycophenolate mofetil (MMF) and methotrexate. Recently, rituximab has been shown to be a very promising drug to be used as maintenance therapy.

Key Points

1. GPA is characterized by granulomatous inflammation, tissue necrosis and vasculitis involving predominantly upper respiratory tract (URT), lungs and kidneys.
2. EGPA is characterized by asthma, peripheral eosinophilia and systemic vasculitis.
3. Assessment of disease activity and damage is important for determining the severity of disease and prognosis of the patient.

REFERENCES

1. Jennette JC, Falk RJ, Bacon PA, Basu N, Cid MC, Ferrario F, et al. 2012 revised International Chapel Hill Consensus Conference Nomenclature of Vasculitides. Arthritis Rheum. 2013; 65:1–11.
2. Ntatsaki E, Watts RA, Scott DG. Epidemiology of ANCA-associated vasculitis. Rheum Dis Clin North Am. 2010; 36(3):447–61.
3. Heckmann M, Holle JU, Arnig L, Knaup S, Hellmich B, Nothnagel M, et al. The Wegener's granulomatosis quantitive trait locus on chromosome 6p21.3 as characterized by tagSNP genotyping. Ann Rheum Dis 2008; 67:972–9.
4. Vaglio A, Martorana D, Maggiore U, Grasselli C, Zanetti A, Pesci A, et al. HLA-DRB4 as a genetic risk factor for Churg-Strauss syndrome. Arthritis Rheum 2007; 56:3159–66.
5. Mahr AD, Neogi T, Merkel PA. Epidemiology of Wegener's granulomatosis: lessons from descriptive studies and analyses of genetic and environmental risk determinants. Clin Exp Rheumatol 2006; 24(2 Suppl 41): S82–91.
6. Wieczorek S, Hellmich B, Arning L, Moosiq F, Lamprecht P, Gross WL et al. Functionally relevant variations of the interleukin-10 gene associated with antineutrophil cytoplasmic antibody-negative Churg-Strauss syndrome, but not with Wegener's granulomatosis. Arthritis Rheum 2008; 58:1839–48.
7. Knight A, Sandin S, Askling J. Risks and relative risks of Wegener's granulomatosis among close relatives of patients with the disease. Arthritis Rheum 2008; 58:302–7.
8. Leavitt RY, Fauci AS, Bloch DA, Michel BA, Hunder GG, Arend WP, et al. The American College of Rheumatology 1990 criteria for the classification of Wegener's granulomatosis. Arthritis Rheum. 1990; 33:1101–7.
9. Masi AT, Hunder GG, Lie JT, Michel BA, Bloch DA, Arend WP, et al. The American College of Rheumatology 1990 criteria for the classification of Churg-Strauss syndrome (allergic granulomatosis and angiitis). Arthritis Rheum. 1990; 33:1094–100.
10. Watts R, Lane S, Hanslik T, Hauser T, Hellmich B, Koldingsnes W, et al. Development and validation of a consensus methodology for the classification of the ANCA-associated vasculitides and polyarteritis nodosa for epidemiological studies. Ann Rheum Dis. 2007; 66:222–7.
11. Sharma A, Naidu GS, Rathi M, Verma R, Modi M, Pinto B, et al. Clinical features and long term outcome of 105 patients of Granulomatosis with Polyangiitis: A single centre experience from north India. Int J Rheum Dis. 2017 Mar 24.

12. Reinhold-Keller E, Beuge N, Latza U, De Groot K, Rudert H, Nolle B, et al. An interdisciplinary approach to the care of patients with Wegner's granulomatosis: Long-term outcome in 155 patients. Arthritis Rheum 2000; 43:1021–32.
13. Comarmond C, Pagnoux C, Khellaf M, Cordier JF, Hamidou M, Viallard JF, et al. Eosinophilic granulomatosis with polyangiitis (Churg-Strauss): clinical characteristics and long-term follow up of the 383 patients enrolled in the French Vasculitis Study Group cohort. Arthritis Rheum 2013; 65:270–281.
14. Guillevin L, Cohen P, Gayraud M, Lhote F, Jarrousse B, Casassus P. Churg-Strauss syndrome. Clinical study and long-term follow-up of 96 patients. Medicine (Baltimore) 1999; 78:26–37.
15. Lanham JG, Elkon KB, Pusey CD, Hughes GR. Systemic vasculitis with asthma and eosinophilia: a clinical approach to the Churg-Strauss syndrome. Medicine (Baltimore) 1984; 63:65–81.
16. Berden AE, Ferrario F, Hagen EC, Jayne DR, Jennette JC, Bajema IM, et al. Histopathologic Classification of ANCA-Associated Glomerulonephritis. J Am Soc Nephrol. 2010; 21:1628–36.
17. Mukhtyar C, Lee R, Brown D, Carruthers D, Dasgupta B, Dubey S, et al. Modification and validation of the Birmingham Vasculitis Activity Score (version 3). Ann Rheum Dis. 2009; 68: 1827–32.
18. Exley AR, Bacon PA, Luqmani RA, Kitas GD, Gordon C, Savage CO, et al. Development and initial validation of the Vasculitis Damage Index for the standardized clinical assessment of damage in the systemic vasculitides. Arthritis Rheum. 1997; 40: 371–80.
19. Mukhtyar C, Guillevin L, Cid MC, Dasgupta B, de Groot K, Gross W, et al. EULAR Recommendations for the Management of Primary Small and Medium Vessel Vasculitis. Ann Rheum Dis. 2009; 68:310–7.

Chapter

51

Henoch-Schönlein Purpura (Immunoglobulin A Vasculitis)

Aarthi Priya T, Ravichandran R

INTRODUCTION

Henoch-Schönlein purpura (HSP) is one of the common type of immune complex mediated vasculitis, primarily involving small blood vessels characterized by IgA deposition. Though it can affect people of any age, most cases occur in children of age 2 to 11 years. In adults, the disease is more severe than in children.

In the 2012, Chapel Hill consensus conference (CHCC) for nomenclature of vasculitis the eponym "Henoch-Schönlein purpura" has been replaced by the name immunoglobulin A vasculitis (IgAV), in recognition of the pathogenic role and pathognomonic deposition of IgA in the skin, gastrointestinal tract, kidney and other organs that is a hallmark feature of this disease.[1] CHCC 2012 defines IgA vasculitis (HSP) as vasculitis with IgA1-dominant immune deposits, affecting small vessels (predominantly capillaries, venules, or arterioles).[1] HSP (IgAV) is a systemic illness with characteristic involvement of skin, gastrointestinal tract, and joints. However, IgAV can occur isolated to skin, analogous to IgA nephropathy without systemic involvement.

Classification Criteria of HSP (IgAV)[2]

The classical features of HSP as originally described by Schönlein and Henoch consists of purpuric rash, arthritis, abnormal urinary sediments, proteinuria, and abdominal pain with bloody diarrhoea.[2] The **American College of Rheumatology (ACR) 1990** criteria for classification of HSP were very simple and easy to apply.[3] There were four criteria *viz* (1) non-thrombocytopenic palpable purpura, (2) disease onset <20 years of age, (3) acute abdominal pain, and (4) biopsy evidence of vasculitis (granulocytes in the walls of arterioles or venules).[3] A patient can be classified as HSP if two of these four criteria are present. It was later realized that the ACR 1990 criteria were not specific for HSP especially since granulocytes in walls of blood vessels could be seen in other vasculitic disorders too such as hypersensitivity vasculitis.

Michel et al proposed another set of classification criteria consisting of: (1) non-thrombocytopenic palpable purpura, (2) disease onset at <20 years of age, (3) bowel angina,(4) gastrointestinal bleeding, (5) haematuria, and (6) no history of medication intake.[4] The presence of three out of these 6 criteria had 87% sensitivity of diagnosis of HSP. These criteria included "no medication intake" in order to exclude hypersensitivity vasculitis.

Helander et al in 1995 further refined the criteria for diagnosis/classification of HSP.[5] These workers emphasized inclusion of presence of IgA immune complex deposits in walls of small vessels as an identifying criterion of HSP to differentiate from other disorders such as urticarial vasculitis or microscopic polyangiitis. The Helander criteria includes—(1) age <20 years or younger, (2) cutaneous IgA deposits, (3) gastrointestinal involvement, (4) upper respiratory prodrome, and (5) mesangioproliferative glomerulonephritis with or without IgA deposits. The presence of 3 out of 5 criteria in a patient with palpable purpura and leukocytoclastic vasculitis (LCV) has a sensitivity of >90% to diagnose HSP.

The latest criteria for diagnosis/classification of HSP are those developed by the EULAR (European League Against Rheumatism), PRINTO (Paediatric Rheumatology International Trials Organization) and PRES (Paediatric Rheumatology European Scociety) and published in 2010.[6]

The ***mandatory entry criteria*** are the ***presence of palpable purpura*** (not thrombocytopenic) or petechiae with lower limb predominance +≥ 1 of the following:

1. Abdominal pain
2. Histopathology: typical LCV with predominant IgA deposits or proliferative glomerulonephritis with predominant IgA deposits
3. Arthritis or arthralgias
4. Renal involvement in the form of proteinuria: >300 mg/24 h or >30 mmol/mg of urine albumin to creatinine ratio on a spot morning sample; and/or haematuria, red blood cell casts: >5 red cells per high power field or ≥2+ on dipstick or red blood cell casts in the urinary sediment).

The EULAR, PRINTO, PRES criteria were 100% sensitive and >87% specific to classify HSP. Compared to the ACR 1990 and Helander criteria, the EULAR/PRINTO/PRES criteria emphasized the importance of LCV with ***predominant IgA deposits***, renal biopsy of glomerulonephritis with ***predominant IgA deposition***, and they also included ***arthritis/ arthralgias.***

Aetiology and Risk Factors

The exact cause of HSP is not known Two third of HSP occur after an upper respiratory tract infection.[7,8] The initial inflammation may be result of immune system responding inappropriately to certain triggers. Infectious triggers include streptococcal infection, varicella, measles, hepatitis A and B. Other triggers include certain foods, drugs, chemicals, insect bites and exposure to cold weather.

HSP occurs predominantly in children and young adults. It is more common in males than in females. It also commonly occurs in winter, autumn, and is rare in summer. HLA B35 and interleukin 1(IL-1) beta gene polymorphisms are associated with severe renal manifestations.[7]

Epidemiology: In North America the annual incidence is estimated at 13.5 cases per 100,000 population whereas in India, the incidence is 20.4 per 100,000 population with highest number of cases in 4–6 years age group.[7] In Taiwan the annual incidence is estimated at 12.9 (11.8–13.4) per 100,000 children.[8]

Clinical Features

The classical triad of purpura, colicky abdominal pain and arthritis is seen in most children, while renal involvement is present in only 30–50%. Other systemic manifestations are rare.

Cutaneous Manifestations

Palpable non-thrombocytopenic purpura occurs in 100% of patients. Common sites are dependent areas of the body such as lower extremities and buttocks. Oedema of feet, hands, scalp, and ears are seen in 20–46%.[7] Other skin lesions include maculopapular rash, petechiae, ecchymoses. Rarely bullous lesions are seen.

Joints

Arthralgia or arthritis are seen in 60–85% of the patients. The knees and ankle joints are commonly involved. The arthritis is transient and self- limiting.

Gastrointestinal System

A colicky type abdominal pain, nausea, vomiting with or without gastrointestinal bleeding is seen in 85% of the patients.[7, 8] Abdominal pain is often periumbilical, or epigastricin location and increases after meals. Endoscopic studies have revealed erosive, haemorrhagic duodenitis. Major haemorrhage occurs in less than 5% and intussusceptions in 2% of children with HSP.

Renal

Renal involvement can occur within 3 months of onset of rash in 10–15% of the patients. Older children above 9 years of age are more prone.[7] Nephritis can range from transient isolated microscopic haematuria to rapidly progressive glomerulonephritis.

IgA Nephropathy and HSP

Thirty percentadult patientsof IgA nephropathy have rash and joint symptoms like HSP. The renal biopsy findings are identical in both conditions. However, HSP is a systemic syndrome with good prognosis unlike IgA nephropathy which is localized to kidney with guarded prognosis.

Genitourinary Manifestations

About 2–35% of children develop acute scrotal swelling due to inflammation and haemorrhage of the scrotal vessels.

Other Systems

CNS involvement is very rare and subtle encephalopathy has been reported. Children may also have impaired lung diffusion capacity. Rarely pulmonary haemorrhage and interstitial pneumonia are seen. Rarely arrhythmias have been reported in adults. The course of the illness in majority of children with early onset disease is monocyclic.[9] But children with HSP will need long-term follow-up since chronicity occurs in about 20% cases.[9]

Adult-onset HSP/IgAV

Onset in adulthood of this form of small of small vessel vasculitis heralds a poorer prognosis than in children. Adults tend to have a greater proportion of renal involvement (up to 85%), which may eventually progress to chronic kidney disease in 30%. Also, the presence of gastrointestinal involvement with persistence of cutaneous rash in an adult with HSP portends the likelihood of eventual renal involvement.

Investigations

HSP is primarily diagnosed clinically by signs and symptoms. Routine laboratory studies are normal unless there is organ involvement. Skin biopsy shows leukocytoclastic vasculitis. Immunofluorescence studies show IgA and C3 deposit on the small vessels of the skin and glomeruli. Serum IgA level is elevated in 50% patients. Urinary podocytes is a new marker for glomerular disease progression and response to treatment.[10]

Management

Treatment is mainly supportive with adequate hydration and monitoring. NSAIDs/analgesics used for joint pain and inflammation. Corticosteroids are used for treatment of painful cutaneous oedema.

Treatment of Abdominal Pain

There is no conclusive evidence that corticosteroids affect the course or outcome of HSP abdominal pain or other gastrointestinal manifestations.

Renal Disease

In majority of children, renal involvement is transient and need not be treated.[11] Risk factors of severe nephritis are adults with renal insufficiency, increasing proteinuria of nephritic range, hypertension, severe histologic findings on biopsy and female gender.[11] Predictors of severe renal disease are increased urinary excretion of 2 tubular proteins: N acetyl βD glucosamine (NAG), α1 microglobulin (β1 MG).[11]

Role of Corticosteroids in HSP Nephritis

Although little is known about disease-modifying treatment of HSP, corticosteroids have been tried owing to their immunosuppressive properties. Four randomized, double-blind, placebo-controlled trials have assessed the effectiveness of corticosteroids in preventing HSP nephritis. Prednisone (1 mg/kg daily for 2 weeks followed by a 2-week weaning period, or 2 mg/kg daily for 1 week plus a 1-week weaning period) resulted in no reduction in the severity of haematuria, proteinuria, or urine protein creatinine ratio.[12–14] A 2009 cochrane meta-analysis that evaluated 3 studies ($n = 569$) determined that prednisone did not prevent the development of kidney disease at 1 month, 3 months, 6 months, or 1 year after the onset of HSP.[15] In a 2012 study, Jauhola and colleagues investigated the long-term renal outcomes of steroids through re-assessing 138 of 171 HSP patients a mean of 7.7 years after their treatment in one of the above-mentioned randomized controlled trials (RCTs).[16] They reported no long-term renal protective effects of prednisone (1 mg/kg daily for 2 weeks plus a 2-week weaning period) versus placebo, determined by urinalysis and blood pressure.[16]

Pediatric therapies for established severe HSP nephritis include corticosteroids, immunosuppressants (azathioprine, mycophenolate mofetil and cyclophosphamide), angiotensin-converting enzyme inhibitors, plasma exchange, and tonsillectomy.[16] In a systematic review, Zaffanello and Fanos evaluated 34 English publications that assessed the effectiveness of these

Chapter

52

Cutaneous Vasculitis and Panniculitis

Lalit Duggal, Nagma Bansal

CUTANEOUS VASCULITIS

Cutaneous vasculitis manifests with dermatologic features such as palpable purpura, urticarial lesions, nodules, ulcers or livedo reticular is due to inflammation of small to medium sized vessels in the skin.[1] It is associated with increased risk of vasculitis of internal organs.

Pathologic Features

Light microscopy: The pathologic findings lie within dermis and subcutaneous tissue. The palpable purpura and urticarial lesions may reveal involvement of post capillary venules (size <50 micron) of the superficial plexus. There is necrosis of endothelial cells, fibrin deposition within vessel wall, extravasation of erythrocytes and neutrophilic infiltration along with leucocytoclasis.

Direct immunoflouroscence (DIF): DIF studies differentiate immune complex mediated disorders from pauci-immune vasculitis. Particularly, these are useful for hypocomplementemic, IgA and lupus vasculitis.

Clinical Features

Palpable purpura: These may be papular, nodular, bullous or ulcerative.[2] These lesions occur more frequently on dependent body areas especially on legs (Fig. 52.1). Occasionally, they may become generalized.

Urticarial lesions: These are usually generalized and differ from common transient urticaria (Fig. 52.2); are long lived, lasting for more than 48 hours, usually resolve with some residual alteration, pain and burning are more prominent features than pruritus and may have associated multisystem involvement.

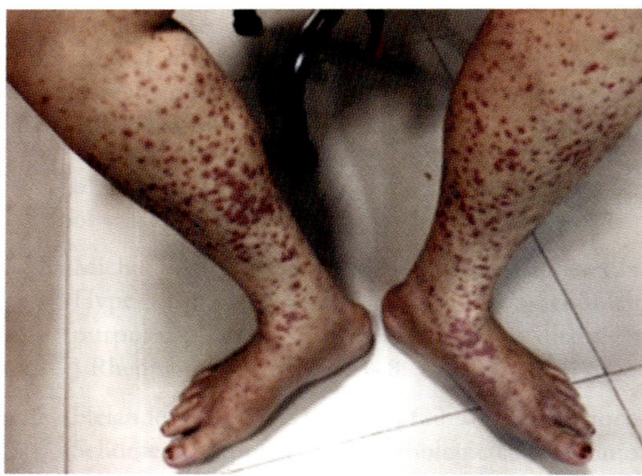

Fig. 52.1: Classical purpuric lesions in dependent parts of a patient with Henoch-Schönlein purpura (IgA vasculitis)

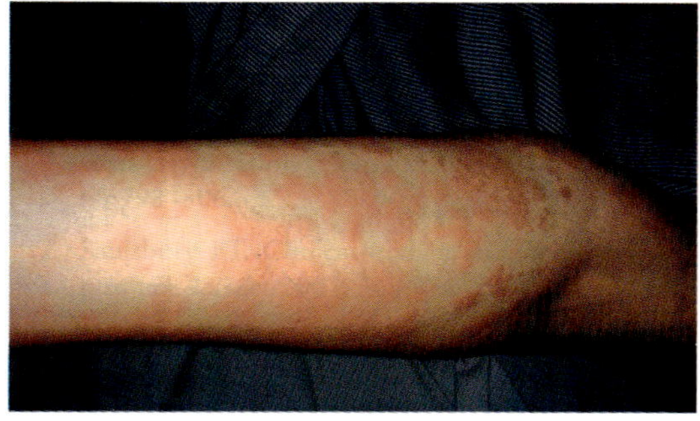

Fig. 52.2: Urticarial lesion in a patient with hypersensitivity vasculitis

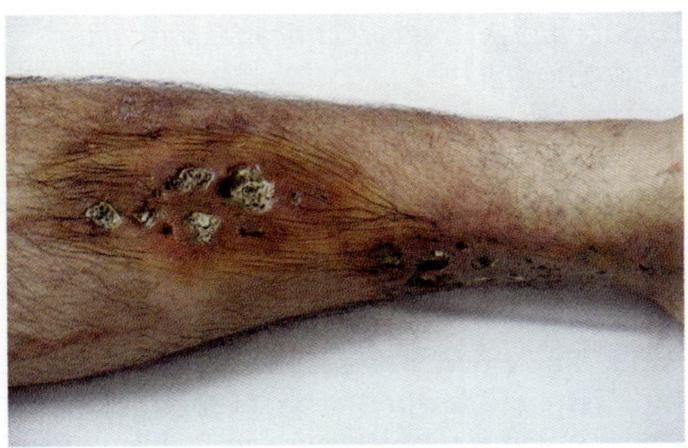

Fig. 52.3: Ulcerative lesions in a patient with cutaneous polyarteritis nodosa

Others are: Vesicles, pustules, superficial ulcerations (Fig. 52.3), macules, splinter haemorrhages and livedo reticularis.

Systemic manifestations: 50% patients may have systemic features.[3] The most common involved organs are gut, kidneys, lungs and musculoskeletal system. Cutaneous vasculitis may be associated with fever, arthralgias, gastrointestinal pain, lymphadenopathy or abnormal urine sediment.

Laboratory Work-up (Fig. 52.4)

All patients with cutaneous vasculitis should undergo a set of investigations which includes complete haemogram, urine examination, acute phase reactants (ESR/CRP), serum creatinine, liver function tests, HCV, HBV, HIV and chest X-ray. Immunological profile should include ANA, RF, anti-Ro and ANCA. Serum cryoglobulins, complement assays, serum protein electrophoresis and blood cultures may be done in selected patients.

Clinical Syndromes (Box 52.1)

Hypersensitivity Vasculitis

Hypersensitivity vasculitis, also known as cutaneous leucocytoclastic angiitis is an immune complex mediated small vessel vasculitis usually in response to some drug exposure or infection.[4] Clinical manifestations are confined to skin with predominant finding of palpable purpura in dependent parts. Characteristic pathologic findings are leucocytoclastic (predominantly eosinophilic) vasculitis with immune complex deposition.[4] The removal of inciting agent helps in resolution, however in 50% cases no inciting agent can be identified. Colchicine, dapsone, low dose steroids and other immunosuppressive agents may be required in resistant cases. Serum sickness is hypersensitivity vasculitis occurring after 1 to 2 weeks of a drug and is associated with systemic symptoms such as fever and arthralgias/arthritis. DRESS (Drug reaction

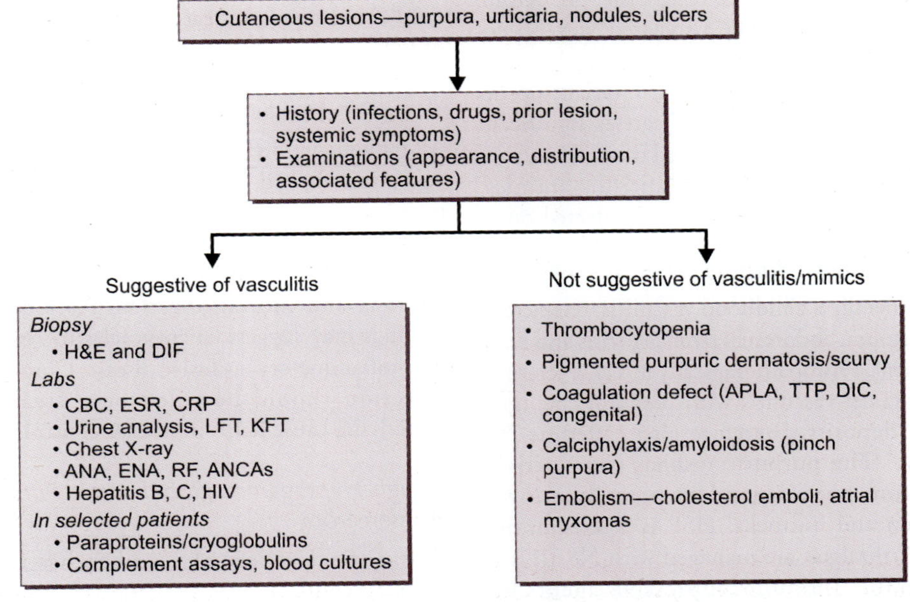

Fig. 52.4: Laboratory work up for cutaneous vasculitis

Box 52.1 Clinical syndromes associated with cutaneous vasculitis

Immune complex mediated

- Hypersensitivity vasculitis
- Henoch-Schnölein purpura (HSP)
- Mixed cryoglobulinaemia
- Urticarial vasculitis (hypocomplementemic, normocomplementemic)
- Erythema elevatum diutinum (EDD)
- Connective tissue diseases associated vasculitis (SLE, rheumatoid arthritis, Sjögren's syndrome, idiopathic inflammatory myositis)

Pauci-immune

- Granulomatosis with polyangiitis (GPA)
- Eosinophilic granulomatosis with polyangiitis (EGPA)
- Microscopic polyangiitis (MPA)

Miscellaneous

- Behçet's disease
- Malignancy associated (Hairy cell leukaemia, lymphoma, etc.)
- Infections (hepatitis C virus, hepatitis B virus, human immunodeficiency virus, subacute bacterial endocarditis, beta-hemolytic streptococci)
- Paraproteinaemia (macroglobulinaemia, cryoglobulinaemia, hyperglobulinaemia)

with eosinophilia and systemic symptoms) is a separate entity with severe drug rash (phenytoin, allopurinol, sulfasalazine) associated with leucocytosis, transaminitis and organ involvement especially kidneys and lungs. This variant occurs 2–4 weeks after exposure and carries significant morbidity/mortality in untreated cases. Treatment requires corticosteroids and anti-histaminics. N-acetyl cysteine may be of help in selected cases.

Henoch-Schönlein Purpura (HSP)/IgA Vasculitis

HSP is generally a childhood vasculitis, characterized by purpura, abdominal pain, arthritis and renal involvement. Histopathologically it is characterized by small vessel vasculitis with predominant IgA1 immune deposits, therefore also called as IgA vasculitis.[5] The purpuric rashhas characteristic distribution over dependent areas like lower extremities and buttocks. HSP is a self-limiting disease. Arthralgias are managed with NSAIDs or paracetamol. Immunosuppressive agents are required for gastrointestinal and renal involvement.

Hypocomplementemic Urticarial Vasculitis Syndrome (HUVS)

The characteristic urticaria like lesions last more than 48 hours, do not blanch when pressed and usually leave post-inflammatory hyper-pigmentation. Burning and pain is more prominent feature than pruritus.[6] The distribution tends to be centripetal involving trunk and proximal extremities.This variety of vasculitis is also associated with systemic features such as fever, malaise, synovitis and extracutaneous organ involvement especially lungs an,.d kidneys. Chronic obstructive pulmonary disease (COPD) and angioedema and laryngealoedema, are severe manifestations, especially in patients with low complement levels.

The major differential is with systemic lupus erythematosus as both may have overlapping features including hypocomplementemia, autoantibodies and interface dermatitis. Urticarial vasculitis with normal complement levels tends to have less systemic involvement. The treatment of HUVS depends upon the manifestations. Skin involvement responds to low-dose corticosteroids, dapsone, hydroxychloroquine and other immunomodulatory drugs. Severe systemic involvement especially nephritis will require high-dose corticosteroids and biologic agents such as anti-tumour necrosis factor alpha agents. In resistant cases, intravenous immunoglobulin may be helpful. COPD is difficult to treat and will influence quality of life.

Vasculitis Associated with Paraproteinaemia

All patients with chronic and recurrent cutaneous vasculitis should be evaluated for presence of paraproteins, including tests for cryoglobulins and immunofixation electrophoresis. The skin lesions may vary from palpable purpura to extensive necrosis and ulcerations. The histopathologic findings may suggest leucocytoclastic vasculitis or non-inflammatory occlusive disease. Chronic HCV infection should be ruled out. Treating the underlying cause helps in disease control.

Benign Hypergammaglobulinemic Purpura of Waldenström

This distinct variety of vasculitis has remitting relapsing course and is associated with raised serum gammaglobulin levels. The purpuric lesions may be on the feet. These tend to increase on prolonged standing. The histopathology of lesions may indicate lymphocytic vasculitis rather than leukocytoclastic, with less vessel wall disruption. The systemic associations are with Sjögren's syndrome and lupus erythematosus with demonstration of anti-Ro antibody. These lesions usually respond to hydroxychloroquine (HCQ) and dapsone. Resistant cases may require rituximab.

Erythema Elevatum Diutinum (EED)

This rare variety of cutaneous vasculitis presents as red yellow papules, plaques or nodules over extensor surfaces such as knees, elbows and hands. The untreated lesions may persist for years and become doughy or hard with time. 20–40% patients may have arthralgias. EED has been found to be associated with various connective tissue disorders, paraproteinaemia especially IgA variant and HIV infection. The biopsy of long-standing lesions reveals concentric perivascular fibrosis.

Paraneoplastic Vasculitis

The paraneoplastic vasculitis is characterized by LCV in patients with lymphoproliferative disorders (Hairy cell leukaemia, lymphoma, etc).[7] The association with solid organ malignancies is rare. However, there are case reports of association with carcinoma of oesophagus. The exact mechanism of vasculitis in malignancy remains unknown. It may be associated with chemotherapy or it may be purely coincidental. Other type of vasculitis such as PAN has been observed with some solid tumours along with LCV. Treatment includes prednisolone and other immunosuppressive agents.

Connective Tissue Disease Associated Vasculitis

Cutaneous vasculitis may be a feature of many connective tissue disorders especially SLE, RA, Sjögren's and MCTD and often signifies active systemic disease. The most common variety is lymphocytic vasculitis, presenting as palpable purpura, urticarial or ulcerative lesions. In SLE, histopathology reveals lymphocytic predominance rather than leucocytoclastis. DIF studies show immune complex deposition (IgG and C3) in and around vessels. The presence of vasculitis in SLE is associated with hypocomplementemia and high titres of ANAs.

Rheumatoid Vasculitis (RV)

Rheumatoid vasculitis classically occurs in patient with seropositive, nodular, long-standing erosive disease, but usually without acute synovitis at time of onset of vasculitis. It affects small to medium sized vessels. It may be indistinguishable from PAN but micro-aneurysms are rare. Lesions may vary from palpable purpura to extensive ulcerations and gangrene. Another manifestation of RV is mononeuritis multiplex. Histopathology of lesions varies from small to medium vessel vasculitis with deposit of immune complexes (IgM, C3, cryoglobins, RF) in vessels. RV is a potentially devastating condition, requiring high-dose glucocorticoids and cyclophosphamide. In resistant cases, TNF inhibitors and rituximab may be of help.

Primary Vasculitic Syndromes with Cutaneous Vasculitis

Pauci-immune vasculitis may be associated with cutaneous lesions especially in EGPA where the figure can go up to 81%. Cutaneous manifestations are palpable purpura, subcutaneous nodules, ulcerations and digital infarcts. Palisaded and granulomatous neutrophilic dermatosis, earlier known as Churg-Strauss granulomas are erythematous to violaceous papules located on extensor surfaces in patients with EGPA. ANCA, especially pANCA, is present in most of the cases with cutaneous vasculitis. The biopsy will reveal leucocytoclastic vasculitis in almost all cases, with rare finding of granulomas in GPA and eosinophils in EGPA. Kawasaki disease (medium vessel vasculitis) in children is associated with desquamation of palms and soles, exanthema in proximal parts and strawberry tongue; histopathology of the lesions fails to show any specific changes. Large vessel vasculitis, especially Takayasu's arteritis, may be associated with cutaneous lesions like tender subcutaneous nodules, livedo reticularis and pyoderma gangrenosum like lesions on legs. The biopsy is suggestive of mixed picture of necrotizing vasculitis with panniculitis. Patients with GCA may rarely have cutaneous manifestation in temporal artery territory. Polyarteritis nodosa PAN and cutaneous PAN are now considered to be two different entities. PAN is a systemic condition and may have skin manifestations in the form of subcutaneous nodules, livedo reticularis and ulcerations.[8] On the contrary, in cutaneous PAN,

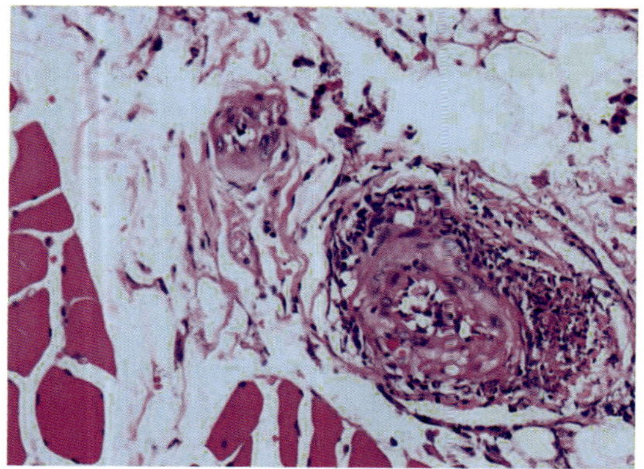

Fig. 52.5: Histopathology of skin lesion in patients with cutaneous PAN-fibrinoid necrosis of a medium artery with extravasation of RBCs

pathology is limited to skin, subcutaneous tissue and underlying nerves and bones. Cutaneous PAN is always associated with painful cutaneous manifestations especially subcutaneous nodules, livedo and ulceration which resolve with hyper-pigmentation. Biopsy of the lesions reveals medium vessel vasculitis with fibrinoid necrosis with or without panniculitis (Fig. 52.5).

Management of cutaneous vasculitis is shown in Fig. 52.6.

PANNICULITIS

These are a group of disorders characterized by inflammation of subcutaneous fat. It presents as subcutaneous nodules and is frequently a sign of systemic disease.[9] Panniculitis may be primary and without an identifiable aetiology (e.g. erythema nodosum) or it may be secondary to some aetiology like infection, vasculitis or connective tissue disease.

Broadly, panniculitis is of four histopathological types:

1. Mostly septal panniculitis without vasculitis
   - Erythema nodosum (EN)
   - Vilanova disease
   - Subacute nodular migratory panniculitis
   - Necrobiosis lipoidica
2. Mostly septal panniculitis with vasculitis
   - Cutaneous PAN
   - Superficial migratory thrombophlebitis
3. Mostly lobular panniculitis without vasculitis
   - Weber-Christian disease (Idiopathic neutrophilic panniculitis)
   - Subcutaneous fat necrosis of newborn
   - Pancreatic panniculitis
   - Post-steroid panniculitis

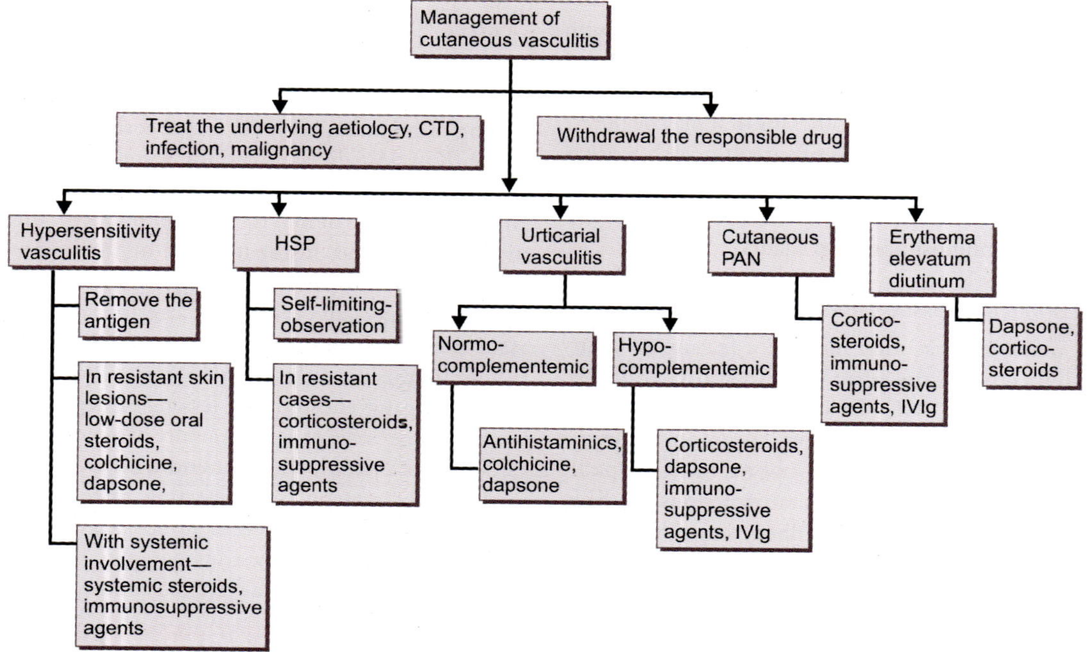

Fig. 52.6 Management of cutaneous vasculitis

- Calcific panniculitis
- α1-anti-trypsin deficiency
- Physical or factitial panniculitis
- Histiocytic cytophagic panniculitis
- Lipodystrophy syndrome
- Connective tissue panniculitis
- Sclerosing panniculitis—lipodermatosclerosis

4. Mostly lobular panniculitis with vasculitis
   - Erythema induratum/nodular vasculitis
   - Small vessel vasculitis known as leukocytoclastic vasculitis
5. Mixed panniculitis
   - Lupus profundus–lupus erythematosus panniculitis (LEP)
   - Erythema nodosum like lesions of Behçet syndrome

Site: The panniculitis usually occurs below the waist area except for SLE, leprosy, tuberculosis, Weber-Christian disease which can manifest above the waist. Shin is the most common location for erythema nodosum whereas calf is predominant site of erythema induratum.

Natural course: The nodules in EN are initially red but on resolution turn blue and completely heal without scarring whereas other lesions breakdown and ulcerate or heal with a scar.

Erythema nodosum: This is the most common form of panniculitis and is the prototype of septal variety without vasculitis.[10] The eruptions are usually acute, self limited and typically presenting as one or more tender erythematosus nodules on anterior aspect of legs (Fig. 52.7). The lesions may be associated with systemic features such as fever, malaise, arthralgia and leucocytosis. These are better palpated than visualized and heal without scar formation. Diagnosis is clinical. The biopsy is suggestive of septal panniculitis without vasculitis. The presence of granulomas suggests sarcoidosis as the aetiology. 50% are idiopathic while the rest are secondary to other aetiologies. (Box 52.2) Idiopathic acute EN is often self-limited. Aspirin and NSAIDs help in alleviating the symptoms. For recurrent and chronic EN, colchicine, oral corticosteroids, HCQ, cyclosporine may be of help.

Subacute nodular migratory panniculitis (Vilanova disease): It is a chronic and recurrent form of EN

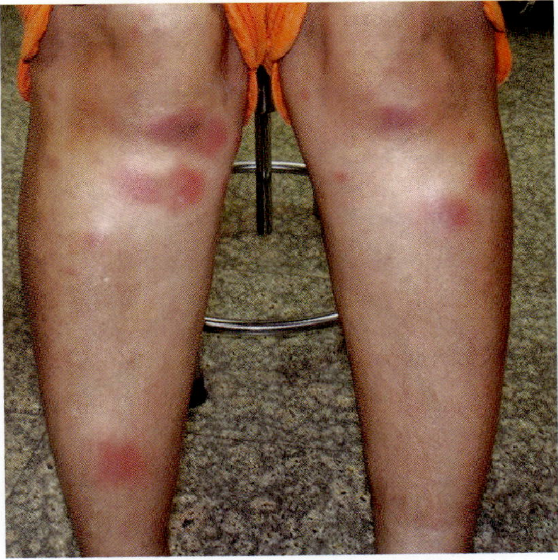

Fig. 52.7: Classical EN lesions

Box 52.2 Secondary aetiologies of erythema nodosum

1. **Infections**
   - Streptococcal infections
   - Upper respiratory tract viral infections
   - Bacterial gastroenteritis (Salmonella, Yersinia)
   - Tuberculosis
   - Coccidiomycosis
   - Histoplasmosis
   - Brucellosis
   - *Chlamydophilia pneumonia*
   - *Chlamydia trachomatis*
   - *Mycoplasma pneumonia*
   - Hepatitis B virus
   - Leprosy
2. **Drugs**
   - Sulfonamides
   - Penicillin
   - Bromides
   - Iodides
   - Oral contraceptive pills
3. **Systemic diseases**
   - Sarcoidosis
   - Inflammatory bowel disease
   - Collagen vascular disorder—SLE, scleroderma, Dermatomyositis
   - Sweet's syndrome
   - Behçet's disease
   - Malignancy-rare
4. **Pregnancy**

found more commonly in middle aged females. In contrast to acute EN, these may be painless. The biopsy of lesion shows granulomatous or fibrotic changes in addition to septal inflammation. The treatment includes colchicine, oral corticosteroids, hydroxychloroquine, cyclosporine and other immunosuppressive agents.

Superficial migratory thrombophlebitis: This is manifested as painful cord like nodular swelling linearly along the superficial veins of the lower limbs due to thrombosis. These may be idiopathic or secondary to other aetiology such as Behçet's syndrome.

Idiopathic lobular panniculitis or Weber-Christian disease (WCD): WCD is an old term. These lesions present as painful cutaneous nodules of unknown aetiology. The biopsy is suggestive of neutrophilic lobular panniculitis. Most of these lesions are now found to have some aetiology or are facticial.

Pancreatic panniculitis: It is a type of lobular panniculitis in patients with pancreatic diseases (pancreatitis, pancreatic carcinoma, post-traumatic and other lesions) associated with polyarthritis and fat necrosis.[11] The release of pancreatic enzymes may play a role in necrosis. The histopathology is suggestive of basophilic alteration of lipocytes.

Post-steroid panniculitis: This variant of lobular panniculitis has been found in children on long-term steroid therapy, after sudden withdrawal of the steroid.[12] These lesions are responsive to re-administration of steroids.

Calcifying panniculitis: It presents as tender erythematous painful nodules in patients with end stage renal diseases due to abnormal calcium and phosphorus balance, leading to calcium deposits in arterial wall (calcific uremic arteriolopathy).[13] Treatment is difficult, renal transplantation may help.

Alpha-1 anti-trypsin deficiency: Lobular panniculitis has been found in patients with alpha-1 anti-trypsin deficiency along with emphysema and cirrhosis.[14] Clinical manifestation consists of painful nodular lesions on trunk and proximal extremities which may lead to ulceration. Alpha-1 anti-trypsin administration may help in treatment along with symptomatic medications.

Physical or factitial panniculitis: It is defined as lobular inflammation in response to tissue injury produced by external trauma or self-injection of various drugs or substances. Besides lobular inflammation, the histopathology may suggest presence of hematoma with abundant neutrophilic infiltrate, with characteristic "swiss cheese pattern".

Histiocytic cytophagic panniculitis: This is a benign chronic condition associated with systemic features like fever, serositis and lymphadenopathy. Histopathologically, there is lobular inflammation of fat along with histiocytic infiltration.[15] This condition may be complicated by haemophagocytic syndrome and responds to oral steroids and cyclosporine. The malignant variant of similar condition is known as subcutaneous panniculitis like T cell lymphoma with little response to oral steroids. The diagnosis is by immunophenotyping for T cell.

Lupus profundus: These lesions are manifested as painful red blue subcutaneous nodules in 1–3% of patients with SLE.[16] These are among the few panniculitis which occur above the waist region typically on face and upper arms apart from thighs and buttocks. The biopsy is suggestive of mixed variety of panniculitis.

Sclerosing panniculitis (lipodermatosclerosis): This condition is a manifestation of venous insufficiency presenting as well-circumscribed, indurated plaques of the lower limb above the ankle. It most commonly occurs in women with history of previous thrombophlebitis and thrombophilia. Biopsy of the lesion is suggestive of lobular panniculitis with sclerosis.[17] Treating the underlying cause and intralesional steroids may be helpful.

EN-like lesions of Behçet's disease: These look like idiopathic EN but differ in having vasculitic component and these heal with residual scarring (Fig. 52.8).

Erythema induratum: It is characterized by granulomatous lobular panniculitis with vasculitis. These are of two types—erythema induratum of Bazin when it is of tubercular origin and nodular vasculitis or erythema induratum of Whitfield when aetiology is multifactorial.[18] These most commonly affect women and typically occur on calf as tender erythematous nodules which may ulcerate.

Erythema nodosum leprosum (ENL): These are type II lepra reactions occurring in borderline leprosy or lepromatous leprosy.[19] These are characterized

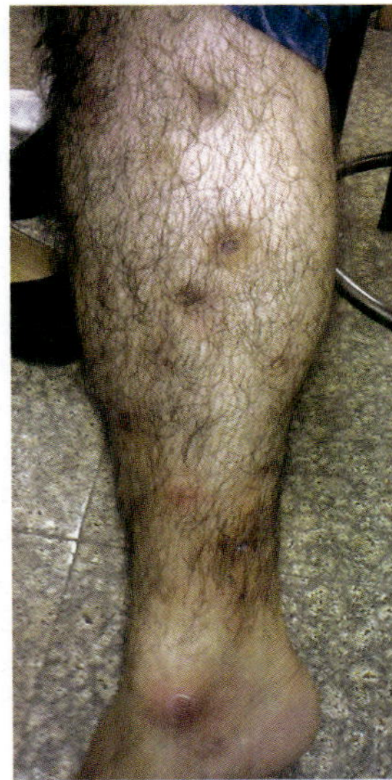

Fig. 52.8: Ulcerative nodular lesions of Behçet's syndrome

by the presence of multiple inflammatory cutaneous nodules (Fig. 52.9) along with systemic symptoms such as fever, malaise, arthritis, iritis, neuritis and lymphadenitis. The biopsy shows mixed panniculitis with vasculitis. The extreme end of type II lepra reaction is known as Lucio's phenomenon with severe ulceration and underlying tissue destruction especially of lower limbs and feet.

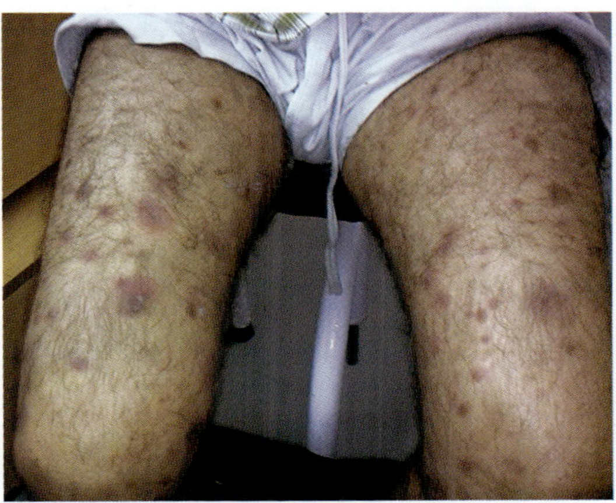

Fig. 52.9: Lesions of erythema nodosum leprosum

Conclusion

Cutaneous vasculitis and panniculitis are two entities that may not necessarily be localised disease.[5] An insight into systemic association of these disorders will perhaps help in achieving correct diagnosis to alleviating the suffering of such patients. Removal of culprit antigen may resolve the condition while mild to aggressive immune suppression may be required for others. The overall outlook depends on early identification of the disorders, and on systemic involvement and its management (Fig. 52.10).

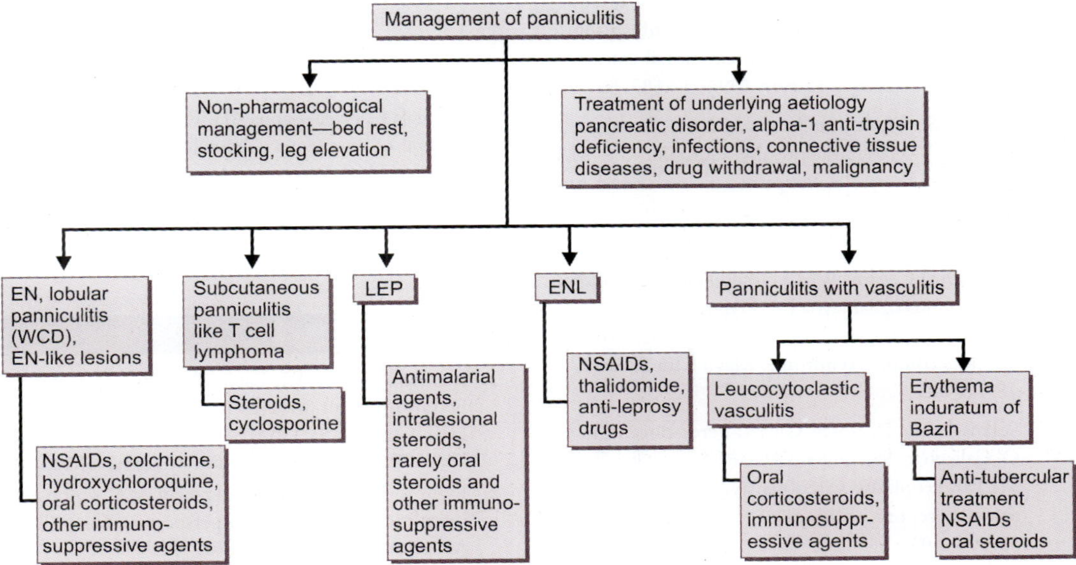

Abbreviations: LEP—lupus erythematosus profundus; ENL—erythema nodosum leprosum.

Fig. 52.10: Management of panniculitis

Key Points

- Cutaneous vasculitis is inflammation of small to medium vessels of various aetiologies and is characterized by cutaneous features such as palpable purpura.
- Panniculitis is inflammation of subcutaneous fat with or without vasculitis and is characterized by subcutaneous nodules.
- Detailed history (drugs, infection and systemic diseases), clinical examination followed by an array of appropriate laboratory tests and lesion biopsy helps in clinching the diagnosis.
- Systemic involvement defines the severity of the condition and immunosuppressive agents are required in most of these conditions.

REFERENCES

1. Russell JP, Weenig RH. Primary cutaneous small vessel vasculitis. Curr Treat Options Cardiovasc Med 2004; 6:139–49.
2. Sams WM Jr, Thorne EG, Small P, Mass MF, McIntosh RM, Stanford RE. Leukocytoclastic vasculitis. Arch Dermatol 1976; 112:219–26.
3. Ekenstam E, Callen JP. Cutaneous leucocytoclastic vasculitis: clinical and laboratory features of 82 patients seen in private practice. Arch Dermatol 1984; 120:484–489.
4. Jennette JC, Falk RJ, Andrassy K, Bacon PA, Churg J, Gross WL, et al: Nomenclature of systemic vasculitides. Proposal of an international consensus conference. Arthritis Rheum 1994; 37:187–92.
5. Blanco R, Martinez-Taboada VM, Rodriguez-Valverde V, García-Fuentes M, González-Gay MA. Henoch-Schönlein Purpura in adulthood and childhood: two different expressions of the same syndrome. Arthritis Rheum 1997; 40:859–64.
6. McDuffie FC, Sams WM Jr, Maldonado JE, Andreini PH, Conn DL, Samayoa EA. Hypocomplementemia with cutaneous vasculitis and arthritis: possible immune complex syndrome. Mayo Clin Proc 1973; 48:340–48.
7. Solans-Laqué R, Bosch-Gil JA, Pérez-Bocanegra C, O' Callaghan AS, Simeon- Aznar CP, Vilardell- Terres M. Paraneoplastic vasculitis in patients with solid tumors: report of 15 cases. J Rheumatol 2008; 35:294–304.
8. Díaz-Pérez JL, De Lagrán ZM, Díaz-Ramón JL, Winkelmann RK. Cutaneous polyarteritis nodosa. Semin Cutan Med Surg 2007; 26:77–86.
9. Requena L, editor. Panniculitis. Dermatol Clin 2008; 26:419–584.
10. Garcia-Porrua C, Gonzalez-Gay MA, Vazquez-Caruncho M, Lopez- Lazaro L, Lueiro M, Fernandez ML, et al. Erythema nodosum: aetiologic and predictive factors in a defined population. Arthritis Rheum 2000; 43:584–92.
11. Dhawan SS, Jimenez-Acosta F, Poppiti RJ Jr, Barkin JS. Subcutaneous fat necrosis associated with pancreatitis: histochemical and electron microscopic findings. Am J Gastroenterol 1990; 85:1025–28.
12. Roenigk HH, Haserick J, Arundell FD. Post steroid panniculitis: report of a case and review of the literature. Arch Dermatol 1964; 90:387–91.
13. Ivker RA, Woosley J, Briggaman RA. Calciphylaxis in three patients with end-stage renal disease. Arch Dermatol 1995; 131:63–8.
14. Smith KC, Su WPD, Pittelkow MR, Winkelmann RK. Clinical and pathologic correlations with 96 patients with panniculitis, including 15 patients with deficient levels of alpha-1-antitrypsin. J Am Acad Dermatol 1989; 21:1192–96.
15. Craig AJ, Cualing H, Thomas G, Lamerson C, Smith R. Cytophagic histiocytic panniculitis—a syndrome associated with benign and malignant panniculitis: case comparison and review of the literature. J Am Acad Dermatol 1998; 39:721–36.
16. Fraga J, García-Díez A. Lupus erythematosus panniculitis. Dermatol Clin 2008; 26:453–463.
17. Requena C Sanmartín O, Requena L. Sclerosing Panniculitis. Dermatol Clin 2008; 26:501–04.
18. Nirmala C, Nagarajappa AH. Erythema Induratum – A type of cutaneous tuberculosis. Indian J Tuberc 2010; 57:160–64.
19. Cuevas J, Rodríguez-Peralto JL, Carrillo R, Contreras Fl. Erythema nodosum leprosum: reactional leprosy. Semin Cutan Med Surg. 2007; 26(2):126–30.

FURTHER READING

1. Jennette JC, Falk RJ, Bacon PA, Basu N, Cid MC, Ferrario F et al. 2012 revised International Chapel Hill Consensus Conference Nomenclature of Vasculitides. Arthritis Rheum. 2013; 65(1):1–11.
2. Pina T, Blanco R, González-Gay MA. Cutaneous vasculitis: a rheumatologist perspective. Curr Allergy Asthma Rep. 2013; 13(5):545–54.

Chapter

53

Behçet's Syndrome

Varun Dhir

Behçet's syndrome is an inflammatory relapsing disease of unknown cause that has a marked geographic variation. It is characterized by a constellation of clinical features including oral and genital ulcerations, eye involvement and other major system involvements.

HISTORY AND EPIDEMIOLOGY

Behçet's syndrome or Behçet's disease is prevalent in countries along the old silk route. As the name suggests this was the route used for silk trading in the olden times and stretched from the east (China) to the Roman Empire around the Mediterranean that now includes many countries in Europe like Turkey. There is high prevalence in countries like Turkey, the middle east, and Japan. Low prevalence is found in the USA and Western Europe (UK). The prevalence in India is also probably low; however, it is definitely not rare.

PATHOGENESIS

The pathogenesis of Behçet's is incompletely understood as of today. The association of HLA-B*5 (initially called HLA-A*5) was reported in Japanese patients in 1973. Subsequently, the broad antigen HLA-B*5 has been found to contain two different specificities—B*51 and B*52 (split antigens). It is now clear that HLA-B*51 is the main HLA serotype associated with Behçet's. Its main allele that has found to be associated is HLA-B*5101. In a meta-analysis, the pooled incidence of HLA-B*5/B*51 was found to be 55 to 63.5% in patients and 16.8 to 21.7% in controls, giving a pooled odds ratio of 5.78 (95% confidence interval from 5 to 6.67).[1] Indeed, HLA B region has been the strongest association in the many genome wide association studies (GWAS) done. In addition, GWAS have found an association with IL-10, STAT4 (downstream in the Th1 pathway), CCR1-CCR3 (chemokine receptors), KLRC4 (killer cell lectin type receptor of unknown function), ERAP1 [ERAP1 encodes endoplasmic reticulum amino-peptidase 1 whose function is to trim the peptides for loading into major histocompatibility complex (MHC) Class I], fucosyltransferase (responsible for synthesis of H antigen expressed on intestinal epithelia and thus on flora) and IL-23R.[2]

The occurrence around a particular route used by traders suggests an infectious aetiology, but none has been conclusively proven. The best postulate is that it is a result of hypersensitivity to an infectious agent. In this regard, heat shock protein-60 (HSP-60) has been found to be over-expressed in Behçet's apthous ulcers and skin lesions. Heat shock proteins are chaperones that help in making proteins fold and are expressed by cells in the face of any stress (not only heat, as the name suggests). Significant sequence homology exists between HSP-60 in humans and HSP-65 in *Mycobacterium tuberculosis* and *Streptococcus* (HSP-65)—around 50%. Further, studies have found that T cells from patients show enhanced reactivity to HSP.[3]

Gamma delta (γδ) T cells are non-conventional T cells found to be expanded in Behçet's in many studies. They have been found to be present in local Behçet's lesions along with HSP65, suggesting that they may be responding to this stimulus.[4] They are thought to be a bridge between the innate and adaptive immune response. They are similar to T cells with respect to T cell receptor (TCR) signaling, but they are non-MHC restricted and have limited diversity. They are thought to recognize non-classical MHC associated lipids and alkyl amines.

Clinical Features and Criteria

The cardinal systems involved by this disease and consequently, the specialists which most commonly encounter them include the skin and mucocutaneous (often present to dermatologists), eye (ophthalmologists), arthritis (rheumatologists), cardiovascular, neurological and gastrointestinal (respective specialities). The spectrum of this disease is vast, ranging from patients having minor (though irritating) features like recurrent apthous ulcers to patients having life-threatening or organ-threatening manifestations like retinitis, brainstem lesions and pulmonary aneurysms.

Muco-cutaneous: Apthous ulcers are the most common manifestation, estimated to occur in nearly all the patients. They are usually the first manifestation in patients. They are indistinguishable from the common apthous ulcers (called canker sores) and occur on the gingival, tongue and buccal mucous membranes. Obviously, their frequency and recurrence is the difference in Behçet's patients from the normal population. They heal without scarring, are usually <10mm (called minor). Sometimes bigger ulcers may occur. The typical lesion is usually round with a sharp erythematous border and covered with a yellowish pseudomembrane. Genital ulcers typically occur on the scrotum (90%) and less frequently on the penis or perianal areas in men. In women they occur usually on the labia majoris, followed by labia minora and then the perianal and femoral area. Vaginal and cervical ulcers are rare. They are larger than oral ulcers and have an irregular border. Genital ulcers usually heal with scarring when large. Skin lesions in Behçet's include erythema nodosum (histopathology showing vasculitis in addition to panniculitis), pseudo-folliculitis, acne-like lesions (face, arms and legs), superficial thrombophlebitis, cutaneous vasculitis and Sweet's syndrome.[2]

Eye manifestations: Ocular involvement is seen in almost half the patients. This is in the form of a chronic relapsing uveitis that is often bilateral and involves the anterior and posterior uveal tracts (posterior uveal inflammation is often associated with retinitis, also called retinal vasculitis). Hypopyon uveitis, where there is a visible layer of pus in the anterior chamber, may occur in 20% and is considered a major clue for the diagnosis of Behçet's when present. Retinal vasculitis presents as exudative and haemorrhagic lesions on the retina and can lead to blindness. This has to be differentiated from viral and sarcoid retinopathy.

Musculoskeletal: Musculoskeletal involvement usually consists of arthralgias or minor arthritis that may be poly-, oligo- or monoarthritis. Knees are the most common joints involved followed by wrists, ankles and elbows. Erosions or deformities are not seen.

Vascular: Vascular lesions occur in the form of deep venous thrombosis (in addition to superficial venous inflammation that can be migratory). Generally, as the venous thrombosis is due to inflammation, the clot is adherent and there is no pulmonary embolism. In case of veins in specific locations, one can get Budd-Chiari syndrome (hepatic venous outflow obstruction). Arterial lesions in the form of aneurysms involving the entire arterial tree including the pulmonary artery. Pulmonary artery (and its branches) aneurysms can be fatal and present with hemoptysis and can appear as non-cavitating masses on the chest radiograph, that can be centrally (main pulmonary artery) or peripherally located (segmental arteries).

Neurologic: Neurologic disease is found in 10% of patients and can be parenchymal (more common—80%) or vascular (less common-20%). The former comprises lesions in the brainstem that present with cerebellar and long tract signs (sensory and motor). On magnetic resonance imaging, there are multiple hyperintense on T2 lesions in brainstem, basal ganglia and cerebellar white matter. Chronic neurologic disease is common in young males and is often progressive. Vascular involvement may occur in the form of dural sinus thrombus.

Intestinal: Bowel involvement may manifest as malena, diarrhoea and perforation. The ileocecal region (same as tuberculosis) is the most commonly involved.

Behçet's Criteria (International Study Group for Behçet's Disease)[5]

This consists of a mandatory criteria that consists of recurrent oral ulcers observed by a physician and occurring at least 3 times in one year;plus two of the following manifestations. These include recurrent genital ulcers, pathergy test—read at 24–48 hours, skin manifestations-erythema nodosum, pseudofolliculitis and aceniform lesions in persons who are post-adolescent and not on steroids and eye lesions-anterior uveitis, posterior uveitis, cells in vitreous, retinal vasculitis. Another set of criteria, the International Criteria for Behçet's Disease (ICBD),[6] provide a quantitiative score for fulfilment of different clinical features scoring two points each for oral ulcers, genital ulcers, ocular lesions and one point for cutaneous involvement, nervous system involvement, arterial or venous pathology and positivity for pathergy test. A score of 4 is required to classify as Behçet's as per these criteria.

The diagnosis of Behçet's requires the coming together of a constellation of clinical features rather than any laboratory test or specific clinical feature. The differential diagnosis of Behçet's is wide depending on the organ system involved. However, common mimics include seronegative spondyarthritis especially reactive arthritis (uveitis, oral ulcers and arthritis), lupus (oral ulcers, arthritis, CNS manifestations, bowel), inflammatory bowel diseases (oral ulcers, bowel involvement, arthritis) and herpes infections (oral and genital ulcers). One must note that rarely, in about 1–2% of patients, a diagnosis of Behçet's disease may be made without the presence or oral ulcers, as per the ICBD criteria.[7] However, such a diagnosis would be most unlikely in the absence of both oral and genital ulcers.

Pathergy and Laboratory Tests

The pathergy test refers to skin hypersensitivity to a needle puncture. There are multiple methods described in literature. Often it can be inferred from asking the patients about needle punctures they received when giving a blood sample. The usual procedure is to use a hypodermic needle (25 G) and give a intradermal prick on the flexor aspect of the forearm. The use of spirit to clean the arm before puncture may be omitted (as per some). Similarly, some inject around 0.1 ml of saline. The reading is done at 48 hours and is classified as positive when there is a pustule formed on the site of injection. As per some variations described the reaction can be graded into erythema with elevation, papule and pustule formation, the latter being the most positive. Again some have described the pustule should be at least 2 mm in diameter. This test is neither sensitive, nor specific for Behçet's, and can be found in other diseases including Sweet's syndrome and pyoderma gangrenosum.

There is no specific laboratory test for Behçet's disease; commonly acute phase reactants and total white cell count may be elevated representing inflammation. The use of genotyping for HLA*B51 is again controversial due to its low sensitivity and is dependent on the population and pre-test probability.

Treatment

Treatment of this disease depends on the major clinical features and clinical situation-whether there is organ or life threatening manifestations.

Colchicine: Colchicine remains the front-line drug used for oral and genital ulcers and works by inhibiting neutrophil function. It is often used at a dose of 0.5 mg twice a day. Although, not safe in pregnancy as per routine classification, there are studies showing it has been continued safely in diseases like familial mediterannean fever. It is used continuously over a long-time to prevent new oral and genital ulcers and prevent new onset anterior and posterior eye inflammation.

Azathioprine: Azathioprine remains one of the front line drugs in Behçet's disease whenever immunosuppressants have to be used like in uveitis, aneurysms, CNS disease. It is often used as a steroid-sparing agent, to bring the dose of steroids down to acceptable levels (7.5 to 10 mg per day prednisolone). The dose of Azathioprine is usually 1.5 to 2.5 mg per kg per day.

Corticosteroids: Corticosteroids remain the prominent anti-inflammatory agents used in Behçet's to control inflammation and limit damage. The way they are used depends on the urgency depending on the risk to the organ (or life). In urgent situations, daily pulses of 500–1000 mg methylprednisolone are used. In case of less urgent situations, high oral doses like 1–1.5 mg per kg per day of prednisolone are used followed by medium (0.5 mg per kg) and low doses (unto 10 mg per day) depending on the clinical situation. Needless to say, it is important to look at blood glucose levels and electrolytes in case of pulse doses, and look out for infections and other corticosteroid adverse effects including cataract, corticosteroid-related fat re-distribution, weight gain, osteoporosis and avascular necrosis on any dose. Corticosteroids ointments can often be used for local control of apthous ulcers, although, oral cavity must be routinely examined for oral thrush.

Cyclosporine: This is an extremely useful and potent immunosuppressant that has a relatively quick action. It is often used for organ-threatening inflammation in Behçets like uveitis and sometimes aneurysms. It is used at a dose of 2.5 to 5 mg per kg per day (often started at a lower dose of 1 mg per kg per day). Due to its tendency to cause hypertension and hypokalaemia, these parameters must be routinely tested as well as drug levels must be monitored. It is also notorious for its tendency to lead to opportunistic infections like pneumocystis. In Behçet's, cyclosporine has a unique tendency to lead to neurological lesions, indistinguishable from neuro-Behçet's. Thus, it is not used in cases of neuro-Behçets or in patients with subclinical lesions (on MRI).

Cyclophosphamide: It is useful and often used as an IV monthly pulse of 15 mg per kg. It is a potent immunosuppressant used for organ-threatening manifestations like retinal vasculitis and posterior uveitis, meningoencephalitis and CNS lesions.

In addition to the above drugs, other less commonly used drugs include thalidomide, pentoxifylline, interferon alpha, sulfasalazine and methotrexate. Thalidomide is used at a dose of 50 to 300 mg per day and must not be used in females unless the risk of the patient getting pregnant is very low as it is highly teratogenic. It is also associated with a time dependant peripheral neuropathy for which monitoring must be done. It is often useful for refractory oral ulcers.

Conclusion

Behçet's disease is a rare systemic inflammatory vasculitis, most commonly seen in Mediterranean countries but also seen in India, characterized by oral and genital ulcers with potential to involve other organ systems such as the vascular and nervous systems. Colchicine forms the mainstay of therapy, with corticosteroids used to tide over acute systemic life-threatening flares and other immunosuppressants like azathioprine or cyclosporine used for steroid-sparing efficacy.

Key Points

1. Behçet's disease is most frequent around countries along the old silk route, highest in Turkey and the Middle-East.
2. Neither pathergy, nor HLA-B*51 nor any laboratory test can diagnose Behçet's; diagnosis requires occurrence of a constellation of clinical features and exclusion of other diseases.
3. Treatment depends on whether there is organ or life-threatening manifestations and includes corticosteroids, immunosuppressants and colchicine.

REFERENCES

1. De Menthon M, Lavalley MP, Maldini C, Guillevin L, Mahr A: HLA-B51/B5 and the risk of Behçet's disease: a systematic review and meta-analysis of case-control genetic association studies. Arthritis Rheum. 2009; 61:1287–96.
2. Takeuchi M, Kastner DL, Remmers EF: The immunogenetics of Behçet's disease: A comprehensive review. J Autoimmun. 2015; 64:137–48.
3. Birtas-Atesoglu E, Inanc N, Yavuz S, Ergun T, Direskeneli H: Serum levels of free heat shock protein 70 and anti-HSP70 are elevated in Behçet's disease. Clin Exp Rheumatol. 2008; 26:S96–8.
4. Hasan MS, Bergmeier LA, Petrushkin H, Fortune F: Gamma Delta (gammadelta) T Cells and Their Involvement in Behçet's Disease. J Immunol Res. 2015; 2015:705831.
5. Criteria for diagnosis of Behçet's disease. International Study Group for Behçet's Disease. Lancet. 1990; 335: 1078–80.
6. International Team for the Revision of the International Criteria for Behçet's Disease. The International Criteria for Behçet's Disease (ICBD): a collaborative study of 27 countries on the sensitivity and specificity of the new criteria. J Eur Acad Dermatol Venereol 2014; 28:338–347.
7. Hatemi G, Yazici Y, Yazici H.Behçet's syndrome. Rheum Dis Clin North Am. 2013 ;39:245–61. doi: 10.1016/j.rdc.2013.02.010.

FURTHER READING

1. Hatemi G, Yazici Y, Yazici H. Behçet's syndrome. Rheum Dis Clin North Am. 2013 ;39:245–61. doi: 10.1016/j.rdc.2013.02.010 (good exposition on the spectrum of Behçet's and its management).
2. International Team for the Revision of the International Criteria for Behçet's Disease. The International Criteria for Behçet's Disease (ICBD): a collaborative study of 27 countries on the sensitivity and specificity of the new criteria. J Eur Acad Dermatol Venereol 2014; 28: 338–347 (Excellent review on different classification criteria for Behçet's disease).
3. Sakane T, Takeno M, Suzuki N, Inaba G. Behçet's disease. N Engl J Med. 1999; 341:1284–91 (Good review article).

CRYSTAL ARTHROPATHY AND OSTEOARTHRITIS

Third National Health and Nutrition Examination Survey. Arthritis Rheum. 2007; 57:109–15.

21. Rodriguez G, Soriano LC, Choi HK. Impact of diabetes against the future risk of developing gout. Ann Rheum Dis. 2010; 69(12):2090–2094.
22. Burns CM, Wortmann RL. Gout therapeutics: new drugs for an old disease. Lancet. 2011; 377:165–77.
23. Zhang W, Doherty M, Bardin T, Pascual E, Barskova V, Conaghan P, et al; EULAR Standing Committee for International Clinical Studies Including Therapeutics. EULAR evidence based recommendations for gout. Part II: Management. Report of a task force of the EULAR Standing Committee for International Clinical Studies Including Therapeutics (ESCISIT). Ann Rheum Dis 2006; 65:1312–24.
24. European Medicines Agency. Summary of product characteristics: adenuric film-coated tablets (online), http://www.medicines.org.uk/emc/medicine/22830/SPC(2012).
25. Sundy JS, Baraf HS, Yood RA, Edwards NL, Gutierrez-Urena SR, Treadwell EL, et al. Efficacy and tolerability of pegloticase for the treatment of chronic gout in patients refractory to conventional treatment: two randomized controlled trials. JAMA 2011; 306: 711–20.

FURTHER READING

1. Neogi T, Jansen TL, Dalbeth N, Fransen J, Schumacher HR, Berendsen D, et al. 2015 Gout Classification Criteria: an American College of Rheumatology/European League Against Rheumatism collaborative initiative. Arthritis Rheumatol. 2015; 67:2557–68.
2. Richette P, Doherty M, Pascual E, Barskova V, Becce F, Castañeda-Sanabria J, et al. 2016 updated EULAR evidence-based recommendations for the management of gout. Ann Rheum Dis. 2017; 76:29–42.
3. Sriranganathan MK, Vinik O, Bombardier C, Edwards CJ. Interventions for tophi in gout. Cochrane Database Syst Rev. 2014; (10):CD010069.
4. van Echteld I, Wechalekar MD, Schlesinger N, Buchbinder R, Aletaha D. Colchicine for acute gout. Cochrane Database Syst Rev. 2014; (8):CD006190.

Chapter

55

Calcium Pyrophosphate Deposition Disease and Others

Jyoti Ranjan Parida

INTRODUCTION

Crystal induced arthropathies include a variety of disorders where inflammation of the joint and soft tissues occurs due to deposition of various crystals. These include mainly uric acid and calcium-containing crystals like calcium pyrophosphate dehydrate (CPPD) and basic calcium phosphate (BCP). Besides, oxalate crystals may accumulate and cause joint inflammation in patients with renal failure whereas cholesterol crystals may get deposited in joints of longstanding rheumatoid arthritis or osteoarthritis.

CPPD DISEASE

CPPD disease is characterized by deposition of calcium pyrophosphate (CPP) crystals mainly in the fibrocartilage and hyaline cartilage of joints leading to inflammation and calcification. This is usually associated with aging, degenerative joint diseases and a variety of genetic and metabolic disorders (Table 55.1).The disease prevalence increases with age and CPP crystals are found in joints of 10–15% of people between 65 and 75 years and 30–50% of people over the age of 85 years.[1]

True prevalence in India is not known due to lack of large studies and may be partly related to lack of awareness among health care professionals.

Table 55.1: Metabolic and endocrine diseases associated with CPPD disease

Strong association	*Possible association*
Hyperparathyroidism	Gout
Hypophosphatasia	Ochronosis
Hypomagnesaemia	Wilson's disease
Haemochromatosis	Hypothyroidism
	X-linked hypophosphatemic rickets
	Familial hypocalciuric hyper-calcemia
	Acromegaly

Pathophysiology (Fig. 55.1)

Formation of CPP Crystals

Formation of CPP crystal is initiated near the surface of chondrocytes and reflects increased levels of either local calcium or inorganic pyrophosphate (iPP) in the milieu. Adenosine triphosphate (ATP) is the main substrate for production of iPP and this hydrolysis is catalyzed by a hormone called nucleoside triphosphate pyrophosphohydrolase (NTPPH). These enzymes are usually associated with the external surface of chondrocytes or present in excess quantity in vesicles produced by collagenase digestion of articular cartilage (Ecto NTPPH). Extracellular iPP can bind to and inhibit growth of basic calcium phosphate (BCP) crystals. Factors that promote the production of iPP are TGF-beta, ascorbate, retinoic acid and thyroid hormone. Tissue non- specific alkaline phosphatase (ALP) hydrolyses iPP to inorganic 2Pi.

Role of ANK Gene

ANK gene polymorphism is associated with hereditary or familial form of CPPD.[2] The ANK gene product is a transmembrane protein that usually serves as an iPP transporter or its regulator.

Pseudo-rheumatoid Arthritis

This group is also known as chronic pyrophosphate arthritis and constitutes less than 5% of CPPD patients. This presents as non-erosive, asynchronous inflammatory arthritis mimicking rheumatoid arthritis (RA). This may be associated with early morning stiffness, symmetrical involvement, flexion contractures and elevated acute phase reactants. Elderly patients may have low titres of rheumatoid factor further making it difficult to differentiate this from RA. Hence, the presence of CPPD crystals in synovial fluid and typical radiographic features suggests CPPD disease, whereas, specific antibodies like anti-CCP antibody and bony erosions are more suggestive of RA. However, the two conditions may overlap and be difficult to differentiate.

Pseudo-neuropathic Joints

This is a severe destructive arthropathy mainly affecting knee joints. The presence of chondrocalcinosis and absence of any neurological disorder aids in the diagnosis of CPPD disease.

Atypical/Periarticular CPPD

CPPD crystals may deposit at sites of tendon insertion, ligament, bursa, joint recesses, etc. and presents with carpal tunnel syndrome, periarthritis, tenosynovitis, and bursitis, etc. Involvement of spine may lead to spinal stenosis, or bony ankylosis resembling diffuse idiopathic skeletal hyperostosis.

Investigations

X-ray

The radiographic hallmarks of CPPD deposition disease are chondrocalcinosis and CPPD arthropathy (osteoarthritic changes in a specific distribution).

Chondrocalcinosis refers to linear or wedge-shaped calcifications and occurs because of deposition of CPPD crystals in fibrous and hyaline cartilage around the joints (Fig. 55.2). It is most frequently seen in the knees, pubic symphysis and wrist joints (triangular cartilage) followed by glenoid and acetabular labra. CPP crystal deposits in hyaline cartilage appear as a radiopaque line parallel to the bone surface.

The arthropathy of CPPD disease radiographically resembles osteoarthritis but the following features may help to differentiate-presence of chondrocalcinosis, normal bone mineralization, uniform joint space loss, subchondral new bone formation with variable osteophytes, more prominent and large subchondral cysts, bilateral distribution and occasional neuropathic changes. Involvement of some particular joints (which are spared by osteoarthritis) like elbow, wrist, ankle and shoulder, etc. increase likelihood of CPPD disease. The most common joints affected by CPPD disease are knees, hands and hips in decreasing order. Other joints that may be involved are acromioclavicular joints, MTP joints and talonavicular joints. In the hand, MCP joints are usually affected whereas IP joints are spared. In foot, arthropathy has a predilection for talonavicular joints.

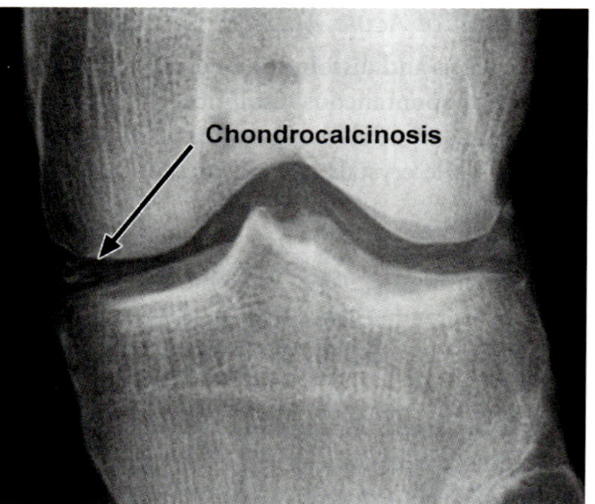

Fig. 55.2: Knee chondrocalcinosis

Calcification due to CPP crystal deposition can occur in bursa, ligaments, and tendons. Linear calcifications involving the Achilles tendon or plantar fascia are often seen.

Spinal involvement is not uncommon and any patient with evidence of multiple levels of degenerative disk disease should be considered for CPPD deposition disease. It may involve any part from cervical through lumbar spine and may be associated with calcification of soft tissues around intervertebral disk space. It may cause atlantoaxial dislocation.

Ultrasonography

Musculoskeletal ultrasound has evolved recently as a promising modality in CPPD disease although more studies are required for validation and determination of its specificity and sensitivity.[7,8]

The following findings in musculoskeletal ultrasound of joint and fibrocartilage suggest presence of CPPD disease:

1. A thin hyperechoic band parallels to bone cortex in hyaline cartilage. Unlike gout where the hyperechoic band is on the superficial part of hyaline cartilage, CPPD crystals deposit within the hyaline cartilage and the band is thin and stippled, in contrast to the smooth pattern of double contour sign in gout (Fig. 55.3).
2. Small rounded hyperechoic amorphous regions with acoustic shadowing in triangular fibrocartilage of wrist and menisci of knees and tendons.
3. Nodular hyperechoic deposits in bursa and joint recesses.
4. Hyperechoic lines of calcifications parallel to Achilles tendon or plantar fascia.

Synovial Fluid

In spite of advances in imaging techniques, demonstration of CPPD crystals by polarized light microscope in tissue or synovial fluid (SF) still remains the gold standard for diagnosis of CPPD disease. These crystals are rhomboid, long or short rods ranging from 2–20 µm in length with absent or weakly positive birefringence (Fig. 55.4).Thus, pseudogout crystals are blue when aligned parallel to the slow ray of the compensator and yellow when they are perpendicular. Because of their small size, cell integrity is very important and fresh synovial fluid samples yields better result. CPPD crystals may be detected in SF of asymptomatic joints and in between inflammatory episodes.

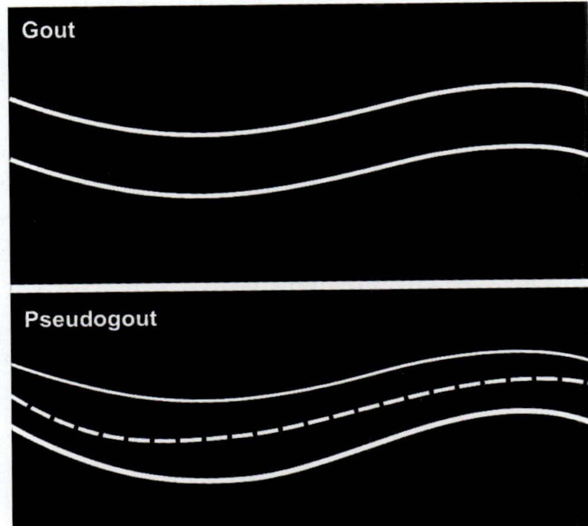

Fig. 55.3: Schematic representation of "double contour sign" of gout and pseudogout in ultrasound

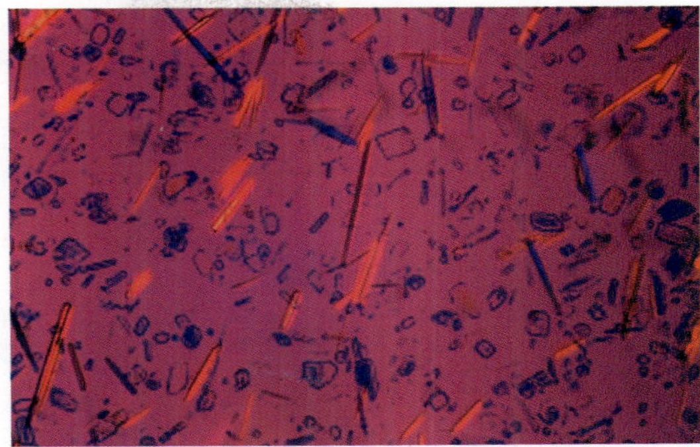

Fig. 55.4: CPPD crystals under polarized microscope rhomboid, long or short rods with absent or weakly positive birefringence

Treatment

Unlike gout, large-scale evidence-based guidelines for treatment of pseudogout are lacking. The treatment approach of pseudogout includes the following:

1. Treatment of acute pseudogout attacks
2. Prophylaxis of recurrent pseudogout attacks
3. Treatment of associated metabolic and endocrine diseases
4. Treatment of chronic CPP inflammatory arthritis

1. Treatment of Acute Pseudogout Attacks

Non-pharmacological treatment for acute attacks includes ice pack application, rest to the affected joint and immobilization. Initial drug treatment for acute pseudogout depends on the number of joints affected and is highlighted in the algorithm (Fig. 55.5).

Intra-articular injection: In case of involvement of one or two joints, long-acting glucocorticoid like Triamcinolone 40 mg with 2% lignocaine for large joints and smaller dose for small joints are preferred. It is very important to rule out any joint infection before intra-articular injection in monoarthritis.

NSAIDs: Usually, short-acting NSAIDs in full anti-inflammatory doses like naproxen (500 mg twice daily), sulindac (200 mg twice daily) or Indomethacin (50 mg thrice daily) are preferred. We prefer naproxen based on its relative cardiac safety in elderly over other NSAIDs. NSAIDs should be avoided in chronic kidney disease, heart failure,

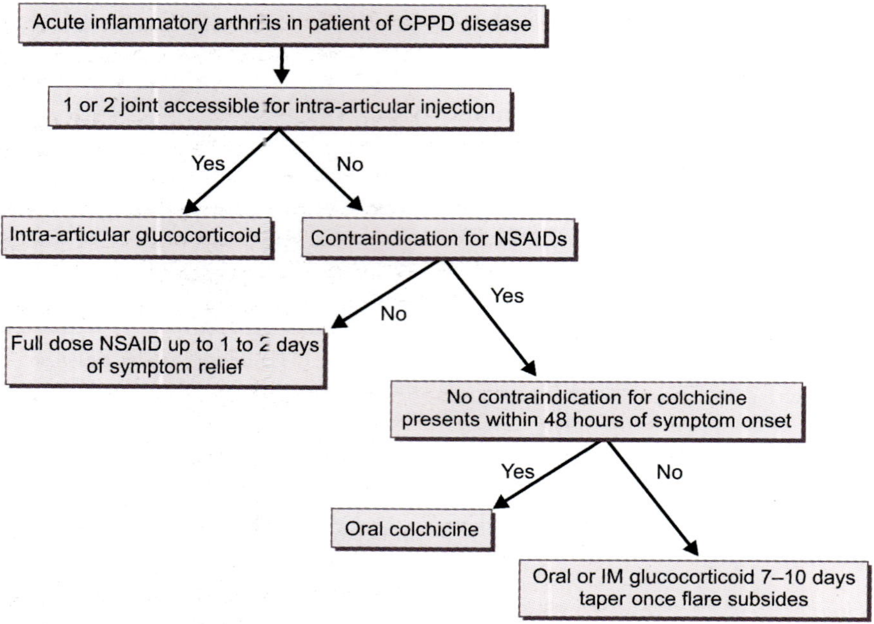

Fig. 55.5: Treatment algorithm of pseudogout

poorly controlled hypertension and duodenal or gastric ulcer. It gives faster relief when started within 48 hours of disease onset and response considered inadequate if significant symptom relief is not achieved within 48–72 hrs. Once symptoms subside, it can be stopped after 1–2 days.

Oral colchicine: For treatment of acute pseudogout, colchicine is effective only if started within 48-hr of disease onset and usually preferred when there is contraindication for NSAID use. Low-dose colchicine (0.5 mg up to three times daily with or without loading dose of 1 mg) than traditional dose (1 mg loading dose followed by 0.5 mg every 2-hr until the development of side-effects) is recommended based on expert opinion. IV colchicine is contraindicated and not used in any condition now.

Oral or parenteral steroid: In patients having any contraindication or inadequate response to NSAIDs and colchicine, oral prednisone 30 to 50 mg once daily or in two divided doses is given until flare resolution begins, and the dose is tapered and stopped over 7 to 10 days. Patients who are unable to take oral steroid, equivalent doses of intramuscular (IM) methylprednisolone injection are given. Although IM ACTH is an alternative option in gout, it is less studied in pseudogout.

2. Prophylaxis of Recurrent Pseudogout Attacks

The main aim of prophylaxis treatment is to reduce the number of future attacks and is offered to patients with three or more attacks of acute pseudogout per year. Colchicine (0.5 mg twice daily) is preferred for this based on its action on cytoskeleton. In case of side-effects like abdominal distress or diarrhea, 0.5 mg once daily is given.

In patients with an inadequate response or intolerance to colchicine, low-dose NSAID like naproxen is used for prophylaxis.

3. Treatment of Associated Metabolic or Endocrine Diseases

It is very important to identify and treat associated metabolic and endocrine diseases although this may not have any effect on the resolution of CPP crystal disease. Oral calcium supplements to suppress PTH in hyperparathyroidism, magnesium supplement (30 meq/day) in hypomagnesemia and appropriate therapy in hemochromatosis are recommended.

4. Treatment of Chronic CPP Inflammatory Arthritis (Pseudo-rheumatoid arthritis)

Oral NSAIDS (in absence of contraindication), colchicine (0.5 mg twice daily) and low-dose glucocorticoid (up to 10 mg of prednisone) are the initial treatment options for patients present with bilateral nonerosive symmetrical polyarthritis. Methotrexate and hydroxychloroquine have been used in refractory diseases with some success.

Newer drugs to block NLRP3 inflammasome - IL1 pathway seems promising but need more trial data before get approval for CPPD disease. Other drugs like probenecid, polyphosphates, transglutaminase 2, phosphocitrate, etc. have been tried to modulate pyrophosphate levels but still not successful.

BASIC CALCIUM PHOSPHATE HYDROXYAPATITE DEPOSITION DISEASE

This disease occurs due to deposition of the mixture of carbonate substituted hydroxyapatite and octacalcium phosphate. The term BCP (Basic calcium phosphate) is used, as these crystals are nonacidic. Because of their small size, they are not detectable by light microscope.

Clinical Feature

The clinical manifestations range from asymptomatic to very destructive arthropathy. Most commonly this involves shoulder joints and first MTP joints. It can also involve other joints like knees, elbow, wrists and digits. In older woman, severe destruction can occur in both shoulders because of BCP crystal deposition causing rotator cuff degeneration and glenohumoral instability, which is known as "Milwaukee shoulder syndrome"[9,10] (Fig. 55.6). It may be associated with large skin hematoma and joint effusion during the acute attack.

It can also cause calcific periarthritis, tendinitis, bursitis, enthesitis and soft tissue calcifications. BCP crystals may present in 30–50% of synovial fluid of osteoarthritis patients and contribute to low-grade inflammation and cartilage destruction.

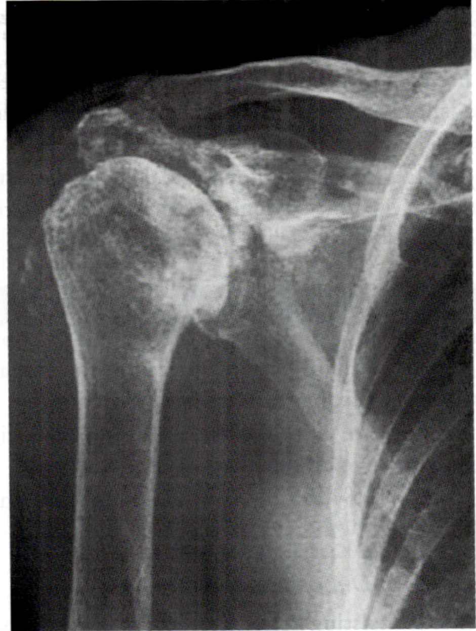

Fig. 55.6: Milwaukee shoulder

Investigations

Synovial fluid: Usually the synovial fluid looks serosanguineous and contains few cells under light microscopy. The crystals may be mistaken as debris or clumps and are not birefringerant under the polarized microscope. A special staining known as "alizarin red S" shows bright red extracellular and intracellular crystals. Newer techniques for crystal identification include electron microscopy with energy dispersive elemental analysis, infrared spectroscopy, ramanmicro spectroscopy, X-ray diffraction, etc. but they need validation before used in routine practice.

Radiology: Intra and/or periarticular calcification with or without erosion along with severe destructive or hypertrophic X-ray changes are characteristic of this disease.

Treatment

Acute attacks of bursitis or arthritis may be self-limiting and resolve within days to weeks. Rest and local application of ice help in relieving pain. Joint aspiration with intra-articular glucocorticoid, NSAIDs, short course of oral colchicine helps in decreasing intensity and duration of attacks. Other agents like local injection of EDTA, subcutaneous anakinra, IVIG, bisphosphonate, etc. have been tried with limited success. In patients with chronic renal failure and undergoing haemodialysis, phosphate-lowering agents are helpful in decreasing crystal load.

OTHER CRYSTAL INDUCED ARTHRITIS

Calcium Oxalate Deposition Disease

It is seen in patients with renal failure receiving hemodialysis or peritoneal dialysis, particularly

OA management. The most commonly used includes:

a. Non-scaffold based approach namely microfracture, autologous cartilage transplantation
b. Scaffold-based technique.

Microfracture (Mf): In this technique the mesenchymal stem cells (MSCs) of bone marrow are implanted to the site of damaged cartilage. Mf repair results in fibrocartilage formation instead of natural hyaline cartilage of joint, which is less robust and are more vulnerable to re-damage by mechanical stress. The mechanically inferior fibrocartilage is prone to failure after less than 12 months. Indications for Mf repair in OA is multifactorial: based on the size, depth and location of the cartilage damage in the joint, as well as the patient's age and body mass index.[12,13] Data of outcomes after Mf repair is available for up to 7 years. The Mf technique is approved by US food and drug administration (FDA). Many clinicians consider Mf technique of cartilage repair to be the gold standard.

Autologous cartilage implantation (ACI): In this technique a full thickness punch biopsy of the cartilage tissue is taken from the low weight bearing joints by an arthroscopic procedure. The tissue collected is expanded in laboratory *in vitro* condition. Once the adequate cell lines are generated these autologous chondrocytes are re-implanted into the debrided cartilage defect by doing second arthroscopy. Benefit of the ACI technique is manifold: less immunogenicity and reduced chances of transmission of pathogenic virus due to use of autologous cell, instead of foreign material. The smaller size of punch biopsy also minimizes complication.[13,14] Disadvantage of ACI includes two major arthroscopic procedures, flap hyperosteosis and dedifferentiation of cells *in vitro*. The maximum follow up data is available for 20 years. Autologous cartilage implantation (ACI) technique is considered gold standard for larger defects.[15,16]

Scaffold based techniques: Scaffolds are materials (e.g. collagen, polyester) that have been engineered to form a new functional tissue for cell/tissue growth in vitro environment. Scaffold based approach has some major advantages over scaffold less method. The chondrocyte implanted in 3D environment of scaffold are less prone to dedifferentiation and thus produces a more natural hyaline like cartilage implant on culture. Matrix-induced autologous chondrocyte implantation (MACI) and Autologous matrix induced chondrogenesis (AMIC) are the two major scaffold based cartilage repair method.

- *Matrix-induced autologous chondrocyte implantation (MACI):* In current clinical practice MACI is the most commonly used technique for scaffold based cartilage repair in OA. In this method the autologous tissue is extracted by arthroscopy and expanded *in vitro*. After 3 days the expanded cells are seeded on porcine-derived mixed collagen (type 1 and type 3) membrane scaffold. These are later re-implanted by another arthroscopic procedure at the site of tissue lesion. Studies have found MACI procedure to have similar outcome to ACI or Mf technique.[17]
- *Autologous matrix induced chondrogenesis (AMIC):* This is the second most common scaffold based technique. This method is similar to Mf, where a mini-arthrotomy is done to place mixed collagen membrane into the defect. The data is limited to small case series only, with maximum follow-up period of 5 years post procedure.[18]

Future research: The human chondrocytes can be cultured in a 3D matrix in *ex vivo* condition over 3–4 weeks to generate healthy articular cartilage implant with better biochemical integrity compared to plain *in vitro* culture. It is well-known that chondrocytes in the joints responds to high mechanical loads (e.g. walking) with increased proliferation and producing more ECM. In these 3D scaffold tissues the mechanical stimulation can be applied in the form of hydrostatic force, shear forces or dynamic compression leading to improved matrix maturation and production of biomechanically robust implants. Transforming growth factor β (TGF-β), particularly TGF-β1 and TGF-β3 are important growth factors used widely in cartilage regeneration in these *ex vivo* studies. Platelet rich plasma (PRP) has been used successfully in these studies, as PRP has been found to be rich source of growth factors. Intra-articular administration of PRP has shown good short-term benefit in some studies.[19]

Table 57.8: Investigational approaches for the management of OA

- Nerve growth factor inhibitors—Tanezumab-phase 3 trials resumed
- Tumour necrosis factor α-inhibitors
- Interleukin 1β inhibitors
- Matrix metalloproteinase inhibitors
- Mitogen activated protein kinase inhibitors
- Inducible nitric oxide inhibition
- Bradykinin B2 receptor antagonist
- Intra-articular bone morphogenic protein 7
- Intra-articular recombinant human fibroblast growth factor
- Calcitonin, bisphosphonates and strontium

Currently the Mf, MACI and ACI technique of cartilage repair and re-implantation are approved by US-FDA for the treatment of cartilage defects. These techniques are still in very nascent stage and only being practiced in research setting. There are very few centers in India who are performing them, due to non-gratifying results as compared to joint arthroplasty.

With the advent of better molecular diagnostics and subsequent discovery of newer pathogenic pathways, many different targets are being explored (Table 57.8). Some of these investigational therapies may come-up into the mainstream in the future.

Key Points

- The management of OA should have holistic approach and person's activity of daily living, occupation and leisure activities should be given due consideration.
- The non-pharmacological management in the form of exercise should be advised to all OA patients irrespective of age, comorbidity, pain severity or disability.
- Pharmacological treatment with NSAIDs/COX-2 inhibitor and opioids should be considered in patients who fail to respond with paracetamol or topical NSAIDs.
- Intra-articular injections with corticosteroid should be considered as an adjunct to core treatments for the relief of moderate to severe pain in people with OA.
- Surgical therapy is the most definitive, efficacious and cost-effective treatment of OA and should be offered after discussion with patient and surgeon, but should not be overly delayed before development of established functional limitation.
- The present cartilage repair technique includes: microfracture, ACI and scaffold based technique; these are still in early stage of development.
- Development of regenerative biological products over the next decade could revolutionize OA care by functionally healing damaged articular cartilage.

Acknowledgement

Figures courtesy of Dr Alok Sud, Professor, Department of Orthopaedics, LHMC, New Delhi and Dr Tankeshwar Baruah, Associate Professor, Central Institute of Orthopaedics, Vardhman Mahavir Medical College, and Safdarjung Hospital, New Delhi.

REFERENCES

1. National Institute for Health and Care Excellence (NICE). Osteoarthritis: care and management Clinical guideline [CG177] Published date: February 2014. Page 11–38. https://www.nice.org.uk/guidance/cg177.
2. Zheng H, Chen C. Body mass index and risk of knee osteoarthritis: systematic review and meta-analysis of prospective studies. BMJ Open. 2015; 5:e007568.
3. Hochberg MC, Altman RD, April KT, Benkhalti M, Guyatt G, McGowan J, et al. American College of Rheumatology 2012 recommendations for the use of non-pharmacologic and pharmacologic therapies in osteoarthritis of the hand, hip, and knee. Arthritis Care Res. 2012; 64:465–74.
4. Jansen MJ, Viechtbauer W, Lenssen AF, Hendriks EJ, de Bie RA. Strength training alone, exercise therapy alone, and exercise therapy with passive manual mobilisation each reduce pain and disability in people with knee osteoarthritis: a systematic review. J Physiother 2011;57(1):11e20. Epub 2011/03/16. http://dx.doi.org/10.1016/s1836–9553(11)70002–9. PubMed PMID: 21402325.
5. Nissen SE, Yeomans ND, Solomon DH, Lüscher TF, Libby P, Husni ME, et al. Cardiovascular Safety of Celecoxib, Naproxen, or Ibuprofen for Arthritis. N Engl J Med. 2016; 375:2519–29.
6. Cooper C, Rannou F, Richette P, Bruyère O, Al-Daghri N, Altman RD, et al. Use of Intra-articular Hyaluronic Acid in the Management of Knee Osteoarthritis in Clinical Practice. Arthritis Care Res. 2017; doi: 10.1002/acr.23204.
7. Thorlund JB, Juhl CB, Roos EM, Lohmander LS. Arthroscopic surgery for degenerative knee: systematic review and meta-analysis of benefits and harms. BMJ. 2015; 350:h2747.

8. McAlindon TE, Bannuru RR, Sullivan MC, Arden NK, Berenbaum F, Bierma-Zeinstra SM, et al. OARSI guidelines for the non-surgical management of knee osteoarthritis. Osteoarthritis Cartilage. 2014; 22: 363–88.
9. Skou ST, Roos EM, Laursen MB, Rathleff MS, Arendt-Nielsen L, Simonsen O, et al. A Randomized, Controlled Trial of Total Knee Replacement. N Engl J Med. 2015; 373:1597–606.
10. Makris EA, Gomoll AH, Malizos KN, Hu JC, Athanasiou KA. Repair and tissue engineering techniques for articular cartilage. Nat Rev Rheumatol. 2015; 11:21–34.
11. Goyal D, Keyhani S, Lee EH, Hui JH. Evidence-based status of microfracture technique: a systematic review of level I and II studies. Arthroscopy. 2013; 29:1579–88. doi: 10.1016/j.arthro.2013.05.027.
12. Gudas R, Gudaitė A, Mickevicius T, Masiulis N, Simonaitytė R, Cekanauskas E, et al. Comparison of osteochondral autologous transplantation, microfracture, or debridement techniques in articular cartilage lesions associated with anterior cruciate ligament injury: a prospective study with a 3-year follow-up. Arthroscopy. 2013; 29:89–97.
13. Saris DB, Vanlauwe J, Victor J, Almqvist KF, Verdonk R, Bellemans J, et al. Treatment of symptomatic cartilage defects of the knee: characterized chondrocyte implantation results in better clinical outcome at 36 months in a randomized trial compared to microfracture. Am J Sports Med. 2009; 37(Suppl. 1):10S–19S.
14. Saris DB, Vanlauwe J, Victor J, Haspl M, Bohnsack M, Fortems Y, et al. Characterized chondrocyte implantation results in better structural repair when treating symptomatic cartilage defects of the knee in a randomized controlled trial versus microfracture. Am J Sports Med. 2008; 36:235–246.
15. Peterson L, Vasiliadis HS, Brittberg M, Lindahl A. Autologous chondrocyte implantation: a long-term follow-up. Am J Sports Med. 2010; 38:1117–1124.
16. Minas T, Von Keudell A, Bryant T, Gomoll AH. The John Insall Award: a minimum 10 year outcome study of autologous chondrocyte implantation. Clin Orthop Relat Res. 2014; 472:41–51.
17. Bartlett W, Skinner JA, Gooding CR, Carrington RW, Flanagan AM, Briggs TW, et al. Autologous chondrocyte implantation versus matrix-induced autologous chondrocyte implantation for osteochondral defects of the knee: a prospective, randomised study. J Bone Joint Surg Br. 2005; 87:640–645.
18. Gille J, Behrens P, Volpi P, de Girolamo L, Resiss E, Zoch W, et al. Outcome of Autologous Matrix Induced Chondrogenesis (AMIC) in cartilage knee surgery: data of the AMIC Registry. Arch Orthop Trauma Surg. 2013; 133:87–93.
19. Filardo G, Kon E, Di Martino A, Di Matteo B, Merli ML, Cenacchi A, et al. Platelet-rich plasma vs hyaluronic acid to treat knee degenerative pathology: study design and preliminary results of a randomized controlled trial. BMC Musculoskelet Disord. 2012; 13:229. doi: 10.1186/1471–2474–13–229.

FURTHER READING

1. Bennell KL, Hunter DJ, Hinman RS. Management of osteoarthritis of the knee. BMJ. 2012; 345:e4934.
2. Nelson AE, Allen KD, Golightly YM, Goode AP, Jordan JM. A systematic review of recommendations and guidelines for the management of osteoarthritis: The chronic osteoarthritis management initiative of the US bone and joint initiative. Semin Arthritis Rheum. 2014; 43:701–12.

OSTEOPOROSIS AND METABOLIC BONE DISEASES

Data from India show that western reference values for BMD may not be applicable to Indians. The Indian Council for Medical Research (ICMR) carried out a large multicenter study to generate an India-specific database which confirmed data from smaller, single-center studies and showed that Indians have lower BMD than Caucasians.[3] Reasons ascribed for lower BMD in Indians include possible genetic differences, nutritional deficiency, and smaller skeletal size. The widespread vitamin D deficiency in India may also have an impact on BMD.[4,5]

EPIDEMIOLOGY

Ageing of the world's population has meant an increase in the number of osteoporotic individuals. Substantial geographic variation has been noted in the incidence of osteoporotic fractures worldwide, with Western populations (North America, Europe and Oceania), reporting increases in hip fracture throughout the second half of the 20th century, with a stabilisation or decline in the last two decades.[6] However, the rates of osteoporotic fracture continue to increase in developing populations, particularly in Asia. The number of individuals aged 50 years or more at high risk of osteoporotic fracture worldwide in 2010 was estimated at 158 million (137 million women and 21 million men) and is set to double by 2040.[7] The highest number of individuals above the fracture threshold was in Asia with more than 11 million men and 73 million women, respectively, comprising 55 and 54% of all men and women aged 50 years identified at risk. Men and women in Europe accounted for 17 and 22% of the global burden, respectively. It has been estimated that 56.2 million major osteoporotic fractures occurred worldwide in 2000 and the numbers rose to 74.2 million in 2010. Accounting for the fact that major fractures constitute only 60–65% of all osteoporotic fractures depending on age, the number of fragility fractures in 2010 is likely to exceed 120 million.[8]

No population based statistics are available from India. The data available are limited.[9] Most studies are hospital based, cross sectional or with a small sample size. Estimates by expert groups like the Osteoporosis Society of India placed the number of osteoporosis patients in India at 26 million approximately (2003 figures), with the numbers projected to increase to 36 million by 2013.[10] The International Osteoporosis Foundation Asia Pacific audit report of 2013 estimated that 50 million people in India are either osteoporotic (T-score lower than –2.5) or have low bone mass (T-score between –1.0 and –2.50).[11]

Hip fractures are commoner in women (~80% of hip fractures) as compared to men. Fracture rates adjusted for age are highest in Scandinavian and North American populations with lower rates in Asian and Latin American populations. Rates seem to be lower in rural areas than in urban areas in any country.[12] Geographical variation in the prevalence and incidence of vertebral fractures is less than that for hip fractures. The prevalence of vertebral fractures in men and women is almost equal, likely due to occupation associated trauma in men. Unlike hip fractures that result primarily from a fall, most vertebral fractures result from routine activities like bending or lifting light objects. Falls account for only one-fourth of vertebral fractures. Vertebral fractures are often clinically silent. However, one vertebral fracture leads to a tenfold increase in risk of subsequent vertebral fractures. The epidemiology of wrist fractures differs from hip and vertebral fractures Most wrist fractures occur in women, 50% of whom are older than 65 years. The incidence in men is low and does not increase much with age.[12]

Detection and treatment before the occurrence of first fracture is uncommon in India. Several factors are responsible for this state of neglect: The long held belief that osteoporosis is a disease of the west and does not exist in India, the notion that osteoporosis is an inevitable consequence of ageing which can neither be treated nor prevented, preoccupation of the health planners with infectious diseases, limited availability of diagnostic facilities and paucity of epidemiologic data.[13] The low life expectancy at birth seen a few decades ago meant that many Indians, in the past, did not live long enough to develop osteoporosis. Osteoporosis was thought to be an exotic disease, seldom suspected, rarely diagnosed. With ageing of the Indian population this has changed. According to 2011 census, the population of India is 1.2 billion of which the aged (60+) population is 7.5%. This translates to more than 90 million elderly in India.

Studies from India have shown that osteoporotic fractures usually occur 10–20 years earlier in Indians compared to Caucasians. One hypothesis advanced to explain the early onset of osteoporosis in Indians is that a dietary deficiency of calcium, beginning early in life, leads to a lower peak bone mass and consequently osteoporosis at an earlier age in Indians. Another contributory factor could be subclinical vitamin D deficiency causing malabsorption of calcium without overt osteomalacia.[14] One interesting observation has been the high proportion of male patients among hip fracture cases in India which is contrary to what is seen in other areas. Lower hospital attendance and health service utilization by women in India, especially the elderly, may be the main reason behind the skewing of data. Vertebral fractures are common in Indians and the prevalence of radiographic vertebral fractures in older adults in Delhi is reported to be 17.9% (18.8% male and 17.1% female); indicating that vertebral fracture prevalence in India is similar to Western populations.[15]

PATHOPHYSIOLOGY

Bone tissue comprises an extracellular matrix within which bone mineral comprising chiefly of hydroxyapatite is deposited. The matrix is composed of type 1 collagen, proteoglycans, and non-collagenous proteins like osteocalcin, osteopontin, etc. Bone cells include osteoclasts, osteoblasts and osteocytes. Osteoclasts are bone resorbing cells derived from hematopoietic cells of monocyte macrophage lineage while osteoblasts are derived from pluripotent stromal stem cells. Osteocytes are terminally differentiated osteoblasts. Osteoporosis is characterized by excessive bone loss due to abnormalities in the bone remodeling cycle, normally a balanced process where resorption by osteoclasts is coupled with bone formation by osteoblasts. Despite the fact that cortical bone comprises 90% of the skeleton, the trabecular (cancellous) bone is more affected by osteoporosis because it has 20 times greater number of remodeling units/volume. The trabeculae decrease in number and size. Both genetic and environmental factors contribute to peak bone mass. Candidate genes linked to BMD include those coding for vitamin D receptor, estrogen receptor, collagen I α1 and LDL receptor-related protein 5 [LRP5]. However, the link between gene polymorphisms and fracture risk is poor.

RANK (receptor activator of NFκB), RANK ligand (RANKL) and OPG are the three key cytokines that regulate osteoclast recruitment and function. Osteoclasts express RANK while osteoblasts express RANKL constitutively on their cell surface. RANKL, a member of the TNF superfamily of ligands and receptors, is essential for the differentiation, activation, and survival of osteoclasts. Osteoprotegerin (OPG), a soluble decoy receptor secreted by osteoblasts, blocks the interaction between RANKL and RANK and serves as a physiological regulator of bone turnover. Bone loss results both from oestrogen deficiency as well as by oestrogen independent, age-related mechanisms like secondary hyperparathyroidism and reduced mechanical loading. Oestrogen deficiency results in a remodeling imbalance, increase in bone turnover and increased osteoclastogenesis. There is increased production of pro-inflammatory cytokines like IL-1 and TNF. TNF stimulates M-CSF (macrophage colony stimulating factor) and RANKL production. RANKL interacts with its cognate receptor RANK to increase osteoclast generation. RANKL and IL-1 also increase osteoclast survival by preventing apoptosis.

Canonical Wnt signalling plays an important role in bone formation. The Wnt signalling pathways are a group of signal transduction pathways that pass signals into a cell through cell surface receptors. The Wnt signalling pathways may be canonical and non-canonical. Canonical pathways involve β-catenin while non-canonical pathway operates independently of β-catenin. The LRP5, a member of the LDL receptor family, is a co-receptor of Wnt located on the osteoblast membrane between two other receptors, Frizzled and Kremen. Wnt signalling involves binding of Wnt to LRP 5/6 and its coreceptor Frizzled resulting in the phosphorylation of LRP 5/6. Axin, which plays a role in destruction of β catenin, binds to this phosphorylated receptor complex and is unable to degrade β-catenin. Accumulation of cytosolic β-catenin leads to nuclear translocation where it activates target gene promoters which result in increased bone mass. The Wnt signalling pathway has antagonists like Dickkopf protein (Dkk) and sclerostin. Sclerostin binding to LRP receptor 5/6 prevents Wnt binding and formation of the Frizzled-LRP complex.[16] Axin remains unphosphorylated and can activate GSK3β (glycogen synthase kinase 3) resulting in degradation of β-catenin. Better understanding of the disease pathophysiology has led to development of several targeted therapies for osteoporosis. Denosumab, a humanized monoclonal antibody to RANKL, mimics the function of OPG and is approved for treatment of osteoporosis. Similarly, inhibitors of sclerostin-like romosozumab and blosozumab are potential therapeutic target. Other agents in the pipeline are cathepsin K inhibitors. Cathepsin K is a tissue-specific cysteine protease expressed by osteoclasts that degrades bone and cartilage matrix proteins, including type 1 collagen.

DIAGNOSIS

Osteoporosis typically produces no symptoms. Often the first manifestation is a low impact fracture, defined as a fracture resulting from trauma equal to or less than fall from standing height. Fractures commonly occur in distal forearm, vertebrae and hips, though any bone can be involved. Vertebral crush fractures lead to spinal deformities, kyphosis (Dowager's hump), loss of height and a protuberant abdomen. Isolated involvement of upper dorsal vertebrae should arouse suspicion of causes other than osteoporosis. Patients may complain of early satiety due to abdominal compression and breathlessness. Hip fractures are associated with excess mortality and are the most dreaded consequence of osteoporosis.

Baseline investigations in a patient with osteoporosis should include blood counts, renal and liver function tests, serum calcium, phosphorus, alkaline phosphatase and urinary calcium excretion. In postmenopausal osteoporosis, these hematological and biochemical investigations are within normal limits and serve to rule out secondary causes.

In absence of tools to measure bone quality, bone mass has become the surrogate marker used to define osteoporosis. Table 58.1 provides a listing of various techniques for assessing bone mass. Techniques like photon absorptiometry, digital X-ray radiogrammetry and single energy X-ray absorptiometry are obsolete. The current gold standard to measure BMD is DXA (dual energy X-ray absorptiometry). The correct terminology as approved by ICSD is DXA not DEXA. DXA can be used for lumbar spine, proximal femur, forearm and even for assessment of total body mineral composition. The radiation risk is negligible. From a clinical stand point it is pertinent to mention that the presence of one or more fragility fractures is sufficient to classify a patient as severe osteoporosis even if the BMD is normal. Peripheral DXA (pDXA) can measure BMD at peripheral sites like heel, phalanges and forearm. It is recommended that pDXA measurements should be interpreted with device-specific upper and lower thresholds.

DXA should ideally be performed at 2 sites: Hip and spine (anteroposterior). The correlation of BMD at different sites is modest and BMD at a specific site is the best predictor of fracture at that particular site. Osteophytes may interfere with BMD measurement in the AP view in the elderly. Lateral spine DXA, though less precise because accurate positioning is difficult, is less affected by spinal degenerative disease. The lumbar spine is the best site for monitoring response to treatment. This is because of greater quantity of trabecular bone that is more sensitive to the changes in BMD that occur in response to drug treatment. Contraindications for bone densitometry include pregnancy, recent gastrointestinal contrast studies and radionuclide tests. DXA results may be inaccurate in presence of osteomalacia, osteoarthritis, vascular calcification, previous fracture, etc.

There are differences in calibration between DXA machines of different companies and an individual patient's bone density reading can differ by as much as 12% when different DXA machines are used. Therefore, whenever possible, follow-up measurements for a given patient should use the same scanning procedure and the same instrument as for the original measurement. Most treatments lead to an increase in BMD of 1–6% over 3 years. The small magnitude of change in response to treatment necessitates that repeat BMD measurements be spaced fairly far apart, usually 2 years or more. An increase in BMD of at least 3–4% is required as "the least significant difference" that exceeds the error of the measurement. Although BMD is useful in guiding decisions to initiate treatment, subsequent changes in BMD do not fully explain reductions in fracture risk. Over reliance solely on BMD should be avoided. Fracture protection benefit may be realized even before BMD gains are detected. No change in BMD should not be taken to imply that therapy is not working

Table 58.1: Various modalities for measurement of bone density

Technique	*Comments*
Conventional skeletal radiography	Insensitive method. Detects loss greater than 30–50% of bone mass.
Dual energy X-ray Absorptiometry (DXA)	Current gold standard. Useful both for axial and appendicular skeleton. Can measure total body mineral content. Drawback: It is a two-dimensional measurement which measures density/area (in grams per square cm). Areal BMD is influenced by bone size. Overestimates fracture risk in individuals with small body frame whose areal BMD is lower. In obese individuals superimposed soft tissue results in elevated BMD readings. Radiation 4 microsievert (chest X-ray radiation is ~20–60 microsievert).
Quantitative computed tomography (QCT)	Provides selective measurement of trabecular and cortical bone and true volumetric bone mineral density. Drawbacks: Cost and high radiation dose (200 microsievert at spine, 1200 microsievert at hip). May be used in very small or large patients or in older individuals with advanced degenerative changes like DISH (diffuse idiopathic skeletal hyper-ostosis).
Peripheral quantitative computed tomography (pQCT)	Used for distal bone (usually radius, tibia). Radiation minimal (3–5 microsievert). Disadvantage is small peripheral regions only can be measured and changes over time are slow. High resolution HR-pQCT is used to study micro-architecture of bone.
Quantitative ultrasound (QUS)	Screening procedure used for cancellous bone in heel and phalanges. Advantages: Low cost, no radiation, portable. Cannot be used for diagnosis. May be able to predict fracture risk, but conflicting evidence for monitoring while on treatment. Drawbacks: Not standardised, operator dependent, poor reproducibility.

because treatment may have prevented bone loss. DXA is also now utilized to pick up vertebral fractures. Vertebral Fracture Assessment (VFA) refers to densitometricspine imaging performed for the purpose of detecting vertebral fractures. This is important because vertebral fracture is the most common osteoporotic fracture and often remains undetected.

Quantitative ultrasound (QUS) is an inexpensive modality of measuring BMD. It may provide some idea about bone quality. The parameters measured include speed of sound (SOS) and broadband ultrasound attenuation (BUA) at the calcaneus. Different systems yield different values which are not comparable. The current role of QUS is as a screening procedure. It cannot be used to diagnosis osteoporosis or make treatment decisions. Low QUS is an independent risk factor for fracture in post-menopausal women over 65 years of age.

OSTEOPOROSIS SCREENING INTERVALS

Recommendations regarding rescreening are limited. Expert groups recommend rescreening in 1 to 2 years if women are at high risk for accelerated bone loss. The Study of Osteoporotic Fractures

Research Group has reported that women with favourable bone density levels at baseline can be retested at greater intervals than previously suggested. Frequent BMD testing is unlikely to improve fracture prediction. Age and baseline BMD are important factors which influence the interval for repeat bone density testing. Gourlay *et al* performed a competing risk analysis, adjusting for oestrogen use and clinical risk factors, in 4957 women aged ≥67 years with normal BMD or osteopenia at baseline.[17] They reported that osteoporosis would develop in less than 10% of older, postmenopausal women during screening intervals that are set at approximately 15 years for women with normal BMD or mild osteopenia (T score greater than –1.50) at the initial assessment, 5 years for women with moderate osteopenia (T score –1.50 to –1.99), and 1 year for women with advanced osteopenia (T score –2.00 to –2.49). Based on this study it has been recommended that DXA be repeated every 2 years in women with advanced osteopenia or if risk factors are present for accelerated bone loss regardless of T score. With moderate osteopenia without risk factors DXA may be repeated in 3 to 5 years. If baseline BMD is normal or mild osteopenia is present with no risk factors, DXA may be repeated in 10 to 15 years.[18]

MEASUREMENT OF BONE QUALITY

Unlike BMD measurements that are well standardized, techniques to measure bone quality *in vivo* are in a state of development. These include quantitative ultrasonography, magnetic resonance (MR) imaging, MR spectroscopy, multidetector CT, and high-resolution peripheral quantitative (HR-pQ) CT. The trabecular bone score (TBS) is a gray-level textural index of bone microarchitecture derived from lumbar spine DXA images using dedicated post-processing software. TBS is a BMD independent predictor of fracture risk. A recent meta-analysis demonstrated that the hazard ratio for major osteoporotic fractures increases by 1.44 per 1 SD decrease in TBS.[19]

WHO SHOULD BE SUBJECTED TO BONE MASS MEASUREMENTS?

Resource constraints do not permit universal screening of all postmenopausal women in most countries. The ISCD recommends screening for osteoporosis in:

- Women aged 65 and older.
- Postmenopausal women under age 65 with risk factors: Medical disorders like RA, patients on long-term corticosteroids (>3 months), current smoking, thinness (BMI <21), parents with history of fragility fracture and previous history of fragility fracture.
- Women during the menopausal transition with clinical risk factors for fracture, such as low body weight, prior fracture, or high-risk medication use.
- Men aged 70 and older.
- Men under age 70 with clinical risk factors for fracture.
- Adults with a fragility fracture.
- Adults with a disease or condition associated with low bone mass or bone loss.
- Adults taking medications associated with low bone mass or bone loss.
- Anyone being considered for pharmacologic therapy.
- Anyone being treated, to monitor treatment effect.
- Anyone not receiving therapy in whom evidence of bone loss would lead to treatment.
- Women discontinuing oestrogen should be considered for bone density testing according to the indications listed above.

BIOCHEMICAL BONE TURNOVER MARKERS (BTMs)

These provide an integrated assessment of global disease activity in contrast to DXA which is regional. Biochemical markers for osteoblastic activity include serum bone-specific alkaline phosphatase, serum osteocalcin and the peptide cleavage products of procollagen synthesis—type I amino terminal propeptide (PINP) and carboxy terminal propeptide of type I procollagen (PICP). Markers for the osteoclastic activity include urinary hydroxyproline, urinary collagen cross links (pyridinoline as well as deoxypyridinoline), urinary cross-linked C telopeptide of type I collagen, urinary cross-linked N telopeptide of type I collagen, urinary hydroxylysine glycosides, serum bone sialoprotein, serum C-terminal pyridinium cross-linked telopeptide domain of type I collagen (S-ICTP), and serum tartarate resistant acid phosphatase

(TRAP). Biochemical markers do not help in diagnosis of osteoporosis or in prediction of fracture risk. These markers are usually employed to monitor response to treatment or adherence.[20] Bone resorption markers decrease relatively faster than bone formation markers. In general, a decrease of 30% after 3 months of treatment with antiresorptives is considered a treatment response. Bone formative agents like teriparatide increase markers of bone formation as well as bone resorption due to physiological coupling of bone formation with resorption. Biochemical assessment has the potential advantage of obviating the need for repeated BMD measurements. A demonstrable fall in these markers may indirectly help to increase patient compliance. Their use in clinical practice, however, is limited by high *in vivo* and assay variability, poor predictive ability in individual patients, and lack of evidence-based thresholds for clinical decision-making. The preferred resorption marker is serum C-terminal telopeptide (S-CTX) and the preferred formation marker is serum carboxy-terminal propeptide of type I collagen (PINP).[21]

The decrease in BTMs compared to pretreatment levels with oral and IV bisphosphonates can range from 30 to 50% and from 40 to 80% with denosumab. Use of a bone resorption marker such as a fasting morning S-CTX may be helpful in evaluating nonresponders with bone loss or fractures on therapy or to identify patients with high bone turnover. An elevated S-CTX level is associated with high bone turnover and could represent malabsorption or poor compliance and the need for evaluation for causes of secondary osteoporosis. BTMs can rise transiently after a recent fracture.[21]

FRACTURE RISK ASSESSMENT

Bone densitometry has several limitations apart from availability, cost, and lack of normative reference values. It does not capture bone quality. Therefore, for a clinician fracture risk assessment assumes greater importance than BMD alone. The WHO in 2008 has come out with an algorithm for 10-year absolute fracture risk called FRAX™ . This can give a 10-year probability of fracture risk even if BMD values are not available. Risk factors included in FRAX™ (www.shef.ac.uk/FRAX) are age, sex, glucocorticoid use, secondary osteoporosis, rheumatoid arthritis, family history, prior fragility fracture, low BMI, current smoking, alcohol consumption >3 units/day and femoral neck BMD. Falls were not included as their risk is unlikely to be modified by a pharmaceutical intervention. The FRAX algorithms are suitable for men and women between 40 and 90 years. Four FRAX™ assessment models have been constructed: The 10-year probability of hip fracture, with and without BMD at the femoral neck, and the 10-year probability of other osteoporotic fractures, with and without BMD at the femoral neck. Other osteoporotic fractures include forearm, clinical spine and humerus. FRAX models are available for several countries including India. Treatment decisions, however, have to be individualized based on the fracture risk and are not provided by FRAX™. It has been suggested that FRAX scores (10-year probability of hip fracture ≥3% or the 10-year probability of major osteoporotic fracture ≥20%) may be used for diagnosis.[22] The United States Preventive Services Task Force (USPSTF) has proposed the use of the FRAX calculator to determine need for screening in women aged 50 to 64 years.[23] If the FRAX 10-year major osteoporotic risk is greater than or equal to 9.3%, which is equivalent to a 65-year-old white woman without risk factors, then the USPSTF recommends screening with DXA. Other screening tools for predicting fracture include the Osteoporosis Risk Assessment Instrument (ORAI), Osteoporosis Self-assessment Tool (OST), Osteoporosis Index of Risk (OSIRIS), and Simple Calculated Risk Estimation Score (SCORE).

Conclusions

Physician awareness is important for detection of osteoporosis before the first fracture. BMD determination by DXA is currently the gold standard for diagnosis. Recent elucidation of pathophysiologic mechanisms has opened vistas for novel therapies.

Key Points

- Osteoporosis has emerged as a major public health problem in India because of an increase in the elderly population.
- DXA is the current gold standard for determining BMD.
- DXA does not capture bone quality.
- The WHO fracture risk assessment tool (FRAX™) aids quantification of fracture risk and can be used even if DXA is not available.

drug treatment. Fracture risk assessment tool (FRAX) is an online fracture risk scoring tool (Fig. 59.1) which provides doctors a way to identify which patients are at risk of having an osteoporotic fracture within the next 10 years (Fig. 59.2), based on specific risk factors. The calculator includes other risk factors over BMD and age. It is country-specific and is used for individuals aged 40 to 90 years, not already receiving treatment. It can be used without bone mineral density (BMD) value, and hence a valuable tool when there is no access to BMD testing. It is advised not to routinely measure BMD without prior fracture-risk assessment using FRAX. FRAX is not suitable for young men and women (age less than 40 years) with secondary causes of osteoporosis.

Following fracture risk assessment using FRAX, the patient is classified as follows:

- Low risk (10% or less)—reassure and reassess in 5 years or less depending on the clinical context.
- Medium risk (10–20%)—measure BMD and recalculate the fracture risk to determine whether an individual's risk lies above or below the intervention threshold.
- High-risk more than 20%—can be considered for treatment without the need for BMD, although BMD measurement may sometimes be appropriate, particularly in younger postmenopausal women.

Intervention Threshold for Treatment

- Women with any prior fragility fracture without the need for further risk assessment.
- Patients with prevalent vertebral fractures without undertaking BMD measurements if these are felt to be inappropriate or impractical.
- Patients with low BMD and a ten-year risk of hip fracture of ≥3% or a ≥20% ten-year risk of a major osteoporosis-related fracture, as assessed with FRAX.
- Post-menopausal women aged 75 and above who have suffered a previous non-hip or non-vertebral fragility fracture should not be initiated on bone-sparing drug treatments unless they are shown to have osteoporosis on dual-energy X-ray absorptiometry (DXA) examination.
- Where both probabilities fall below the treatment threshold, a further assessment is recommended in 5 years or less depending on the clinical context.
- Given the heterogeneity of Indian scenario, intervention thresholds, and management may need to be individualized.

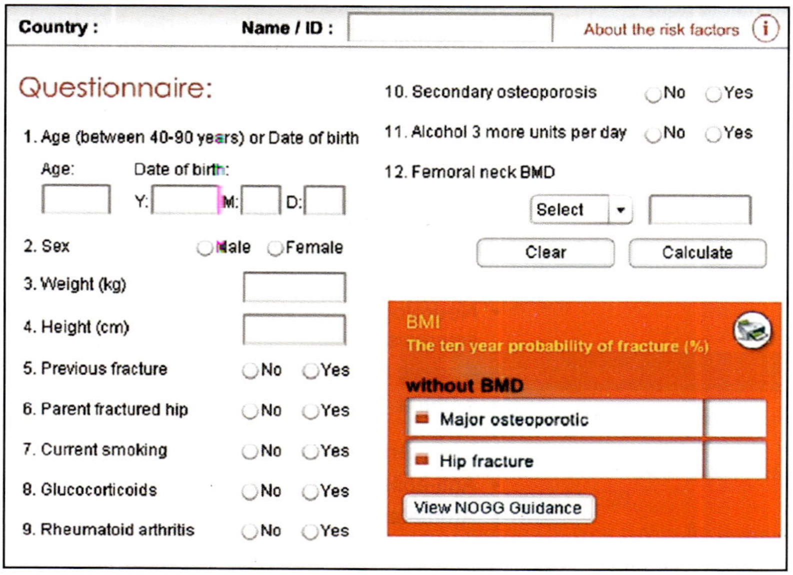

Country : Name / ID : About the risk factors

Questionnaire:

1. Age (between 40-90 years) or Date of birth
Age: Date of birth: Y: M: D:
2. Sex ○Male ○Female
3. Weight (kg)
4. Height (cm)
5. Previous fracture ○No ○Yes
6. Parent fractured hip ○No ○Yes
7. Current smoking ○No ○Yes
8. Glucocorticoids ○No ○Yes
9. Rheumatoid arthritis ○No ○Yes
10. Secondary osteoporosis ○No ○Yes
11. Alcohol 3 more units per day ○No ○Yes
12. Femoral neck BMD
Select
Clear Calculate

BMI
The ten year probability of fracture (%)
without BMD
Major osteoporotic
Hip fracture
View NOGG Guidance

Fig. 59.1: Screen page for the input of FRAX variables (version 3.2.http://www.shef.ac.uk/FRAX). The output is the 10-year probability of a major osteoporotic and hip fracture

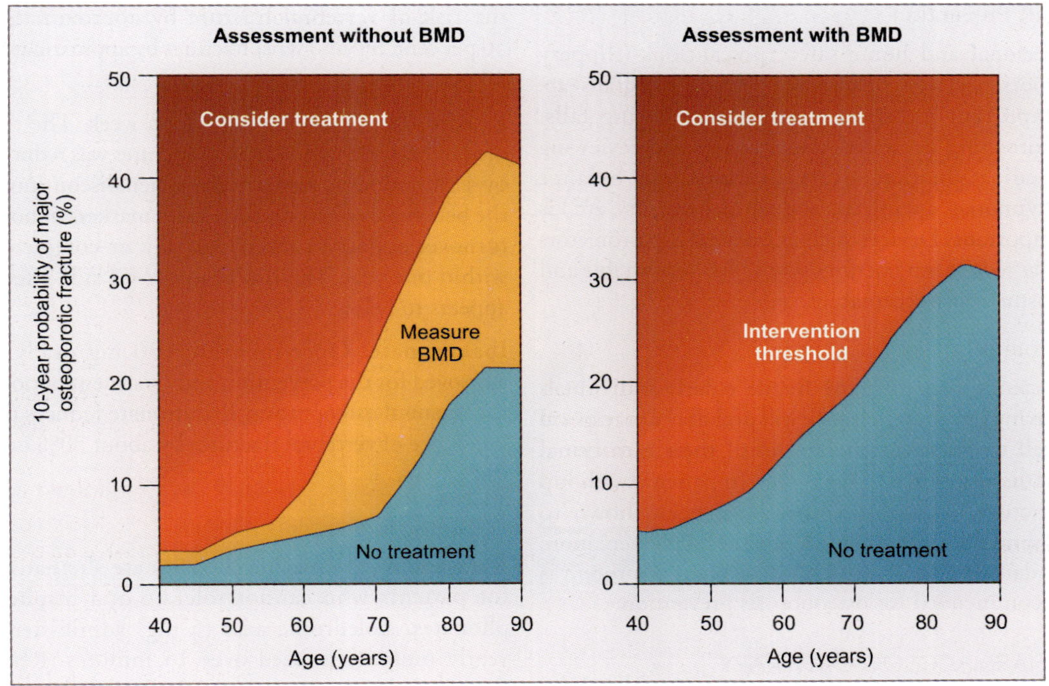

Fig. 59.2: Assessment and intervention thresholds based on the 10-year risk of major fracture

USE OF NON-PHARMACOLOGIC THERAPY[2]

Lifestyle Modification

Exercise, reducing caffeine and alcohol intake, quitting smoking are all lifestyle modifications that can optimize peak bone mass, minimize bone loss and ultimately prevent fractures.

Physical Activity[3]

Exercise improves bone strength, muscular performance, joint flexibility and balance to decrease the risk of falls and fractures. It produces a small increase in bone density (0.5–3%) and decreases the frequency of hip fractures in the elder population. Exercise such as walking, jogging and stair climbing involving bone loading promotes increase in bone mass/bone density in healthy sedentary post-menopausal women. Exercise programs that include weight bearing activity three or four times per week for 45 minutes or weight lifting two or three times per week for 20–30 minutes are beneficial. Bicycling, water aerobics and chair exercises may help maintain bone density in patients with restricted activity. Static weight-bearing exercise, (single-leg standing) may be considered to slow decline of hip BMD. Swimming is not considered as weight-bearing exercise.

Smoking[4]

Smoking is associated with decreased sex hormone concentrations, early menopause, increased bone turnover markers, decreased calcium absorption and reduced bone mineral density. All smokers should be encouraged to quit smoking.

Alcohol

Alcohol appears to have an effect on osteoblasts, slowing bone turnover. Those consuming more than two drinks per day (3.5 units) are at a higher risk of hip fracture and should be considered for fracture-risk assessment. They should be advised to reduce their alcohol intake to (<21 units/week in men, <14 units per week in women). Alcoholics are more susceptible to osteoporosis because of poor nutrition, impaired calcium and vitamin D metabolism.

Caffeine-containing Foods and Beverages

Excessive caffeine consumption may be associated with increased calcium excretion, higher rates of bone loss and modest increases in fracture risk.

Fall Prevention

Personal and home safety precautions (slippery floors, obstacles, insufficient lighting, handrails) are important measures aimed at preventing falls. Correcting decreased visual acuity and reviewing medication that alters alertness and balance (hypnotics, anxiolytics and antidepressants, etc.) is important. Devices such as external hip protectors that reduce hip fractures should be comfortable and cosmetically acceptable.

Protein Intake

Osteoporosis in many fragile elderly individuals having low protein intake will probably not respond well to pharmacotherapy until their nutritional status is repaired. This is seen in patients with hip fracture, whose outcomes have been shown to improve dramatically with protein supplementation. A daily intake of 1 g/kg body weight of protein is recommended for osteoporosis prevention.

PHARMACOLOGIC THERAPY

Bisphosphonates[5]

Bisphosphonates inhibit bone resorption, raise bone density, and reduce fracture risk.[5] They have a half-life >10 years. Nitrogen-containing bisphosphonates bind to hydroxyapatite and block the enzyme farnesyl pyrophosphate synthase in bone tissue and inhibit osteoclast-mediated bone resorption. Alendronate, risedronate, ibandronate and zoledronic acid are approved both the treatment and prevention of osteoporosis in post-menopausal women and for men with osteoporosis. All approved bisphosphonates reduce spine fracture risk. All except ibandronate reduce hip fracture risk. In general the risk of vertebral fracture is reduced by approximately 50 per cent, hip and wrist fractures by 30 per cent. Concern about side-effects is the main reason for poor adherence to therapy. Bisphosphonates should be taken on an empty stomach with a 30–60 minute post-dose fast. Calcium supplements should not be taken within two hours of bisphosphonates. Not recommended if e GFR <35 ml.

Oral Bisphosphonates

Alendronate: Orally 70 mg weekly once for both treatment and prevention in post-menopausal women and for men with osteoporosis. It reduces the risk of vertebral fracture by approximately 50 per cent, hip and wrist fractures by approximately 30 per cent.

Risedronate: Orally 35 mg once a week. The risk of vertebral and non-vertebral fractures was reduced by 41% and 39%, respectively. When discontinued the beneficial effects on BMD and markers of bone turnover appear to revert partially or completely within one year. The risk of upper GI side-effects appears to below.

Ibandronate: Once-monthly 150 mg orally is approved for the prevention and treatment of post-menopausal osteoporosis. Ibandronate reduces the incidence of vertebral fractures by about 50% over three years.

IV Bisphosphonates

Intravenous (IV) bisphosphonates are alternatives for patients who cannot tolerate oral bisphosphonates. Zoledronic acid (5 mg) administered yearly must be infused over 15 minutes. Renal function should be assessed before and is not recommended if e-GFR <35 ml. Zoledronic acid is associated with myalgia, fever, and headache. It is important to adequately hydrate before the infusion. Zoledronic acid reduces the incidence of vertebral fractures by 70% hip fractures by 40%. IV Ibandronate 3 mg once every three months given as a 30 second intravenous injection.

Drug Holiday[6]

It has been shown that bone loss and bone turnover increase within 12–18 months of discontinuing treatment with either oral or parenteral bisphosphonates, suggesting that the persistence of the skeletal effects of retained bisphosphonates may be short-lived in real life compared to the clinical trial setting. Oral bisphosphonate is recommended for 5 years. At the end of 5 years reassess fracture risk with DXA/FRAX and consider drug holiday in appropriate patients. Bisphosphonates may be stopped in low-risk women (e.g. no history of fracture and T-scores better than –2.5) as long as they are followed by BMD and assessment of risk factors. For women at highest risk for fracture, continue up to 10 years, as BMD and fracture benefits are maintained with no increased risk of adverse events. One study showed one dose of zoledronic acid maintained BMD up to 2 years. Therefore, three intravenous doses may be given in 5 years. FRAX may also be helpful in deciding on the duration of the 'drug holiday'.

Bisphosphonates and Mortality[7, 8]

There is good evidence that bisphosphonates decrease mortality. In one study IV zoledronic acid after hip fracture decreased mortality by 28%. Those patients on bisphosphonates at the time admitted to ICU followed for a year had lower mortality than those not on it. All other variables were looked at and none found to explain this.

SIDE EFFECTS OF BISPHOSPHONATES[9]

Gastrointestinal

Reflux, pill esophagitis, oesophageal ulcers and rarely oesophageal stricture. Use of a nonsteroidal anti-inflammatory drug increases the risk. Bisphosphonates should not be used in patients with Barrett's oesophagus and in patients with motility disorders.

Acute phase reactions mostly with first dose of iv zoledronic acid. Flu-like symptoms: low grade fever, myalgia and arthralgia can occur within 24 to 72 hours of intravenous infusion. Paracetomol relieves the symptoms.

Hypocalcaemia may be transient and asymptomatic. This can be severe in vitamin D deficient individuals. Vitamin D deficiency should be treated prior to bisphosphonate infusions. Level of 30 ng/ml is advised for patients with osteoporosis.

Uveitis rare, idiosyncratic and cannot be predicted. Some first-time users of bisphosphonates had an increased risk of uveitis and scleritis. These all occurred within one week of starting zoledronic acid. Usually, respond to prednisolone and most patients do well.

Cancers

There is currently no definitive evidence of bisphosphonates causing oesophageal cancer.

Musculoskeletal

Some patients have severe musculoskeletal pain. This is rare and may sometimes not resolve completely with stopping the bisphosphonate.

Atrial Fibrillation

The risk of AF from oral bisphosphonates is small. Care should be taken when giving IV bisphosphonates for patients with underlying or history ofheart disease.

Osteonecrosis of the Jaw[10]

Osteonecrosis of the jaw (ONJ) is defined as a greater than eight-week history of exposed or necrotic bone, poor healing after extraction in the maxillofacial region which is not due to irradiation of the jaw. ONJ has been reported with all four bisphosphonates but rare (1/100,000 cases). The incidence is less with an oral bisphosphonate and four times higher in individuals with risk factors (given below), receiving a parenteral bisphosphonate. It is more common in cancer patients. The mechanism is unclear, probably multifactorial.

Risk factors include dental extractions, dental implants, poorly fitting dentures, steroids, smoking, pre-existing dental disease, cancer and anti-cancer therapy, oral bone manipulating surgery and duration of exposure to bisphosphonate treatment.

Advise: Good oral hygiene and regular dental visits. Start bisphosphonate after dental treatment. If patient is already on a bisphosphonate and needs dental procedure, go ahead and do it, no evidence that stopping is going to help. Bisphosphonates have a long terminal half-life stay in bone for years.

Atypical Femur Fractures[11]

Atypical femoral fractures are an uncommon type of stress fracture that occurs with little or no trauma. They have been associated with bisphosphonate and denosumab treatments. Prolonged therapy can lead to over suppression of bone turnover and increased skeletal fragility. While the ultimate strength of the bone improves, its toughness decreases by as much as 20% due to the accumulation of micro damage and lack of effective remodelling within the bone. The decrease in toughness may lead to failure in areas with high tensile forces, such as the sub trochanteric and diaphyseal regions of the femur. Many atypical femur fractures were seen in younger women with osteopenia. Therefore, use FRAX to see if patient really needs therapy.

These patients fracture first and then fall. The mechanism is not clear, but both treatment duration and recent exposure to the drugs are associated with this risk, patients tended to have a higher rate of chronic disease and most had T-scores >–2.5. Glucocorticoids and PPIs were recognised as important risk factors. Geometry of femur important: varus angle of shaft and neck, bowing of femur may increase the risk of AFF by increasing the tension on the lateral aspect of the femur. East Asian ethnicities are at higher risk. The risk has been estimated to double with treatment longer than 3 years, appears to decline with discontinuation of the drug, hence the potential benefit of a "drug holiday".

Awareness of this possibility is necessary so that symptomatic patients (mid-thigh pain) can be appropriately investigated. X-rays show cortical thickening and scalloping of the lateral cortex in the diaphyseal region of the femur prior to fracture. MRI shows bone oedema in this area. Prophylactic fixation of insufficiency fractures is recommended in patients with a history of long-term bisphosphonate use, persistent thigh pain and radiographic changes. Rehabilitation after fixation may be long. These patients may not need any treatment for osteoporosis as the bisphosphonates have a long half-life. Teriparatide or denosumab may be options if necessary. The overall benefit of bisphosphonates in preventing hip fractures greatly exceeds the risks of causing atypical fracture. Bisphosphonate should be discontinued in patients with suspected atypical fracture while they are being evaluated. Although drug holidays have been recommended, they have not been proven to prevent these fractures.

ANABOLIC THERAPY FOR OSTEOPOROSIS

Teriparatide[12]

Teriparatide is recommended to prevent vertebral and non-vertebral fractures in both men and post-menopausal women who cannot use other osteoporosis medications.

Teriparatide is a bone forming agent which increases trabecular bone mass and also reconnectivenss of bone architecture. It not only increases bone formation but also increases bone resorption. Therefore, it has a remodelling stimulating rather than a pure anabolic effect. It activates the Wnt signal pathway which is the major anabolic pathway in osteoblasts and inhibits endogenous sclerostin is an inhibitor of this pathway. PTH upregulates RANK ligand causing bone resorption.

1–34 Teriparatide is produced using recombinant DNA technology. Intermittent administration of PTH results in an increase of the number and activity of osteoblasts, leading to an increase in bone mass and in an improvement in skeletal architecture.

The effects are maximal at trabecular bone sites such as the spine. BMD gains occur in the first few months, although anti fracture efficacy is evident only after six months of treatment. BMD changes with PTH begin to level off after 18 months. Increases in bone formation occur more quickly than bone resorption, leading to marked increase in lumbar BMD. Teriparatide decreases vertebral fracturerisk by 65 to 75% and non-vertebral fragility fractures by 53%. The percentage of vertebral fracture reduction that could be attributed to the improvement in spine BMD was only 30 to 40 per cent. The non-BMD determinants of bone strength could include changes in bone geometry, microarchitecture, remodelling rate and/or collagen/mineral matrix properties. Teriparatide does not significantly accelerate fracture repair. Teriparatide increases BMD in women previously treated with bisphosphonates, although the improvement may be less than in women not previously exposed to bisphosphonates. After PTH treatment is discontinued, an antiresorptive preferably a bisphosphonate should be used to preserve or increase gains in BMD acquired with PTH. It is indicated for steroid induced osteoporosis, where a study has shown that teriparatide more effectively increases bone density and substantially reduces vertebral fracture risk by 90% compared to alendronate.

The combination of teriparatide with denosumab causes a faster and greater increases in BMD. Combination of teriparatide with bisphosphonates has not been found to be advantageous and can reduces the anabolic effect of teriparatide.[13] Recent studies show increased cortical porosity in radius and tibia. Denosumab can be given following teriparatide. Teriparatide alone should not be given after denosumab. Teriparatide can be added to denosumab. Biosimilar PTH is available in India.

Dose: 20 mcg/day administered subcutaneously as daily injections.

Side-effects: Orthostatic hypotension, nausea, myalgia and arthralgia, headache and hypercalcaemia. Serum uric acid increases with teriparatide and may precipitate an attack of gout in some. Patients with uric acid levels greater than 7.5 mg/dl probably should not be treated with PTH until their uric acid is controlled. Teriparatide is not recommended for use beyond 2 years due to lack of safety and efficacy data. Osteosarcoma was observed in rats exposed to teriparatide at much higher doses.

Abaloparatide[14]

Abaloparatide is a novel peptide that engages the parathyroid hormone receptor (PTH1 receptor) While teriparatide interacts with both subtypes of PTH receptors, abaloparatide selectively activates the type 1 receptor. As a result there is less stimulation of bone resorption than teriparatide. Therefore, a larger dose can be given. Abaloperide can be administered without causing any disturbance in calcium levels which is the dose limiting problem with teriparatide. There is no need for refrigeration. Abaloparatide has completed phase-3 development for potential use as a daily self-administered SC injection.

Denosumab[15]

Denosumab is a humanized monoclonal antibody against receptor activator of nuclear factor kappa-β (RANK) ligand hence preventing its binding to RANK receptors on osteoclasts. Osteoblasts produce a protein called RANK ligand, which bind to the receptors found of osteoclasts causing them to fuse, mature and attach to bone. After menopause decrease oestrogen leads to increase in RANK ligand which triggers increase osteoclast activity, leading to increased bone resorption.

Denosumab is recommended for the primary and secondaryprevention in post-menopausal women at increased risk of fractures who are unable to comply with, have intolerance or contraindication to bisphosphonates. It is also indicated for treatment of osteoporosis in men at high risk of fracture, prevention of bone loss in women with breast cancer treated with aromatase inhibitors, men with prostate cancer on gonadotropin releasing hormone agonists. It is contraindicated in patients with hypocalcaemia. There is no nephrotoxity.

Denosumab improves bone mineral density of the lumbar spine and hips, and reduces the incidence of vertebral fractures by about 68%, hip fractures by about 40% and non-vertebral fractures by about 20% over 3 years. The reduction in vertebral fracture noted with denosumab is similar to teriparatide and intravenous zoledronic acid and greater than that reported for alendronate. When denosumab treatment is stopped there is arapid drop in BMD and alternative agents should be considered to maintain BMD. Treatment cessation is associated with rebound increase in bone turnover markers and increased risk for new vertebral fractures. Denosumab can be given after discontinuation of bisphosphonates.[31] Denosumab can be given after teriparatide, whereas teriparatide should not be given after denosumab but can be added to denosumab.

Dose: 60 mg subcutaneous injection every six months.

Adverse effects of Denosumab: Back pains, limb pains, other musculoskeletal pain, hypercholesterolaemia, cystitis may occur. Others reported include cellulitis, pancreatitis, ONJ (two cases) and atypical femur fractures.[16,17] No increase in susceptibility to infection was found in patients with rheumatoid arthritis on anti-TNF and denosumab.

Raloxifene[18]

Raloxifene is a selective oestrogen receptor modulator (SERM). Raloxifene reduces the risk of vertebral fractures by 30%. It may be considered as a treatment option when other treatments are contraindicated or unsuitable.

Adverse effects: Increased risk of deep vein thrombosis, pulmonary embolism and stroke.

Dose: 60 mg once daily.

HORMONE REPLACEMENT THERAPY (HRT)

HRT prevents fractures in post-menopausal women. The risk of stroke, venous thromboembolism, cardiovascular disease and cancer is increased in older women and with longer term therapy. Hormone replacement therapy may be considered for the prevention of fractures in younger post-menopausal women as the balance of benefits and risks may be more favourable than in those aged over 60 years. In each case the overall risks should be assessed before starting treatment.

Strontium Ranelate[19]

Strontium ranelate may be considered in post-menopausal women and adult men at high-risk of fracture when other treatments are contraindicated. Strontium ranelate has an inhibitory effect on

REFERENCES

1. Kanis JA, Hans D, Cooper C, Baim S, Bilezikian JP, Binkley N, et al. Task Force of the FRAX Initiative. Interpretation and use of FRAX in clinical practice. Osteoporos Int. 2011; 22:2395–411.
2. Weaver CM, Alexander DD, Boushey CJ, Dawson-Hughes B, Lappe JM, LeBoff MS, et al. Calcium plus vitamin D supplementation and risk of fractures: an updated meta-analysis from the National Osteoporosis Foundation. Osteoporos Int. 2016; 27:367–76.
3. Howe TE, Shea B, Dawson LJ, Downie F, Murray A, Ross C, et al. Exercise for preventing and treating osteoporosis in postmenopausal women. Cochrane Database Syst Rev. 2011 Jul 6;(7):CD000333.
4. Cornuz J, Feskanich D, Willett WC, Colditz GA. Smoking, smoking cessation, and risk of hip fracture in women. Am J Med. 1999; 106:311–4.
5. Cosman F, de Beur SJ, LeBoff MS, Lewiecki EM, Tanner B, Randall S, et al. Osteoporosis International, Clinician's Guide to Prevention and Treatment of Osteoporosis Osteoporos Int. 2014; 25:2359–81.
6. Adler RA, El-Hajj Fuleihan G, Bauer DC, Camacho PM, Clarke BL, et al. Managing Osteoporosis in Patients on Long-term Bisphosphonate Treatment: Report of a Task Force of the American Society for Bone and MineralResearch. J Bone Miner Res. 2016; 31:16–35.
7. Lyles KW, Colón-Emeric CS, Magaziner JS, Adachi JD, Pieper CF, Mautalen C, et al. HORIZON Recurrent Fracture Trial. Zoledronic acid and clinical fractures and mortality after hip fracture. N Engl J Med. 2007; 357:1799–809.
8. Lee P, Ng C, Slattery A, Nair P, Eisman JA, Center JR. Preadmission Bisphosphonate and Mortality in Critically Ill Patients. J Clin Endocrinol Metab. 2016; 101:1945–53.
9. Suresh E, Pazianas M, Abrahamsen B. Safety issues with bisphosphonate therapy for osteoporosis. Rheumatology (Oxford). 2014; 53:19–31.
10. Khosla S, Burr D, Cauley J, Dempster DW, Ebeling PR, Felsenberg D, et al. American Society for Bone and Mineral Research. Bisphosphonate-associated osteonecrosis of the jaw: report of a task force of the American Society for Bone and Mineral Research. J Bone Miner Res. 2007; 22:1479–91.
11. Adler RA. Bisphosphonates and atypical femoral fractures. Curr Opin Endocrinol Diabetes Obes. 2016; 23:430–434.
12. Neer RM, Arnaud CD, Zanchetta JR, Prince R, Gaich GA, Reginster JY, et al. Effect of parathyroid hormone (1–34) on fractures and bone mineral density in postmenopausal women with osteoporosis. N Engl J Med. 2001; 344:1434–41.
13. Cosman F, Eriksen EF, Recknor C, Miller PD, Guañabens N, Kasperk C, et al. Effects of intravenous zoledronic acid plus subcutaneous teriparatide [rhPTH(1-34)] in post-menopausal osteoporosis. J Bone Miner Res. 2011; 26:503–11.
14. Miller PD, Hattersley G, Riis BJ, Williams GC, Lau E, Russo LA, et al. ACTIVE Study Investigators. Effect of Abaloparatide vs Placebo on New Vertebral Fractures in Postmenopausal Women With Osteoporosis: A Randomized Clinical Trial. JAMA. 2016; 316:722–33.
15. Cummings SR, San Martin J, McClung MR, Siris ES, Eastell R, Reid IR, et al. FREEDOM Trial. Denosumab for prevention of fractures in postmenopausal women with osteoporosis. N Engl J Med. 2009; 36:756–65.
16. Anastasilakis AD, Toulis KA, Goulis DG, Polyzos SA, Delaroudis S, Giomisi A, et al. Efficacy and safety of denosumab in postmenopausal women with osteopenia or osteoporosis: a systematic review and a meta-analysis. Horm Metab Res. 2009; 41:721–9.
17. Curtis JR, Xie F, Yun H, Saag KG, Chen L, Delzell E. Risk of hospitalized infection among rheumatoid arthritis patients concurrently treated with a biologic agent and denosumab. Arthritis Rheumatol. 2015; 67:1456–64.
18. Delmas PD, Genant HK, Crans GG, Stock JL, Wong M, Siris E, et al. Severity of prevalent vertebral fractures and the risk of subsequent vertebral and nonvertebral fractures: results from the MORE trial. Bone. 2003; 33:522–32.
19. Reginster JY, Seeman E, De Vernejoul MC, Adami S, Compston J, Phenekos C, et al. Strontium ranelate reduces the risk of nonvertebral fractures in post-menopausal women with osteoporosis: Treatment of Peripheral Osteoporosis (TROPOS) study. J Clin Endocrinol Metab. 2005; 90:2816–22.
20. Chesnut CH 3rd, Silverman S, Andriano K, Genant H, Gimona A, Harris S, et al. A randomized trial of nasal spray salmon calcitonin in postmenopausal women with established osteoporosis: the prevent recurrence of osteoporotic fractures study. PROOF Study Group. Am J Med. 2000; 109:267–76.
21. McClung MR, Grauer A, Boonen S, Bolognese MA, Brown JP, Diez-Perez A, et al. Romosozumab in postmenopausal women with low bone mineral density. N Engl J Med. 2014; 370:412–20.
22. Nair R, Maseeh A. Vitamin D: The "sunshine" vitamin. J Pharmacol Pharmacother. 2012; 3:118–26.
23. Cauley JA, Cawthon PM, Peters KE, Cummings SR, Ensrud KE, Bauer DC, et al. Osteoporotic Fractures in Men (MrOS) Study Research Group. Risk Factors for Hip Fracture in Older Men: The Osteoporotic Fractures in Men Study (MrOS). J Bone Miner Res. 2016; 3:1810–19.

FURTHER READING

1. Black DM, Rosen CJ. Clinical Practice. Postmenopausal Osteoporosis. N Engl JMed. 2016; 374:254–62.
2. Compston J, Cooper A, Cooper C, Gittoes N, Gregson C, Harvey N, et al. National Osteoporosis Guideline Group (NOGG). UK clinical guideline for theprevention and treatment of osteoporosis. Arch Osteoporos. 2017; 12(1):43.

Chapter

60

Glucocorticoid Induced Osteoporosis

Molly Mary Thabah, Jeet Patel

INTRODUCTION

Glucocorticoids are widely used therapeutic agents in rheumatology practice. They form the backbone in the management of rheumatoid arthritis (RA), systemic lupus erythematosus (SLE), and almost all forms of vasculitis.[1] They also form the mainstay of treatment of asthma, and inflammatory bowel disease. One of the most prevalent and serious consequences of treatment with glucocorticoids is glucocorticoid induced osteoporosis (GIOP), which is also the most common cause of secondary osteoporosis at large. GIOP is associated with significant morbidity because of resultant fractures.

The association of glucorticoid excess due to Cushing's syndrome and fractures became obvious even before Phillip Hench and William Kendall were awarded the Nobel Prize in 1950 for their discovery of the beneficial effects of "substance E" on a patient with rheumatoid arthritis.[1,2] Since then there is plenty of evidence of the deleterious effects of therapeutic glucocorticoids on bone resulting in osteoporosis and fractures.[3]

Data obtained from the UK general practice research database (GPRD) which is representative of the UK general population show that 0.9% of the population are prescribed oral corticosteroid.[4] The highest users are individuals in the age group of 70–79 years of age, and the most common disease for which corticosteroid was most likely to be used long term was arthritis, followed by chronic obstructive pulmonary disease (COPD). Between the years 1989 to 2008 there has been a rising trend for long-term glucocorticoid therapy prescriptions in the UK.[5] The increase prescriptions for long-term glucocorticoid therapy was more likely for patients with RA, and polymyalgia rheumatica/giant cell arteritis, but the trend is declining or stable for other diseases like asthma, COPD, and ulcerative (UC). In Germany as many as 53 to 80% patients with RA receive systemic glucocorticoids along with disease modifying antirheumatic drugs (DMARDs) as part of the management of RA.[6]

PATHOGENESIS OF GIOP

Glucocorticoids causes osteoporosis by directly affecting all three type of bone cells—the osteoblast, the osteoclast, and the osteocyte. While the function of the osteoblast is to form bone, the osteoclasts are responsible for bone resorption, both together they perform bone remodeling. Bone remodeling is a process whereby old bone is being replaced by new bone, resulting in the "complete regeneration of the adult skeleton every 10 years".[7] The osteoclast and osteoblast function as a team forming the basic multicellular unit (BMU) to carry out remodeling. The BMUs tunnel through the cortical bone and across trabecular bone creating a trench, with the osteoclast removing bone, then they leave the resorption site. Into this newly excavated area the osteoblasts move in and deposit new bone, which later become mineralised. Normally bone mass is preserved due to a tight balance between bone resorption and bone formation. In osteoporosis, this process becomes disturbed with either increase in bone resorption as seen in postmenopausal osteoporosis or decrease in bone formation as seen in involutionary osteoporosis. In GIOP the mechanism of osteoporosis is an initial relative increase in bone resorption followed by stable decrease in bone formation.

The lifespan of an osteoblast is 3 months and that of an osteoclast is 2 weeks. Soon after eroding bone to a certain extent the osteoclasts die by apoptosis, similarly almost 70% osteoblasts also undergo apoptosis. The remaining osteoblasts either turn into osteocytes or they become cells that line the newly formed bone (lining cells). The function of the osteocyte was relatively unexplored till recently where it is hypothesized that it has a role in bone remodeling by its capacity to sense micro-injury followed by recruitment of osteoblasts and osteoclast to repair injured bone.[8] Osteocytes are actually more abundant in number than osteoblasts.[7,8] They are embedded throughout the mineralized bone matrix, they communicate with one another as well as with lining cells via their dendritic processes that run along the canaliculi.[8] Whenever the osteocyte sense injury to the bone, they send signals to the lining cells through their dendritic process, which in turn cause the osteoclast precursors to home to the injured area and start repair process.

The key mechanism of GIOP is direct action of glucocorticoid on the osteoblasts by decreasing their number, promotion of osteoblasts and osteocyte apoptosis.[9] The characteristic histological changes of bone exposed to glucocorticoids consist of thinning of trabeculae, increase trabecular space, decreased osteoid width, *in situ* death of portions of bone, with markedly reduced bone formation rate and reduced bone turnover.[9] Much of these changes are convincingly explained by increased osteoblasts apoptosis.[9] *In vivo* experiments where wild type mice and mice with transgenic expression of 11β-hydroxysteroid dehydrogenase type 2 (an enzyme that inactivates glucocorticoid) on osteoblasts were exposed to excess glucocorticoids, it was shown that there was equivalent bone loss in both type of mice.[10] There was an increase in osteoblast and osteocyte apoptosis that occurred in wild type mice that was prevented in transgenic mice. Consistent with this, by histology it was found that osteoblasts, osteoid area, and bone formation rate was higher in glucocorticoid treated transgenic mice. Bone strength was also reduced in the wild type mice, but was preserved in transgenic mice. Therefore, it was concluded that increased osteoblast and osteocyte apoptosis (death) plays a key role in causing GIOP.

Using a similar experimental approach, when wild-type mice and mice with transgenic expression of 11β-hydroxysteroid dehydrogenase type 2 on osteoclasts, were exposed to excess glucocorticoids for 7 days, there was decreased number of cancellous osteoclasts in transgenic mice but not in wild-type mice, suggesting that excess glucocorticoids do not reduce osteoclasts number.[11] Rather in this experiment and one previous other the survival of osteoclasts are increased, which explains early rapid bone loss characteristic of GIOP.[12]

Other Important Factors Involved in the Pathogenesis of GIOP

Canonical Wnt signalling pathway involving β-catenin is important in generation of osteoblasts and bone formation (reader is referred to the chapter on epidemiology, pathogenesis, and diagnosis of osteoporosis). Co-receptors for Wnt signalling include the LRP5 (a member of the LDL receptor family) which is located on the osteoblast membrane, between two other receptors Frizzled and Kremen, the phosphorylation of which leads to cytosolic accumulation of β-catenin, activation of target gene promoters which result in increased bone mass. The antagonists of Wnt signalling included Dickkopf-1 (DKK-1) and sclerostin.

Glucocorticoids enhances the expression of DKK-1, suppresses bone morphogenetic proteins (BMP) which are critical to osteoblast differentiation.[13–15]

Another mechanism of inhibiting Wnt signalling by glucocorticoids is by oxidative stress, attenuating AKT phosphorylation, and increasing activation of forkhead box (Fox) O transcription factor family.[16] They also increase PPAR-γ which induces terminal adipocyte differentiation while suppressing osteoblast differentiation thus contributing to reduced number of osteoblasts and bone formation.[17]

Sclerostin, an inhibitor of Wnt signaling, and bone formation has emerged as an important therapeutic target in the treatment of osteoporosis. It is now convincingly shown in murine models that glucocorticoids cause increased expression of sclerostin in wild type mice leading to reduced bone formation and reduced bone strength, whereas sclerostin deficiency leads to activation of Wnt/β-catenin signaling and high bone mass as a result of increased bone formation (with unchanged resorption).[18]

Impact of Glucocorticoids on BMD and Fracture Risk

Bone loss in GIOP is rapid, the onset is soon after glucocorticoids are initiated.[19] In a longitudinal, clinical, and histomorphometric study on 23 patients requiring long-term glucocorticoid therapy, where biochemical and bone biopsies were carried out before and during treatment with 10–25 mg of prednisone, it was shown that total bone volume on iliac crest biopsies decreased up to 27%. This decreased occurred with first 5–7 months of therapy.[20] On subsequent biopsies (including after 12 months of treatment) the changes were minor. It was noted in the bone biopsies that there was significant lack of osteoblastic activity which also confirmed that GIOP is caused by decreased osteoblastic activity due to increased osteoblastic death.

Bone mineral density (BMD) decreases rapidly by 2–4.5% after just 6 months of therapy followed by a much slower decline.[19,20] This initial rapid decline in BMD is attributed to accelerated bone resorption, and a slower, later phase of bone loss to inadequate bone formation.

Both lumbar spine and hip BMD is lower in glucocorticoid users compared to controls.[21] Bone loss occurs in both cortical and trabecular bone, but is greatest in trabecular bone. Hence it not surprising that in GIOP most fractures occur in the vertebral bodies as they have the most trabecular bone.

A meta-analysis of 66 studies on BMD measurements in 2891 glucocorticoid users, also confirmed the finding that bone loss occurs with the first months of starting glucocorticoid therapy.[21] In this meta-analysis there were 23 papers which evaluated fractures, and since the GPRD study is the largest study to date to evaluate risk of fractures among glucocorticoid users it was analyzed separately.[3,4] The relative rate (RR) of overall fracture in glucocorticoid users was 1.33 in the GPRD study and 1.91 in the other studies.[21] The RR for vertebral fractures was 2.60 and 2.86 respectively in the GPRD study and the other studies; while risk of hip fractures was increased by 60% in the GPRD study and 101% in the 23 other studies.

A strong relationship exists between daily glucocorticoid dose and risk of fracture;[22,23] the higher the dose the greater the risk.[21,22] The GPRD study showed that excess fracture risk was stable when the dose of prednisolone was <5 mg but increased for higher dose.[3,21] Even a dose as low as 2.5 mg of prednisolone taken for a long time carries an increased risk of fracture.[3] Another study showed that patients who use daily prednisolone dose of 20 mg had a 60% higher non-vertebral fracture rate compared to controls.[21]

There is also a strong relationship between cumulative glucocorticoid dose and decreases in spine and hip BMD. A dose as low as 7.5 mg of prednisolone daily is associated with significant reductions in BMD.[24,25]

According to the GPRD study, the onset of fracture risk is rapid. The risk of non-vertebral fractures increased by 54% in the first year of therapy in patients using 7.5 mg or more of prednisolone per day.[3] Increased risk for vertebral fractures is also present. There is a high rate of occurrence of new vertebral fracture among glucocorticoid users within first year of glucocorticoid therapy.[26,27]

Fortunately, as rapidly as bone loss occurs the fracture risks also rapidly return to baseline soon after glucocorticoid therapy is stopped. In the GPRD study, most of the excess risk of fracture disappeared within 1 year of stopping therapy. This effect was most pronounced for vertebral fractures, but was also apparent for hip.[3,21] There is also evidence that after stopping glucocorticoids the BMD returns to baseline,[24] and the BMD of a non-glucocorticoid user is comparable to a prior-glucocorticoid user.[28] But at the same time even though the BMD of non-glucocorticoid user and prior glucocorticoid user is comparable, prior glucocorticoid users still stand a higher risk for fracture.[29]

There is evidence that suggests the risk and rate of fracture is higher in GIOP patients when compared to idiopathic osteoporosis,[30] or post-menopausal osteoporosis at a similar level of BMD[22] although some studies show otherwise.[31] It is possible that glucocorticoids not only reduce the BMD but it induces micro-architectural changes of bone resulting in poor bone quality.[32] Hence fractures in GIOP occur independent of the level of BMD. This disparity between BMD and fracture makes BMD measurement alone inadequate to assess fracture risk in GIOP. This poor bone quality is explained by increased apoptosis of the osteocyte, a cell which is recently recognize to play an important role in determining bone strength.[7,8]

Risk Factors for GIOP

The risk factors for GIOP listed in Table 60.1 include among others higher doses of glucocorticoids, longer duration of therapy, higher cumulative dose, and continuous use of glucocorticoids. Continuous treatment with oral 10 mg/day of prednisone for more than 90 days was associated with a 7-fold increase in hip fractures and a 17-fold increase in vertebral fractures.[33] Data from the GPRD show that intermittent use of high dose glucocorticoids (daily dose ≤15 mg of prednisone) or cumulative dose less than 1 gram is associated with small increased risk for fracture.[34] In contrast the intermittent use of multiple high dose (≥15 mg prednisolone) or cumulative exposure of more than 1 gm is associated with substantial fracture risk.

Another important risk factor for GIOP is the activity of 11β-HSD. The enzyme 11β-HSD has two isoenzymes 11β-HSD1 and 11β-HSD2. The isoenzyme 11β-HSD1 catalyzes conversion of inactive glucocorticoids to the active form (cortisone to cortisol), while 11β-HSD2 is inactivator of glucocorticoids. Increased expression of 11β-HSD1in osteoblasts facilitates local synthesis of active glucocorticoids with consequent effects on osteoblastic proliferation. It is shown that the activity of 11β-HSD1 increases with age and on exposure to glucocorticoids.[35]

It seems that natural and synthetic glucocorticoids differ in their vulnerability to 11β-HSD, with dexamethasone causing more osteoporosis than prednisone possibly because it is resistant to inactivation by 11β-HSD2.[36,37]

Underlying inflammation and disease are all important contributors to GIOP. RA, spondyloarthritis (SpA), SLE, GCA/PMR are all associated with bone loss which is in inflammation driven. RA in particular, the bone resorption is mediated by RANK-RANK ligand osteoprotegerin network, whereas the Wnt-βcatenin signalling pathway is important in SpA.[38] Effective control of disease activity by biologic DMARDs and small molecules have led to slowing down of radiographic progression and osteoporosis.[39]

Table 60.1: Risk factors for glucocorticoid induced osteoporosis (GIOP) and fractures

Risk factors
1. Advanced age, low BMI, postmenopausal status, low BMD, physical inactivity, smoking, frequent falls
2. Prolonged duration of glucocorticoid treatment
3. Higher doses (e.g. daily dose of >10 mg prednisolone (or equivalent) for more than 90 days)
4. Intermittent, multiple doses amounting to cumulative dose >1 gm/year
5. Prevalent fractures
6. Underlying disease and inflammation, for example, rheumatoid arthritis, ankylosing spondylitis, juvenile idiopathic arthritis.
7. Increase activity of 11βHSD1
8. Glucocorticoid receptor genotype

Abbreviations: BMD, bone mineral density; BMI, body mass index; 11βHSD1, 11- beta hydroxy steroid dehydrogenase

Clinical and Laboratory Evaluation for GIOP

Clinical: Any patient receiving long-term glucocorticoids in any dose is at risk for osteoporosis. There are no clinical symptoms and signs of osteoporosis nor GIOP, until the occurrence of a fracture. The most common site for fracture is the vertebral body, as the spine is mostly composed of cancellous bone. The only hint of vertebral fracture would be a loss in height or change in posture. New onset back pain can sometimes be present, however, most vertebral fractures are asymptomatic and is detected incidentally on plain radiographs done for other reasons.

Patients who are prescribed glucocorticoids need to be explained of the side effects which include osteoporosis, osteonecrosis, weight gain, hyperglycemia, susceptibility to infection, electrolyte abnormalities among others.[37] Before a patient is prescribed anti-resorptive therapy such as a bisphosphonate it is pertinent to obtain a baseline serum calcium, phosphorus, serum alkaline phosphatase, and 25-hydroxy vitamin D (vitamin D) levels to avoid drug induced hypocalcaemia. Vitamin D deficiency is particularly common in patients with systemic inflammatory rheumatic diseases, plus glucocorticoid therapy is independently associated with vitamin D deficiency. Vitamin D deficiency could also be due to inadequate intake and impaired synthesis due to poor sun exposure.

Bone turnover markers: Bone turnover markers are not helpful in the diagnosis nor in the follow-up of GIOP per se, as the pathogenesis of GIOP is due to a low turnover of bone mass.[37] But if another disease process such as hyperparathyroidism is suspected, then it might be useful to obtain them.

DXA: Assessments of the BMD by dual energy absorptiometry (DXA) alone is not sufficient to predict fracture.[37] The risk for fracture among glucocorticoid users appears is in the initial 3 months of starting glucocorticoid therapy, even before significant change in BMD occurs.[21,23,37] This rapid onset of fracture risk is because of the direct effect on the bone quality mediated by osteocyte apoptosis, and this poor bone quality is not captured by DXA.[37] BMD measurements by DXA are useful for follow-up.

FRAX® is a WHO recommended fracture risk assessment tool used to calculate the ten-year probability of fracture. It is useful in identifying individuals who have a high risk for fracture, will benefit from therapy in postmenopausal and senile osteoporosis. Fracture risk assessment by FRAX® in GIOP may be underestimated because it includes glucocorticoid use as a dichotomous variable (Yes or No). Glucocorticoid use is entered yes if the patient is currently on oral glucocorticoids or has been exposed to at least ≥5 mg prednisolone (or equivalent) for 3 months. FRAX® does not include the actual dose of prednisolone or cumulative dose or the total duration of intake of glucocorticoid. Moreover, it includes femoral neck BMD (g/cm^2) while in GIOP the vertebral fractures are more prevalent. When glucocorticoid dose are entered as a risk factor into FRAX, what one gets is the risk for a moderate dose of prednisolone 2.5 to 7.5 mg. For those taking prednisolone of 7.5 mg or more FRAX adjustment is required as follows: For hip fracture risk multiply the risk value given by FRAX by 1.20. Major osteoporotic fractures risk multiply the risk value by 1.15.

TREATMENT OF GIOP

General Measures

At the start of glucocorticoid therapy all patients need to be given elemental calcium 800–1000 mg/day and vitamin D 600–800 units/day. An effort must be made to prescribe glucocorticoids at the minimum effective dose, treat the underlying disease effectively, and to use glucocorticoid sparing agent wherever possible. Patients also need to undergo aerobic exercise training, avoid immobilization, so that further osteoporosis is prevented.

Bisphosphonates

Specific treatments include anti-resorptive treatment with bisphosphonates (Table 60.2). Alendronate, risedronate, etidronate, and zolendronate are all effective and approved agents for prevention and treatment of GIOP.[26,40–43]

A meta-analysis of 9 studies comprising a total of 1134 patients investigating the efficacy of alendronate in preventing GIOP in patients with rheumatic diseases, showed that alendronate can increase the BMD at the lumbar spine and hip, but neither vertebral nor non-vertebral fractures can be prevented.[40] A single 5 mg yearly infusion of zolendronate is comparable and probably more effective compared to risedronate in treating GIOP and preventing loss in BMD.[42] This strategy of using yearly zolendronate is particularly useful as adherence to bisphosphonates is poor.

Table 60.2: Pharmacological treatment of glucocorticoid induced osteoporosis (GIOP)

Drug	*Dose*	*Route of administration*	*Current status*
Alendronate	70 mg once a week	Oral	First line therapy in GIOP
Risedronate	35 mg once a week	Oral	
Zolendronate	5 mg once a year	Intravenous infusion	When patient is not tolerating oral therapy or non-compliance to oral therapy
Teriparatide	20 mcg once a daily	Subcutaneous injections	Second line treatment
Denosumab	60 mg twice a year	Subcutaneous injections	Undergoing phase III trials in GIOP

Pooled data from 27 RCTs including 3075 patients who had received a bisphosphonate for prevention or for treatment of GIOP, show that bisphosphonates are able to reduce the risk of vertebral fractures, and can stabilize or increase BMD at the lumbar spine and femoral neck when compared to controls.[43] More specifically, risedronate, etidronate, and teriparatide are associated with decreased vertebral fractures.[44]

Based on currently available evidence, experts agree that bisphosphonates should only be used in the first 2 years of glucocorticoid therapy. This is because there is insufficient data to recommend long-term use of bisphosphonates.

Teriparatide

Teriparatide, a recombinant human parathyroid hormone (1–34) is an anabolic agent and it stimulates new bone formation by directly causing osteoblastogenesis, and reduces osteocyte apoptosis, thereby directly targeting the key pathogenetic mechanism of GIOP.[45] In an 18-month, double blind RCT, teriparatide was found to increase the lumbar spine and total hip BMD much more than an increase caused by alendronate.[45]

Teriparatide also causes new bone formation as shown in a recent RCT which compared the effect of alendronate and teriparatide on the trabecular bone score (TBS) among patients who were on chronic glucocorticoid therapy.[46] The TBS derived from reanalysis of DXA scans is a good measure of trabecular bone volume, number of trabeculae and their connectivity. A high TBS indicates that the bone microarchitecture is strong and fracture resistant. In this trial involving 214 patients in the alendronate group and 214 patients in the teriparatide group followed for 36 months, TBS increased by 3.7% in the teriparatide arm. Increases in the lumbar spine BMD was also higher in the teriparatide arm.

The major problems with teriparatide is the cost, daily subcutaneous injections, need for refrigeration of the drug, and occurrence of sporadic hypercalcemia. Even with these limitations teriparatide has a clear role in patients who have established osteoporosis, and have a significantly high risk for fractures. It is also suggested that therapy with teriparatide should be followed by bisphosphonate therapy. Denosumab a human monoclonal antibody to RANKL is currently undergoing phase III clinical trials.

FRACTURE RISK ASSESSMENT FOR PATIENTS ON GLUCOCORTICOIDS

Initial risk assessment: The 2017 ACR guidelines for management of GIOP suggests that clinical assessment for fracture risk in all age groups should be done within 6 months of starting glucocorticoid therapy and repeated yearly.[47]

In adults <40 years in absence of osteoporotic fracture or other significant risk factors BMD testing is not required. But if there is prior fracture and other significant risk factors such as malnutrition, weight loss or low body weight, hypogonadism, secondary hyperparathyroidism, family history of hip fracture, alcohol use, etc. then they have to undergo BMD testing.

In adults >40 years FRAX with glucocorticoid dose correction and BMD is to be done with 6 months of glucocorticoid treatment.

In the clinical setting glucocorticoid users can be stratified into those having high fracture risk, moderate fracture risk, and low fracture risk based on history of osteoporotic fracture, BMD results, and FRAX scores.

ADULTS MORE THAN 40 YEARS

High fracture risk

- Past history of osteoporotic fracture
- Men aged more than 50 years and postmenopausal women with BMD T score <–2.5 at spine or hip
- Glucocorticoid dose adjusted FRAX 10 years risk of major osteoporotic fracture >20% and/or hip fracture >3%.

Moderate fracture risk

- Glucocorticoid adjusted FRAX 10 years risk of major osteoporotic fracture 10–19% and/or hip fracture >1% and <3%.

Low fracture risk

- Glucocorticoid adjusted FRAX 10 years risk of major osteoporotic fracture <10% and/or hip fracture <1%.

ADULTS LESS THAN 40 YEARS

High fracture risk

- Past history of osteoporotic fracture.

Moderate fracture risk: Only if continuing prednisolone >7.5 mg per day for >6 months

- Very low bone mass with Z score < – 3 at hip or spine
- Rapid bone loss (>10%/year at the hip or spine)

Low fracture risk: None of the above apart from on glucocorticoid treatment.

TREATMENT RECOMMENDATIONS[47]

To quit smoking, limit alcohol intake, and exercise. Oral calcium 800–1000 mg/day, vitamin D 600–800 units/day. Both for men and women greater or less than 40 years, with moderate and high fracture risk oral bisphosphonate is given. If oral bisphosphonate is not tolerated then bisphosphonate by intravenous (IV) route is given, if inappropriate then teriparatide or denosumab or raloxifene.

SPECIAL POPULATIONS

For women of child bearing age with moderate to high fracture risk, not planning pregnancy now or might consider pregnancy after their osteoporosis treatment treat with oral bisphosphonate, if bisphosphonate is contraindicated or inappropriate give teriparatide which has a short duration of action in case of pregnancy.

Patients who have solid organ transplants who are continuing glucocorticoid treatment same recommendation as other adult in the same age group if eGFR >30. Those who have renal transplant must be evaluated for metabolic bone disease. Avoid Denosumab as data not yet available on infections in these patients on multiple immunosuppressants.

For children: Optimize calcium and vitamin D intake. But those who have sustained a fracture and are continuing to receive prednisolone >0.1 mg/kg for more than 3 months must be given oral bisphosphonates or IV if oral not tolerated.

Other special group: Age >30 years, those on >30 mg/with and cumulative dose >5 grams/year, treat with oral bisphosphonate.

Follow Up

- Adults more than 40 years, a baseline FRAX with BMD (glucocorticoid included as risk factor with dose adjusted). Patients who are not treated either because they are at low risk or refused treatment, BMD repeated after one year if patient is on high dose glucocorticoid or else once in 2 years.
- Those receiving active treatment repeat BMD every 2 years if there is concern of poor drug adherence, poor absorption or continuing high glucocorticoid use.
- BMD repeated after 2 years of completion of osteoporosis therapy if patient continues use of glucocorticoids.

GIOP is the most common cause of secondary osteoporosis. Bone loss and consequent fracture risk occur soon after initiation of glucocorticoids. No dose is considered safe for osteoporosis. The clinician must be aware of this condition as occurrence of osteoporotic fractures increases the morbidity and cost of treatment.

Key Points

1. Bone loss in GIOP is rapid, the onset is soon after glucocorticoids are initiated. Bone loss is greatest in the first year of glucocorticoid use, maximum in trabecular bone, hence greatest risk in vertebra.
2. The risk of glucocorticoid-induced fractures is strongly related to the daily and also the cumulative dose.
3. Fracture rates drop quickly by one year when steroids are stopped and completely by 3 years.
4. The main mechanism of bone loss is osteoblast apoptosis.
5. Accelerated osteocyte apoptosis contribute to poor quality bone and risk of fracture is higher in GIOP patients when compared to idiopathic osteoporosis, or post-menopausal osteoporosis at a similar level of BMD.
6. GC users are stratified into high fracture risk, moderate fracture risk, and low fracture risk based on history of osteoporotic fracture, BMD results, and FRAX scores.
7. Both for men and women greater or less than 40 years, with moderate and high fracture risk oral bisphosphonate is given.
8. Alendronate, risedronate, etidronate, and zolendronate are approved for prevention and treatment of GIOP.

REFERENCES

1. van der Goes MC, Jacobs JW, Bijlsma JW. The value of glucocorticoid co-therapy in different rheumatic diseases-positive and adverse effects. Arthritis Res Ther. 2014;16 Suppl 2:S2 [Internet] Cited on Jan 2017. Available on http://arthritis-research.com/content/16/S2/S2.
2. Kirwan JR, Bálint G, Szebenyi B. Anniversary: 50 years of glucocorticoid treatment in rheumatoid arthritis. Rheumatology (Oxford). 1999; 38:100–2.
3. van Staa TP, Leufkens HG, Abenhaim L, Zhang B, Cooper C. Use of oral corticosteroids and risk of fractures. J Bone Miner Res. 2000; 15:993–1000.
4. van Staa TP, Leufkens HG, Abenhaim L, Begaud B, Zhang B, Cooper C. Use of oral corticosteroids in the United Kingdom. QJM. 2000; 93:105–11.
5. Fardet L, Petersen I, Nazareth I. Prevalence of long-term oral glucocorticoid prescriptions in the UK over the past 20 years. Rheumatology (Oxford). 2011; 50: 1982–90.
6. Thiele K, Buttgereit F, Huscher D, Zink A; German Collaborative Arthritis Centres. Current use of glucocorticoids in patients with rheumatoid arthritis in Germany. Arthritis Rheum. 2005; 53:740–7.
7. Manolagas SC, Weinstein RS. New developments in the pathogenesis and treatment of steroid-induced osteoporosis. J Bone Miner Res. 1999; 14:1061–6.
8. Manolagas SC. Corticosteroids and fractures: a close encounter of the third cell kind. J Bone Miner Res. 2000; 15:1001–5.
9. Weinstein RS, Jilka RL, Parfitt AM Manolagas SC. Inhibition of osteoblastogenesis and promotion of apoptosis of osteoblasts and osteocytes by glucocorticoids. Potential mechanisms of their deleterious effects on bone. J Clin Invest. 1998 102:274–82.
10. O'Brien CA, Jia D, Plotkin LI, Bellido T, Powers CC, Stewart SA, et al. Glucocorticoids act directly on osteoblasts and osteocytes to induce their apoptosis and reduce bone formation and strength. Endocrinology. 2004; 145:1835–41.
11. Jia D, O'Brien CA, Stewart SA, Manolagas SC, Weinstein RS. Glucocorticoids act directly on osteoclasts to increase their life span and reduce bone density. Endocrinology. 2006; 147:5592–9.
12. Weinstein RS, Chen JR, Powers CC, Stewart SA, Landes RD, Bellido T, et al. Promotion of osteoclast survival and antagonism of bisphosphonate-induced osteoclast apoptosis by glucocorticoids. J Clin Invest. 2002; 109:1041–8.
13. Ohnaka K, Tanabe M, Kawate H, Nawata H, Takayanagi R. Glucocorticoid suppresses the canonical Wnt signal in cultured human osteoblasts. BiochemBiophys Res Commun. 2005; 329:177–81.
14. Ohnaka K, Taniguchi H, Kawate H, Nawata H, Takayanagi R. Glucocorticoid enhances the expression of dickkopf-1 in human osteoblasts: novel mechanism of glucocorticoid-induced osteoporosis. Biochem Biophys Res Commun. 2004; 318:259–64.
15. Hayashi K, Yamaguchi T, Yano S, Kanazawa I, Yamauchi M, Yamamoto M, et al. BMP/Wnt antagonists are upregulated by dexamethasone in osteoblasts and reversed by alendronate and PTH: potential therapeutic targets for glucocorticoid-induced osteoporosis. BiochemBiophys Res Commun. 2009; 379:261–6.
16. Almeida M, Han L, Ambrogini E, Weinstein RS, Manolagas SC. Glucocorticoids and tumor necrosis factor α increase oxidative stress and suppress Wnt protein signaling in osteoblasts. J Biol Chem. 2011 Dec 30; 286:44326–35.
17. Lecka-Czernik B, Gubrij I, Moerman EJ, Kajkenova O, Lipschitz DA, Manolagas SC, et al. Inhibition of Osf2/Cbfa1 expression and terminal osteoblast differentiation by PPAR gamma 2. J Cell Biochem. 1999; 74:357–71.
18. Sato AY, Cregor M, Delgado-Calle J, Condon KW, Allen MR, Peacock M, et al. Protection From Glucocorticoid-Induced Osteoporosis by Anti-Catabolic Signaling in the Absence of Sost/Sclerostin. J Bone Miner Res. 2016; 31:1791–802.
19. Pearce G, Tabensky DA, Delmas PD, Baker HW, Seeman E. Corticosteroid-induced bone loss in men. J Clin Endocrinol Metab. 1998; 83:801–6.
20. LoCascio V, Bonucci E, Imbimbo B, Ballanti P, Adami S, Milani S, et al. Bone loss in response to long-term glucocorticoid therapy. BoneMiner. 1990; 8:39–51.
21. van Staa TP, Leufkens HG, Cooper C. The epidemiology of corticosteroid-induced osteoporosis: a meta-analysis. Osteoporos Int. 2002; 13:777–87.
22. van Staa TP, Leufkens HGM, Abenhaim L, Zhang B, Cooper C. Fractures and oral corticosteroids: relationship to daily and cumulative dose. Rheumatol 2000; 39:1383–9.
23. van Staa TP, Laan RF, Barton IP, Cohen S, Reid DM, Cooper C. Bone density threshold and other predictors of vertebral fracture in patients receiving oral glucocorticoid therapy. Arthritis Rheum. 2003; 48:3224–9.
24. Laan RFJM, van Riel PLCM, van de Putte LBA, van Erning LJTO, van't Hof MA, Lemmens JAM. Low-dose prednisone induces rapid reversible axial bone loss in patients with rheumatoid arthritis. Ann Intern Med 1993; 119:963–8.

25. McKenzie R, Reynolds JC, O'Fallon A, Dale J, Deloria M, Blackwelder W, et al. Decreased bone mineral density during low dose glucocorticoid administration in a randomised-placebo controlled trial. J Rheumatol 2000; 27:2222–5.

26. Cohen S, Levy RM, Keller M, Boling E, Emkey RD, Greenwald M, et al. Risedronate therapy prevents corticosteroids-induced bone loss. Arthritis Rheum 1999; 42:2309–18.

27. Adachi J, Bensen WG, Brown J, Hanley D, Hodsman A, Josse R, et al. Intermittent etidronate therapy to prevent corticosteroid-induced osteoporosis. N Engl J Med 1997; 337:382–7.

28. Hall GM, Spector TD, Griffin AJ, Jawad ASM, Hall ML, Doyle DV. The effect of rheumatoid arthritis and steroid therapy on bone density in postmenopausal women. Arthritis Rheum 1993; 36:1510–6.

29. Kanis JA, Johansson H, Oden A, Johnell O, de Laet C, Melton III LJ, et al. A meta-analysis of prior corticosteroid use and fracture risk. J Bone Miner Res. 2004; 19:893–9.

30. Luengo M, Picado C, Del Rio L, Guanabens N, Montserrat JM, Setoain J. Vertebral fractures in steroid dependant asthma and involutional osteoporosis: a comparative study. Thorax 1991; 46:803–6.

31. Selby PL, Halsey JP, Adams KR, Klimiuk P, Knight SM, Pal B, et al. Corticosteroids do not alter the threshold for vertebral fracture. J Bone Miner Res. 2000; 15:952–6.

32. Chappard D, Legrand E, Basle MF, Fromont P, Racineux JL, Rebel A, et al. Altered trabecular architecture induced by corticosteroids: a bone histomorphometric study. J Bone Miner Res 1996; 11:676–85.

33. Steinbuch M, Youket TE, Cohen S. Oral glucocorticoid use is associated with an increased risk of fracture. Osteoporos Int. 2004; 15:323–28.

34. De Vries F, Bracke M, Leufkens HG, Lammers JW, Cooper C, Van Staa TP. Fracture risk with intermittent high-dose oral glucocorticoid therapy. Arthritis Rheum. 2007; 56:208–14.

35. Cooper MS, Rabbitt EH, Goddard PE, Bartlett WA, Hewison M, Stewart PM.Osteoblastic 11beta-hydroxysteroid dehydrogenase type 1 activity increases with age and glucocorticoid exposure. J Bone Miner Res. 2002; 17:979–86.

36. Weinstein RS. Glucocorticoid-induced osteonecrosis. Endocrine. 2012; 41:183–90.

37. Weinstein RS. Glucocorticoid-induced osteoporosis and osteonecrosis. Endocrinol MetabClin North Am. 2012; 41:595–611.

38. Szentpétery Á, Horváth Á, Gulyás K, Pethö Z, Bhattoa HP, Szántó S, et al. Effects of targeted therapies on the bone in arthritides. Autoimmun Rev. 2017; 16:313–20.

39. Haugeberg G, Helgetveit KB, Førre Ø, Garen T, Sommerseth H, Prøven A. Generalized bone loss in early rheumatoid arthritis patients followed for ten years in the biologic treatment era. BMC Musculoskelet Disord. 2014; 15:289.

40. Kan SL, Yuan ZF, Li Y, Ai J, Xu H, Sun JC, et al. Alendronate prevents glucocorticoid-induced osteoporosis in patients with rheumatic diseases: A meta-analysis. Medicine (Baltimore). 2016;95(25):e3990 [internet] cited on Jan 2017. Available at https://www.ncbi.nlm.nih.gov/pmc/articles/PMC4998340/

41. Reid DM, Hughes RA, Laan RF, et al. Efficacy and safety of daily risedronate in the treatment of corticosteroid-induced osteoporosis in men and women: a randomized trial. J Bone Miner Res. 2000; 15:1006–13.

42. Reid DM, Devogelaer JP, Saag K, Roux C, Lau CS, Reginster JY, HORIZON investigators. Zoledronic acid and risedronate in the prevention and treatment of glucocorticoid-induced osteoporosis (HORIZON): a multicentre, double-blind,double-dummy, randomised controlled trial. Lancet. 2009; 373:1253–63.

43. Allen CS, Yeung JH, Vandermeer B, Homik J. Bisphosphonates for steroid-induced osteoporosis. Cochrane Database Syst Rev. 2016;10:CD001347. [internet] cited on jan 2017. Available from http://onlinelibrary.wiley.com/doi/10.1002/14651858.CD001347.pub2/abstract.

44. Amiche MA, Albaum JM, Tadrous M, Pechlivanoglou P, Lévesque LE, Adachi JD, et al. Efficacy of osteoporosis pharmacotherapies in preventing fracture among oral glucocorticoid users: a network meta-analysis. Osteoporos Int. 2016; 27:1989–98.

45. Saag KG, Shane E, Boonen S, Marín F, Donley DW, Taylor KA, et al. Teriparatide or alendronate in glucocorticoid-induced osteoporosis. N Engl J Med. 2007; 357:2028–39.

46. Saag KG, Agnusdei D, Hans D, Kohlmeier LA, Krohn KD, Leib ES, et al. Trabecular bone score in patients with chronic glucocorticoid therapy-induced osteoporosis treated with alendronate or teriparatide. Arthritis Rheumatol. 2016; 68(9):2122–8.

47. Buckley L, Guyatt G, Fink HA, Cannon M, Grossman J, Hansen KE, et al. 2017 American College of Rheumatology guideline for the prevention and treatment of glucocorticoid-induced osteoporosis. Arthritis Rheumatol. 2017; 69:1521–37.

FURTHER READING

1. Buckley L, Guyatt G, Fink HA, Cannon M, Grossman J, Hansen KE, et al. 2017 American College of Rheumatology guideline for the prevention and treatment of Glucocorticoid-Induced Osteoporosis. Arthritis Rheumatol. 2017; 69(8):1521–37.
2. Krishnamurthy V, Sharma A, Aggarwal A, Kumar U, Amin S, Rao URK et al. Indian rheumatology association guidelines for management of glucocorticoid-induced osteoporosis. Indian J Rheumatol 2011; 6:68–75
3. van Staa TP, Leufkens HG, Abenhaim L, Zhang B, Cooper C. Use of oral corticosteroids and risk of fractures. J Bone Miner Res. 2000; 15(6):993–1000.
4. van Staa TP, Leufkens HG, Abenhaim L, Begaud B, Zhang B, Cooper C. Use of oral corticosteroids in the United Kingdom. QJM. 2000; 93(2):105–11.

Chapter

61

Principles and Interpretation of DXA Scan for the Rheumatologist

Puneet Mashru, Aswin Nair, Nikhil Gupta, Debashish Danda

INTRODUCTION

Osteoporosis is defined as a skeletal disorder characterized by compromised bone strength predisposing a person to an increased risk of fracture.

Over time, various devices have been used to assess bone mineral density (BMD) and other markers of fracture risk and bone strength. These can be divided into central and peripheral devices.

Peripheral Devices

These are devices, which use forearm, heel, finger, tibia, etc. for fracture risk assessment.[1] They are useful in assessing for fracture prediction and help in identifying patients who require a central DXA scan. Treatment may be initiated in high-risk patients based on validated peripheral measurements if central DXA facilities are not available. The drawbacks of these technologies are that they have lower sensitivity compared to DXA and they are not useful for monitoring response to therapy.

1. **Peripheral dual energy X-ray absorptiometry (pDXA):** pDXA uses the principle of dual energy to measure BMD at peripheral sites. Of the various areas, only the forearm (distal 33% of radius) can be used to diagnose osteoporosis.[2] pDXA of the forearm can be used to assess global and vertebral fragility fracture risk in postmenopausal women.
2. **Peripheral quantitative computer tomography (pQCT):** This is done on a pQCT scanner that measures volumetric BMD (g/cm^3) of the forearm. It is portable and easier to use with lower radiation compared to QCT. pQCT predicts hip fracture in postmenopausal women.
3. **Quantitative ultrasound (QUS):** QUS uses ultrasound to assess fracture risk. It can be used at multiple sites like tibia, finger but only the heel has been validated. This technology does not measure BMD but measures propagation velocity (SOS) and signal attenuation (BUA). Validated heel QUS predict global, vertebral and hip fracture risk in postmenopausal women independent of central DXA BMD.

Central Devices

These are devices that use spine and hip for fracture risk assessment.

1. **Quantitative computer tomography (QCT):** BMD assessment of the spine (L1–L3) or femur can be carried out on any commercial CT scanner with the appropriate software. This tool has been pivotal in understanding the pathophysiology and response of various drugs in osteoporosis. QCT offers the advantage of measuring volumetric BMD (g/cm^3), which is more accurate compared to areal BMD (g/cm^2) measured by DXA in individuals with small frame. However, an important disadvantage of QCT is the higher radiation dose.[2]
2. **Dual energy X-ray absorptiometry (DXA):** The DXA scanner uses the principle of dual energy, i.e. two different energy peaks produced by the X-ray tube (30–50 and >70 keV),[3] to differentiate bone from soft tissue. This modality has been approved by the Food and Drug Administration (FDA) for clinical use in 1988 and currently the gold standard to measure BMD. The WHO

classification for diagnosis of osteoporosis, osteopenia and normal BMD is based primarily on reference data obtained by DXA. It has been used in multiple epidemiological and pharmaceutical trials in osteoporosis. A distinct advantage of this modality is that the exposure to radiation is trivial.

DXA scan has limitations as mentioned below

a. Being a two-dimensional measurement (g/cm2) fracture risk in small body framed individuals would be overestimated. Likewise in obese individuals fracture risk would be underestimated due to superimposed soft tissue.[4]
b. DXA may be inaccurate in patients with degenerative disease, aortic calcification, post fracture etc.
c. DXA cannot differentiate between osteoporosis and other causes of low BMD like osteomalacia, multiple myeloma, renal bone diseases etc.

INDICATIONS AND CONTRAINDICATIONS FOR BONE MINERAL DENSITY (BMD) TESTING

If BMD assessment is not done when indicated, an opportunity to treat osteoporosis may be lost and might result in fracture. Paradoxically, when a DXA scan is done without valid indications, it would lead to wasteful use of resources and may lead to inappropriate treatment. Multiple guidelines are available by several societies. The International Society of Clinical Densitometry (ISCD) indications are mentioned in Table 61.1.[2]

Contraindications for Central DXA include pregnancy, patients recently undergone nuclear medicine test or GI contrast studies, extensive orthopedic instrumentation and very obese individuals exceeding the manufacturer's weight limit.

Which Skeletal Sites to use for BMD Measurement?[2,5]

Table 61.2 lists the sites of the body for measurements of BMD, which differs in adults and children.

DXA Scan Interpretation

A stepwise approach in interpreting the DXA scan will help in avoiding common pitfalls (Table 61.3).

Table 61.1: ISCD indications for BMD testing

- Women aged ≥65 years and men aged ≥70 years
- Postmenopausal women <65 years and men <70 years old with clinical risk factors for fracture (patients on corticosteroids > 3 months, rheumatoid arthritis, h/o fractures, low BMI, smoking other medical conditions and drugs associated with decrease in BMD.)
- Women during menopausal transition with clinical fracture risk factors (low body weight, prior fracture or high-risk medication use)
- Adults with a fragility fracture
- Adults with a disease or condition associated with low bone mass or bone loss
- Adults taking medications associated with low bone mass or bone loss
- Anyone being considered for pharmacologic therapy
- Anyone being treated, to monitor treatment effect
- Anyone not receiving therapy in whom evidence of bone loss would lead to treatment
- Women discontinuing estrogen should be considered for BMD testing according to the indications listed above

BMD: Bone mineral density

Table 61.2: Which Skeletal sites to use for BMD measurement?[2,5]

Adults	*Pediatric/Adolescent*
1. PA Spine	1. PA Spine
2. Hip	2. Total body less head (TBLH)
3. Forearm: Recommended in patients where hip or spine is invalid, patients with hyperparathyroidism or in obese patients exceeding the weight limit of DXA machine.	3. Hip is not preferred due to variability in skeletal development

Check Demographics

The ISCD recommends using Caucasian female normative database for women of all ethnic groups. As the fracture risk for males and females is similar at the same BMD values, ISCD recommends the use of female database for males.[6] Z-score calculation depends on correct input of race, age and sex. Thus incorrect input of these demographics will lead to false reports.

Table 61.3: Stepwise interpretation of DXA scans

Steps in interpreting DXA scan
I. Check Demographics—name, age, sex, race
II. Spine
a. Check scan acquisition—image should include part of T12 vertebra up to part of L5 vertebra with both iliac crest visible
b. Positioning—spine centred and straight
c. Correctly labeled vertebrae with intervertebral markers in the intervertebral disc spaces
d. Verify Lateral margins
e. Note BMD progression L1 to L4
f. Look for artifacts and exclude abnormal vertebrae if there is more than 1 SD difference in T-score between adjacent vertebrae
g. ROI*: Ideally should include L1–L4 spine (at least 2 vertebrae required for interpretation)
III. Femur
a. Check scan acquisition—image should include acetabulum, greater trochanter and extend below lesser trochanter
b. Positioning—straight with a little or none of the lesser trochanter visible.
c. Femoral neck box—should include only femoral neck
d. Verify lateral margins
e. Look for artifacts
f. ROI*: Total hip/femoral neck values—choose whichever has a lower T-score
IV. Forearm
a. Check scan acquisition—distal cortex of radius and ulna should be visible
b. Positioning- forearm should be centred and straight
c. Verify lateral margins
d. Look for artifacts
e. ROI*: 33% Radius
V. Comparison with previous DXA: Compare BMD and same ROI
VI. Final report: Appropriate use of T/Z scores

**ROI:* Region of interest (The anatomical area imaged, analyzed and for which bone mineral density values are calculated.); SD: Standard deviation

Spine BMD

Scan Acquisition and Positioning

Patient positioning is a very important step to obtain a valid BMD. This step is dependent on the technologist doing the scan. Correct spine acquisition should include part of T12 vertebra above up to part of L5 vertebra below with both iliac crest visible (Fig. 61.1a). The spine should be centred and straight (Fig. 61.2a). Rotation should be avoided, as it would lead to falsely lower BMD values.[7]

Spine Scan Analysis

a. **Correct labelling of vertebrae:** Labelling of the vertebra should be confirmed by clinician, as the DXA software does it automatically. Labeling of vertebrae should be done from bottom up with the L4–L5 disc margin being at the superior margin of iliac crest. The shape of the vertebrae will also help in correctly identifying the vertebra. L1 to L3 appears as U-shaped, L4 as X-shaped and L5 shaped like a bow tie.
b. **Correct placement of intervertebral markers:** The DXA software automatically places intervertebral markers in the intervertebral disc spaces (Fig. 61.3a), though at times the placement may be incorrect (Fig. 61.3b). The radiologist can use the histogram aid (Fig. 61.3c), which is a part of the DXA software, to correctly place the intervertebral markers.
c. **Verify properly marked out lateral margins:** The vertebral borders are automatically identified and marked out by the machine (Fig. 61.4a). At times, the lateral margin might be inadequately marked out which can lead to falsely low BMD values.

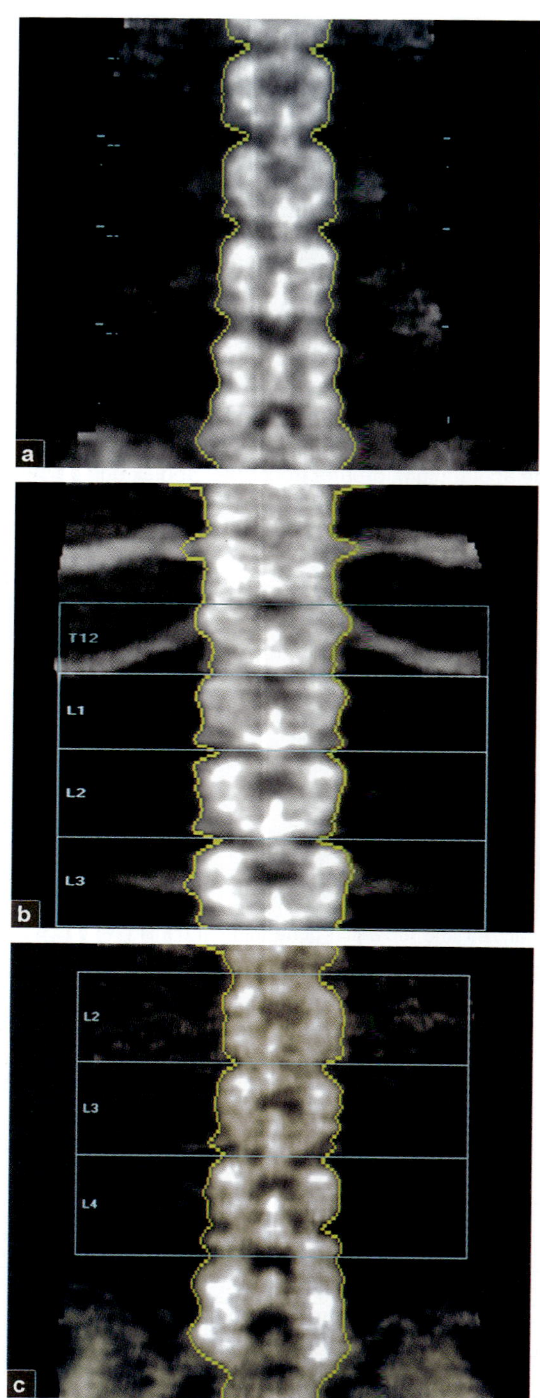

Fig. 61.1: (a) Correct spine acquisition—part of T12 to L5 vertebra seen with both iliac crest visible; Incorrect spine acquisition (b) started too high, L4 vertebra cannot be seen; (c) started too low, L1 vertebra cannot be seen

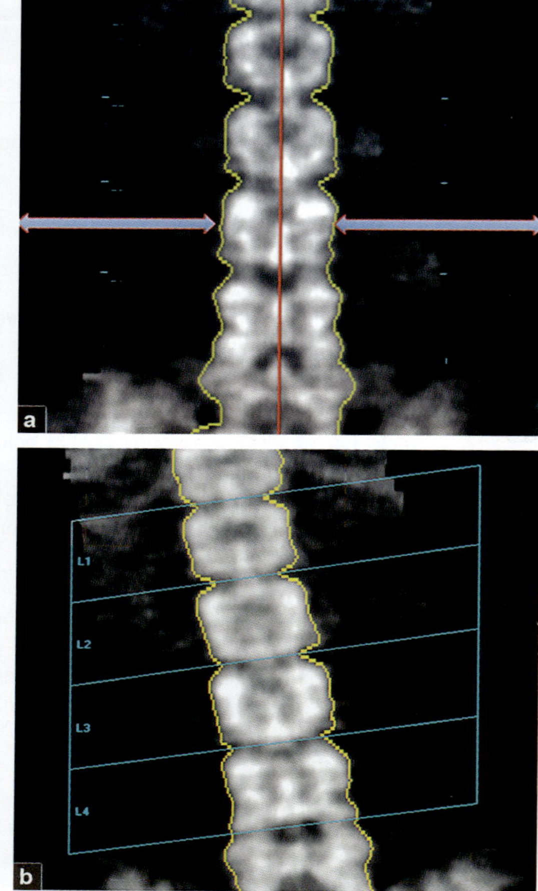

Fig. 61.2: (a) Correct positioning; (b) Incorrect positioning: Spine is not straight

d. **Look at BMD progression and T scores:** BMD values should progressively increase from L1 to L4, however, L3 vertebral BMD may be more than L4 (Fig. 61.5a). Presence of more than 1 SD difference in the T-scores between two vertebrae should alert the clinician to look for artifacts like vertebral fracture, osteosclerosis, etc. The abnormal vertebra should be excluded from BMD analysis (Fig. 61.5b).

Artifacts

It is important to look for artifacts as they can either falsely increase or decrease BMD values. Osteophytes and osteochondrosis can lead to falsely elevated BMD values (up to14%).[8] Falsely elevated BMD values can be seen in advanced stages of

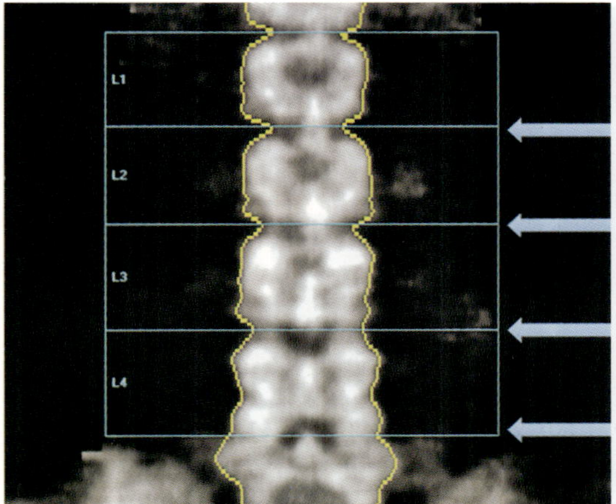

Fig. 61.3: (a) Correctly labeled vertebrae and placed intervertebral markers

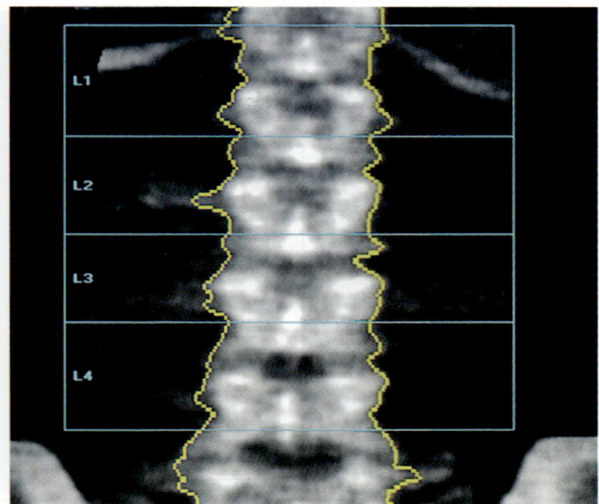

Fig. 61.3: (b) Intervertebral markers not placed in intervertebral disc space

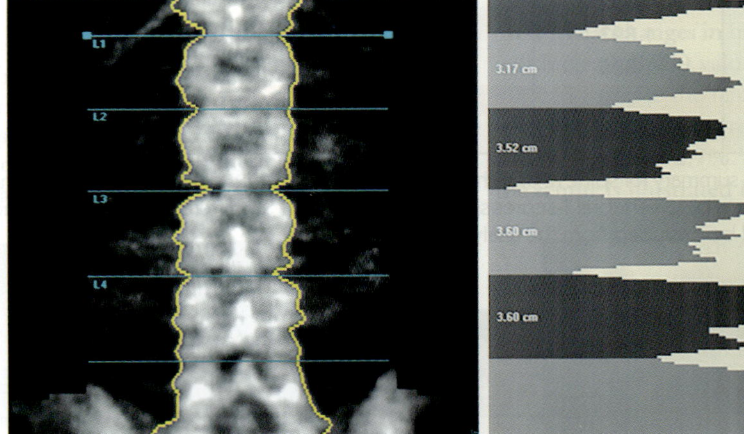

Fig. 61.3: (c) Histogram aid used by the radiologist to correctly place the intervertebral markers

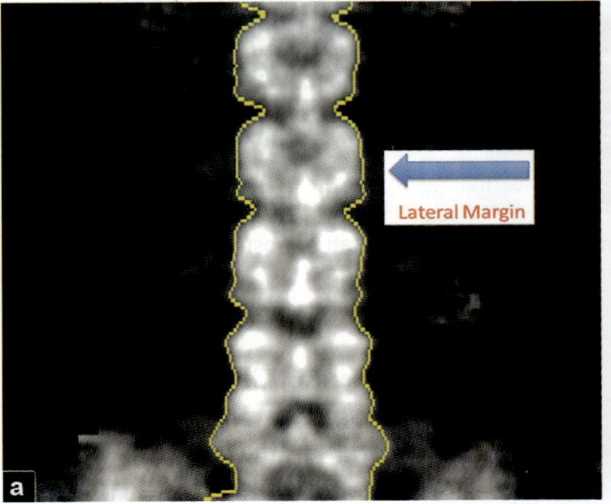

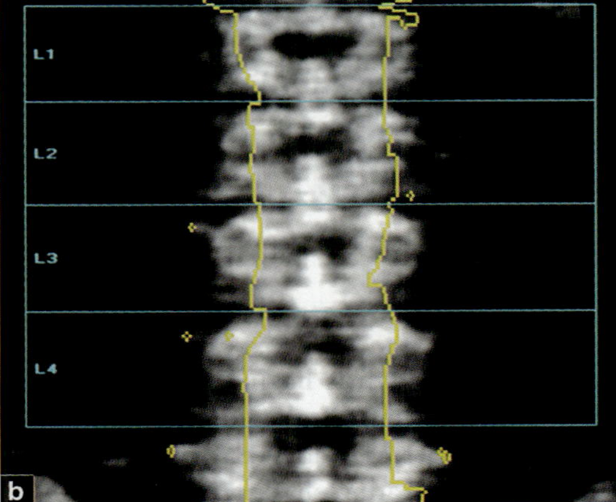

Fig. 61.4: (a) Lateral margins correctly outlined which appears as a yellow margin over the vertebrae; (b) Lateral margins incorrectly outlined

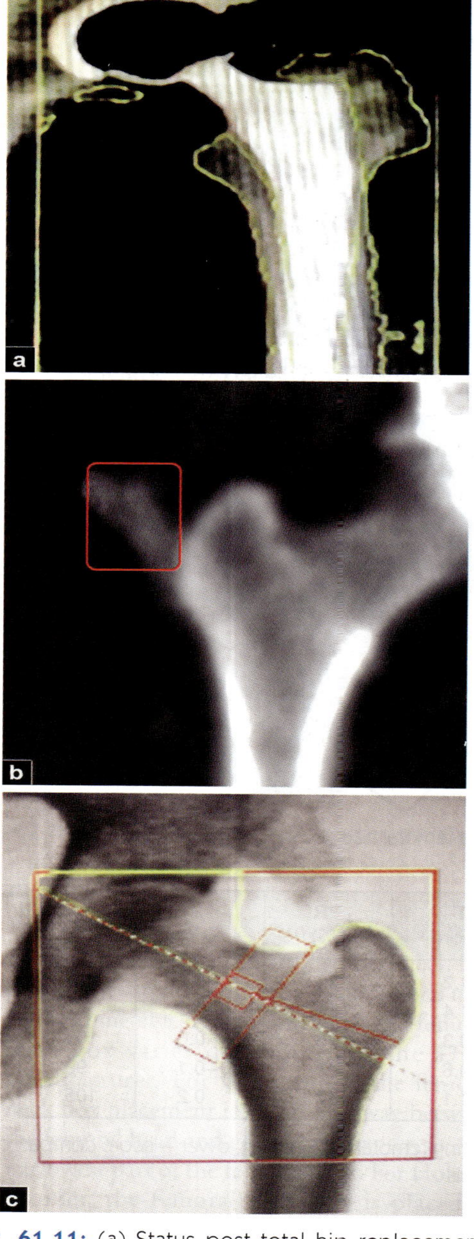

Fig. 61.11: (a) Status post total hip replacement; (b) Patient's finger over the greater trochar ter; (c) Enthesitis related calcification in a case of ankylosing spondylitis

Forearm BMD

The non-dominant forearm should be used for measurement. Difference in BMD between the dominant and non-dominant arm has been found to vary between 6 and 9%.[11]

Scan Acquisition Positioning and ROI

The forearm should be centred with the ulna and radius being aligned straight. The distal cortex of radius and ulna should be visible (Fig. 61.12a). ISCD recommends the use of distal one-third radius (33% radius) for measuring forearm BMD.

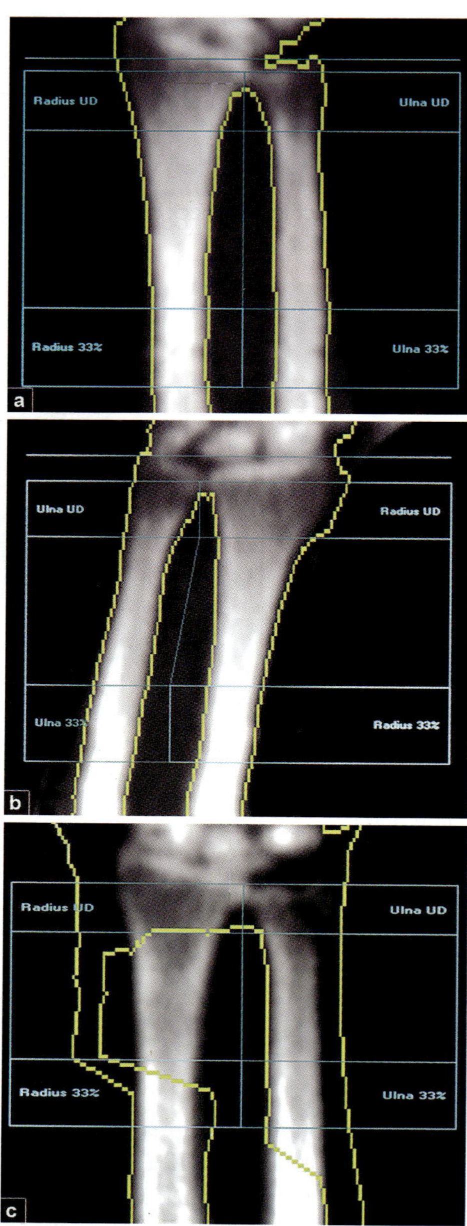

Fig. 61.12: (a) Correct forearm acquisition and positioning with lateral margins correctly outlined; (b) Incorrect positioning: Forearm not straight; (c) Lateral margins incorrectly outlined by the DXA software

Artifacts

Forearm scans usually have fewer artifacts, as they are less prone to degenerative changes as compared to the spine. A common artifact encountered is due to movement (Fig. 61.13a), which occurs in up to 20% of the scans leading on to underestimation of BMD.[11] Other artifacts include use of bangles (Fig. 61.13b), wristbands, etc.

When to Repeat a DXA Scan?

DXA scan should be repeated 1–2 years after initiation of therapy and subsequently every 2 years. In patients with normal T-score or upper low bone mass without risk factors, DXA can be repeated at longer intervals.[13]

Monitoring with DXA

DXA scan is commonly used to monitor BMD as it has good precision. When comparing serial DXA scans, one must use change in BMD (g/cm^2) values and not T-scores for monitoring. On repeating a DXA scan, it is mandatory to repeat the scan at the same centre and to compare the same ROI. DXA scan can be repeated on another machine, if it has been cross calibrated with the machine on which the first scan was done.[2]

Which Skeletal Sites to use for Monitoring Change in BMD?

PA spine is the preferred site since it is most responsive to therapy. If the spine cannot be used for the assessment due to significant degenerative changes, extensive orthopedic instrumentation, etc. the total hip is preferred. Forearm is not preferred since it does not respond well to treatment.[14]

When do We Say the Change in BMD is Significant?

The least significant change (LSC) for each machine should be calculated, i.e. the measurement error in BMD contributed by the machine and the technologist. Serial BMD measurements beyond the LSC value ought to be considered as a significant change.

Example of correctly comparing a scan:

Baseline spine BMD—0.981 g/cm^2

Repeat spine BMD—0.932 g/cm^2

Difference—0.049 g/cm^2

LSC at spine—0.031 g/cm^2

Does it exceed LSC?: Yes, therefore indicating the change in BMD is significant.

It should be reported as % change (±) LSC. Thus (0.049/0981) × 100 = loss of 4.9% ± LSC = –4.9% ± 3.1%

Examples of Common Pitfalls in Serial DXA Interpretation

Case 1: A 82-year-old female patient's serial DXA shows a drop in T-scores compared to her previous scan (Fig. 61.14a).

Case 2: A 54-year-old postmenopausal female patient's serial spine BMD measurements showed significant increase of 4.9% in BMD from 0.786 in 2014 to 0.825 g/cm^2 in 2015 (Fig. 61.14b).

Case 3: A 64-year-old female on Alendronate since 4 years showed a significant increase in BMD from 0.820 in 2013 to 0.865 in 2016 (Fig. 61.14c).

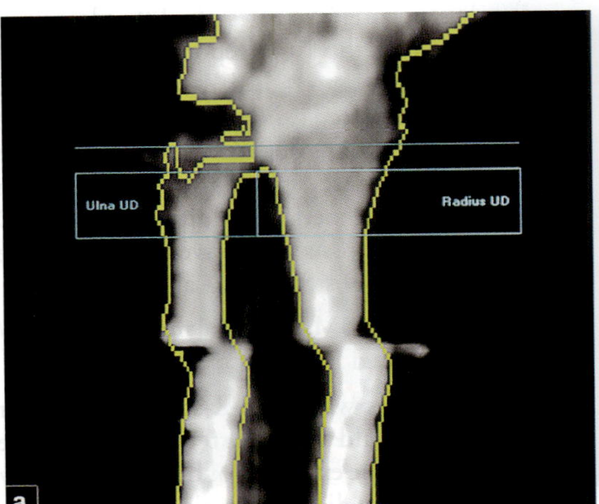

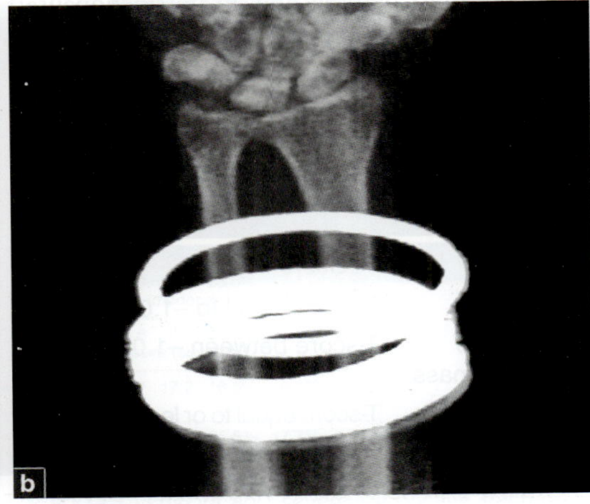

Fig. 61.13: (a) Movement artifact; (b) Bangle artifact

may bring about change in diagnostic classification and treatment.

VFA is indicated in individuals with a T-score less than –1.0 with one or more of the following, men ≥ age 80 years or women age ≥ 70 years; historical height loss >4 cm (>1.5 inches); self-reported but undocumented prior vertebral fracture and glucocorticoid therapy equivalent to ≥ 5 mg of prednisone or equivalent per day for ≥ 3 months.

Further imaging is recommended in the presence of any of the following: Two or more mild deformities without any moderate or severe deformities, equivocal fractures, T7–L4 vertebra not identified clearly, vertebral lesions which cannot be attributed as benign etiologies, vertebral deformities with history of malignancy and presence of lytic and sclerotic changes.[2]

Advantages of VFA

a. Sensitivity of 92% in detecting grade 2 and grade 3 vertebral fracture with a negative predictive value of 98%.[17]
b. Convenient for the patient as it can be done as part of the DXA study.
c. Lesser exposure to radiation dose compared to X-ray thoracolumbar spine
d. No parallax distortion especially of the lumbar spine as compared to X-rays.

Disadvantage

Image resolution is modest which may make it difficult to outline the vertebral borders and it is not reliable in identifying other spinal pathologies.

Trabecular Bone Score (TBS)

TBS is a novel DXA software that utilizes gray-level texture analysis of the PA spine image to indirectly assess the skeletal microarchitecture. It has been recognized as an independent risk factor for fracture, meant to complement routine DXA scan. A higher TBS correlates with better skeletal microarchitecture, while a lower TBS correlates with weaker skeletal microarchitecture. This score has been found to be lower in postmenopausal women and men with previous fragility fractures compared to those without fragility fractures. Diabetic patients and patients on glucocorticoids have lower TBS, which would explain the increased fracture risk despite higher BMD values.[18] Notably, degenerative changes have no impact on TBS score unlike BMD. TBS values alone are not meant to determine treatment recommendations in patients. It should be used in association with FRAX and BMD to adjust FRAX probability of fracture in postmenopausal women and older men.[3]

Key Points

- DXA is the gold standard for diagnosing osteoporosis.
- Looking at the DXA images and interpreting a DXA scan in a stepwise manner will reduce common pitfalls.
- Use BMD and not T-scores for comparison with previous DXA scans and know the LSC value.
- Use of VFA improves fracture risk assessment and management of osteoporosis.
- TBS is a novel upcoming tool for fracture assessment that is particularly useful in patients with diabetes and on glucocorticoids.

REFERENCES

1. Eis SR, Lewiecki EM. Peripheral bone densitometry: Clinical applications. Arq Bras Endocrinol Metabol. 2006; 50(4):596–602.
2. 2015 ISCD Official Positions - Adult - International Society for Clinical Densitometry (ISCD) [Internet]. [cited 2017 Mar 31]. Available from: http://www.iscd.org/official positions/2015-iscd-official-positions-adult/
3. Link TM. Osteoporosis Imaging: State of the Art and Advanced Imaging. Radiology. 2012; 263 (1):3–17.
4. Rexhepi S, Bahtiri E, Rexhepi M, Sahatciu-Meka V, Rexhepi B. Association of Body Weight and Body Mass Index with Bone Mineral Density in Women and Men from Kosovo. Mater Sociomed. 2015; 27(4):2 59–62.
5. 2013 ISCD Official Positions - Pediatric - International Society for Clinical Densitometry (ISCD) [Internet]. [cited 2017 Mar 31]. Available from: http://www.iscd.org/official-positions/2013-iscd-official-positions-pediatric/
6. Binkley N, Adler R, Bilezikian JP. Osteoporosis Diagnosis in Men: The T-score Controversy Revisited. Curr Osteoporos Rep. 2014; 12(4):403–9.
7. Girardi FP, Parvataneni HK, Sandhu HS, Cammisa FP Jr, Grewal H, Schneider R, Lane JM. Correlation between vertebral body rotation and two-dimensional vertebral bone density measurement. Osteoporos Int. 2001; 12(9):738–40.

8. Rand T, Seidl G, Kainberger F, Resch A, Hittmair K, Schneider B, et al. Impact of spinal degenerative changes on the evaluation of bone mineral density with dual energy X-ray absorptiometry (DXA). Calcif Tissue Int. 1997; 60(5):430–3.
9. Mullaji AB, Upadhyay SS, Ho EK. Bone mineral density in ankylosing spondylitis. DEXA comparison of control subjects with mild and advanced cases. J Bone Joint Surg Br. 1994; 76(4):660–5.
10. Goh JC, Low SL, Bose K. Effect of femoral rotation on bone mineral density measurements with dual energy X-ray absorptiometry. Calcif Tissue Int. 1995; 57(5):340–3.
11. Garg MK, Kharb S. Dual energy X-ray absorptiometry: Pitfalls in measurement and interpretation of bone mineral density. Indian J Endocrinol Metab 2013; 17(2):203–10.
12. Berntsen GK, Tollan A, Magnus JH, Søgaard AJ, Ringberg T, Fønnebø V. The Tromsø Study: artifacts in forearm bone densitometry—prevalence and effect. Osteoporos Int 1999; 10(5):425–32.
13. Cosman F, de Beur SJ, LeBoff MS, Lewiecki EM, Tanner B, Randall S, et al. National Osteoporosis Foundation. Clinician's Guide to Prevention and Treatment of Osteoporosis. Osteoporos Int. 2014; 25(10):2359–81.
14. Lenchik L, Kiebzak GM, Blunt BA; International Society for Clinical Densitometry Position Development Panel and Scientific Advisory Committee. What is the role of serial bone mineral density measurements in patient management? J Clin Densitom. 2002; 5 Suppl: S29–38.
15. New Delhi: ICMR: Published by Director General; 2010. Population based reference standards of Peak Bone Mineral Density of Indian males and females—an ICMR multi-center task force study; pp. 1–24.
16. Shetty S, Kapoor N, Naik D, Asha HS, Thomas N, Paul TV. The impact of the Hologic vs the ICMR database in diagnosis of osteoporosis among South Indian subjects. Clin Endocrinol (Oxf). 2014; 81(4): 519–22.
17. Rea JA, Li J, Blake GM, Steiger P, Genant HK, Fogelman I. Visual assessment of vertebral deformity by X-ray absorptiometry: a highly predictive method to exclude vertebral deformity. Osteoporos Int. 2000; 11(8):660–8.
18. Harvey NC, Glüer CC, Binkley N, McCloskey EV, Brandi ML, Cooper C, et al. Trabecular bone score (TBS) as a new complementary approach for osteoporosis evaluation in clinical practice. Bone. 2015; 78:216–24.

FURTHER READING

1. 2015 ISCD Official Positions - Adult - International Society for Clinical Densitometry (ISCD) [Internet]. [cited 2017 Mar 31]. Available from: http://www.iscd.org/official-positions/2015-iscd-official-positions-adult/
2. Garg MK, Kharb S. Dual energy X-ray absorptiometry: Pitfalls in measurement and interpretation of bone mineral density. Indian J Endocrinol Metab. 2013; 17(2):203–10.

Chapter

62

Metabolic Bone Diseases

Arup Kumar Kundu, Shyamashis Das

INTRODUCTION

Metabolic bone diseases are a heterogeneous group of disorders which affects the strength of the bones. Usually, these disorders occur due to abnormalities of minerals (calcium, magnesium, and phosphorus), vitamin D, and bone mass or bone structure. Common manifestations are altered serum calcium level and/or skeletal failure. Worldwide the most common metabolic bone disease is osteoporosis, essentially a disease of elderly, can lead to fragility fracture if untreated.[1] Other common disorders are vitamin D deficiency leading to osteomalacia or rickets, renal osteodystrophy, parathyroid disorders, Paget's disease, bone disease due to gastrointestinal and hepatic disorders, and chemotherapy-induced bone loss. These disorders are generally reversible when underlying abnormalities are appropriately treated, and should be distinguished from genetic bone disorders which occur due to abnormalities in certain signaling cascades or specific cell types.

Structure, Function and Remodeling of Bone

The outer part of each bone has dense skeletal tissue known as cortical bone, responsible for skeleton's mechanical strength. At the ends of long bones and within the vertebrae, a fine network of bone tissue is found, known as trabecular or cancellous bone. Trabecular bone has close contact with the bone marrow and is largely responsible for the metabolic role of bone.

Structure of bone consists of extracellular matrix and cellular components. The integrity of the extracellular matrix is maintained throughout life by continuous remodeling of it by the cellular components. Type 1 collagen, the principal matrix protein, forms a fibrillar architecture by cross-linking of the precursor procollagen, and is responsible for tensile strength of the bone. There are non-collagenous matrix proteins also, including glycoproteins and proteoglycans which form the ground substance. Calcium and phosphate containing hydroxyapatite-like crystals are deposited between the collagen fibrils to provide rigidity of the bone.[2] Mesenchymal-derived osteoblast lineage (i.e. osteoblasts and osteocytes) and hematopoietic precursor derived osteoclasts constitute the cellular elements of bone. Osteoblasts produce organic matrix and responsible for new bone formation, whereas osteoclasts cause bone resorption by local secretion of acid hydrolase into the resorption lacuna.

The orderly sequence of bone resorption and formation is known as bone remodelling, which is a continuous process influenced by several hormones, cytokines, vitamin D and ageing. Osteoblastic activity is stimulated by parathyroid hormone (PTH), vitamin D, insulin like growth factors (IGF-1), bone morphogenic proteins, and proteins of Wnts signaling pathway. Bone resorption by osteoclasts is accelerated by RANK (receptor activator of nuclear factor kappa-B) ligand, M-CSF (macrophage colony-stimulating factor), IL-1, IL-6, and many other proinflammatory cytokines. Other important chemical mediators in bone remodelling include IL-33 (produced by osteoblasts and inhibits osteoclast formation) and cardiotrophin-1 (produced by osteoclasts and stimulates osteoblasts).[3, 4]

Other than providing the endoskeletal support to our body, the major function of bones is to act as a reservoir of calcium. Many essential cellular functions are dependent on extracellular calcium concentration. The critical level of calcium in the extracellular fluid is maintained by release of calcium from bone.

Calcium Regulatory Hormones

Three hormones play important role in calcium homeostasis—vitamin D, PTH, and calcitonin. Vitamin D, a hormone having molecular structure similar to steroid hormones, is either obtained from food or produced in the skin from 7-dehydrocholesterol by photochemical cleavage following exposure to sunlight. The active form of vitamin D, 1,25-dihydroxyvitamin D, is produced by hydroxylation in the liver and kidney by the enzymes 25-hydroxylase and 1-α- hydroxylase, respectively. In kidney 1-α-hydroxylase is stimulated by PTH, low phosphate concentration, growth hormone, etc. 1, 25-dihydroxy vitamin D maintains serum calcium and phosphate concentration by promoting calcium and phosphate absorption from the proximal gut and stimulating calcium mobilization from bone.[5]

PTH regulates extracellular calcium concentration by stimulating calcium mobilization from bone by indirectly increasing osteoclast activity, promoting calcium reabsorption in the proximal convoluted tubules of kidney, increasing phosphate excretion in renal tubules, increasing conversion of 25-hydroxyvitamin D to 1, 25-dihydroxy vitamin D in the kidney. As PTH receptors are abundant on osteoblasts but not on osteoclasts, the PTH mediated bone resorption by osteoclasts occurs indirectly through cytokines secreted by osteoblasts. Intermittent pulsatile administration of PTH causes overall anabolic effect on bone turnover through PTH-1 receptor.[6]

Calcitonin can inhibit bone resorption by its direct action on osteoclasts. However, as a whole, the role of calcitonin in bone metabolism appears to be small.

Growth hormone increases the level of IGF-1 in the circulation and in the micro-environment of bone, and thus it influences the skeletal growth. Steroid hormones like glucocorticoid, oestrogen and androgen can also influence bone metabolism. Although bone cells have oestrogen and androgen receptors, it is difficult to demonstrate their direct effect on development and maintenance of bone.

Like IGF-1, insulin also has some anabolic effects on bone growth. It has been observed that patients with type 2 diabetes mellitus are associated with modest increases in bone mineral density for age; and studies on markers of bone metabolism showed that insulin stimulates osteoblastic bone formation and decreases bone resorption. Thyroid hormones are necessary for skeletal maturation and attainment of peak bone mass. Hypothyroidism causes impaired skeletal development and growth retardation, whereas thyrotoxicosis results in accelerated bone maturation, decreased bone mass and an increased risk of osteoporotic fracture.

Metabolic bone diseases may be classified as below:

- Osteoporosis
- Vitamin D deficiency leading to osteomalacia or rickets
- Renal osteodystrophy
- Parathyroid disorders
- Paget's disease of bone
- Bone disease due to gastrointestinal and hepatic disorders
- Chemotherapy-induced bone loss
- Heritable metabolic skeletal disorder

Osteoporosis

Osteoporosis, which is defined as decrease in the quantity of bone per unit volume with normal mineralization leading to compromising its mechanical function, is increasingly being recognized as a public health problem in most of the countries including India. Osteoporosis is discussed further in Chapters 58, 59, and 60.

Vitamin D Deficiency States—Rickets and Osteomalacia

Vitamin D is crucial for maintaining calcium and phosphorus level in blood. Deficiency of vitamin D leads to decreased level of active metabolite 1, 25-dihydroxy vitamin D in serum results in lowering of the calcium-phosphate product to below normal leading to deficient mineralization of bones. Rickets and osteomalacia occurs due to defective mineralization of newly formed bone matrix or osteoid.

Rickets occurs when mineralization defect during skeletal growth results in impaired epiphyseal growth in children leading to soft and fragile bones with typical bony deformities. Osteomalacia refers to defective mineralization of the mature skeleton in adults. In adults with vitamin D deficiency, unmineralized osteoid accumulates on bone surfaces. This results in compromise of the mechanical strength of the skeleton, leading to increased susceptibility to fractures.[7] The causes of rickets and osteomalacia are given in Table 62.1.

Impaired skeletal growth and bony deformities are two main clinical features of rickets. Characteristic bony abnormalities are bowing of long bones, rib deformities (rachitic rosary, Harrison's groove), frontal bossing, craniotabes, increased kyphosis and lordosis of thoracolumbar spine, and delayed eruption of permanent teeth. Children with rickets may have proximal muscle weakness, hypotonia and symptoms of hypocalcaemia. Clinical features of osteomalacia in adults are relatively non-specific; the characteristic features are widespread bone pain, bony tenderness, muscle pain and proximal myopathy.

Table 62.1: Causes of rickets and osteomalacia

Causes
Vitamin D deficiency
• Diet deficient in vitamin D and reduced sunlight exposure
• Increased vitamin D requirements in childhood
Vitamin D malabsorption
• Small bowel malabsorption syndrome
• Chronic pancreatic insufficiency
Impaired vitamin D metabolism
• Chronic renal failure
• Chronic liver failure
• Congenital renal 1-α- hydroxylase deficiency (vitamin D-dependent rickets)
Hypophosphatemia
• X-linked hypophosphataemic vitamin D-dependent rickets (renal tubular defects in phosphate handling)
• Fanconi's syndrome
Drugs
• Antiepileptics (e.g. phenytoin, barbiturates)
• Fluoride therapy
• Bisphosphonates
Miscellaneous
• Type 1 (distal) renal tubular acidosis
• Hypophosphatasia (inherited alkaline phosphatase deficiency)
• Malignancy

Biochemical abnormalities in rickets and osteomalacia are decreased serum phosphate and elevated alkaline phosphatase concentration (Table 62.2). Measurement of 25-hydroxyvitamin provides confirmatory evidence of vitamin D deficiency. In case of children with rickets having normal vitamin D levels, evaluation for urinary phosphate wasting, Fanconi's syndrome and other tubular disorders should be done. Elevated levels of PTH in vitamin D deficiency state suggest secondary hyperparathyroidism.

Bone mineral density (BMD) measured by DXA scan may show osteopenia or even osteoporosis, which is usually correctable by treatment with vitamin D and calcium. Radiographs in rickets show widened irregular epiphysis, and fraying, cupping, and splaying of the metaphysis. Characteristic radiographic feature in osteomalacia is pseudofractures or 'Looser zones', considered as a type of insufficiency fracture, most commonly seen in the pubic rami, medial proximal femur, lateral scapula, proximal ulna and ribs. 'Looser zones' are a band of transverse lucencies traversing partially through a bone, usually at right angles to the involved cortex. They are often symmetrical and have sclerotic irregular margins. Iliac crest bone biopsy may be considered in case of diagnostic difficulty. Histologically the bone biopsy shows widening of the layer of osteoid, poorly mineralized new bone formation and areas of hypomineralization.

Vitamin D deficiency is defined as a serum 25-hydroxyvitamin D concentration below 20 ng/ml. However, to maximize intestinal calcium absorption and minimize circulating PTH levels, maintaining serum 25-hydroxyvitamin D level above 30 ng/ml is desired. To treat vitamin D deficiency, vitamin D 60,000 IU/week for 8 to 12 weeks along with oral calcium supplementation is recommended. Thereafter, vitamin D can be supplemented with 60000 IU/month or 2000 IU/day. Recommendation of daily calcium intake varies with age: 800 mg, 1300 mg, 1000 mg, and 1200 mg for age groups <12 years, 13–18 years, 19–50 years and >50 years respectively. Vitamin D deficiency resulting from dietary deficiency or malabsorption responds very well to vitamin D replacement.[8,9]

Table 62.2: Typical blood biochemical features in metabolic bone disease

Disorders	*Calcium*	*Phosphate*	*Alkaline phosphatase*	*PTH*
Osteoporosis	Normal	Normal	Normal	Normal
Rickets/osteomalacia	Normal or decreased	Decreased	Increased	Increased
Hyperparathyroidism	Increased	Decreased	Increased	Increased
Malignancy	Increased	Normal or decreased	Normal or increased	Decreased

Renal Osteodystrophy

Renal osteodystrophy is a heterogeneous group of metabolic bone disorders (MBD) associated with chronic kidney disease. These disorders are termed together as chronic kidney disease—mineral and bone disorder (CKD-MBD). The defects in mineral metabolism begin even in early stage of CKD. The earliest form of CKD-MBD is secondary hyperparathyroidism which may be defined by quantitative histomorphometry of bone or elevated 1–84 intact PTH in serum. Double tetracycline-labeled quantitative bone histomorphometry is considered "gold standard" for defining the specific forms of renal osteodystrophy. CKD-MBD carries greater risk of fractures for change in bone quality. There are four main types of CKD-MBD according to bone histomorphometry.[10]

1. Osteitis fibrosa cystica: It results from increased bone turnover due to secondary hyperparathyroidism. It frequently occurs when glomerular filtration declines to less than 60 mL/minute leading to a series of abnormalities that initiate and maintain increased PTH secretion. The principal contributory factors to secondary hyperparathyroidism are: phosphate retention, decreased free calcium level, decreased 1, 25-dihydroxy vitamin D level, reduced expression of vitamin D receptors and calcium-sensing receptors.
2. Adynamic bone disease: Here, the bone turnover is low, which represents the major bone lesion in peritoneal dialysis and hemodialysis patients. Bone turnover is characteristically reduced associated with reduced activity of both osteoblasts and osteoclasts. In contrast to osteomalacia, both the new bone formation and its subsequent mineralization are subnormal, resulting in no increase in osteoid formation as seen in osteomalacia. Bone biopsy shows normal or decreased osteoid volume, and reduced numbers of osteoblasts and osteoclasts. Although aluminum deposition may cause this disorder, the principal factor underlying adynamic bone disease is excessive suppression of PTH release, induced by the relatively high doses of vitamin D analogues and possibly of calcium-based phosphate binders. In animal studies, administration of the non-calcium based phosphate binder could reverse adynamic bone disease. Adynamic bone disease (as determined by bone biopsy or intact serum PTH <100 pg/mL) should be treated to allow PTH secretion to rise by using non-calcium-based phosphate binders rather than calcium-based phosphate binders, vitamin D and a low dialysate calcium concentration.[11]
3. Osteomalacia: Here, like adynamic bone disease, bone turnover is low but it is associated with an increased volume of unmineralized osteoid. Vitamin D deficiency possibly plays an important role in the development of this. However, earlier this disorder was thought to be primarily due to aluminum toxicity from aluminum-containing antacids used as phosphate binders. Aluminium is known to inhibit bone mineralization. The mineralization lag time is prolonged in osteomalacia (>100 days; whereas <35 days in healthy individuals and patients with pure osteitis fibrosa). Treatment of aluminum-induced osteomalacia consists of cessation of aluminum-containing antacids and aluminum removal from the body by chelators.
4. Mixed uremic osteodystrophy: Features of both high and low bone turnover are observed (overlapping features of the above three). This

Key Points

- Metabolic bone disease is a heterogeneous group of disorders which include osteoporosis, osteomalacia/rickets, renal osteodystrophy, parathyroid disorders, Paget's disease, bone diseases due to gastrointestinal disorders and chemotherapy-induced bone loss.
- Metabolic bone disease occurs due to abnormalities of calcium, phosphorus, vitamin D or bone structure. Vitamin D, PTH and calcitonin are the hormones which play important role in calcium homeostasis.
- Osteoporosis is due to reduced BMD, disrupted microarchitecture of bones, and alteration of the amount and variety of non-collagenous proteins in bone.
- Rickets and osteomalacia occur due to defective mineralization of newly formed bone matrix as a consequence of vitamin D deficiency.
- Treatment varies diversely between the disorders in this group but early diagnosis and multidisciplinary approach can potentially cure some of them, and prevent complications in others.

REFERENCES

1. Holroyd C, Dennison E, Cooper C. Epidemiology of osteoporosis and classification. In: Hochberg MC, Silman AJ, Smolen JS, Weinblatt ME, Weisman MH, editors. Rheumatology, 6th edition. Philadelphia, PA: Mosby; 2015. pp 1633–40.
2. Dempster DW. Bone microarchitecture and strength. Osteoporos Int 2003;14 (Suppl 5):54–6.
3. Sims NA, Walsh NC. Intercellular cross-talk among bone cells: new factors and pathways. Curr Osteoporos Rep. 2012; 10:109–17.
4. Raggatt LJ, Partridge NC. Cellular and molecular mechanisms of bone remodeling. J Biol Chem 2010; 285:25103–8.
5. Holick MF, Garabedian M. Vitamin D: photobiology, metabolism, mechanism of action, and clinical applications. In: Favus MJ, editor. Primer on the metabolic bone diseases and disorders of mineral metabolism. 6th ed. Washington, DC: American Society of Bone and Mineral Research; 2006. p.129–37.
6. Neer RM, Arnaud CD, Zanchetta JR, Prince R, Gaich GA, Regin-ster JY, et al. Effect of parathyroid hormone (1–34) on fractures and bone mineral density in postmenopausal women with osteoporosis. N Engl J Med 2001; 344:1434–41.
7. Holick MF. Resurrection of vitamin D deficiency and rickets. J Clin Invest 2006; 116:2062–72.
8. Wagner CL, Greer FR. Section on Breastfeeding and Committee on Nutrition. Prevention of rickets and vitamin D deficiency in infants, children, and adolescents. Pediatrics 2008; 122:1142–52.
9. Standing Committee on the Scientific Evaluation of Dietary Reference Intakes Food and Nutrition Board, Institute of Medicine. Dietary reference intakes for calcium, phosphorus, magnesium, vitamin D and fluoride. Washington, DC: National Academy Press; 1999.
10. Martin KJ, González EA. Metabolic bone disease in chronic kidney disease. J Am Soc Nephrol 2007; 18: 875–85.
11. Malluche HH, Porter DS, Monier-Faugere MC, et al. Differences in bone quality in low-and high-turnover renal osteodystrophy. J Am Soc Nephrol 2012; 23: 525–32.
12. KDIGO clinical practice guidelines for the diagnosis, evaluation, prevention, and treatment of chronic kidney disease-mineral and bone disorder (CKD-MBD). Kidney Int 2009; 76 (Suppl 113):S1–S130.
13. Reid IR, Lyles K, Su G, Brown JP, Walsh JP, del Pino-Montes J, et al. A single infusion of zoledronic acid produces sustained remissions in Paget disease: data to 6.5 years. J Bone Miner Res 2011; 26:2261–70.
14. Brufsky AM. Cancer treatment-induced bone loss: pathophysiology and clinical perspectives. Oncologist 2008; 13(2):187–95.
15. Skowronska-Jozwiak E, lorenc RS. Metabolic bone disease in children: etiology and treatment options. Treat Endocrinol. 2006; 5(5):297–318.

FURTHER READING

1. Holick MF. Vitamin D deficiency. N Engl J Med 2007; 357:266–81.
2. Lane NE. Metabolic bone disease. In: Firestein GS, Budd RC, Gabriel SE, Mc Innes IB, O'Dell JR, editors. Kelley and Firestein's Textbook of Rheumatology, 10th edition. Philadelphia, PA: Elsevier; 2017. pp. 1730–50.

Section

X

JUVENILE IDIOPATHIC ARTHRITIS

Chapter

63

Classification and Pathogenesis of Juvenile Idiopathic Arthritis

Sanat Phatak, Amita Aggarwal

Juvenile idiopathic arthritis (JIA) is a blanket term referring to a heterogenous group of childhood joint disorders which share: (a) onset prior to the age of 16 years; (b) chronicity-persistence for more than 6 weeks and (c) unknown aetiology.

CLASSIFICATION OF JIA

Chronic arthritides in children have been known for more than a century and were described in detail by Frederick Still. Early attempts at classifying these disorders were made in early eighties. The disorder was termed 'juvenile rheumatoid arthritis' or JRA by the American Rheumatology Association (ARA) classification and 'juvenile chronic arthritis' or JCA by the EULAR classification. ARA divided JRA into three classes: Systemic JRA, polyarticular JRA, and pauciarticular JRA. There were some important differences between these two classification systems, chiefly differing symptom durations needed and the absence of the spondyloarthropathy spectrum of disorders in the ARA classification. Moreover, 'JRA' was a subset of JCA in the EULAR classification referring to only rheumatoid factor positive polyarticular disease. These discrepancies in terminologies led to much confusion and precluded direct comparison between the two.

To resolve this issue, a task force was set-up under the aegis of the International league of associations of rheumatology (ILAR) to classify chronic childhood arthritides. The ILAR classification was first released in 1994 with an aim to segregate patients into homogenous and mutually exclusive groups, suitable for research purposes such as inclusion into clinical trials.[1] This approach was expected to guide proper management since the natural history of each subset differs. A cut-off of 16 years for inclusion into the JIA umbrella was decided arbitrarily—it has no biological basis. The classification has subsequently been revised and the second revised version is currently in use in which exclusion criteria were added (Table 63.1).[2,3]

Classification into JIA categories is an exercise in recognizing patterns. Therefore, the ILAR classification is useful in stereotypical cases. Since JIA is a diagnosis of exclusion other possibilities must be diligently excluded. In addition these must also be considered when presentations do not fit into any category easily (Table 63.2).

In recent times, JIA classification is being rethought.[4] Some forms of JIA are mere extensions of their adult counterparts e.g. polyarticular JIA resembles rheumatoid arthritis, enthesitis related arthritis (ERA) resembles spondyloarthropathy and systemic JIA (sJIA) resembles adult onset Still's disease. On the other hand, oligoarticular JIA is a well-defined disorder peculiar to children with no real adult counterpart. It has also been suggested that the presence of serologic markers, such as antinuclear antibody (ANA), rheumatoid factor (RF) and anti-citrullinated peptide antibody are more likely to form homogeneous groups rather than groups based on joint counts.[5] In addition joint involvement evolves overtime thus this arbitrary division of 4 or less or more than 4 does not define homogeneous groups. The inclusion of family history of psoriasis is another difficult criterion to ascertain. Perhaps a new classification considering these issues is

Table 63.1: ILAR classification of JIA

General definition of JIA: Arthritis of unknown aetiology that begins before the sixteenth birthday and persists for at least 6 week; other known conditions are excluded.

Category	*Definition*	*Exclusions*
Oligoarthritis persistent oligoarthritis: Affecting ≤4 joints throughout the disease course (Extended oligoarthritis: affecting a total of >4 (joints after the first (6 months of disease)	Arthritis affecting 1–4 joints during the first 6 months of disease	1–5
RF-negative polyarthritis	1. Arthritis affecting ≥ 5 joints during the first 6 months of disease and 2. Test for RF is negative	1, 5
RF-positive polyarthritis	1. Arthritis affecting ≥5 joints during the first 6 month of disease, and 2. ≥2 positive RF tests (as routinely defined in an accredited laboratory), at least 3 months apart during the first 6 month of disease	1–3, 5
Psoriatic arthritis	1. Arthritis and psoriasis, or 2. Arthritis and at least 2 of the following: Dactylitis Nailpitting (minimum of 2 pits on1or more (nails at anytime) or onycholysis 3. Psoriasis in a first-degree relative	2–5
Enthesitis related arthritis	1. Arthritis and enthesitis, or 2. Arthritis or enthesitis,with at least 2 of the following: The presence of or a history of sacroiliac joint tenderness and/or inflammatory lumbosacral pain The presence of HLA-B27 Onset of arthritis in a male >6 yr of age Acute (symptomatic) anterior uveitis History of ankylosing spondylitis, ERA, sacroiliitis with inflammatory bowel disease, reactive arthritis, or acute anterior uveitis in a first-degree relative	1, 4, 5
Systemic JIA	Arthritis in 1 or more joints with, or preceded by, fever of at least 2 weeks' duration that is documented to be daily and quotidian (fever that rises to ≥39°C once a day and returns to ≤37°C between fever peaks) for at least 3 days, and accompanied by 1 or more of the following: 1. Evanescent (nonfixed) erythematous rash 2. Generalized lymphnode enlargement 3. Hepatomegaly and/or splenomegaly 4. Serositis	1–4
Undifferentiated arthritis	Arthritis that fulfills criteria in no category or in ≤2 of the above categories	

1. Psoriasis or a history of psoriasis in the patient or first-degree relative
2. Arthritis in an HLA-B27–positive male beginning after the sixth birthday
3. Ankylosing spondylitis, ERA, sacroiliitis, with inflammatory bowel disease, reactive arthritis, acute anterior uveitis, or a history of one of these disorders in a first-degree relative
4. The presence of IgMRF on at least 2 occasions at least 3 months apart
5. The presence of systemic JIA in the patient

Table 63.2: Differential diagnosis of JIA

Type of JIA	*Differentials*	*Clues towards diagnosis*
Systemic JIA	Hematologic malignancy (ALL, Lymphoma)	Can present as polyarthritis, monoarthritis or fever Up to one third have musculoskeletal symptoms Night pains, blood cytopenias may be clues; ANA and RF do not help
	Juvenile SLE, connective tissue disease	Older girls with ANA positivity–SLE more likely Other features of organ involvement, blood cytopenias, hemolysis
	Kawasaki disease	First episode with prolonged fever. Desquamation, lymphadenopathy
	Periodic fever syndromes	Evolution of specific symptoms (mucosal inflammation, peritonitis) give clues
Oligoarticular JIA	Infectious arthritis	More painful than JIA Local warmth, systemic features prominent Tubercular arthritis an important differential in resource limited settings like India
	Foreign body synovitis	Can present as chronic monoarthritis without systemic features
	Bleeding disorders	Hemophilia has severe pain and early deformity especially if left unnoticed
	Benign local tumours	Night dull aching pains in benign tumours, bony swelling
	Orthopaedic problems	Trauma related effusions are rare in children, therefore other causes should be ruled out Trauma, meniscal injury, slipped femoral epiphysis, Osgood Schlatter disease
	Benign disorders: 'Growing pains' hypermobility	Transient effusions in hypermobility
Polyarthritis	Infectious arthritis	Gonococcal arthritis can resemble adolescent age group, skin rash
	CTD, SLE	Associated with other features: Systemic, skin
	Hematologic malignancy	May present as only arthritis; signs out of proportion to symptoms

required. Further the explosion of knowledge in genomics and proteomics may also help us to define the categories better Albeit only for research purposes. Till we have that, the current classification is good and systematic approach to classify a patient with JIA should be followed (Fig. 63.1):

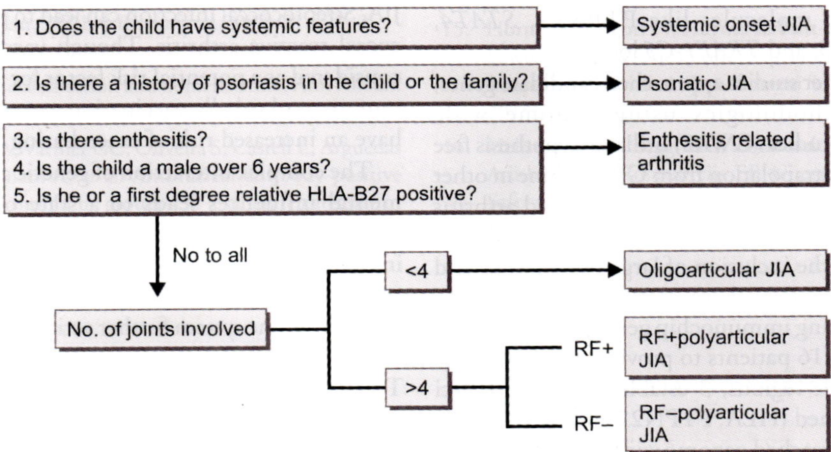

Fig. 63.1: Classification of systemic approach to JIA

Chapter

64

Clinical Features of Juvenile Idiopathic Arthritis

Sathish Kumar

Clinical manifestations of juvenile idiopathic arthritis (JIA) can be divided into articular and extra-articular.

1. ARTICULAR MANIFESTATIONS

Arthritic joint exhibits a number of cardinal signs of inflammation like swelling, erythema, heat, pain, and loss of function (Fig. 64.1). Children with arthritis may not complain of pain while at rest but active or passive motion of joints typically elicits pain. Joint tenderness is usually maximal at the joint line or just over the hypertrophied inflamed synovium. Swelling of joint may be due to periarticular soft tissue oedema, from intra-articular effusion, or from hypertrophy of the synovial membrane. Large synovial cysts are an unusual complication. These may occur in the popliteal space (Baker cyst) which can rupture into adjacent muscles or dissect into the calf. Consider Baker cyst rupture if there is sudden sharp pain and swelling in the calf, followed by crescentic ecchymoses about the malleoli. In JIA, inflamed joints are often warm but usually not erythematous. If erythematous think of septic arthritis or acute rheumatic fever or other reactive arthritides.

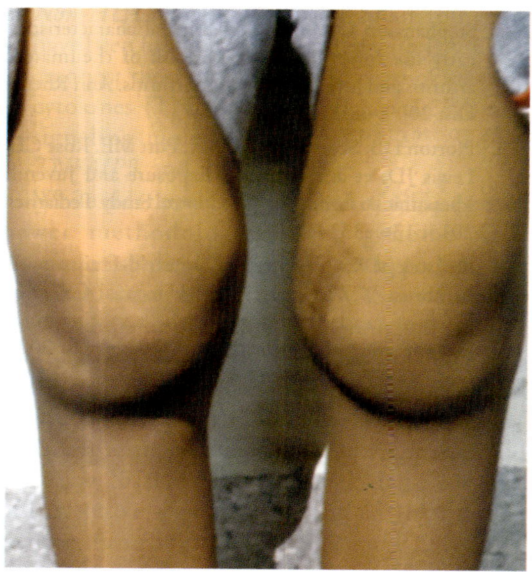

Fig. 64.1: Bilateral knee effusion in a child with oligoarthritis

Morning stiffness and gelling following inactivity especially in morning are common manifestations of joint inflammation.[1] But young children infrequently describe these symptoms in various ways according to their developmental age. Often young children do not complain of pain and instead refuse to use the affected joint entirely or assume a posture of guarding the joints.

Joint Distribution

Both large and small joints can be affected in JIA. Based on number of joints and site of involvement, JIA is classified to oligoarthritis and polyarthritis.[2]

Oligoarticular disease affects up to 4 joints at presentation, with the knee joints mostly affected, followed by the ankles. This subtype almost never affects the hips, and rarely the smaller joints of the hands and feet. Oligoarticular disease is characterized by asymmetric large joint arthritis, early onset (before 6 years of age), female predilection, high frequency of positive ANAs, and a high risk of asymptomatic anterior uveitis.

Polyarticular disease is defined as the presence of arthritis in 5 or more joints during the first 6 months of disease (Fig. 64.2). The arthritis is usually symmetrical and usually involves the large and small joints of the hands and feet. Axial skeleton including the cervical spine and the temporomandibular joints may also be involved. Two subtype of polyarticular disease includes both RF-negative and RF-positive diseases. Both types affect girls more frequently than boys. Arthritis predominantly affecting the joints of the lower extremity including hip and sacroiliac joints characterizes enthesitis related arthritis (ERA) JIA. Psoriatic arthritis tends to be asymmetrical and involves both large and small joints, sometimes including the distal interphalangeal joints.

The temporomandibular (TMJ) joint and the cervical, thoracic and lumbar spine can be involved in children with JIA.[3] TM joint is usually affected in polyarticular disease. JIA often affects the cervical spine, and the most common changes in the upper cervical spine are anterior atlantoaxial subluxation and impaction.

Enthesitis

Enthesitis is a distinct pathologic feature of ERA and psoriatic arthritis. Enthesitis is defined as inflammation of an enthesis, which is a site where a tendon, ligament, or joint capsule attaches to bone. Enthesitis in ERA is often symmetric and largely affects the lower limbs.[4] The most commonly affected entheses were the patellar ligament insertion at the inferior pole of the patella, plantar fascial insertion at the calcaneus (Fig. 64.3), and the Achilles tendon insertion at the calcaneus followed by hip extensor insertion at the greater trochanter was the most common site involved. Enthesitis does not go parallel with disease activity or respond as well to therapy as arthritis.

Dactylitis

Dactylitis or 'sausage-like digit' is defined as a diffuse swelling of a digit. Dactylitis is one of the hallmark features of the spondyloarthropathies especially psoriatic arthritis.[5] Fingers and toes are involved and are presented with swelling, slight redness, and deformity. Dactylitis has been thought to be a result of the concomitant swelling and inflammation of the flexor tendon sheaths of the metacarpophalangeal, metatarsophalangeal, or interphalangeal joints.

Tenosynovitis

Tenosynovitis is quite commonly present in children with JIA. Usually it is not a striking or isolated clinical complaint. The most common sites are the extensor tendon sheaths on the dorsum of the hand, the extensor sheaths over the dorsum of the foot, and those of the posterior tibial tendon and the peroneus longus and brevis tendons around the ankle. Carpal tunnel syndrome is uncommon in children with involvement of the wrists.

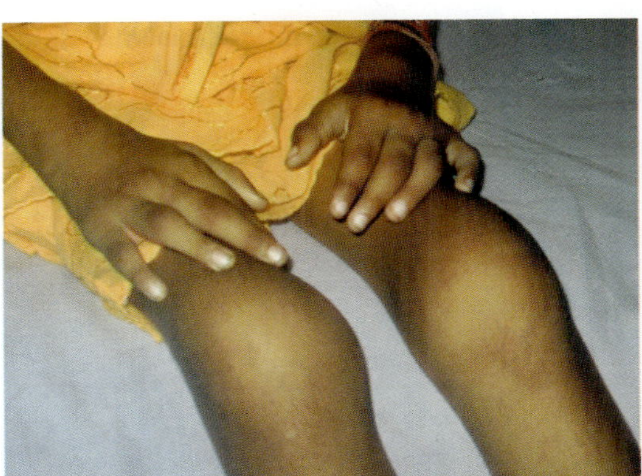

Fig. 64.2: Polyarthritis involving all PIP and MCP joints and both knee effusion in a child with polyarticular RF positive JIA

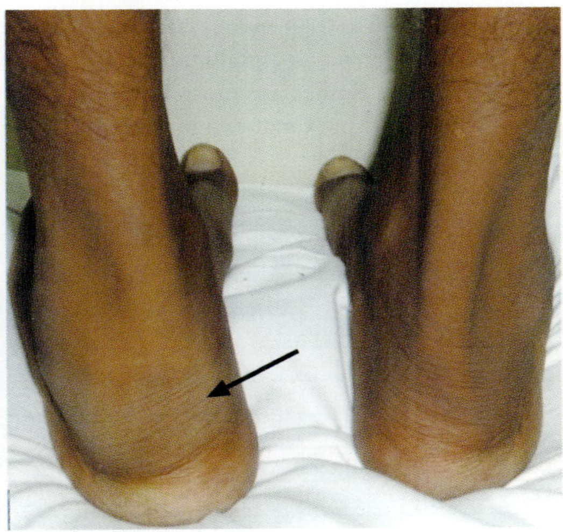

Fig. 64.3: Enthesitis of left tendo Achilles in a child with enthesitis related arthritis (ERA)

Constitutional Symptoms

Child with JIA can have anorexia, weight loss and growth failure. Significant fatigue is a common symptom in children with polyarticular or systemic disease, especially at onset and during disease flares. Sleep disturbance and night pain contribute to fatigue.[6]

2. EXTRA-ARTICULAR MANIFESTATIONS

Extra-articular manifestations are more common in systemic type JIA (sJIA). Systemic symptoms like classical fever and rash may precede overt arthritis.[7] The arthritis associated with systemic onset disease is usually polyarticular affecting both large and small joints. Asymmetric, oligoarticular arthritis is less common.

Classical systemic arthritis fever typically rises to 39°C or higher on a daily or twice-daily basis, followed by a rapid return to the baseline temperature or below (Fig. 64.4). Although this quotidian pattern is highly suggestive of systemic onset disease, all children may not present this fever pattern. Fever may occur at any time of the day, but characteristically presents in the late afternoon to evening in conjunction with the rash. During febrile period, an affected child commonly appears ill but then appears well when the fever breaks.

Classical rash (Fig. 64.5) of systemic arthritis is accompanied with fever. The rash is evanescent (usually comes and goes with the fever spikes) and consists of discrete, circumscribed, salmon-pink macules (2 mm to10 mm in size) that may be surrounded by a ring of pallor or may develop central clearing. Rashes are most common occur on the trunk and proximal extremities, including the axilla and inguinal areas. But can also develop on the face, palms, or soles of affected individuals. The rash last for up to a few hours and leave no residua. Individual lesions may be elicited by rubbing and/or scratching the skin (Koebner phenomenon), by a hot bath, or by psychological stress. The rash is occasionally pruritic but is never purpuric.

Pericarditis and pericardial effusions are especially common in children with systemic onset disease.[8] Pericarditis may precede the development of arthritis or may occur at any time during the course of disease. Pericarditis tends to occur in older children, but it is not related to sex, age at onset, or severity of joint disease. Most pericardial effusions are either asymptomatic or present with dyspnea or precordial pain that may be transferred to the back, shoulder, or neck. Examination will reveal diminished heart sounds, tachycardia, cardiomegaly and a pericardial friction rub, usually at the left lower sternal border. Pneumonitis or pleural effusions may also occur with carditis or may be asymptomatic. Rheumatoid nodules that are described in adult rheumatoid arthritis are rare in childhood.

Systemic onset disease can also present with lymphadenopathy or splenomegaly either alone or in combination. Marked symmetric lymphadenopathy is particularly common in the anterior cervical, axillary, and inguinal areas. Splenomegaly

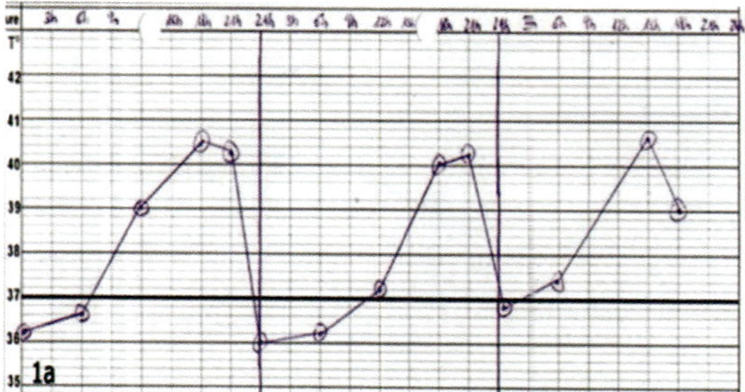

Fig. 64.4: Classical quotidian fever pattern in a child with systemic JIA

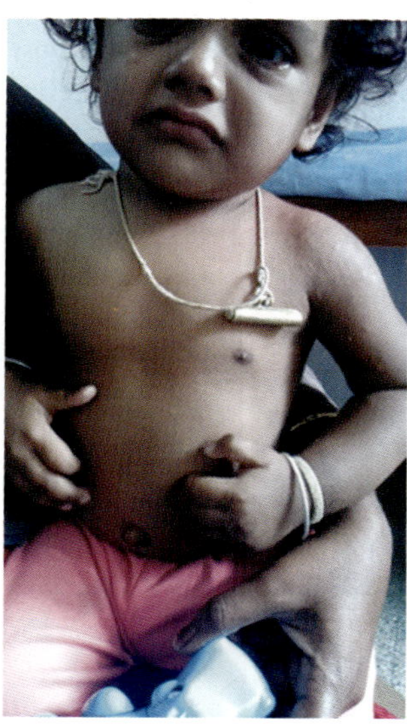

Fig. 64.5: Classical evanescent rash during febrile period of systemic arthritis in 16 months old child

is generally most prominent within the first years after onset of systemic onset disease. Hepatomegaly is less common than splenomegaly.

Macrophage activation syndrome (MAS) is a rare but life-threatening complication of systemic onset disease which is characterized by demonstration of histiophagocytosis in bone marrow.[9] The main manifestations of MAS include fever, hepatosplenomegaly, lymphadenopathy, severe cytopenias, serious liver disease, and disseminated intravascular coagulation. Consider MAS in a child with sJIA when presents with continuous fever, rash, liver dysfunction and shock.

Growth abnormalities—General and Local

Children with JIA have abnormalities of physical growth and development. Linear growth is retarded during periods of active systemic disease.[10] Growth retardation was associated with glucocorticoid administration, nutritional status, bone mineral density and early disease onset. Growth suppression in JIA is related to high levels of proinflammatory cytokines, such as IL6, IL1β, and TNF-α, seen in children with active arthritis. Levels of growth hormone and IGF-I and II may be impaired.

Localized growth disturbances lead to leg length discrepancy and micrognathia. In inflamed joints, there is an accelerated development of ossification centers of the long bones and premature fusion of the physis. The result may be either overgrowth of the affected limb or ultimately premature fusion of the involved epiphyses, resulting in diminished length. Knee arthritis causes accelerated growth and epiphyseal maturation and discrepancy of leg lengths results. A difference of greater than 1 cm is probably significant. Apparent leg length in equality may also result from pelvic rotation and scoliosis. Marked alterations of facial morphology (e.g. micrognathia, retrognathia) occur due to TM joint damage (Fig. 64.6).

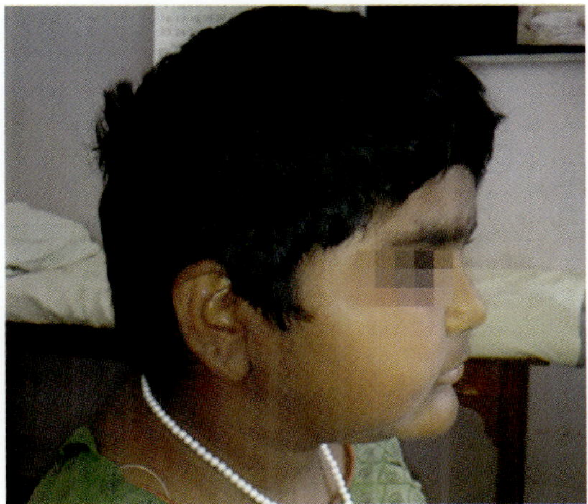

Fig. 64.6: A 12 years old girl with polyarthritis with bilateral TM joint involvement and retrognathia

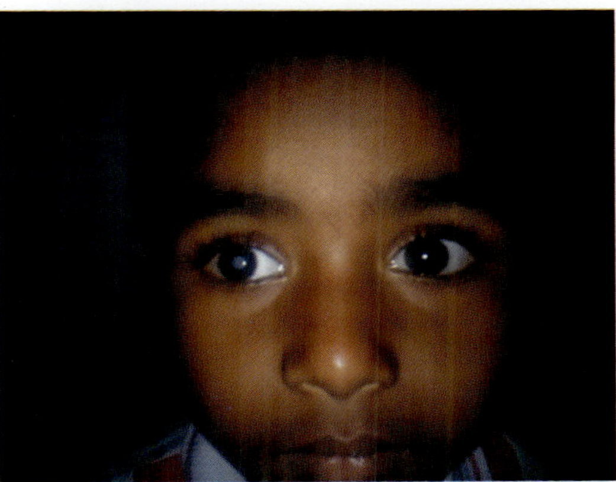

Fig. 64.7: A 5 years old boy with oligoarthritis with ANA positivity with bilateral uveitis with right eye cataract and left eye synechiae

3. UVEITIS

Uveitis associated with JIA is usually chronic, anterior, nongranulomatous uveitis. The only known independent risk factor for developing uveitis is a positive ANA test. Onset is typically insidious and often entirely asymptomatic, although up to one-half of affected children have some symptoms attributable to uveitis (e.g. pain, redness, headache, photophobia, and change in vision) later in the course of their disease. Uveitis may be present at the time of diagnosis, may develop during the course of JIA, or may be an initial manifestation of JIA that is usually detected in the course of routine ophthalmologic examination. JIA patients should be screened routinely to prevent delay in diagnosis of uveitis. Complications of uveitis include posterior synechiae, cataracts, band keratopathy, glaucoma and visual impairment (Fig. 64.7).

Conclusion

Juvenile idiopathic arthritis presents with varied articular and extra-articular clinical features. Clinical features at presentation and during follow up identifies subtypes. It is important to recognize the disease and to treat early, before soft-tissue deformities and joint damage become irreversible.

Key Points

- Juvenile idiopathic arthritis is not a single disease, but a term that encompasses all forms of arthritis that begin before the age of 16 years, persist for more than 6 weeks.
- The majority of children with JIA present with stiffness, joint swelling, limp or functional impairment, with pain either not apparent or not verbalised.
- More frequently in children, parents notice abnormalities in an uncomplaining child such as clumsiness, a change in mood or avoidance of activities or play.
- Joint swelling is not the only important clinical finding—care must be taken to look for other physical signs to diagnose JIA especially sJIA.
- A thorough history and examination leads to most diagnoses of JIA. Investigations are used to confirm suspicions only.

REFERENCES

1. Chronic arthritis in childhood. Petty RE, Cassidy JT. In: Ross Petty R, Laxer R, Lindsley C, Wedderburn L (eds.) Textbook of Pediatric Rheumatology. 6th Edition. Philadelphia, PA: Elsevier Saunders; 2011. p. 221–25.
2. Ravelli A, Martini A. Juvenile idiopathic arthritis. Lancet. 2007; 369:767–78.
3. Cannizzaro E, Schroeder S, Müller LM, Kellenberger CJ, Saurenmann R. Temporomandibular joint involvement in children with juvenile idiopathic arthritis. J Rheumatol. 2011; 38:510–5.
4. Aggarwal A, Misra DP. Enthesitis-related arthritis. Clin Rheumatol. 2015; 34:1839–46.
5. Ravelli A, Consolaro A, Schiappapietra B, Martini A. The conundrum of juvenile psoriatic arthritis. Clin Exp Rheumatol. 2015; 33:S40–3.
6. Coulson EJ, Hanson HJ, Foster HE. What does an adult rheumatologist need to know about juvenile idiopathic arthritis? Rheumatology (Oxford). 2014; 53:2155–66.
7. Cimaz R. Systemic-onset juvenile idiopathic arthritis. Autoimmun Rev. 2016; 15:931–4.
8. Martini A. Systemic juvenile idiopathic arthritis. Autoimmun Rev. 2012; 12:56–9.
9. Kumar S. Systemic Juvenile Idiopathic Arthritis: Diagnosis and Management. Indian J Pediatr. 2016; 83:322–7.
10. Giancane G, Consolaro A, Lanni S, Davì S, Schiappapietra B, Ravelli A. Juvenile Idiopathic Arthritis: Diagnosis and Treatment. Rheumatol Ther. 2016; 3:187–20.

FURTHER READING

1. Kahn PJ. Juvenile idiopathic arthritis—what the clinician needs to know. Bull Hosp Jt Dis (2013). 2013; 71:194–9.
2. Macaubas C, Nguyen K, Milojevic D, Park JL, Mellins ED. Oligoarticular and polyarticular JIA: epidemiology and pathogenesis. Nat Rev Rheumatol. 2009; 5:616–26.

Chapter

65

Management of Juvenile Idiopathic Arthritis

Sujatha Sawhney

INTRODUCTION

Juvenile idiopathic arthritis (JIA) is the commonest rheumatologic condition that afflicts children. This condition has many subcategories of arthritis included under its umbrella and the presentation of these patients is thus varied.[1]

The clinical spectrum of patients is protean and may be

- A preschool female toddler who has a painless swollen knee joint noted by a vigilant parent: oligoarticular JIA (oJIA)
- A 10-year-old male child with high spiking fevers, an evanescent rash that peaks at the spike of the fever and is accompanied by swollen wrists: Systemic onset JIA (SoJIA or sJIA)
- A three-year-old child with symmetric arthritis of the wrists, small joints of the hands and the knees: Rheumatoid factor negative polyarticular JIA (RF negative polyarticular JIA)
- A 15-year-old girl with symmetric small joint arthritis of the hands, wrists and elbows: Rheumatoid factor positive polyarticular JIA (RF positive polyaticular JIA)
- A 9-year-old boy with pain and swelling of an ankle joint, the first metatarsophalangeal joint and buttock pain: Enthesitis related arthritis (ERA)
- A 12-year-old girl with a swollen wrist and dactylitis of two digits with a history of psoriasis: Psoriatic arthritis.

Clearly children with such varied clinical presentations should be and indeed are treated differently.

It is important to understand that the JIA criteria are to *classify* children with idiopathic arthritis are not *diagnostic* criteria. The diagnosis still remains one of exclusion, such that each subcategory of the disease has different differential diagnoses to be considered depending upon the age, sex, joint distribution, extra-articular features, analysis of the complete blood count, etc.[2]

MANAGEMENT ISSUES IN JIA

This chapter will be discussed under the following headings:

- Principles of management of JIA
- Objective assessment
- Medications available
- Management of various types of JIA
- Specific pediatric issues
  - Growth
  - Deformities
  - Osteoporosis
  - Uveitis
  - When should methotrexate be stopped?
  - Schooling
  - Psychological issues, compliance
  - Prognosis
- Transition of the adolescent with JIA to adult care
- Conclusions

PRINCIPLES OF MANAGEMENT OF JIA

The aims of good management of JIA are conveniently divided into short and long-term goals.

- DMARDs: Disease modifying anti-rheumatic drugs
- BRMs: Biologic reaction modifiers
- ASCT: Autologous stem cell transplantation

In the eighties and nineties children with JIA had a poor long-term outcome because of late recognition and a few pharmacologic choices. The three therapeutic advances in the last decade that have made a substantial difference to the quality of life and outcome of children with JIA are:

- Intra-articular injections of triamcinalone hexacetonide
- Weekly methotrexate
- Use of twice weekly etanercept and the availability of several other BRMs.

NSAIDs

The choice of NSAIDs is empiric; clinical trials of NSAIDs by the PRCSG (pediatric rheumatology collaborative study group) in the USA concluded that 65% of the children who were going to respond would do so in the first 4 weeks. If a child does not respond, it is logical to use an NSAID from a chemical class different to the one used earlier. 25–33% of patients with oligoarthritis show a significant response to NSAIDs alone. Aspirin is not used commonly any longer for the following reasons—the need for frequent dosing, greater frequency of liver enzyme abnormality, and the association of Reyes syndrome.[17,18] Only ibuprofen and nimesulide are available in the liquid form in India. In the vast majority of children who tolerate non-selective COX inhibitors well, there is no added advantage of using a COX-2 inhibitor. With the availability of biologics and routine use of methotrexate, the number of patients' currently using NSAIDs has reduced over the years. On a review of charts in a US centre in 2003, 53% used an anti-inflammatory dose compared to 35% in 2008.[19]

Steroids

These are very potent drugs and can be used in JIA in many ways: Oral, intravenous and also intra-articular. The use of systemic steroids is limited primarily because of their well-known side effect profile with long-term use. Oral low dose steroids are used sometimes to help patients with severe polyarticular JIA. Both IV and oral steroids are used to treat disease flares in SoJIA.[9] It is important to use the lowest possible dose for a short interval and not let the high disease activity be masked by long-term steroids.[20,21] Specific indications for the use of systemic steroids include: aggressive uveitis, systemic features or macrophage activation syndrome (MAS) in SoJIA, the presence of poor prognostic features and erosive disease in children with polyarticular disease.[22]

Triamcinalone hexacetonide is the preferred intra-articular steroid preparation due to the long-term and predominantly local effects, with a reported median duration of improvement for a period of 74 weeks. In 60% of patients the response persists for at least 6 months, in 45% there is no inflammation for at least one year. Early and continued use of intra-articular steroid has been recently shown to be associated with less leg length discrepancy in young children with oligoarticular JIA.[23] This agent is not available in India, where we use triamcinalone acetonide.

Side effects of intra-articular steroids are very few. It is a very safe procedure that may need appropriate sedation in a child under 7 or 8 years of age. Problems after a joint injection are:

- Periarticular subcutaneous atrophy
- Asymptomatic calcification

There is no report of it being associated with any joint or cartilage damage.[24,25]

Disease Modifying Anti-rheumatic Drugs

Methotrexate

Introduction: Has been established as a treatment for JIA in 1992 after a controlled trial showed efficacy at 10 mg/m^2.[26] This drug is the first choice across most centres in the world. In addition to being a folate antagonist, the effect is mediated via an increase in adenosine, a potent anti-inflammatory mediator.[22]

Dose and route: The route can be oral or injectable, and the safety profile is very good. Most children tolerate this medicine very well and in younger children doses at 1 mg/kg/week are commonly used. The maximum dose is usually restricted to 25–30 mg/week. Patients respond within three months, very occasionally is the response delayed beyond this. There is recent data to support the use of higher dose of methotrexate as published recently in the TREAT trial where a weekly dose of up to 40 mg was given to children with polyarticular JIA.[15] The poorest response with methotrexate is in patients with SoJIA. Folic acid at a dose of 1 mg/day is often given to children receiving methotrexate and appears to lessen the toxicity and increase the tolerability of the drug.

Side effects, precautions and monitoring: Precautions to be exercised while the patient is on methotrexate are as follows: No live vaccination, avoidance of alcohol, and no pregnancy while on the drug. Regular monitoring of hemoglobin, white cell count, platelets and liver function tests every 10–12 weeks is mandatory. Many patients develop nausea and emesis a few years after taking methotrexate; this is because of accumulation of polyglutamated methotrexate intracellularly.[27]

Leflunomide

Leflunomide is another DMARD that has been recently been shown to be as effective as methotrexate. It can be used alone or in combination with other DMARDs.

Dose: Between 5 and 20 mg per day depending upon the size of the child.

Precautions: The female patients who start this drug should be strictly counseled against pregnancy as this is highly teratogenic. Some children may develop diarrhoea while on it. The liver function and complete blood count needs to be checked every 10–12 weeks.[28] It can be used an alternative to the child who is intolerant of methotrexate.[22]

Sulfasalazine

This medication is effective in children with enthesitis related arthritis and interestingly *contraindicated* in the active phase of systemic onset JIA, where it has been reported to cause macrophage activation syndrome.

It is a combination of sulphapyridine and 5 amino salicylic acid, and thus has a cross reactivity with other sulpha containing compounds.

Dose and route: It is orally administered in two divided doses per day. The starting dose is 10 mg/kg/per day, built up to 50 mg/kg/day.

Side effects and precautions: This drug can cause side effects in 11–31% of patients: Nausea, rash, elevated liver enzymes and cytopenias. As with methotrexate the complete blood counts and liver enzymes need monitoring every 8 weeks while the patient is on sulphasalazine.[29]

Hydroxychloroquine

An anti-malarial, it has been used in children with JIA since 1951. It is a mild but safe DMARD. There is only one placebo controlled trial that has shown the drug to be effective in reducing pain on movement.[30]

Dose and route: In children the dose is 6 mg/kg/day, given as a single daily dose.

Side effects and monitoring: Ophthalmic screening every six months to one year is recommended in children who take this drug. This is to screen for retinal toxicity. It can cause a "grey tone change" to the skin colour in Asian patients. It may rarely cause nausea, dyspepsia and abdominal discomfort in some children.[8]

Other DMARDs

Only a brief mention will be made of these agents.

Cyclophosphamide

It is a powerful alkylating agent and a place only for the treatment of severe SoJIA where it has been used in combination with IV methyl prednisone boluses.[31]

Cyclosporine

This drug inhibits the early phase of T cell activation and production of several cytokines. It has been combined with methotrexate, shown to reduce fever in SoJIA, be effective for uveitis and specifically for treatment of macrophage activation syndrome. It is used only occasionally in clinical practice. The use is limited by the side effect profile which includes increased facial hair, gingivitis, hypertension and nephrotoxicity.[32]

Thalidomide

It has both anti-TNF and anti-angiogenic properties. It has a place in the treatment of refractory SoJIA. Used in a dose of up to 5 mg/kg/day it has been shown to be effective in >50% of patients with refractory SoJIA. Two important side effects are somnolence and peripheral neuropathy; the latter is irreversible and needs to be monitored.[33]

BIOLOGICS

There are many cellular and molecular mechanisms that participate in inflammation in patients with juvenile arthritis. Specific biologic agents have been developed that can target one or more steps involved in the immune response.[34]

Interference with cytokines: T cell and monocyte derived cytokines are responsible for clinical features seen in JIA such as arthritis, enthesitis, and for systemic features like fever. Cytokines are thus important targets for therapeutic manipulation. The cytokines that can be manipulated in clinical practice are those that have an important role to play in both the initiation and perpetuation of the rheumatoid process. The three main cytokines that can be manipulated are:

- Tumor necrosis factor alpha
- IL-1
- IL-6

TNF Inhibitor (TNFi) Agents

Agents that block TNF-alpha are excellent tools to reduce disease activity, improve function and even retard or reverse structural damage. Three TNFi available are etarnercept, infliximab, and adalimumab. The use, dose and side effects of each are detailed in Table 65.1. Double blind, randomized controlled studies have shown all three agents to be an effective therapy in patients with JIA. Currently these drugs are given to patients who not respond to an adequate trial with methotrexate.[35–38]

Generic issues with anti-TNF agents: Safety issues are a concern because of the ubiquitous role of TNF.

- To date the only consistent adverse event seen with etanercept has been injection site reactions. There should be caution, however, with using etanercept in patients with a serious infection, or recurrent infections or patients with untreated or latent tuberculosis.
- There has been a recent controversy about the role of TNFi and malignancies. The FDA reported that there is evidence that treatment with TNFi in children may increase the risk of malignancy. However, the cases were confounded by the potential risk of malignancy associated with underlying illnesses and the use of concomitant immunosuppressants; therefore, a clear causal relationship could not be established. The FDA had 48 cases of malignancy in children treated with TNFi, half were lymphomas.[39] The German Biologics registry identified five of 1290 patients to have a malignancy over an eight-year observation period. All patients had been

Table 65.1: Anti-TNF agents for JIA

Name	*Etanercept*	*Infliximab*	*Adalimumab*
Chemical content	Fully humanized	Mouse chimeric	Fully humanized
Chemical structure	P-75 soluble TNF receptor protein fused to human Fc region of human IgG1	Chimeric IgG1 anti-TNF alpha antibodies—mouse antibodies and constant region of human antibody	Recombinant human IgG1 Monoclonal antibody
Binds to	Soluble TNF alpha	Soluble TNF-alpha and membrane bound TNF alpha	Soluble TNF-alpha and membrane bound TNF-alpha
Half life	Four days	Weeks	2 weeks
Route	Subcutaneous	Intra-venous (IV)	Subcutaneous
Dose	0.4 mg/kg/dose/twice weekly or 0.8 mg/kg once weekly	3–6 mg/kg/dose	24 mg/m^2
Frequency	By weekly or weekly	As 0, 2, 6 weeks and then every eight weeks	Every 2 weeks
Methotrexate use	Optional	Recommended (to prevent anti-infliximab antibodies)	Optional
Risk of tuberculosis	+	++	++

exposed to multiple cytotoxics in addition to TNF blockade.[40] Thus it is important that patients on TNFi be followed up carefully and screened appropriately over the long term. Some rare neurological and hematological events have also been noted.

- Active tuberculosis (TB) may develop soon after initiation of treatment with infliximab. Thus before prescribing the drug physicians should screen patients for latent tuberculosis infection or disease. Screening is usually with X-ray chest, purified protein derivative (PPD) and computerized tomography (CT) chest if needed. The risk of TB is more with infliximab than with etanercept as it binds to both the soluble and membrane bound TNF.[41,42]
- Infliximab a mouse chimeric antibody, is known to give infusion reactions and must be given in a hospital setting with adequate monitoring facilities. Though in the western world their use is open ended, in India these agents are prohibitively expensive and are used to reduce disease burden and are very seldom continued for years on end. The rough cost of etanercept therapy for a 30 kg child is 4.16 lakh rupees for one year, for infliximab is 3.24 lakh rupees per year + incidental hospital charges for nine admissions in the first year. Adalimumab (originator) is not yet available in India.
- Data from the German registry has shown that the combination therapy with etanercept and methotrexate is far superior to use of etanercept alone. The likelihood of achieving a PedACR70 increased with combination therapy with an odds ratio of 2.1 (95% CI 1.2 to 3.5). The added advantage of using the combination of methotrexate and a TNFi is the reduced development of anti-drug antibodies against the TNFi.[43]
- Non-response: If a patient does not respond to the 1st TNF, there are three options: Switch the TNFi, change to tocilizumab, or abatacept.

ANAKINRA

This is interleukin-1 receptor antagonist (IL-1Ra). It is a naturally occurring acute phase anti-inflammatory protein part of the IL-1 supergene family. It is an important physiologic regulator of IL-1 induced inflammatory activity. Anakinra is a human recombinant form of IL-1 Ra that is produced by recombinant technology in *E. coli.* It is given daily by subcutaneous injection. In pediatrics it is used in a dose of 0.1 mg/kg/day and it has been shown to be effective in etanercept resistant SoJIA patients and also as a first line therapy for macrophage activation syndrome (MAS).[44] This is currently unavailable in India.

IL-6 Receptor Antagonist

There is evidence that IL-6 is an important driver of the clinical features of SoJIA. Tocilizumab is a genetically engineered humanized monoclonal antibody directed toward IL-6 receptor. There is evidence for the use of tocilizumab for both SoJIA and polyarticular JIA. This drug has recently received FDA approval in children with SoJIA in USA. It is used in a dose of 12 mg/kg in children with a weight <30 kg and 8 mg/kg in children with weight >30 kg. The systemic features respond within a few weeks, arthritis abates more slowly. The drug is given two weekly for the child with SoJIA and 4 weekly for polyarticular JIA.[45,46]

Biosimilars

These are generic biological drugs intended to have the same efficacy as the innovator and are being used in many countries due to their low cost. However, as these are high molecular weight proteins with complex structures and produced in biological systems they are different from chemical generic compounds. Minor differences in post-translational modifications (e.g. glycosylation, deamidation), microheterogenicity (difference in charge), etc. may lead to difference in pharmacokinetics, pharmacodynamics, efficacy and immunogenicity. No trial of biosimilars is available for children with JIA. In India multiple biosimilars are available but rigorous well-planned studies are lacking. Their clear role as regards switching from innovator to biosimilar, use as substitutes as is done for usual drugs, etc. is still to be determined.

Tolerance Induction

Trials of oral type II collagen have been tried in adults and children, as there is documented autoimmunity to type II collagen in both adult and juvenile arthritis. Studies in humans have shown variable results, but with advances in other areas this has not been further studied.

Inhibition of MHC/Antigen/T Cell Receptor (TCR) Interaction

This is also called the tri-molecular complex. If the initiating antigen was known an immunization program could be developed to prevent disease. Another possible route of attack is to block the MHC sites by anti-MHC antibodies. This is currently under study.

Inhibition of Cellular Function and Cell to Cell Interaction

T cells are critical in initiating the rheumatoid process in both children and adults. Another promising approach has been to block co-stimulatory interactions between T and B cells, for example, by inhibiting B7 pathway with CTLA-4Ig. Initial results in adult patients look promising.[34] A selective T cell co-stimulation modulator abatacept has been shown to be very effective in children with polyarticular JIA both in the short term and long term, up to 21 months.[47,48]

B Cell Blockade

The anti-CD20 molecule rituximab is very effective in adult rheumatoid arthritis. There is no class I evidence for its use in JIA, however, there are recent reports of the effectiveness of rituximab in both polyarticular JIA and SoJIA which need to be replicated in other centres.[49,50]

Stem Cell Transplantation in JIA

Autologous haemopoietic stem-cell transplantation (AHSCT) has been described as a possible treatment for severe autoimmune disease refractory to conventional treatment. The first four children with severe forms of juvenile idiopathic arthritis to receive AHSCT were reported in 1999, Collaborative European trails with strict entry criteria and pre-transplant conditioning are on going. Patient selection is critical prior to undertaking this modality of treatment where the mortality has now come down from 14 to 5%.[51]

MANAGEMENT OF VARIOUS SUBTYPES OF JIA[3, 8, 9, 34, 52]

Oligoarticular JIA

Mild disease can be treated with NSAIDs or intra-articular steroids. With functional limitations both modalities are started together. If only NSAIDs are used alone, marked improvement should occur in 4 weeks and resolution of the involved joints occurs in 3 to 4 months. Once the patient is in clinical remission on drugs (6 months) the NSAIDs may be discontinued.

If intra-articular steroids are used, resolution occurs in a week. Adequate, age appropriate sedation administered by a skilled team is a prerequisite.

DMARDs are generally not recommended for this group, although occasionally methotrexate may be useful for difficult to control uveitis or when damage to a critical joint threatens function (wrist/hip). If the disease extends, methotrexate should be considered for the treatment. Oligoarticular JIA is not always a benign disease. Persistent elevation of ESR, longstanding anaemia, or difficult to control arthritis and or uveitis are all indications for aggressive treatment/DMARD use.[53]

Polyarticular JIA

The patient with mild disease can be treated with NSAIDs and hydroxychloroquine. The vast majority of children will need methotrexate. Low dose steroids which are tapered over a few months are very useful in many patients and the benefits outweigh the risks. TNF antagonists are reserved for the methotrexate resistant patients.

Systemic Onset JIA (SoJIA)

The patients who have this disease vary significantly from each other. There are some children with raging fevers and barely discernable arthritis, others with severe arthritis and then a group that have a life-threatening condition called macrophage activation syndrome where the child with SoJIA develops high grade continuous fever, a marked drop in the platelets and ESR and features of coagulopathy. A child with very mild disease may improve with NSAIDs alone over a period of a few weeks. A vast majority of the children will require steroids that can be given orally at a dose of 1–2 mg/kg/day slowly tapered off over a period of months or even IV methylprednisolone at a dose of 30 mg/kg/day, maximum 500 mg given daily for 2–3 days. Methotrexate has a good effect on the arthritis but its role in reducing the systemic features of this condition is not well understood.

Thalidomide may be used in the resistant patient where it controls both the arthritis and the systemic features. The resistant patients are a challenge to manage and strategies that are beneficial include use of pulsed methyl prednisolone, thalidomide and biologics. The biologic agents used in this subcategory are anakinra and toculizumab; TNF blockers have the least efficacy in this subcategory. Of note, anti-TNF agents have the least efficacy in this subcategory and have no significant effect on the systemic features.

Cyclosporine has a dramatic effect in children with macrophage activation syndrome and may also be used to control systemic features in the difficult to control patients.[22]

Enthesitis Related Arthritis (ERA)

The components of this disease that need attention are:

- Acute anterior uveitis
- Peripheral joint disease
- Enthesitis
- Axial joint disease

There is convincing evidence that TNF plays a pivotal role in inflammation in these patients. Of late, there is data on the role of IL17/IL 23 in the pathogenesis of adults with ankylosing spondylitis (AS). The discovery of this pathogenic pathway leads to the development of an IL-17 inhibitor, Secukinumab, which has recently been approved by the US FDA for treatment of active AS.[54,55] If the patient has < 4 peripheral joints without critical joint involvement such as hip disease they may be treated with NSAIDs and intra-articular steroids. With hip joint disease most children would get methotrexate as for polyarticular disease. Sulfasalazine is also effective in these patients. With axial disease the child would need regular NSAIDs for weeks together. If there is no improvement in three months TNFi should be used. It should be noted that there are several adult studies that have confirmed the lack of response of axial disease to traditional DMARDs. Acute uveitis may be treated with topical steroids, and in recurrent uveitis the child will benefit with long-term methotrexate or infliximab or adalimumab.[22,56] Of note, abatacept, rituximab and tocilizumab have no role to play in the management of ERA.

Psoriatic Arthritis

Asymmetric involvement of small joints especially the DIP joint, and dactylitis is characteristic of psoriatic arthritis. Significant nail pitting often precedes arthritis. The skin and joint disease may not always follow the same course. In addition to local skin treatment NSAIDs/intra-articular steroids are used for localized disease involving a few joints, and methotrexate is used for aggressive disease involving multiple joints.[57]

American College of Rheumatology (ACR) Guidelines for the Management of JIA

The ACR has chosen to describe the pathways of care for five groups of children: Oligoarticular disease, polyarticular disease, SoJIA with systemic features, SoJIA with articular disease and finally, the child with axial disease. Each category of disease has been divided into low, moderate and high disease activity and poor prognostic markers have been listed for each category. The key points that have not been covered in the ACR recommendations are:

- The indications for systemic glucocorticoids for the treatment of synovitis were not considered, owing to a lack of published evidence.
- The treatment of uveitis, enthesitis, and macrophage activation syndrome was not considered.

Role of folic acid, weaning of methotrexate/biologic response modifiers (BRM)/abnormal liver function tests have not been discussed. The guidelines are heavily weighted in favour of use of biologics within 3–6 months of ongoing disease in most categories which is not practical for the majority of Indian children with JIA. Each category is stratified per the prognostic features with details available on the application of these recommendations for the practicing clinician.[58]

SPECIFIC PAEDIATRIC ISSUES

Growth

Significant chronic disease in childhood often impacts the growth in children and thus all children with JIA should be placed on a growth chart and periodic assessment of their height velocity should be calculated. Not just the disease but the drugs used to control severe inflammation such as steroids can permanently stunt the height of children. The best strategy to maximize growth is aggressive

as per standard definitions recently validated for use for JIA patients.

Strategies likely to help children with JIA in our country are:

- Formal inclusion of pediatric rheumatology core curriculum both at the graduate and post-graduate levels in medical schools in India.
- Increased awareness amongst general pediatricians and orthopedic surgeons about the prevalence of JIA in India and the urgency of early referral of these patients to specialist centres.
- Establishment of several tertiary level centres with good skill and expertise to direct care for these children.[52]

Key Points

1. JIA is a group of heterogeneous diseases; it remains a diagnosis of exclusion.
2. Each subcategory of the disease should be managed per the age of the child, number of joints involved and the criticality of involved joints.
3. The disease should be assessed objectively and regularly assessed with the aim of achieving clinical remission in shortest possible time.
4. Early diagnosis and appropriate management by a multidisciplinary team gives the best possible outcome.

REFERENCES

1. Petty RE, Southwood TR, Baum J, Bhettay E, Glass DN, et al. Revision of the proposed classification criteria for juvenile idiopathic arthritis: Durban, 1997 [see comments]. J Rheumatol. 1998; 25(10):1991–4.
2. Sawhney S. Juvenile idiopathic arthritis: Classification, clinical features, and management. Indian J Rheumatol 2012; 7:11–21.
3. Cassidy JT, Petty RE. Chronic Arthritis in Childhood. In Textbook of Pediatric Rheumatology. 5th edition. Philadelphia, PA: Elsevier Saunders; 2005. pp. 206–60.
4. Sawhney S. Spectrum of juvenile idiopathic arthritis in a tertiary care centre in Northern India. Indian J Rheumatol 2006; 14:3–6.
5. Wallace CA. On beyond methotrexate treatment of severe juvenile rheumatoid arthritis. Clin Exp Rheumatol. 1999; 17:499–504.
6. Rossi F, Di Dia F, Galipo O, Pistorio A, Valle M, Magni-Manzoni S, et al. Use of the Sharp and Larsen scoring methods in the assessment of radiographic progression in juvenile idiopathic arthritis. Arthritis Rheum. 2006 15; 55:717–23.
7. Zhao Y, Wallace C. Judicious use of biologicals in juvenile idiopathic arthritis. Curr Rheumatol Rep. 2014; 16:454.
8. Wallace CA. Current management of juvenile idiopathic arthritis. Best Pract Res Clin Rheumatol. 2006 20:279–300.
9. Ravelli A, Martini A. Juvenile idiopathic arthritis. Lancet. 2007; 369:767–78.
10. Giannini EH, Ruperto N, Ravelli A, Lovell DJ, Felson DT, Martini A. Preliminary definition of improvement in juvenile arthritis. Arthritis Rheum 1997; 40:1202–9.
11. Giannini EH, Lovell DJ, Felson DT, Goldsmith CH. Preliminary core set of outcome variables for use in JRA clinical trials. Arthritis Rheum 1994; 37:Suppl: S428.
12. Consolaro A, Ruperto N, Bazso A, Pistorio A, Magni-Manzoni S, Filocamo G, et al. for the Paediatric Rheumatology International Trials Organisation. Development and validation of a composite disease activity score for juvenile idiopathic arthritis. Arthritis Rheum 2009; 61:658–66.
13. Consolaro A, Bracciolini G, Ruperto N, Pistorio A, Magni-Manzoni S, Malattia C, et al. Remission, minimal disease activity and acceptable symptom state in juvenile idiopathic arthritis. Defining criteria based on the juvenile arthritis disease activity score. Arthritis Rheum 2012; 64:2366–74.
14. Wallace CA, Ruperto N, Giannini E; Childhood Arthritis and Rheumatology Research Alliance; Pediatric Rheumatology International Trials Organization; Pediatric Rheumatology Collaborative Study Group. Preliminary criteria for clinical remission for select categories of juvenile idiopathic arthritis. J Rheumatol. 2004; 31(11):2290–4.
15. Wallace CA, Giannini EH, Spalding SJ, Hashkes PJ, O'Neil KM, Zeft AS, et al. Trial of early aggressive therapy in polyarticular juvenile idiopathic arthritis. Arthritis Rheum 2012; 64:2012–21.
16. Collado P, Malattia C. Imaging in paediatric rheumatology: Is it time for imaging? Best Pract Res Clin Rheumatol. 2016; 30:720–35.
17. Kvien TK. Hoyeraal HM, Sandstad B. Naproxen and acetylastliclic acid in the treatment of pauciarticular and polyarticular juvenile rheumatoid arthritis: assessment of tolerance and efficacy in a single center 24-weeks double-blind parallel study. Scand J Rheumatol. 1984; 13:342–50.
18. Giannini EH, Cawkwell GD. Drug treatment in children with juvenile rheuamatoid arthritis. Pediatr Clin North Am. 1995; 42:1099–125.

19. Kochar R, Walsh K, Jain A, Spalding S, Hashkes P. Decreased use of non-steroidal anti-inflammatory drugs for the treatment of juvenile idiopathic arthritis in the era of modern aggressive treatment. Rheumatol Int 2012; 32:3055–60.
20. Kahn P. Juvenile idiopathic arthritis—an update on pharmacotherapy. Bull NYU Hosp J Dis. 2011; 69: 264–76.
21. Beresford MW, Baildam EM. New advances in the management of juvenile idiopathic arthritis—1: non-biological therapy. Arch Dis Child Educ Pract Ed 2009; 94:144–50.
22. Blazina S, Markelj G, Avramovic MZ, Toplak N, Avcin T. Management of Juvenile Idiopathic Arthritis: A Clinical Guide. Paediatr Drugs. 2016; 18:397–412.
23. Sherry DD. What's new in the diagnosis and treatment of juvenile rheumatoid arthritis. J Pediatr Orthop. 2000; 20:419–20.
24. Padeh S, Passwell JH. Intra-articular corticosteroid injections in the management of children with chronic arthritis. Arthritis Rheum. 1998; 41: 1210–14.
25. Hagelberg S, Magnusson B, Jenner G, Andersson U. Do frequent corticosteroid injections in the knee cause cartilage damage in juvenile chronic arthritis? Long-term follow-up with MRI. J Rheumatol. 2000; 27 (suppl 58):95.
26. Giannini EH, Brewer EJ, Kuzmina N, Shaikov A, Maximov A, Vorontsov I, et al. Methotrexate in resistant juvenile rheumatoid arthritis. Results of the USA–USSR double-blind, placebo-controlled trial. The Pediatric Rheumatology Collaborative Study Group and the Cooperative Children's Study Group. N Engl J Med. 1992; 326:1043–9.
27. Wallace CA. The use of methotrexate in childhood rheumatic diseases. Arthritis Rheum. 1998; 3:381–91.
28. Silverman E, Mouy R, Spiegel L Jung LK, Saurenmann PK, et al. Leflunomide or methotrexate for Juvenile Rheumatoid Arthritis. N Engl J Med. 2005; 352:1655–66.
29. Burgos-Vargas R, Vazquez-Mellado J, Pacheco-Tena C, Hernandez-Garduno A, Goycochea-Robles MV. A 26-week randomised, double blind, placebo controlled exploratory study of sulfasalazine in juvenile onset spondyloarthropathies. Ann Rheum Dis. 2002; 61:941–2.
30. Brewer EJ, Giannini EH, Kuzmina N, Alekseev L. Penicillamine and hydroxychloroquine in the treatment of severe juvenile rheumatoid arthritis. Results of the USA–USSR double-blind placebo-controlled trial. N Engl J Med. 1986; 314:1269–76.
31. Shaikov AV, Maximov AA, Speransky AI, Lovell DJ, Giannini EH, Solovyev SK. Repetitive use of pulse therapy with methylprednisolone and cyclophosphamide in addition to oral methotrexate in children with systemic juvenile rheumatoid arthritis—preliminary results of a long-term study. J Rheumatol. 1992; 19:612–6.
32. Mouy R, Stephan JL, Pillet P, Haddad E, Hubert P, Prieur AM. Efficacy of cyclosporine A in the treatment of macrophage activation syndrome in juvenile arthritis: report of five cases. J Pediatr 1996; 129:750–4.
33. Lehman TJ, Schechter SJ, Sundel RP, Oliveira SK, Huttenlocher A, Onel KB. Thalidomide for severe systemic onset juvenile rheumatoid arthritis: A multicenter study. J Pediatr. 2004; 145:856–7.
34. Laxer RM. Pharmacology and Drug Therapy. In Cassidy JT, Petty RE, Editors. Textbook of Pediatric Rheumatology. 5th edition. Philadelphia, PA: Elsevier Saunders; 2005. pp. 76–141.
35. Schmeling H, Mathony K, John V, Keysser G, Burdach S, Horneff G. A combination of etanercept and methotrexate for the treatment of refractory juvenile idiopathic arthritis: a pilot study. Ann Rheum Dis. 2001; 60:410–2.
36. Lovell DJ, Giannini EH, Reiff A, Cawkwell GD, Silvermann ED, Nocton JJ, et al. Etanercept in Children with Polyarticular Juvenile Rheumatoid Arthritis. N Engl J Med 2000; 342:763–9.
37. Lovell DJ, Ruperto N, Goodman S, Reiff A, Jung L, Jarosova K, et al; Pediatric Rheumatology Collaborative Study Group; Pediatric Rheumatology International Trials Organisation. Adalimumab with or without methotrexate in juvenile rheumatoid arthritis. N Engl J Med. 2008; 359:810–20.
38. Gerloni V, Pontikaki I, Gattinara M, Desiati F, Lupi E, Lurati A, et al: Efficacy of repeated intravenous infusions of an anti-tumor necrosis factor alpha monoclonal antibody, infliximab, in the persistently active, refractory juvenile idiopathic arthritis. Results of an open-label prospective study. Arthritis Rheum 2005; 52:548–53.
39. Diak P, Siegel J, La Grenade L, Choi L, Lemery S, McMahon A. Tumor necrosis factor α blockers and malignancy in children: Forty-eight cases reported to the food and drug administration. Arthritis Rheum 2010; 62:2517–24.
40. Horneff G, Foeldvari I, Minden K, Moebius D, Hospach T. Report on malignancies in the German juvenile idiopathic arthritis registry. Rheumatology (Oxford) 2011; 50:230–6.
41. Keane J, Gershon S, Wise RP, Mirabile-Levens E, Kasznica J, Schwieterman WD, et al. Tuberculosis associated with infliximab, a tumor necrosis factor alpha-neutralizing agent. N Engl J Med. 2001; 345:1098–104.

42. Botsios C. Safety of tumour necrosis factor and interleukin-1 blocking agents in rheumatic diseases. Autoimmun Rev. 2005; 4:162–70.

43. Horneff G, De Bock F, Foeldvari I, Girschick HJ, Michels H, Moebius D, et al. Safety and efficacy of combination of etanercept and methotrexate compared to treatment with etanercept only in patients with juvenile idiopathic arthritis (JIA): preliminary data from the German JIA Registry. Ann Rheum Dis. 2009; 68:519–25.

44. Ohlsson V, Baildam E, Foster H, Jandial S, Pain C, Strike H, et al. Anakinra treatment for systemic onset juvenile idiopathic arthritis (SOJIA). Rheumatology (Oxford). 2008; 47:555–6.

45. Yokota S, Imagawa T, Mori M, Miyamae T, Aihara Y, Takei S, et al. Efficacy and safety of tocilizumab in patients with systemic-onset juvenile idiopathic arthritis: a randomised, double-blind, placebo-controlled, withdrawal phase III trial. Lancet 2008; 371:998–1006.

46. Imagawa T, Yokota S, Mori M, Miyamae T, Takei S, Imanaka H, et al. Safety and efficacy of tocilizumab, an anti-IL-6-receptor monoclonal antibody, in patients with polyarticular-course juvenile idiopathic arthritis. Mod Rheumatol 2012; 22(1):109–15.

47. Ruperto N, Lovell DJ, Quartier P, Paz E, Rubio-Perez N, Silva CA, et al. Abatacept in children with juvenile idiopathic arthritis: a randomised, double-blind, placebo-controlled withdrawal trial. Lancet 2008; 372:383–91.

48. Ruperto N, Lovell DJ, Quartier P, Paz E, Rubio-Pérez N, Silva CA, et al. Long-term safety and efficacy of abatacept in children with juvenile idiopathic arthritis. Arthritis Rheum 2010; 62:1792–802.

49. Kasher-Meron M, Uziel Y, Amital H. Successful treatment with B-cell depleting therapy for refractory systemic onset juvenile idiopathic arthritis: a case report. Rheumatology (Oxford) 2009; 48:445–46.

50. Alexeeva E, Valieva S, Bzarova T, Semikina E, Isaeva K, Lisitsyn A, et al. Efficacy and safety of repeat courses of rituximab treatment in patients with severe refractory juvenile idiopathic arthritis. Clin Rheumatol 2011; 30:1163–72.

51. Wulffraat NM, Vastert B, Tyndall A. Treatment of refractory autoimmune diseases with autologous stem cell transplantation: focus on juvenile idiopathic arthritis. Bone Marrow Transplant. 2005; 35 Suppl 1:S27–9.

52. Sawhney S. Recent advances in management of JIA. In Dutta A, Sachdeva A, Editors. Advances in Pediatrics. 1st edition. New Delhi: Jaypee Publishers; 2006. pp. 944–53.

53. Lehman TJA. Oligoarticular JIA: Is it a benign disease? Pediatric Rheumatol Online J. 2005; 3:144–46.

54. Braun J, Baraliakos X, Kiltz U. Secukinumab (AIN457) in the treatment of ankylosing spondylitis. Expert Opin Biol Ther. 2016;16:711–22.

55. Sieper J, Poddubnyy D. Axial spondyloarthritis. Lancet. 2017; 390:73–84.

56. Petty RE, Malleson P. Spondyloarthropathies of childhood. Pediatr Clin North Am 1986; 33:1079–96.

57. Woo P, Wedderburn LR. Juvenile chronic arthritis. Lancet 1998; 351:969–73.

58. Beukelman T, Patkar NM, Saag KG, Tolleson-Rinehart S, Cron RQ, DeWitt EM, et al. American college of rheumatology recommendations for the treatment of juvenile idiopathic arthritis: initiation and safety monitoring of therapeutic agents for the treatment of arthritis and systemic features. Arthritis Care Res. 2011; 63:465–482.

59. Rabinovich CE. Bone mineral status in juvenile rheumatoid arthritis. J Rheumatol. 2000; 27 Suppl 58:34–7.

60. Bianchi ML, Cimaz R, Bardare M, Zulian F, Lepore L, Boncompagni A, et al. Efficacy and safety issues of alendronate for the treatment of osteoporosis in diffuse connective diseases in children. Arthritis Rheum. 2000; 43:1960–66.

61. Weiss AH, Wallace CA, Sherry DD. Methotrexate for resistant chronic uveitis in children with juvenile rheumatoid arthritis. J Pediatr. 1998; 133(2):266–8.

62. Wells JM, Smith JR. Uveitis in juvenile idiopathic arthritis: recent therapeutic advances. Ophthalmic Res. 2015; 54:124–7.

63. Clarke SL, Sen ES, Ramanan AV. Juvenile idiopathic arthritis-associated uveitis. Pediatr Rheumatol Online J. 2016; 14:27.

64. Holzinger D, Frosch M, Kastrup A, Prince FHM, Otten MH, Van Suijlekom-Smit LWA, et al. The Toll-like receptor 4 agonist MRP8/14 protein complex is a sensitive indicator for disease activity and predicts relapses in systemic-onset juvenile idiopathic arthritis. Ann Rheum Dis 2012; 71:974–80.

65. Ravelli A, Martini A. Early predictors of outcome in juvenile idiopathic arthritis. Clin Exp Rheumatol. 2003; 21(5 Suppl 31):S89–93.

66. Spiegel LR, Schneider R, Lang BA, Birdi N, Silverman ED, Laxer RM, et al. Early predictors of poor functional outcome in systemic-onset juvenile rheumatoid arthritis: a multicenter cohort study. Arthritis Rheum. 2000; 43:2402–9.

67. Aggarwal A, Agarwal V, Danda D, Misra R. Outcome in juvenile rheumatoid arthritis in India. Indian Pediatr 2004; 41(2):180–4.

68. Athreya BH. A general approach to management of children with rheumatic diseases. In: Cassidy JT, Petty RE. Textbook of Pediatric Rheumatology. Philadelphia: WB Saunders; 200. pp. 190–211.
69. Zak M , Pedersen FK. Juvenile chronic arthritis into adulthood: a long-term follow up study. Rheumatology 2000; 39:198–204.
70. Hashkes PJ Laxer RM. Medical treatment of juvenile arthritis. JAMA, 2005; 294:1671–84.

FURTHER READING

1. Blazina Š, Markelj G, Avramoviè MZ, Toplak N, Avèin T. Management of juvenile idiopathic arthritis: a clinical guide. Paediatr Drugs. 2016; 18: 397–412.
2. Pagnini I, Bertini F, Cimaz R. Difficult-to-treat juvenile idiopathic arthritis: current and future options. Paediatr Drugs. 2016; 18:101–8.

INFECTIONS AND ARTHRITIS

Common Modes of Infection

- Bacterial infection from distant site, e.g. infective endocarditis, meningitis
- Direct inoculation during surgery or joint injection
- Complications of procedures like sterno-clavicular joint infection in case of subclavian vein catheterisation
- Septic arthritis of hip from femoral venepuncture.[5]

Clinical Features[6–8]

- Bacterial arthritis classically presents with acute onset fever, joint pain, swelling and limited range of motion.
- Usually occurs in a single joint, most commonly of the lower extremity. Knee is the most common joint involved followed by hip and then ankle.
- Polyarticular involvement is seen in around 10 per cent of cases and are more common in young age, immunosuppressed and patients suffering from rheumatoid arthritis.[9]
- Injection drug users have a predilection to develop bacterial arthritis in axial joints, such as the sternoclavicular or sternomanubrial joint.[1, 3]

Diagnosis

a. Clinical

- Acute onset monoarthritis of lower limb especially knee joint should always raise the suspicion of joint infection (Fig. 66.1).
- In addition to pain there will be limitation in range of motion.
- Systemic response like fever will be present in immunocompetent patient.

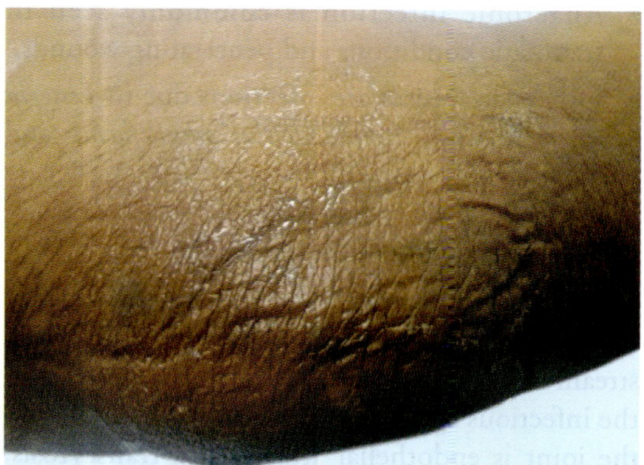

Fig. 66.1: Septic arthritis right knee with peau-de-orange appearance of overlying skin

b. Laboratory Evaluation

- Complete blood count (CBC) will show polymorphonuclear leucocytosis.
- Erythrocyte sedimentation rate (ESR) and C-reactive protein (CRP) levels are raised and help in ascertaining response to treatment.
- Blood culture should be advised in patient with profound systemic response and is positive in around 30% cases.
- Synovial fluid analysis should be performed as soon as possible when bacterial arthritis is suspected. The identification of organisms in synovial fluid is the primary criterion for the diagnosis of bacterial arthritis. Fluid should be examined for crystals also to rule out gout or pseudogout. Typically, the fluid is purulent with leukocyte count more than 50000/mm^3. Gramstain and culture should be done in all cases. In special cases acid-fast bacilli stain and culture, polymerase chain reaction are also undertaken.

c. Imaging

- Plain radiograph is an important adjunct to the physical examination. It is imperative to compare the involved extremity with the opposite healthy side. Widening of joint space, increased opacity within the joint, sublaxation, erosion of subchondral bone are a few important findings. It helps in ruling out osteomyelitis, erosions due to gout and fracture if any.
- Ultrasonography is of not much importance except identifying and quantifying joint effusions which are located deep like hip joint.
- Magnetic resonance imaging is not advised frequently in septic arthritis. Its use is mainly to evaluate possible concomitant osteomyelitis.

Differential Diagnosis

1. Crystal-induced Arthritis

- History of episodic monoarthritis of acute onset involving fist metatarsophalangeal joint
- Risk factors for crystal induced arthritis (gout) are hypertension, dyslipidaemia, diabetes mellitus
- Synovial fluid analysis demonstrating monosodium urate crystals of gout or calcium pyrophosphate dihydrate crystals of pseudogout
- One must be cautious that septic arthritis may co-exist with a flare of crystal-induced arthritis.

2. Reactive Arthritis

- Young individual with antecedent history of urinary or gut infection
- Asymmetric oligoarthritis of lower limb
- Extra-articular skin lesions like circinate balanitis over glans penis or keratoderma blenorrhagica over sole of foot may be present in patients with Reiter's syndrome.

3. Rheumatoid Arthritis

- Though rare rheumatoid arthritis may sometimes present as mono-arthritis
- Serological tests like rheumatoid factor and anti-cyclic citrullinated protein positivity will clinch the diagnosis of RA
- Gramstain and culture negative synovial fluid

4. Lyme disease

- Acute monoarthritis with characteristic rash, fever, and migratory arthralgias.
- History of travel to an endemic area
- Immunoglobulin G (IgG) antibodies to *Borrelia burgdorferi* positive

Septic Arthrits in Children

- In neonates and young infants it may manifest primarily as septicaemia,cellulitis, or fever without any localising sign of infection.[4]
- The signs and symptoms can be subtle
- More than one joint may be involved
- Causative organisms are different (Group B Streptococci, *N. gonorrhoeae*, and gram-negative bacilli in addition to *S. aureus*)
- Bacterial arthritis at times may represent as a complication of osteomyelitis

Management

Septic arthritis should be managed on the following principles[6–8]

- Antibiotic therapy
- Joint lavage or arthrocentesis
- Rapid mobilisation of the affected joint

Antibiotic Therapy

Diagnostic and therapeutic joint aspiration is the keystone in the management of septic arthritis. However, one need not wait to initiate antibiotic therapy. The initial choice of antimicrobial regimens is generally based on coverage of the most likely organisms to cause infection. Regimen may be streamlined to a single agent once antimicrobial susceptibility data are available. There is no role for intra-articular antibiotics. Choice of antibiotics is given in Table 66.1.

Joint Drainage

There is a requirement of joint drainage to drain out the pus from affected joint. Options for

Table 66.1: Antibiotics in septic arthritis

Organisms	*Antibiotics*	*Duration of therapy*
Gram-positive cocci	Vancomycin (30 mg/kg in 24 hr 2 divided doses)	2 wk followed by switch over to oral antibiotics for another 2 wk
Gram-negative bacilli	Ceftriaxone (2 g intravenously once daily) or Cefotaxime (2 g IV every eight hours) or Ceftazidime (1 to 2 g intravenously every eight hours)	Same
Pseudomonas aeruginosa	Ceftazidime plus aminoglycoside	Same
Methicillin-resistant *S. aureus* (MRSA)	Vancomycin (30 mg/kg in 24 hr 2 divided doses, Daptomycin (6 mg/kg/day IV) Linezolid (600 mg by mouth [PO] or IV twice daily) Clindamycin (600 mg PO or IV three times daily)	Same
Cephalosporin allergic patients	Ciprofloxacin (400 mg IV every 12 hours or 500 to 750 mg orally twice daily) plus aminoglycoside	Same

drainage include needle aspiration, arthroscopic drainage, or arthrotomy (open surgical drainage). Knee joint can be drained by large bore needle. Decrease in joint temperature, reduction in leucocyte count and amelioration of pain with increase mobility signifies adequate drainage. Drainage of bigger joints like hip and shoulder should be done by arthroscopy or arthrotomy. A few studies have shown that arthroscopy with large volume irrigation is better than arthrotomy. To conclude septic arthritis is a medical emergency which needs urgent confirmation of diagnosis and treatment in order to avoid permanent joint damage.

Key Points

- Septic arthritis is a common cause of acute onset monoarthritis due to bacterial infection.
- Bacteraemia is more likely to localize in a joint with pre-existing arthritis. Patients with rheumatoid arthritis, gout, pseudogout and prosthetic joints appear to be more prone to bacterial arthritis;
- Organisms such as *S. aureus* and Streptococci are the most common to cause joint infections. Infection due to gram-negative bacilli is generally seen following trauma or in patients with severe underlying immunosuppression.
- Common joint involved is the knee joint followed by wrists, ankles, and hip joint respectively.
- Isolation of bacteria in the synovial fluid is the definitive diagnostic test. But leukocyte count, blood culture are also important. Crystals need to be ruled out.
- Treatment consists of antibiotic therapy and joint drainage. It should be initiated pending results of gramstain and can be modified after the result of gramstain and culture sensitivity. The duration of therapy is around three to four weeks.

REFERENCES

1. Mathews CJ, Coakley G. Septic arthritis: current diagnostic and therapeutic algorithm. Curr Opin Rheumatol 2008; 20:457.
2. Kaandorp CJ, Krijnen P, Moens HJ, et al. The outcome of bacterial arthritis: a prospective community based study. Arthritis Rheum 1997; 40:884.
3. Nade S. Septic arthritis. Best Pract Res Clin Rheumatol 2003; 17:183.
4. Petty RE. Septic arthritis and osteomyelitis in children. Curr Opin Rheumatol. 1990; 2:616–21
5. Fromm SE, Toohey JS. Septic arthritis of the hip in an adult following repeated femoral venipuncture. Orthopedics 1996; 19:1047.
6. Bhojani K, Kalke S. Septic arthritis including tuberculosis. In: Rao URK, editor. Manual of Rheumatology. Indian Rheumatology Association. Mumbai: National Book Depot, 2014; pp. 424–32.
7. Krishnamurthy V. Infection and arthritis. In: Rao URK, editor. Manual of Rheumatology. Indian Rheumatology Assoc. Mumbai: National Book Depot; 1999. pp. 131–140.
8. Joshi VR. Joint infection. J Gen Med 2004; 16:31–32.
9. Dubost JJ, Fis I, Denis P, Lopitaux R, Soubrier M, Ristori JM, et al. Polyarticular septic arthritis. Medicine (Baltimore). 1993; 72:296–310.

FURTHER READING

1. Wang DA, Tambyah PA.Septic arthritis in immunocompetent and immunosuppressed hosts. Best Pract Res Clin Rheumatol. 2015; 29(2):275–89.
2. Coakley G, Mathews C, Field M, Jones A, Kingsley G, Walker D, et al. BSR & BHPR, BOA, RCGP and BSAC guidelines for management of the hot swollen joint in adults. Rheumatology (Oxford). 2006; 45: 1039–41.

Chapter

67

Musculoskeletal Tuberculosis

Arun Hegde, Krishnan Shanmuganandan

INTRODUCTION

Musculoskeletal tuberculosis (TB) accounts for 1–3% of all forms of TB and 10–15% of extra-pulmonary cases.[1] Several clinical patterns of musculoskeletal TB have been described, including spondylitis, peripheral arthritis, reactive arthritis or Poncet disease, osteomyelitis, and soft tissue abscesses. In this chapter, we review the clinical features, diagnosis and treatment of the various forms of musculoskeletal TB.

TB SPONDYLITIS

Tuberculous spondylitis, also known as Pott's disease is named after Percivall Pott, and is one of the oldest described diseases of humankind. Tuberculous spondylitis is the most common manifestation of musculoskeletal tuberculosis, accounting for approximately 40–50% of cases and has the potential to cause serious morbidity and severe deformities.

Demographics

Although some series have found that Pott's disease does not have a sexual predilection, the disease is more common in males with male-to-female ratio of 1.5–2:1. It occurs primarily in adults, although in developing countries involvement in younger children and older adults predominates.[2]

Pathophysiology

Pott's disease is usually secondary to haematogenous dissemination from a primary focus in the lung or lymph nodes. There are two types, the more common paradiscal type which involves destruction adjacent to the endplates of two or more vertebral bodies and spreads along the arteries, and the central type, that spreads along the Bateson's plexus of veins. Thoracic vertebrae are most commonly involved followed by the lumbar and cervical vertebrae. The basic lesion comprises a combination of osteomyelitis and arthritis. The anterior vertebra body adjacent to the subchondral plate is usually affected. In adults, disc disease is secondary to spread of infection from a vertebral body whereas in children, the disc is the primary site.

Clinical Features

Pott's disease is usually preceded by constitutional symptoms in form of fever and weight loss. The average duration of symptoms prior to diagnosis is 4 months,[3] but it can be considerably longer because of the non-specific nature of the chronic back pain. Back pain can be constant, dull aching or radicular in nature, and usually experienced for weeks prior to seeking treatment. Neurologic abnormalities can result from spinal cord compression, and include paraplegia, paresis, impaired sensation, nerve root pain, and cauda equina syndrome. Patients manifest with symptoms of cord or root compression. Cervical spine tuberculosis can present as neck pain, quadriparesis, or retropharyngeal abscess resulting in dysphagia, respiratory distress and hoarseness. Sequelae of Pott's spine include vertebral collapse, kyphosis and spinal cord compression due to narrowing of the spinal canal by abscesses, granulation tissue, or direct dural invasion. Kyphotic deformity is seen usually with thoracic spine involvement. Cold abscesses resulting from the spread of infection to adjacent soft tissues can present as psoas abscesses in the region of lumbar spine or femoral trigone.

Diagnosis

The diagnosis of Pott's spine is primarily clinical, with support from the laboratory and radiological modalities. Laboratory studies used in the diagnosis include tuberculin skin test (TST) with positive results in 84–95% of patients. The erythrocyte sedimentation rate (ESR) is generally markedly elevated up to >100 mm/hour. Confirmation of the diagnosis ideally comes from microbiologic studies. Bone tissue or abscess samples obtained are positive for acid fast bacilli (AFB) in up to 50% of cases.[4]

Plain radiographs are the first line of radiological investigation and the earliest changes include narrowing of joint space and indistinct paradiscal changes of the vertebral bodies. Paravertebral abscesses can result from collection of tuberculous granulation tissue presenting as a fusiform radio dense shadow in the thoracic spine called the 'bird nest appearance'. Vertebral bodies can develop erosions along their anterior borders producing a scalloped appearance called 'aneurysmal phenomenon'.[5] Computerised tomography (CT) provides earlier and better delineation of irregular lytic lesions, sclerosis, disc collapse, and disruption of bone circumference as compared to plain radiographs. Magnetic resonance imaging (MRI) is the very good for evaluating disc space infection, osteomyelitis of the spine and neural compression. Figures 67.1 to 67.3 show MRI imaging findings in the lumbosacral, thoracic and cervical spine respectively in a case of Pott's disease.

Treatment

Management consists of administration of anti-tubercular therapy (ATT) for a duration of 9–12 months as per the World Health Organization (WHO) 2010 guideline for treatment of

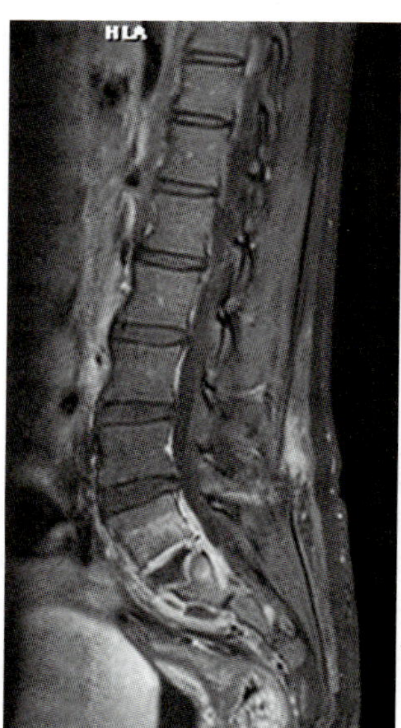

Fig. 67.1: Contrast enhanced T1 sagittal MRI image of lumbar spine shows non-enhancing fluid component at L5–S1 intervertebral disc associated with abnormal marrow enhancement of L5 and S1 vertebral body. There is prevertebral and epidural soft tissue component also. The radiological picture is suggestive of Pott's spine

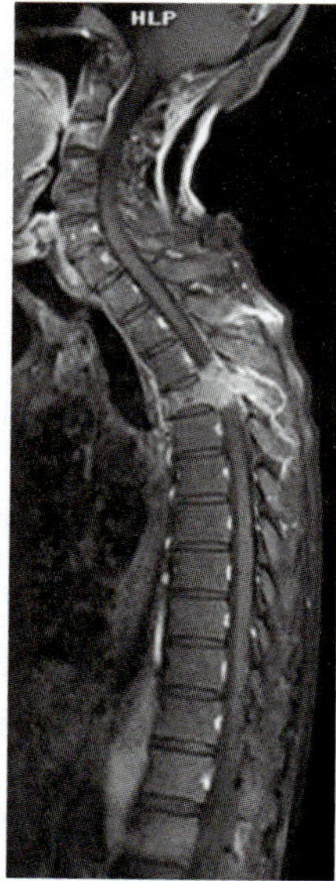

Fig. 67.2: Contrast enhanced T1 sagittal MRI of dorsal spine shows collapse of D4 vertebral body with soft tissue element in the spinal canal suggestive of Pott's spine

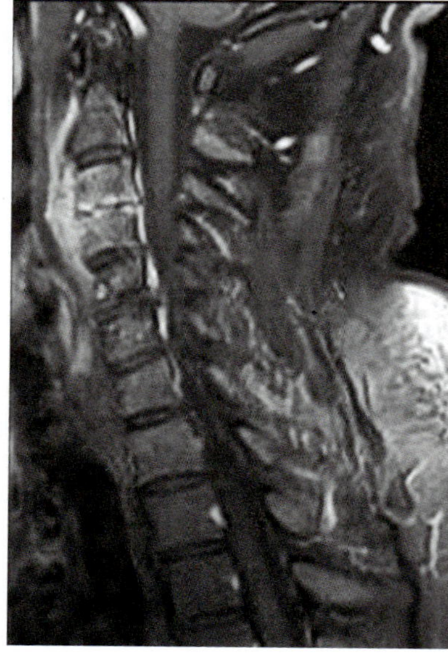

Fig. 67.3: Contrast enhanced T1 sagittal image of the cervical spine shows destruction of C3 and C4 intervertebral disc with abnormal marrow enhancement suggestive of Pott's spine

Table 67.1: Dosages of first-line anti tubercular drugs (Adapted from WHO (2010) treatment of tuberculosis guidelines, 4th edition)

Drug	*Recommended dose per day*	
	Adults: Dose and range (mg/kg body weight)	*Children: Dose and range (mg/kg body weight)*
Isoniazid	5 (4–6)	12 (10–15)
Rifampicin	10 (8–12)	15 (10–20)
Pyrazinamide	25 (20–30)	35 (30–40)
Ethambutol	15 (15–20)	20 (15–25)
Streptomycin	15 (12–18)	15 (12–18)

tuberculosis.[5] Therapy consists of initial intensive phase of 8 weeks of oral daily treatment with isoniazid (H), rifampicin (R), pyrazinamide (Z), and ethambutol (E) or streptomycin (S), followed by 28 to 36 weeks of treatment with isoniazid plus rifampicin. Pyridoxine 25 to 50 mg daily is added to regimes containing isoniazid. The doses of the various first-line anti-tubercular drugs used is given in Table 67.1. The latest Government of India RNTCP (Revised National Tuberculosis Control Programme) recommends administering daily fixed dose combinations of first-line anti-tubercular drugs (HRZE) as intensive phase for the first two months, however during the continuation phase (CP) only pyrazinamide is to be stopped whereas the other three drugs (HRE) would continue in daily doses for at least the next 16 weeks.[6] It leaves the option of extending the duration of CP beyond 16 weeks to the clinical decision of the treating physician.

Surgical treatment of Pott's spine is reserved for neurologic deficits such as acute neurologic deterioration, paraparesis, and paraplegia, progressive spinal deformity as defined by collapse and destruction of more than 50% of the vertebral body, spinal instability, progressive kyphosis (while on medical management) and large paraspinal abscess.

TUBERCULOSIS OF THE SACROILIAC JOINT

Approximately 10% of musculoskeletal tuberculosis involves the sacroiliac (SI) joints.[7] The diagnosis of sacroiliac joint tuberculosis often gets missed owing to vague symptoms at presentation, and inaccessibility of the joint when the patient is examined in the supine position. Sometimes it may present as a psoas abscess and may not be diagnosed until it presents with spontaneous drainage in the groin.

Clinical Features

Patients often present with buttock pain, persistent low back pain and difficulty in walking. Constitutional symptoms in form of weight loss, poor appetite and night sweats are invariably present. Sacroiliac pain can present as referred pain in groin, posterior thigh, and knee. Musculoskeletal examination reveals tenderness over the affected sacroiliac joint along with positive pelvic compression test, FABER stress tests and the Gaenslen's test.[8] There may be signs of femoral or sciatic nerve irritation if the distended anterior joint capsule comes in contact with the lumbosacral plexus.

Diagnosis

Laboratory investigations reveal an elevated ESR in almost all patients. Plain radiographs of SI joints reveal haziness initially, followed by joint space widening and sclerosis of the joint margins. Unilateral sacroilitis, along with the absence of additional manifestations in the spine, are most typical in tuberculosis and help to differentiate it from ankylosing spondylitis (Fig. 67.4). CT scans

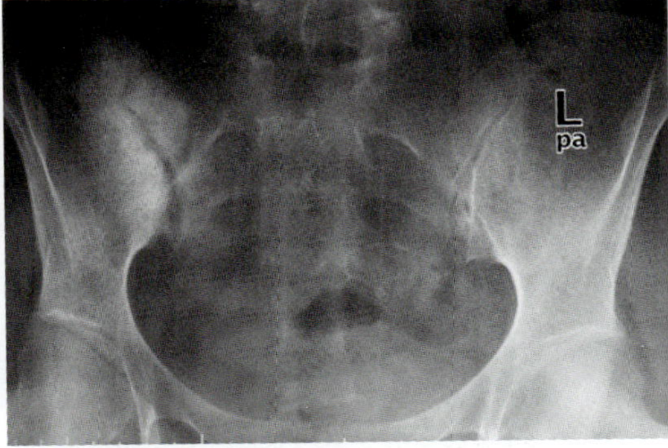

Fig. 67.4: Plain radiograph of pelvis AP view shows reduction in right sacroiliac joint space with articular margin irregularity and sclerosis suggestive of unilateral sacroilitis most likely TB

show joint space widening, sclerosis of the joint margins and sequestra within the joint. MRI is the best radiological modality in early disease, showing capsular distension, associated bone oedema and destruction. Bone scans are helpful when the diagnosis is not clear. A diagnostic aspiration or closed needle biopsy of the sacroiliac joint is appropriate when the disease is in its early stages, with high AFB positivity ranging from 40 to 60%.[9]

Treatment

ATT is the treatment of choice. Surgical management in form of curettage and arthrodesis is carried out when joint instability is anticipated.

TUBERCULOUS PERIPHERAL ARTHRITIS

Tubercular peripheral arthritis is usually monoarticular, chronic, and insidious but involvement of multiple sites can also be seen in 15% of cases. Any large joint can be affected, including hips (15%), knees (15%), and ribs (5%).[10] Tubercular peripheral arthritis usually starts as osteomyelitis in the growth plate of bones and then spreads into the joint spaces. Peripheral tuberculous arthritis tends to involve children and young adults in endemic areas.

Clinical Features

Most common symptoms include local pain, swelling and restriction of movement of the affected joint. Regional muscle wasting and deformity may also develop. Systemic symptoms like fever, weight loss and night sweats may or may not be present during active TB arthritis. Evidence of active pulmonary TB is seldom seen in individuals with tubercular arthritis at the time of diagnosis.

Diagnosis

The diagnosis of tuberculous arthritis is often delayed on an average by 16–19 months.[11] Confirmation of diagnosis requires the demonstration of AFB from any fluid or tissue. Different culture mediums, such as Lowenstein-Jensen medium, as well as both radiometric and non-radiometric techniques, can be used for confirmation of diagnosis. The newer methods are capable of providing quicker results.

Radiographic features appear late and are usually noted 2 to 5 months after disease onset. The classical triad of radiologic characteristics of TB arthritis is juxta articular osteoporosis, peripheral osseous erosion and gradual narrowing of intra-articular space, also known as Phemister triad. The joint space is relatively preserved in early TB arthritis. Bone scan findings are non-specific in form of increased uptake. The MRI findings in tubercular arthritis comprise synovitis, synovial effusion, bony erosions (either central or peripheral), pannus, abscess, and bone chips (Figs 67.5 and 67.6).

Synovial fluid analysis reveals a turbid, non-haemorrhagic fluid with elevated white blood cell count (10,000–20,000 cells/mL) with predominant polymorphs. Synovial fluid cultures are positive

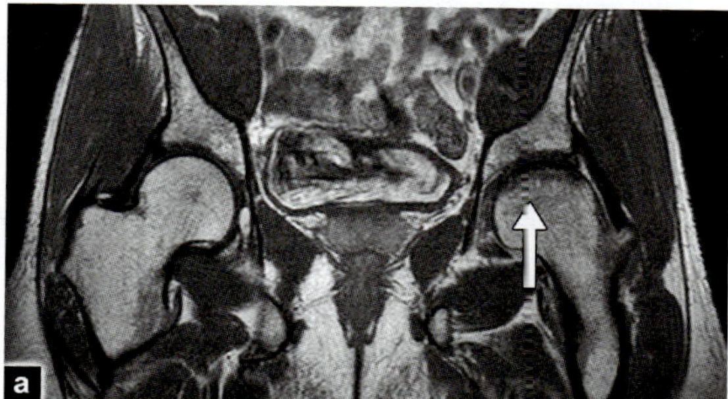

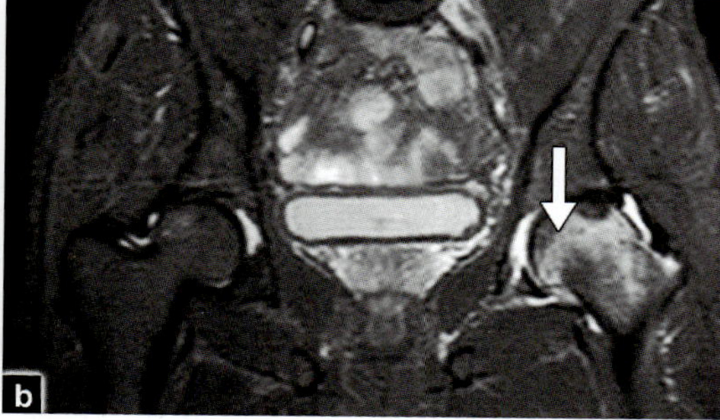

Fig. 67.5: Multiple erosions are seen in the superior and lateral aspects of left femoral head with minimal flattening of the contour. Multiple areas of altered signal intensity are seen in the left femoral head, neck and proximal shaft, which appear hypointense on T1W image (thin arrow) in (a) Marrow edema is seen in the left femoral head (thick arrow) neck, proximal shaft and left acetabulum which appears hyperintense on STIR image in (b)

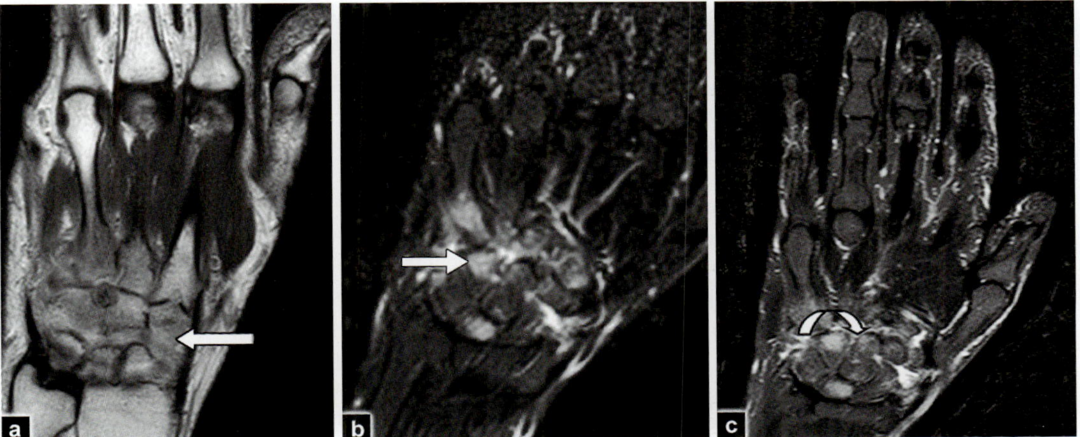

Fig. 67.6: MRI of left wrist joint show diffuse soft tissue thickening (thick white arrow) in wrist joint around the carpal bones, adjacent bases of metacarpals extending to surround some of the extensor tendons of the hand (a). Minimal synovial effusion (curved arrow) seen in some of the intercarpal joint spaces (c). There is reduction of intercarpal and carpo-metacarpal joint spaces. A few well defined bony erosions are seen involving some of the carpal bones.
(a) T1 weighted non-fat saturated sequence showing hypointense signal of the synovium
(b) T2 Fat suppressed sequence showing hyperintense signal of the thickened synovium, carpal bones and bases of metacarpals suggestive of marrow oedema (thin arrow)
(c) Post-contrast T1 Fat suppressed sequence shows abnormal enhancement of the soft tissue in the wrist and marrow of carpal bones

in roughly 20–40% of cases.[12] The gold standard of diagnosis remains synovial biopsy, with positive findings seen in 80% of cases including lymphocytes, caseating granulomas, and giant cells.[13]

Treatment

The treatment of tuberculous arthritis consists of ATT, which when constituted early can result in near-complete resolution and preservation of function. The duration of ATT should be for at least 9–12 months, but children and immuno-compromised patients require a longer duration of treatment. The role of operative treatment is limited to performing a biopsy, joint debridement, drainage of abscess and synovectomy.

REACTIVE ARTHRITIS (PONCET DISEASE)

Poncet's disease is classically a "polyarthritis associated with visceral tuberculosis in which there is no evidence of bacteriologic involvement of the joints themselves." It was first reported by Antonin Poncet in 1897. Poncet's disease is a non-destructive and parainfective polyarthritis, occurring in patients with active tuberculosis, which resolves completely on anti-tuberculosis therapy. It is more prevalent in the younger population and is slightly more common in women.

Pathogenesis

The pathogenesis of Poncet's disease is still poorly understood and the following theories have been proposed.

a. **Immune mediated:** A hypersensitive immune response to tubercular protein produces a reaction in the joint spaces, similar to that produced in the skin in erythema nodosum.
b. **Genetic:** A genetic predisposition might also be seen, since Poncet's disease occurs only in a small proportion of patients with active tuberculosis. In one patient with Poncet's disease the HLA haplotype was DR4.[14]
c. **Molecular mimicry:** Since mycobacteria and their antigens are known to be arthritogenic, molecular mimicry between mycobacterial antigens and host tissues and the resulting immunological cross reactivity may also play a role. In this regard an antigenic similarity, between a fraction of *M. tuberculosis* and human cartilage has also been shown.[15]

Clinical Features

Clinical features comprise fever, malaise, and polyarthritis of large joints. The arthritis is acute to subacute at onset and tends to involve the large joints symmetrically. Small joint involvement is also seen. Although often described as a polyarthritis, recent literature review reveals Poncet's disease to be more often a pauciarticular, symmetrical, predominant large joint arthritis. In a recent case series of 23 patients from India the ankle joint is the most common joint involved followed by the knee.[16]

Diagnosis

The diagnosis is clinicalas there are no standard criteria and is considered after excluding other potential causes of arthritis in a patient with active tuberculosis. Recently certain criteria have been proposed by Sharma *et al* however they would require validation in prospective studies.[16] The tuberculin test is frequently positive and the extrapulmonary site of tuberculosis is usually seen in lymph nodes.

Treatment

Poncet's disease responds well to non-steroidal anti-inflammatory drugs (NSAIDs) and resolves completely on anti-tuberculous therapy.

TUBERCULOUS OSTEOMYELITIS

Tuberculous osteomyelitis is a rare entity comprising less than 5% of all cases of osteoarticular tuberculosis. Symptom duration varies from days to months and co-existing visceral disease is uncommon. Although any bone may be involved, the femur and tibia are the most commonly involved, followed by short bones such as metacarpals, metatarsals, and phalanges.[17] Because of a low index of suspicion for its diagnosis, early lesions may be neglected. Bony involvement is secondary to haematogenous dissemination of mycobacteria in the diaphysis.

Clinical Features

The duration of symptoms ranges from days to months and coexisting tuberculosis is uncommon. Clinical features comprise mild pain and swelling of bone, with slight warmth and tenderness, and overlying boggy swelling of the soft tissues. Enlargement of regional lymph nodes and the presence of an abscess or sinus are also of great significance. Notable features of a tuberculous sinus include bluish discolouration at the periphery, undermined edges, serosanguinous discharge, matted draining lymph nodes and fixation to the bone.

Diagnosis

The lesion may be metaphyseal or diaphyseal and may penetrate the physis or extend into an adjacent site. The most common presentation is a solitary lytic lesion with a sclerotic rim. Bone sequestration is rarely seen. Spina ventosa a spindle shaped expansion with multiple layers of subperiosteal new bone, occurs in the small bones of the hands and feet. If plain radiographs are normal more sensitive investigations such as MRI and CT are required to detect and localise lesions. Biopsy is required to confirm diagnosis.

Treatment

ATT remains the mainstay of treatment and judicious surgical intervention may help to promote early healing.

OTHER FORMS OF MUSCULOSKELETAL TB

Primary tuberculous pyomyositis, tuberculous bursitis, and tuberculous tenosynovitis are rare entities constituting 1% of skeletal tuberculosis.[18]

Tuberculous tenosynovitis involves most commonly the tendon sheaths of the hand and wrist. Tuberculous bursitis occurs most commonly around the hip and involves the greater trochanteric bursa. Primary tuberculous pyomyositis and tenosynovitis of the tendons of the ankle and foot are rare and are seldom reported in literature. Figure 67.7 depicts tubercular dactylitis of left index finger.

TUBERCULOSIS RELATED TO TREATMENT WITH ANTI-TUMOUR NECROSIS FACTOR ALPHA (TNF-α) BIOLOGIC AGENTS

Over the last two decades anti-TNF-α agents have been successfully used in the management of chronic inflammatory disorders. Both clinical trial and registry data showed that patients treated with these biologics have significantly increased risk of reactivation of latent TB infection (LTBI), owing to the fact that TNF plays a role in the maintenance of granuloma integrity.[19]

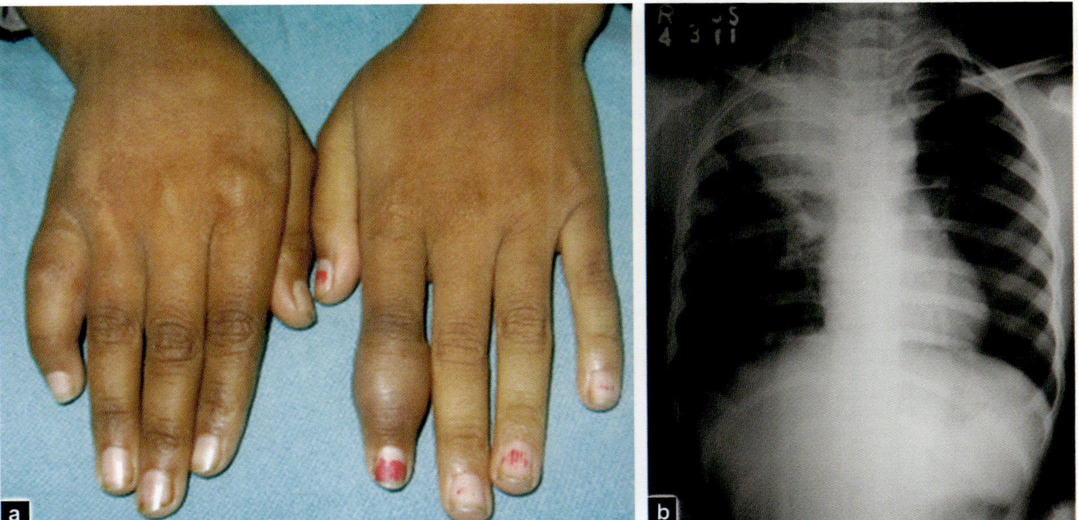

Fig. 67.7: (a) Tuberculous dactylitis of the left index finger. (b) The corresponding chest radiograph PA view shows right upper lobe fibrosis

Reactivation of latent TB usually occurs concomitantly with initiation of anti-TNF therapy and has a rapidly progressive course. The most common manifestation of TB following anti-TNF therapy is occurrence of extra-pulmonary forms of TB. The risk of TB differs from one anti-TNF agent to another. For example, monoclonal antibodies like infliximab and adalimumab have a markedly higher risk of TB reactivation when compared to soluble TNF receptor etanercept.[20]

Diagnosis

Since *M. tuberculosis* bacilli are difficult to identify in patients with latent TB, detection of an adaptive type I cellular immune response is the current diagnostic method to detect latent TB. Two techniques available to detect this response are the in vivo tuberculin skin test (TST) and the *ex vivo* IFN-γ release assays (IGRAs). TST has its limitations because of high false-positive rates especially in a TB endemic country like India, infection with environmental mycobacteria, and because of false positive results caused by Bacille Calmette-Guérin (BCG) vaccination, and the failure to discriminate true negative responses from anergy which occurs due to malnutrition or severe immunodeficiency.

IGRA tests work on the principle of detection of released IFN-γ in patients' blood samples after stimulation in vitro with *M. tuberculosis* specific antigens such as ESAT6, culture filtrate protein 10 or TB7.[7] antigen (absent in all strains of BCG vaccines and in most other non-tuberculous mycobacteria). IFN-γ is released from activated T cells post-stimulation and is subsequently measured using either enzyme linked immunospot assay (ELISPOT), marketed as T-SPOT.TB or by quantifying the IFN-γ released into the supernatant using an ELISA, marketed as QuantiFERON-TB Gold. In tube unlike TST, a negative IGRA test can differentiate anergy from a true negative result. Owing to the use of specific antigens, IGRA tests have higher specificity than TST, especially in populations with a high prevalence of BCG vaccination. Malaviya et al developed an algorithm to detect LTBI in patients with inflammatory rheumatic diseases receiving anti-TNF therapy using a combination of clinical history, TST, IGRAs, and CT chest (Fig. 67.8). Using a combination of these methods to screen for TB the authors reported that tuberculosis flare reduced from 18 to 4.6% in patients taking TNF alpha inhibitors.

Treatment of LTBI in Patients Planned for TNF Inhibitors

The treatment of latent TB as per the TBNET consensus statement (2010) includes either 9–12 months of isoniazid as monotherapy[21] or three months of isoniazid and rifampicin. The initiation

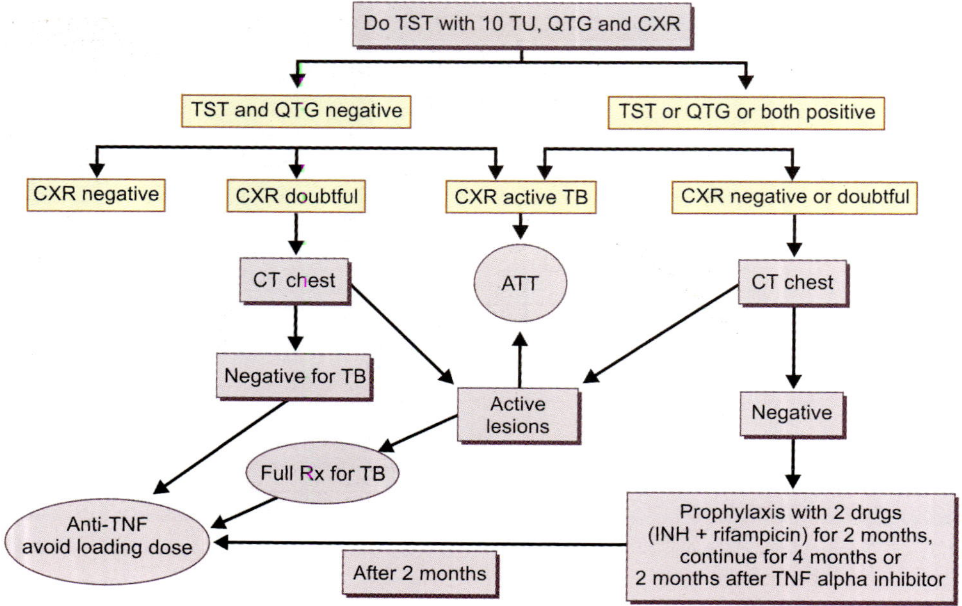

Fig. 67.8: Protocol recommended for the screening of latent tuberculosis before commencement of anti-TNF therapy. (*Abbreviations*: TST—tuberculin sensitivity test, QTG—Quantiferon TB Gold, CXR—Chest X-ray) (*Adapted from* Malviya et al. Preventing tuberculosis flare in patients with inflammatory rheumatic diseases receiving tumour necrosis factor-α inhibitors in India—An audit report. J Rheumatol. 2009; 36(7):1414–2010)

of anti-TNF treatment is generally recommended after 3–4 weeks of LTBI treatment. The occurrence of TB cannot be completely prevented during the anti-TNF therapy despite LTBI treatment. For this reason, the possibility of TB development in the course of treatment always needs to be taken into account. When diagnosed with active TB, ATT should be initiated and anti-TNF therapy should be stopped. The treatment period for active TB is identical to that of ordinary TB patients. In the Indian context if LTBI is detected, Malaviya et al recommend prophylaxis with combination of Isoniazid and Rifampicin for 4 months or till 2 months after the last dose of anti-TNF therapy. Anti-TNF can be initiated 2 months after completion TB prophylaxis treatment.[22]

Conclusion

Musculoskeletal TB is one of the important forms of extra-pulmonary TB and can have significant consequence if not recognised and treated early on. Common sites include spine and the weight bearing joints. Diagnosis is frequently missed and a high degree of clinical suspicion along with the radiological, microbiologic and biopsy findings are important for diagnosis. Conventional ATT remains the mainstay of treatment.

Key Points

1. Musculoskeletal TB accounts for 1–3% of all forms of TB and 10–15% of extra-pulmonary tuberculosis.
2. Most common manifestation of musculoskeletal tuberculosis is tuberculous spondylitis also known as Pott's disease accounting for 40–50% of cases.
3. Tuberculous peripheral arthritis accounts for 15% of cases of musculoskeletal tuberculosis. Evidence of active pulmonary TB is seldom seen in individuals with tubercular arthritis at the time of diagnosis.
4. Poncet's disease is a form of reactive arthritis in the presence of extra-articular tuberculosis presenting either as oligoarthritis or polyarthritis.
5. Tuberculous osteomyelitis is a rare entity comprising of 2–3% of osteoarticular tuberculosis.
6. In the era of increasing use of anti-TNF agents, detection of LTBI and its appropriate treatment is a must to prevent reactivation of tuberculosis.

REFERENCES

1. Malaviya AN, Kotwal PP. Arthritis associated with tuberculosis. Best Pract Res Clin Rheumatol 2003; 17(2):319–343.
2. Moon MS, Kim SS, Lee BJ, Moon JL. Spinal tuberculosis in children: Retrospective analysis of 124 patients. Indian J Orthop. 2012; 46(2):1508.
3. Pertuiset E, Beaudreuil J, Lioté F, Horusitzky A, Kemiche F, Richette P, et al. Spinal tuberculosis in adults. A study of 103 cases in a developed country, 1980–1994. Medicine (Baltimore) 1999; 78(5): 309–20.
4. Chauhan A, Gupta BB. Spinal tuberculosis.JIACM 2007; 8(1):110–4.
5. World Health Organization. 2010. Treatment of tuberculosis guidelines. 4th ed. [internet] Cited May 2017. Available from http://www.who.int/tb/publications/2010/9789241547833/en/.
6. Government of India. Central Tuberculosis Division. Directorate General of Health Services. Technical and Operational Guidelines for TB Control in India 2016. [Internet]. Cited on May 2017. Available from http://tbcindia.gov.in/index1.php?lang=1&level=3&sublinkid=4611&lid=3221.
7. Tuli SM.Tuberculosis of the Skeletal System (Bones, Joints, Spine and Bursal Sheaths). New Delhi: Jaypee Brothers Publication; 2010:3–15.
8. Prakash J. Sacroiliac tuberculosis—A neglected differential in refractory low back pain—Our series of 35 patients. J Clin Orthop Trauma. 2014; 5(3): 146–153.
9. Ramlakhan RJS, Govender S. Sacroiliac Joint tuberculosis. Int Orthop 2007; 31(1):121–124.
10. García-Arias M, Pérez-Esteban S, CastañedaS. Septic Arthritis and Tuberculosis Arthritis. J Arthritis 2012; 1:102.doi:10.4172/2167–7921.1000102.
11. Yao DC, Sartoris DJ. Musculoskeletal tuberculosis. Radiol Clin North Am 1995; 33(4):679–89.
12. Verettas D1, Kazakos C, Tilkeridis C, Dermon A, Petrou H, Galanis V. Polymerase chain reaction for the detection of *Mycobacterium tuberculosis* in synovial fluid, tissue samples, bone marrow aspirate and peripheral blood. Acta Orthop Belg. 2003; 69(5): 396–9.
13. Pigrau-Serralach C, Rodriguez-Pardo D. Bone and Joint tuberculosis. Euro Spine J 2013; 22(Suppl 4): 556–566.
14. Ames PRJ, Capasso G, Testa V, Maffulli N, Tartora M, Gaeta GB. Chronic tuberculous rheumatism (Poncet's disease) in a gymnast. British J Rheumatol 1990; 29:72–74.
15. Holoshitz J, Klajman A, Drucker I, Lapidot Z, Yaretzky A, Frenkel A, et al. T lymphocytes of patients with rheumatoid arthritis patients show augmented reactivity to a fraction of mycobacteria cross reactive with cartilage. Lancet 1986; 328(8502): 305–309.
16. Sharma A, Pinto B, Dogra S, Sharma K, Goyal P, Sagar V, et al. A case series and review of Poncet's disease, and the utility of current diagnostic criteria. Int J Rheum Dis. 2016; 19(10):1010–1017.
17. Hoschberg MC, Silman AJ, Smolen SJ, Weinblatt ME, Weisman MH. Rheumatology. Philadelphia: Elsevier Mosby; 2015.
18. Abdelwahab IF, Bianchi S, Martinloi C, Kein M, Hermann G. Atypical extraspinal musculoskeletal tuberculosis in immunocompetent patients. Can Assoc Radiol J 2006; 57(5):278–86.
19. Harris J, Keane J. How tumour necrosis factor blockers interfere with tuberculosis immunity. Clin Exp Immunol 2011; 161(1):1–9.
20. Keystone EC, Papp KA, Wobeser W. Challenges in diagnosing latent tuberculosis infection in patients treated with tumor necrosis factor antagonists. J Rheumatol 2011; 38:1234e43.
21. Solovic I, Sester M, GomezReino JJ et al. The risk of tuberculosis related to TNF antagonist therapies: a TBNETconsensus statement. Eur Respir J. 2010; 36(5):1185–206.
22. Malaviya AN, Kapoor S, Garg S, Rawat R, Shankar S, Nagpal S et al. Preventing tuberculosis flare in patients with inflammatory rheumatic diseases receiving tumour necrosis factor-alpha inhibitors in India—An audit report. J Rheumatol. 2009; 36(7):1414–2010.

FURTHER READING

1. Bhargava DA, Malaviya AN, Kumar A. Tuberculous Rheumatism(Poncets disease)—A case series. Indian J Tub 1998; 45:215.
2. Ramlakan RJ, Govender S. Sacroiliac joint tuberculosis. Int Orthop 2007; 31(1):121–4.
3. Tseng C, Huang RM, Chen KT.Tuberculosis Arthritis: Epidemiology, Diagnosis, Treatment. Clin Res Foot Ankle 2014; 2:131.

Chapter

68

Arthritis in Leprosy

Latika Gupta, Anupam Wakhlu, Vikas Agarwal

INTRODUCTION

Leprosy is a chronic granulomatous infectious disease caused by *Mycobacterium leprae* with skin and nerves being predominantly affected. As of 1985 leprosy was endemic in 122 countries with a global prevalence of 12 per 10,000. After the world health assembly accepted the goal of elimination of leprosy as a public health problem (prevalence rate of less than one case per 10,000 persons) by 2000 AD,[1] widespread use of multi-drug treatment (MDT) has reduced the disease burden dramatically. Leprosy has been eliminated from 108 out of the 122 endemic countries. In 2002, 70% of the world's registered leprosy patients were in India. The World Health Organization's April 2005 report suggested that the South-East Asia (SEA) region was the only area with prevalence above the elimination goal. In the SEA region, India alone represents 80% and 88% of prevalent and incident cases respectively. Although the prevalence of leprosy in India was 0.84 per 10,000 according to the June 2006 report; nine states/union territories of the country have not been able to achieve the elimination target of the World Health Assembly till date.[2] India is considered to be the epicentre of the problem, where 260,000 of the 408,000 people diagnosed across the world in 2004 reside.[3] The states of Uttar Pradesh and Bihar have the highest number of cases (Fig. 68.1).

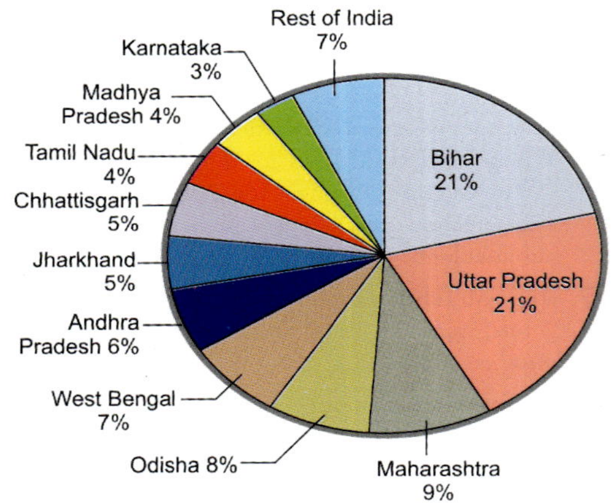

Fig. 68.1: Registered prevalence per state, India. April 2003

Epidemiology

The frequency of arthritis in leprosy is difficult to estimate because the prevalence varies from 1 to 78%.[4–6] Arthritis was found in a miniscule 1% (27 patients) cases, out of 2500 Indian leprosy patients at a dermatology department.[7] Cossermelli observed arthritis as the first manifestation of leprosy in 12 out of 44 patients with arthritis[8] while Gibson reported arthralgia/arthritis as the first manifestation of leprosy in 39% (12/31) of patients in lepra reaction.[9] However, in another series of 18 unselected patients, 14 had arthritis,[10] whereas Atkin et al reported that one-third of 66 consecutive leprosy outpatients had an inflammatory arthritis.[5] A recent review has suggested that rheumatic symptoms occur in as many as 75% of leprosy patients and are the third most common reason for hospital admission.[11] In a recent report from Nagpur, authors have reported rheumatic manifestation in 61.4% patients (54.3% having arthritis and 17.4% having soft tissue rheumatism) in a rheumatology clinic.[12] Thus, wide variability in the reported prevalence suggests that documentation of rheumatologic signs and symptoms depends on the specialty in which the patient is being treated and rheumatic manifestations of leprosy as a whole is often under-reported by non-rheumatology services. Therefore, it is a felt need for awareness of the condition among rheumatologists, as joint involvement may be the only manifestation in some cases.

Lepromatous manifestations are most often seen in males across all series. These are more common in people from rural areas although some studies have found equal prevalence in rural and urban areas. Although rheumatologic manifestations can be seen in all types of leprosy, these are more common in the lepromatous variety.

Clinical Features

The classical clinical presentation of leprosy is in the form of hypo- or anesthetic and anhidrotic macules, patches or plaques or widespread papulonodular lesions, with or without paresthesias or loss of nerve function in the distribution of mononeuropathy or mononeuritis multiplex/polyneuropathy and blindness.

Rheumatological manifestations of leprosy are less common/less commonly recognized than the classical dermatological and neurological presentation. Joint involvement has been recognized as early as 600 BC in Chinese literature.[13] For the sake of pathophysiologic clarity, arthritis or rheumatological manifestations in leprosy can be divided into three broad categories. (a) Arthritis as a consequence of direct infiltration of synovium (b) Reactional states with arthritis or rheumatological manifestations (c) Manifestations arising as a consequence of complications such as peripheral neuropathy, e.g. (Charcot's or neuropathic joints). Though neuropathic joints and septic arthritis are well-known types of articular involvement in leprosy, inflammatory arthritis is much more common during the reactional states. The pattern of arthritis may vary from monoarticular involvement to symmetric polyarticular disease mimicking RA and or oligoarticular involvement with/without sacroiliitis or enthesitis mimicking seronegative spondyloarthropathy.

Charcot's arthropathy is the classic rheumatologic manifestation of leprosy often described in textbooks; it is usually seen in long-standing cases with neurologic involvement. Although it is deemed to be prevalent in one in 10 cases, data from rheumatology units has yielded a surprisingly lower number.[14] Characteristically, these patients have severely damaged joints with a little pain. Ankle and knee are most commonly involved. Findings may range from subluxation, dislocation or pathologic fractures to complete joint destruction.

Till recently, inflammatory arthritis was not recognized as an important feature of leprosy.[15] However, since the first report (1964) by Lele et al, inflammatory arthritis has been recognized as a characteristic presentation of leprosy, especially during reactional states. They described 13 cases of leprosy that developed acute painful symmetric polyarthritis involving hand joints and mimicked rheumatoid arthritis (RA).[7] In 6 cases, synovial biopsy showed granulomatous reaction without evidence of *M. leprae* infiltration. Five years later, Modi et al. reported 27 cases of leprosy (16 lepromatous, 6 borderline, 3 tuberculoid and 2 pure neuritic) presenting with similar polyarthritis mimicking RA from South India. Of these 27 cases, 21 had lepra reaction. Synovial biopsy (23 cases) was reported as non-specific in 10, healed scarred lesions in 9; however, in 4 cases, *M. leprae* could be demonstrated. In 6 cases of arthritis with no reactional state, therapy with dapsone aggravated the arthritis and led to involvement of the newer joints.[16] Similar cases were reported from different parts of the globe (Table 68.1). Gibson et al reported the presence of arthritis in 20 out of 31 (65%) patients with lepra reaction.[9] Arthritis in type 1 (12/20) and type 2 reactions (8/20) could not be differentiated. Lele and Modi observed involvement of small joints of hands in 22 (81%) cases and most were in reaction. The arthritis was diagnosed within a month of onset in most patients.[16] Thus various forms of inflammatory arthritis have been described in association with leprosy over the years.

Reactional states are systemic inflammatory complications arising from immunologic phenomenon, seen in 30–50% of all leprosy patients. Type 1 Lepra reaction most commonly affects patients with borderline disease and is called upgrading reaction when induced by anti-lepromatous therapy. Rarely, it can signify deterioration in patient's immunity when it occurs spontaneously; it is then called

Table 68.1: Reaction associated acute arthritis

Author, year	*No. of patients*	*Type of arthritis*	*Joints involved*	*Synovial histology*
Lele et al. 1965[7]	13	Acute symmetric polyarthritis	Wrists, PIP, knee	6, granuloma + No AFB
Modi et al. 1969[16]	27	Acute symmetric polyarthritis	Wrists, PIP, ankle	AFB + (4)
Chandrasekaran et al. 1993[15]	7	Acute symmetric polyarthritis	MCP, PIP, ankle, knee	ND
Gibson et al. 1994[9]	20	Acute symmetric polyarthritis	Wrist, MCP, PIP, ankle, MTP	AFB – (2)

MCP, metacarpophalangeal; PIP, proximal interphalangeal; ND, not done

downgrading reaction. Type 2 Lepra reaction, on the other hand, affects lepromatous leprosy (LL) patients or those with borderline leprosy (BL). It is also called erythema nodosum leprosum (ENL). ENL has been more commonly described from rheumatology units than type 1 reaction. The cytokine profile is typically Th2 type (increased IL-6, IL-10). It usually occurs after first 6 months of the anti-leprosy therapy, although leprosy presenting with ENL as the first manifestation is well documented. Arthritis in type 1, reversal reaction, has similar presentation like ENL. However, the underlying pathogenic process is exaggerated CD4+ cell mediated immune response to *M. leprae* antigens. The cytokine profile is typically Th1 type (increased TNF-α, IFN-γ, IL-2, IL-4). It commonly occurs in the first six months of antileprosy therapy.

Unique to Indian subcontinent is the spontaneous onset of lepra reactions that has been reported recently.[17] This further adds to the diagnostic dilemma. Reactional state associated arthritis usually begins abruptly and is associated with constitutional symptoms, fever and weight loss, and other features of reaction, new/worsening of cutaneous lesions or paresthesias. It characteristically involves small joints of hands and wrist and may mimic RA. It may last for a few weeks to months. Patients may have repeated episodes of such type of arthritis (Box 68.1). Occasionally patients with reactional states may present with insidious onset chronic symmetric polyarthritis mimicking RA.

A type 2 reaction is marked by the sudden appearance of numerous painful nodules due to panniculitis. Some of these lesions can ulcerate and exude pus. The patients with ENL look ill; they usually have high fever, lymphadenopathy, orchitis, iridocyclitis, muscle tenderness and arthritis. The natural course of the condition is 1–2 weeks, but recurrences are common.

Box 68.1 Characteristics of arthritis in reactional states

- May precede, accompany or follow the reaction
- Acute in onset
- Severe pain, morning stiffness and disability
- Parallel the duration and course of reaction
- Responds to antileprosy and anti-inflammatory therapy especially glucocorticoids
- Mostly resolve without sequelae
- Synovial biopsies do not show *M. leprae*
- Radiographs do not show erosions

Another presentation reported during reactional states is swollen hand and feet syndrome. In 1980, Albert and colleagues from California reported rheumatic manifestations on 15 patients with leprosy requiring hospitalization. Authors described 3 cases with acute onset painful oedema over the dorsum of the hands with marked restriction of the movements and nodules along the extensor tendons. Biopsy of the nodules showed granulomatous reaction with positivity for *M. leprae*. All these 3 patients responded to antileprosy and glucocorticoid therapy.[6] At our centre, we have seen ten such cases in last 10 years (Fig. 68.2).

In 1987[18] and 1989,[5] Atkins et al reported insidious onset chronic symmetric polyarthritis mimicking RA but not associated with lepra reaction. Reaction independent chronic RA-like

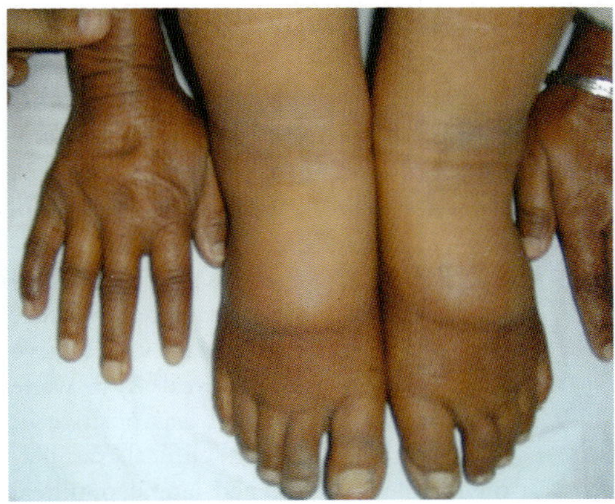

Fig. 68.2: Swollen hand and feet syndrome

arthritis was described first among patients from Papua New Guinea[18] where 31 of the 55 patients had chronic inflammatory polyarthritis. Later, in another study from Egyptian leprosy colony,[5] 20 of 33 patients were reported to have reaction independent chronic symmetric polyarthritis. The largest series of reaction independent arthritis, 39 out of 44 patients, was reported by Cossermalli et al.[8] Arthritis in majority of their patients was chronic (mean duration 11 years) and RA-like in distribution (Box 68.2). Leprosy was diagnosed more than 10 years ago in most of their patients and was currently inactive in 19 patients. There was no difference in the prevalence of arthritis among patients with and without active leprosy. Chronic RA-like presentations are the most likely to be misdiagnosed. Atkins et al. has reported boutonnière and swan neck deformities in their patients.[5] Though these patients had considerable relief with antileprosy therapy, their arthritis never resolved completely. A few of the patients had erosive

Box 68.2 Characteristics of chronic arthritis in leprosy

- Insidious onset
- Seen in all forms of leprosy
- Usually polyarticular and symmetric
- May be asymmetric and oligoarticular
- Spontaneous exacerbations and remissions
- Responds to antileprosy therapy but rarely resolves completely
- Radiographs may show erosions
- Synovial biopsy may show *M. leprae*/granuloma

arthritis just like RA (Table 68.2). Some of them respond to disease modifying antirheumatic drugs such as hydroxychloroquine and sulfasalazine.

Another presentation reported by Cossermalli et al. was sacroiliitis (seen in 64% of their patients) and tenosynovitis.[8] Prasad et al. have also described articular manifestations akin to spondyloarthropathy. In this study one-third of those with arthritic manifestation had asymmetric lower limb oligoarthritis. Sacroiliitis is not a common manifestation, although Massina et al. had an astonishing prevalence of 63% of their 55 cases. Most had bilateral sacroiliitis although it did not correlated with lower back pain. Lepromatous leprosy is most common subtype associated with this manifestation.

Leprosy can some times lead to vasculitic feature leading to diagnostic confusion. This happens in cases with high bacillary load where the vascular endothelium is teeming with bacilli. Popularly known as "lucio leprosy" as "lucio" means 'to glow' in Mexican. Skin of the patient is diffusely infiltrated so the natural wrinkles are obliterated, imparting a shiny hue. This variety is extremely rare in Asia, most cases being described from Mexico

Table 68.2: Reaction independent chronic arthritis

Author, year	*No. of patients*	*Type of arthritis*	*Joints involved*	*Radiology*
Atkins et al 1987[17]	31	Symmetrical polyarthritis	Wrists, MCP, PIP, knee, MTP	ND
Atkins et al 1989[5]	20	Symmetrical polyarthritis	Wrists, MCP, PIP, knee, MTP	Erosive disease (n = 12)
Cossermelli et al 1998[8]	39	Symmetrical polyarthritis	Wrists, MCP, PIP, knee, MTP	Erosive disease

MCP, metacarpophalangeal; PIP, proximal interphalangeal; MTP, metatarsophalangeal; ND, not done and Costa Rica. It is the one of most severe forms of leprosy and can be fatal at times.

Albeit rare, cryoglobulinemia can be another cause for vasculitis in a leprosy patient without lucio phenomenon. Although cryoglobulins have been described in up to 40% cases of leprosy, vasculitis is rare and limited to case reports. The cryoglobulins in these cases are polyclonal IgG and IgM.

Leprosy is a great masquerader and true to its name varied manifestations resembling different connective tissue diseases have been described. Biopsy proven cases of leprosy can present with heliotrope rash, muscle weakness and elevated muscle enzymes; oral ulcers, malar rash and photosensitivity, or a picture completely resembling scleroderma with Raynaud's, pitting scars and skin thickening. Other rheumatological manifestations like enthesitis and isolated tenosynovitis[19] (Fig. 68.3) have also been reported.

Recently, we have reported five patients (18%) presenting with arthritis, tenosynovitis with/without paresthesias and thickened nerves.[20] These belong to pure neuritic subtypes of leprosy, as they do not have cutaneous manifestations. Nerve biopsy in all the cases was confirmatory. Pure neuritic (PN) form of leprosy (no skin manifestations) is common in India (4.6%) and Nepal (8.7%).[21] Three of our patients with RA-like presentation were on disease-modifying drugs. Patients with PN leprosy lack skin manifestations and the only clue may be from the presence of tenosynovitis and neuropathy. At times, patients present with only tenosynovitis in the form of swollen hands and feet syndrome (HFS). Some series from Argentina have described up to 66% of their patients. This may be a feature of pure neuritic leprosy, or less commonly a part of generalized Lepra reaction. Presence of thickened or tender nerves can clinch the diagnosis. Ulnar and common peroneal nerve thickening should always be look for; at times the radial, supraclavicular, grater auricular, facial can also be thickened.

Some case scenarios observed by the authors exemplify the capability of leprosy to mimic presentations of common rheumatological disorders.

1. *Reactive arthritis like:* A young male presented with isolated acute onset inflammatory knee arthritis 2 weeks after a diarrheal illness. Inflammatory parameters were raised. Everything pointed to a reactive arthritis till synovial fluid reveals AFB. Wade Fite stain for lepra bacilli was positive in synovial fluid. Subsequent clinical examination revealed a thickened lateral popliteal nerve. The diagnosis was confirmed by a sural nerve biopsy. Patient recovered completely with antileprosy therapy.
2. *Chronic monoarthritis like:* A 25 years female presented with progressively increasing pain in the left elbow for the last 5–6 months. There were no systemic complaints. Examination revealed tenderness in the medial aspect of the elbow joint with swelling and marked tenderness in the posteromedial lower arm region. Ultrasound (USG) examination revealed an anechoic linear collection extending from the cubital tunnel to the lower part of the arm as described above (Fig. 68.4). A tubercular collection was suspected in the elbow and arm region. Attempts to aspirate this collection led to radiating paresthesias in the ulnar nerve distribution. A very thick ulnar nerve swelling was then suspected. Repeat USG examination revealed an ulnar nerve with cross sectional area of 22 mm^2 (upper normal of 7.5 mm^2) and diameter of 49 mm (Fig. 68.4). Further examination revealed

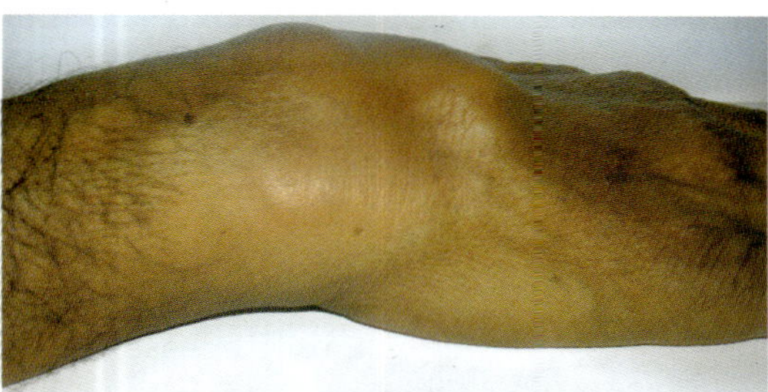

Fig. 68.3: Isolated tenosynovitis of extensor tendons of the hand

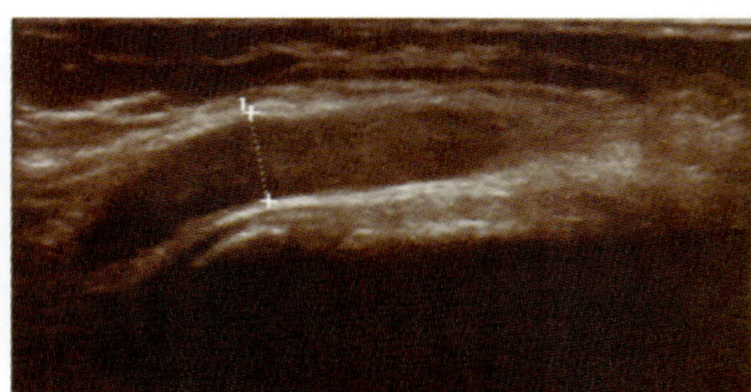

Fig. 68.4: Ultrasound examination revealing a markedly thickened ulnar nerve mimicking localized collection

25% sensory loss in the nerve distribution and a thickened lateral popliteal nerve. Sural nerve biopsy confirmed leprosy. Patient responded dramatically to initiation of steroids with antileprosy therapy.

3. *Sarcoid like:* A 35-year male presented with a 3-month history of ankle arthritis and recurrent typical attacks of erythema nodosum (Fig. 68.5). He denied any sensory loss. Radiograph of chest did not reveal any evidence of lymphadenopathy and there was no evidence of pulmonary/extra-pulmonary tuberculosis at that time. Biopsy of the subcutaneous nodules revealed evidence of both septal and nodular panniculitis, which led to a suspicion of leprosy. Bilateral lateral popliteal nerves were thickened without definite sensory loss. There were no skin lesions. Sural nerve biopsy confirmed leprosy and the patient completely recovered with therapy.
4. *Relapsing polychondritis like:* A 40-year-old woman presented with complaints of recurrent episodes of pain in the ear, throat and nose intermittently for the last 3 years. These episodes used to respond well to NSAIDs and steroids given by local practitioners. Ear involvement was limited to bilateral erythematous painful involvement of the pinna (Fig. 68.6). Pain in the throat region was limited to tenderness of the thyroid cartilage and tracheal rings. There was also some change in the quality of voice. Nasal involvement was characterized by tenderness of the cartilaginous nasal septum. A tentative diagnosis of relapsing polychondritis was entertained. However, closer examination of the ear revealed involvement of the cartilage of the ear as well as the pinna, which is typically spared in RP. A thickened great auricular nerve prompted a search for leprosy. Slit skin smear was positive for lepra bacilli. The patient responded well to multibacillary treatment along with NSAIDs and steroids.
5. *Psoriatic arthritis like:* A 25-year-old male presented with a seronegative asymmetrical small and large joint polyarthritis of 9-month duration. There was no history of back pain and no skin lesions of psoriasis. Nails appeared dystrophic and brittle (Fig. 68.7). On enquiry, the patient revealed that he had actually chewed the nails off but had not felt much pain. Note that the nails are

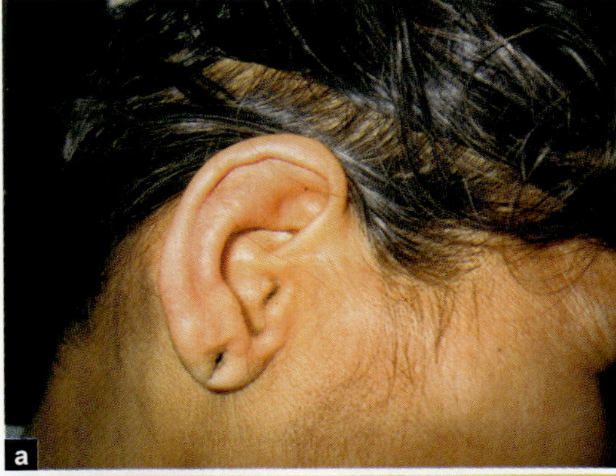

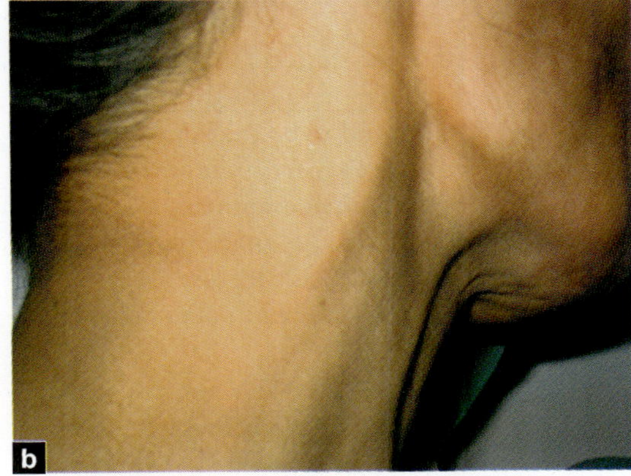

Fig. 68.6: Auricular chondritis (a) with thickened right greater auricular nerve, (b) mimicking relapsing polychondritis

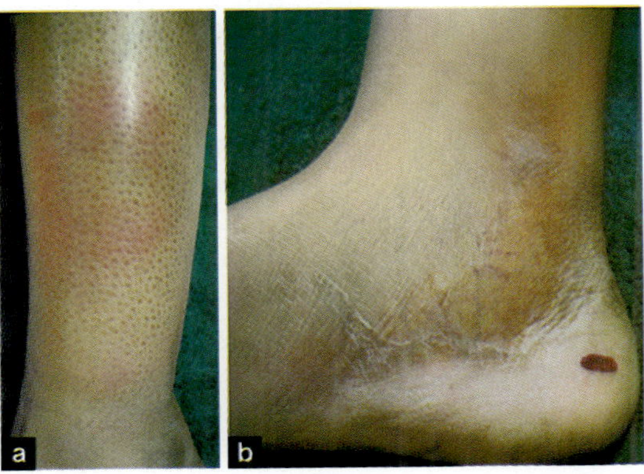

Fig. 68.5a and b: (a) Erythema nodosum lesions with (b) ankle arthritis mimicking sarcoidosis

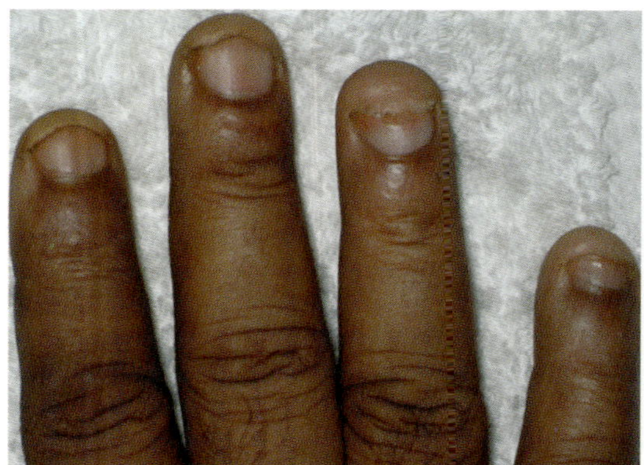

Fig. 68.7: Dystrophic nails

involved in the ring and little finger, which is in the ulnar nerve distribution. Ulnar nerve was found to be thickened and there was significant sensory loss in that distribution. Abdominal examination revealed a hypohidrotic, hypo-anaesthetic, hairless patch consistent with leprosy. Skin biopsy confirmed the same. The patient responded well to therapy.

6. *Vasculitis like:* A fifty-year female presented with ulcerative, superficially gangrenous and vasculitic lesions (Fig. 68.8) of the lower limbs for the last one month. There was associated pyrexia, asymmetrical oligoarthritis and anorexia. She also had loss of peripheral sensations in the lower limbs. Skin biopsy revealed ulcerated dermis and epidermis with presence of foamy macrophages and evidence of vasculitis. Wade Fite stain was positive for lepra bacilli. The patient was diagnosed as a case of Lucio phenomenon and was treated with MDT and steroids. Patient showed gradual improvement.

Diagnosis

Clinically, attributing arthritis in a patient of leprosy with characteristic neuro-cutaneous lesions is easy; however, sometimes arthritis may be the presenting complaint. In such cases careful attention to skin involvement; hypo/anaesthetic macules or patches or erythematous macules, nodules, with history of paresthesias, thickened and tender peripheral nerves and or sensory and motor neuropathy may help in suggesting the diagnosis of leprosy. Histopathologically, synovium in chronic polyarthritis patients may demonstrate granuloma, epithelioid cells and *M leprae*. Pernambuco et al[22] described 25 cases of arthritis from whom 32 synovial fluid samples, 20 synovial and three bone biopsies were obtained to provide a spectrum of change from non-inflammatory to turbid effusions and a synovium which was usually inflammatory and contained mycobacteria in nine samples taken from six patients.

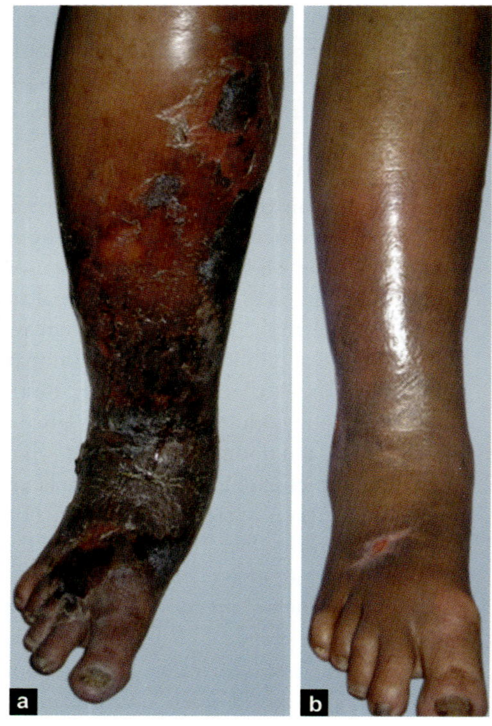

Fig. 68.8: Lucio phenomenon mimicking necrotizing vasculitis (before and after MDT)

Radiological features of leprosy related arthritis has been studied in different series. Juxta-articular erosions were seen in 12 (43%) out of 28 available radiographs.[5] Modi and Lele reported erosions in only 3 out of 27 patients with acute arthritis and leprosy.[16] Periosteitis, bone resorption and deformed joints have been reported.[13] Cossermaeli et al reported Grade 1 to 3 sacroiliitis in 35 out of 55 radiographs.[8]

Recently, a new serological test for detection of antibodies to the *M. leprae* specific PGL-I has been developed. These antibodies are present in 90% of patients with untreated lepromatous disease, but only 40–50% of patients with paucibacillary disease, and 1–5% of healthy controls.[23] Skin

Table 68.3: Characteristics of different forms of arthritis in leprosy

	Onset	*Symmetric*	*Oligo/ polyarthritis*	*Response to*		*Joint erosions*	*M. leprae in synovium*
				Antimicrobial	*Glucocorticoid*		
Lepra reaction	Acute	Yes	Yes	Poor	Good	No	No
Chronic arthritis	Insidious	Yes	Yes	Good	Not much	Yes	Yes
Swollen hand	Acute	Yes	Yes	Poor	Good	No	No
Pure* neuritic	Acute to insidious	Yes	Yes	Yes	Yes	No	No

**M. leprae can only be demonstrated in the nerve biopsy. May present with tenosynovitis alone.*

smears, taken to detect intradermal acid-fast bacilli, have high specificity, but low sensitivity, as up to 70% of all leprosy patients may be smear negative.[24]

Treatment

The first-line drugs against leprosy are rifampicin, clofazimine, and dapsone. According to World Health Organization (WHO), all patients should receive a multidrug combination with monthly supervision. For paucibacillary disease Rifampicin 600 mg once monthly and Dapsone 100 mg daily for 6 months and for multibacillary leprosy, rifampicin 600 mg and clofazimine 300 mg once a month along with clofazimine 50 mg and dapsone 100 mg daily for 24 months and for paucibacillary single lesion, rifampicin 600 mg, ofloxacin 400 mg, and minocycline 100 mg single dose is enough.[25]

The treatment of reactions associated arthritis is aimed at controlling acute inflammation, easing pain, and reversing nerve damage. Corticosteroids along with multidrug therapy regimen form the backbone of management of reactions. Prednisolone administration should be started at the dose of 1 mg/kg/day and gradually tapered by 5 mg every 2–4 weeks depending upon the improvement in the signs and symptoms. Patients with borderline-tuberculoid reactions commonly need corticosteroids for 3–4 months, whereas those with borderline-lepromatous reactions may need treatment for 6 months. Recovery is less in patients with pre-existing impairment of nerve function or with chronic or recurrent reactions. Sometimes type 2 reaction (erythema nodosum leprosum, ENL) may not respond to MDT and prednisolone; in such cases, clofazimine can be used at 300 mg daily for several months or thalidomide (TNF-blocker) (400 mg daily) is better than steroids and is the drug of choice for young men with severe ENL. Pentoxifylline, which inhibits TNF-production, has been used to treat ENL but was inferior to both thalidomide and steroids. Monoclonal antibodies targeting the TNF-molecule seem logical choice for therapy of severe refractory ENL but needs more critical evaluation before it can be recommended routinely.[23]

Conclusion

Leprosy, by itself, as a consequence of its reactional states or as a consequence of its complications may mimic completely a number of rheumatological disorders, in addition to the classical forms of articular involvement already described. It requires a high index of suspicion coupled with good clinical examination to clinch the diagnosis.

Arthritis, tenosynovitis, may be the presenting manifestations and may mimic RA or spondyloarthropathy. Increased awareness about articular manifestations and careful attention to cutaneous lesions and thickened peripheral nerves may help in clinching the diagnosis of leprosy.

REFERENCES

1. World Health Assembly. Leprosy resolution WHA 44.9, Forty-fourth World Health Assembly, May 13, 1991.
2. The World Health Organization—India website information on leprosy (Accessed June 21, 2006, at http://www.whoindia.org/EN/Section3/Section122_1213.htm)

3. Making progress towards leprosy elimination. Lancet 2006; 367:276.
4. Bonvoisin B, Martin JM, Bouvier M, Bocquet B, Boulliat J, Duivon JP. Articular manifestations in leprosy. Sem Hop 1983; 59:302–05.
5. Atkin SL, el-Ghobarey A, Kamel M, Owen JP, Dick WC. Clinical and laboratory studies of arthritis in leprosy. Br Med J. 1989; 298:1423–5.
6. Albert DA, Weisman MH, Kaplan R. The rheumatic manifestations of leprosy (Hansen disease). Medicine 1980; 59:442–8.
7. Lele RD. Sainani GS, Sharma KD. Leprosy presenting as rheumatoid arthritis. J Assoc Physicians India 1965; 13:275–7.
8. Cossermelli-Messina W, Festa Neto C, Cossermelli W. Articular inflammatory manifestations in patients with different forms of leprosy. J Rheumatol 1998; 25:111–19.
9. Gibson T, Ahsan Q, Hussein K. Arthritis of leprosy. British Journal of Rheumatology 1994; 33:963–966.
10. Alcocer J, Herrera R, Lavelle C, Guidino J, Fraga A. Inflammatory arthropathy in leprosy. Arthritis Rheum 1979; 22:587.
11. Pernambuco J, Cossermelli-Messina W. Rheumatic manifestations of leprosy: clinical aspects. J Rheumatol 1993; 20:897–898.
12. Vengadakrishnan K, Saraswat PK Mathur PC. A study of rhematological manifestations of leprosy. Ind J Dermatol Venereol Leprol 2004; 70:76–78.
13. Messner RP. Arthritis due to mycobacteria, fungi and parasites. In Koopman WJ, McCarty DJ, editors. Arthritis and Allied conditions: a textbook of rheumatology. Baltimore, MD: Williams and Wilkins; 1997. P.2305–20.
14. Horibe S, Tada K, Nagano J. Neuroarthropathy of the foot in leprosy. J Bone Joint Surg Br 1988; 70:481–85.
15. Chandrasekaran A. Infection and arthritis, our experience, a retrospective and prospective study. J Indian Rheumatol Assoc 1993; 1:7–20.
16. Modi TH, Lele RD. Acute joint manifestations in leprosy. J Assoc Physicians India 1969; 17:247–54.
17. Prasad S, Misra R, Aggarwal A, Lawrence A, Haroon N, Wakhlu A, et al. Leprosy revealed in Rheumatology clinic: a case series. Int J Rheum Dis 2013; 16:129–33.
18. Atkin SL, Welbury RR, Stanfield E, Beavis D, Iwais B, Dick WC. Clinical and laboratory studies of inflammatory polyarthritis in patients with leprosy in Papua New Guinea. Ann Rheum Dis 1987; 46:688–90.
19. Agarwal V, Wakhlu A, Aggarwal A, Pal L, Misra RN. Tenosynovitis as the presenting manifestation of Leprosy. J Indian Rheumatol Assoc 2002; 10:69–70.
20. Haroon N, Agarwal V, Aggarwal A, Kumari N, Krishnani N, Misra R. Arthritis as presenting manifestation of pure neuritic leprosy-a rheumatologist's dilemma. Rheumatology (Oxford). 2007; 46:653–56.
21. Mahajan PM, Jogaikar DG, Mehta JM. A study of pure neuritic leprosy: clinical experience. Indian J Lepr 1996; 68:137–41.
22. Pernambuco J, Opromolla D, Tolentino M, Fleury R. Arthritis in lepromatous Hansen's disease. Int J Lepr 1979; 47:353–354.
23. Britton WJ, Lockwood DNJ. Leprosy. Lancet 2004; 363:1209–19.
24. Lockwood DN. Leprosy elimination: a virtual phenomenon or a reality? BMJ 2002; 324:1516–18.
25. WHO. Expert Committee on Leprosy, 7th Report, 1998:1–43.

MCQs

1. **Which is the commonest form of arthritis seen in leprosy**
   a. Charcot's arthropathy
   b. Septic
   c. Reaction associated oligo- or poly-arthritis
   d. Swollen hand and feet syndrome
2. **Following type of arthritis is responsive to glucocorticoids**
   a. Reaction independent chronic erosive arthritis
   b. Reaction associated acute arthritis
   c. Swollen hand and feet syndrome
   d. Both b and c
3. **Erosive arthritis in leprosy is commonly seen in**
   a. Reaction associated arthritis
   b. Reaction independent arthritis
   c. Charcot's arthritis
   d. Swollen hand and feet syndrome
4. **Arthritis in pure neuritic leprosy**
   a. May present acutely or insidiously
   b. Does not occur
   c. Always present with paresthesias
   d. May respond to NSAIDs alone
5. **Arthritis in leprosy may mimic**
   a. Rheumatoid arthritis
   b. Seroegative spondyloarthropathy
   c. Sarcoidosis
   d. All of the above
6. **To confirm the diagnosis of leprosy associated arthritis**
   a. Demonstration of *M. leprae* is a must in the synovium
   b. Combination of the clinical and radiological features are enough
   c. Demonstration of *M. leprae* at any site, other than the joint is enough
   d. Combination of clinical, histological and radiological features may help

Chapter

69

Brucella Arthritis

Yojana Gokhale

INTRODUCTION

Osteoarticular brucellosis is the most frequent, localized complication of brucellosis, a worldwide zoonosis. Due to severe rheumatism associated with this febrile illness or due to osteoarticular complications of the disease, these patients can present to rheumatologist. Rheumatologists should be well-versed with clinical and radiological features of brucellosis as well as its lab diagnosis.

Epidemiology

Brucellosis is a common zoonotic disease of worldwide distribution. Only 17 countries in the world are declared brucella free by the World Health Organization. It is an important economic problem and a serious health hazard in Gulf countries, Mediterranean region, Indian subcontinents, Mexico and South-Central America. The disease is transmitted to man by consumption of contaminated milk or milk products or contact with infected cattle. In some parts of India 19–39% buffalos are seropositive for brucellosis.[1, 2] With non-availability of pasteurized milk in many parts of the country and farming being the main occupation in rural India both these modes of transmission are possible. Many residents of urban slums return to their villages during the monsoon where they are exposed to domesticated animals and can acquire the infection. Infection can also be acquired during foreign travel (Gulf countries) or consumption of soft Mexican cheese. The organisms enter human body through skin, conjunctiva, gut or respiratory mucosa. Brucellosis is reported from all parts of India, Mumbai,[3] Rajasthan,[4] Hyderabad,[5] Kerala.[6]

Microbiology

Brucellae are gram-negative aerobic, cocco-bacilli. They are fastidious organisms, have long incubation period (4–6 weeks), grow well on Castaneda biphasic (solid and liquid) medium. There are four subtypes which cause human infection. *B. melitensis* (sheep, goat, camel), *B. abortus* (cow, buffalo, camel, horses), *B. suis* (pigs), *B. cannis* (dogs). *B. melitensisis* most virulent. Brucellae are killed by boiling or pasteurization of milk but survive in biological material for long periods at low temperature. Brucellosis is a sexually transmitted disease in cattle, so a single infected bison can infect the whole herd. The infected animal should be culled to protect the herd. This is not done for obvious reasons and the infected animal keeps spreading the infection to man through infected milk and also to the rest of the herd. In cattle it causes recurrent abortion.

Immunity and Pathogenesis

Exposure to brucellosis generates both humoral and cell-mediated immune response. Antibodies promote clearance of extra-cellular brucellae. IgM antibodies appear by one week of acquiring the infection and persist for 9–12 months. The IgG antibodies appear by 4 weeks and persist beyond 18–24 months. Blocking antibodies appear in over 50% patients by 3–6 months of acquiring the infection. They are responsible for false-negative standard 'Tube agglutination test' by 3–6 months time.[7]

reported in 13–52% patients of brucellosis. Acute sacroilitis due to brucellosis occurs in the early febrile stage of infection. It presents as unilateral pain over the sacroiliac joint, fever with chills and sweats. The sacroiliac maneuvers are very painful, as also the hip movements and the straight leg raising test. It lasts for a few weeks to resolve spontaneously. All acute unilateral sacroiliitis should be investigated for brucellosis. In few cases opposite sacroiliac joint may subsequently get affected. Sacroiliitis may be associated with lumbar spondylitis with low back pain simulating ankylosing spondylitis. Brucella organisms may be cultured from sacroiliac joint aspirate.

Differential Diagnosis of Brucella Spondylitis and Sacroiliitis

1. **Spinal tuberculosis:** Clinically as well as radiologically spinal tuberculosis can mimic brucella arthritis and brucella spondylitis. Brucellosis has predilection for the lumbar spine where as tuberculosis affects the dorsal spine more often. Spinal deformities are common in tuberculosis. Positive brucella serology is useful in making the diagnosis of brucella spodilitis.

2. **Prolapsed intervertebral disc:** Patients of brucella spondylitis experience low back pain radiating to leg and straight leg raising test can be positive. The radiograph may be normal in early brucella spondylitis. An abnormal bone scan and positive brucella serology can aid in the diagnosis.

3. **Degenerative lumbar spondylosis:** Anterior osteophytes in the elderly patients' radiograph may simulate healed old brucella spondylitis. Such changes on lateral radiograph are common in elderly patients in endemic areas.

4. **Ankylosing spondylitis:** Sacroiltis and osteophytes due to healed superior end plate lesion may mimic ankylosing spondylitis.

5. **Malignancy or secondary deposits:** Back pain, abnormal tracer uptake in bones and the spine on bone scan and abnormal marrow signals on the MRI may be mistaken for malignancies like myeloma or secondaries.

6. Staphylococcal osteomyelitis of the spine.

Brucella Arthritis[8,11]

Arthralgia is reported in over 80% patients of brucellosis. Bone scans in patients of brucellosis with musculoskeletal symptoms show abnormal uptake at multiple sites. Arthritis may be monoarthritis usually affecting the large joints or migratory polyarthritis. The synovial fluid may grow Brucella. The radiographs may be normal in early cases. Bone scans would show increased tracer uptake. Septic and destructive arthritis of one joint may occur due to brucellosis. If there is delay in diagnosis destructive changes and radiological abnormalities, such as joint space narrowing occur. Brucella arthritis is easily diagnosed in countries where brucellosis is endemic. It is more common in children than adults. There are not much reports on brucellar arthritis of joints other than sacroiliac joint from India.

Tenosynovitis, bursitis and osteomyelitis due to brucellosis are reported.

Treatment of Osteoarticular Brucellosis[8]

Principles

1. Always use combination therapy (2 or 3 drugs)
2. Duration of therapy should be 6–12 weeks or more
3. Paraspinal abscesses do respond to medical line of treatment. Surgery should be reserved for patients with neurological deficit or relapse due to abscess.[8]

Anti-Brucella Drugs

Doxycycline 100 mg bid, rifampicin 900 mg daily, ciprofloxacin 500 gm bid, co-trimoxazole 10 mg/ kg of TMP (i.e. 3 tablets bid), Inj. streptomycin 1 gm daily for 3 weeks. Ceftriaxone is also effective against brucellae.

Relapse

It is known to occur in 20% cases. Its diagnosis is clinical. One can document rising titers of antibodies. Treatment is like fresh attack.

Prevention

Using pasteurized milk is the best method of prevention. For occupations where brucellosis is occupational hazard use of gloves masks, gowns should be promoted. Animal disease can be prevented by culling infected cattle and immunization of the newborn and the uninfected animals.

Key Points

- Diagnosis of brucella spondylitis and sacroiliitis should be considered in patients with chronic backache with history of animal contact or consumption of raw milk or milk products or travel to Gulf countries and in all patients with acute unilateral sacroiliitis.
- IgM, IgG antibodies towards brucella by ELISA are highly sensitive and specific.
- Combination anti-brucella drugs (2 or 3) should be used for 6–12 weeks for treatment.

REFERENCES

1. Brahmabhatt MN, Varasada RN, Bhong CD, Nayak JB. Seroprevelance of Brucella Spp. in buffalos in the central Gujrat region of India. Buffalo Bulletin 2009; 28:73–75.
2. Chauhan HC, Chandel BS, Shah NM. Seroprevalence of brucellosis in buffaloes of Gujarat. Indian Vet J 2000; 77:1105–6.
3. Gokhale YA, Ambardekar AG, Bhasin A, Patil M, Tillu A, Kamath J, et al, Brucella spondylitis and sacroiliitis in the general population in Mumbai. J Assoc Physicians India 2003; 51:659–66.
4. Kochar DK, Agarwal N, Jain N, Sharma BV, Rastogi A, Meena CB. Clinical profile of Neurobrucellosis- a report of 12 cases from Bikaner (north-west India), J Assoc Physicians India 2000; 48: 376–80.
5. Gokhle YA, Bichile LS, Gogate A, Tillu AV, Zamre. Brucella spondylitis: an important treatable cause of low backache. J Assoc Physicians India. 1999; 47: 384–8.
6. Deepak S, Bronson SG, Sibi, Joseph W, Thomas M. Brucella isolated from bone marrow. J Assoc Physicians India 2003; 51:717–18.
7. Ariza J, Pellicer T, Pallarés R, Fos A, Gudiol F. Specific antibody profile in human brucellosis. Clin Infect Dis 1992; 14:131–40.
8. Madkour MM. Bone and joint brucellosis. In Madkour MM. Brucellosis. London: Butterworth; 1981.pp. 90–104.
9. Maleknejad P, Hashemi FB, Fatollahzadeh B, Jafari S, Peeri Dogaheh H. Direct urease test and acridine orange staining on BACTEC blood culture for rapid presumptive diagnosis of brucellosis. Iranian J Publ Health 2005; 34: 52–55.
10. Yagupsky P, Paled N, Press J. Use of BACTEC 9240 Blood Culture System for Detection of Brucella melitensis in Synovial Fluid. J Clin Microbiol 2001; 39:738–9.
11. Pascual E. Brucella arthritis. In: Isenberg DA, Madison PJ, Woo P, Klars D, Breedveld FC, editors. Oxford Text book of Rheumatology. Oxford, UK: Oxford University Press; 2004.pp. 937–44.

FURTHER READING

1. Brucellosis in Humans and animals, e-book, free download, by World Health Organization. Accessed June 2017. Available at http://www.who.int/csr/resources/publications/deliberate/WHO_CDS_EPR_2006_7/en/
2. Madkour MM. Madkour's Brucellosis. Berlin, Heidelberg: Springer; 2001.
3. Mantur BG, Amarnath SK. Brucellosis in India—a review. J Biosci 2008; 33: 539–47.
4. Pappas G, Akritidis N, Bosilkovski M, Tsianos E. Brucellosis. N Engl J Med. 2005; 352: 2325–36.

Chapter

70

Rheumatic Manifestations of Human Immunodeficiency Virus Infection

Ajay Wanchu, Marcia Friedman

INTRODUCTION

Musculoskeletal disorders are common with HIV infection and the prevalence increases in the late stages of disease. These disorders can result in significant morbidity and may impair the patient's quality of life. Various musculoskeletal syndromes that can affect HIV-infected patients include reactive arthritis, infective arthritis, diffuse infiltrative lymphocytic syndrome (DILS) and myositis. These manifestations are categorized in Table 70.1. The approximate incidence of various rheumatic diseases in HIV infected patients is mentioned in Table 70.2.

Table 70.1: Classification of rheumatic disorders in HIV-infected patients

Manifestations occurring anytime in the course of HIV disease

- HIV-associated arthralgias and arthritis including crystal induced arthritis
- HIV-associated seronegative arthritides, including reactive, psoriatic undifferentiated spondyloarthritis
- Diffuse infiltrative lymphocytic syndrome (DILS)
- Vasculitis
- Myopathy
- Bone disorders

Manifestations of advanced HIV disease (result of CD4 helper T-cell dysfunction)

- Septic arthritis
- Osteomyelitis
- Pyomyositis

Manifestations associated with HIV treatment

- Osteoporosis
- Avascular necrosis (AVN)

Manifestations associated with immune reconstitution syndrome (IRIS)

- Rheumatoid arthritis (RA)
- Systemic lupus erythematosus (SLE)
- Sarcoidosis

Several important factors need to be taken into consideration when approaching an HIV-infected patient with a musculoskeletal syndrome. All musculoskeletal syndromes that occur in non-HIV infected patients can occur in HIV-infected patients.[1] HIV infection can alter clinical presentation and course of a pre-existing rheumatic disorder. Thus, reactive arthritis, inflammatory arthritis, or musculoskeletal infections may be more severe in their presentation and course in HIV-infected persons. In other instances, musculoskeletal

Table 70.2: Relative frequency HIV disease manifestations[59]

Disease manifestation	*Percentage*
Arthralgia	5.5
Septic arthris	1
Osteomyelitis	0.9
Avascular necrosis	0.7
SLE	0.3
Psoriasis	0.2
Rheumatoid arthritis	0.1
Polymyositis	0.1
Scleroderma	0.1
Psoriatic arthritis, reactive arthritis	0
Ankylosing spondylitis, Sjögren's syndrome	0

infections may be more insidious and subtle in onset. The chances of an opportunistic infection as a cause for a musculoskeletal complaint depend on the stage of the HIV disease. During early stages (CD4 count above 300), opportunistic infections are unlikely, although resistance to common bacterial pathogens may be reduced. For the most part, diagnostic tests and treatment regimens for musculoskeletal syndromes are similar to non-HIV infected patients, except that the HIV-related medications may result in side effects and interactions that affect differential diagnosis and limit therapeutic options.

MANIFESTATIONS OCCURRING ANY TIME IN THE COURSE OF HIV DISEASE

Several rheumatic manifestations occur throughout the course of HIV disease. These include seronegative spondyloarthritides, DILS, vasculitis, and myopathy. The development of some specific rheumatic disorders and paucity of others in HIV infected patients is consistent with the subgroup of these disorders as being immune-mediated.

HIV-associated Arthralgia

The frequency of otherwise unexplained arthralgia was originally reported by Berman et al to occur in approximately 40% of patients.[2] Usual sites of involvement are knees, shoulders and elbows. Most individuals presenting with arthralgia alone do not develop inflammatory joint disease and they are usually seronegative for rheumatoid factor and ANA. Course is usually self-limited within 6 weeks of onset of joint symptoms. The most appropriate treatment, along with non-narcotic analgesics such as paracetamol, is reassurance. Recent evaluations of arthralgia in HIV positive patients have shown similar rates of joint pain.[3]

HIV-associated Arthritis

Arthritis has been reported to occur in more than 10% of patients in some reports.[4] It was initially described as oligoarthritis, predominantly affecting lower extremities that tended to be self-limiting, lasting less than 6 weeks. It can be distinguished from HIV associated reactive arthritis in that there is no mucocutaneous involvement or enthesopathy, and there is no association with HLA-B27 or any other known genetic factor. The main distinguishing features between HIV-associated arthritis and reactive arthritis are presented in Table 70.3. Synovial fluid cultures are typically sterile. A debate is going regarding the causal role of HIV virus in these infections. Evidence for a direct role of HIV includes the isolation of HIV virus DNA and HIV antigen from synovial fluid samples.[5] X-rays of the affected joints are usually normal. Treatment of this disorder includes nonsteroidal anti-inflammatory drugs (NSAIDs) and, in more severe cases, low-dose glucocorticoids. Rarely these patients may require hydroxychloroquine (HCQ) or sulfasalazine (SSZ).

Spondyloarthritides

Prevalence of seronegative spondyloarthritides (including reactive arthritis, psoriatic arthritis and undifferentiated arthritis) has increased in last few years and is reported to be 0.4 to 10%. Clinical features are similar to non-HIV patients but with less uveitis and axial skeleton involvement. They also have frequent extra-articular manifestations like conjunctivitis, urethritis, circinate balanitis and keratoderma blennorrhagicum. Treatment usually includes HAART, SSZ, and tumour necrosis factor (TNF) alpha inhibitor therapy as described below.

Reactive Arthritis

Reactive arthritis (formerly known as Reiter's syndrome), the first rheumatic syndrome reported

Table 70.3: Contrasting features of HIV-associated arthritis and reactive arthritis

Feature	*HIV arthritis*	*HIV-associated reactive arthritis*
Joint involvement	Variable	Asymmetric oligoarthritis
Enthesopathy	Absent	Present
Mucocutaneous involvement	Absent	Present
Synovial fluid white blood cells	500–2000/μL	2000–10,000/μL
Organisms in synovium	? HIV	Chlamydia
HLA-B27	Absent	Present
Course	Self-limited	Chronic/relapsing

in patients with HIV infection, can be a severe illness, but whether it occurs more frequently in HIV-infected patients than in the general population is controversial. Two small retrospective studies of HIV-infected patients reported a prevalence of reactive arthritis syndrome between 5 and 10%, which is 100 to 200 times higher than the expected prevalence in the general population.[6,7] Another study of more than 1,000 homosexual men found the frequency of reactive arthritis to be only 0.3 to 0.5%, and there was no difference in the frequency between HIV-positive and HIV-negative subjects.[8] It has been hypothesized that areas where HIV infection is more likely sexually acquired may show a higher rate of reactive arthritis than regions such as Spain where HIV infection is more often acquired through IV drug abuse.[9] The overall rate of HLA-B27 positivity in HIV patients with reactive arthritis is similar to non-HIV infected patients (63–75% in HIV patients compared to 70–80% in non-HIV infeted patients); however, while 80–90% of Caucasian patients with HIV and reactive arthritis are HLA-B27 positive, in African patients this gene is rarely seen.[5]

The presentation of reactive arthritis in HIV positive patients is similar to that of non-HIV infected patients, with the exception of a lower rate of uveitis and axial involvement. This arthritis mainly involves the lower extremities especially the knees. Frank synovitis is less common but may occur at the ankle, subtalar, metatarsophalangeal, and interphalangeal joints of the feet. Synovitis of the upper extremities is uncommon but may lead to contractures and joint fusion. Reactive arthritis is usually accompanied by enthesitis (sausage toes or fingers, Achilles tendinitis, and plantar fasciitis). Multi-digit dactylitis, especially in the upper extremities, is common and may be relatively painless. Enthesopathy may occur at the medial and lateral epicondyles, rotator cuff, or flexor tendons of the digits. Even though radiographs may reveal sacroiliitis, clinical sacroiliitis and axial involvement are uncommon.

Mucocutaneous features such as keratoderma blennorrhagicum and circinate balanitis are common. Psoriasiform skin rashes are also common and can be extensive, such that it can be difficult to distinguish HIV-associated reactive arthritis from psoriatic arthritis.[5,10] Urethritis occurs as often as it does in HIV-negative reactive arthritis. Constitutional features of Reiter's syndrome, including weight loss, malaise, lymphadenopathy, and diarrhoea, are difficult to distinguish from features of adrenal HIV disease. Clinicians should recommend HIV testing for patients with Reiter's syndrome whose behaviour puts them at an increased risk for HIV infection. It is important to differentiate this entity from HIV-associated arthritis. The major differentiating features are shown in Table 70.3.

The treatment is similar to that of HIV-negative patients with reactive arthritis. NSAIDs are the mainstay of treatment. Indomethacin is often recommended, not only for its efficacy, but also for its unique inhibition of HIV replication observed *in vitro*.[5] Phenylbutazone is rarely used now because of its high side effect profile, but can be useful in refractory cases. Patients frequently have an inadequate response to NSAIDs alone, and these drugs are frequently used in combination with second line drugs. SSZ has been shown to be effective in some studies at doses of 1.5 to 2 g/day. Methotrexate (MTX) was initially thought to be contraindicated because of early reports of *Pneumocystis carinii* pneumonia and other opportunistic infections. Later data have shown that with careful monitoring of HIV viral loads, cell counts, and CD4 counts>100/mm^3, methotrexate can be a safe and effective treatment for reactive arthritis and psoriatic arthritis occurring in the setting of HIV infection. HCQ has also been reported to be useful not only in treating HIV-associated reactive arthritis but also in reducing HIV replication as evidenced in various *in vitro* and *in vivo* studies.[11]

Despite fears of worsening immunosuppression, biologics such as TNF-alpha blockers have been used with surprisingly good safety profiles. The largest case series to date reports on 23 patients with HIV and a variety of rheumatic diseases who used TNF-inhibitors, these patients had rates of infection comparable to rates observed in larger databases.[12] Other smaller case series have found similar safety profiles in well controlled HIV positive patients with adequate CD4 counts, suppressed viral loads, and careful clinical monitoring. However, until larger studies are available, these agents should be used with extreme caution and only in patients with otherwise refractory HIV-associated reactive arthritis and spondyloarthropathy.

Psoriasis and Psoriatic Arthritis

Psoriatic arthritis with or without psoriasis occurs in HIV-infected persons with prevalence probably being the same as that in non-HIV infected persons (1 to 2%), though the severity of the HIV-associated psoriasis and psoriatic arthritis tends to be worrisome.[13] The foot and ankle are the commonest and most severe sites of inflammation. Intense enthesopathy and dactylitis, especially in the feet, usually accompany the arthritis. The enthesopathy can be a major cause of disability. Frank synovitis and synovial effusions are less frequent, but can occur at the ankle and subtalar, metatarso-phalangeal, and interphalangeal joints of the feet. Sacroiliac and spine involvement may also occur. The radiologic appearance of these joints may mimic classic psoriatic arthritis, with "pencil-in-cup" deformities and osteolysis, even in the absence of frank psoriasis.

Nail involvement is a common presenting symptom of arthritis, especially in the distal interphalangeal joints of the hands and feet. Many patients with psoriatic skin manifestations or onycholysis have only these findings and do not meet the criteria for the diagnosis of psoriatic arthritis. Nail involvement occurs in most patients who present with inflammatory articular symptoms. There is a high clinical correlation between skin and joint involvement, and joints may develop erosive changes and crippling deformities. Patients with HIV infection and psoriatic arthritis fall into one of the two patterns of disease: Either the articular disease is sustained and aggressive progressing to joint erosions, or it is characterized by mild and intermittent joint involvement.

The treatment is similar to that for non-HIV infected patients. Numerous reports have also indicated that antiretroviral treatment may be effective. NSAIDs, such as indomethacin (75 to 50 mg per day), can be used initially for the joint symptoms, but results have not been encouraging. There are few data on second-line agents such as gold, MTX, SSZ, and azathioprine. There are reports that phenylbutazone (100 mg 3 times per day) is an effective drug for this arthritis, and neutropenia has not been a problem even in patients receiving zidovudine concurrently. Patients in whom psoriatic skin or joint disease does not respond to NSAIDs may be treated with phenylbutazone (100 mg 3 times a day) or SSZ (1 to 2 g per day). Etretinate may also be helpful. The use of psoralen and pulsed UV-A (PUVA) phototherapy has helped the skin and joints of some HIV-infected individuals with psoriatic arthritis. As above, biologics such as TNF inhibitors are often avoided due to concern of further immunosuppression, however, more recent data suggests that with proper monitoring these drugs may have a better safety profile than previously thought.

Undifferentiated Spondyloarthropathy

This is a frequently encountered problem where there are features of reactive arthritis or psoriatic arthritis in patients who do not otherwise develop full-blown disease, such as enthesopathy (plantar fasciitis, Achilles tendinitis). The treatment is symptomatic (NSAIDs, intralesional corticosteroid injections, when indicated), and SSZ in patients with more extensive disease.

Gout

Up to 42% of HIV patients may have hyperuricemia. Annual incidence of gout in this population is 0.5%. Uncontrolled HIV replication and rapid cell turnover could be related to hyperuricemia.[14] Other possible mechanisms could be mitochondrial toxicity from some anti-retrovirals like stavudine and didanosine.

Diffuse Infiltrative Lymphocytic Syndrome (DILS)

DILS is a syndrome unique to HIV infection that superficially resembles Sjögren's syndrome and is characterized by massive parotid enlargement and xerostomia. It is a CD8+ T-cell infiltration disease that most commonly involves the salivary glands but can involve multiple organs. It can appear at any stage of HIV disease (at any CD4 level) and the prevalence varies between 3 and 7.8% in different series.[15] The prevalence of this condition appears to be decreasing with the widespread use of HAART therapy.[16]

The most common presenting symptoms of this disease is salivary gland enlargement with or without sicca symptoms. The mean duration between the discovery of HIV seropositivity and onset of symptoms in one study was 3.4 years. Submandibular and lacrimal gland enlargement often occurs together and this enlargement is accompanied by sicca symptoms in more than 60% patients. In addition to glandular manifestations, there are several extraglandular features in these patients (Table 70.4). Significantly, the reported frequency of lymphocytic interstitial pneumonitis (LIP) varies between one-third of patients and nearly two-thirds of patients with DILS.[15] Since the introduction of protease inhibitors, this complication occurs much less frequently. Extra-glandular manifestations like renal tubular acidosis, lymphoma, polymyositis, and neurologic involvement are usually associated with a poor prognosis. The major features distinguishing this disorder from Sjögren's syndrome and another mimic in the form of Hepatitis C are shown in Table 70.5.

The primary pathogenic association is reflected as a distinct host immune response in people with the HLA-DR5 phenotype.[17] The association with HLA-DRB1 alleles expressing the ILEDE amino acid sequence in the third diversity region, usually HLA-DRB1*1102, DRB1*1301, and DRB1*1302 has been associated with delayed progression to AIDS among patients with DILS. This is due to delay in the evolution of the HIV-1 virus from the less aggressive M-tropic strain to the more rapidly replicating T-tropic strain by a more effective CD8 lymphocyte response. Pathologically, there is focal sialadenitis, similar to that seen in Sjögren's syndrome, although there is less destruction of the salivary glands.

Table 70.4: Extra-glandular manifestations among patients with DILS[34]

Neurological
- 7th cranial nerve palsy
- Aseptic lymphocytic meningitis
- Peripheral neuropathy

Pulmonary
- Lymphocytic interstitial pneumonitis (LIP)

Gastrointestinal
- Lymphocytic hepatitis

Renal
- Renal tubular acidosis
- Interstitial nephritis

Hematological
- Lymphoma

Musculoskeletal
- Peripheral arthritis
- Polymyositis

To establish the diagnosis the suggested criteria are: The patient (1) must be HIV-seropositive, (2) must have bilateral salivary gland enlargement or xerostomia persisting for more than 6 months, and (3) must have salivary or lacrimal gland lymphocytic infiltration as confirmed by histology orGa scintigraphy, in the absence of granulomatous or neoplastic enlargement.[17] CD8+ lymphocytes constitute most of the inflammatory infiltrate, unlike that seen in primary (non-HIV-associated)

Table 70.5: Differences between DILS and Sjögren's syndrome[16]

Variables	*Sjögren's syndrome*	*DILS*	*Hepatitis C*
Sicca	Present	Present	Present
Glandular manifestations	Moderate parotid enlargement	Moderate to severe parotid enlargement	Mild to moderate parotid enlargement
Extraglandular manifestations	Mainly pulmonary, gastrointestinal, renal and neurologic involvement	Mainly musculoskeletal, pulmonary, gastro-intestinal and neurologic involvement	Mainly gastrointestinal and musculoskeletal involvement
Infiltrating lymphocytic	CD4	CD8	CD4
Phenotype autoantibodies	High frequency of RF, ANA, anti-Ro/SSA and anti-La/SSB	Low frequency of RA factor, ANA and anti-Ro mostly absent	High frequency of RA factor and ANA, anti-Ro/SSA present, though less frequently than primary SS
HLA association	B8, DR2, DR3 and DR4	B45, –B49, –B50 DR11 (DRS) and DRw6 (DRBI*1301*1302)	DR11 (DRS)

Sjögren's syndrome. Also, the SSA and SSB antibodies are usually negative, another feature which can help in distinguishing between the two diseases. It has been suggested that all patients with painless parotid gland enlargement and sicca symptoms should be tested for HIV.

Management of these patients depends upon the extent of the manifestations. Treatment or optimization of the underlying HIV infection with anti-retrovirals is also advised and may improve or resolve their tissue infiltration.[15] If glandular swelling does not worry patients and the sicca symptoms are mild or even absent, then, reassurance and observation are enough. Pilocarpine, in doses of 5–10 mg thrice daily, may be needed for sicca symptoms. In addition, regular dental care is required. One report suggests that moderate doses of corticosteroids (up to 30 to 40 mg of prednisolone per day) are useful in treating both the glandular swelling and sicca symptoms of DILS without adversely affecting the frequency of opportunistic infections, raising the viral loads or depressing the CD4 counts.[18] The effect is usually transient. LIP may need higher doses of corticosteroids (up to 60 mg/day of prednisolone), sometimes for extended periods.

Vasculitis

Every type of vasculitis, including small, medium and large vessel vasculitis, has been described in patients with HIV infection.[19] Patients presenting with vasculitis should be tested routinely for HIV, hepatitis B and hepatitis C infection. In one study 23% patients with symptomatic HIV disease had vasculitis.[20] These include hypersensitivity vasculitis, polyarteritis nodosa, Henoch-Schönlein purpura, Kawasaki's disease both in children and adults, giant cell arteritis, granulomatosis with polyangiitis (formerly known as Wegener's granulomatosis), cryoglobulinemic vasculitis and primary angiitis of the central nervous system (PACNS) and infectious vasculitides.[21]

Polyarteritis nodosa (PAN): PAN seems to be the most common form of vasculitis in HIV infected individuals. HIV associated PAN is not typically associated with hepatitis B or C viruses, and does not typically have severe systemic sequelae (Table 70.6).[22]

Cerebral vasculitis or primary angiitis of central nervous system (PACNS): Among HIV patients with stroke, PACNS is a cause for cerebrovascular disease about 13% of the times. It can occur at anytime during the course of HIV disease, immediately after HIV infection, or when CD4 count falls below 200 copies/µl.

Cryoglobulinemic vasculitis: Prevalence of cryoglobulinemic vasculitis (mostly Type II and III mixed cryoglobulins[23] in HIV positive patients ranges from 1 to 27% but it is much more frequent with HIV and hepatitis C co-infection; indeed it is unclear whether HIV and cryoglobulinemic vasculitis are causally or coincidentally related. Whether HIV infection itself plays a role in cryoglobulinemia itself is also unclear; in one report cryoblubulins were not significantly associated with HIV while in other studies cryoglobulins were detected in 17–26% of HIV patients.[24] In patients with HIV and hepatitis C co-infection, cryoglobulinemic vasculitis presents with palpable purpura, mononeuritis multiplex, arthralgias, and nephritic syndrome. It responds well to hepatitis C treatment with ribavirin and interferon. Rituximab has also been successfully used in the treatment of cryoglobulinemic vasculitis with or without hepatitis C.[25,26]

Table 70.6: Differences between different forms of polyarteritis nodosa[22]

	Classical PAN	*Non-classical PAN*	*Non-classical PAN*
Association	HBV	HIV	HCV
Symptoms	Waxing, waning	Acute	Acute
Organs involved	Renal, GI and CNS	Skin, joints	Skin and others like HBV
Response to therapy	Fair (steroids and cytotoxic)	Excellent (mostly steroids only; HAART)	Good (Rituxan); relapses common; Peg IFN + Ribavarin
Severity	Severe	Mild	Moderate
Relapses	Common	Uncommon	Very common
Prognosis	Poor (35% mortality)	Excellent	Fair (10% mortality)

It remains to be seen whether rituximab can be used in cryoglobulinemic vasculitis associated with HIV with or without HCV co-infection

Infectious vasculitis: Vasculitis resulting from cytomegalovirus (CMV) infection can result in prominent gastrointestinal and cutaneous manifestations and its inclusion bodies can be demonstrated by microscopy. In such cases, antiviral agents like ganciclovir or foscarnet should be started and any concomitant immunosuppressive therapy reduced. Any drugs that can cause hypersensitivity vasculitis must be withdrawn. Life-threatening complications affecting the lungs, kidneys or the nervous system should be treated with steroids and immunosuppressants. Antiretroviral drugs may help, and additional therapy in the form of prophylaxis for herpes zoster and Pneumocystis pneumonia must be instituted.

Myopathy

Skeletal muscle involvement may occur at all stages of HIV infection and may be classified as follows: (1) HIV-associated myopathies and related conditions; (2) muscle complications of antiretroviral therapy; (3) opportunistic infections and tumour infiltrations of skeletal muscle; and (4) rhabdomyolysis.[27,28] In one series of skeletal muscle biopsies performed in adult patients with HIV infection, at autopsy there was atrophy in most and inflammatory infiltrates, necrosis and infection in few.[29]

Polymyositis: HIV-associated polymyositis typically occurs early in HIV infection. The usual presentation is that of a subacute, progressive proximal muscle weakness occurring in the patient with elevated creatine kinase. Auto-antibodies like anti-Jo-1 and anti-Mi-2 are typically absent. Electromyography shows myopathic motor unit potentials with early recruitment and full interference patterns, fibrillation potentials, positive sharp waves, and complex repetitive discharges suggesting an irritative process. Microscopic examination of muscle biopsies reveals interstitial inflammatory infiltrates of variable intensity associated with degenerating and regenerating myofibrils, like those seen in polymyositis without HIV-1.[30] Concomitant vasculitis is rare. As seen in patients who are HIV seronegative, the predominant cell populations are CD8+ T cells and macrophages invading or surrounding healthy muscle fibres in the endomysial distribution. The etiopathogenesis of the disorder is not clear. It is controversial whether the virus directly leads to inflammatory myopathy. The treatment of HIV-associated polymyositis is similar to that for other inflammatory myopathies. Both creatine kinase and muscle weakness usually respond to glucocorticoids.[31] Refractory cases might need immunosuppressive agents such as methotrexate or azathioprine. These should, however, be used with caution with careful monitoring of the patient's clinical status, CD4+ counts, and HIV viral load. Concomitant DILS may be present in up to half the patients. Dermatomyositis, inclusion body myositis and nemaline rod myopathy[32] are also seen very rarely in association with HIV.

Rhabdomyolysis: HIV can also be complicated by rhabdomyolysis. This can be caused by HIV infection itself, as a consequence of interaction between protease inhibitors and statins (cytochrome p450 mediated interaction leading to high levels of statins), or as an adverse effects of other drugs like integrase inhibitors (raltegravir) or from hypersensitivity to abacavir.

HIV wasting syndrome and other forms of muscle disease: HIV wasting syndrome should be distinguished from other causes of generalised weight loss in HIV patients from infections, lactic acidosis, etc. Muscle biopsy may show diffuse atrophy or type II neurogenic atrophy without much inflammation.

Other forms of muscle disease in HIV-positive individuals have included myalgias, myasthenic syndrome, chronic fatigue, and fibromyalgia. HIV associated myasthenia is not always associated with antibodies to acetylcholine receptors and could be a transient phenomenon. Treatment is based on antiretroviral therapy, combined with, if necessary, pyridostgimine, prednisone, or intravenous immunoglobulin.[33]

HAART related myopathy: With the introduction of antiretroviral drugs in the treatment of HIV infection, especially zidovudine (AZT), there have been reports of a diffuse myopathy.[34] It is characterized by insidious onset of myalgias, muscle tenderness, and proximal muscle weakness after a mean duration of 11 months therapy with the drug. It tends to be dose-related, being associated with mitochondrial dysfunction. Both clinically, by EMG and by muscle biopsy, it is difficult to distinguish it from HIV-associated polymyositis, although inflammatory infiltrates tend to be much less severe and sometimes absent in patients receiving AZT. Ragged red fibres may be seen suggesting mitochondrial dysfunction. A rare life-threatening syndrome with severe hepatic steatosis, lactic acidosis, mitochondrial myopathy and pancreatitis resembling Reye's syndrome in children, has been described with stavudine in combination another nucleoside analogue and a protease inhibitor. Consequently, in any HIV-infected individual who presents with an elevated CK, especially when symptoms of myalgia or muscle weakness are present, the drug should be discontinued for 4 weeks and the patient re-evaluated before EMG or muscle biopsies are undertaken.

Infectious myositis (pyomyositis): Pyomyositis can be an important cause of muscle problems in patients with HIV infection.[35] Organisms implicated have included *Staphylococcus aureus, Salmonella enteritidis,* Microsporum, and Toxoplasma. Their management is on lines similar to HIV uninfected individuals. Extra nodal non-Hodgkin's lymphoma has been described in HIV patients and presents with rapidly growing muscle mass with fever. Histopathological examination shows a lymphomatous proliferation destroying the muscle tissue. Magnetic resonance imaging can be useful in distinguishing lymphomatous involvement of muscle from other causes of limb swelling in HIV patients. Kaposi sarcoma can also involve the muscle.

MANIFESTATIONS OF ADVANCED HIV DISEASE

During later stages of infection patients are predisposed to a several infectious disorders. These might be either localized or systemic in nature and may primarily or secondarily involve the musculoskeletal system. The profile of organisms infecting the joints can also vary with high-risk behavior patterns or underlying conditions associated with HIV infection. Thus, intravenous drug users are more likely to be infected with *Staphylococcus aureus, Pseudomonas aeruginosa,* and *Candida albicans.*

As described in the previous section, musculoskeletal infections include septic arthritis, osteomyelitis and pyomyositis. Data is conflicting regarding the incidence of septic arthritis among the HIV infected. There is no good data to show that bacterial infections of bones or joints due to the usual infective agents, such as *S. aureus,* are more frequent in patients with HIV infection.[36] A large retrospective case series of 4023 HIV-infected patients studied between 1985 and 1996 found that *S. aureus* was the commonest infective agent isolated.[37] The incidence was not related to CD4 counts suggesting that parenteral drug use rather than HIV infection *per se* predisposed to the joint infections. Even so, at CD4 counts dropping below 100/μl, atypical mycobacteria can be isolated from some joints. Most common among these are *M. hemophilum* and *M. kansasii.* Studies from Africa have shown that septic arthritis occurred much less frequently than HIV-associated aseptic arthritis.[38] On the other hand, there are studies to show an increased incidence of septic arthritis among HIV-infected haemophiliacs compared with those who are uninfected.[39] Management of these conditions is similar to that of HIV uninfected individuals. Some form of joint drainage may be required.

Osteomyelitis typically occurs at very low CD4 counts. The most common organism remains *Staphylococcus aureus.* It can often be polymicrobial. Moreover, after instituting HAART, pre-existing quiescent mycobacteria like *M. avium* intracellulare can cause symptomatic osteomyelitis.

Other infections of relevance include those by atypical mycobacteria and fungi. Among the former the common species implicated are *Mycobacterium* avium intracellulare complex (MAI), *M. kansasii, M. haemophilum, M. terrae,* and *M. fortuitum.* M. haemophilum is implicated in more than half of the cases, and *M. kansasii* accounts for an additional 25%. Other lesions seen in about half the patients are cutaneous nodules, ulcers, and draining sinus tracts.[39] These infections occur late in the course of HIV, usually when the CD4+ T lymphocyte count is less than 100/μL. Clarithromycin has been reported as effective in most atypical mycobacterial bone and joint infections along with standard antituberculous therapy.

At CD4 counts less than 100/μL there is a high risk of development of fungal infections, which can have oligoarticular or polyarticular arthritis and may also result in tendon sheath effusion. These infections may be especially difficult to eradicate and may need chronic suppressive antifungal therapy.

Table 70.8: Consequences of B cell defects related to rheumatic defects

Polyclonal hypergammaglobulinemia
Circulating immune complexes
Anti-neutrophil cytoplasmic antibodies
Autoantibodies
• Antinuclear antibodies
• IgG and IgA rheumatoid factors
• Anti-phospholipids
• Anti-platelet
• Anti-RBC
• Anti-lymphocyte surface antigen
• Anti-neuronal cells
• Anti-neurotransmitters
• Anti-myelin associated proteins

patients with AIDS. Anti-β2 glycoprotein-1antibody is seen in up to 47% patients, whereas lupus anticoagulant is generally not seen. Despite the presence of these antibodies, the full picture of anti-phospholipid syndrome is not more common in HIV patients than general population.[58] Antinuclear neutrophilic cytoplasmic antibodies (ANCAs), both c-ANCA and p-ANCA, have also been detected in the serum of HIV-positive individuals, in frequencies of up to 43% by ELISA and 18 percent by indirect immunofluorescence. In most instances these serological abnormalities represent false positives that are rarely of clinical significance.

Summary

An entire spectrum of rheumatic diseases occurs in HIV-positive individuals, with most of them being typical of what is seen in non-HIV populations. Others may be particular to the setting of HIV. (DILS, HIV-associated arthritis). Treatment of the HIV infection is part of the management of many of these disorders. Antiretroviral agents improve the clinical symptoms of HIV-associated rheumatic diseases. Some drugs used in the treatment of the rheumatic diseases (indomethacin, hydroxychloroquine) have antiretroviral properties too. With the availability of highly active anti-retroviral agents the effects that these agents have on the presentation, course, and outcome of the HIV-associated rheumatic diseases remain to be determined.

Key Points

1. Most of the rheumatic diseases seen in patients with HIV are similar to what is seen in non-HIV populations, though their presentation may at times be more severe.
2. DILS, HIV-associated arthritis, IRIS, and possibly reactive arthritis represent inflammatory disorders particular to HIV.
3. Hepatitis C, gonorrhoea, and chlamydia may be comorbid conditions in the HIV population, and can increase the risk of rheumatic diseases such as cryoglobulinemic vasculitis, HCV arthritis, reactive arthritis, and gonococcal arthritis.
4. Infections, including atypical organisms and fungi, must always be considered high on the differential when evaluating an arthritic or rheumatic complaint in a patient with late or poorly controlled HIV infection.
5. Anti-retroviral treatment may cause side effects relevant to the musculoskeletal system including osteoporosis, osteomalacia, and avascular necrosis.
6. Non-biologic DMARDs can generally be used in well-controlled HIV; biologic DMARDs should be used with extreme caution in HIV populations, though early data suggests that in well-controlled HIV the use of TNF inhibitors may have a safety profile similar to the general population.

REFERENCES

1. Wanchu A. Rheumatic manifestations of HIV infection. APLAR Journal of Rheumatology. 2004; 7(2):189–91.
2. Berman A, Espinoza LR, Diaz JD, Aguilar JL, Rolando T, Vasey FB, et al. Rheumatic manifestations of human immunodeficiency virus infection. Am J Med. 1988; 85:59–64.
3. Ogdie A, Pang WG, Forde KA, Samir BD, Mulugeta L, Chang KM, et al. Prevalence and risk factors for patient-reported joint pain among patients with HIV/hepatitis C coinfection, hepatitis C monoinfection, and HIV monoinfection. BMC Musculoskelet Disord. 2015; 16:93.
4. Louthrenoo W. Rheumatic manifestations of human immunodeficiency virus infection. Curr Opin Rheumatol. 2008; 20(1):92–9.
5. Adizie T, Moots RJ, Hodkinson B, French N, Adebajo AO. Inflammatory arthritis in HIV positive patients: A practical guide. BMC Infect Dis. 2016; 16:100.
6. Fischer H. Reiter's syndrome and the acquired immune deficiency syndrome. Arthritis Rheum. 1985;4.

7. Winchester R. AIDS and the rheumatic diseases. Bull Rheum Dis. 1990; 39(5):1–10.
8. Hochberg MC, Fox R, Nelson KE, Saah A. HIV infection is not associated with Reiter's syndrome: data from the Johns Hopkins Multicenter AIDS Cohort Study. AIDS. 1990; 4:1149–51.
9. Walker-Bone K, Doherty E, Sanyal K, Churchill D. Assessment and management of musculoskeletal disorders among patients living with HIV. Rheumatology (Oxford). 2016.
10. Reveille JD, Williams FM. Infection and musculoskeletal conditions: Rheumatologic complications of HIV infection. Best Pract Res Clin Rheumatol. 2006; 20:1159–79.
11. Sperber K, Kalb TH, Stecher VJ, Banerjee R, Mayer L. Inhibition of human immunodeficiency virus type 1 replication by hydroxychloroquine in T cells and monocytes. AIDS Res Hum Retroviruses. 1993; 9:91–8.
12. Wangsiricharoen S, Ligon C, Gedmintas L, Dehrab A, Tungsiripat M, Bingham C 3rd, et al. The Rates of Serious Infections in HIV-infected Patients Who Received Tumor Necrosis Factor (TNF)-alpha Inhibitor Therapy for Concomitant Autoimmune Diseases. Arthritis Care Res (Hoboken). 2016.
13. Espinoza LR, Berman A, Vasey FB, Cahalin C, Nelson R, Germain BF. Psoriatic arthritis and acquired immunodeficiency syndrome. Arthritis Rheum. 1988; 31:1034–40.
14. Manfredi R, Mastroianni A, Coronado OV, Chiodo F. Hyperuricemia and progression of HIV disease. J Acquir Immune Defic Syndr Hum Retrovirol. 1996; 12:318–9.
15. Ghrenassia E, Martis N, Boyer J, Burel-Vandenbos F, Mekinian A, Coppo P. The diffuse infiltrative lymphocytosis syndrome (DILS). A comprehensive review. J Autoimmun. 2015; 59:19–25.
16. Basu D, Williams FM, Ahn CW, Reveille JD. Changing spectrum of the diffuse infiltrative lymphocytosis syndrome. Arthritis Rheum. 2006; 55: 466–72.
17. Itescu S, Winchester R. Diffuse infiltrative lymphocytosis syndrome: a disorder occurring in human immunodeficiency virus-1 infection that may present as a sicca syndrome. Rheum Dis Clin North Am. 1992; 18:683–97.
18. Williams FM. Short-term moderate dose prednisone therapy in the diffuse infiltrative lymphocytosis syndrome: a prospective study. Arthritis Rheum. 1997; 40(S92).
19. Kaye BR. Rheumatologic manifestations of infection with human immunodeficiency virus (HIV). Ann Intern Med. 1989; 111:158–67.
20. Gherardi R, Belec L, Mhiri C, Gray F, Lescs MC, Sobel A, et al. The spectrum of vasculitis in human immunodeficiency virus-infected patients. A clinicopathologic evaluation. Arthritis Rheum. 1993; 36:1164–74.
21. Pagnoux C, Cohen P, Guillevin L. Vasculitides secondary to infections. Clin Exp Rheumatol. 2006; 24(2 Suppl 41):S71–81.
22. Patel N, Patel N, Khan T, Patel N, Espinoza LR. HIV infection and clinical spectrum of associated vasculitides. Curr Rheumatol Rep. 2011; 13:506–12.
23. Dimitrakopoulos AN, Kordossis T, Hatzakis A, Moutsopoulos HM. Mixed cryoglobulinemia in HIV-1 infection: the role of HIV–1. Ann Intern Med. 1999; 130:226–30.
24. Saadoun D, Aaron L, Resche-Rigon M, Pialoux G, Piette JC, Cacoub P, et al. Cryoglobulinaemia vasculitis in patients coinfected with HIV and hepatitis C virus. AIDS. 2006; 20:871–7.
25. Sneller MC, Hu Z, Langford CA. A randomized controlled trial of rituximab following failure of antiviral therapy for hepatitis C virus-associated cryoglobulinemic vasculitis. Arthritis Rheum. 2012; 64:835–42.
26. De Vita S, Quartuccio L, Isola M, Mazzaro C, Scaini P, Lenzi M, et al. A randomized controlled trial of rituximab for the treatment of severe cryoglobulinemic vasculitis. Arthritis Rheum. 2012; 64:843–53.
27. Authier FJ, Chariot P, Gherardi RK. Skeletal muscle involvement in human immunodeficiency virus (HIV)-infected patients in the era of highly active antiretroviral therapy (HAART). Muscle Nerve. 2005; 32:247–60.
28. Zell SC, Nielsen S. Clinical correlates to muscle biopsy findings in HIV patients experiencing fatigue: a case series. J Int Assoc Physicians AIDS Care (Chic). 2002; 1:90–4.
29. Wrzolek MA, Sher JH, Kozlowski PB, Rao C. Skeletal muscle pathology in AIDS: an autopsy study. Muscle Nerve. 1990; 13:508–15.
30. Hiniker A, Daniels BH, Margeta M. T-Cell-Mediated Inflammatory Myopathies in HIV-Positive Individuals: A Histologic Study of 19 Cases. J Neuropathol Exp Neurol. 2016; 75:239–45.
31. Johnson RW, Williams FM, Kazi S, Dimachkie MM, Reveille JD. Human immunodeficiency virus-associated polymyositis: a longitudinal study of outcome. Arthritis Rheum. 2003; 49:172–8.
32. Dalakas MC, Pezeshkpour GH, Flaherty M. Progressive nemaline (rod) myopathy associated with HIV infection. N Engl J Med. 1987; 317:1602–3.
33. Strong J, Zochodne DW. Seronegative myasthenia gravis and human immunodeficiency virus infection: response to intravenous gamma globulin and prednisone. Can J Neurol Sci. 1998; 25:254–6.
34. Dalakas MC, Illa I, Pezeshkpour GH, Laukaitis JP, Cohen B, Griffin JL. Mitochondrial myopathy caused by long-term zidovudine therapy. N Engl J Med. 1990; 322:1098–105.
35. Al-Tawfiq JA, Sarosi GA, Cushing HE. Pyomyositis in the acquired immunodeficiency syndrome. South Med J. 2000; 93:330–4.

36. Hughes RA, Rowe IF, Shanson D, Keat AC. Septic bone, joint and muscle lesions associated with human immunodeficiency virus infection. Br J Rheumatol. 1992; 31:381–8.
37. Vassilopoulos D, Chalasani P, Jurado RL, Workowski K, Agudelo CA. Musculoskeletal infections in patients with human immunodeficiency virus infection. Medicine (Baltimore). 1997; 76:284–94.
38. Blanche P, Taelman H, Saraux A, Bogaerts J, Clerinx J, Batungwanayo J, et al. Acute arthritis and human immunodeficiency virus infection in Rwanda. J Rheumatol. 1993; 20:2123–7.
39. Hirsch R, Miller SM, Kazi S, Cate TR, Reveille JD. Human immunodeficiency virus-associated atypical mycobacterial skeletal infections. Semin Arthritis Rheum. 1996; 25:347–56.
40. Ghanem KG, Vassilopoulos D, Gebo KA. Evaluation of Rheumatic Complaints in Patients with HIV. In: Imboden JB, Hellmann DB, Stone JH, editors. CURRENT Diagnosis & Treatment: Rheumatology, 3e. New York, NY: The McGraw-Hill Companies; 2013.
41. Mehsen-Cetre N, Cazanave C. Osteoarticular manifestations associated with HIV infection. Joint Bone Spine. 2016.
42. Brown TT, Qaqish RB. Antiretroviral therapy and the prevalence of osteopenia and osteoporosis: a meta-analytic review. AIDS. 2006; 20:2165–74.
43. Shiau S, Broun EC, Arpadi SM, Yin MT. Incident fractures in HIV-infected individuals: a systematic review and meta-analysis. AIDS. 2013; 27:1949–57.
44. Parsonage MJ, Wilkins EG, Snowden N, Issa BG, Savage MW. The development of hypophosphataemic osteomalacia with myopathy in two patients with HIV infection receiving tenofovir therapy. HIV Med. 2005; 6:341–6.
45. French MA. HIV/AIDS: Immune reconstitution inflammatory syndrome: a reappraisal. Clin Infect Dis. 2009; 48:101–7.
46. Foulon G, Wislez M, Naccache JM, Blanc FX, Rabbat A, Israel-Biet D, et al. Sarcoidosis in HIV-infected patients in the era of highly active antiretroviral therapy. Clin Infect Dis. 2004; 38:418–25.
47. Bijlsma JW, Derksen RW, Huber-Bruning O, Borleffs JC. Does AIDS 'cure' rheumatoid arthritis? Ann Rheum Dis. 1988; 47:350–1.
48. Azeroual A, Harmouche H, Benjilali L, Mezalek ZT, Adnaoui M, Aouni M, et al. Rheumatoid arthritis associated to HIV infection. Eur J Intern Med. 2008; 19:e34–5.
49. Muller-Ladner U, Kriegsmann J, Gay RE, Koopman WJ, Gay S, Chatham WW. Progressive joint destruction in a human immunodeficiency virus-infected patient with rheumatoid arthritis. Arthritis Rheum. 1995; 38:1328–32.
50. du Toit R, Whitelaw D, Taljaard JJ, du Plessis L, Esser M. Lack of specificity of anticyclic citrullinated peptide antibodies in advanced human immunodeficiency virus infection. J Rheumatol. 2011; 38:1055–60.
51. Bambery P, Deodhar SD, Malhotra HS, Sehgal S. Blood transfusion related HBV and HIV infection in a patient with SLE. Lupus. 1993; 2:203–5.
52. Wanchu A, Sud A, Singh S, Bambery P. Human immunodeficiency virus infection in a patient with systemic lupus erythematosus. J Assoc Physicians India. 2003; 51:1102–4.
53. Yen YF, Chuang PH, Jen IA, Chen M, Lan YC, Liu YL, et al. Incidence of autoimmune diseases in a nationwide HIV/AIDS patient cohort in Taiwan, 2000–2012. Ann Rheum Dis. 2016.
54. Hazarika I, Chakravarty BP, Dutta S, Mahanta N. Emergence of manifestations of HIV infection in a case of systemic lupus erythematosus following treatment with IV cyclophosphamide. Clin Rheumatol. 2006; 25:98–100.
55. Hentges F, Hoffmann A, Oliveira de Araujo F, Hemmer R. Prolonged clinically asymptomatic evolution after HIV-1 infection is marked by the absence of complement C4 null alleles at the MHC. Clin Exp Immunol. 1992; 88:237–42.
56. Jindal R, Solomon M, Burrows L. False positive tests for HIV in a woman with lupus and renal failure. N Engl J Med. 1993; 328:1281–2.
57. De Milito A. B lymphocyte dysfunctions in HIV infection. Curr HIV Res. 2004; 2:11–21.
58. Loizou S, Singh S, Wypkema E, Asherson RA. Anticardiolipin, anti-beta(2)-glycoprotein I and antiprothrombin antibodies in black South African patients with infectious disease. Ann Rheum Dis. 2003; 62:1106–11.
59. Yao Q, Frank M, Glynn M, Altman RD. Rheumatic manifestations in HIV-1 infected in-patients and literature review. Clin Exp Rheumatol. 2008; 26: 799–806.

FURTHER READING

1. Fox C, Walker-Bone K. Evolving spectrum of HIV-associated rheumatic syndromes. Best Pract Res ClinRheumatol. 2015; 29:244–58.
2. Vassilopoulos D, Chalasani P, Jurado RL, Workowski K, Agudelo CA. Musculoskeletal infections in patients with human immunodeficiency virus infection. Medicine (Baltimore). 1997; 76:284–94.
3. Walker-Bone K, Doherty E, Sanyal K, Churchill D. Assessment and management of musculoskeletal disorders among patients living with HIV. Rheumatology (Oxford). 2016, Dec 24.

Chapter

71

Viral Arthritis

Rudra Prasad Goswami, Parasar Ghosh

INTRODUCTION

Polyarthralgia and myalgia are common clinical manifestations of most viral infections, however significant rheumatic manifestations during or after the course of such infections are relatively rare. However acute infections with hepatitis B virus (HBV), Parvovirus B19, alphaviruses, and chronic infections with human T-cell lymphotropic virus (HTLV)-1, hepatitis C virus (HCV), and HIV virus causes well described musculoskeletal manifestations. Post-viral arthritis like syndromes is recently recognized.[1–3]

ALPHAVIRUS

Viruses belonging to the genus Alphavirus prominently cause arthritis (Tables 71.1 and 71.2). The Aalphavirus is a genus comprising single stranded RNA virus and belongs to the family Togaviridae. These Alpha viruses are arthropod-borne among which chikungunya virus (CHIKV) is clinically most relevant in the Indian context, while other viruses (Table 71.1) although present worldwide are not very common in India. Clinical infection with alphaviruses is usually acute onset, producing fever, rash, arthralgia/arthritis and myalgias, which may eventually evolve into chronic arthralgia/arthritis in a proportion of affected individuals.

Chikungunya Virus (CHIKV)

CHIKV was first isolated in 1952 from the blood of a febrile patient in Tanzania and since then, has been reported almost worldwide. The first Indian outbreak of CHIKV infection was reported in 1963 in Kolkata followed by outbreaks in Chennai, Puducherry, and Vellore in 1964 and in Visakhapatnam in 1965.[4, 5] The period between 2004 though 2011 witnessed a global epidemic of CHIKV; the epidemic started from Kenya, spread throughout the Indian Ocean, crossed over to India and southeast Asia.[6] This outbreak was linked to transmission via the *Aedes albopictus* mosquito vector that has a wide geographical distribution.[3] It is estimated that 2 million people are affected with CHIKV infection, with India alone having almost 1.3 million infected individuals.[7]

O'nyong'nyong Virus (ONNV)

The prevalence of ONNV is restricted to Africa, it was first isolated from anopheline mosquitoes and human serum in the 1959 epidemic in East Africa.[8] The virus was quiescent for 3 decades in East Africa from the 1960s to 1990s, when it was resurfaced in Uganda in the 1990s.[8, 9]

Sindbis Virus

Sindbis virus was isolated from Culex mosquitoes in the Sindbis district Egypt in 1995, but unlike the ONNV it has spread from East Africa to both southern Africa and to Europe. In Europe it is known by various names such as Ockelbo disease (Sweden), Pogosta disease (Finland), and Karelian fever in Russia.[10] The virus is also prevalent in Malaysia, Philippines, China, Papua New Guinea, and Australia.[11]

Table 71.1: Alphaviruses associated with rheumatic diseases[2–3, 34]

Virus	*First description*	*Vector*	*Primary host*	*Endemicity*
Ross River virus	1928, New South Wales, Australia	Aedes; Mansonia uniformis; Culex annulirostris	Rodents, marsupials, domestic animals	Australia, Papua New Guinea, Solomon Islands and other Pacific Islands
Sindbis virus	1952, Sindbis village, near Cairo, Egypt	Aedes; Culiseta; Culex	Birds (Grouse and Passerines)	Endemic in north Europe; outbreaks in Finland, Sweden, Norway and Russia (late summer and early autumn)
Chikungunya virus	1952, Newala, Tanzania	Aedes species; Mansonia africana	Baboons, monkeys, Scotophilus bat species	Africa, Asia with documented cases in Europe, USA and Oceania
Mayaro virus	1954, Trinidad and Tobago	Haemagogus janthinomys	Marmosets	Northern South America (Bolivia, Brazil, Peru)
O'nyong-nyong virus	1959, Northern Uganda	*Anopheles funestus*, *A. gambiae*	[b]	Africa (rare epidemics)
Barmah Forest virus	1974, Barmah forest, Australia	Aedes vigilax; Culex annulirostris	Likely to be marsupials (possums, kangaroos, etc.)[a]	Annual epidemics in Australia

[a]Definitive evidence lacking.
[b]No non-human host have been identified yet, but human to human transmission does not seem to occur.
Modified and reused with permission from Assunção-Miranda I, Cruz-Oliveira C, Da Poian AT. Molecular mechanisms involved in the pathogenesis of *Alphavirus* induced arthritis. Biomed Res Int. 2013;2013: 973516 under creative commons license.

Ross River Virus (RRV) and Mayaro Virus

RRV is endemic in Australia, Papua New Guinea, and Pacific Islands and remains the most common arboviral disease in humans in Australia.[12] Mayaro virus is endemic in South America.[13]

Clinical Features of the Alphaviral Arthritides

Acute infection: CHIKV disease is characterized by an abrupt onset of high grade fever with rigor. Polyarthralgia or polyarthritis may occur beginning two to five days after the onset of fever. The pattern of arthralgia is bilaterally symmetrical involving both small (interphalangeal joints and wrists) and large joints (shoulders and knees). Axial involvement may present in up to 50% of involved persons.[1–4,15,16]

Skin rash follows fever by 2–3 days, usually in the form of a pruritic maculopapular rash involving trunks and limbs and rarely face.[17,18] Patients also commonly have fatigue, generalised weakness, headache, nausea, and vomiting. Conjunctivitis often occurs and is characteristic. However, unlike dengue fever, retro-orbital pain does not occur.

Haematological findings include leukopaenia with relative lymphocytosis; occasional thrombocytopaenia and raised acute phase reactants. Unlike dengue fever, bleeding is rare.[19] Diagnosis in this phase of illness is virus isolation by PCR. The virus can be cultured from blood using C6/36 (mosquito cell line) or Vero (monkey kidney cell line) cells.[3]

Chronic phase: Although the exact prevalence of chronic rheumatological disease caused by Alphavirus is not well-known, nevertheless the alphaviruses often cause a chronic, episodic, often debilitating, polyarthralgia/polyarthritis associated

Table 71.2: Alphavirus associated rheumatological clinical symptomatology[3]

Virus	*Occurrence*	*Incubation period*	*Asymptomatic (%)*	*Major symptomatology (%)*				*Other characteristic feature(s)*
				Fever	*Rash*	*Myalgia*	*Arthralgia*	
Chikungunya virus	Large epidemics, common in India	1–2 weeks	15	90	50	90	>95	Tenosynovitis; conjunctivitis
Ross River virus	Australia, Pacific Islands	1–3 weeks	50–75	40	50	40	>80	Fatigue, photophobia, lymphadenopathy and rarely encephalitis
Barmah Forest virus	Australia	1 week	[a]	50	40	>50	>70	Lethargy and headache
Sindbis virus	Russia, Scandinavia	1 week	[b]	25	90	50	95	Headache
O'nyong- nyong virus	Africa	1 week	[c]	90	80	70	>60	Pruritis, lymphadenopathy, conjunctivitis, rarely bleeding
Mayaro virus	South America	1–2 weeks	8	100	40	75	>50	Oedema, retro-orbital pain, rarely bleeding

[a]A large number of asymptomatic cases likely similar to Ross River virus, but definite estimates are not available.

[b]Approximately one-third may asymptomatic, however not very well-defined.

[c]0.6% population seroprevalence of IgM antibody against Sindbis virus was identified from Finland; asymptomatic cases especially in children are not uncommon but definite estimates not present.

Modified and reused with permission from Suhrbier A, Jaffar-Bandjee M-C, Gasque P. Arthritogenic Alphaviruses—An Overview. Nat Rev Rheumatol. 2012;8:420–9.

with fatigue. Protracted course of prominent arthralgia is seen especially with Sindbis virus, RRV, and CHIKV.[20] CHIKV tenosynovitis involving dorsal part of wrists and finger joints is a characteristic feature.[21] The prevalence of CHIKV chronic rheumatism widely varies from as low as 1.6% to as high as 57%.[21–26]

There are many studies from India on CHIKV chronic arthritides (Table 71.3).[23–29] Knees, ankles, and the distal interphalangeal (DIP) joints are the common sites of arthritis along with proximal interphalangeal (PIP) and metacarpal phalangeal (MCP) joints. Data from prospectively conducted Indian studies suggest that chronicity is not more than 5%. Many of these patients are actually rheumatoid factor (RF) and/or anti-citrullinated peptide antibodies (ACPA) positive raising the doubt that these cases are actually unmasked rheumatoid arthritis given the especially high community prevalence of RA.

A meta-analysis of post-CHIKV chronic rheumatism, wherein chronic rheumatism was defined as: "arthritis, arthralgia lasting >2 months from acute attack, both relapsing or lingering and without a history of previous rheumatologic disease or musculoskeletal symptoms" was recently published.[30] Rheumatoid arthritis, unspecific or post-viral arthritis, and seronegative spondylitis were included in the arthritis group. Post-viral polyarthralgia, fibromyalgia, chronic articular pain, and frozen shoulder or plantar fasciitis were included in the musculoskeletal symptoms". Overall the prevalence of chronic rheumatism was 40% (95% confidence interval 31–49%) and that of chronic arthritis was 13.6% (95% CI: 9–18%). However, the prevalence in pooled data from the India showed a much lesser prevalence of chronic rheumatism of 27% (95% CI: 15–38).[30]

Community based studies from Australia reported 50–57% prevalence of chronic arthralgia with RRV infection and similar rate for Pogosta were reported from Finland.[20, 31]

The predictors of CHIKV chronic rheumatism were identified in a study from the Reunion Island (Sisoko et al.)[21] among patients who had acute CHIKV followed up for a mean of 15 months, where 57% (84/147) subjects had self-reported rheumatic symptoms. Among these 63% (53/84) had persistent symptoms and 37% had recurrent symptoms. The predictors of chronicity identified was age more than 45 years at initial presentation (OR 3.9, 95% CI: 1.7–9.7), severe initial presentation (OR 4.8, 95% CI: 1.9–12.1) and presence of osteoarthritis (OR 2.9, 95% CI: 1.1–7.4).[22] Whether these persistent symptoms are true auto-immunity or older male with worsening osteoarthritis is difficult to differentiate. Moreover in this study the pattern of joint involvement even arthritis, tenosynovitis and arthralgia, these types of basic distinctions are not categorized.

A follow-up of TELECHIK study which is 2-year follow-up of 346 patients with CHIKV infection found that 111 (32.1%) individuals reported relapsing chronic symptoms, and 150 (43.3%) complained of lingering (persistent) chronic symptoms.[32] Determinants of relapsing remitting symptomatology were: age 45–59 years (OR: 2.9, 95% CI: 1.0–8.6), or greater than 60 years (OR: 10.4, 95% CI: 3.5–31.1), severe initial involvement at presentation (OR: 3.6, 95% CI: 1.5–8.2), and CHIKV-specific IgG titers (OR: 3.2, 95% CI: 1.8–5.5, per one unit increase). Determinants for persistent symptoms were: age 45–59 years (OR: 6.4, 95% CI: 1.8–22.1), >60 years (OR: 22.3, 95% CI: 6.3–78.1), severe initial involvement (OR: 5.5, 95% CI: 2.2–13.8) and CHIKV-specific IgG titers (OR: 6.2, 95% CI: 2.8–13.2, per one unit increase).[32]

Immune Alterations in CHIKV Infections

Following entry, the virus resides within skin fibroblasts and thereafter upon entering bloodstream it induces a type I interferon response. Thereafter, it spreads to various sites including joints and skeletal muscles. Mononuclear cells infiltrate these sites and CHIKV can persist within these cells providing an ongoing source of antigenic stimulation.

During acute stages of CHIKV infection numerous cytokines are elevated: type 1 cytokines (IFN-α, IL-1, IL-2, IL-6, IL-15, G-CSF, GM-CSF, RANTES), type 2 cytokine (IL-4) and others (CXCL-9, CCL-3, eotaxin/CCL-11), some of which has been associated with severity of the disease such as IL-1β, IL-6, and RANTES.[34] By the second week however a shift to Th2 cytokine profile is observed (elevated IL-4 and eotaxin).[33]

Table 71.3: Indian studies on Chikungunya virus associated arthritis

Studies	*Type of study*	*n (chronic rheumatism)/n (total)[a]*	*Follow-up duration (months)*	*Mean age (years)*	*RF/CCP*	*Pattern of joint involvement*	*Others*
Gauri,[25] 2016 Rajasthan	Prospective	34/50	60	36	70% (among those with arthritis)	MCP, ankle and knee most common sites (% not given)	Risk factors for chronicity: Thrombocytopaenia, past history of arthritis and duration of fever > 10 days
Chaaithanya[27] 2014 Karnataka	Prospective	9/203	36	58	Not tested	Intercarpal and meta-carpal joints	
Chopra,[24] 2012 Maharashtra	Prospective	24/509	24	45		11.39% had ankle pain	
Mathew,[28] 2011 Kerala	Retrospective	437/1396	15	48	0.1%	83.3% had knee involvement	Rheumatoid arthritis ($n = 6$) and spondyloarthritis ($n = 11$) were rare
Chopra and Vanugopalan,[27] 2011 Maharashtra	Retrospective	172/212	12			KJ: 78.9%, SJ: 44.5%, AJ: 36%, EJ and hand joints: 31%	Inflammatory arthritis in 15%, none fulfiled ACR EULAR criteria
Ganu,[23] 2011 Maharashtra	Prospective	16/625	18	50	2/16 RF +, 9/16 CCP +, and 6 both negative	Wrists, MCPs, PIPs, elbows, shoulders, knees, ankles and MTPs were most commonly affected; DIP joints were involved in 12.5%.	14 patients treated with methotrexate
Paul,[29] 2011 Kerala	Prospective	30/134	18	?		AJ 63%, KJ 55%, PIP 52%, MCP 50%, DIP 49%; tenosynovitis in 5%	All were given HCQs, and only 4% required the medication for more than 9 months; Low dose steroids were used in 9%
Manimunda,[22] 2010 Karnataka	Prospective	94/203	10	35	20 patients tested, only 1 was CCP +	27% had knee joint involvement	34/94 fulfiled ACR/EULAR 2010 criteria for rheumatoid arthritis

a n (chronic rheumatism) = number of patients with chronic rheumatism; *n* (total) = total sample size

Abbreviations used: ACR: American College of Rheumatology; AJ: Ankle joint; CCP: anti-cyclic citrullinated protein antibody; DIP: distal interphalangeal joint; EULAR: European League against Rheumatism; HCQ: Hydroxychloroquine; KJ: knee joint; MCP: metacarpophalangeal joint; PIP: proximal interphalangeal joint; RF: Rheumatoid factor

Increased levels of MCP-1, TNF-α and IFN-γ are associated with joint symptoms.[34] During chronic stages of the disease persistent arthralgia is associated with CXCL10 and CXCL9 and with more severe disease at 6 months post-infection. Cytokines like IFN-α, IL-6, IL-8, etc. have been demonstrated within the synovium in chronic CHIKV arthralgia. Viral persistence has been associated with expression of IFN alpha, IL-10, monocyte chemotactic protein 1 (MCP-1 or CCL2), and proinflammatory cytokines.[35]

CHIK virus may also cause bone erosion mimicking RA probably mediated by MCP-1, RANTES and CCL2.[34]

Specific polymorphisms in human leukocyte antigen (HLA) have also been observed in alphavirus-induced arthritis. Rheumatoid arthritis associated alleles HLA-DRB1*01 and HLA-DRB1*04 have been variably associated with Chikungunya virus and Sindbis virus arthritis patients. Some of these patients later developed auto-antibodies like anti cyclic cirtrullinated antibodies and even antinuclear antibodies.[33]

Treatment of Chronic Chikungunya Arthritis

CHIKV rheumatic symptoms is managed with non-steroidal anti-inflammatory drugs (NSAIDs) or short tapering course of oral glucocorticoids in moderate to low doses. In those patients who develop chronic arthritis, chloroquine (CQ), Hydroxychloroquine (HCQ) and disease-modifying anti-rheumatic drugs (DMARDs) have tried.

Acute Rheumatic Symptoms (Arthritis)

Ravichandran and Manian used ribavirin 200 mg twice daily for 7 days in crippling post-CHIKV lower limb arthritis in 10 patients. Faster resolution of joint symptoms was observed in those treated with ribavirin and 70% could discontinue NSAIDs.[36]

Chloroquine has been tried in acute and sub-acute CHIKV arthritis. The basic assumption is that CHIKV enters cells through endosomal vesicles and CQ and HCQ prevents effective endosomal vesicle acidification and therefore has a theoretical chance to reduce proper viral processing.[37] This was recently shown in a Vero-cell based model of CHIKV infection.[37]

However, in a double-blind placebo-controlled randomized trial from the Reunion Islands of 54 patients with acute Chikugunya infection, CQ (600 mg at day 1, 600 mg at days 2 and 3, and 300 mg at days 4 and 5) had no effect on prevention of arthritis at 6 months.[38] But the dosing of CQ and short duration of the therapeutic intervention may have confounded the results.

Some other trials raised significant question on usefulness of CQ therapy in early CHIKV infection. For example, a parallel group randomised controlled trial from Pune, India compared the effectiveness of CQ 250 mg/day versus Meloxicam 7.5 mg/day. There were no significant efficacy differences between the meloxicam group and the CQ group but in both groups improved significantly.[27]

Chronic Arthritis

An open label pilot study on the efficacy of chloroquine phosphate in 10 patients with chronic CHIKV arthritis over a period of 20 weeks of therapy, reported significant improvement in Ritchie articular index and morning stiffness with 70% improvement of patient global assessment improvement.[39]

Ganu et al. studied the efficacy of sulfasalazine (SSZ) and methotrexate (MTX) in16 Chikungunya IgM positive patients having arthritis lasting more than 3 months in spite of NSAIDs and HCQ. They reported good response in 71% and 12% in SSZ+MTX versus SSZ alone. However, most of their patients were anti CCP antibody positive (9/16) and none of their synovial biopsy samples had detectable CHIKV RNA.[23]

The usefulness of combination DMARDs in chronic persistent post CHIKV polyarthritis was demonstrated in a recent randomised controlled trial reported by Ravindran et al. from Calicut India.[40] They recruited 139 patients with persistent arthritis for >1 year after the initial CHIKV infection. Those patients who were already on HCQ and had active arthritis were randomized to receive either fixed-dose combination therapy (MTX, SSZ, and HCQ) or to continue with HCQ only. Both groups received oral prednisolone up to 6 weeks. Assessments at every 4 weeks were carried out for primary efficacy (disease activity score; DAS ESR 28) and secondary efficacies, HAQ-Indian version and pain VAS 100 mm. Seventy-two patients were finally randomized (37 combination therapy and 35 monotherapy). This study provide evidence of superiority of combination DMARD therapy over HCQ monotherapy.[40]

PARVOVIRAL ARTHRITIC SYNDROMES

Parvovirus B19 is a member of *Erythrovirus* genus and Parvoviridae family. It replicates in the red blood cell (RBC) precursors after binding to its receptor, the blood group P antigen, which is also expressed by other cells such as megakaryocytes, endothelial cells, hepatocytes, smooth muscle cells, placental trophoblasts and cardiac cells. A co-receptor is generally required for infectivity such as $\alpha5\beta1$ integrin. In addition, B19 can bind specifically to Ku80 auto-antigen expressed on various immune cell types including B and T-cells and also on bone marrow stromal cells. Binding to Ku80 in co-operation with the blood group antigen P and the $\alpha5\beta1$ integrin as a co-receptor helps in efficient cellular entry.[41, 42]

Parvovirus B19 infections occur in outbreaks and are transmitted by aerosols, although parenteral routes ie blood or blood product contamination. Vertical transmission and solid organ or haemato-poietic stem cell transplantation has also been documented. The incubation period is 1 to 2 weeks (up to 4 weeks). During outbreaks, the household transmission rate of parvovirus B19 can be as high as 50%.[42] The clinical outcomes of B19 infection depend on age of the patient as well as the immune status.

In children, infection with B19 causes erythema infectiosum or fifth disease. The disease begins with mild prodrome consisting of fever, malaise headache, and myalgia. This is followed later by appearance of a exanthematous rash on the cheeks which is a "slapped cheek" appearance. A diffuse maculopapular rash appear later which may be pruritic gradually spreading to the extremities.[1, 2, 42]

In adults, infection with Parvovirus B19 is mostly asymptomatic (25 to 70% cases). Symptomatic cases manifest flu-like symptoms. The natural history of B19 infection varies from an early viremic phase to antibody response phase. The viremic phase is generally asymptomatic and patients actively shed the virus in nasal and oropharyngeal secretions. Profound articulocytosis coincides with peak viremia and may be prolonged. In the second week cytopenias can develop, even pancytopenia.[42] Infection during early pregnancy results in the syndrome of fetal hydrops.[43]

In the immunocompromised individual ongoing replication of the virus occurs inside erythroblasts, leading to pure red cell aplasia. Transient aplastic crisis can occur in patients with hemolytic anaemia especially sickle cell anaemia and also in beta-thalassemia, hereditary spherocytosis, and malaria, etc.[42,43]

A number of auto-antibodies such as RF, ANA, anti-DNA anti-ENA, anti-phospholipid antibodies, and others may be transiently positive.[1, 2]

Joint symptoms develop 2 weeks after clinical onset, coinciding with development of IgM antibody. Acute B19-related joint involvement is more common in adults than children. Arthralgia and arthritis can be seen in up to 80% adults. In children, the clinical course is typically "slapped-cheek" appearance followed by asymmetric mono to oligoarthritis mostly involving the knee joints.

In adults, this rash is seen in <50% and may be very subtle in those who actually develop it and lasts for a shorter period (<4 days). The pattern of joint involvement in adults may resemble RA with polyarthritis involving PIP, MCP joints, followed by the knees, wrists, and ankles associated with early morning stiffness.[1,2,42] Added to this clinical presentation the transient presence of RF and auto-antibodies may confuse the clinician. The joints symptoms usually resolve within 2 weeks except for a small subset of patients. In cohorts of acute arthritis routine serology and PCR for B19 is rarely positive, simply reflecting the rarity of B19 as a cause of acute onset arthritis.[42]

Parvovirus B19 has been implicated as a trigger in RA, systemic lupus erythematosus, systemic juvenile idiopathic arthritis, and adult onset Still's disease, remitting seronegative symmetrical synovitis with pitting oedema (RS3PE) syndrome, and papular-purpuric 'gloves and Socks syndrome. It has also been associated with fibromyalgia and chronic fatigue syndrome but definitive evidence is lacking.[1, 2, 42] Some studies linked HLA-DR4 with severity of B19 arthritis.[44]

The laboratory detection of parvovirus B19 infection depends on identification of specific IgM anti-B19 antibodies reactive against viral capsid protein, most importantly VP2 protein or NS1 with

23. Ganu MA, Ganu AS. Post-chikungunya chronic arthritis: our experience with DMARDs over two year follow up. J Assoc Physicians India 2011; 59:83–6.
24. Chopra A, Anuradha V, Ghorpade R, Saluja M. Acute Chikungunya and persistent musculoskeletal pain following the 2006 Indian epidemic: a 2-year prospective rural community study. Epidemiol Infect. 2012; 140:842–50.
25. Gauri LA, Thaned A, Fatima Q, Yadav H, Singh A, Jaipal HP, Chaudhary A. Clinical Spectrum of Chikungunya in Bikaner (North Western India) in 2006 and Follow up of Patients for Five Years. J Assoc Physicians India. 2016; 64:22–25.
26. Chaaithanya IK, Muruganandam N, Raghuraj U, Sugunan AP, Rajesh R, Anwesh M, et al. Chronic inflammatory arthritis with persisting bony erosions in patients following chikungunya infection. Indian J Med Res 2014; 140:142–5.
27. Chopra A, Venugopalan A. Persistent rheumatic musculoskeletal pain and disorders at one year post-chikungunya epidemic in south Maharashtra: a rural community based observational study with special focus on native persistent rheumatic musculoskeletal cases and selected cytokine expression. Indian J Rheumatol 2011; 6 Suppl:5–11.
28. Mathew AJ, Goyal V, George E, Thekkemuriyil DV, Jayakumar B, Chopra A. Rheumatic-musculoskeletal pain and disorders in a naive group of individuals 15 months following a Chikungunya viral epidemic in south India: a population based observational study. Int J Clin Pract 2011; 65:1306–12.
29. Paul BJ, PMoni SP, Pannarkady G, Thachil EJ. Clinical profile and long-term sequelae of Chikungunya fever. Indian J Rheumatol 2011; 6:12–19.
30. Rodríguez-Morales AJ, Cardona-Ospina JA, Fernanda Urbano-Garzón S, Sebastian Hurtado-Zapata J. Prevalence of Post-Chikungunya Infection Chronic Inflammatory Arthritis: A Systematic Review and Meta-Analysis. Arthritis Care Res (Hoboken). 2016; 68:1849–58.
31. Westley-Wise VJ, Beard JR, Sladden TJ, Dunn TM, Simpson J. Ross River virus infection on the North Coast of New South Wales. Aust N Z J Public Health. 1996; 20:87–92.
32. Gérardin P, Fianu A, Michault A, Mussard C, Boussaïd K, Rollot O, Grivard P, Kassab S, Bouquillard E, Borgherini G, Gaüzère BA, Malvy D, Bréart G, Favier F. Predictors of Chikungunya rheumatism: a prognostic survey ancillary to the TELECHIK cohort study. Arthritis Res Ther. 2013; 15:R9.
33. Assunção-Miranda I, Cruz-Oliveira C, Da Poian AT. Molecular mechanisms involved in the pathogenesis of alphavirus-induced arthritis. Biomed Res Int. 2013; 2013:973516.
34. Rulli NE, Rolph MS, Srikiatkhachorn A, Anantapreecha S, Guglielmotti A, Mahalingam S. Protection from arthritis and myositis in a mouse model of acute chikungunya virus disease by bindarit, an inhibitor of monocyte chemotactic protein-1 synthesis. J Infect Dis 2011; 204:1026–30.
35. Kelvin AA, Banner D, Silvi G, Moro ML, Spataro N, Gaibani P, et al. Inflammatory cytokine expression is associated with chikungunya virus resolution and symptom severity. PLoS Negl Trop Dis 2011; 5(8): e1279.
36. Ravichandran R, Manian M. Ribavirin therapy for Chikungunya arthritis. J Infect Dev Ctries. 2008; 2: 140–2.
37. Khan M, Santhosh SR, Tiwari M, Lakshmana Rao PV, Parida M. Assessment of in vitro prophylactic and therapeutic efficacy of chloroquine against Chikungunya virus in vero cells. J Med Virol. 2010; 82:817–24.
38. De Lamballerie X, Boisson V, Reynier JC, Enault S, Charrel RN, Flahault A, et al. On chikungunya acute infection and chloroquine treatment. Vector Borne Zoonotic Dis. 2008; 8:837–9.
39. Brighton SW. Chloroquine phosphate treatment of chronic Chikungunya arthritis. An open pilot study. S Afr Med J. 1984; 66:217–8.
40. Ravindran V, Alias G. Efficacy of combination DMARD therapy vs. hydroxychloroquine monotherapy in chronic persistent chikungunya arthritis: a 24-week randomized controlled open label study. Clin Rheumatol 2017; 36:1335–40.
41. Munakata Y, Saito-Ito T, Kumura-Ishii K, et al. Ku80 autoantigen as a cellular coreceptor for human parvovirus B19 infection. Blood 2005; 106:3449–56.
42. Colmegna I, Alberts-Grill N. Parvovirus B19: its role in chronic arthritis. Rheum Dis Clin North Am. 2009; 35:95–110.
43. Florea AV, Ionescu DN, Melhem MF. Parvovirus B19 infection in the immunocompromised host. Arch Pathol Lab Med 2007; 131:799–804.
44. Kerr JR, Mattey DL, Thomson W, et al. Association of symptomatic acute human parvovirus B19 infection with human leukocyte antigen class I and II alleles. J Infect Dis 2002; 186:447–52.
45. Peterlana D, Puccetti A, Corrocher R, Lunardi C. Serologic and molecular detection of human Parvovirus B19 infection. Clin Chim Acta 2006; 372:14–23.
46. Vassilopoulos D, Calabrese LH. Management of rheumatic disease with comorbid HBV or HCV infection. Nat Rev Rheumatol 2012; 8:348–57.
47. Cacoub P, Commarmond C, Sadoun D, Desbois AC. Hepatitis C Virus Infection and Rheumatic Diseases: The Impact of Direct-Acting Antiviral Agents. Rheum Dis Clin North Am. 2017; 43:123–32.

FURTHER READING

1. Calabrese LH, Naides SJ. Viral arthritis. Infect Dis Clin North Am 2005; 19:963–80.

MISCELLANEOUS

Chapter

72

Relapsing Polychondritis

Sundeep Grover

INTRODUCTION

Relapsing polychondritis (RP) is a rare systemic connective tissue disease characterized by recurrent bouts of inflammation, followed by degeneration and deformation of the cartilage of the ears, nose, larynx and the tracheobronchial tree.[1–6] These cases also have other systemic manifestations like arthritis, ocular inflammation, audiovestibular involvement, skin lesions, heart valve incompetence and vasculitis.

EPIDEMIOLOGY

RP chiefly affects middle aged adults, with a slight female preponderance. Cases have been reported in the very young and elderly too.[7] In the US, the estimated annual incidence of RP is $3.5/10^6$ population.[3]

Genetic studies have identified a relationship between RP and human leukocyte antigen (HLA)-DR4. The frequency of HLA-DR4 in patients with RP is up to 56% in patient group and 26% in the healthy controls.[8]

PATHOPHYSIOLOGY

There is probably an autoimmune response against certain unidentified cartilage antigens with resultant cartilage matrix destruction by proteases including proteases released by chondrocytes undergoing apoptosis.[4, 9–11]

The triggering factor for cartilage destruction can be chemical, toxic, and infectious agents or direct trauma. RP cases have been reported to occur after trauma to the pinna[12] and intravenous administration of an indefinite substance that may have a direct toxic effect on the cartilage.[13] This suggests that there may be a direct association with trauma and the triggering of autoimmune phenomena.

Evidence which supports a pathogenic role for the humoral and cellular immune systems in occurrence of RP are as follows:[14]

1. About 30% of patients with RP have another autoimmune disease as shown in Table 72.1.
2. The damaged cartilage is infiltrated by CD4+ T cells and plasma cells and contains immune deposits.
3. Autoantibodies against collagen type II are detectable in over 30% of patients of RP.
4. A T cell response specific of peptides found in collagen type II (which constitutes 95% of all cartilage collagen) is found in some patients.
5. Over half the patients with RP carry the HLA-DR4 antigen.
6. High dose glucocorticoid therapy is effective in most patients of RP.
7. Symptoms resembling those of RP can be induced in animal models by injecting type II collagen.

It can be hypothesized that a damage to the cartilage containing chondrocyte epitopes leads to cytokine release and local inflammation following autoantibody production in an inherently susceptible host.[15]

Table 72.1: Systemic diseases associated with relapsing polychondritis

Disease group	*Subtypes*
Connective tissue diseases	Systemic lupus erythematosus, Sjögren's syndrome, systemic sclerosis, rheumatoid arthritis, juvenile idiopathic arthritis
Spondyloarthropathies	Ankylosing spondylitis, reactive arthritis, psoriatic arthritis
Systemic vasculitides	ANCA-associated vasculitis, Takayasu arteritis, Behçet's disease, polyarteritis nodosa, cryoglobulinemia
Hematologic diseases	Myelodysplastic syndrome, lymphoma
Gastrointestinal diseases	Inflammatory bowel disease, primary biliary cirrhosis
Dermatologic diseases	Sweet's syndrome, dermatitis herpetiformis, lichen planus, psoriasis
Other inflammatory diseases	Retroperitoneal fibrosis, meningoencephalitis, familial Mediterranean fever, sarcoidosis, autoimmune thyroid disease

CLINICAL FEATURES

Inflammation of the cartilage (chondritis) is the hallmark clinical feature of RP. The initial manifestation is usually articular in the form of arthralgias. Frank chondritis is usually absent in more than half of the cases in the initial stages of the disease resulting in a diagnostic delay of months to sometime years. The development and subsequent correct identification of clinical chondritis provides the diagnosis in patients evaluated previously for unexplained articular, audio vestibular or cutaneous manifestations; sometimes patients presenting with unexplained prolonged fever, weight loss and asthenia.

Auricular chondritis is specific for RP, if infection and trauma are ruled out. Auricular inflammation is seen in 20% of patients at presentation and in 90% of cases at some point over entire course of the disease. Pinna is the most common site of involvement in patient with chondritis. One or both ears may be affected. A swollen, red, painful and tender pinna sparing the ear lobe is characteristic (Fig. 72.1). There may be recurrent self-limiting episodes causing the cartilage to collapse and leading to cauliflower ear deformity seen in about 10% patients.

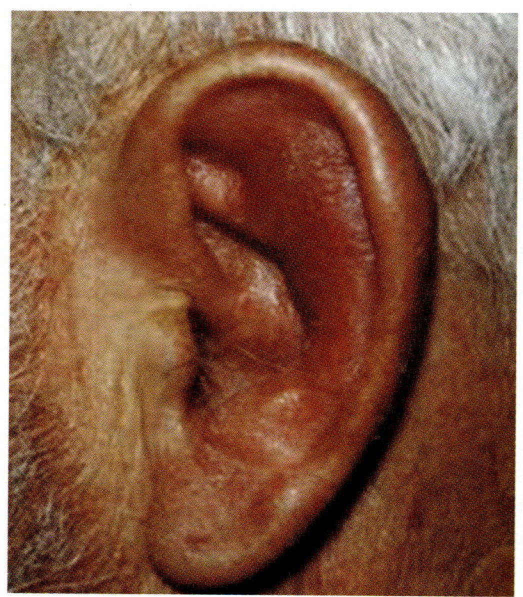

Fig. 72.1: Pinna involvement in relapsing polychondritis

Nasal chondritis is seen in 15% patients at presentation and 65% develop it at some point of time over the disease course. It involves the bridge of the nose and secondary atrophy of the same leads to a saddle nose deformity.

Chondritis of the larynx and tracheobronchial tree is seen in 10% patients at presentation and in 50% over the course of the disease. It is seen more commonly in females and is an important cause of mortality in RP patients. Laryngeal involvement presents as pain, dysphonia or transient aphonia. Recurrent inflammation leads to laryngomalacia or laryngeal stenosis sometimes needing an emergency tracheostomy as a life-saving measure. Tracheobronchial involvement is potentially the most severe manifestation of RP and often presents with expiratory dyspnoea, pain, cough and recurrent severe lower respiratory tract infections.[5,16,17]

Ocular inflammation in RP can affect any part of the eye and may affect 20 to 60% of the RP cases. The most common ocular manifestations are episcleritis (Fig. 72.2), peripheral ulcerative keratitis, scleritis and uveitis.[18,19]

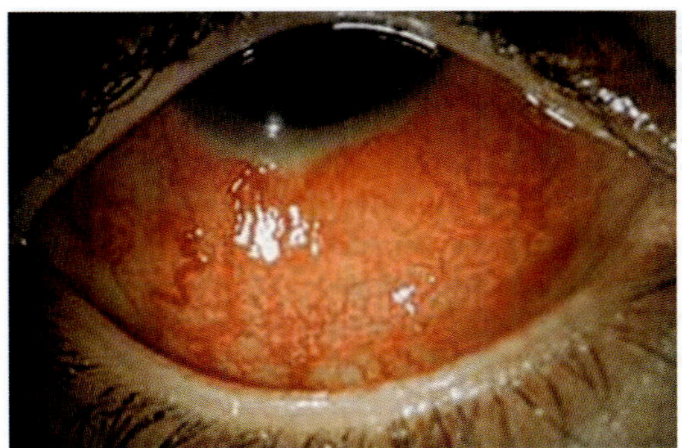

Fig. 72.2: Episcleritis in a patient with relapsing polychondritis

Arthritis is present in 33% of the patients with RP at presentation and is eventually observed in 50–75% of the patients with RP.[20] It characteristically affects the manubriosternal, sternoclavicular, and costochondral joints. Any joint may be involved, but the frequently affected peripheral joints are the metacarpophalangeal, proximal interphalangeal, knees, and wrist joints. Usually, arthritis is either polyarthritis or oligoarthritis with or without synovitis and may last from weeks to months. It has an episodic, self-remitting, asymmetric, nonerosive, and nondeforming course.[21]

Approximately 10% of the patients with RP have clinically remarkable valvular disease mostly affecting aortic or mitral valves. Additionally, cardiac involvement may be associated with pericarditis, heart blocks, and myocardial infarction. Valvular involvement can occur any time during the disease course.[22] Progression of the disease is often insidious; as a result, echocardiography should be periodically performed to determine valvular dysfunction.

Renal disease arises in a minority of RP cases. Kidney pathologies reported to occur in RP cases include immunoglobulin A (IgA) nephropathy, tubulointerstitial nephritis, and glomerulonephritis.[5]

In addition to audiovestibular manifestations, neurological involvement in RP appears in approximately 3% of the cases. The most common neurologic features are cranial neuropathies of the second, sixth, seventh, and eighth nerves. However, miscellaneous neurological conditions can occur such as hemiplegia, seizures, organic brain syndrome, dementia, and cerebral dysfunction.[23]

Several nonspecific skin lesions may develop in patients with RP. About 36% of patients have skin lesions such as oral ulcers, papules, purpura, and nodules. Skin biopsy studies show nonspecific findings, including cutaneous leucocytoclastic vasculitis, small vessel thrombosis, and panniculitis.[24]

RP presents in combination with myelodysplastic syndrome (MDS) in a significant number of cases and can be considered a paraneoplastic manifestation of MDS. Older patients with newly diagnosed RP should undergo a hematologic evaluation to look for MDS.[25]

Associated Diseases

As many as one-third of the patients with RP have a concomitant disease including systemic vasculitis, dermatologic or hematologic disease, or other systemic rheumatic disease (Table 72.1).[26] These diseases may precede RP, occur after the diagnosis of RP or present simultaneously with RP.

DIAGNOSIS

Relapsing polychondritis has no pathognomonic clinical, radiological, and histopathological features. The diagnosis is established by a combination of clinical manifestations, supportive laboratory and radiological tests and if possible, a biopsy from an involved cartilaginous site.

The McAdam's criteria consist of a set of 6 clinical features, diagnosis of RP requires the presence of three or more of the clinical findings summarized in Table 72.2.[1] These criteria were modified by Damiani et al. who suggested a new set of criteria which include histological features and therapeutic responses (Table 72.2).[27]

There is no specific laboratory test for RP, but measurement of acute phase reactants ESR, and/or C-Reactive protein (CRP) can be helpful similar to that in other autoimmune diseases. Approximately 10% of the patients may have peripheral eosinophilia that suggests vasculitis as a potential differential diagnosis.

Particularly when there is an overlap with other rheumatologic disorders, patients with RP may have a positive serology such as rheumatoid factor, antineutrophil cytoplasmic antibody (ANCA), antinuclear antibody (ANA), and false-positive venereal disease research laboratory (VDRL) tests.[5,26]

3. Luthra HS. Relapsing Polychondritis. In: Klippel JH, Dieppe PA, Editors. Rheumatology. St Louis: Mosby; 1998. p.1–4.
4. Trentham DE, Le CH. Relapsing polychondritis. Ann Intern Med 1998; 129:114–22.
5. Lahmer T, Treiber M, von Werder A, Foerger F, Knopf A, Heemann U, et al. Relapsing polychondritis: an autoimmune disease with many faces. Autoimmun Rev 2010; 9:540–6.
6. Mathew SD, Battafarano DF, Morris MJ. Relapsing polychondritis in the Department of Defence population and review of the literature. Semin Arthritis Rheum 2012; 42:70–83.
7. Belot A, Duquesne A, Job-Deslandre C, Costedoat-Chalumeau N, Boudjemaa S, Wechsler B, et al. Pediatric-onset relapsing polychondritis:case series and systematic review. J Pediatric 2010; 156:484–9.
8. Zeuner M, Straub RH, RauhG, Albert ED, Scholmerich J, Lang B. Relapsing polychondritis: clinical and immunogenetic analysis of 62 patients. J Rheumatol 1997; 24:96–101.
9. Ouchi N, Uzuki Kamataki A, Miura Y, Sawai T. Cartilage destruction is partly inducted by the internal proteolytic enzymes and apoptotic phenomenon of chondrocytes in relapsing polychondritis. J Rheumatol 2011; 38:730–7.
10. Arnaud L, Mathian A, Haroche J, Gorochov G, Amoura Z. Pathogenesis of relapsing polychondritis: a 2013 update. Autoimmun Rev 2014; 38:730–7.
11. Frisenda S, Perricone C, Valesini G. Cartilage as a target of autoimmunity; a thin layer. Autoimmun Rev 2014; 12;591–8.
12. Alissa H, Kadanoff R, Adams E. Does mechanical insult to cartilage trigger relapsing polychondritis? Scand J Rheumatol 2001; 30:311.
13. Berger R. Polychondritis resulting from intravenous substance abuse. Am J Med 1988; 85:415–7.
14. Puéchal X, Terrier B, Mouthon L, Costedoat-Chalumeau N, Guillevin L, Le Jeunne C. Relapsing polychondritis. Joint Bone Spine. 2014; 81:118–24.
15. Stabler T, Piette JC, Chevalier X, Marini-Portugal A, Kraus VB. Serum cytokine profiles in relapsing polychondritis suggest monocyte/macrophage activation. Arthritis Rheum 2004; 50:3663–7.
16. Letko E, Zafirakis P, Baltatzis S, Voudouri A, Livir-Rallatos C, Foster CS. Relapsing polychondritis:a clinical review. Semin Arthritis Rheum 2002; 31: 384–95.
17. Tillie-Leblond I, Wallaert B, Leblond D, Salez F, Perez T, Remy-Jardin M, et al. Respiratory involvement in relapsing polychondritis. Clinical, functional, endoscopic, and radiographic evaluations. Medicine (Baltimore) 1998; 77:168–76.
18. Isaak BL, Liesegang TJ, Michet CJ. Ocular and systemic findings in relapsing polychondritis. Ophthalmology 1986; 93:681–9.
19. Yu EN, Jurkunas U, Rubin PA, Baltatzis S, Foster CS. Obliterative microangiopathy presenting as chronic conjunctivitis in a patient with relapsing polychondritis. Cornea 2006; 25:621.
20. Arkin CR, Masi AT. Relapsing polychondritis: review of current status and case report. Semin Arthritis Rheum 1975; 5:41–62.
21. Sharma A, Gnanapandithan K, Sharma K, Sharma S. Relapsing polychondritis: a review. Clin Rheumatol. 2013; 32:1575–83.
22. Lang-Lazdunski L, Hvass U, Paillole C, Pansard Y, Langlois J. Cardiac valve replacement in relapsing polychondritis. A review. J Heart Valve Dis 1995; 4: 227–35.
23. Sundaram MB, Rajput AH. Nervous system complications of relapsing polychondritis. Neurology 1983; 33:513–5.
24. Frances C, el Rassi R, Laporte JL, Rybojad M, Papo T, Piette JC. Dermatologic manifestations of relapsing polychondritis. A study of 200 cases at a single center. Medicine (Baltimore) 2001; 80:173–9.
25. Chopra R, Chaudhary N, Kay J. Relapsing polychondritis. Rheum Dis Clin N Am 2013; 39: 263–276.
26. Emmungil H, Aydýn SZ. Relapsing polychondritis. Eur J Rheumatol 2015; 2(4):155–9.
27. Damiani JM, Levine HL. Relapsing polychondritis-report of ten cases. Laryngoscope 1979; 89:929–46.
28. Heman-Ackah YD, Remley KB, Goding GS Jr. A new role for magnetic resonance imaging in the diagnosis of laryngeal relapsing polychondritis. Head Neck 1999; 21:484–9.
29. Yamashita H, Takahashi H, Kubota K, Ueda Y, Ozaki T, Yorifuji H, et al. Utility of fluorodeoxyglucose positron emission tomography/computed tomography for early diagnosis and evaluation of disease active ty of relapsing polychondritis: a case series and literature review. Rheumatology(Oxford) 2014; 53:1482–90.
30. Cutolo M, Seriolo B, Pizzorni C, Secchi ME, Soldano S, Paolino S, et al. Use of glucocorticoids and risk of infections. Autoimmun Rev 2008; 8:153–5.
31. Stewart KA, Mazenic DJ. Pulse intravenous cyclophosphamide for kidney disease in relapsing polychondritis. J Rheumatol 1992; 19:498–500.
32. Kemta Lekpa F, Kraus VB, Chevalier X. Biologics in relapsing polychondritis: a literature review. Semin Arthritis Rheum 2012; 41:712–9.

33. Dooms C, De Keukeleire T, Janssens A, Carron K. Performance of fully covered self expanding metallic stents in benign airway strictures. Respiration 2009; 77:420–6.

34. Tsunezuka Y, Sato H, Shimizu H. Tracheobronchial involvement in relapsing polychondritis. Respiration 2000; 67:320–2.

35. Dion J, Costedoat-Chalumeau N, Sène D, Cohen-Bittan J, Leroux G, Dion C, et al. Relapsing Polychondritis Can Be Characterized by Three Different Clinical Phenotypes: Analysis of a Recent Series of 142 Patients. Arthritis Rheumatol. 2016; 68:2992–3001.

FURTHER READING

1. Hazra N, Dregan A, Charlton J, Gulliford MC, D'Cruz DP. Incidence and mortality of relapsing polychondritis in the UK: a population-based cohort study. Rheumatology (Oxford). 2015; 54:2181–7.
2. Kent PD, Michet CJ Jr, Luthra HS. Relapsing polychondritis. Curr Opin Rheumatol. 2004; 16:56–61.
3. Mathian A, Miyara M, Cohen-Aubart F, Haroche J, Hie M, Pha M, et al. Relapsing polychondritis: A 2016 update on clinical features, diagnostic tools, treatment and biological drug use. Best Pract Res Clin Rheumatol. 2016; 30:316–33.

Chapter

73

Adult Onset Still's Disease

Durga Prasanna Misra, Vir Singh Negi

INTRODUCTION

Adult onset Still's disease (AOSD) refers to a symptom complex of high spiking fever, evanescent rash and arthritis with markedly elevated acute phase response. First described by Bywaters in 1971, this uncommon disease is remarkable for its severity that remains a diagnostic and therapeutic challenge for rheumatologists.[1] It shares many features in common with the systemic onset variant of juvenile idiopathic arthritis (sJIA) which is seen in the pediatric age group.

AETIOPATHOGENESIS

The presentation of AOSD with fever and rash mimics that of many infectious diseases. As of yet unidentified infectious triggers have been proposed to induce attacks of AOSD. Proposed triggers include viruses, such as measles virus, rubella virus, cytomegalovirus, Ebstein-Barr virus and bacteria such as *Borrelia burgdorferi* and *Mycoplasma pneumoniae.*[1] However, no definite agent has been implicated as of yet as a trigger of AOSD. Various genes, both HLA and non-HLA (summarized in Table 73.1), have been implicated to predispose towards AOSD.[1,2]

Table 73.1: Genetics of AOSD[1,2]

Gene	*Influence on AOSD*
HLA Bw35	Lesser severity of AOSD
HLA DR4, B17, B18, B35,	Susceptibility towards AOSD
HLA DRw6	Predisposition towards shoulder/hip joint involvement
HLA DRB1*1501, DRB1*1201	Predispose towards chronic polyarthritis
HLA DQB1*1602	Predisposes towards both systemic and chronic polyarthritis phenotypes
HLA DRB1*14	Associates with monocyclic systemic variant of AOSD
HLA DRB1*04	May be protective against AOSD
Polymorphisms in IL-18, MIF	Predispose to AOSD

AOSD: Adult onset still's disease; HLA: Human leucocyte antigen; IL: Interleukin; MIF: Macrophage migratory inhibitory factor

Numerous abnormalities in the immune system characterize AOSD.[1,2] Innate immune cells such as macrophages and neutrophils in the peripheral blood of patients with AOSD display an activated phenotype. Higher levels of the chemokine CXCL8, macrophage colony stimulating factor (M-CSF), macrophage migratory inhibiting factor (MIF), interferon gamma (IFN-γ) and intracellular adhesion molecule 1 (ICAM-1) are found in these patients as a consequence of activated macrophages and neutrophils. The natural killer (NK) cells, one of the functions of which is to exert a suppressive action on macrophages, are defective. This contributes towards uncontrolled macrophage activation, which is one of the mechanisms responsible for predisposition towards macrophage activation syndrome (MAS) in these individuals. Abnormalities in the TLR7-induced activation of Myd88 is implicated as one of the mechanisms driving neutrophil activation in AOSD. Activation of the inflammasome in dendritic cells drives secretion of pro-inflammatory IL-1β and IL-18. AOSD is also associated with increased circulating TNF-α and IL-6. The activation of adaptive immune system is evidenced by Th-1 cell (secreting IFN-γ and soluble IL-2 receptor, which activate macrophages) and Th-17 cell (secreting IL-17, which recruits neutrophils) activation in AOSD.

Is AOSD an Autoinflammatory Syndrome?

Considering the shared phenotype of rash, serositis and arthritis with absence of autoantibodies or autoreactive T lymphocytes and responsiveness to IL-1β-targeted therapy, an argument has emerged that AOSD and its childhood counterpart sJIA are in fact autoinflammatory (periodic fever) syndromes. However, the absence of a single predisposing gene, presence of a characteristic pattern of fever and tendency to cause destructive arthritis are points against this argument.[1]

CLINICAL FEATURES

AOSD is a rare disease, with estimated incidence of 1.6 to 4 per million per year and prevalence ranging from 1 to 34 per million population.[1] It is seen most commonly in the 4th to 5th decade with no particular gender predilection. The presentation may be predominantly with systemic features such as fever and weight loss (systemic form—60%), which in turn may be monocyclic with a single episode which never recurs, or a polycyclic course with recurrent flares of disease (each approximately 30%). The remaining 40% have a chronic disease with polyarthritis as the predominant feature, with or without systemic features.[1]

Fever is the most characteristic feature of AOSD, almost universally seen. It has a typical pattern of high-spiking temperature which resolves during the course of the day and may reach a subnormal body temperature in between spikes. The patient often feels quite unwell during the spikes of fever, and completely normal when there is no fever. Another characteristic feature is the transient evanescent maculopapular eruption over the trunk and limbs, which may be difficult to identify in patients with a darker tone of skin colour. Other rarely described cutaneous features include persistent plaque-like lesions, urticarial lesions or bullous lesions.[3] However, it is imperative to note that **the presence of a fixed rash should raise serious doubts about the diagnosis of AOSD in the mind of the clinician** and prompt a thorough search for plausible alternative diagnoses, as discussed below.

Arthritis, often involving large joints such as hips, shoulders and knees, also affecting the wrists and ankles, is a major cause of disability in AOSD. To start with, it may be oligoarticular. In established AOSD, the arthritis usually follows a symmetric polyarticular pattern. Synovial fluid aspirate from affected joints may appear thick and purulent due to intense neutrophilic infiltrate, mimicking septic arthritis. Unless promptly treated, the arthritis is erosive and leaves behind sequelae such as ankylosis (most commonly in the wrist) and deformity. Involvement of the axial skeleton is commonly seen in the neck, which may result in ankylosis or atlanto-axial subluxation. Serositis in the form of exudative pleural and/or pericardial effusions may be seen in up to half the patients, which may leave behind sequelae such as constrictive pericarditis and pleural thickening. Hepatomegaly, splenomegaly and lymphadenopathy are often encountered as a part of reticulonodular hyperplasia. Patients often complain of sore throat, which is attributable to a non-infective purulent pharyngitis as part of the spectrum of AOSD.

Laboratory abnormalities include striking elevations of the erythrocyte sedimentation rate (ESR) and serum C-reactive protein (CRP). A peripheral blood picture of neutrophilic leucocytosis is almost universal, often accompanied by thrombocytosis and anaemia of chronic disease. Very high levels of serum ferritin (often greater than 1000 ng/mL) are seen in AOSD, akin to that seen in MAS (which can complicate AOSD), catastrophic antiphospholipid syndrome (CAPS) and septic shock, all of which are now considered to constitute the hyper-ferritinemic syndrome.[4] Glycosylated ferritin, which normally constitutes more than 50% of the total ferritin, is reduced to less than 20% in AOSD. Serum transaminases may be deranged in the context of active disease. Plain radiographs of the chest may show pleural or pericardial effusions; fleeting parenchymal lung shadows may also be evident. Biopsy of affected lymph nodes, often done to rule out differential diagnosis of hematologic malignancy, may show reactive hyperplasia.

Autoantibodies such as antinuclear antibody (ANA) and rheumatoid factor (RF) are undetectable in AOSD. Levels of serum procalcitonin have been reported to be elevated in AOSD and thus may not be useful to distinguish AOSD from sepsis.[5] Positron emission tomography-computerized tomography (PET-CT), often done in these patients as part of the work-up of pyrexia of unknown origin (PUO), shows uptake in long bones as well as enlarged lymph nodes, spleen and affected joints.[6]

Literature on AOSD from Indian centers is scarce, with only small case series available in the published literature.[7] In general, patients from India have lower prevalence of skin rash (possibly due to difficulty in identifying the rash due to dark colour of skin) and sore throat, and up to a quarter of patients with arthritis may have joint destruction.

DIAGNOSIS

It cannot be over-emphasized that **AOSD is a diagnosis of exclusion.** Similar presentation may be due to infections (such as tuberculosis, HIV, brucellosis or infective endocarditis), malignancies (like leukemias, lymphomas, renal cell carcinoma), connective tissue diseases (SLE, inflammatory myositis) and inflammatory bowel diseases. These must be excluded first by a thorough history and clinical examination, followed by investigations, including multiple blood and urine cultures, echocardiography (including transesophageal echocardiography if routine transthoracic echocardiography is normal), serology for HIV, bone marrow (including culture for tuberculosis and fungal organisms), chest X-ray, ultrasound abdomen and pelvis and contrast-enhanced computerized tomography of chest and abdomen (to identify focus of occult infection, including tuberculosis). Usually, patients with AOSD are negative for ANA and RF. Once plausible alternative explanations for the symptom complex have been excluded, the clinician can be satisfied with a diagnosis of AOSD (Fig. 73.1).

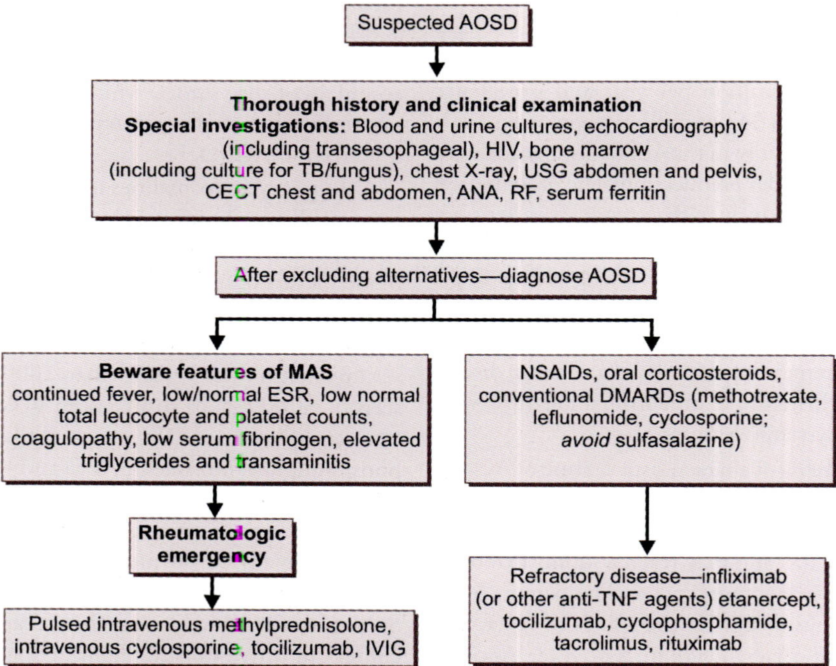

Fig. 73.1: Proposed management algorithm for Indian patients with AOSD

ANA: Antinuclear antibody; AOSD: Adult onset still's disease; CECT: Contrast-enhanced computerized tomography; DMARD: Disease-modifying anti-rheumatic drug; ESR: Erythrocyte sedimentation rate; HIV: Human immunodeficiency virus; IVIG: Intravenous immunoglobulin; MAS: Macrophage activation syndrome; NSAID: Non-steroidal anti-inflammatory drugs; RF: Rheumatoid factor; TB: Tuberculosis; TNF: Tumour necrosis factor α; USG: Ultrasound

Various classification criteria of differing sensitivity and specificity have been proposed for AOSD.[8] These are represented in Table 73.2. It is worthwhile noting that the single test of glycosylated ferritin less than 20% is highly specific (93%) for AOSD, but is limited by its low sensitivity (43%) and the lack of widespread availability of this test in an Indian scenario. Also, the levels of glycosylated ferritin do not normalize for months even after attaining remission of AOSD.[7,8]

COMPLICATIONS OF AOSD

AOSD can be associated with life-threatening complications.[9] The most important of these is the macrophage activation syndrome (MAS), affecting 12–14% patients, which may occur *per se* in the context of active AOSD, or triggered by infections (including *Salmonella typhi*) or drugs such as sulfasalazine or anti-IL-1β therapy. MAS is characterized by a change in the nature of fever from intermittent to a high grade persistent fever. This may be accompanied by bleeding tendencies, including mucosal, skin and internal organ bleeds, resulting from both the characteristic coagulopathy along with thrombocytopenia. Other lineages are also affected, with leucopenia (neutropenia) and anaemia being common accompaniments. Elevation of transaminases, serum triglycerides and hypofibrinogenemia (resulting in normal or low ESR) are characteristic. Serum ferritin is usually markedly elevated, often greater than 10000 ng/mL. Levels of soluble IL-2 receptor α (CD25) and soluble CD 163 are increased in the peripheral blood, reflecting uncontrolled activation of immune cells. Although demonstration of hemophagocytosis on bone marrow is diagnostic of MAS, one should not wait for this before instituting therapy for MAS when the clinical situation is highly suggestive. Change in pattern of fever of a patient with AOSD, or active disease in the context of normal ESR and lack of elevation of platelet count or total leucocyte count should lead to a suspicion of MAS, which is one of the very few rheumatologic emergencies. Early institution of high-intensity immunosuppressive therapy (with daily pulses of methylprednisolone, often accompanied with intravenous cyclosporine) is imperative to arrest the cytokine storm that is characteristic of MAS. Anti-interleukin 1β therapy and anti-interleukin 6 therapy have also been anecdotally reported to be effective in MAS in the context of AOSD.

Other rare but severe complications associated with AOSD include disseminated intravascular

Table 73.2: Classification criteria for AOSD[7]

	Major criteria	*Minor criteria*	*Requirement*	*Sensitivity/Specificity*
Yamaguchi	• Intermittent fever ≥39°C for at least one week • Typical rash • Arthralgia of at least 2 weeks duration • Leucocytosis (>10,000/mm³, with >80% neutrophils)	• Sore throat • Lymphadenopathy/ splenomegaly • Abnormal liver functions • Negative test for ANA and RF	5 criteria, at least 2 of which are major criteria	96.2%/92.1%
Fautrell	• Intermittent fever ≥39°C • Transient rash • Arthralgia • Pharyngitis (>80% neutrophils in peripheral smear) • Glycosylated ferritin ≤20%	• Maculopapular rash • Leucocytosis (≥10,000/mm³)	4 major criteria *OR* 3 major criteria with both minor criteria	80.6%/98.5%

ANA: Antinuclear antibody; RF: Rheumatoid factor

coagulation (DIC), which may occur in the presence of MAS or *de novo* in AOSD. DIC in AOSD has been reported to be complicated with myocarditis, which can be life-threatening. Thrombotic thrombocytopenic purpura (TTP) is another rare complication with AOSD, characterized by schistocytes in the peripheral blood and requiring plasma exchange with intensification of immunosuppression. AOSD may rarely be complicated by diffuse alveolar haemorrhage or pulmonary hypertension, both of which necessitate the intensification of immunosuppression along with supportive therapy. Long duration of AOSD can predispose to systemic AA amyloidosis which often affects the kidney and requires heightened immunosuppression with cyclophosphamide, anti-IL-1β targeted therapy or tocilizumab.

MANAGEMENT

Non-steroidal anti-inflammatory drugs (NSAIDs) are often used for symptomatic relief as the patient with AOSD is undergoing diagnostic evaluation, however, once diagnosis is established, additional therapy is warranted. It is usual to start patients on oral corticosteroid (equivalent to 1 mg/kg prednisolone).[1] Additional maintenance therapy with conventional disease-modifying antirheumatic drugs (DMARDs), usually methotrexate (up to 25 mg/week) is useful for both systemic and articular manifestations and helps to taper the dose of steroid with reduction of long-term toxicity due to the same. Failure of methotrexate warrants use of other DMARDs such as leflunomide or cyclosporine.[10,11] Sulfasalazine should not be used for the fear of precipitating MAS.

Biologic DMARDs are emerging as therapeutic options in AOSD refractory to conventional DMARDs.[12] Anti-interleukin 6 targeted therapy with tocilizumab is useful for systemic as well as articular manifestations. Etanercept and infliximab have been reported to be useful in about a third of patients with refractory disease. Considering the role of inflammasome activation with increased production of IL-1β in the pathogenesis of AOSD, therapies directly targeting IL-1β such as anakinra, canakinumab and rilonacept have shown a lot of promise in AOSD and sJIA. The requirement for daily injections with anakinra is a limiting factor, however, the monthly dosing schedule of canakinumab addresses this limitation. The utility of IL-1β targeted therapies in an Indian scenario is limited by their unavailability in the Indian market. Other biologic agents are also costly, most often out of the reach of Indian patients. Hence, therapies like monthly cyclophosphamide or tacrolimus, while not having gained widespread acceptance the world over, may have a niche in our patients with difficult-to-treat AOSD.[13,14] Anecdotal reports suggest the usefulness of rituximab, which may be an option in our patients in view of lesser cost compared to other biologics.[15] Figure 73.1 is a proposed algorithm for the management of AOSD in an Indian scenario.

Prognosis

When AOSD follows a monocyclic course, the disease tends to wane in a few months and the long-term outlook is good. The polycyclic systemic and polyarthritis variants of AOSD have variable prognosis. Certain factors have been identified to be associated with poor prognosis, such as hip or shoulder involvement, evanescent rash, polyarthritis, pleuritis, interstitial pneumonitis, hyperferritinemia and lack of response of fever to corticosteroid therapy.[9]

Conclusion

AOSD is a rare but potentially life-threatening rheumatic disease characterized by high spiking fever, evanescent rash and arthritis with markedly elevated acute phase reactants and hyperferritinemia. Infective, neoplastic and connective tissue diseases as cause of these symptoms should be ruled out before making a diagnosis of AOSD. Management involves use of oral corticosteroids along with conventional DMARDs. Refractory patients are candidates for biologic therapy targeting IL-1β, IL-6 or TNF. MAS is a serious life-threatening complication of AOSD, suspected in the presence of persistent fever with normal ESR and normal or falling total leucocyte and platelet counts, that mandates immediate intensive immunosuppressive therapy.

Key Points

- AOSD is characterized by quotidian fever, evanescent rash, arthritis with marked elevation of ESR, CRP and serum ferritin.
- AOSD is a diagnosis of exclusion.
- Management entails use of corticosteroids along with conventional or biologic DMARDs.

REFERENCES

1. Gerfaud-Valentin M, Jamilloux Y, Iwaz J, Seve P. Adult-onset Still's disease. Autoimmun Rev 2014; 13:708–22.
2. Mavragani CP, Spyridakis EG, Koutsilieris M. Adult-Onset Still's Disease: From Pathophysiology to Targeted Therapies. Int J Inflam 2012; 2012:879020.
3. Cozzi A, Papagrigoraki A, Biasi D, Colato C, Girolomoni G. Cutaneous manifestations of adult-onset Still's disease: a case report and review of literature. Clin Rheumatol 2016; 35:1377–82.
4. Rosario C, Zandman-Goddard G, Meyron-Holtz EG, D'Cruz DP, Shoenfeld Y. The hyperferritinemic syndrome: macrophage activation syndrome, Still's disease, septic shock and catastrophic anti-phospholipid syndrome. BMC Med 2013; 11:185.
5. Shaikh MM, Hermans LE, van Laar JM. Is serum procalcitonin measurement a useful addition to a rheumatologist's repertoire? A review of its diagnostic role in systemic inflammatory diseases and joint infections. Rheumatology (Oxford) 2015; 54:231–40.
6. Yamashita H, Kubota K, Mimori A. Clinical value of whole-body PET/CT in patients with active rheumatic diseases. Arthritis Res Ther 2014; 16:423.
7. Reddy Munagala VV, Misra R, Agarwal V, Lawrence A, Aggarwal A. Adult onset Still's disease: experience from a tertiary care rheumatology unit. Int J Rheum Dis 2012; 15:e136–41.
8. Mahroum N, Mahagna H, Amital H. Diagnosis and classification of adult Still's disease. J Autoimmun 2014; 48–49:34–7.
9. Efthimiou P, Kadavath S, Mehta B. Life-threatening complications of adult-onset Still's disease. Clin Rheumatol 2014; 33:305–14.
10. Jamilloux Y, Gerfaud-Valentin M, Henry T, Seve P. Treatment of adult-onset Still's disease: a review. Ther Clin Risk Manag 2015; 11:33–43.
11. Castaneda S, Blanco R, Gonzalez-Gay MA. Adult-onset Still's disease: Advances in the treatment. Best Pract Res Clin Rheumatol 2016; 30:222–38.
12. Al-Homood IA. Biologic treatments for adult-onset Still's disease. Rheumatology (Oxford) 2014; 53:32–8.
13. Tsuji Y, Iwanaga N, Adachi A, Tsunozaki K, Izumi Y, Moriwaki Y,et al. Successful Treatment with Intravenous Cyclophosphamide for Refractory Adult-Onset Still's Disease. Case Rep Rheumatol 2015; 2015:163952.
14. Nakamura H, Odani T, Shimizu Y, Takeda T, Kikuchi H. Usefulness of tacrolimus for refractory adult-onset Still's disease: Report of six cases. Mod Rheumatol 2014:1–5.
15. Belfeki N, Smiti Khanfir M, Said F, Said F, Hamzaoui A, Ben Salem T, et al. Successful treatment of refractory adult onset Still's disease with rituximab. Reumatismo 2016; 68:159–62.

FURTHER READING

1. Al-Homood IA. Biologic treatments for adult-onset Still's disease. Rheumatology (Oxford) 2014; 53:32–8.
2. Efthimiou P, Kadavath S, Mehta B. Life-threatening complications of adult-onset Still's disease.Clin Rheumatol 2014; 33:305–14.
3. Gerfaud-Valentin M, Jamilloux Y, Iwaz J, Seve P. Adult-onset Still's disease. Autoimmun Rev 2014; 13:708–22.

Chapter

74

Sarcoidosis

S Aparna Reddy, Alladi Mohan

INTRODUCTION

Sarcoidosis is a multisystem granulomatous disorder, with causative agent not proven till date. The available evidence suggests that individuals with a susceptible genotype, when exposed to one or more potential causative antigens may develop the disease. Sarcoidosis is characterized by a sustained inflammatory response that ensues after the disease initiation resulting in granuloma formation. Sarcoidosis is a heterogeneous disease, with a varied clinical presentation. Individuals with sarcoidosis can have a spectrum of clinical manifestations ranging from being an incidental finding on the chest radiograph to a potentially life-threatening disease, requiring aggressive therapy.

HISTORY

Jonathan Hutchinson,[1] Caesar Boeck, are credited with the initial descriptions of sarcoidosis. The milestones in the discovery of condition named sarcoidosis, to the establishment of World Association of Sarcoidosis and other Granulomatous Disorders (WASOG) by the sarcoidologists all over the world is shown in Fig. 74.1.[1]

EPIDEMIOLOGY

Global Scenario

Sarcoidosis can affect individuals of all races, and ethnic groups. Reliable epidemiological data on sarcoidosis are lacking as clear-cut diagnostic or classification criteria are not available. The peak incidence is seen in second and third decade of life.[2] Prevalence of sarcoidosis appears to be higher in the Scandinavian countries.[3] The incidence of sarcoidosis peaks around fourth decade in African-Americans, and compared to other ethnic groups, sarcoidosis is more chronic and fatal in this ethnic group.[4] The variation in annual incidence can be

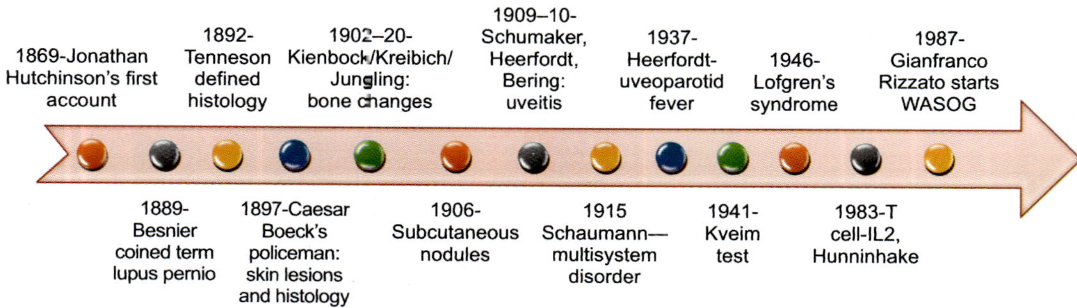

Fig. 74.1: Milestones in sarcoidosis resesarch
IL-2—interleukin-2, WASOG—World Association of Sarcoidosis and other Granulomatous disorders

as low as seen as 0.73 per 100,000 in Japanese men to as high as 71 per 100,000 in African-American women.[4,5] Another population based study done in US showed incidence of 10 persons per 100,000 per year, with majority having intrathoracic involvement, and the overall mortality was same as general population.[6]

Sarcoidosis in India

Few studies have documented the burden of sarcoidosis in India. In Eastern India, sarcoidosis constituted 10–12 per 1000 new registrations annually at a respiratory unit at Kolkata; and 0.2 per 100,000 population in the metropolitan areas of Kolkata.[7] Sarcoidosis constituted 61.2 per 100,000 new cases seen in Vallabhbhai Patel Chest Institute at Delhi during the period 1957–1974.[8,9] Published reports suggest that sarcoidosis affects all ethnic groups in India.[9–15] Table 74.1 shows comparison of the clinical presentation, laboratory abnormalities, and radiographic findings in Indian patients with sarcoidosis with observations documented in studies from the West.[9–22] However, these figures appear to be underestimates and the real burden of sarcoidosis in India is not known.[9]

AETIOLOGY OF SARCOIDOSIS

Sarcoidosis results from a complex interaction between environmental, immunological, and genetic factors of individual.

Table 74.1: Comparison of clinical profile, laboratory characteristics, and radiological findings in Indian patients with sarcoidosis with observations from other parts of the world

Variable	*Studies from India (n = 765) (9–15) (%)*	*Studies from other parts of the world (n = 10003) (16–22) (%)*
Female gender	43–71	57–61
Under 40 years	25–70	70–86
Thoracic involvement	61–97	88–99
Ocular involvement	08–40	4–27
Erythema nodosum	02–20	11–34
Other skin lesions	10–24	
Parotid enlargement	03–15	0.5–6
Neurological involvement	01–11	0.3–9
Fever	35–54	
Constitutional symptoms	14–57	
Arthralgias/arthritis	18–35	30–40
Cardiac involvement	0–12	03
Peripheral lymphadenopathy	19–42	8–27
Hepatomegaly	14–42	12
Splenomegaly	02–27	0.3–10
Radiologic stage at presentation		
0	01–03	4–13
I	45–62	28–65
II	30–34	17–31
III	07–18	7–13
Kveim positive	45–96	73–84
Tuberculin skin test negative	59–97	55–70
Hypercalcaemia	18–40	0.7–18
Hypercalciuria	10–49	2–50

Source: References 9–22

Environmental Factors

Environmental exposure to irritants such as smoke from burning wooden firewood, pollen grains, exposure to inorganic particles, insecticides, several mycobacterial antigens, may act as precipitating factors for the cause of sarcoidosis.[23, 24] In an observational study,[25] during the 5 years after the World Trade Center (WTC) disaster,[26] patients with pathologic evidence consistent with new-onset sarcoidosis were detected among New York City's fire-fighters who were exposed to WTC "dust"; 13 of these had occurred during the first year. The incidence rate of sarcoidosis during the first year after exposure to WTC dust (86/100,000) had significantly increased compared to the incidence rate of 15/100,000 observed during the preceding 15 years.[25] In a recent study,[26] 9 of the 50 fire-fighters who had developed sarcoidosis after the WTC disaster; and two other fire-fighters earlier diagnosed to have sarcoidosis had presented with refractory arthritis and required additional disease-modifying antirheumatic drugs (DMARDS) for adequate control of their joint manifestations unlike those who were not exposed to the dust. These observations suggest a role for environmental triggers in the causation of sarcoidosis. A higher incidence of sarcoidosis in non-tobacco smokers has been described in some studies;[15, 27] however, this association merits further study.

Genetic Factors

Sarcoidosis is thought to develop as a consequence of a persistent and exaggerated granulomatous response in a genetically predisposed individual to an environmental factor. Published data suggest that concordance rates are higher in monozygotic compared to dizygotic twins. In A Case-Control Etiologic Sarcoidosis Study (ACCESS)[28] a five-fold occurrence of sarcoidosis in siblings or parents was observed in sarcoidosis cases compared to control subjects.

Following the reporting of associations between class I human leucocyte antigen (HLA) B8 antigens and acute sarcoidosis, other HLA class II antigens, encoded by HLA-DRB1 and DQB1 alleles, have been found to be associated with sarcoidosis.[29] HLA-DQB1*0201 and HLA-DRB1*0301 have been reported to be markers associated with good prognosis in patients with acute disease.[29] Another susceptibility gene which has been well-studied is butyrophilin-like 2 (BTNL2) gene; several other susceptibility loci related to interleukin-23 (IL-23)/interleukin-12 (IL-12) signalling pathways have also been studied.[29]

Infections and Sarcoidosis

In a recent review,[30] the authors had reported that there was no evidence for sarcoidosis being an infectious disease but suggested that sarcoidosis develops as a result of an increased immune response to "molecular patterns of killed and partially degraded mycobacterium and propionibacteria". A more recent meta-analysis[31] which evaluated the association of infectious agents with sarcoidosis, however, revealed an aetiological link between Propionibacterium acnes (odds ratio [OR] 18.8, 95% confidence intervals [CI] 12.62–28.01), mycobacteria (OR 6.8, 95% CI 12.62–28.01) and sarcoidosis. In this meta-analysis,[31] no association between Borrelia, human herpes virus-8, *Rickettsia helvetica*, *Chlamydia pneumoniae*, Epstein-barr virus, retrovirus, and sarcoidosis was noted. The pathogenetic role of mycobacterial antigens and *Propionibacterium acnes* in the causation of sarcoidosis is shown in Table 74.2.[31] The association between infection and sarcoidosis merits further research.

IMMUNOPATHOGENESIS

The immunopathogenesis of sarcoidosis involves a close interaction of both innate and adaptive immune mechanisms. The key histopathologic hallmark of sarcoidosis is the presence of "epithelioid-cell-rich granulomas that are non-necrotising" (Fig. 74.2). Formation of granuloma occurs in four stages such as initiation, accumulation, effector phase and resolution.[32] In the lung, the granulomas are found in a distinctly lymphangitic distribution along the bronchovascular bundles of the pulmonary tree. The ongoing granuloma formation in various organs leads to the diverse manifestations in sarcoidosis.

One proposed hypothesis for the chronicity of the disease is, that sarcoidosis is triggered by an exaggerated type 1 helper T-cell (Th1) response to microbial and tissue antigens which are associated with the abnormal aggregation of serum amyloid A within granulomas, thereby leading to progressive

Table 74.2: Pathogenetic role of mycobacterial antigens and *Propionibacterium acne* in the causation of sarcoidosis

Causative agents	*Pathogenic antigen*	*Mechanism involved in granuloma formation*
Mycobacterial antigens	Early secreted antigenic target 6 kDa (ESAT-6) Superoxide dismutase A (Sod A) *M. tuberculosis* catalase-peroxidase (mKatG) *M. tuberculosis* heat-shock protein-70 (Mtb-hsp) *M. tuberculosis* mycolyltransferase antigen	Activates TLR2-dependent immune responses
Propionibacterium acne	Propionibacterial DNA and RNA	Propionibacterium-induced granuloma formation depends on TLR9 and TLR2

TLR—Toll-like receptor; DNA—Deoxyribo nucleic acid; RNA—Ribo nucleic acid
Source: Reference 31

chronic granulomatous inflammation in the absence of detectable ongoing infection.[32]

The levels of inflammatory mediators determine the stage of progression versus resolution. For example, high tumour necrosis factor-alpha (TNF-α) levels are observed in progressive disease, whereas, fibrotic disease has high transforming growth factor-beta (TGF-β) levels.[32] This mainly results from a tilting of balance from Th1 to type 2 helper T-cell (Th2) responses.[32, 33] Usually, a check in the proliferative capacity is made by the upregulation of programmed death-1 in the T-effector (T_{eff}) cells.[34] Apart from TGF-β, other mediators of fibrosis include interleukin-5 (IL-5), interleukin-7 (IL-7), matrix metalloproteinases (MMPs) within are responsible for myofibroblast differentiation, fibroblast expansion, accumulation of extra-cellular matrix and alteration of micro-architecture of organs.[35]

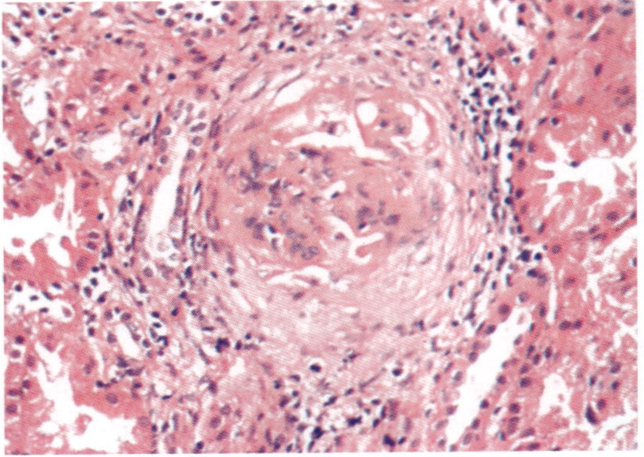

Fig. 74.2: Sarcoid granuloma. Photomicrograph showing the granuloma organized with a central core consisting of macrophages, foreign body giant cells, epithelioid cells and encircled by lymphocytes (Haematoxylin and eosin X 400)

In the resolution phase of sarcoidosis the balance tilts towards regulatory T-cells (T_{reg}), leading to production of interleukin-2 (IL-2) production, thereby strongly inhibiting T-cell proliferation. This leads a condition of "immune paradox", where, in spite of extensive local inflammation, anergy may develop, as shown by inadequate immune response to tuberculin.[36] These details are summarized in Table 74.3.[37, 38]

CLINICAL PRESENTATION

Sarcoidosis is a systemic disease, and the organ involvement can be due to direct tissue infiltration (e.g. sarcoidosis of lung, and heart), or can occur as an epiphenomenon of varied intense inflammation manifesting as erythema nodosum or resulting in hypercalcaemia. In majority of series, the pulmonary involvement is seen in 90–95% of cases.[9–22, 39]

Constitutional Symptoms

As with any granulomatous disorder, patients with sarcoidosis present with constitutional symptoms such as fever, fatigue, malaise, weight loss. In India sarcoidosis is an uncommon, but important cause of pyrexia of unknown origin.[9–15, 40]

Pulmonary Sarcoidosis

Among the various organ systems that are involved, pulmonary involvement, is by far the most common, accounting for about 90% of the patients.[41] Patients with pulmonary sarcoidosis present with

Table 74.3: Immunopathogenesis of sarcoidosis

Pathogenetic mechanism	*Cells, cell receptors and inflammatory mediators involved*	*Immunological role*
Innate immunity	Monocytes	Increased expression of surface Fcγ receptors (CD16, CD32, CD64), complement receptor 1 (CD35), and β-integrin CD11b
	Macrophages	Detect antigens via pattern recognition receptors, e.g. TLRs, then internalise and process the antigen. Present processed antigen to T-cell receptor in presence of co-stimulatory molecules (CD80, CD83, CD86). Differentiate into epithelioid cells, which gain secretory capability, lose phagocytic capacity and fuse to form multinucleate giant cells
	S100A8 (calprotectin) S100A9 (calgranulin B)	Bind to TLR4 and RAGE and initiate an immune response including NF-kB activation and induction of cytokine release
	Serum amyloid A	Activates NFkB via TLR2
Adaptive immunity	Dendritic cells	Mature myeloid dendritic cells activate naive T-cells
	BTNL2	Inhibits T-cell activation and induction of T_{reg}
Myeloid cells	Neutrophil	Phagocytizing and killing of micro-organisms in acute sarcoidosis Elevated neutrophil/lymphocyte ratio has prognostic value
	CD14+ CD16+ monocyte	Basis for the constant replacement of granuloma with myeloid cells
	M1 alveolar macrophages	Secrete IL-12 and IL-18, which increase Th1 differentiation at the granulomatous site. Release TNF, which is required for the induction and maintenance of granuloma Enhanced TLR-9 expression High levels of CXCL10. Elevation of two innate host defense signaling pathways: (i) Fcγ-receptor mediated phagocytosis; and (ii) clathrin mediated endocytosis
Lymphoid cells	CD4+ Th1	Secrete IFN-γ, IL-2, CXCR3, monocyte chemotactic protein-1, macrophage inflammatory protein-1, granulocyte-macrophage colony stimulating factor Upregulation of JAK-STAT signaling pathways, such as, IL-12, IL-18, CCL2, CXCL9,and CXCL10
	Th17 cells	Secretes IL-17A and IL-17F Decreases expression of CTLA-4
	Hybrid T-cell (Th17.1)	Features of both Th17 and Th1 cells
	T_{reg}	Decreased expression of CTLA-4
Fibrotic response	Alveolar macrophages	Release profibrotic chemokine CCL18 in chronic sarcoidosis Secretes IL-4, which is necessary for B-cell development and activation and downregulates the release of proinflammatory cytokines
	Th2 lymphocytes	Synthesize fibronectin,CCL18, upregulation of TGF-β, increase ratio of CD8+ to CD4+ cells

BTLN-2—butyrophilin-like 2; CD = cluster differentiation; CCL—chemokine ligand; CXCL—C-X-C motif ligand; CTLA-4—cytotoxic T-lymphocyte antigen-4; IFN-γ—interferon-γ; IL—interleukin; JAK-STAT—Janus kinase signal transducer and activator of transcription; NF-kB = nuclear factor kappa B; RAGE = receptor for advanced glycation endproducts; Th—T-helper; T_{reg}—regulatory T cells; TGF-β—transforming growth factor-β; TLR—toll-like receptor; TNF—tumour necrosis factor

Source: References 37, 38

dyspnoea, dry cough, chest discomfort, and wheezing. Other uncommon manifestations include involvement of the airways (larynx, trachea, and bronchi) resulting in airway obstruction and bronchiectasis, a clinical presentation sometimes mimicking bronchial asthma. The other rare manifestations include pleural involvement in form of pleural plaques, nodules, effusion, pneumothoraces and haemoptysis, clubbing among others.

Most often sarcoidosis is detected incidentally on the chest radiograph, that may show intrathoracic lymphadenopathy and pulmonary infiltrates (Fig. 74.3). The staging system for pulmonary sacroidosis based on chest radiograph is shown in Table 74.4.[42] The major limitation of this system is that the stages do not represent chronological evolution of the disease.[42] Patients with stages I, II have up to 90% and 70% spontaneous remissions, respectively, whereas in stage III, 90% can have progressive disease, and stage IV indicates end-stage lung disease.[42] These patients can have both restrictive as well as obstructive lung disease, mixed restrictive and obstructive pattern. Airway hyperreactivity is also common seen.[43] The clinical course of sarcoidosis is variable, approximately two-thirds of patients progress to remission within one year of diagnosis, remaining one-third have either persistent or recurrent disease.[44]

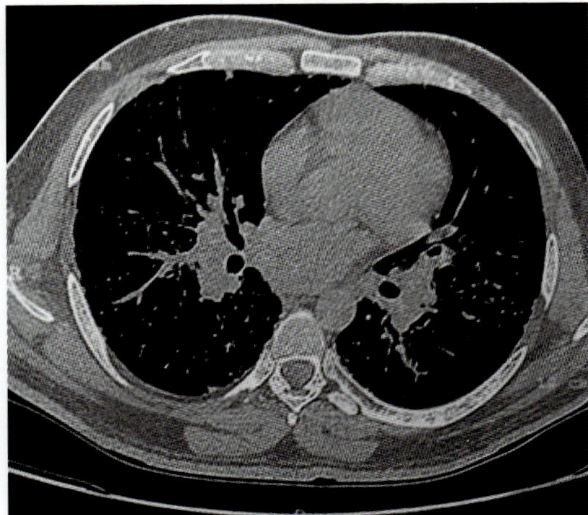

Fig. 74.3: Pulmonary sarcoidosis. CECT chest (mediastinal window) showing nodular interstitial septal thickening with ill-defined nodules in both lungs, predominantly in peribronchvascular and subpleural location. Non-necrotic enlarged lymph nodes are evident in mediastinum and hilar regions

Table 74.4: Chest radiograph staging in sarcoidosis

Stage	*Chest radiograph findings*
Stage 0	Normal
Stage 1	Bilateral hilar lymphadenopathy without infiltration
Stage 2	Bilateral hilar lymphadenopathy with infiltration
Stage 3	Infiltration alone
Stage 4	Fibrotic bands, bullae, hilar retraction, bronchiectasis, and diaphragmatic tenting

Source: Reference 42

In the pulmonary fibrosis the distribution of inflammation is in upper and middle lobe predominantly with thin linear bands, dense fibrotic masses, and cystic lung disease.[45] In patients with pulmonary fibrosis, potential complications, such as, pulmonary hypertension, pulmonary aspergillosis syndromes (including aspergilloma and chronic aspergillus infection), and opportunistic infections can occur.

Cutaneous Sarcoidosis

Skin involvement occurs in 25–35% of patients with sarcoidosis.[46] The manifestations can occur as isolated lesions or in crops of macules, papules, and plaques. The anatomical sites commonly involved are nape of the neck and upper back, extremities, and trunk. The skin lesions have been classified as specific and non-specific, based on the presence of non-caseating granulomas on histopathological examination. Specific lesions include macules, papules, plaques, nodules, lupus pernio, scar infiltration, alopecia, ulcerative lesions, and hypopigmentation among others. The most common non-specific lesion is erythema nodosum. Other cutaneous manifestations include metastatic calcification, prurigo, erythema multiforme, digital clubbing, and Sweet's syndrome.[47]

Erythema nodosum is a non-specific panniculitis and is considered to be the hallmark of acute sarcoidosis. Patients with erythema nodosum present with elevated, red, and tender subcutaneous nodules typically on the anterior aspects of the legs. Sometimes it accompanies other manifestations of Lofgren's syndrome (detailed below), and resolves spontaneously within weeks.

Lupus pernio, represents chronic sarcoidosis has cosmetic relevance as it can cause disfigurement of face due to the erosion of underlying cartilage and bone. The sites commonly involved are nose, cheeks, lips, and ears.[46] Skin biopsy shows non-caseating granulomas.

Ocular Sarcoidosis

Eye involvement is seen in 25–80% of patients with sarcoidosis.[48] Majority of the patients (65%) present with anterior uveitis. Chronic anterior uveitis, is more common than acute anterior uveitis. Posterior-segment involvement is seen in nearly 30% of the patients.[48] In about 10–15% of patients both the anterior and posterior segments are involved. Criteria for the diagnosis of ocular sarcoidosis are shown in Table 74.5.[49] Other less common eye manifestations in sarcoidosis include conjunctival follicles, enlargement of lacrimal gland, keratoconjunctivitis sicca, dacryocystitis, and retinal vasculitis.

Hepatic and Splenic Sarcoidosis

Liver involvement is twice as common in African-Americans compared to white Americans, and among them 60% present with constitutional symptoms such as fever, loss of appetite and weight loss, hepatomegaly and elevated hepatic transaminases is evident.[50, 51] Splenic involvement is usually asymptomatic. Sometimes individuals manifest

Table 74.5: Diagnosis of ocular sarcoidosis

Seven signs in the diagnosis of ocular sarcoidosis

i. Slit lamp examination revealing "mutton-fat" keratic precipitates, or small granulomatous keratic precipitates such as "soap sud" like debries or and or iris nodules (Koeppe/Busacca)
ii. Trabecular meshwork nodules and or tent-shaped peripheral anterior synechiae
iii. Vitreous opacities displaying "snowballs" or "strings of pearls"
iv. Multiple chorioretinal peripheral lesions (active and/or atrophic)
v. Posterior uveitis in form of nodular and/or segmental periphlebitis along the veins "candle wax drippings" and/or retinal macroaneurysm in an inflamed eye
vi. Optic disc nodules or granuloma(s) and/or solitary choroidal nodule
vii. Bilaterality

Laboratory investigations suggested for the diagnosis ocular sarcoidosis in patients having the above mentioned intraocular signs

i. Negative tuberculin skin test in a BCG-vaccinated patient or in a patient having had a positive tuberculin skin test previously
ii. Elevated serum angiotensin converting enzyme levels and/or elevated serum lysozyme
iii. Chest radiograph revealing bilateral hilar lymphadenopathy
iv. Abnormal liver enzyme tests
v. Chest CT in patients with a negative chest radiograph result

After excluding other possible causes of uveitis, four levels of certainty for the diagnosis of ocular sarcoidosis were recommended:

- Definite ocular sarcoidosis (Level 1)
  - Biopsy supported diagnosis which is compatible with uveitis
- Presumed ocular sarcoidosis (Level 2)
  - Biopsy not done but chest radiograph showing bilateral hilar lymphadenopathy associated with a compatible uveitis
- Probable ocular sarcoidosis (Level 3)
  - Biopsy not done and the chest radiograph does not show bilateral hilar lymphadenopathy but 3 of the above listed intraocular signs and 2 positive laboratory tests are present
- Possible ocular sarcoidosis (Level 4)
  - Lung biopsy has been done and the result was negative but at least 4 of the above ocular signs and 2 positive laboratory investigations are present

BCG—bacille Calmette-Guerin; CT—computed tomography
Source: Reference 49

constitutional symptoms and marked splenomegaly may be present in up to 6% of cases (Fig. 74.4).[51] Hepatomegaly and splenomegaly can be identified on ultrasonography, computed tomography.[51] Up to 65% of patients have non-caseating granulomas of the liver.[51]

Lymphadenopathy

Sarcoidosis presenting as peripheral lymphadenopathy has been observed in 8–44% of patients.[9–22,41,53] The lymph nodes commonly affected are cervical (posterior triangle than anterior triangle of the neck), axillary, epitrochlear, and inguinal.[9–22, 53] Hilar lymphadenopathy is one of the components of acute sarcoidosis (Lofgren's syndrome). Sometimes as unusual presentation in form of intrabdominal lymphadenopathy involving the portahepatis, retroperitoneal and preaortic and paraaortic nodes.[9–22, 41, 53]

Sarcoidosis of Musculoskeletal System

Bone and joint involvement in sarcoidosis can be widespread. The musculoskeletal involvement can be acute or chronic and can range from being asymptomatic to presentation with frank arthritis.

Sarcoid Arthropathy and Periarthropathy

Joint involvement in sarcoidosis can be inflammatory and non-inflammatory. Acute inflammatory arthritis with frank joint swelling can involve any joint. With the widespread use of magnetic resonance imaging (MRI) and ultrasonography, the ankle swelling which was previously attributed to ankle synovitis is now considered to be either periarthritis or tenosynovitis.[54] The criteria proposed by Visser et al[55] Table 74.6 for the diagnosis of arthritis in sarcoidosis have a sensitivity of 93% and specificity of 99%.

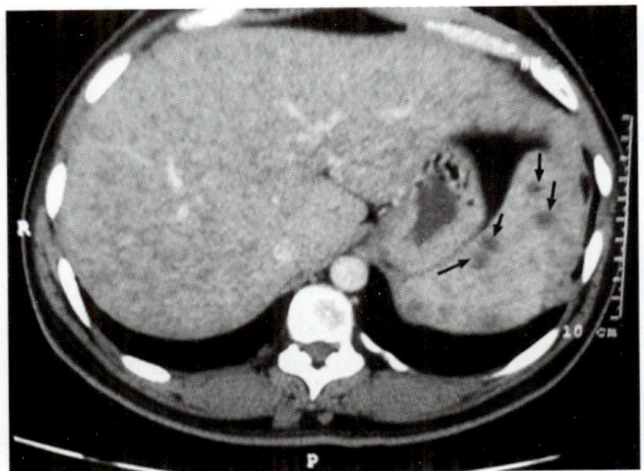

Fig. 74.4: Splenic sarcoidosis. CECT abdomen showing mild splenomegaly, multiple tiny hypodense nodules splenic parenchyma (arrows)

The triad of acute arthritis with bilateral hilar lymphadenopathy and erythema nodosum is referred to as Lofgren's syndrome.[56] While Lofgren's syndrome is considered to be a common rheumatological manifestation of sarcoidosis by rheumatologists,[57] it is considered to be a rare/uncommon manifestation of sarcoidosis in India.[58,59] Acute sarcoidosis usually resolves within in weeks of onset, and once it resolves it usually does not recur.

Chronic granulomatous infiltrative arthritis is much less common in sarcodosis. Granulomatous inflammation can affect periarticular structure resulting in bursitis, enthesopathies, carpal tunnel syndrome, dactylitis, nodular tenosynovitis, and sacroiliitis.

Osseous Sarcoidosis

The patients, with osseous sarcoidosis present with focal pain, tenderness, swelling, and erythema overlying the affected bone. The common sites of involvement are phalanges of the hands and feet.[60] Radiographs of phalanges show conventional "lacy" trabecular pattern with osteolysis and "punched out" cystic lesions.[60]

Sarcoid Myopathy

Three patterns of sarcoid myopathy have been described. These include chronic myopathy, nodular myopathy, and acute myositis. Chronic sarcoid myopathy is the most common among these. Patients with chronic sarcoid myopathy present with relatively painless but insidiously progressive muscle weakness. This condition is often

Table 74.6: Visser's criteria for sarcoidosis arthritis*

Criteria
Age < 40 years
Erythema nodosum
Symmetric ankle arthritis
Symptom duration < 2 months

*Presence of 3 of the 4 characteristics establishes the diagnosis (sensitivity 93%, specificity 99%)

Source: Reference 55

mistaken for drug-induced myopathy but muscle biopsy can demonstrate infiltrating granulomatous inflammation. Acute myositis presents with pain due to acute oedema of the involved muscle.[61] Nodular myopathy presents with palpable, localized tumour-like nodular swellings within the muscle.

Hypercalcaemia and Renal Disease

Hypercalciuria occurs in 40% of patients with sarcoidosis, hypercalcaemia and renal calculi are observed in 11% and 10% patients respectively.[62] Dysregulated production of 1, 25-dihydroxy vitamin D1, 25 (OH) D3 by activated macrophages present in granulomas is thought to be mechanism underlying the occurrence of hypercalcaemia in sarcoidosis. Occurrence of hypercalcaemia or hypercalciuria in the presence of normal serum parathyroid hormone levels, normal or increased serum calcitriol, and low serum ergocalcitriol is regarded as a "highly probable" criterion for the diagnosis of sarcoidosis.[62] Vitamin D supplementation theoretically can precipitate the granulomatous inflammation. In a recent randomized, placebo-controlled clinical trial of vitamin D supplementation on surrogate measures of skeletal health in patients with sarcoidosis who had serum 25(OH)D3 levels less than 50 nmol/L, it was reported that only one of the 27 patients studied developed hypercalcaemia. The authors found no differences between the vitamin D supplementation and placebo groups in terms of the albumin-adjusted serum calcium, 24-hour urine calcium, parathyroid hormone levels, bone turnover markers or bone mineral density.[63] Renal sarcoidosis can manifest as hypercalcaemia, hypercalciuria, tubulo-interstitial nephritis and nephrocalcinosis.[64] Rarely renal sarcoidosis can present as pseudotumour (Fig. 74.5).[65]

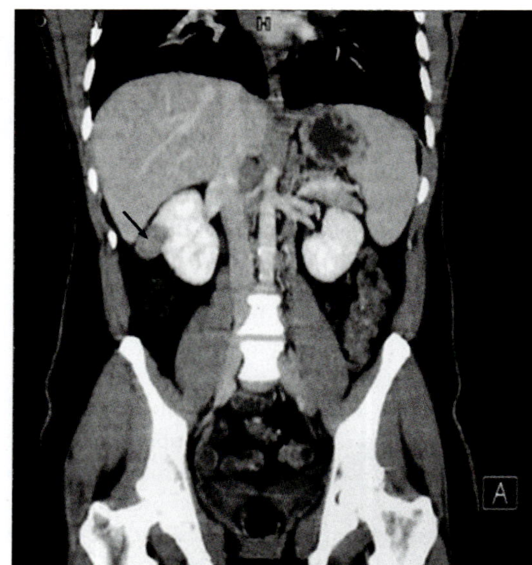

Fig. 74.5: Renal sarcoidosis presenting as pseudotumour. CECT abdomen showing a 35 × 25 mm exophytic enhancing mass lesion in the anterior midpolar region of the right kidney

Cardiac Sarcoidosis

Cardiac sarcoidosis has been less frequently documented in clinical (5%) series as compared to autopsy series (25%).[53, 62] The sarcoid granulomas are commonly seen in left ventricular free wall followed by interventricular septa involving the conduction system. Hence patients with cardiac sarcoidosis can present with cardiomyopathy, arrhythmias. The patchy nature of cardiac involvement decreases the yield of endocardial biopsy. Imaging modalities such as cardiac MRI with gadolinium enhancement and positron-emission tomography (PET)-computed tomography (CT); and PET-MRI are valuable aids in the diagnosis of myocardial sarcoidosis.[66]

Ventricular tachycardia (VT) due to cardiac sarcoidosis can be fatal and is difficult to distinguish from idiopathic ventricular tachycardia. Pleomorphism and cycle length variation during VT have been found to be useful in distinguishing VTs due to cardiac sarcoidosis from idiopathic VT. Table 74.7[53, 66, 67] summarizes the screening and suggested evaluation of patients with cardiac sarcoidosis.

Neurosarcoidosis

Neurologic manifestations are evident in about 10% of patients with sarcoidosis; however, autopsy studies reveal neurological involvement in about 25% patients with sarcoidosis.[68] Facial nerve palsy which may be bilateral and recurrent has been commonly described; headache, ataxia, cognitive dysfunction, weakness, and seizures are other neurological manifestations.[68] Involvement of the hypothalamus, pituitary gland can present as

Table 74.7: Diagnosis of and screening for cardiac sarcoidosis

Diagnosis of cardiac sarcoidosis

Histopathological diagnosis

Presence of non-caseating granuloma on histopathological examination of myocardial tissue with no alternative cause identified (including negative organismal stains if applicable)

Clinical diagnosis

Probable* cardiac sarcoidosis is present if:

i. there is histopathological diagnosis of extra-cardiac sarcoidosis; and
ii. one or more of the following is present:
   a. cardiomyopathy or heart block responsive to corticosteroid and/or immunomodulatory therapy
   b. unexplained reduced left ventricular function (ejection fraction <40%)
   c. unexplained spontaneous or inducible sustained ventricular tachycardia
   d. Mobitz type II second-degree or third-degree atrioventricular heart block
   e. characteristic findings of patchy uptake on dedicated PET imaging
   f. delayed enhancement on cMRI
   g. positive gallium scanning
   h. unexplained defect on perfusion scintigraphy or SPECT
   i. T2-prolongation on cMRI
iii. Other causes for the cardiac manifestation(s) have been reasonably excluded

Screening for cardiac involvement in patients with sarcoidosis

Class I recommendations:

i. Assessment for history of unexplained syncope, pre-syncope, or significant palpitations
ii. A baseline 12-lead ECG

Class IIa recommendations:

i. Echocardiography
ii. Advanced cardiac imaging with cMRI or PET in patients with one or more positive findings on clinical history, ECG, or echocardiography

cMRI—cardiac magnetic resonance imaging; ECG—electrocardiogram; PET— positron emission tomography; SPECT—single photon emission computerized tomography; ECG—electrocardiography

Source: References 53, 66, 67

* 'Probable involvement' is considered adequate to establish a clinical diagnosis of cardiac sarcoidosis

endocrine disorders such as syndrome of inappropriate anti-diuretic hormone, diabetes insipidus, galactorrhoea, and amenorrhoea. Spinal cord involvement can result in acute transverse myelitis. The peripheral nerve involvement is caused by granulomatous infiltration, extrinsic compression, or vasculitic ischaemia of large nerve fibres. In patients with neurosarcodosis, MRI reveals leptomeningeal thickening and enhancement, most typically at the base of the brain. In 50% of patients small fibre neuropathy (SFN) has been reported.[69] Autonomic dysfunction and sensory symptoms such as upper extremity pain, muscle cramps, abnormal temperature sensation, hyperesthesias have also been observed.

Haematologic Abnormalities

The most common haematologic abnormality in sarcoidosis is anaemia. It is usually mild, and presumed to be anaemia of chronic disease. However, study bone marrow infiltration has been documented in 27% of sarcoidosis patients with anaemia.[70] Other haematologic manifestations are leucopenia, especially lymphopenia, due bone marrow infiltration, redistribution of lymphocytes to active disease sites, and hypersplenism. Peripheral lymphadenopathy and splenomegaly is seen in about one-third of the patients.[71]

TNF-inhibitor-induced Sarcoidosis-like Illness

TNF-α is main cytokine responsible for sarcoid granuloma, TNF inhibitors have been used in the treatment of sarcoidosis. But these agents have the paradoxical effect of inducing new-onset granulomatous disease resembling sarcoidosis. The sarcoid-like reaction may be due to upregulation of interferon-α through TNF and IL-12 inhibition.[72] The individual with TNF polymorphisms may also develop sarcoidosis-like reaction. Among them etanercept, infliximab, adalimumab, certolizumab, have been reported as causative agents.[73–75] Similarly, patients treated with tocilizumab, anakinra may develop drug-induced sarcoidosis.[76]

SARCOIDOSIS IN CHILDREN

Sarcoidosis is relatively rare in children as compared to adults.[77] Familial juvenile systemic granulomatosis, (Blau's syndrome) an autosomal dominant, autoinflammatory syndrome characterized by the clinical triad of granulomatous recurrent uveitis, dermatitis and symmetric arthritis. The gene responsible has been identified in the caspase recruitment domain gene CARD15/NOD2.[78] The same findings were reported in another multicentre, observational study on articular, ophthalmological, therapeutic and radiological data in Blau's syndrome patients.[78]

SARCOIDOSIS IN THE ELDERLY

While sarcoidosis occurs usually in the second or third decade of life approximately 30% of cases occur in elderly patients.[119] Elderly-onset sarcoidosis (EOS) is defined as the occurrence of sarcoidosis in people aged 65 years or more.[79] Fatigue, uveitis and specific skin lesions are more common compared to younger patients, while erythema nodosum and other pulmonary abnormalities are less frequent. A granulomatous reaction in the elderly, should raise the clinical suspicion of tuberculosis, neoplasia, granulomatosis with polyangiitis, sarcoid like reaction due to interferon and TNF-α blockers and sarcoidosis. The treatment is empiric, based on clinical condition of the patient. However, the treatment decisions are difficult due to the comorbidities and increased risk of toxicities from treatments, particularly corticosteroids.[79]

SARCOIDOSIS AND AUTOIMMUNE DISORDERS

There is a apparent similarity between sarcoidosis and autoimmune diseases based on the immunological profile. However, dissimilarities in terms of clinical presentation and therapeutic response are present.[80] In sarcoidosis lungs are the most commonly affected whereas, the musculoskeletal and dermatologic abnormalities are predominantly seen in other autoimmune disorders.[80] It has been observed that psoriasis was associated with a disease severity-dependent increased risk of developing sarcoidosis.[81]

Biomarkers

There is a strong need for biomarkers that facilitate the diagnostic process for a better characterization of patient's phenotypes and their prognosis or to indicate the specific targets of effective treatments.[82] The various biomarkers and their significance has been shown in Table 74.8.[82]

APPROACH TO MANAGEMENT

Sarcoidosis has variable presentation and hence a clear cut rigid management algorithm is difficult to postulate. The assessment of a patient with suspected sarcoidosis, evaluation and subsequent follow-up details are shown in Table 74.9.[83] Modern imaging modalities, such as, PET-CT and PET-MRI have shown potential for defining the active disease extent and assessing active disease on follow-up. However, imaging findings are not specific for sarcoidosis and only indicate the site(s) and extent of involvement. The diagnosis of sarcoidosis requires the identification of two organ system sites of involvement. It also requires exclusion of the many diseases that sarcoidosis can mimic sarcoidosis as shown in Table 74.10.[83]

Diagnosis and Classification Criteria

Till date there is no classification criteria for sarcoidosis, as it is a heterogenous disease. The American Thoracic Society, the European Respiratory Society and the World Association of Sarcoidosis and Other Granulomatous Disorders suggested diagnosis based on following criteria:[84] (i) a compatible clinical picture; (ii) demonstration of non-caseating granulomas on histology; and (iii) exclusion of other conditions presenting similarly. Some of the conditions that should be differentiated from sarcoidosis are shown in Table 74.10.[83]

Table 74.8: Biomarkers of sarcoidosis and their clinical significance

Biomarker	*Source of origin*	*Clinical significance*
ACE	Monocyte-macrophage, endothelial cells	Acute stage, levels influenced by polymorphisms
sIL-2R	Lymphocyte	Disease severity, extrapulmonary organ involvement
SAA	Monocyte-macrophage	Higher level in tissue and serum in sarcoidosis
BALF β-1-antitrypsin	Predominantly produced by hepatocytes	Downregulated only in patients without LS; associated with spontaneous resolution
BALF protocadherin-2 precursor	Sarcoidosis BALF proteome, using (SELDI-TOF MS), determines the protein patterns in BALF	Cell adhesion; upregulated in sarcoidosis across all studied phenotypes
Chitotriosidase	Monocyte-macrophage	Disease progression
BALF tenascin-C	Keratinocytes, fibroblasts	Fibrosis and ECM associated; levels correlated with infiltrates on chest radiographs in sarcoidosis
IL-17RC	Lymphocyte	Elevated expression in retinal tissues
TGF-β1	Fibroblasts	Fibrosis and ECM related; associated with pulmonary fibrosis
miR-34a and miR-144	Mononuclear cells	NF-kB signalling and activation of the T-cells

ACE—angiotensin-converting enzyme; BALF—bronchoalveolar lavage fluid; ECM—extracellular matrix; IL-17RC—interleukin 17 receptor C; sIL-2R—soluble interleukin-2 receptor; SAA—serum amyloid A; LS—Lofgren's syndrome; SELDI-TOF MS—surface-enhanced laser desorption/ionization-time-of-flight mass spectroscopy; miR—microRNA; NF-kβ—Nuclear factor-κappa B

Source: References 82

Table 74.9: Approach to management of sarcoidosis

History (environmental, occupational, family history) and physical examination
Baseline investigations:
Complete haemogram including ESR
Liver function test and serum calcium
Urinalysis (including urinary calcium)
CRP
TST/IGRA
Chest radiograph
Pulmonary function testing
Electrocardiogram
Echocardiography
Biopsy of affected organ, with special stains and culture of the specimen
Assessment of extent of disease and staging of disease severity: Further diagnostic laboratory and radiologic testing and referrals to subspecialists are directed by the results of the initial assessment in order to establish the extent of organ system involvement
Imaging
- Conventional radiography
- Ultrasonography
- CT
- MRI (including cMRI)
- PET-CT, PET-MRI

Treatment
Follow-up and monitoring every 6–12 months
Clinical, laboratory evaluation

ESR—erythrocyte sedimentation rate; CRP—C-reactive protein; TST—tuberculin skin test; IGRA—interferon-gamma release assays; CT—computed tomography; MRI—magnetic resonance imaging; cMRI—cardiac MRI; PET—positron emission tomography

Source: Reference 83

Table 74.10: Some of the conditions that mimick sarcoidosis

Category	*Conditions*
Inflammatory and immunologic diseases	Sjögren's syndrome Psoriatic arthritis and other inflammatory arthritides Primary biliary cirrhosis Granulomatous hepatitis Granulomatosis with polyangiitis (Wegener's disease) and other vasculitides Crohn's disease Combined variable immunodeficiency Giant cell myocarditis Vogt-Koyanagi-Harada syndrome
Infections	Tuberculosis Atypical mycobacterial infections Spirochetal diseases (e.g. Lyme disease, syphilis) Brucellosis Histoplasmosis and other fungal infections Cat-scratch fever
Neoplasms and lymph node diseases	Non-Hodgkin's lymphoma Hodgkin's disease Metastatic malignancies Kikuchi-Fujimoto disease Castleman's disease
Exposures, drugs, and toxins	Chronic beryllium disease Pneumoconiosis (e.g. talc, aluminium) Hypersensitivity pneumonitis TNF inhibitors

TNF—tumour necrosis factor
Source: Reference 83

Assessment of Quality of Life

In evaluating a patient with sarcoidosis, assessing the health-related quality of life (HRQL) is important. HRQL has various domains including emotional, cognitive, spiritual, physical, and social. Based on the extent of involvement of these domains, the decisions for disease management are taken. Certain HRQL which are common to other diseases, are also used in sarcoidosis patients, such as, the Short Form-36 (SF-36), World Health Organization–Quality of Life 100 (WHOQOL-100) and WHOQOL-BREF.[85, 86] The HRQL measures specific for sarcoidosis are the Sarcoidosis Health Questionnaire (SHQ),[87] Kings Sarcoidosis Questionnaire (KCQ), and Sarcoidosis Assessment Tool (SAT).[88] These measures have both organ specific as well as modules focusing on medication effects.

St George Respiratory Questionnaire (SGRQ) is a useful measure for assessing lung involvement. Though SGRQ is not sarcoidosis specific, it has shown good correlation with other sarcoidosis specific patient reported outcomes (PROs).[88]

Fatigue, a common symptom in sarcoidosis, is assessed by Fatigue Assessment Scale (FAS),[89] the Functional Assessment of Chronic Illness Therapy Fatigue (FACIT-F), and the Patient Reported Outcomes Measurement Information Systems (PROMIS) Fatigue Instrument (PFI), and the SAT fatigue subscale. For depression the Beck Depression Inventory and Center for Epidemiologic Studies-Depression Scale have been used. Cutaneous Sarcoidosis Activity and Morphology Instrument (CSAMI) and Sarcoidosis Activity and Severity Index (SASI) have been used to assess cutaneous sarcoidosis disease severity.[90]

Severe Sarcoidosis

Respiratory failure is the most frequent cause of death in sarcoidosis, except in Japanese patients in whom cardiac involvement is common.[5, 91] Pulmonary hypertension is an independent predictor of mortality in sarcoidosis, irrespective of specific organ involvement.[92–94]

Table 74.11: Treatment of sarcoidosis

Organ system	*Indication for immunosupression*	*First-line*	*Second-line*	*Third-line*
Pulmonary	Symptomatic pulmonary disease (cough, wheeze, dyspnoea) with infiltrates Progressive deterioration of pulmonary function (FEV_1, FVC < 70%) Significant upper respiratory tract sarcoidosis	Oral CS* Inhaled CS	Methotrexate Azathioprine Leflunomide Hydroxychloroquine	Infliximab Adalimumab
Ocular	Posterior or intermediate uveitis Anterior uveitis refractory to topical therapy or with toxicities from topical therapy Optic neuritis	Oral CS Topical CS	Methotrexate Azathioprine Leflunamide	Infliximab Adalimumab
Cutaneous Lupus pernio Plaques, nodules Erythema nodosum	Disfiguring lesions	Oral CS* NSAIDs for EN (Lofgren's syndrome)	Methotrexate Azathioprine Leflunomide Hydroxychloroquine Mycophenolate mofetil	
Liver	Impaired synthetic function Hyperbilirubinaemia Progressive increase of transaminase levels Portal hypertension	Oral CS*	Methotrexate Azathioprine Leflunomide Hydroxychloroquine Mycophenolate mofetil	
Splenic	Pain or early satiety caused by enlargement Cytopenias caused by hypersplenism	Oral CS* Azathioprine	Methotrexate	
Cardiac	Second-degree or third-degree conduction block Ventricular dysrhythmias Cardiomyopathy	Oral CS*	Methotrexate Azathioprine Leflunomide Mycophenolate mofetil	Infliximab Adalimumab Cyclophosphamide

(*Contd.*)

Table 74.11: Treatment of sarcoidosis (*Contd.*)

Organ system	*Indication for immunosupression*	*First-line*	*Second-line*	*Third-line*
Neurosarcoidosis	Intracerebral or spinal cord involvement Cranial nerve palsies Pituitary sarcoidosis Granulomatous peripheral nerve disease	Severe cases pulse, i.v. methylprednisolone 500–1,000 mg/day	Methotrexate Azathioprine Leflunamide Cyclosporine Mycophenolate mofetil	Infliximab Adalimumab Cyclophosphamide Rituximab
Small fibre neuropathy	Disabling paresthesias	Pregabalin, amitryptiline	Infliximab Adalimumab Intravenous immunoglobulin	
Haematological	Cytopenia lymphadenopathy	Oral CS*	Methotrexate Azathioprine	Infliximab Adalimumab
Renal Hypercalciuria Hypercalcaemia	Significant hypercalcaemia/ hypercalciuria nephrolithiasis	Oral CS*	Hydroxychloroquine Azathioprine Mycophenolate mofetil	Infliximab
Sarcoidosis associated fatigue		Methyl phenidate Armodafinil Cognitive behavioural therapy Rehabilitation	Infliximab	
Musculoskeletal	Symptomatic myositis Significant disabling arthritis Granulomatous arthritis	Oral CS* NSAIDS	Methotrexate Hydroxy chloroquine Azathioprine Leflunomide	

* for details, see text

CS—corticosteroids; FEV_1—forced expiratory volume in the first second; FVC—forced vital capacity; NSAIDs—non-steroidal anti-inflammatory drugs

Source: References 62, 96

Sudden death can result from cardiac arrhythmias.[95] A left ventricular ejection fraction (LVEF) less than 50% is a contributor to morbidity and mortality in heart disease, and this has also been proven in sarcoidosis patients.[95] Late gadolinium enhancement (LGE) on cardiac magnetic resonance, a marker of myocardial fibrosis that is typically focal, involving the subepicardial or midmyocardial walls of the left ventricle, is an independent predictor of mortality and cardiac adverse events.[95] Recently, an extent of LGE greater than 20% of the myocardial mass has been linked to a higher incidence of cardiac events, even after adjustment for the degree of left ventricular impairment, which correlates with the extent of myocardial fibrosis.[95]

Treatment

The clinical course for sarcoidosis is variable. In about half the cases, spontaneous resolution has been observed within 2–5 years; thereafter, remission is less likely. Sarcoidosis of two years or less duration is acute, while disease duration greater than three to five years is chronic sarcoidosis. Disease progressing despite treatment has been called refractory sarcoidosis. Till date uniform evidence-based treatment protocol and data on clinical outcomes are lacking in sarcoidosis. Hence, treatment of sarciodosis is largely individual patient oriented.

The treatment decisions are based on the following:

i. Presence of incapacitating constitutional symptoms;
ii. High disease activity, progressive disease, presence of severe organ dysfunction or irreversible damage; and
iii. Risk of death.

The treatment options are listed in Table 74.11.[62,96] Oral glucocorticoids constitute the first-line of treatment. Oral prednisolone is administered in a dosage 20–40 mg daily for 6–12 weeks, with the dosage reduced thereafter.[62] In life-threatening sarcoidosis, such as, cardiac sarcoidosis, neurosarcoidosis or severe eye involvement, an initial dose of 1 mg/kg daily is indicated.[96] Most workers advocate a minimum of 12 months of maintenance therapy to prevent relapse. However, some workers suggest that treatment for six months may be adequate. The dosage is adjusted as per the clinical response. Patients with chronic sarcoidosis might require low-dose treatment for many years. In patients with mild skin involvement, antimalarial drugs or tetracyclines might be preferred over corticosteroids.[96] Third-line therapies (e.g. infliximab, adalimumab) are indicated for refractory sarcoidosis. Several clinical scoring systems have been proposed for assessing disease extent and activity,[97] but their long-term benefits merit further study.

Prognosis

Usually, patients with sarcoidosis have good prognosis, and most of them achieve remission in 2–5 years time. The mortality rate is less then 5%.[98] The common causes of morbidity in patients with sarcoidosis are due to organ damage secondary to granulomatous inflammation, treatment related complications, and psychological burden if the disease. The poor prognostic factors which increase the rate of mortality and mobidity and mandate an aggressive therapy include;[98]

i. age over 40 years;
ii. African-American ethnicity;
iii. lupus pernio/cutaneous sarcoidosis;
iv. requirement with high dose steroids;
v. extrapulmonary involvement such as cardiac, neurological (except isolated cranial nerve palsy)
vi. hypercalcemia; and
vii. pulmonary features such as stage 3–4 chest radiograph, pulmonary hypertension,[99] significant lung function impairment leading to moderate to severe dyspnoea, broncho-alveolar fluid neutrophilia.

Key Points

- Sarcoidosis multisystem granulomatous disorder that is seen in all parts of the world.
- Genetic predisposition and exposure to yet unknown transmissible agent(s) and/or environmental factors are considered as aetiological agents for sarcoidosi.
- Consistent with multisystem nature of the disease, patients with sarcoidosis present themselves to various specialities; pulmonary involvement is by far the most common seen in about 90% of the patients.
- The diagnosis of sarcoidosis is based on a compatible clinical and/or radiological picture, histopathological evidence of non-caseating granulomas in tissue biopsy specimens and exclusion of other diseases capable of producing similar clinical or histopathological appearances.
- Corticosteroids remain the mainstay of treatment and should be considered in symptomatic patients with evidence of radiographic or pulmonary function deterioration; steroid-sparing alternative treatments may be beneficial in certain situations.

REFERENCES

1. James DG, Sharma OP. From Hutchinson to now: a historical glimpse. Curr Opin Pulm Med 2002; 8: 416–23.
2. Rybicki BA, Major M, Popovich J Jr, Maliarik MJ, Iannuzzi MC. Racial differences in sarcoidosis incidence: a 5-year study in a health maintenance organization. Am J Epidemiol 1997; 145:234–41.
3. James DG, Hosoda Y. Epidemiology. In : James DG, editor. Sarcoidosis and other granulomatous disorders. Los Angeles: Marcel Dekker; 1994 p. 729–43.
4. Cozier YC, Berman JS, Palmer JR, Boggs DA, Serlin DM, Rosenberg L. Sarcoidosis in black women in the United States: data from the Black Women's Health Study. Chest 2011; 139:144–50.
5. Morimoto T, Azuma A, Abe S, Usuki J, Kudoh S, Sugisaki K, et al. Epidemiology of sarcoidosis in Japan. Eur Respir J 2008; 31:372–9.
6. Ungprasert P, Carmona EM, Utz JP, Ryu JH, Crowson CS, Matteson EL. Epidemiology of Sarcoidosis 1946–2013: A Population-Based Study. Mayo Clin Proc 2016; 91:183–8.
7. Gupta SK. Epidemiology of sarcoidosis in India. In: Proceedings of symposium. VP Chest Institute, Delhi. 2002; 27–31.
8. Hosoda Y, Kosuda T, Yamamoto M, Hongo O, Mochizumi H, Mikami R, et al : A cooperative study of sarcoidosis in Asia and Africa: descriptive epidemiology. Ann N Y Acad Sci 1976; 278:347–54.
9. Sharma SK, Mohan A. Sarcoidosis: global scenario & Indian perspective. Indian J Med Res 2002; 116: 221–47.
10. Gupta SK, Gupta S. Sarcoidosis in India: a review of 125 biopsy proven cases. Sarcoidosis 1990; 7: 43–9.
11. Bambery P, Behera D, Gupta AK, Kaur U, Jindal SK, Deodhar SD, et al. Sarcoidosis in north India: the clinical profile of 40 patients. Sarcoidosis 1987; 4:155–8.
12. Kumar R, Goel N, Gaur SN. Sarcoidosis in north Indian population: a retrospective study. Indian J Chest Dis Allied Sci 2012; 54:99–104.
13. Jindal SK, Gupta D, Aggarwal AN. Sarcoidosis in India:practical issues and difficulties in diagnosis and management. Sarcoidosis Vasc Diffuse Lung Dis 2002; 19:176–84.
14. Rosha D, Panda BN. Sarocidosis in India-an evaluation of 60 cases. Lung India 2000; 18:70–3.
15. Kahsyap S, Kumar M, Pal LS, Saini V. Clinical profile of sarcoidosis in Himachal Pradesh: Northern India. J Assoc Physicians India 1998; 45:574–6.
16. Breyne H, James DG. A tale of two cities. Sarcoidosis 1985; 2:122–3.
17. Djurie B, Handi L, Vezendi S et al. Sarcoidosis in six European cities. In: Williams JW, Davies BH, editors. Sarcoidosis. Cardiff: Alpha Omega 1980.
18. James DG, Neville E, Siltzbach LE. A world wide review of sarcoidosis. Ann N Y Acad Sci 1976; 278:321–34.
19. Hosoya S, Kataoka M, Nakata Y, Maeda T, Nishizaki H, Hioka T, et al. Clinical features of 125 patients with sarcoidosis:Okayama University Hospital review of a recent 10-year period. Acta Med Okayama 1992; 46:31–6.
20. Loddenkemper R, Kloppenborg A, Schoenfeld N, Grosser H, Costabel U. Clinical findings in 715 patients with newly detected pulmonary sarcoidosis–results of a cooperative study in former West Germany and Switzerland. WATL Study Group. Wissenschaftliche Arbeitsgemeinschaft für die Therapie von Lungenkrankheitan. Sarcoidosis Vasc Diffuse Lung Dis 1998; 15:178–82.
21. Judson MA, Boan AD, Lackland DT. The clinical course of sarcoidosis:presentation, diagnosis, and treatment in a large white and black cohort in the United States. Sarcoidosis Vasc Diffuse Lung Dis 2012; 29:119–27.
22. Jayakrishnan B, Al-Busaidi N, Al-Lawati A, George J, Al-Rawas OA, Al-Mahrouqi Y, et al. Clinical features of Sarcoidosis in Oman: a report from the Middle East region. Sarcoidosis Vasc Diffuse Lung Dis 2016; 33:201–8.
23. Newman LS, Rose CS, Bresnitz EA, Rossman MD, Barnard J, Frederick M, et al; ACCESS Research Group. A case control etiologic study of sarcoidosis: environmental and occupational risk factors. Am J Respir Crit Care Med 2004; 170:1324–30.
24. Song Z, Marzilli L, Greenlee BM, Chen ES, Silver RF, Askin FB, et al. Mycobacterial catalase-peroxidase is a tissue antigen and target of the adaptive immune response in systemic sarcoidosis. J Exp Med 2005; 201:755–67.
25. Izbicki G, Chavko R, Banauch GI, Weiden MD, Berger KI, Aldrich TK, et al. World Trade Center "sarcoid-like" granulomatous pulmonary disease in New York City Fire Department rescue workers. Chest 2007; 131:1414–23.
26. Loupasakis K, Berman J, Jaber N, Zeig-Owens R, Webber MP, Glaser MS, et al. Refractory sarcoid arthritis in World Trade Center-exposed New York City firefighters: a case series. J Clin Rheumatol 2015; 21:19–23.
27. Valeyre D, Soler P, Clerici C, Pre J, Battesti JP, Georges R, et al. Smoking and pulmonary sarcoidosis: effect of cigarette smoking on prevalence, clinical manifestations, alveolitis, and evolution of the disease. Thorax 1988; 43:516–24.

28. Rybicki BA, Iannuzzi MC, Frederick MM, Thompson BW, Ros sman MD, Bresnitz EA, et al; ACCESS Research Group. Familial aggregation of sarcoidosis. A case-control etiologic study of sarcoidosis (ACCESS). Am J Respir Crit Care Med 2001; 164:2085–91.
29. Fingerlin TE, Hamzeh N, Maier LA. Genetics of Sarcoidosis. Clin Chest Med 2015; 36:569–84.
30. Valeyre D, Prasse A, Nunes H, Uzunhan Y, Brillet PY, Müller-Quernheim J. Sarcoidosis. Lancet 2014; 383:1155–67.
31. Esteves T, Aparicio G, Garcia-Patos V. Is there any association between sarcoidosis and infectious agents—a systematic review and meta-analysis? BMC Pulm Med 2016; 16:165.
32. Iannuzzi MC, Fontana JR. Sarcoidosis: clinical presentation, immunopathogenesis, and therapeutics. JAMA 2011; 305:391–9.
33. Patterson KC, Hogarth K, Husain AN, Sperling AI, Niewold TB. The clinical and immunologic features of pulmonary fibrosis in sarcoidosis. Transl Res 2012; 160:321–31.
34. Braun NA, Celada LJ, Herazo-Maya JD, Abraham S, Shaginurova G, Sevin CM, et al. Blockade of the programmed death-1 pathway restores sarcoidosis CD4(+) T-cell proliferative capacity. Am J Respir Crit Care Med 2014; 190:560–71.
35. Patterson KC, Franek BS, Müller-Quernheim J, Sperling AI, Sweiss NJ, Niewold TB. Circulating cytokines in sarcoidosis: phenotype-specific alterations for fibrotic and non-fibrotic pulmonary disease. Cytokine 2013; 61:906–11.
36. Miyara M, Amoura Z, Parizot C, Badoual C, Dorgham K, Trad S, et al. The immune paradox of sarcoidosis and regulatory T cells. J Exp Med 2006; 203:359–70.
37. Bonham CA, Strek ME, Patterson KC. From granuloma to fibrosis: sarcoidosis associated pulmonary fibrosis. Curr Opin Pulm Med 2016;22: 484–91.
38. Sakthivel P, Bruder D. Mechanism of granuloma formation in sarcoidosis. Curr Opin Hematol 2017; 24:59–65.
39. Chappell AG, Cheung WY, Hutchings HA. Sarcoidosis: a long-term follow up study. Sarcoidosis Vasc Diffuse Lung Dis 2000; 17:167–73.
40. Gupta SK. Sarcoidosis : a journey through 50 years. Indian J Chest Dis Allied Sci 2002; 44:247–53.
41. Sharma SK, Mohan A. Sarcoidosis in India: Not so rare! J Indian Acad Clin Med 2004; 5:12–21.
42. Scadding JG. Prognosis of intrathoracic sarcoidosis in England. A review of 136 cases after five years' observation. Br Med J 1961; 2:1165–72.
43. Shorr AF, Torrington KG, Hnatiuk OW. Endobronchial involvement and airway hyperreactivity in patients with sarcoidosis. Chest 2001; 120:881–6.
44. Kouranos V, Jacob J, Wells AU. Severe Sarcoidosis. Clin Chest Med 2015; 36:715–26.
45. Nishino M, Lee KS, Itoh H, Hatabu H. The spectrum of pulmonary sarcoidosis: variations of high-resolution CT findings and clues for specific diagnosis. Eur J Radiol 2010; 73:66–73.
46. Yanardag H, Pamuk ON, Pamuk GE. Lupus pernio in sarcoidosis: clinical features and treatment outcomes of 14 patients. J Clin Rheumatol 2003; 9: 72–6.
47. Fernandez-Faith E, McDonnell J. Cutaneous sarcoidosis: differential diagnosis. Clin Dermatol 2007; 25:276–87.
48. Bonfioli AA, Orefice F. Sarcoidosis. Semin Ophthalmol 2005; 20:177–82.
49. Herbort CP, Rao NA, Mochizuki M; members of Scientific Committee of First International Workshop on Ocular Sarcoidosis.. International criteria for the diagnosis of ocular sarcoidosis: results of the first International Workshop on Ocular Sarcoidosis (IWOS). Ocul Immunol Inflamm 2009; 17:160–9.
50. Judson MA. Hepatic, splenic, and gastrointestinal involvement with sarcoidosis. Semin Respir Crit Care Med 2002; 23:529–41.
51. Warshauer DM, Lee JK. Imaging manifestations of abdominal sarcoidosis. AJR Am J Roentgenol 2004; 182:15–28.
52. Modaresi Esfeh J, Culver D, Plesec T, John B. Clinical presentation and protocol for management of hepatic sarcoidosis. Expert Rev Gastroenterol Hepatol 2015; 9:349–58.
53. Statement on sarcoidosis. Joint Statement of the American Thoracic Society (ATS), the European Respiratory Society (ERS) and the World Association of Sarcoidosis and Other Granulomatous Disorders (WASOG). Am J RespirCrit Care Med 1999; 160: 736–55.
54. Le Bras E, Ehrenstein B, Fleck M, Hartung W. Evaluation of ankle swelling due to Lofgren's syndrome: a pilot study using B-mode and power Doppler ultrasonography. Arthritis Care Res (Hoboken) 2014; 66:318–22.
55. Visser H, Vos K, Zanelli E, Verduyn W, Schreuder GM, Speyer I, et al. Sarcoid arthritis: clinical characteristics, diagnostic aspects, and risk factors. Ann Rheum Dis 2002; 61:499–504.
56. Lofgren S. Primary pulmonary sarcoidosis. II. Clinical course and prognosis. Acta Med Scand 1953; 145: 465–74.

Chapter

75

Amyloidosis

Shankar Subramanian, Manjit Ahlawat

INTRODUCTION

First described by Rudolph Virchow in 1854, Amyloidosis is a disease of protein misfolding, characterised by extracellular deposition of amyloid fibrils that affects normal tissue function.[1] Amyloid fibrils, that are around 10 nm in diameter, are characterized by insoluble, rigid and an unbranching β pleated sheet structure. On basis of organ/tissue involved amyloidosis can be classified as localized (fibril deposition limited to one site or organ) or systemic. This review will confine itself to systemic amyloidosis.

EPIDEMIOLOGY

A rare disease, the incidence of systemic amyloidosis increases with age. Current classification is based on chemical characteristic of the precursor protein. About 30 different proteins have been identified that cause amyloidosis, the most common ones being AL, SAA, ATTR and Aβ_2M amyloidosis (Table 75.1). AL amyloidosis is the commonest type in western world while AA is common in developing nations where chronic inflammatory diseases like tuberculosis are still endemic. Kidney is the most common organ involved in amyloidosis followed closely by heart.[1, 2]

PATHOGENESIS

Amyloid formation occurs when a protein fails to acquire its physiological and functional tertiary structure. This misfolded protein interacts with some molecules (serum amyloid P-component, heparan sulphate) and assembles to form fibrils (Fig. 75.1). These accumulate in extracellular space

Table 75.1: Features of common forms of systemic amyloidosis

Type	*Incidence*	*Protein*	*Association*	*Median survival*
AL	10 per million/year	Amyloid light chain	2% of monoclonal B-cell or plasma cell dyscrasia Overt myeloma is rare	6–12 months
AA	2 per million/year	Serum amyloid A protein	Chronic inflammation like TB. Occurs 1–17 years later	3–4 years
ATTR*	Rare	Transthyretin	>120 gene mutations reported, familial amyloid polyneuropathy (FAP) most common	10 years
Aβ2M		β2 microglobulin	ESRD patients on dialysis for 6–10 years	

*Acquired ATTR amyloidosis presents later than hereditary ATTR. Senile amyloidosis now called ATTRwt (wild-type ATTR amyloidosis) is seen in > 10% of people aged > 80 yr in autopsy studies

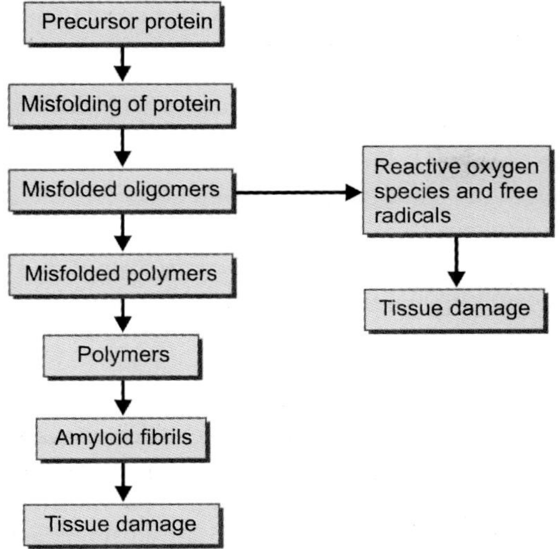

Fig. 75.1: Flowchart highlighting the pathogenesis of amyloidosis

and lead to tissue damage and organ dysfunction. The fibril precursor proteins are commonly synthesized in liver or bone marrow and by unidentified mechanisms trapped at different locations.[3–5] Factors that enhance propensity of a protein to form amyloid fibril include:

a. **Precursor protein concentration:** Under normal conditions a mixture of many types of light chains are generated, some of which may be amyloidogenic in nature. However, due to their heterogeneous nature and low concentration, risk of aggregation is low. In AL amyloidosis a single type of light chain is produced due to clonal expansion of plasma cells. Thus, a critical concentration for amyloid fibril formation may be reached. In AA amyloidosis, chronic inflammation leads to increased concentration of acute phase reactant SAA while in ESRD the concentration of large molecular weight $A\beta_2M$ rises as it cannot be excreted effectively.
b. **Propensity of a protein for mis-folding and aggregation:** Certain proteins like monoclonal immunolglobulin light chains, wild type transthyretin and lipoprotein AL and inherited proteins seen in various hereditary amyloidosis are inherently unstable and amyloidogenic in nature.
c. **Proteolytic remodelling:** Protein mis-folding and proteolytic cleavage may expose amyloidogenic amino acid regions that are normally hidden within folded portion of proteins. For example, cleavage of membrane protein 2B by protease furin leads to release of amyloid β peptides.
d. **Role of other components:** Serum amyloid P-component is a circulating glycoprotein of the pentraxin family and found in all types of amyloid fibrils. It partially protects amyloid fibrils from proteolytic degradation. Heparan sulphate accelerates transition of stable native state of SAA to mis-folded amyloidogenic state. It also interacts with amyloidogenic immunoglobulins, transthyretin and amyloid β protein and accelerates fibril formation.
e. **Aging and amyloidosis:** With advancing age there is inadequate control of protein synthesis and deterioration of proteostasis which increases chances of mis-folded proteins to escape from quality control system of cells and leads to formation of aberrantly folded proteins.
f. **Kinetics of fibril formation:** *In vitro* studies have shown that fibril formation follows process of nucleated growth, usually seen in crystallization. Nucleus generation is slow and is the rate limiting step. After lag phase elongation phase commences, in which fibril formation proceeds by addition of mis-folded proteins to the end of initial aggregates.
g. **Organ tropism:** Various factors such as local protein concentration, tissue specific proteolytic enzymes, local pH, interaction with local collagen, etc. have been implicated in organ tropism.
h. **Tissue damage:** Oligomers are more toxic than mature fibrils and thought to affect structural integrity of cell membrane. They cause cell damage by formation of reactive oxygen species and by disturbance to cellular ion homeostasis and may cause cell death by apoptosis or other mechanisms.

CLINICAL FEATURES

Clinical picture mainly depends on organ involved. Kidneys and heart are commonly involved while CNS involvement is rare (except in ATTR amyloidosis).

Renal involvement characteristically manifests as nephrotic range proteinuria. About 50% AA amyloidosis may have ESRD along with nephrotic syndrome. Nephrotic syndrome with bilateral enlarged kidneys should always have amyloidosis as a differential diagnosis.[3,4]

Cardiac involvement is the most common cause of mortality in amyloidosis. Besides myocardium, the valvular tissue and small vessels supplying the myocardium can be involved. Amyloid cardiomyopathy has a restrictive physiology and is characterized by thickening of interventricular septum, normal size left ventricular cavity and features of diastolic dysfunction. Involvement of conduction system can cause atrial and ventricular arrhythmias. Angina and myocardial ischaemia are rare and usually results from small vessel involvement.[6]

Peripheral neuropathy affects the small fibres and can be predominantly painful (AL) or present with loss of superficial sensations (ATTR). Carpal tunnel syndrome is common. Autonomic neuropathy manifests with diarrhoea, constipation, gastroparesis, impotence or severe postural hypotension that may be debilitating.[5]

Hepatic manifestations are usually benign and typically consist of hepatomegaly with isolated increase in serum alkaline phosphate. Liver failure is rare. Splenic involvement is common but rarely symptomatic.

Skin manifestations include papules, nodules and characteristic periorbital haematoma. Lesions are usually localized to trunk and face.

Osteoarticular manifestations (Aβ$_2$M amyloidosis) include symmetrical polyarthropathy affecting shoulders, wrists, small joints of hand and knees. Spondyloarthropathy and cord compression can also be seen.[7]

GI tract involvement is common but often asymptomatic. Common symptoms are occult bleeding and altered bowel habits. More severe complications include malabsorption, hollow viscous perforation or intestinal on struction. Macroglossia is a characteristic feature of AL amyloidosis.

Infiltration of exocrine glands causes sicca symptoms and endocrine glands may lead to adrenal or thyroid deficiencies.

Cerebral amyloid angiopathy and leptomeningeal amyloidosis are the main features of hereditary ATTR amyloidosis but amyloid angiopathy can be seen with other types of amyloidosis also. It can present as cerebral infarction, haemorrhage, epilepsy, ataxia or dementia.[8]

DIAGNOSIS

Amyloidosis is frequently asymptomatic in early stages and clinical manifestations occur fairly late, so diagnosis relies on a very high index of suspicion. Any patient with symptoms consistent with amyloidosis and no alternative explanation should be investigated thoroughly.[4,5,9] Diagnosis of amyloidosis involves the following steps:

a. **Proving presence in tissue:** Amyloidosis is a tissue based diagnosis and can be demonstrated by characteristic apple green birefringence under polarized light with congo red staining. Abdominal fat pad biopsy is relatively simple and under ideal conditions has up to 90% sensitivity and 100% specificity for AL, AA and hereditary ATTR amyloidosis. Rectal tissue and bone marrow have sensitivity of 80% and 60% respectively. Biopsy from involved tissue has the highest sensitivity (≈ 100%) but to avoid the biopsy of vital organs, abdominal fat pad biopsy, rectum, bone marrow or salivary gland biopsy is preferred. Figure 75.2 shows renal amyloidosis with Congo red staining.
b. **Typing of amyloidosis:** Typing of amyloid is usually done by immunohistochemistry of biopsy sample and is very important as prognosis and treatment differs markedly between different types of amyloidosis. Immunohistochemistry is usually sufficient for AA amyloidosis but is less reliable in ATTR and AL amyloidosis. Proteomic analysis by mass spectrometry is emerging as an important new method.
c. **Determining organ involvement:** Thorough clinical history and clinical examination may point towards organs involved. Serum amyloid P component (SAP) has universal presence in amyloidosis and radio labelled SAP scintigraphy can be used to determine visceral organ involvement and has a high sensitivity for AA and AL amyloidosis (up to 90%). Additionally, it can determine total body amyloid load and monitor progression.

   Gold standard for cardiac amyloidosis is endomyocardial biopsy but is an invasive procedure. ECG abnormalities are common in amyloidosis and low QRS voltage is seen in up to 70% case

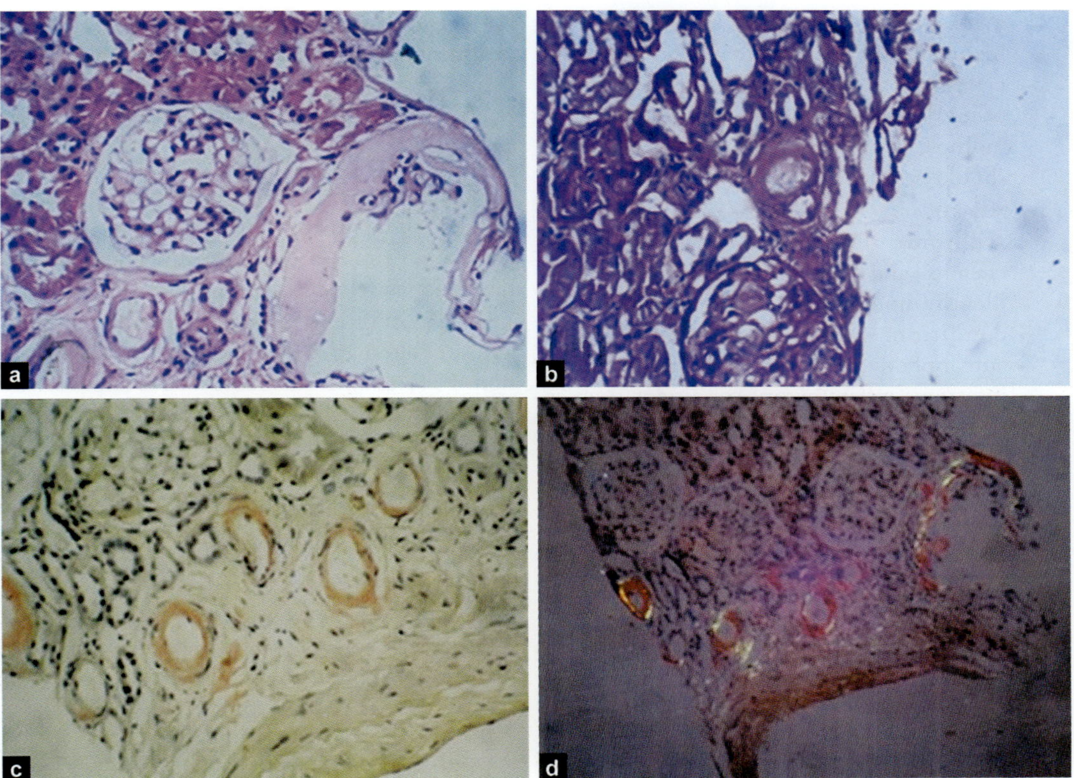

Fig. 75.2: (a) Photomicrograph of kidney biopsy showing deposition of pale homogenous extracellular eosinophilic material in the vessel walls (H and E, X 40). (b) The material is only weakly PAS positive (PAS, X 40). (c) The material is congophilic (Congo red, X 40) and (d) shows apple green birefringence on polarised microscopy (Congo red, polarised light, X 40)

of AL amyloidosis. Imaging is the corner stone for evaluation of suspected cardiac amyloidosis. Echocardiography may demonstrate LV thickness, right ventricular free wall and biatrial enlargement and diffuse valvular thickness. LV ejection fraction usually remains preserved. In advanced cases atrial stand still with intramural thrombosis can be seen. Low voltage ECG with echocardiography demonstrating increased myocardial thickness is highly suggestive of cardiac amyloidosis.

Cardiac CEMRI is emerging as a promising diagnostic modality. IVS and ventricular wall thickening with associated diffuse and delayed gadolinium enhancement is highly suggestive of cardiac amyloidosis. Novel MRI techniques such as MR relaxometery can help in characterization of physical and dynamic properties of cardiac tissue.

Use of cardiac biomarkers such as cardiac troponin T and I (cTnT, cTnI) and brain natriuretic peptide (BNP) has had significant impact on diagnosis and management of AL amyloidosis. Prognostic scoring systems have been developed using free light chain concentration, cTnT levels and NT-proBNP levels.

Demonstration of proteinuria (>0.5 g/d) and enlarged kidneys (USG) can be used for confirmation of kidney involvement along with a positive biopsy from some other tissue.

d. Characterization of underlying disease:

- *AL amyloidosis*: Serum and urine protein electrophoresis can be used to demonstrate paraproteinaemia. Serum free light chain assay (sFLC assay) is used for quantification of free light chain load and for monitoring of progress and treatment response. Bone marrow studies (aspiration and biopsy) with histopathological, cryptogenetic and flow cytometric analysis can be used for diagnosis of underlying plasma cell dyscrasia. If multiple myeloma is suspected skeletal survey should be done.

- *AA amyloidosis*: Identification of clinical syndrome and various serological, biochemical and imaging studies can be done to identify underlying chronic inflammatory condition. Acute phase reactants and autoantibody profile is used to diagnose non-infectious inflammatory conditions such as rheumatoid arthritis, SLE or juvenile inflammatory arthritis. Chronic infection such as tuberculosis can be diagnosed with the help of imaging and microscopic studies.
- *ATTR amyloidosis*: DNA analysis needs to be done to demonstrate TTR mutation.

TREATMENT

Current treatment is based on the precursor-product concept. This concept relies on the principle that break in supply of precursor amyloid protein stops further growth and deposition of amyloid (Table 75.2). Importance of early diagnosis and treatment cannot be over emphasized.

Goals of treatment amyloidosis are

1. Eliminate production or accumulation of amyloidogenic precursor protein
2. Supporting affected organs

AL amyloidosis: Treatment of AL amyloidosis relies on the principle to return abnormally increased levels of serum free light chains to normal level by suppressing or eradicating underlying plasma cell dyscrasia.[10] Thus, treatment of primary amyloidosis is basically treatment of underlying plasma cell or lymphoproliferative disorder. Treatment regimens are adapted from multiple myeloma treatment and are usually administered by hematologists. Common treatment regimens include regimens based on Bortezomib (proteasome inhibitor), Melphalan, Thalidomide and Lenalidomide. The most effective therapy amongst eligible patients is high dose Melphalan followed by autologous stem cell transplantion. Treatment is highly individualized and depends on various factors such as age, organ dysfunction, pace of disease progression, performance status of patient and regime toxicity.

AA amyloidosis: Like primary amyloidosis aim of treatment in secondary amyloidosis (AA amyloidosis) is sustained suppression of precursor protein SAA by treating underlying chronic inflammatory condition.[11]

ATTR amyloidosis: Liver transplant is the current standard of treatment for FAP but is not viable for familial amyloid cardiomyopathy, familial leptomeningeal amyloidosis and wild type ATTR.[8]

In view of various limitations of liver transplant, a host of alternative therapies are under investigation. TTR tetramer stabilizers such as diflunisal and tafamids act by stabilizing native tetramer structure. Tetramer dissociation is necessary for ATTR amyloid fibril formation. Gene therapy especially gene silencing therapy using small interfering RNAs have shown promising results.

Aβ2M amyloidosis: Currently only effective treatment available for Aβ2M amyloidosis is renal transplant but there is some research where use of high-flux membranes and adsorption columns in hemodialysis have been studied.

Supportive/Symptomatic Treatment

Heart Failure

Loop diuretics are the drugs of choice. Other drugs such as β-blockers, ACE inhibitors/ARB antagonists or digitalis are either ineffective or even sometimes dangerous.[6] Amiodarone is the first-line therapy for arrhythmias and can be used for prophylaxis if any complex ventricular arrhythmias are detected on Holter study.

Kidney

In nephrotic syndrome fluid retention is managed with diuretics esp loop diuretics.

Hemodialysis/peritoneal dialysis is indicated for end stage renal disease.[4,10] In selected patients who have achieved persistent hematological remission for a minimum of 1 year renal transplant can be offered.

Autonomic Neuropathy

Disabling autonomic dysfunction and postural hypotension associated with autonomic neuropathy can be managed with use of support stockings along with α agonist midodrine. Dose of midodrine is 2.5 mg TID and can be increased gradually up to

Table 75.2: Organ affection in various common types of systemic amyloidosis

Type of amyloidosis	*Kidney**	*Heart*	*Liver*	*Nerves*	*GI*	*Others*
AL amyloidosis	+++	++	+	+ both peripheral neuropathy autonomic neuropathy	+	Skin Lungs Spleen Locomotor macroglossia
AA amyloidosis	+++	++			+	Endocrine glands spleen
ATTR (mutant) amyloidosis	+	+++		+	+	Leptomeninges eyes
ATTR (wild-type) amyloidosis	++	++		Carpel tunnel syndrome		
Aβ2M amyloidosis					+	Osteoarticular tissue

*In certain types of amyloidosis like ALECT2 and AApoAII, Kidney might be the only organ involved

10 mg TID. Fludrocortisone can also be tried, but because of fluid retention it is poorly tolerated especially if there is associated amyloid heart or renal disease.[3]

Conclusion

Amyloidosis is a rare but potentially devastating disease, which can occur due to deposition of amyloid fibrils of differing types in different body tissues. Diagnosis relies on demonstration of amyloid in various tissues. Cardiac and renal involvement are associated with adverse prognosis. Management is focused on therapies to reduce load of the abnormal protein in the circulation, attempt dissolution of the deposited protein from the tissues as well as symptomatic therapy based on the organ involved.

Key Points

- Amyloidosis is a rare disease of protein mis-folding
- Tissue damage occurs due to extracellular deposition of amyloid fibrils
- AL, SAA, ATTR and $A\beta_2$Mare the most common amyloidosis
- Diagnosis is essentially tissue based
- Treatment comprises of managing the underlying disease
- Prognosis is uniformly poor without definitive treatment

REFERENCES

1. Hazenberg BPC. Amyloidosis. A clinical overview. Rheum Dis Clin North Am. 2013; 39(2):323–45.
2. Chugh KS, Datta BN, Singhal PC, Jain SK, Sakhuja V, Dash SC. Pattern of renal amyloidosis in Indian patients. Postgrad Med J 1981; 57(663):31–5.
3. Gillmore JD, Hawkins PN. Pathophysiology and treatment of systemic amyloidosis. Nat Rev Nephrol 2013; 9(10):574–86.
4. Palladini G, Merlini G. Systemic amyloidoses: What an internist should know. Eur J Intern Med 2013; 24(8):729–39.
5. Pettersson T, Konttinen YT. Amyloidosis-recent developments. Semin Arthritis Rheum 2010; 39(5): 356–68.
6. Patel KS, Hawkins PN. Cardiac amyloidosis: where are we today? J Intern Med. 2015; 278(2):126–44.
7. M'Bappé P, Grateau G. Osteo-articular manifestations of amyloidosis. Best Pract Res Clin Rheumatol 2012; 26:459–75
8. Sekijima Y. Transthyretin (ATTR) amyloidosis: clinical spectrum, molecular pathogenesis and disease-modifying treatments. J Neurol Neurosurg Psychiatry. 2015; 86(9):1036–43.
9. Siakallis L, Tziakouri-Shiakalli C, Georgiades CS. Amyloidosis: review and imaging findings. Semin Ultrasound CT MR. 2014; 35(3):225–39.
10. Kastritis E, Dimopoulos MA. Recent advances in the management of AL Amyloidosis. Br J Haematol. 2016; 172(2):170–86.
11. Westermark GT, Fändrich M, Westermark P. AA amyloidosis: pathogenesis and targeted therapy. Annu Rev Pathol 2015; 10:321–44.

FURTHER READING

1. Baker KR and Rice L.The Amyloidoses: Clinical Features, Diagnosis and Treatment. Methodist Debakey Cardiovasc J. 2012; 8(3):3–7.
2. Wechalekar AD, Gillmore JD, Hawkins PN. Systemic amyloidosis. Lancet. 2016; 387(10038):2641–54.

Chapter

76

Fibromyalgia

Sakir Ahmed, Able Lawrence

INTRODUCTION

Fibromyalgia is a chronic disease characterized by widespread pain, fatigue and varying degrees of other overlapping functional symptoms. It has a prevalence of 1–2% in the general population though community based surveys have estimates of around 5%.[1–3] It predominantly affects women and incurs considerable distress at the personal level, and economic and social burden at societal level. Early descriptions of "neurasthenia" and "fibrositis" drew considerable scepticism and was believed to be psychogenic until Moldofsky et al. reported the association with tender points and widespread pain with characteristic sleep abnormalities and paved the way for the 1990 ACR classification criteria.[4–6] However, there were practical challenges in the use of tender points in the diagnosis as it required considerable practice, had poor reproducibility, and reliability in the hands of non-experts besides recognition of "fibromyalgianess" as a continuous spectrum rather than a discrete entity motivating the development of 2010/2016 criteria.[7]

AETIOLOGY AND PATHOGENESIS

Fibromyalgia (FM) tends to show familial aggregation suggesting genetic risk factors. Genes that alter the function of neurotransmitter function demonstrate the strongest association with fibromyalgia. While COMT and TAAR1 alter catecholamine signalling, GABRB3 mutations alter the sensitivity of inhibitory GABA receptors.[8] On the other hand, mutations in SCN9A that affect pain transmission in the dorsal root ganglia has also been implicated in severe fibromyalgia.[9] However, these studies have not been replicated in independent studies. The ability of serotoninergic and noradrenergic drugs (SSRI, SNRI, SNNRIs) to modify FM has implicated serotonin (5HT) and norepinephrine (NA) in the pathogenesis. Role of ketamine and dextromethorphan has implicated NMDA receptors.[10] Fibromyalgia patients have increased opioids peptides in their CSF while PET has demonstrated decreased µ-opioid receptor binding. Thus, opioid receptors have high baseline occupancy in fibromyalgia.[11] Therefore, opioid drugs may actually be harmful in patients with fibromyalgia.

Quantitative sensory testing in patients with fibromyalgia shows diminished sensory function in a significant proportion of patients and quantitative histological and functional studies have shown decreased unmyelinated sensory fibres in the skin of a subset of patients with fibromyalgia.[12] While the aetiology of these neurophysiological abnormalities is unclear, electron microscopic studies have shown ultrastructural defects in mitochondria of fibromyalgia patients[13] and metabolomics studies have shown altered tryptophan catabolism in FM as well as altered lysophosphocholine platelet activating factor (PAF) receptor interactions.[14] Their role in the pathogenesis is unknown.

Although prevalence of obesity is increased in patients with fibromyalgia, causality is controversial as patients tend to be less physically active. The association between obstructive sleep apnoea and symptoms indistinguishable from fibromyalgia suggests possible role of nocturnal sympathetic overactivity playing a role in this subgroup. Sleep abnormalities were the earliest biomarker for fibromyalgia with the characteristic finding being increased alpha intrusions during slow delta wave sleep. Restless legs syndrome and chronic fatigue syndrome overlap considerably with fibromyalgia and are associated with decreased sleep efficiency and increased sleep latency respectively.[15]

Fibromyalgia is a disorder of central pain. There are defects in descending inhibitory pathways as well as increased ascending pain transmission in the spinal cord. There is attenuated "diffuse noxious inhibitory control (DNIC)" that controls whole body analgesic after a noxious stimulus.[16] Functional neuroimaging has shown increased pain processing in superior temporal gyrus, sensory cortices I and II, as well as insula and putamen. Magnetic resonance spectrometry has shown increased glutamate in insular cortex.[17]

Certain individuals have inherent difficulty in expressing their thoughts and emotions. This is termed alexithymia. Presence of alexithymia is predictive of chronic pain irrespective of current pain status.[18] Acute and chronic stress are frequently implicated by patients before the onset of fibromyalgia. Physical trauma like whiplash injury as well as emotional stress like post-traumatic stress disorder (PTSD), and child abuse predispose to fibromyalgia. Secondary fibromyalgia can arise due to sustained peripheral pain signalling as in inflammatory arthritis, osteoarthritis, etc. Infectious agents like hepatitis C, Ebstein-Barr virus, Parvovirus, Borrelia (Lyme disease) and recently Chikungunya virus have been implicated.

Wind-up phenomenon is increased in fibromyalgia. Wind-up is the temporal summation of pain on repetitive stimulation of peripheral nociceptors afferents. Treatment of peripheral pain generators like myofascial points often lead to improvement in FM. Chronic pain can result in central sensitisation which result from altered pain modulators namely enkephalins and receptors like $\alpha 2\delta$, Calcium channel NK1 receptor and NMDA receptor. Recently the role of unmyelinated nerve fibres in initiating and perpetuating inflammation has been recognised. When inflammation arises in nerves and microglia because of neuronal activation without direct external stimuli, it is called neurogenic neuroinflammation.[10] Rodents exposed to stress in "social defeat" models have been shown to develop immune inflammation in the brain symptoms consistent with fibromyalgia and depression.[19] This demonstrates how stress can initiate and perpetuate fibromyalgia like disease. The peripheral component of neuroinflammation is involves molecules like TRPA1 (transient receptor potential A1), TRPV1 (transient receptor potential vanniloid 1), Toll like receptor (TLRs), NLRs (NOD like receptor), CGRP (calcitonin gene related peptide), etc.[10] Hyper-excitable C-type nociceptors and increase peptidergic innervation of AV shunts have been demonstrated in FM, leading credence to the neuroinflammation theory.

CLINICAL FEATURES

The main complaint of patients with fibromyalgia is persistent and widespread pain. It may start regionally and then spread to involve other body areas. The degree of pain is usually rated by patients as more severe than the pain of rheumatoid arthritis. Early morning stiffness is common but not necessarily present. Body swelling is a common complaint but objective swelling is rare. But when observed in patients, it is attributable to vasomotor phenomena.[10] Diverse sleep abnormalities and sleep disorders are common in fibromyalgia. Patients have non-restorative sleep and disturbance in sleep is associated with further worsening of pain symptoms. The characteristic polysomnographic finding in patients with fibromyalgia is alpha intrusions in slow delta wave sleep. Increased sleep latency and poor sleep efficiency may also be seen in subsets patients.[15]

Fatigue is a common symptom in fibromyalgia and overlaps with chronic fatigue syndrome. Patients generally have poor tolerance to physical activities including exercise. Pathologic fatigue can be distinguished from sleep deprivation by sleep latency which is decreased in latter and increased in patients with chronic fatigue syndrome. Depression is common among patients attending referral centres but epidemiological studies have failed to show association of depression with fibromyalgia. Indeed, depression is a common comorbidity in most chronic diseases. Clinical features of fibromyalgia are summarized in Table 76.1.

Table 76.1: Features of fibromyalgia
1. Generalised body pain
2. Distress
3. Fatigue
4. Sleep disturbance
5. Allodynia
6. Dysthesia
7. Complaint of "swelling"
8. Change in skin colour and temperature
9. Dermatographism*
10. Raynaud's phenomenon*
*Rare in Indian population

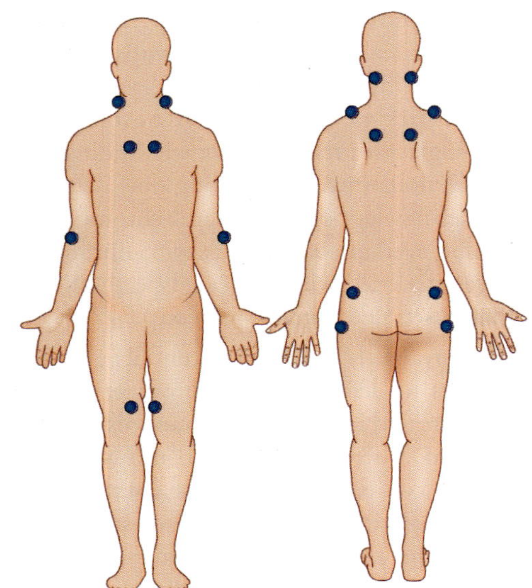

Fig. 76.1: Tender points required for 1990 classification criteria

DIAGNOSIS

Fibromyalgia should be considered in patients with ***persistent and widespread*** pain for at least 3 months. The 1990 criteria required at least 11 out of 18 tender points (Fig. 76.1).[6] While the original 1990 criteria remains valid and reliable in experienced hands, it was found to have drawbacks. In primary care, physicians are often unwilling or untrained to look for tender points. Therefore, an attempt to shift the focus from the expert to the primary physician was made with the preliminary diagnostic criteria of 2010 which was further modified in 2016.[6,19] The ACR 2010 criteria depends on two scales, (a) Widespread pain index (WPI) and (b) Symptoms severity scale (SS). A minimum WPI of 7 with SS of 5, or WPI of 3–6 and SS of at least 9 are required for diagnosis (Fig. 76.2). The 2010 ACR criteria as

Widespread pain (in last 7 days) (✓)			
R Jaw		L Jaw	
R shoulder		L shoulder	
R upper arm		L upper arm	
R lower arm		L lower arm	
Neck			
Chest			
Abdomen			
Upper back			
Lower back			
R hip or buttock		L hip or buttock	
R upper leg		L upper leg	
R lower leg		L lower leg	
WPI = (Total no of ticks)			

Symptom severity	
Severity of each symptom during the last ***7 days***	Score*
Fatigue	
Trouble thinking or remembering	
Waking up tired	
* Scoring: 0 no problem; 1 slight problem; 2 moderate problem; 3 severe problem	
Tick (✓) if present in last ***6 months***	
Pain or abdominal cramp	
Depression	
Headache	
SS = (sum of symptom severity plus no of ticks)	

Fig. 76.2: 2016 modification of the 2010 American College of Rheumatology (ACR) diagnostic criteria for fibromyalgia. WPI, widespread pain index; SS, symptom severity

originally proposed was applicable only after exclusion of all other co-morbid conditions. However, the 2016 modification no longer require any exclusion but emphasised the requirement of persistent and widespread pain for at least 3 months. The 2016 criteria defined widespread pain as pain present in at least four out of five regions (the four quadrants and spine).[20] The diagnosis of fibromyalgia can be made in parallel with any other diagnosis (Table 76.2). It should be kept in mind that the 2016 criteria have not superseded or invalidated the 1990 criteria. Thus, both criteria should complement one another. The 2010/2016 criteria were based on the realisation that fibromyalgia forms a continuous spectrum from the normal to the very severe and Poly Symptomatic Distress Scale (= WPI + SS) gives an index of fibromyalgia severity, i.e. "fibromyalgianess".

A major caveat in the diagnosis of fibromyalgia is the spectrum of overlapping disorders that can both mimic fibromyalgia as well as co-exist with it. Some of these disorders share common pathophysiology and have been grouped together as "functional somatic syndromes (FSS)" or "central sensitization syndrome". Thus, it is important for the physician diagnosing fibromyalgia to be familiar with and screen for these disorders even among patients with established fibromyalgia.

Table 76.2: Functional somatic syndrome

Conditions mimicking and/or overlapping fibromyalgia
A. Inflammatory conditions
1. Seronegative spondyloarthropathy
2. Inflammatory myositis
3. Rheumatoid arthritis
4. SLE
B. Mechanical syndromes
1. Hypermobility syndrome
2. Degenerative disk disease
3. Myofascial pain
4. Osteoarthritis
5. Temporomandibular joint syndrome
6. Facet arthropathy
C. Functional disorders
1. Dysmenorrhoea
2. Chronic fatigue syndrome
3. Functional bowel disorder syndrome
4. Hyperventilation syndrome
5. Interstitial cystitis
6. Irritable bowel syndrome
7. Migraine headache
8. Periodical Limb Movement Syndrome (PLMS)
9. Vestibular disorders
D. Psychiatric disorders
1. Mood disorders
2. Stress disorders (PTSD)
3. Anxiety disorders

The minimum investigations required in a patient of fibromyalgia are haemoglobin, acute phase reactants (ESR, CRP), serum calcium and phosphate, alkaline phosphatase and thyroid function tests. Depending on clinical presentation, Creatinine kinase (CPK), X-ray or MRI of sacroiliac joints, HIV, etc. may also be required. Polysomnography may be done in selected patients to diagnose sleep disorders. Today ultrasound is an important tool to pick-up inflammatory arthritis in doubtful cases.

MANAGEMENT

Patients with fibromyalgia require multidisciplinary care (Table 76.3). Patient education reassurance forms the cornerstone as patients who are usually anxious and worried about serious medical conditions. Cognitive behavioural therapy and mindfulness meditation would help reduce maladaptive behavioural patterns. Patients should be encouraged to increased physical activity and exercise and obese patients should reduce weight.

Table 76.3: Management options

Management
A. Non-pharmacological
1. Patient education
2. Exercise and physiotherapy
3. Cognitive behaviour therapy
B. Pharmacological
1. Anticonvulsants: Pregabalin, gabapentin
2. TCA: Amitriptyline, cyclobenzaprine
3. SNRIs: Duloxetine, milnacipran
4. 5HT3 antagonist: Tropisetron
C. Others
1. Mindfulness techniques
2. Tai chi
3. Yoga
4. Acupuncture

Abbreviations: TCA, tricyclic antidepressants; SNRIs, serotonin and norepinephrine reuptake inhibitors; 5HT3, 5-hydroxytryptamine 3 (serotonin)

Chapter

77

Joint Hypermobility Syndrome

Bidyut Kumar Das

INTRODUCTION

Hypermobility of the joints denotes an increased range of joint movement. It has been recognized since the time of Hippocrates and is commonly appreciated as a contributing factor to excellence, for certain professions like gymnastics, musicians, and ballet dancing. Joint mobility is pronounced at birth but declines thereafter. It is greater in females and people of Asian or African descent. Hypermobility syndrome, also now known as joint hypermobility syndrome (JHS) is a common disorder. Originally defined as the occurrence of symptoms in otherwise healthy hypermobile individuals. It is now seen as a multisystemic disorder of connective tissue that is responsible for chronic widespread pain (often mislabelled fibromyalgia), gastrointestinal dysmotility, anxiety and phobic states, and dysautonomias, thus adding additional symptoms to the musculoskeletal sequelae of joint pain and instability. Most authorities regard it as being the same as the hypermobility type of Ehlers-Danlos syndrome (EDS-HT), formerly EDS type III. Unfortunately, JHS still remains a less understood entity.

JHS shows an autosomal dominant inheritance with variable penetrance. It is three times more common in females than in males. Hormonal influences may play a role, although the mechanisms have not been identified. The precise genetic defect underlying JHS is unknown. It is presumed that mutation in the collagen gene, affects the cross-linking structure of collagen resulting in laxity of the tissue.[1] Current studies have focussed on Tenascin X (TNX), a putative gene, whose mutation has been linked to hypermobility in experimental animal model and also to the hypermobile phenotype of EDS.[2]

CLINICAL PRESENTATION

Musculoskeletal

JHS may manifest as (a) soft tissue injury or overuse lesion affecting a ligament, tendon, muscle, enthesis, or joint; (b) recurrent joint instability, subluxation, or dislocation; (c) chronic non-inflammatory joint or spinal pain without structural abnormality; and (d) secondary osteoarthritic or spondylotic changes in peripheral or spinal joints, respectively. The differentiating feature of patients with JHS from other aetiology of hypermobility is the range and frequency of the manifold varieties of acute and chronic painful episodes occurring over the course of their lives, with increase in intensity of pain over time. Avoidance of painful movement as a strategy to avoid pain (described as kinesiophobia) inevitably leads to muscle deconditioning, the development of adverse postures, and loss of function, which leads to poor quality of life.

Extra-articular Manifestations

- Thin skin with paper thin scars
- Stretch marks (striae atrophicae)
- Hiatal hernias
- Gastroesophageal reflux
- Abdominal wall hernias
- Uterine or rectal prolapsed

- Rectocele or cystocele in parous women
- Varicose veins
- Mitral valve prolapse
- Spontaneous pneumothorax
- Psychiatric disturbances like depression, anxiety, phobia and panic attacks.[3]
- Autonomic disturbance, including presyncope, syncope, and palpitations as a consequence of orthostatic hypotension, orthostatic intolerance, and postural orthostatic tachycardia syndrome.[4]
- Gastrointestinal symptoms such as bloating, reflux, constipation, and diarrhoea.
- Pan intestinal dysmotility, which may result from laxity of intestinal connective tissue.[5]

DIAGNOSIS

There is no biochemical or genetic marker at present to confirm the diagnosis. It is entirely clinical and based on certain validated criteria. The first quantitative system for analyzing joint hypermobility was suggested in 1964 which was subsequently modified in 1973 and widely accepted as Beighton score (Table 77.1).[6] The scoring system rests on the ability of the individual to perform certain passive manoeuvres in nine determined sites resulting in a nine point scoring system. There is no single diagnostic test for hypermobility. Figure 77.1a–e demonstrates the manoeuvres to be performed at the designated joints. The major limitations are that it is an "all or none" criterion that gives no indication of severity. Because only five body sites are sampled, hypermobility may be missed, especially the pauciarticular variety. Hypermobility may be uncovered by a person's response to a validated 5-point questionnaire. Two affirmative answers would suggest an 85% chance of being hypermobile (Table 77.2).[7]

Table 77.1: Beighton hypermobility score

Assessment site	*Right*	*Left*
Hyperextension of 5th MCP joint (Fig. 77.1a)	1	1
Thumb touching the forearm (Fig. 77.1b)	1	1
Hyperextension of elbow >10^0 (Fig. 77.1c)	1	1
Hyperextension of knee joint >10^0 (Fig. 77.1d)	1	1
Hand touching the floor with full palms with knees extended (Fig. 77.1e)	1	
Maximum possible score	9	
Hypermobility is present if total score ≥4		

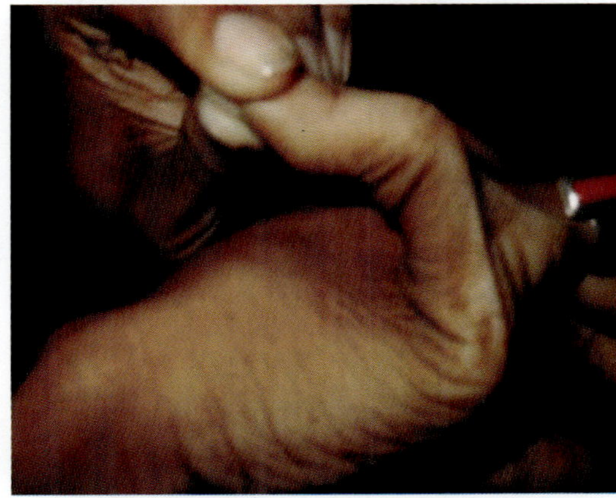

Fig. 77.1a: Passive dorsiflexion of the little finger beyond 90°

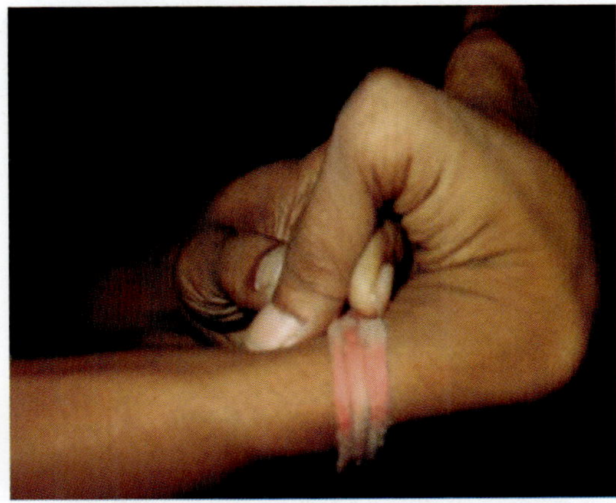

Fig. 77.1b: Passive apposition of the thumb to the flexor aspect of the forearm

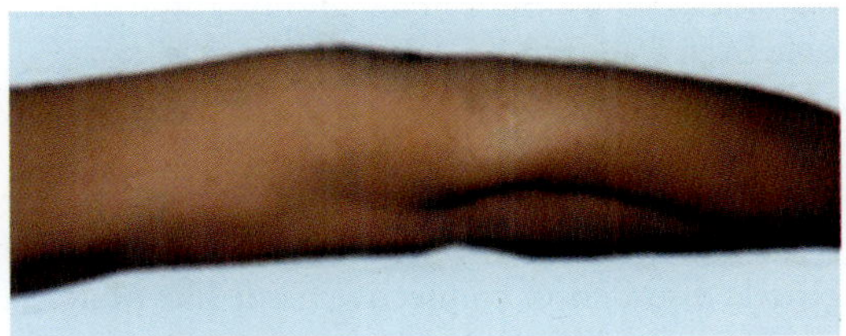

Fig. 77.1c: Hyperextension of the elbow beyond 10°

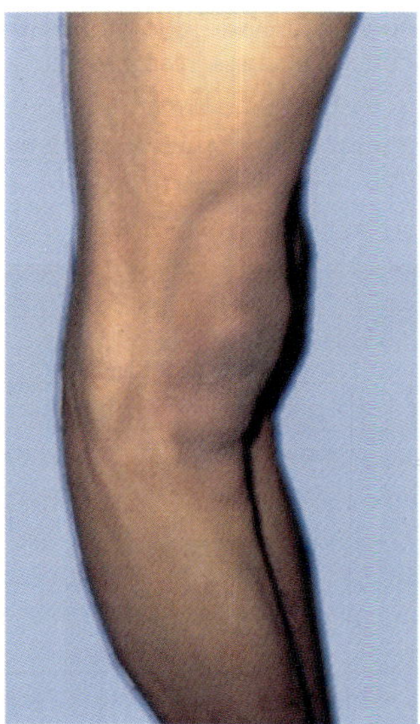

Fig. 77.1d: Hyperextension of the knee beyond 10°

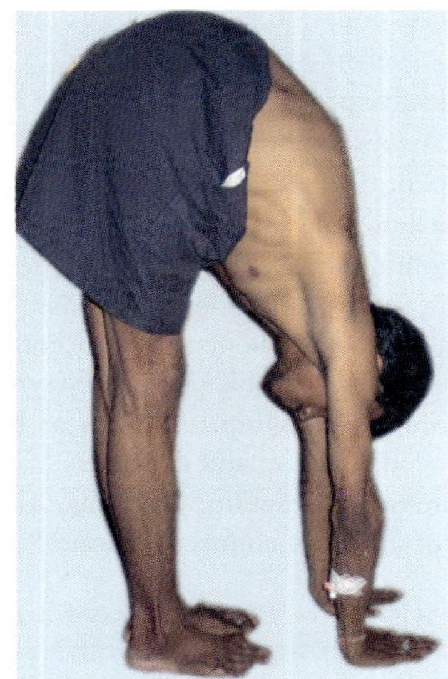

Fig. 77.1e: Forward flexion of the trunk with the knees straight, with palms resting on the floor

Table 77.2: Five-part questionnaire for identifying hypermobility

- Can you now (or could you ever) place your hands flat on the floor without bending your knees?
- Can you now (or could you ever) bend your thumb to touch your forearm?
- As a child did you amuse your friends by contorting your body into strange shapes or could you do the splits?
- As a child or teenager did your shoulder or knee cap dislocate on more than one occasion?
- Do you consider yourself double-jointed?

The diagnosis of JHS is based on clinical assessment of signs and symptoms included under the revised Brighton criteria 1998 comprising of major and minor points (Table 77.3).[8] The reproducibility of this diagnostic criteria is at present unknown. However, in the hands of experienced examiners, both Beighton test and Brighton criteria have been found to be reproducible.

DIFFERENTIAL DIAGNOSIS

JHS has to be distinguished from more severe genetic disorders of connective tissue like Marfan's syndrome, EDS and osteogenesis imperfecta (Table 77.4). There are no reliable laboratory tests

Table 77.3: Revised diagnostic criteria for benign joint hypermobility syndrome (Brighton 1998)

Major criteria

A Beighton score more than or equal to 4

Arthralgia for longer than 3 months in 4 joints or more

Minor criteria

Beighton score of 1 to 3 (0, 1, 2, or 3 if aged more than 50 years)

Arthralgia >3 months in 1 to 3 joints or back pain >3 months, spondylosis, spondylolysis, or spondylolisthesis

Dislocation or subluxation in more than one joint, or in one joint on more than one occasion

Soft tissue rheumatism >3 lesions (epicondylitis, tenosynovitis, bursitis)

Marfanoid habitus (tall, slim, arm span: height ratio >1.03 or upper: lower segment ratio <0.89)

Abnormal skin: striae, hyperextensibility of skin, papyraceous scars, thin skin

Eye signs: drooping eyelids, myopia or anti-mongoloid slant

Varicose veins, hernia, uterine/rectal prolapse

- *Benign Joint hypermobility syndrome (BJHS) is diagnosed if there are 2 major OR 1 major + 1 minor OR 4 minor criteria. Two minor criteria sufficed if an affected first degree relative is present.*
- *Criteria major 1 and minor 2, as well as major 2 and minor 2 are mutually exclusive.*
- *BJHS is excluded if Ehlers-Danlos (other than type III or hypermobility type) and Marfan syndrome are present.*

Table 77.4: Other genetic connective tissue disorders to be considered in the differential diagnosis of joint hypermobility syndrome

Site	*Marfan's syndrome*	*Ehlers-Danlos syndrome*	*Osteogenesis imperfecta*
Joints	Hypermobility	Hypermobility	Hypermobility
Skeleton	Marfanoid habitus	Osteoporosis	Osteoporosis ++
Skin	Hyperextensibility	Hyperextensibility +	Hyperextensibility
Eyes	Ectopia lentis		Blue sclera
Vasculature	Aortic dilatation Mitral valve prolapse	Mitral valve prolapse Intracranial aneurysm	Mitral valve prolapse

that can establish the diagnosis of these disorders. But there are well enumerated clinical diagnostic criteria for Marfan's syndrome, Ehlers-Danlos syndrome and JHS.

Certain acquired diseases may present with polyarticular laxity of joints, namely, acromegaly, hyperparathyroidism, rheumatic fever (Jaccoud's arthropathy) and chronic alcoholism. They can be clearly identified from history and clinical examination.

MANAGEMENT

JHS is a difficult condition to treat. Multidisciplinary approach to treatment is recommended. Simple analgesics and anti-inflammatory drugs play an important role in the management of acute and acute-on-chronic pain from soft tissue injuries and degenerative disease. The majority of patients, however, report that these agents are of little benefit in the control of chronic pain. As with other chronic pain syndromes both serotonergic and noradrenergic agents can be used, although no randomized controlled trials of the use of these agents for JHS have been conducted. Local injection of corticosteroid may also be of value in treating isolated soft tissue lesions, corticosteroid inhibits fibroblast formation, and this may add to the risk for delayed or poor healing of soft tissue already inherent in JHS. So, judicious use of corticosteroids is warranted.[9]

Conventional physiotherapy has often been deemed unhelpful (or counter productive) by patients with JHS. However, a new approach has been developed in which physiotherapeutic methods are used to build-up the muscle strength responsible for joint and core stability, restore diminished proprioceptive acuity, use of mobilizing techniques.[10] Finally, educating the patient about JHS and working with clinical psychologists to enable them to manage chronic pain by various techniques like pacing and relaxation are very helpful in long-term management.

Key Points

- The joint hypermobility syndrome (JHS) is the most common disorder among the hereditary disorders of connective tissue, affecting a subset of the 10 to 20 per cent of the general population.
- The major clinical features of the JHS include symptoms and findings related to the musculoskeletal system and skin.
- Systemic features are often present, including chronic widespread pain, fatigue, autonomic dysfunction, and gastrointestinal dysmotility.
- The diagnosis of JHS is made clinically, based upon the medical history and physical examination, using the Brighton 1998 criteria for JHS.
- The differential diagnosis for JHS includes the conditions that have generalized joint hypermobility as a clinical feature, particularly Marfan syndrome and EDS, other than EDS-HM.
- Treatment of patients with JHS should be individualized. Patients should receive education regarding the nature of the condition, and pain management should employ a multidisciplinary approach.
- The management of systemic manifestations associated with JHS is generally the same or very similar to usual management for these conditions in patients who do not have JHS.

REFERENCES

1. Henney AM, Brotherton DH, Child AH, et al. Segregation analysis of collagen genes in two families with joint hypermobility syndrome. Br J Rheumatol 1992; 31:169–74.

2. Zweers M, Hakim AJ, Grahame R, Schalkwijk J. Tenascin-X deficiency and haploinsufficiency as a cause of generalized joint hypermobility. Arthritis Rheum 2004; 50:2742–9.

3. Gratacós M, Nadal M, Martin-Santos R, Pujana MA, Gago J, Armengol L, et al. A polymorphic genomic duplication on human chromosome 15 is a susceptibility factor for panic and phobic disorders. Cell 2001; 106:367–79.

4. Gazit Y, Nahir AM, Grahame R, Jacob G. Dysautonomia in the hypermobility syndrome. Am J Med 2003; 115:33–40.

5. Zarate N, Farmer AD, Grahame R, Mohammed SD, Knowles CH, Scott SM, et al. Unexplained gastrointestinal symptoms and joint hypermobility: is connective tissue the missing link? Neurogastroenterol Motil 2010; 22(3): 252–e78.

6. Beighton PH, Solomon L, Soskolne CL. Articular mobility in an African population. Ann Rheum Dis 1973; 32:413–7.

7. Hakim AJ, Grahame R. A simple questionnaire to detect hypermobility: an adjunct to the assessment of patients with diffuse musculoskeletal pain. Int J Clin Pract 2003; 57:163–6.

8. Grahame R, Bird HA, Child A. The revised (Brighton 1998) criteria for the diagnosis of benign joint hypermobility syndrome (BJHS). J Rheumatol 2000; 27:1777.

9. Harvey W, Grahame R, Panayi GS. Effects of steroid hormones on human fibroblasts *in vitro*. II. Antagonism by androgens of cortisol-induced inhibition. Ann Rheum Dis 1976; 35:148–51.

10. Ferrell WR, Tennant N, Sturrock RD, Ashton L, Creed G, Brydson G, et al. Amelioration of symptoms by enhancement of proprioception in patients with joint hypermobility syndrome. Arthritis Rheum 2004; 50:3323–8.

Chapter

78

Cardiovascular Disease in Autoimmune Rheumatic Diseases

Vishad Viswanath

INTRODUCTION

Aided by the advances in immunology and genetics, the last two decades have witnessed a paradigm shift in the management of autoimmune diseases. On the diagnostic front, early detection of autoimmune diseases, even preclinical disease, has become possible increasing scope for early intervention. Therapeutically, there has been an ever expanding armamentarium of medications for safer and better disease control. As a result, the mortality and morbidity related to both disease activity and long-term treatment has reduced. The improved longevity in autoimmune diseases has not been without consequences. Therein, a novel pattern has emerged where cardiovascular diseases (CVD) are now the leading cause of mortality and morbidity in autoimmune rheumatic diseases.

Epidemiology of Cardiovascular Disease in Autoimmune Rheumatic Diseases

A higher risk of CVD (defined as any of myocardial infarction, angina, coronary intervention, stroke or peripheral vascular disease) is being increasingly identified in patients with autoimmune diseases. The most well-studied among these is the association between rheumatoid arthritis (RA) and CVD. Patients with RA have two fold increased risk for developing myocardial infarction (MI) in comparison to general population. This risk for CVD in RA is identical to that seen in type II diabetes. In addition, patients with RA have an increased propensity for silent MI, recurrent MI and increased case fatality rate following an MI. While atherosclerosis is the most important cause of CVD in RA, other causes such as autonomic dysfunction, micro-vascular disease, and non-ischaemic heart disease are also contributory. Among the risk factors for CVD include raised ESR, persistent synovitis, extra-articular manifestations, elevated RF and anti-CCP. Indeed this risk of CVD in RA is in excess of that can be predicted by the traditional cardiovascular risk factors.[1]

CVDs are the major cause for late mortality in SLE as well. A bimodal pattern of mortality has been known in SLE since 1970s with deaths early in the disease course (first peak) occurring due to disease activity and infections, while late mortality (second peak) is predominantly due to CVD. In fact, there is 2–10 fold increased risk of MI in SLE with almost 50 times increased risk in women aged 35–44 years. Poor prognosis and higher chance for reinfarction are seen in patients with SLE post angioplasty. Endothelial dysfunction as demonstrated by a reduction in flow mediated vasodilatation, reduced coronary reserve as documented by reduced coronary dilatation to adenosine, myocardial perfusion defects demonstrated by SPECT and increased carotid plaque burden shown by carotid USG are all more common in SLE patients as compared to controls. Higher disease activity, presence of lupus nephritis, anti-cardiolipin antibody positivity, corticosteroid usage (>20 mg/day) and increased disease duration are all associated with higher CVD risk in SLE.[2]

Studies in ankylosing spondylitis and psoriatic arthritis also demonstrate higher CVD risk with risk ratios approaching that seen in RA. Higher CVD risk or endothelial dysfunction has also been established in scleroderma. Increased CVD related complications are seen in vasculitic diseases as well. In addition to the direct injury to the arterial wall and resultant vessel wall abnormality, traditional risk factors as hypertension, insulin resistance and corticosteroid use contributes to the CVD risk in vasculitis. Coronary arteritis and consequent ischaemic heart disease is well-known in vasculitic disorders as Takayasu's arteritis, polyarteritis nodosa and Kawasaki disease. The association between gout and CVD is well-known, with nearly 60% increased risk of CVD and mortality due to CVD in patients with gout.[3]

PRECLINICAL AUTOIMMUNITY AND ATHEROSCLEROSIS[4]

With widespread availability of autoantibody testing it is not uncommon to detect auto-antibodies in apparently asymptomatic individuals. In many autoimmune diseases, appearance of auto-antibodies have been known to predate onset of clinical disease. Such a situation is considered as a state of preclinical autoimmunity. Preclinical autoimmunity is being increasingly recognized as a risk factor for CVD. The association between anti cardiolipin antibodies and CVD is well-known. Not only is there increased risk for thrombosis, anti-cardiolipin antibodies cause endothelial dysfunction and vessel stenosis as well. Population studies have shown an association between a higher risk for CVD in patients with Rheumatoid factor positivity without joint symptoms. Other antibodies as ANA and anti-CCP in clinically asymptomatic patients have also been recognized as a risk factor for CVD. In a recent study, in patients with ST elevation MI without RA, having anti-CCP antibody positivity higher long-term mortality was reported [HR-3.1 (CI 1.4–7.1)].[5] It has been hypothesized that the anti-CCP antibodies get incorporated into the atheromatous plaque driving further inflammation.

INFLAMMATION AND ATHEROSCLEROSIS

The role of inflammation in pathogenesis of atherosclerosis and incident CVD has been in focus for last two decades. Rather than being a process of passive lipid accumulation and eventual vessel wall occlusion, both innate and adaptive immune systems are key players in atherosclerosis. Hence, it is not surprising that autoimmune diseases, characterized by profound systemic inflammation have a higher incidence of atherosclerosis.

Endothelial Dysfucnction and Atherosclerosis[6,7]

Endothelial dysfunction is one of the earliest events leading to atherosclerosis. Normal vessel wall endothelium has anticoagulant and anti-inflammatory properties and modulates the vessel tone producing vasodilator mediators as Nitric oxide. However, a dysfunctional endothelium over expresses pro coagulant and pro-inflammatory molecules like clotting factors, cytokines and immune adhesion molecules and generates reduced amounts of nitric oxide.[6]

Traditional CVD risk factors as hypertension, insulin resistance and smoking as well as inflammatory mediators as TNF-α and IL-1 can induce endothelial dysfunction. The pro inflammatory and pro coagulant mediators produced from the dysfunctional endothelium induce leucocyte chemotaxis and enhance endothelial cell permeability. The chemo-attracted leucocytes; predominantly macrophages and lymphocytes; along with small lipid molecules enter into the subendothelial layer by a process aided by local expression of adhesion molecules. Upon entry into the sub endothelial layer, the monocytes differentiate into macrophages. The LDL component of lipids become oxidized and are taken up by macrophages. These lipid laden macrophages are called foam cells and secrete pro-inflammatory cytokines as TNF-alpha and IL-1 (Fig. 78.1).

The inflammatory milieu in the subendothelial layers of the blood vessel leads to subendothelial smooth muscle proliferation and neovascularization which along with the foam cells form the atheromatous plaque. The new blood vessels formed in the plaque are leaky and fragile. As the plaque grows, it becomes increasingly complex and the core becomes relatively hypoxic. The foam cells in the core undergo apoptosis leading to deposition of its lipid component thereby forming a soft lipid core. In the early stages, the thrombogenic content in the plaque is prevented from coming in contact with the blood stream by a fibrous cap. However, as the local inflammation progresses, production of

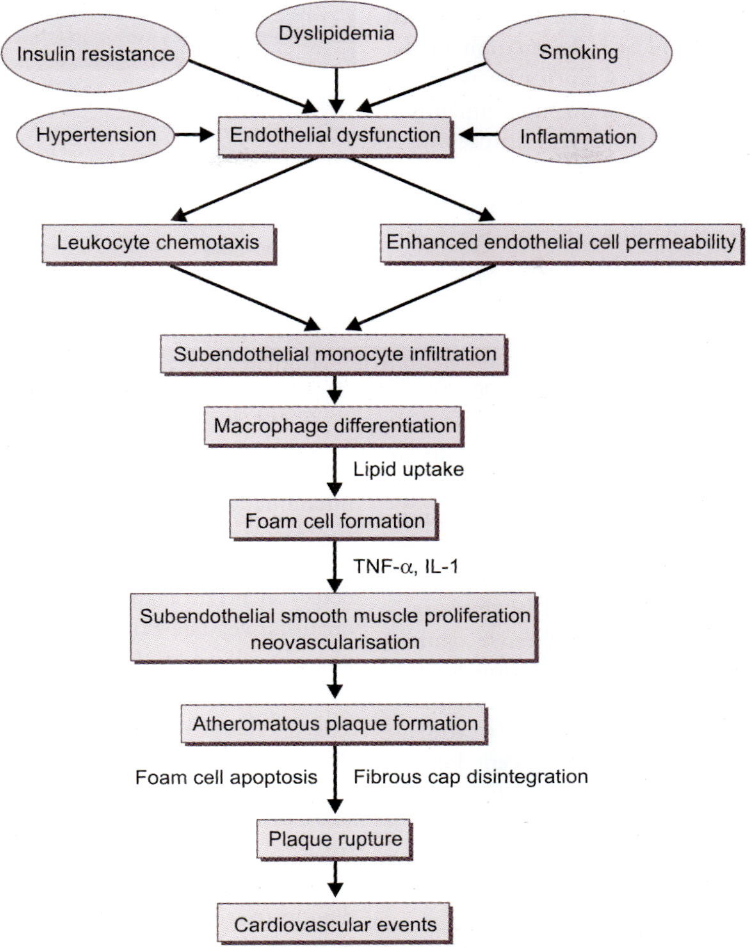

Fig. 78.1: Pathogenesis of cardiovascular diseases (CVD) in rheumatoid arthritis (RA)

enzymes like MMP-3 digests the collagen in the fibrous cap. In addition the fragile vessels can hemorrhage into the lesion leading to change in size and morphology of the plaque. The outcome of this process is rupture of the plaque and its exposure to the circulation. Rupture of the atheromatous plaque has been associated with cardiovascular events. At the time of rupture, the plaque is rich in inflammatory cells as macrophages, Th1 lymphocytes and neutrophils.

Role of Chronic Inflammation in Atherosclerosis

Chronic inflammation has been known to modulate traditional as well as nontraditional CVD risk factors. Endothelial dysfunction has been recognized even in the incipient stages of inflammatory disease as RA, SLE, systemic sclerosis, ankylosing spondylits (AS) and psoriatic arthritis even in the absence of traditional risk factors. High levels of pro-inflammatory mediators as TNF-α, IL-1 and Interferon-γ are seen in these diseases and are known to produce endothelial dysfunction. In addition, these mediators also lead to increased expression of TLRs and pro-inflammatory mediators as MMP 3 and MCP-1. Indeed the process of atherosclerosis recapitulates many events in synovial inflammation in RA.[7]

RA patients have more plaque burden than general population and the burden is higher in those with active RA. Atherosclerosis in RA progresses most rapidly within the first 6 years of diagnosis and the risk of CV morbidity starts to increase from

7–10 years of onset. Imaging studies have confirmed the presence of subclinical inflammation in the blood vessels of RA patients. Culprit atherosclerotic lesions in patients with RA show more inflammation than stenosis. Histological studies have demonstrated that the atherosclerotic plaques in patients with RA are unstable and more prone to rupture compared to those in non-RA CVD subjects. Not surprisingly, RA patients with active disease have more unstable plaques than those with well-controlled disease.

Shared Risk Factors between Inflammatory Disease and CVD

Certain genetic and environmental risk factors are common to RA and CVD. Genome wide association studies (GWAS) have identified the shared epitope (HLA DRB1) to be a risk factor for both RA and CVD. Another gene, TNF -304, polymorphism is seen in both RA and CVD. Similarly, smoking, a major risk factor for CVD, has also been recognized as one of the strongest known risk factors for RA. Not only are smokers predisposed to RA, but they are prone to be poorly responsive to treatment; predisposing to higher disease activity. Periodontitis, especially due to *Porphyromonas gingivalis,* has been recently recognized as a risk factor for anti- CCP positivity and RA. Emerging data shows that the same may be a risk factor of CVD as well.

Role of Traditional Risk Factors in Atherosclerosis[7,8]

Traditional CVD risk factors as insulin resistance, hypertension, and metabolic syndrome are more common in RA. The severity of insulin resistance correlates with disease activity and treating RA has been shown to bring down the insulin resistance. An interesting relation has been shown between body mass index and CVD in RA. Although visceral obesity increases insulin resistance and incident CVD, patients with active RA are leaner than those with well-controlled disease and are more susceptible to CVD (Table 78.1).

Dyslipidemia, albeit in a non-traditional pattern is seen in RA. Patients with active RA have low levels of both LDL and HDL cholesterol. This so called lipid paradox, where patients with active RA tend to have lower LDL and total cholesterol but a higher risk for CVD, is believed to be due to higher catabolic rate of cholesterol esters in the setting of inflammation. The lipid paradox is reversible with treatment; best demonstrated with Janus kinase inhibitors and anti-IL- 6 monoclonal antibody. The defect in HDL is appears to be qualitative than quantitative. The HDL fraction in RA is rich in phosphorylated HDL which has a pro atherogenic effect by causing oxidation of LDL and promoting foam cell formation. A number of HDL associated proteins also undergo alterations in RA and other inflammatory states resulting in impairment of HDL mediated cholesterol efflux from the peripheral tissues.

Table 78.1: Risk factors for atherosclerotic CVD in autoimmune rheumatic diseases

Non-traditional	*Traditional*
• Disease activity	• Age
• Genetics-HLA-DRB1 epitope, TNF 304 polymorphism	• Hypertension • Diabetes mellitus • Smoking
• Periodontitis	• Insulin resistance
• Anti-cardiolipin antibody	• Metabolic syndrome
• Nephritis	• Dyslipidemia
• Drugs: Corticosteroid, NSAIDs	
• Vasculitis	

An increased risk for hypertension has been demonstrated in RA in some studies. This could be the result of using medication like corticosteroids as well as higher peripheral vascular resistance due to inflammation.

Atherosclerosis in SLE

In SLE, in addition to endothelial dysfunction, presence of anti-cardiolipin antibodies, nephritis and use of corticosteroids predispose to CVD. Interferon alpha, a major player in SLE induces macrophages and foam cell formation. Among the multitude of antibodies in SLE, antibodies to lipoproteins lead to increased formation of oxidized LDL. These lead to acceleration of atherosclerosis. Activated platelets, leucocytes and endothelial cells release a variety of exosomes and micro-particles in the setting of inflammatory rheumatic diseases. These exosomes packaged with auto-antigens, cytokines and pro-coagulant factors may accelerate atherosclerosis.

ASSESSMENT OF CVD[9]

CVD risk is underestimated in patients with autoimmune rheumatic diseases. A wide variety of risk scores as the Framingham score are used to assess the risk of CVD in an individual. But these scores do not take into account the increased incidence of CVD in autoimmune diseases. More recent risk scores as QRISK 2 have intrinsically accounted for increased risk of CVD in RA. EULAR recommend that a multiplication factor of 1.5 be used to conventional risk algorithm to estimate the risk of CVD in patients with RA. Another very useful screening tool for CVD risk assessment in inflammatory rheumatic diseases is carotid Doppler. Presence of carotid plaques and carotid intima medial thickness correlate with CVD risk in these patients. EULAR recommends CVD assessment to be done in inflammatory rheumatic diseases at least once every 5 years. They also recommend a baseline Lipid profile to be done for all patients with inflammatory joint diseases, ideally at a time when the disease is in remission or at least when the disease is stable. A total cholestrol /HDL ratio is considered to be the best predictor of CVD than individual lipid components. Angiopoetin 2 and osteoprotogerin receptor in serum are emerging biomarkers for endothelial dysfunction and CVD in RA. However, they are not yet used except in clinical studies. Newer techniques such as coronary calcium scoring have not yet been put into practice in patients with autoimmune diseases.

MANAGEMENT OF CVD IN RHEUMATIC DISEASES[9–11]

The risk for CVD in autoimmune diseases is often underrated during the course of treatment. There are no specific guidelines for risk stratification or target levels for conventional cardiovascular factors in inflammatory rheumatic diseases. The existing guidelines in traditional CVD is currently extrapolated in most clinical settings. Concern about the underlying rheumatic disease is often a hindrance to timely and aggressive therapy for patients with acute CVD event; increasing the morbidity and mortality associated with it. Early data suggests that early and tight control of disease activity is important in mitigating CVD risk in inflammatory rheumatic diseases. However, whether mere clinical remission will suffice in reducing CVD risk is unknown. It is possible that even subclinical inflammation can drive CVD in inflammatory joint diseases and a higher target as suppression of TNFα and other pro-inflammatory cytokines may be required to effectively normalize the CVD risk.

Among DMARDs, methotrexate usage has been effective in reducing the CVD risk in RA. Methotrexate has been shown to reverse the dyslipidemia associated with RA and *in vitro* studies have shown efficacy in inhibition of foam cell formation. In fact a prospective NIH funded study is evaluating the effect of low dose methotrexate in high risk post MI patients for CVD prevention in patients without inflammatory rheumatic diseases. Hydroxy chloroquine use in SLE has been associated with reduced plaque burden, improved lipid profile, lesser risk of thromboembolism and overall increased survival. Mycophenolate Mofetil in experimental animal models have shown to attenuate atherogeneseis, a finding that was not observed even with statins. But the role of other conventional DMARDs in CVD reduction is still not established.

Early evidence suggest that biological therapy especially anti-TNF agents in all probability have a role in reducing the risk, but convincing data is yet to emerge. Lipid profile seems to improve with use of anti-TNF therapy. The use of anti IL-6 agents tocilizumab is associated with an increase in LDL and total cholesterol, but its influence on CVD is still unclear. JAK kinase inhibitor, tofacitinib has also been found have profound effect on lipid levels in RA patients. A 15–20% increase in total cholesterol and HDL levels has been demonstrated with 1–3 months of initiation of tofacitinib and the change in lipid levels correlate with fall in CRP.

Another area of uncertainty is the role of corticosteroids in CVD in inflammatory diseases. That corticosteroid use is associated with CVD is well-known as their use can cause hypertension and insulin resistance, but it is possible that their usage in diseases as RA might actually lead to earlier suppression of inflammation reducing the CVD risk. In any case attempt should be made to minimize the dose and duration of steroid therapy. The use of NSAIDs is a proven risk factor for CVD in general population and hence their use should be to the minimum possible extent.

The efficacy of established management strategy to reduce traditional cardiovascular risk factors in RA patients need to be further explored. For example it has been shown that patients with RA are less likely to respond to lipid lowering therapy with statin than non RA controls. The role of structured exercise therapy and smoking cessation in RA should be emphasized. Control of the traditional CVD risk factors are hypertension, diabetes mellitus and dyslipidemia is important. A lower level of target LDL cholesterol (<100 mg/dl) has been proposed in patients with SLE. But early reports of use of statins in SLE have not shown any reduction in plaque density or improvement in endothelial dysfunction.

Conclusions

Cardiovascular diseases are emerging as major co morbidities in autoimmune rheumatic diseases. The link between CVD and inflammation is still very intriguing. Both traditional and non-traditional risk factors seem to contribute to the heightened risk for CVD in autoimmune diseases. More research is needed to define targets to minimize the CVD risk.

Key Points

- Higher incidence of CVD is seen in most autoimmune rheumatic diseases
- Disease activity is a major risk factor for CVD in autoimmune rheumatic diseases
- Even preclinical autoimmune disease appears to be a risk factor for CVD
- Regular screening for CVD is recommended in autoimmune rheumatic disease
- The role of DMARDs in modifying CVD risk also needs to be considered

REFERENCES

1. Prasad M, Hermann J, Gabriel SE, Weyand CM, Mulvagh S, Mankad R, et al. Cardiorheumatology: cardiac involvement in systemic rheumatic disease. Nat Rev Cardiol. 2015; 12:168–76.
2. Teknodu MG, Kravvariti E, Konstantonis G, tentolouris N, Sfikaskis PP, Protogeru A. Subclinical atherosclerosis in Systemic Lupus Erythematosus: Comparable risk with Diabetes mellitus and Rheumatoid Arthritis. Autoimmun Rev 2017; 16: 308–12.
3. Mason JC, Liby P. Cardiovascular disease in patients with chronic inflammation: mechanisms underlying premature cardiovascular events in rheumatologic conditions. Eur Heart J. 2015; 36: 482–9.
4. Majka DS, Chang RW. Is preclinical autoimmunity benign? The case of cardiovascular disease. Rheum Dis Clin North Am. 2014; 40:659–68.
5. Skeoch S, Bruce IN. Atherosclerosis in rheumatoid arthritis: is it all about inflammation. Nat Rev Rheumatol. 2015; 11:390–400.
6. Hermans MP J, van der Velden D, Montero Cabezas JM, Putter H, Huizinga TWJ, Kuiper J, et al. Long-term mortality in patients with ST-segment elevation myocardial infarction is associated with anti-citrullinated protein antibodies. Int J Cardiol. 2017; 240:20–24.
7. Nurmohamed MT, Heslinga M, Kitas GD. Cardiovascular comorbidity in rheumatic diseases. Nat Rev Rheumatol. 2 015; 11:693–704.
8. Baghdadi LR, Woodman RJ, Shanahan EM, Mangoni AA. The impact of traditional cardiovascular risk factors on cardiovascular outcomes in patients with rheumatoid arthritis: a systematic review and meta-analysis. PLoS One. 2015; 10:e0117952.
9. Agca R, Heslinga SC, Rollefstad S, Heslinga M, McInnes IB, Peters MJ, et al. EULAR recommendations for cardiovascular disease risk management in patients with rheumatoid arthritis and other forms of in?ammatory joint disorders: 2015/2016 update Ann Rheum Dis 2017; 76:17–28.
10. Castaneda S, Nurmohammed MT, Gonzalez-Gay MA. Cardiovascular disease in inflammatory rheumatic diseases. Best Pract Res Clin Rheumatol. 2016; 30:851–69.
11. Charles Schoeman C, Gonzalez-Gay MA, Kaplan I, Boy M, Geier J, Luo Z, et al. Effects of tofacitinib and other DMARDs on lipid profile in rheumatoid arthritis: implications for the rheumatologist. Semin Arthritis Rheum. 2016; 46:71–80.

Chapter

79

Rheumatic Manifestations of Systemic Diseases

Liza Rajasekhar

INTRODUCTION

Pattern recognition is an important tool in the clinical diagnosis of rheumatic diseases. In a dedicated rheumatology practice it is important to be constantly aware of symptoms or signs in a patient which are at variance from the usual pattern of rheumatic diseases. The usual categories of systemic diseases which may be present or associated with rheumatic manifestations include:

1. Endocrine disorders
2. Infections
3. Malignancies
4. Inflammatory bowel disease
5. Hematologic disorders

Endocrine Disorders

1. **Thyroid disorders:** Both hyper and hypothyroidism are associated with rheumatic manifestations. Carpal tunnel syndrome, proximal myopathy are reversible conditions seen with hypothyroidism. Thyroid muscle disorder may be associated with cramps and especially in the elderly may resemble polymyalgia rheumatica. Rarely there may be an arthritis resembling seronegative rheumatoid arthritis.

   In hyperthyroidism proximal myopathy may occur. A thyroid acropachy is often described which usually occurs in the context of dermopathy and ophthalmopathy. It resembles pulmonary hypertrophic arthropathy in that patients have clubbing of fingers and toes with a adistinct radiologic pattern of sub-periosteal speculated, frothy, or lacy appearance, different from the laminal periosteal proliferation of classic pulmonary osteoarthropathy.
2. **Diabetes mellitus**
   1. The rheumatic manifestations of diabetes mellitus often correlate with sustained severe hyperglycaemia and predict macro-vascular complications like atherosclerotic cardiac or central nervous system disease. Trigger fingers, flexor tenosynovits and Dupuytren's contracture have all been reported and relate to diabetes triggered fibrotic changes in the tendons and palmar fascia.
   2. Disorganization of the joints commonly known as neuropathic joint is another rheumatic manifestation of uncontrolled diabetes. It is believed to arise from increased effect of matrix metalloproteinases in the chondrocytes. This occurs due to increased ligation of receptors for advanced glycation end products. Loss of sympathetic tone in diabetes, leading to increased blood flow and increased osteoblastic activity also contributes. Radiologically this manifests as localized osteopenia followed by subchondral osteolysis and later cartilage and bone disorganization.Unlike syphilis where the knee is commonly involved, diabetic neuropathic joints are usually the ankle and small joints of the feet.
   3. Patients with Type 1 DM may have an increased risk of RA since they share a common risk gene the 620 W PTPN22. This association is more strong for anti-CCP positive RA.

However, universally it is noted that axial and sacroiliac involvement is rare. Criteria for the diagnosis of Poncet's disease have been proposed.[3] These criteria require the presence of inflammatory non-erosive arthritis of no known aetiology in a patient with a concurrent diagnosis of extra-articular tuberculosis or a complete response to anti-tubercular therapy supported by either Mantoux positivity or presence of anyone of the allergic manifestations of hypersensitivity to tubercular proteins like erythema nodosum, papulonecrotic tuberculids, phylectenular conjuctivitis.[4] Since it is often difficult to isolate tubercle bacilli from suspected lesions, many clinicians believe that what is passed off as Poncet's is actually lack of sensitive tests to diagnose tuberculosis of the synovium. However, it is important to keep in mind this diagnosis since tuberculosis is endemic in India. There is also need for more well documented cases of Poncet's arthritis for validating these criteria.

The mechanism of Poncet's arthritis is an immune reaction to the tubercle bacilli. The synovial fluid is non-inflammatory and tubercle bacilli cannot be cultured from these fluids.

b. Tuberculosis also can affect the skeletal system by direct microbial growth and injury and this is classically seen in spinal tuberculosis or Pott's spine. Tubercular synovitis commonly affects the wrist and is radiologically associated with carpal osteopenia with erosions involving multiple contiguous carpal bones unlike rheumatoid arthritis which is also associated with carpal osteopenia but the erosions are seen on one or more non-contiguous carpal bones.

c. Erythema nodosum a common finding in many rheumatic diseases is also seen in tuberculosis. Another cutaneous finding of tuberculosis which may resemble erythema nodosum is an erythematous lesion found on the posterior surface of the legs with a histology of panniculits with vasculitis. This is called erythema induratum of Bazin.

3. **Leprosy:** Musculoskeletal manifestations are the third most common clinical manifestation of leprosy after skin and neurological manifestations. Patients generally have oligoarthritis. A symmetric non-erosive polyarthritis even in the absence of the lepra conversion reactions has been reported.[5] Rheumatoid factor maybe positive in up to one-third of patients with leprosy especially lepromatous leprosy.

4. **Hepatitis C infection:** Hepatitis C infection is associated with many extra hepatic manifestations which include arthralgia, myalgias, vasculitis, sicca symptoms, peripheral neuropathy, other autoimmune manifestations and malignancy. The cause is due to the lymphotrophic nature of the virus resulting in immune dysregulation.[2]

Malignancy

The common rheumatic syndromes associated with malignancies include dermatomyositis, vasculitis and scleroderma. Occasionally, stiff man syndrome occurs. These occur either as a result of antibodies to auto-antigens or cancer related antigens or as paraneoplactic manifestations. In patients with dermatomyositis older than 50 years, a significant proportion may have an underlying malignancy which is diagnosed with in two years of myositis onset.

1. **Lung cancer:** Small cell lung cancer can cause rheumatic manifestations however neurological manifestations like myasthenia or Eaton Lambert syndrome are more common than rheumatic manifestations.[8]

2. **Multiple myeloma:** An elderly patient presenting with diffuse osteopenia and bone pains, with raised ESR, renal dysfunction may have an underlying plasma cell dyscrasia.

3. **Lymphoma:** 2% patients can have musculoskeletal-skeletal manifestations.

4. **Leukaemias:** About 20% have some musculoskeletal manifestations at presentation. The musculoskeletal manifestations of leukaemia includes arthralgia, arthritis, myalgias occurring in different combinations.[6] A third of these patients may not have any other clinical manifestation at presentation. The typical patient is a young male with mono- or

oligoarthritis/arthralgia/myalgia. Polyarthritis is extremely rare. The hip joint is most commonly involved. Early morning stiffness is rare, hence significant early morning stiffness is an important pointer to another diagnosis especially JIA. The radiographs and ultrasound are usually normal. Pain out of proportion to clinical findings, severe anaemia and cytopenias maybe a clue to underlying leukaemias. Elevated LDH maybe another clue to underlying malignancy, though LDH is often elevated as an acute phase reactant even in rheumatic diseases. The extent of elevation of LDH is more in malignancy than in JIA. Strong ANA positivity maybe found in less than a fifth of children with acute lymphoblastic leukaemias, though it is much less common in patients with lymphoma. In up to one-third of cases the suspicion of malignancy may find resonance in radiologic abnormalities. These may be osteoporosis, related vertebral wedge fractures or permeating lyric lesions.

Hematologic Disorders

1. **Sickle cell disorder:** Children with sickle cell disorder especially homozygous sickle cell disease often have articular manifestations in the form of arthralgia, arthritis and often excruciatingly painful dactylitis. The oligo or polyarthralgia maybe be associated with joint or finger swellings which lasts for 2 to 3 weeks. The joint maybe hot and a differential diagnosis of trauma, septic arthritis or occasionally gouty arthritis is considered. The synovial fluid is non-inflammatory in nature. Avascular necrosis of the bone is another important rheumatic manifestation of sickle cell disease and should be considered in the work-up of any patient with avascular necrosis of bone of unknown aetiology. Presence of hemolytic anaemia would be an important clue.

2. **Thalassemia:** Patients with thalassemia who survive to adulthood due to iron chelating therapy often have osseous abnormalities. These include coarsening of trabeculae, osteoporosis and medullary widening giving the bones a square appearance. Fractures can happen.

3. **Hemochromatosis:** Is a genetic disorder where excessive intestinal absorption of iron results in tissue damage causing diabetes, liver disease and cardiomyopathy. Acute and chronic arthopathy can also occur. Typically seen in a middle aged male patient. Patient may have episodes of acute synovitis of large joints due to calcium pyrophosphate dihydrate deposition. They may also develop a chronic arthropathy of the small joints with features of secondary osteoarthritis like joint space narrowing, small cysts and osteophytes, characteristically described as beak like in the second and third metacarpophalangeal joints.

4. **Haemophilia:** In patients with hemophilia, hemarthrosis manifesting as acute synovitis may announce the underlying deficiency of factor VIII or IX. Joint involvement may also proceed to chronic synovitis as a result of the synovial reaction to the iron released from the blood and over a period of time to degenerative arthritis. This is usually seen in the weight bearing joints, the knee elbows and ankles. The first episode usually occurs after the child starts walking.

Inflammatory Bowel Disease

Up to one-third of patients with inflammatory bowel disease (IBD) have musculoskeletal manifestations. The most common rheumatic manifestation is inflammatory arthritis. The pattern is usually a peripheral arthritis with or without a sacroiliitis or spondylitis.[8] Other rheumatic manifestations of IBD included well-defined diseases like Sjögren's syndrome, Takayasu's arteritis. Minor manifestations like aphthous ulcers and neutrophilic dermatoses are often reported.

Conclusion

A large number of systemic diseases present with rheumatic complains. Hence when one sees a patient with rheumatic complaint in the rheumatology clinic or in an internal medicine practice, it is imperative to probe for history of other systemic illness. For example, the presence of knee joint swelling in a young male should prompt one to look for haemophilia; arthralgias and myalgias could be the manifestation of viral hepatitis. Paraneoplastic syndromes could present with lupus like features or RA like features or scleroderma like features. Therefore, it is important to be aware of symptoms or signs which do not fit in the usual pattern of rheumatic diseases.

Table 80.1: Rheumatological emergencies (disease-wise) (*contd.*)

- Acute nephritis
- Hypertensive crisis
- Pulmonary renal syndrome

Systemic sclerosis and mixed connective tissue disease
- Digital vasculitis and ischaemia
- Scleroderma renal crisis

Inflammatory myositis (poly/dermatomyositis)
- Respiratory failure

Crystal-induced arthropathies
- Acute gout
- Acute interstitial nephritis

Arthritis related to infections
- Septic arthritis
- Reactive arthritis

Osteoporosis
- Fracture

Miscellaneous disorders
- Haemophilic arthropathy
- Rupture of Baker's cyst

occlusion of blood vessel represents common basis for most emergencies in rheumatology. Vasculitis and its consequences may be primary or it may be secondary to rheumatic diseases such as rheumatoid arthritis (RA), systemic lupus erythematosus (SLE), and non-rheumatic diseases like sarcoidosis, malignancy and severe bacterial infections.[3] Medium vessel vasculitis like polyarteritis nodosa (PAN), anti-neutrophil cytoplasmic antibody (ANCA) associated vasculitis can present with mesenteric ischemia or gangrene. Treatment should be started after ruling out infection. In organ-threatening or life-threatening vasculitis aggressive immunosuppression using intravenous (IV) methyl prednisolone, IV immunoglobulin, cyclophosphamide, mycophenolate mofetil or rituximab is required.[4]

LUPUS FLARE

Lupus flare is defined as an increase in disease activity requiring a change or intensification of treatment and is seen in 27–66% of patients.[5] Flare of existing lupus can be due to stress, exposure to sunlight, infection, withdrawal of steroid and pregnancy. The precipitating factor of a flare should be identified and treated accordingly. A patient with flare of lupus can present with features of increased skin rash, arthritis, serositis, vasculitic, renal or neuropsychiatric manifestations. Severe hematologic manifestations in the form of cytopenias are also seen. Prompt diagnosis and treatment is required in flares involving cardiovascular, renal and central nervous system.[6] Features seen in infection differentiating it from a flare are: (i) Fever with chills, (ii) Leukocytosis and raised C-reactive protein, (iii) high C3, C4 complement levels. However, it is to be remembered that sometimes infection and a flare can coexist making the differentiation difficult. Best treatment of lupus flares is by withdrawal of the precipitating factor and increasing the dose of glucocorticoids. Life-threatening emergencies in SLE and their response to glucocorticoids are listed in Table 80.2.

SCLERODERMA RENAL CRISIS

Scleroderma renal crisis (SRC) constitutes one of the medical emergencies in rheumatology.[7] SRC most often occurs in patients with diffuse cutaneous systemic sclerosis especially in the first few years of disease. It is more often seen in winters and is associated with use of high dose doses of steroids. Most of the patients have high blood pressure but normotensive crisis may also be seen (10%). Renal

Table 80.2: Life-threatening manifestations of SLE and their response to glucocorticoids

Manifestations usually responsive to high doses of glucocorticoids
- Subacute cutaneous lupus
- Acute polyarthritis
- Polyserositis
- Myocarditis
- Lupus pneumonitis
- Glomerulonephritis
- Haemolytic anaemia
- Thrombocytopenia
- Diffuse CNS syndrome

Manifestations not often responsive to glucocorticoids
- Thrombosis
- Scarred end stage renal disease
- Membranous glomerulonephritis
- Psychosis not related to SLE

involvement is characterized by acute onset of renal failure, abrupt onset of moderate to marked hypertension and urine sediment that is usually normal or reveals only mild proteinuria with a few cells or casts. In addition to acute hypertension there is rising serum creatinine which may not become normal. SRC is thought to be due to severe Raynaud's phenomenon of renal arteries. Risk factors for SRC are diffuse skin disease, unexplained anaemia, new cardiac event (congestive heart failure, pericardial effusion), high dose steroid, anti-RNA polymerase III antibody. The other clinical features of SRC apart from accelerated hypertension, and renal failure are headache, seizures, thrombocytopenia, microangiopathic hemolytic anaemia, encephalopathy, pulmonary edema and hypertensive retinopathy. It is fatal if not diagnosed and managed promptly. ACE inhibitors are the drugs of choice in the management of SRC hypertensive crisis. The mortality of SRC has reduced considerably in recent years due to early and aggresive use of ACE inhibitors. But still 40% patients require dialysis and mortality at 5 years is 30 to 40%. Careful management of hypertension by avoiding intravenous nitroprusside or labetalol, tapering and withdrawal of steroids to avoid addisonian crisis is needed. Renal transplant may also be needed in patients with irreversible kidney damage.

CATASTROPHIC ANTIPHOSPHOLIPID SYNDROME

Catastrophic antiphospholipid syndrome is a devastating emergency. It usually presents as multiorgan failure.[8,9] Antiphospholipid syndrome (APS) can be primary (more than 90%) or secondary to connective tissue diseases, malignancy and infections. Pulmonary, gastrointestinal, cardiac, renal and neurological involvement is common. If three or more organs are involved, it is called catastrophic antiphospholipid syndrome. Precipitating factors are surgery, infections, oral contraceptive use and withdrawal of anticoagulants. Antiphospholipid antibody, lupus anticoagulant or anti-beta2 glycoprotein1 antibody can be positive. Patients with antiphospholipid antibodies may present as emergencies owing to vascular occlusion in both arterial and venous system. Unexplained sudden limb ischaemia, acute pulmonary thromboembolism, acute myocardial infarction, acute cerebrovascular accident, acute bowel infarction or intestinal ischaemia, acute Budd-Chiari syndrome, epilepsy, severe haemolytic anaemia and thrombocytopenia, abortion and other obstetric complications may be the presenting feature. Stroke at a young age is one of the commonest manifestations of antiphospholipid syndrome.[10] SLE, RA and other rheumatological entities may have antiphospholipid antibodies with thrombotic complications superadded on the background features which characterize their specific diagnosis. Catastrophic vascular occlusion is usually sudden, diagnostically confusing and life threatening. Despite adequate treatment mortality is high.

PULMONARY RENAL SYNDROME

Pulmonary renal syndrome is a combination of diffuse alveolar haemorrhage and rapidly progressive glomerulonephritis.[11] It was described by Goodpasture. There are three types: Type I is antibody mediated and is related to anti-glomerular basement membrane antibodies, type II is immune complex mediated and is related to SLE, type III is pauci immune mediated and is related to ANCA associated vasculitis [most common].[12] Signs and symptoms include dyspnoea, cough, fever, haemoptysis, peripheral oedema and hematuria. Treatment of alveolar hemorrhage and rapidly progressive glomerulonephritis consists of pulse methylprednisolone, plasma exchange, and cyclophosphamide. The underlying autoimmune disorder should be identified and treated. In addition to immunosuppresive therapy supportive treatment consists of oxygen therapy, monitoring of vital signs, and ventilation if required.

MACROPHAGE ACTIVATION SYNDROME

Macrophage activation syndrome is life-threatening complication of rheumatic diseases. It is more frequently seen in systemic onset juvenile idiopathic arthritis, systemic lupus erythematosus and other connective tissue disorders.[13] It has a high mortality (70%) and is mostly under diagnosed (30%). The condition is triggered by viral infection especially Ebstein-Barr, cytomegalovirus and drugs such as asprin, antiretroviral drugs, sulphasalazine, methotrexate and TNF blockers like etanercept and anti IL 1 receptor therapy (anakinra). It is characterized by non- remitting high fever of more than one week, pancytopenia, hepatosplenomegaly, lymphadenopathy, increased liver enzymes, high triglycerides, high lactic dehydrogenase, low sodium levels and abnormal coagulation profile. Treatment is mainly supportive and should be started early to avoid organ damage. It comprises transfusion of fresh frozen plasma and antibiotics in the presence of infection. Parental high dose corticosteroids and cyclosporine A have been reported to be useful. Uses of IV immunoglobulins, cyclophosphamide, plasma exchange and etoposide have shown conflicting results.

ACUTE CARDIAC EMERGENCIES

Pericarditis is common in SLE but massive pericardial effusion presenting with cardiac tamponade is rare. Once diagnosed, pericardiocentesis along with steroid therapy should be given. Tuberculous pericarditis must be ruled out. Acute myocarditis with congestive heart failure require intensive therapy. The cornerstone in management include glucocorticoids, immunosuppressive therapy with cyclophosphamide and IV immonoglobulins if required. Acute myocardial infarction may be the presenting feature of antiphospholipid antibody syndrome, PAN and SLE. Besides routine therapy, anticoagulants are important to save these patients. Anticoagulation is also required in cardiomyopathy to prevent thrombus formation and thromboembolism.

Endocarditis may be encountered in some patients with SLE (Libman Sacks endocarditis) and in APS, with sterile blood culture. Embolism from left atrial clot may occur resulting in ischaemia in other organs. Anticoagulation is needed in these patients. If infection is suspected, full course antibiotic therapy should be given guided by blood culture and sensitivity tests.

NEUROLOGICAL EMERGENCIES

Commonest neurologic presentations are due to vasculitis that may manifest as encephalopathy, focal or multifocal CNS disturbances, hemorrhagic or ischaemic stroke and myelitis. Cranial neuropathies single or multiple may occur. Characteristic rheumatologic disorders associated with such presentation are Behçet's disease, SLE and PAN while they are distinctly rare in RA and SSA. Corticosteroid and other immunosuppressive agents are useful as the basic pathogenic mechanism in CNS damage is immune mediated. Pulse therapy may be needed in serious cases. However, in certain subset of patients with APS cortical micro infarcts due to thrombotic occlusions occur and anticoagulant therapy is indicated. Stenting of the occluded vessel may also be needed. Sometimes vasculitic and thrombotic process may coexist and in acute stages it is safer to treat both. If seizure is the presenting feature, antiepileptic drugs have to be employed. Acute psychiatric illness may be the presenting feature of lupus cerebral vasculitis where anti psychotic drugs are used besides steroids and immunosuppression. However, if patients with SLE on steroid therapy develop abnormal behaviour, steroid induced psychosis has to be considered. Steroid psychosis usually develops within a week of starting high dose steroids.

OCULAR EMERGENCIES

Ischaemic optic neuropathy due to vasculitis may be observed in temporal arteritis, PAN and SLE while retinal vasculitis is observed in Behçet's syndrome. Immediate steroid therapy using high dose prednisone [1 mg/kg/day] may save vision in the vast majority of subjects. Painful red eye due to conjunctivitis or iridocyclitis may be encountered in seronegative spondyloarthropathies (SpA) and Behçet's syndrome. Granulomatous scleritis is observed in rheumatoid arthritis. Iridocyclitis is a major problem in children with pauciarticular arthritis and positive antinuclear antibodies. It can lead to blindness from secondary glaucoma, cataract formation and band keratopathy. These patients should be handled by taking help from ophthalmologists. Use of local drugs like mydriatics and corticosteroid eye drops beside systemic steroids are recommended. Uveitis usually responds to local steroids and systemic steroids may not be needed. Scleromalacia perforans is an extremely rare complication of RA; it requires urgent medical and surgical intervention.

EMERGENCIES INVOLVING PERIPHERAL JOINTS AND SPINE

Septic Arthritis

Septic arthritis is a medical emergency that is associated with high mortality and joint destruction or deformity if left untreated. It is associated with high spiking fever, swelling and redness of the involved joint which is usually a large joint. The presentation is usually monoarticular involving large joints; polyarticular involvement is also seen, e.g. gonococcal arthritis and associated with active bacteremia. Septic arthritis is particularly observed in prosthetic joints, elderly patients, diabetics, chronically debilitated or immunocompromised subjects or RA. Usual bacterial pathogen is Staphylococcus or Streptococcus but a wide variety of microbes may be involved like Gram-negative bacteria and anaerobes, mycobacteria or even fungi.

Synovial fluid gram stain and cultures will be helpful in identifying the organism. Management involves immediate implementation of appropriate antibiotics given parenterally for the first 2 weeks, followed by oral antibiotics for at least a further 4 to 6 weeks. Usually prolonged antibiotic regime is required in RA patients where there is high chance of sequestration of organism due to altered anatomy and bone erosions. Joint drainage done by closed needle aspiration is needed as long as effusion is detectable. Tissue debris of clots may be removed by sterile saline irrigation. Septic joint is one of the most preventable and successfully treatable rheumatologic emergencies.

Acute Arthritis

A patient may land in the emergency for acute arthritis, which can be a new onset symptom or it could be a flare of arthritis of RA. If the clinical presentation is acute monoarthritis then septic arthritis, gout, or trauma should be ruled out. In case of acute oligo or polyarthritis the causes could be viral arthritis, reactive arthritis, acute rheumatic fever, HIV, or disseminated gonococcal infection etc. A good history will be able to differentiate most of these conditions.

Septic arthritis has to be ruled out. Synovial fluid must be aspirated, sent for culture and empirical antibiotics should be started in order to save the joint. Treatment of gout consists of non steroidal antiinflammatory drugs (NSAIDs). If it is a flare of RA management with NSAIDs, low dose steroids may be required.

Acute Gout

Acute gout is one the commonest rheumatological emergencies. The patient, usually a male, presents with acute pain in one of the lower limb joints, commonly first metatarsophalangeal joint. Presentation is usually with monoarthritis or oligoarthritis. It is common in older individuals, mostly men. The acute stage is managed with NSAIDs, colchicine and corticosteroids. Allopurinol should not be given in acute phase of gout.

Atlantoaxial and Sub-axial Subluxation

Atlantoaxial subluxation a late complication of RA results from damage to bones, ligaments and cartilages caused by invasive rheumatoid synovial tissue. Proliferating synovial tissue with its associated inflammation can weaken or damage the transverse ligament and erode or fracture the odontoid process which can lead to subluxation. It can be precipitated by whiplash injury or trauma to the neck. Furthermore, destruction of the lateral atlanto axial facet, atlanto occipital joints and their associated ligaments contribute to the neurologic complications of rheumatoid cervical spine.

Cervical spine involvement (atlantoaxial subluxation) can result in potentially life threatening spinal cord compression. In addition to direct compression of nervous system structures, signs and symptoms of vascular insufficiency to neural tissues can be evident, e.g. vertebrobasilar insufficiency. MRI play a prominent role in identifying the patients with impending neural compression who will require prompt therapeutic intervention.

Atlantoaxial instability can be conservatively managed by using Philadelphia collar. Surgical treatment consists of fusion with or without decompression. In RA patients care should be taken during routine surgery to avoid forceful neck movements which can result in dislocation, fracture and death.

Acute Synovial Rupture

Acute synovial rupture of knee was classically seen in RA patients and some patients with osteoarthritis having Baker's cyst in the popliteal fossa and patients with early knee effusion. Rarely wrist, shoulder and elbow joints may also be involved. Rupture of popliteal cyst must be carefully distinguished from deep vein thrombosis as erroneous use of anticoagulants in the former will lead to haemorrhage into the joints and calf leading to secondary muscle necrosis with subsequent contracture.

Diagnosis is mainly clinical and may be confirmed by an arthrogram or ultrasonogram. Management includes rest to the joints, along with intraarticular corticosteroid injection to prevent fluid reaccumulation. Early mobilization and muscle strengthening exercises are needed when acute symptoms subside.

Avascular Necrosis

Avascular necrosis may present with acute excruciating pain in the joint. It is commonly seen in patients on long-term steroids and SLE. It can also be seen in antiphospholipid syndrome. Treatment is adequate analgesia, bone grafting or joint replacement.

Low Backache

Low backache can present as a medical emergency.[14]

The first task of the clinician should be to rule out medical emergencies such as cauda equina syndrome, epidural abscess, rupture of aortic aneurysm or dissection of aorta. It is important to know if the pain is inflammatory in nature. The clinician should look for saddle anaesthesia, bladder/bowel involvement, asymmetric loss of reflexes, pulse inequality, hypotension, or circulatory instability. Adequate use of NSAIDs is required for pain relief. In acute flare of spondyloarthropathy, increase in dose of DMARDs, addition of another DMARD or biologic response modifiers may be needed. Neurological compromise or features of cord compression warrants urgent CT myelogram or MRI and neurosurgical intervention.

Osteoporotic vertebral collapse can present with acute backache with or without neurological deficit. Pain management is by NSAIDs; salmon calcitonin nasal spray is also used. In some patients vertebroplasty may be needed.

EMERGENCIES DUE TO IATROGENIC COMPLICATIONS

Non-steroidal anti-inflammatory drugs are the most widely used pharmacologic agents in rheumatological practice. Their major drawbacks include gastrointestinal side effects leading to hemorrhage and the recently recognized enteropathy. The risk for gastrointestinal bleeding is cumulative and dose-related. Serious gastrointestinal side effects due to NSAIDs can be prevented by concomitant use of proton pump inhibitors, oral PGE2 analogue misoprostol or H_2-receptor antagonists.

There are several other complications associated with the use of NSAIDs that may pose as serious emergencies listed in Table 80.3.

Acute adrenal insufficiency may be precipitated by corticosteroid withdrawal in patients who have received prednisolone in doses as low as 3 mg/day for a prolonged duration or within 7 to 28 days with doses of 20 to 30 mg/day. Recovery of the hypothalamic-pituitary-adrenal axis may take as long as 1 year. One must be particularly cautious about this condition during the stress of intercurrent illness or surgery and adequate corticosteroid coverage is necessary.

Exfoliative dermatitis and Stevens-Johnson syndrome are the dreaded complications with gold, penicillamine, sulphasalazine and other DMARDs. Long-term use of DMARDs can cause complications

Table 80.3: Emergencies arising as complications of drug therapy

Nonsteroidal anti-inflammatory drugs
- Acute gastritis
- Perforated/bleeding peptic ulcer
- Acute enteritis
- Analgesic nephropathy
- Acute interstitial nephritis
- Hypersensitivity reactions
- Thrombocytopenia
- Pancytopenia
- Stevens-Johnson syndrome
- Erythema multiforme
- Toxic epidermal necrolysis
- Precipitation of asthma

Glucocorticoids
- Psychosis
- Acute Addisonian crisis due to withdrawal

Disease modifying anti-rheumatic drugs
- Hypersensitivity reactions
- Thrombocytopenia
- Increased transaminases
- Aplastic anaemia
- Exfoliative dermatitis
- Stevens-Johnson syndrome

Biologic response modifiers
- Infusion reactions
- Infections

like sepsis, neutropenia, anaemia, thrombocytopenia, etc. Haematological toxicity is the most serious complication of DMARDs and immunosuppressive agents; hence regular blood monitoring is required. Profound leucopoenia may lead to fatal infections while thrombocytopenia may lead to fatal bleeding. Treatment of such conditions includes appropriate supportive measures, blood component therapy and bone marrow stimulation with growth factors. Methylprednisolone in bolus doses may be life saving. A rare dangerous side effect of methotrexate is interstitial pneumonitis or "methotrexate lung".

A clinical diagnosis in a patient with rheumatological emergency can be made by thorough history, clinical examination, and laboratory workup should be done as warranted by the clinical condition.

GENERAL PRINCIPLES IN MANAGEMENT OF RHEUMATOLOGICAL EMERGENCIES

The primary aim of treatment is to save life and prevent irreversible organ damage by instituting suitable and aggressive therapy. It is useful to have a team approach involving anaesthetist, nephrologists, cardiologist and other relevant specialists. Basic emergency therapy common to other diseases should be given. Before starting aggressive treatment with high dose of glucocorticosteroid, infections must be excluded and co-morbid conditions should also be considered.

OTHER APPROPRIATE THERAPY

Anticoagulation

Patient coming with complications consequent to arterial and venous occlusions as in primary or secondary antiphospholipid syndromes require effective anticoagulant therapy with IV heparin followed by oral anticoagulation. Acute pulmonary vascular disease with pulmonary hypertension owing to vasculitis and thrombosis requires full dose of anticoagulation therapy.

New therapies for aggressive SLE include IV gamma globulins, mycophenolate mofetil and immunoablation with autologous stem cell transplantation. B-lymphocyte depletion by rituximab has shown promising results.[15]

Antihypertensive Drug Therapy

Hypertensive crisis common is lupus nephritis which warrants not only adequate control of blood pressure but immunosuppressive therapy also. Systemic sclerosis and PAN are other important causes of hypertensive crises. Angiotensin converting enzyme inhibitors are particularly valuable in SLE, systemic sclerosis. Calcium channel blocker, post-synaptic alpha blockers and diuretics are also helpful. Intravenous sodium nitroprusside or nitroglycerine may be given in accelerated/ malignant hypertension to prevent target organ damage.

Key Points

- Rheumatological emergencies can be life threatening and should be treated quickly and aggressively.
- A team approach is needed involving rheumatologist and many other specialities depending on the type of emergency.
- Treatment is aimed at preventing mortality and organ damage.
- Before starting treatment infection is to be ruled out and co-morbidities must be taken into account.

REFERENCES

1. Kumar A, Marwaha V, Grover R. Emergencies in rheumatology. J Ind Med Assoc. 2003; 101:520–526.
2. Rao URK, Shantaram V. Rheumatolological emergencies. J Assoc Physicians India 2006; 54 (Suppl):58–61.
3. Gonzalez-Gay MA, Garcia-Porrua C. Epidemiology of the vasculitides. Rheum Dis Clin N Am. 2001; 27:729–49.
4. Yates M, Watts RA, Bajema IM, Cid MC, Crestani B, Hauser T, et al. EULAR/ERA-EDTA recommendations for the management of ANCA-associated vasculitis. Ann Rheum Dis. 2016; 75:1583–1594.
5. Sprangers B, Monahan M, Appel GB. Diagnosis and treatment of lupus nephritis flares—an update. Nat Rev Nephrol. 2012; 8:709–17.
6. Bichile LS, Chewoolkar VC. Lupus Flare: how to diagnose and treat. Medicine update .2011; 281–286.
7. Steen VD. Scleroderma renal crisis. Rheum Dis Clin North Am. 2003; 29:315–333.
8. Cervera R, Asherson RA, Font J. Catastrophic Antiphospholipid Syndrome. Rheum Dis Clin North Am. 2006; 32:575–90.

9. Asherson R, Cervera R, de Groot PG, Erkan D, Boffa MC, Piette JC, et al. Catastrophic antiphospholipid syndrome: international consensus statement on classification criteria and treatment guidelines. Lupus. 2003; 12:530–4.
10. Nagaraja D, Christopher R, Manjari T. Anti-cardiolipin antibodies in ischemic stroke in the young: Indian experience. J Neurol Sci. 1997; 159:137–142.
11. Lee RW, D'Cruz DP. Pulmonary renal vasculitis syndromes. Autoimmun Rev. 2010; 9:657–660.
12. Talarico R, Barsotti S, Elefante E, Baldini C, Tani C, Mosca M. Systemic vasculitis and the lung. Curr Opin Rheumatol. 2017; 29:45–50.
13. Ravelli A. Macrophage activation syndrome. *Curr Opin Rheumatol.* 2002; 14:548–52.
14. Handa R , Aggarwal P , Wali J P . Rheumatological emergencies in clinical practice. J Ind Acad Clin Med. 2000; 5:135–141.
15. Vigna-Perez M, Hernández-Castro B, Paredes-Saharopulos O, Portales-Pérez D, Baranda L, Abud-Mendoza C, et al. Clinical and immunological effect of Rituximab in patients with lupus nephritis refractory to conventional therapy a pilot study. Arthritis Res Ther. 2006; 8:R83.

FURTHER READING

1. Gutiérrez-González LA. Rheumatologic emergencies. Clin Rheumatol. 2015; 34:2011–9
2. Mandell BF. Three rheumatologic emergencies: a sore toe, a cough, hypertension. Cleve Clin J Med. 2005; 72(1):50–6.
3. Slobodin G, Hussein A, Rozenbaum M, Rosner I. The emergency room in systemic rheumatic diseases. Emerg Med J. 2006; 23:667–71.

ANSWER KEY: MULTIPLE CHOICE QUESTIONS

Chapter 6

1. b	2. d	3. c	4. b	5. d	6. c	7. c	8. c
9. b	10. d	11. d	12. c	13. d	14. b	15. b	

Chapter 22

1. c	2. d	3. d	4. b	5. b	6. c	7. d	8. a
9. c	10. c						

Chapter 68

1. c	2. d	3. c	4. a	5. d	6. d